AF714301

Michael Grossman

# Forensic Chemistry Fundamentals

## Also of interest

*Analytical Chemistry.*
*Principles and Practice*
Victor Angelo Soffiantini, 2021
ISBN 978-3-11-072119-5, e-ISBN (PDF) 978-3-11-072120-1

*Practical Chemistry.*
*Vol. 3: Instrumental Analysis and Quantitative Analytical Chemistry*
Kasaye Bahiru Tola and Girma Selale Geleta, 2025
ISBN 978-3-11-170221-6; e-ISBN 978-3-11-170226-1

*Analytical Methods in Chemical Analysis.*
*An Introduction*
Edited by: Shikha Kaushik and Banty Kumar, 2023
ISBN: 978-3-11-079480-9; e-ISBN 978-3-11-079481-6

*General and Inorganic Chemistry.*
*Theoretical Foundations, Main Group Elements, Transition Metals, Lanthanides, and Actinides*
Irena Kostova, 2025
ISBN: 978-3-11-171222-2; e-ISBN 978-3-11-171224-6

*Chemical Analysis in Cultural Heritage*
Edited by: Luigia Sabbatini and Inez Dorothé van der Werf, 2020
ISBN: 978-3-11-045641-7; e-ISBN 978-3-11-045753-7

Michael Grossman

# Forensic Chemistry Fundamentals

Forensic Analytical Chemistry, Instrumental Forensic Analysis, Forensic Report Writing, Criminalistics, Locard's Principle

2nd, Completely Revised and Extended Edition

DE GRUYTER

**Author**
Dr. Michael Grossman,
Barrister and Solicitor of the Ontario Bar
Toronto, Ontario,
Canada. M5W 1P3
d441267@yahoo.ca
j903154@gmail.com

ISBN 978-3-11-914970-9
e-ISBN (PDF) 978-3-11-161855-5
e-ISBN (EPUB) 978-3-11-161860-9

**Library of Congress Control Number: 2025944465**

**Bibliographic information published by the Deutsche Nationalbibliothek**
The Deutsche Nationalbibliothek lists this publication in the Deutsche Nationalbibliografie; detailed bibliographic data are available on the Internet at http://dnb.dnb.de.


Cover image: ilbusca/DigitalVision Vectors/Getty Images
Typesetting: Integra Software Services Pvt. Ltd.

www.degruyterbrill.com
Questions about General Product Safety Regulation:
productsafety@degruyterbrill.com

In memory of my uncle – and with thanks for his encouragement –

Dr. Jonathan Grossman
1915–2010
of New York City and Washington, D.C.

American labor historian, educator, businessman and Jewish community participant

# Preface

This second edition is an update for cited references; as well as for some re-organization of topic presentation and format; and to better advise some readers about avoiding some of the more detailed scientific treatments. And, some important topics are expanded on. Otherwise, the format and range of topics are largely the same, but with an intention to better explain. And, of course, continued editing for typos and errors. The Chapter section numbering, although not identical, remains much the same.

An Introduction follows this Preface to try to explain in more detail about my approach – as author – as to how this book is constructed. This Introduction tries to explain those aspects of approach that are not-quite too directly related to the forensic chemistry chapters immediately following – but hopefully would be helpful to the reader. However, some readers might find convenience in delaying reading – or avoiding – it.

The intention remains to try to explain about science in legal processes and law in the application of science – developed from fundamental concepts, and to try to present in a simple and understandable way. Readers with limited science backgrounds might safely only skim, for bare concept, if not skip over, the more detailed scientific explanations, without too much loss of meaning of the main theme text.

The science discussed centres around analytical chemistry, in the context of the legal systems historically derived from, and still related to, the *Common Law of England* – of Canada, USA, UK and the Commonwealth. It is hoped that the various aspects of science and law in this book's several chapters can come together to equip the reader with a good essential understanding of what forensic chemistry is about.

This book derives from my practice of criminal law defence, with particular interest in forensic science.

In retrospect, this book is a kind of a law practice notebook of forensics that, at various times, I contemplated and often used.

Hopefully, this book would be useful to lawyers, students and judges for when the law involves science, and also to scientists, engineers, health practitioners and students, for when they would be concerned with legal matters. Also, importantly, it is hoped that this book would be of interest to citizens, who should be the ultimate critics and overseers of public policy and governance.

This book is essentially designed to explain concepts of both science and law – with theory, simple examples, to try to convey an appreciation of the essentials.

Thanks to the editorial personnel at De Gruyter, Berlin, for essential help in the preparation of this book – especially Jessika Kischke, Stella Müller and Dr. Christene Smith, and Janani Muneeswaran, Pakkamudiyanpet, Pondicherry, India. And, to my friend Christine Dick, Toronto, for her help. (All errors are, however, mine.)

M.G.
Toronto, Canada,
August 2025.

https://doi.org/10.1515/9783111618555-202

# Copyright permissions

Acknowledgements[1] and thanks for copyright permissions for extracts and illustrations:

- Edmund Cooper Estate Trust, *The Overman Culture*, 1971, 1974.[2]
- Taylor and Francis Group, Dr. Carl Meyer, ed., *EXPERT WITNESSING. . .*, CRC Press, 1999.[3]
- And1mu, "EM-Wave," Wikimedia Commons, 2016.[4]
- Furfur, "Hand drawing," Wikimedia Commons, 2017.[5]
- Sidney Harris Chemistry Cartoon; New Haven, Connecticut.[6]
- Dr. Nick Kim, "Analytical Chemists in the Wild West", "*Okay – who put my lunch through the mass spectrometer . . .?,*" "*. . . peer review . . .*"; Wellington, New Zealand.[7]
- Robert Silverberg, *A Time of Changes*, 1971, Agberg, Oakland, California.[8]
- *Vanderbilt Law Review*; and Professor Pamela R. Metzger, Dedman School of Law, Director of the Deason Family Criminal Justice Reform Center, Southern Methodist University, Dallas, Texas 75275; "Cheating the Constitution," Tulane University, 2006. [9]

**1** Copyright acknowledgements:
**2** Edmund Cooper. Introduction.
**3** T&F. Dr. Carl Meyer. Chapter 1.
**4** And1mu, "EM-Wave," Wikimedia Commons. Chapter 4.
**5** Furfur, "Hand drawing," Wikimedia Commons. Chapter 4.
**6** Sidney Harris Chemistry Cartoon. "www.CartoonStock.com". Chapter 5.
**7** Nick Kim. Cartoons. Chapters 2, 6 and 8.
**8** Robert Silverberg, *A Time of Changes*. Chapter 9.
**9** Professor Pamela R. Metzger, *Vanderbilt Law Review*. Chapter 10.

https://doi.org/10.1515/9783111618555-203

# Contents

# About the author

**Michael Grossman**, B.S., Ph.D., LL.B., Barrister & Solicitor, of the Ontario Bar, is a chemist[1] from before becoming a lawyer – practising criminal law defence in Toronto, Canada,[2] and a member of the American Chemical Society and of the ACS Division of Chemistry and the Law.[3] He has previously published *The Law of Occupational Health and Safety in Ontario*.[4]

1 Ph.D., University of New Brunswick, Fredericton, Canada, 1972, physical chemistry – photochemistry. B.S., chemistry, Rensselaer Polytechnic Institute, Troy, New York, 1967. High school diploma, Baltimore Polytechnic Institute, 1962.

2 LL.B., UNB Law School, 1980. Called to the Bar of Ontario, 1986. Criminal law defence, Toronto, since 1987; sole practice since ~2000.

3 ACS member since 1968; ACS CHAL since 1981.

4 *The Law of OH&S in Ontario, Second Edition*, Butterworths, Toronto and Vancouver, 1994, ISBN 0-409-90414-7 (1st ed., 1988, ISBN 0-409-80938-1).

https://doi.org/10.1515/9783111618555-205

# Introduction

*The scientist has a lot of experience with ignorance and doubt and uncertainty . . .*

[Richard Feynman, American physicist] [1, 2, 3][1]

[. . .]

*"Child," said the Queen, "what do you want to do when you grow up?"*
*"I want to understand," said Michael, almost without thinking.*
*"You want to understand?"*
*"Yes, your Majesty."*
*"What do you want to understand?"*
*"Everything," said Michael desperately. "I want to understand about people, I want to understand about machines . . . I – I want to know everything there is to know."*
*The Queen patted his head. "You want too much child. You want far too much."*

[. . .]

[Edmund Cooper, *The Overman Culture*, 1971, 1974] [4, 5][2]

## 1.1 This second edition

As indicated in the Preface, this second edition updates the first.[3] This Introduction is immediately followed by chapters more directly related to forensic chemistry and more detailed scientific treatments.[4]

## 1.2 Intent

The intention of this second edition remains much the same as the First, with a similar style of presentation – including that the main text can be read for essential meaning – with only limited reference to notes and references [6, 7, 8, 9, 10, 11, 12, 13, 14, 15, 16, 17, 18, 19],[5,6,7]

1 Introductory quotes. Introduction 1.2.
2 Edmund Cooper Estate Trust copyright acknowledgement. Introduction top of first page.
3 First ed., December 2021. ISBN 978-3-11-071878-2
4 Some footnoted as: Too much scientific technical detail here? Introduction 1.1 and 2.1
5 See Preface.
6 US *First* Amendment. Introduction top of first page; Introduction 4.3 and Chapter 10.3.
7 *Grundgesetz für die Bundesrepublik Deutschland* Artikel 5. Introduction top of first page; Introduction 4.3; Chapter 10.3.

https://doi.org/10.1515/9783111618555-001

## 1.3 Wherefrom; why; audience

Throughout my practice of criminal law defence,[8] I have been involved with forensics as a chemist, and chemistry prompted my involvement in workplace health and safety law (part of labour law) as an early area of interest.

In the past years, I have tried to explain forensic chemistry in various written formats. Various scientific issues and my attempted explanations have appeared incidental to legal issues in written court pleadings, or in other presentations or publications [20] or as my own notes. This book is to try to give a comprehensive general treatment of forensic chemistry's fundamental concepts[9] with examples; and with many endnotes to help document what is presented and for further information that some readers might want.[10]

In retrospect, this book is a kind of a law-practice notebook [21],[11,12,13] with forensics that, at various times, I contemplated and sometimes used – some-fewer times with some success – before the courts, including plea bargains that might not have otherwise happened. But even when not successful as before the courts, in whatever way, some presentations were successful in that they helped me better understand the science and the law and maybe helped the judge understand it too – notwithstanding an adverse ruling. This is especially true for the very essentials of forensic science that may not obviously show up in court presentations. And, importantly, inadequacies in past presenting might be improved on, if not prevented from happening again.

## 2.1 An explain book

This book is designed to explain[14,15] concepts of both science and law by way of presenting theory, with reference to simple examples [22].[16] Hopefully, readers would

8 See Preface.

9 Forensic analytical chemistry; chemistry more generally. Otherwise than forensic – useful for understanding; and perhaps a few digressions. Introduction 2.4.

10 The endnotes [numeric] are intended for data of direct and relevant interest; sometimes at length; also included are some (maybe too many and too diversional) of peripheral and allegorical interest. The footnotes [numeric] are intended only for data of direct and relevant interest; kept brief; and are usually for cross-referencing of topics within this book.

11 As a kind of law practice notebook. Author's biases. Introduction 1.2; Chapter 6.

12 Author's biases. Introduction 1.2, 4.1 and 4.3; Chapters 1, 4, 10 and 11.

13 Disclaimers. Error. Not legal advice. Commercial products. Introduction 1.3 and 4.2; Chapters 1, 5, 6 and 9.

14 See Preface.

15 An explain book. Not encyclopaedic. References. Introduction 2 and 3.1; Chapter 6.

16 Endnotes and footnotes. Introduction 1.2, 1.3 and 2.1.

gain an appreciation of what forensic chemistry is about, along with important underlying concepts, and a knowledge of where to look for more details and how to understand them. Explanations of concepts of forensic methods are dealt with in Chapter 6.[17]

I have tried to develop explanations carefully from essential concepts. I hope this would be especially helpful to readers with limited science or law backgrounds; what may be important to them may seem tedious and trite to the more experienced, who may want to skip through some of it.[18] But, the more experienced might also see it as an important introductory formalism.

Readers with limited science backgrounds might safely only skim, for bare concept, if not skip over,[19] the more detailed scientific explanations, without too much loss of meaning of the main text theme.

Also, explanations developed from essential concepts risk including too much tedious detail for the sake of comprehensiveness. Some readers may also want to skim or skip through some of this and go on to more directly related forensic matters.

## 2.2 *Gedankenexperiment*

Some of the examples used to help explain theory are cited from real sources and also sometimes from fictional literature [23].[20] Many examples are as *Gedankenexperiment* [24, 25, 26, 27, 28, 29][21] – thought experiments. As used in this book [30, 31, 32], these are fictitious examples concocted[22] to illustrate the concepts that are attempted to be explained.

Thought experiments are for things and movements imagined as real – that could be real – but a description of how they logically must work is sufficient to consider the concepts. Thought descriptions are convenient; practical construction would be difficult and expensive – some so much so as to be effectively impossible in practice, or actually impossible; but are important in principle.

That parts of a *Gedankenexperiment* may seem improbable – or beyond – perhaps over-concocted – should not detract too much from its usefulness as a tool of explanation – convenient for both author and reader. *Gedankenexperiment* also allows for convenient alternate variations for further explanation.

*Gedankenexperiment* is a concept borrowed from physics. Some famous examples (details beyond the scope of this book):

**17** Concept. Scientific methods. Introduction 2.1; Chapter 6.

**18** Footnoted as: Too much scientific technical detail here? Introduction 1.1 and 2.1.

**19** Footnoted as: Too much scientific technical detail here? Introduction 1.1 and 2.1.

**20** Fictional literature. Christie. Poe. Introduction 2.2; Chapters 5 and 8.

**21** *Gedankenexperiment*. Introduction 2.2.

**22** Fictional literature. Ethylene glycol. CAS 107-21-1. Introduction 2.2; Chapters 4 and 7.

- The clock in the photon-emitting box [33, 34, 35, 36, 37, 38, 39, 40] of Niels Bohr and Albert Einstein
- The cat that is/is not alive in another box [41, 42, 43, 44, 45, 46, 47, 48][23] of Erwin Schrödinger
- A *Jeans* cube [49, 50, 51, 52]
- The stone dropped from a moving railway carriage, measured with two clocks in different places of observation, of Albert Einstein [53, 54]
- The concept of EM wave propagation, from James Clerk Maxwell, considered by Einstein when developing relativity theory [55]

## 2.3 Ink analysis

In this book, ink analysis, for examination of suspected forged documents, is used for main examples,[24] chosen over the more recently popularly modern DNA[56, 57][25] analysis and proteomics [58, 59, 60], which would unnecessarily complicate the presentation of essential forensic principles with more complex chemistry.[26]

However, document examination – of ink and otherwise – has a popularity of its own – with interesting historical stories – for famous fakes such as

- Howard Hughes "Mormon will" [61, 62]
- Hitler diary [63, 64]
- JFK-Marilyn Monroe papers [65, 66]
- Letter forgeries of Lee Israel [67, 68, 69, 70]
- The Vinland map [71, 72, 73]

Perhaps a most famous of document examinations involved the

- *bordereau* and *petit bleu* [74, 75], and the letters forged by Colonel Hubert Henry, [76] offered as evidence against Captain[27] Alfred Dreyfus – the Dreyfus Affair of the Third Republic of France [77, 78, 79, 80, 81, 82, 83, 84]:[28,29,30]

23 Schrödinger's cat. Introduction 2.2; Chapter 4.

24 Variations of Ink analytical chemistry – as examples – *Gedankenexperiment* – continue throughout this book. Introduction 2.2 and 2.3.

25 DNA. Importance in forensic science. Electrophoresis. Introduction 2.3; Chapter 6.

26 However, this Second edition has expanded explanation for DNA.

27 Promotion to brigadier general ~130 years later. See endnote.

28 Captain Alfred Dreyfus. *bordereau, petit bleu.* Colonel Hubert Henry. Introduction 2.3; Chapter 1.

29 Ethics, lawyers, experts, *etc.* Introduction 2.3; Chapters 1, 5, 8 and 9.

30 Dreyfus. Zola. *J'accuse*. Introduction 2; Chapter 8.

> *. . . J'accuse les trois experts en écritures, les sieurs Belhomme, Varinard et Couard, d'avoir fait des rapports mensongers et frauduleux, à moins qu'un examen médical ne les déclare atteints d'une maladie de la vue et du jugement . . .*
>
> [Émile Zola, "J'ACCUSE . . . !" Paris, *L'Aurore*, 13 January 1898] [85, 86, 87][31]

And, there are other famous documents of disputed or unresolved origins such as
- from the Woodstock typewriter of the Alger Hiss trial [88, 89];[32]
- a Ten Commandments [90] tablet [91, 92]; and
- assuming it is a kind of document – the Shroud of Turin [93, 94, 95, 96, 97, 98, 99, 100, 101][33] – perhaps a most difficult and controversial forensic problem imaginable (neither analysis nor conclusion attempted here).

## 2.4 View of science

This book offers my view of forensic science[34]; others may view it differently and validly in their own ways. I concentrate on what I think most important, and hopefully know well-enough to be able to explain to others for the overall understanding. Topics of lesser importance, in this context, but relevant, are explained in less detail or by references to other literature or are simply mentioned with references.

Although discussions and examples in this book are essentially directed towards forensic science concepts – with specialization in analytical chemistry – many of those concepts and practices are shared more generally so that chemistry and other science concepts are often referred to – with implications towards forensic analytical chemistry [102].[35,36,37]

---

**31** = ". . . I accuse the three handwriting experts, . . . B . . ., V . . . and C . . ., of submitting reports that were deceitful and fraudulent, unless a medical examination finds them to be suffering from a condition that impairs their eyesight and judgement. . . ."

**32** Woodstock typewriter. Alger Hiss. Introduction 2.3; Chapter 2.

**33** Shroud of Turin. Carbon-14. Introduction 2; Chapters 3 and 7.

**34** Author's biases. Introduction 1.2, 2, 4.1 and 4.3; Chapters 10 and 11.

**35** Forensic analytical chemistry; chemistry more generally. Otherwise than forensic – useful for understanding; and perhaps a few digressions. Introduction 2.4.

**36** Forensic science. Introduction 2; Chapters 1, 2 and 5.

**37** Analytical chemistry. Expanded definition – more detail and example. Introduction 2; Chapters 1, 2 and 5.

## 3.1 Not encyclopaedic

This book is intended neither as an encyclopaedic[38] reference nor as a handbook for specific methods; nor for hands-on directions for lab work. The explanations and examples presented here are intended to explain concept – hopefully helpful in understanding any forensic issue – and hopefully to give an understanding of forensics in general overall concept. However, many literature citations[39] and examples are included.

## 3.2 SI and CAS

Writing about science can be difficult-enough without the added confusion of historical units and naming systems. From at least the time of the French revolution, there have been organized efforts for standardized international scientific descriptors – the origin of the metric system.[40]

The editorial convention adopted for this book is *Le Système International* [103]. SI definitions and conventions should be assumed. Generally, this is the modern internationally used metric system [104, 105, 106, 107]. This is generally consistent with the Canada *Weights and Measures Act* [108].

Further, chemical substances are described according to the terminology and systems of the Chemical Abstracts Service of the American Chemical Society, along with molecular structural diagrams; or at least with reference to CAS and molecular structural diagrams. Again, generally, this is the modern internationally used system [109, 110].[41,42]

However, other historical systems persist [111, 112, 113, 114], and sometimes remain more convenient, perhaps because everyone is so used to them; these others must, however, be relatable to SI and CAS.

## 4.1 Bias

Within the context of my hopeful intention to present reasonable discussions, this book necessarily reflects my biases [115]:[43,44,45]

---

**38** An explain book. Not encyclopaedic. References. Introduction 2 and 3.1; Chapter 6.
**39** Endnotes and footnotes. Introduction 1.2, 2.1 and 3.1.
**40** SI CAS Introduction 3.2; Chapters 3 and 4.
**41** Chemical names as used in this book. Introduction 3.2; Chapter 4.
**42** Molecular structural drawing. CAS. Chapters 4.4 and 4.6.
**43** Author's biases. Introduction 1.2, 2, 4.1 and 4.3; Chapters 10 and 11.
**44** re criticism of Health Canada, Judges *et al.* Introduction 4.1 and 4.3; Chapters 10 and 11.
**45** Criticisms in a *free and democratic society*: Introduction 4.1, endnote.

- As a chemist, from before becoming a lawyer, I look for and raise scientific issues in my client's cases. I think this is quite reasonable, although Crown counsel whom I encounter, and judges before whom I appear, often do not seem to concur[46] – sometimes it seems to get on their nerves.[47,48] Too few judges,[49] Crown counsel – and defence lawyers too – in my experience – show enthusiasm in trying to understand the world according to the concepts of physical science. Nor do they seem enthusiastic about understanding science as important beyond the legal issues before them – as an important aspect of human culture and civilization. Nor, sadly, does there appear the joy in trying to understand science as apparent with Isaac Asimov, George Gamow and Richard Feynman and their readers [116, 117, 118, 119, 120, 121, 122, 123, 124, 125, 126, 127].[50]
- My law practice of advocacy as defence counsel – my work environment[51] – at least raises what lawyers might call a "reasonable apprehension of bias". The work of prosecutors and law enforcement investigators, and judges, is no less important, valid or valuable. I have tried to give a balanced overview; readers should scrutinize and criticize this book on that basis.
- Topics dealt with in this book are largely in the context of analytical chemistry for criminal law and workplace health and safety law.
- Clear, transparent and formal reporting[52] of the science invoked in legal process is of paramount importance, as is the quality assurance process.[53] To obfuscate is wrong.[54]
- Forensic scientists should always strive to do good science, be prepared to accept when they fall short, and always continue to try to do better.[55]

---

**46** ". . . no merit to the Applicant's / defence assertion . . ." *O'Connor*. Introduction 4.1; Chapter 10; Chapter 10 Appendix.

**47** Other descriptions: "fishing expeditions", "speculative", "fanciful", "disruptive", "unmeritorious", "obstructive", "time-consuming", "unnecessary", "judicial resources", "no credible grounds", "relevance" and "fundamentally flawed".

**48** Appendix: Law case – Old City Hall, Toronto, 2018. Introduction 4.1; Chapter 10; Chapter 10 Appendix.

**49** *Daubert*. Judicial unfamiliarity with science. Introduction 4.1; Chapters 1 and 10.

**50** Asimov. Gamow. Feynman. Introduction 4; Chapter 9.

**51** Advocacy. Introduction 4.1; Chapters 1 and 8.

**52** Lab documentation. Introduction 4.1, Chapter 5.

**53** Government enquiries: Introduction 4, Chapters 8 and 9.
- Grange re Susan Nelles, 1984;
- Griffiths, re Gemma Ramlal, Health Canada, 2001;
- Goudge re Dr. Smith, Hospital for Sick Children, 2008;
- Lang re "Motherisk", Dr. Koren Hospital for Sick Children, 2015;
- Gillese re Ontario Long-Term Care Homes murders, 2019.

**54** Obfuscation. Introduction 4; Chapter 10.

**55** Science. Introduction 4; Chapter 2.

- In view of the danger of *quod volimus credimus libenter* [128],[56,57] a forensic scientist's work in a specific case should, in concept, be separated from the narrative being theorized for the related human story. This separation would be more difficult for the scientist involved in investigation rather than for the later evaluation of results. As best as practicable, the scientist should be blind[58] to the related human story to relieve a reasonable apprehension of bias. For example [129], the forensic analytical chemist, who happens to hold war-on-drugs warrior opinions, should not know that the analysis is for prosecution of a well-known drug dealer who deserves to go to jail; nor should the forensic analytical chemist, who happens to hold anti-war-on-drugs opinions, know that the analysis is for the victim of an unfair police sting, who has no criminal record at all.
- In a *free and democratic society* [130],[59] fairly operated good process – in good faith – should be expected, with the state as fully accountable to its citizens. Accountability means transparency of process so that the citizenry can scrutinize it. The workings of government should be clearly visible. Not only should forensic scientists try to do good science, but also be publicly accountable. Effectively, public accountability is a fundamental part of doing good science.[60] To obfuscate is wrong, for democracy as well as for science and engineering.[61]
- Judges are bound by precedent authority [131]; scientists must not be.[62] Although reliance by scientists on previously published works is necessary, and necessarily convenient in a practical sense, and high regard for previous scientists' work is appropriate, and other scientists deserve deference [132]; every science declaration is always, in some way, provisional and should never be absolved or excused from explanation. This should be contrasted as for judges and engineers, whose practical decisions are often calendar-deadline-directed – and funding-directed – with limited opportunity to go back for corrective reconsiderations.[63]
- *Supremacy of Parliament* [133] is a legal concept in a *free and democratic society*.[64] Parliament, the Governor-in-Counsel, and the courts command the law as applied to citizens; what to do when they bungle science is a problem.
- Science is explainable [134, 135]:[65]
  - Science conclusions are to be accepted, in whole or part, or not, with explanations, given in good faith, [136] based on factual (experimental) observations

56 *Quod volimus credimus libenter*. Latin. Introduction 4.1; Chapter 11 top of first page
57 ≈ believe what we want to believe
58 *Justitia*. Blind. Introduction 4.1; Chapters 5 and 8.
59 *Free and democratic society*. Canadian constitution. *Charter*. Introduction 4.1; Chapter 5.
60 Confidential *verses* public. Introduction 4; Chapter 5.8 and Chapter 8.
61 Obfuscation. Introduction 4; Chapter 10.
62 Advocacy. Introduction 4; Chapters 1 and 8.
63 Judges, scientists, calendar-deadlines. Introduction 4.1; Chapter 1.3.
64 *Free and democratic society*. Canadian constitution. *Charter*. Introduction 4.1; Chapter 5.
65 Explanation. Method validation. Introduction 4.3; Chapters 5.3.2 and 10.3.

and measurements. Scientific explanations must make rational sense – supportable with mathematical expression; and scientists must address reasonable questions rather than rely solely on authority. The opinion of a purported scientist who would rely on authority at the expense of explanation should be avoided.[66] Science is *nullius in verba.* [137, 138, 139, 140, 141, 142, 143, 144][67] *Ipse dixit* [145, 146, 147, 148, 149][68] is not good enough. Unexplained science is bad science,[69] for that reason alone.[70]
  - Science is necessarily a public process because explanations must be available to potential critics, who inconveniently and embarrassingly, might sometimes turn out to be correct [150].[71]
  - In general, **explaining science should be regarded as a serious on-going conversation – in good faith** – both as written publication as well as spoken conversation amongst scientists and others. Citizen participation in scientific discussion should be seen as necessary.
- However, sometimes satisfactory explanations remain elusive; the science may not be advanced well-enough [151]. And maybe we will never know about some things, but scientists should be encouraged to keep trying [152].
- That justice is so often so elusive is no reason to stop striving. This is for good citizenship.

I hope readers would see these biases as consistent with usually accepted ideas of science, justice and democratic government that I try to express in the first nine chapters of this book.

## 4.2 Accuracy attempted

While I strive for accuracy, and correct presentation, I cannot guarantee it [153, 154, 155].[72,73,74] I present what I write as reasonably reliable, based on, hopefully, sensible explanation and interpretation of searchable reference material. Readers should criticize this book on that basis.

---

**66** Science and explanation. Introduction 4.1; Chapter 1.
**67** *Nullius in verba* ≈ We do not take anybody's word for it. Introduction 4.3; Chapters 2, 5 and 8.
**68** *ipse dixit* ≈ he himself said it – an assertion without proof. Introduction 4.1.
**69** Science – good, bad and junk. Introduction 4.3; Chapters 1, 2, 5, 8, 10 and 11.
**70** Reasonable doubt. Science explanation. Introduction 4.1; Chapters 1.1.1, 10.7, 10.8 and 11.11.
**71** Semmelweis. Handwashing. Counting measurement. Introduction 4.3; Chapters 3 and 8.
**72** Disclaimers. Error. Not legal advice. Commercial products, *etc.* Introduction 1.3 and 4.2; Chapters 1, 5, 6 and 9.
**73** Cited literature. Accuracy. Introduction 2 and 4.3; Chapter 2.
**74** Error. Introduction 4.2; Chapter 8.

## 4.3 Additional bias

However, I also have additional biases,[75,76] from my litigation, over the years:[77,78,79,80]

- Concerning the quality of science done by a Canadian Federal Government Agency,[81] in relation to the war on drugs of abuse and recreation, and how federal criminal prosecution uses legislation to interfere with how the science is done, interpreted and reported.[82] And concerning a serious quality assurance failure not adequately addressed by the *Griffiths* Report (2001).[83,84]
- And, concerning science and determining, within Canadian law, drug-impaired driving.[85,86,87,88]

Because these biases would be rather more controversial, the remaining two chapters, as essays, are set off at the end of this book and labelled as "biased opinion – polemic".[89,90] Readers are invited to criticize on that basis.[91]

## 5.1 On to Chapter 1

On to Chapter 1, *et seq.* – immediately following this Introduction – for topics more directly related to forensic chemistry.[92]

M.G.
Toronto, Canada,
August 2025

75 US *First* Amendment. Introduction top of first page; Introduction 4.3; Chapter 10.3.

76 *Grundgesetz für die Bundesrepublik Deutschland* Artikel 5. Introduction top of first page; Introduction 4.3; Chapter 10.3.

77 Legislative process – policy making. Introduction 4.3; Chapters 1, 10 and 11.

78 Author's biases. Introduction 1.2, 2, 4.1 and 4.3; Chapters 10 and 11. (And also to some extent in Chapters 1.5.3 and 6.20).

79 re criticism of Health Canada, Judges *et al.* Introduction 4.1 and 4.3; Chapters 10 and 11.

80 *Proper scientific report. CD&S Act.* Health Canada. Introduction 4.3; Chapters 5, 8, 9, 10 and 11.

81 Health Canada. Introduction 4.3; Chapter 10.

82 Health Canada. DAS. Ethics . Introduction 4.3; Chapters 5 and 10.

83 Fraud. Griffiths Report, 2001. QA failure. Gemma Ramlal, 2001. Annie Doukhan, 2012. Introduction 4.3; Chapters 8–10.

84 War on drugs. *CD&A Act.* Mass (grams) measurement. *W&M Act.* Introduction 4. Chapters 3, 6, 8 and 10.

85 DRE. Scientific measurement. Introduction 4.3; Chapter 11.

86 Drug-impaired driving. DRE. Introduction 4.3; Chapter 11.

87 Chart – legislation. Introduction 4.3; Chapter 11.

88 Legislative process – policy making. Introduction 4.2; Chapters 1, 10 and 11.

89 Polemic. Health Canada. Introduction 4; Chapters 9 and 10.

90 Some science controversary remains. Introduction 4; Chapters 1, 2, 10 and 11.

91 Essay-biased opinion – polemic. Introduction 4.3; Chapters 10 and 11.

92 See Introduction 1.1.

## Notes

1 Timothy Ferris, Astronomical Notebook/"Minds and Matter," *The New Yorker*, 15 May 1995, pages 46–50, at pages 46 and 47.

2 https://www.goodreads.com/author/quotes/1429989.Richard_P_Feynman
https://www.goodreads.com/quotes/899021-the-scientist-has-a-lot-of-experience-with-igno rance-and
http://laserstars.org/bio/Feynman.html
https://en.wikiquote.org/wiki/Richard_Feynman
https://calteches.library.caltech.edu/1575/1/Science.pdf https://calteches.library.caltech.edu/1575/1/Sci ence.pdf
https://stepsandleaps.wordpress.com/2013/03/15/the-roles-of-ignorance-uncertainty-and-doubt/

3 See also: Richard P. Feynman, *The Meaning of It All / Thoughts of a Citizen Scientist*, Perseus Books Group, Cambridge, Massachusetts 02412; Basic Books, New York City, 10016-8810; ISBN-13: 978-0-465-02394-3; ISBN-10: 0-465-02394-0; 1963, 1998; "THE UNCERTAINTY OF SCIENCE" [Chapter] 1, page 1, *et seq.*

4 Edmond Cooper, *The Overman Culture*, Coronet Books, Hodder Paperbacks Ltd., London, SCIENCE FICTION; 1971, 1974; ISBN 0 340 17,860 4; at pages 40 and 41.

5 For use of this quote here: Permission granted by the Edmund Cooper Estate Trust.

6 This book's **intentions:**

7 An important intention is that the explanations within this book's main text can be read for essential meaning. Readers can defer the many included endnotes – of literature references, including commentaries, cross-references and digressions – until later – or much later, or read them hardly at all. Some readers may find some endnotes helpful for further reference or explanation or interesting; or less so; or not helpful. And, some readers may find some of the endnotes too digressive, tedious, peripheral, obscure or too author-self-indulgent, to bother with.

8 The endnotes are intended for data of direct and relevant interest; also included are some – maybe too many and too digressive – of peripheral and allegorical interest. The footnotes are intended only for data of direct and relevant interest and are usually for cross-referencing of topics within this book.

9 There are very many literature citations, other references and commentaries found in the endnotes. These are intended to be helpful for further explanation and study or fodder for criticism. But the uninterrupted reading of the main text of this book is intended to make sense on its own, notwithstanding these endnotes.

10 The brief introductory quotes, from easily searchable, if not well-known sources, used throughout this book are intended to try to connect some this book's concepts – in an attempted literary sort of way – with a broader culture. Although the connections may be somewhat tenuous, allegorical and perhaps sometimes ironic, quirky, fanciful, attempted humorous or out of context, they would illustrate that some of this book's concepts may also have some meaning elsewhere. All this without the quotes necessarily having fully logical meanings; and that other writings and interpretations of those quoted might be problematic. Generally, the quotes begin each chapter or book part, but also may be found within their components, or as actual parts of the chapter's narrative. Quotes may be as published texts, photos, artwork, cartoons, illustrations and drawings.

11 **Caution alert:** Readers may find this endnote group kind-of tedious and maybe avoidable:

12 Literature citations are intended, variously, to refer to
- comprehensive reference works;
- supporting reference works (sometimes introduced with "See:" or "See, for example:");
- specific examples.

Scientific, law and other literatures are referred to; and sometimes news reports, *etc.* about the science and law. Cited references do not reflect an exhaustive literature search – rather they are typically one or a few references from prominent sources as lead-in to the topic, with other references cited therein.

**13** Cross-references within this book may appear as among chapter footnotes and endnotes – mostly as footnotes.

**14** Citations appearing in footnotes and endnotes are to documents with published and searchable data. Internet locations or links are often provided and sometimes only Internet links. But Internet links – quite convenient – can go stale (not work anymore – for a variety of reasons) faster than the more traditional print references (and sometimes perhaps so fast that they might be regarded as fragile). Readers might then try to use data from a stale link, or topic keywords, to search *de novo.* (The author's experience from preparing this book's second edition, from the first edition, suggests that finding and correcting stale Internet links can be a kind of adventure of itself, partly because not only can a link just simply not work anymore but also can lead puzzlingly elsewhere. And, copying links by cutting and pasting is additional opportunity for error.)

**15** This second edition includes several Internet links to **YouTube** – those that are thought to be relevant, helpful and well-explained – and those that seem to have arguable meaning rather than nonsense and with titles that actually describe content. But as seems with YouTube more generally, some cautions are suggested – including:

- YouTube has its own terms of reference, restrictions and protocols.
- Some come with advertisements irrelevant to the topic linked.
- There is often scant literature reference including for artwork or diagrams. And artwork and diagrams sometimes seem to have only a casual, non-specific, relevance.
- There may have been YouTube editing since citation in this book.

**16** As a kind of author's notebook this book would include two kinds of biases:

- Some scientific methods – of lesser general application – and with less direct involvement with forensics or analytical chemistry – are presented (perhaps somewhat self-indulgently) because they arose from the author's law practice; or other experience or were otherwise of interest.
- Some topics, considered of lesser importance for this book, are presented, only very briefly, for mention, with references to other sources. This brevity may also reflect the bias of the author's limited knowledge of these topics.

**17** Quotes appearing in this book from elsewhere have been considered in terms of copyright status:

- Some, while still under copyright, are with permission of the copyright holder.
- Some would not – or no longer – be covered by copyright restrictions and now be in the *public domain.* https://en.wikipedia.org/wiki/Public_domain
- Some, while still under copyright are as *fair dealing* (Canada, UK, *etc.*) or *fair use* (USA) exceptions – with sometimes-difficult-to-determine criteria that allow brief quotation for purposes of education, criticism, review, *etc.* – when relevant to concepts being presented or argued within this book. With this book published as a textbook it has an educational character. Also, this book, including comment on public policy issues, dealt with in established literature – including newspapers and public government and judicial documents – has a critical and review character. It should also be noted that public policy comment has US *First Amendment*, and similar elsewhere, freedom of expression protection. In regard to *fair dealing* and *fair use*, the explicit statement of source, and brevity of the quote, are important. Also important is an expectation of no competition to the copyright holder's own sales efforts.
- https://en.wikipedia.org/wiki/Fair_dealing_in_Canadian_copyright_law https://en.wikipedia.org/wiki/Copyright_Act_of_1976 https://en.wikipedia.org/wiki/Fair_use

- That some brief text quotations would appear to have several respectable sources would suggest that copyright might not strictly apply.
- Quotation, briefly, of case law – court judgments – is a common practice for academic legal writing and is considered here as *fair dealing* and *fair use*. And similar for legislation, and for government reports. Important for these considerations is relevance to a legal or public policy issue and brevity of the quote.
- Quotation, briefly, from academic journal articles, academic texts and published standards is also considered here as *fair dealing* and *fair use*, depending on the context. Important for this is relevance to a legal or public policy issue and brevity of the quote.
- Quotation, briefly, of index lists from other publications, for purpose of discussion of concept, is considered here as *fair dealing* and *fair use*. Important for this is relevance to legal or public policy issues and brevity of the quote.
- Quotation, briefly, from published news and magazine items is also considered here as *fair dealing* and *fair use*, depending on the context. Important for this is relevance to a legal or public policy issue and brevity of the quote.
- Artwork is in a special category – all appearing here are with permission; or otherwise allowed by copyright law.
- (Some text quotes and artwork that the author would like to have used do not appear in this book (although some literature references are given) because of license expense, or because the actual copyright holder could not be adequately identified, contact data were too obscure or would appear to require too much heavy-duty bureaucratic process.) (For example, in Chapter 6, identifying the present copyright holder of the Leo O'Mealia drawings of Sherlock Holmes, Bell Syndicate (?), *Boston Globe*, 1930, was too difficult.)
- Molecular structural diagrams and technical drawings in this book that are by the author of this book are therefore without other copyright. This includes most of these diagrams and drawings.
- Copying in this book from Wikimedia Commons is as according to its terms, although those terms and identifying the author can be puzzling. https://commons.wikimedia.org/wiki/Main_Page

**18** US *First Amendmentt*
https://constitution.congress.gov/constitution/amendment-1/

**19** *Grundgesetz für die Bundesrepublik Deutschland* Artikel 5
https://www.gesetze-im-internet.de/gg/BJNR000010949.html
https://www.gesetze-im-internet.de/englisch_gg/englisch_gg.html#p0035
https://www.btg-bestellservice.de/pdf/80201000.pdf
https://digitalcommons.law.umaryland.edu/cgi/viewcontent.cgi?params=/context/mlr/article/2740/&path_info=48_2_247_free.pdf

**20** Such as related to my participations with the Division of Chemistry of the Law of the American Chemical Society. And with the Workers' Health and Safety Legal Clinic, Toronto, funded by Legal Aid Ontario.

**21** Sometimes, but not often, commercial products might be mentioned; if so see: Disclaimers. Error. Not legal advice. Commercial products. Introduction 1.3 and 4.2; Chapters 1, 5, 6 and 9.

**22** When literary fiction is quoted, or referred to, as example, it should be taken in that context; not as for actual modern advice. See the actual forensic science literature cited nearby.

**23** See also: P.D. James, *Talking About Detective Fiction*; Vintage Books, New York City, 2011, 2009. ISBN 978-0-307-74313-8.

**24** George Gamow, *Thirty Years that Shook Physics/The Story of Quantum Theory*, Anchor Books, Doubleday and Company, Inc., Garden City, New York, 1966; pages 13–15.

**25** Gamow, page 13. https://en.wikipedia.org/wiki/Thought_experiment

**26** Ursula K. Le Guin, "Schrödinger's Cat," *The Compass Rose*, Short Stories, BANTAM PAPER-BACK, Toronto, . . ., 1983; ISBN 0-553-23512-5; from *Universe 5*, 1974.
**27** Le Guin, page 46.
**28** https://www.merriam-webster.com/dictionary/gedankenexperiment
https://en.wikipedia.org/wiki/Thought_experiment
**29** See John Gribbin, *In Search of Schrödinger's Cat/Quantum physics and reality*, Bantam Books, New York City, 10036; 1984; ISBN 978-0-553-34253-6.
**30** – with the meaning of *Gedankenexperiment*, and grammatical use, perhaps stretched a bit here.
**31** – another word that might be included with this concept – often used in the context of legal literature: hypothetical.
**32** perhaps the use of this term in this book is noted too repetitively – but the author wants to try to make sure that readers are clearly informed as to when an example is *Gedankenexperiment* or from real and literature-reported observations. And similar for published fictional literature when used as example.
**33** Gribbin, pages 178–181.
**34** George Gamow, *Thirty Years that Shook Physics/The Story of Quantum Theory*, Anchor Books, Doubleday and Company, Inc., Garden City, New York, 1966; page 115, Fig. 25. "*Fig. 25. Bohr's ideal experiment which disproved Einstein's statement that the relation* $\Delta E\ \Delta t \geq h$ *is wrong.*" Illustrations by George Gamow. The Science Study Series edition: 1966. Copyright 1966 Educational Services Incorporated.
http://calteches.library.caltech.edu/2459/
https://www.pdfdrive.com/thirty-years-that-shook-physics-the-story-of-quantum-theory-d187158637.html
https://en.wikipedia.org/wiki/George_Gamow
**35** https://penguinrandomhouseeducation.com/book/?isbn=9780553342536
**36** https://en.wikipedia.org/wiki/Einstein%27s_thought_experiments#/media/File:Einstein's_light_box.svg
https://en.wikipedia.org/wiki/Einstein%27s_thought_experiments#
**37** Gribbin, page 179.
**38** Prokaryotic Caspase Homolog, "Einstein's light box," 2018; ". . . licensed under the Creative Commons Attribution-Share Alike 4.0 International license." https://upload.wikimedia.org/wikipedia/commons/e/e3/Einstein%27s_light_box.svg
**39** File:Einstein's light box.svg From Wikimedia Commons, the free media repository:
https://commons.wikimedia.org/wiki/File:Einstein%27s_light_box.svg
https://en.wikipedia.org/wiki/Einstein%27s_thought_experiments#
**40** Gribbin, pages 203–208.
**41** Ursula K. Le Guin, "Schrödinger's Cat," *The Compass Rose*, Short Stories, Bantam Paperback, Toronto, . . ., 1983; ISBN 0-553-23512-5; from *Universe 5*, 1974.
**42** *Gedankenexperiment*. Le Guin, page 46.
**43** https://penguinrandomhouseeducation.com/book/?isbn=9780553342536
**44** https://en.wikipedia.org/wiki/Schr%C3%B6dinger%27s_cat
https://en.wikipedia.org/wiki/Schr%C3%B6dinger%27s_cat#/media/File:Schrodingers_cat. svg
**45** Gribbin, cover.
**46** See John Gribbin, *In Search of Schrödinger's Cat / Quantum physics and reality*, Bantam Books, New York City, 10036; 1984; ISBN 978-0-553-34253-6; pages 203–208.
**47** File:Schrodingers cat.svg From Wikimedia Commons, the free media repository Author: Dhat-field
https://commons.wikimedia.org/wiki/File:Schrodingers_cat.svg
https://en.wikipedia.org/wiki/Schr%C3%B6dinger%27s_cat
**48** George Gamow, *Thirty Years that Shook Physics/The Story of Quantum Theory*, Anchor Books, Doubleday and Company, Inc., Garden City, New York, 1966.

**49** Gamow, pages 13, 14, 15.
**50** Sir James Jeans https://en.wikipedia.org/wiki/James_Jeans
https://www.britannica.com/biography/James-Jeans
http://applet-magic.com/rayleighjeans.htm
https://chem.libretexts.org/Bookshelves/Physical_and_Theoretical_Chemistry_Textbook_Maps/Supplemental_Modules_(Physical_and_Theoretical_Chemistry)
https://chem.libretexts.org/Bookshelves/Physical_and_Theoretical_Chemistry_Textbook_Maps/Supplemental_Modules_(Physical_and_Theoretical_Chemistry)/Quantum_Mechanics/02._Fundamental_Concepts_of_Quantum_Mechanics/Deriving_the_Rayleigh-Jeans_Radiation_Law
**51** *Jeans* cube – Gamow, page 13.
**52** See: Albert Einstein, *Relativity/The Special and General Theory* [in German 1916]/Translated. . . [English, 1920]; Penguin Books, New York City, 10014; 2006; ISBN 0-014-30982-2. [written, 1916, for an audience ". . . interested in the theory, . . . but . . . not conversant the mathematical apparatus of theoretical physics. . .".]
**53** Einstein, pages 13 and 14.
**54** DNA. Importance in forensic science. Electrophoresis. Introduction 2.3; Chapter 6.
**55** John D. Norton, "Chasing a Beam of Light: / Einstein's Most Famous Thought Experiment," Department of History and Philosophy of Science, University of Pittsburgh. "Copyright John D. Norton, December 2004. Rev. February 15, 2005. Reformatted April 14, 2005 on a transatlantic flight returning to Pittsburgh from the Israel Academy of Science and Humanities conference on Einstein".
https://sites.pitt.edu/~jdnorton/Goodies/Chasing_the_light/
https://sites.pitt.edu/~jdnorton/jdnorton.html
https://en.wikipedia.org/wiki/John_D._Norton
https://www.hps.pitt.edu/people/john-d-norton
**56** The intention is not to diminish the importance of DNA forensic science – it is very important. Rather, for this book, ink chemistry is thought to be a better example because of its simplicity. This also somewhat reflects an author's bias because of greater knowledge about ink chemistry.
**57** Sam Knight, "Hidden Traces/*How historical manuscripts are giving up their secrets," The New Yorker, Annals of Science*, 26 November 2018, pages 38→45.
**58** "*Annals of Science* November 26, 2018 Issue/Do Proteins Hold the Key to the Past?/*New methods are allowing a group of scientists to re-examine the world's libraries and archives, in search of the hidden lives of authors./*By Sam Knight/November 19, 2018." https://www.newyorker.com/magazine/2018/11/26/do-proteins-hold-the-key-to-the-past
**59** https://pmc.ncbi.nlm.nih.gov/articles/PMC8473418/
**60** https://en.wikipedia.org/wiki/Proteomics
**61** "*Melvin and Howard,*" 1980; Jonathan Demme, Director; with Jason Robards, Paul LeMat, Mary Steenburgen.
https://en.wikipedia.org/wiki/Melvin_and_Howard
https://en.wikipedia.org/wiki/Melvin_Dummar
**62** Katharine Q. Seelye, "*Melvin Dummar, 74. Dies; Claimed Howard Hughes Left Him Millions,*" *The New York Times*, Obituaries, 13 December 2018, page B14.
**63** Robert Harris, *Selling Hitler/the story of the Hitler diaries*, Faber and Faber, London, 1986; ISBN 0-571-14726-7.
**64** Clay Risen, "*Gerd Heidemann, Journalist Duped by Fake Hitler Diaries, Dies at 93,*" *New York Times*, Obituaries, 13 December 2024.
https://www.nytimes.com/2024/12/13/business/media/gerd-heidemann-dead.html
**65** See:

- David Samuels, "DEPT. OF DISPUTATION/FAKES/*Who forged the J.F.K.-Marilyn Monroe papers?*", *The New Yorker*, 03 November 1997, pages 62–75.

- Benjamin Weiser, "Charges of Forgery and Betrayal/Friendship and Trust Lie at Heart of Case Challenging Validity of Kennedy Papers," *The New York Times*, Monday, 12 April 1999, pages B1 and B4.

**66** https://en.wikipedia.org/wiki/John_F._Kennedy_document_hoax
https://www.latimes.com/archives/la-xpm-1997-sep-26-mn-36463-story.html

**67** Kathryn Hughes, "Fakes and fortunes: has the time come to forgive literary forger Lee Israel?," *The Guardian*, [2019].
https://www.theguardian.com/books/2019/jan/14/fakes-and-fortunes-has-the-time-come-to-forgive-literary-forger-lee-israel-

**68** Lee Israel, *Can You Ever Forgive Me?: Memoirs of a Literary Forger*, Simon and Schuster; Reprint, 2015; ISBN-10: 9781416588689; ISBN-13: 978-1416588689
https://www.amazon.ca/Can-You-Ever-Forgive-Me/dp/141658868X

**69** Richard Brody, *The New Yorker*, AT THE MOVIES, [*Can You Ever Forgive Me?*], 22 October 2018, page 16.

**70** https://en.wikipedia.org/wiki/Lee_Israel
https://en.wikipedia.org/wiki/Can_You_Ever_Forgive_Me%3F

**71** Mike Cummings, "Analysis unlocks secret of the Vinland Map – it's a fake" / "Yale conservators and conservation scientists have found that the map, once hailed as the earliest depiction of the New World, is awash in 20th-century ink," *Yale News*, 1 September 2021.
https://news.yale.edu/2021/09/01/analysis-unlocks-secret-vinland-map-its-fake

**72** https://www.smithsonianmag.com/history/medieval-map-of-north-america-identified-as-20th-century-forgery-180978751/
https://www.nytimes.com/2021/09/30/us/yale-vinland-map-fake.html

**73** https://en.wikipedia.org/wiki/Vinland_Map
https://news.yale.edu/2018/02/28/yale-putting-high-tech-tests-its-controversial-vinland-map

**74** *Petit bleu* of the French Post Office, used for the Paris *pneumatic* tube.

**75** Shirer, page 40, et seq.

**76** See: William L. Shirer, *The Collapse of the Third Republic/An Inquiry into the Fall of France in 1940*, POCKET BOOK EDITION, Simon and Schuster of Canada, Ltd., Richmond Hill, Ontario; 1969, 1971; Standard Book Number 72-91306; Book One/3 "The Dreyfus Affair/1894-1906," page 26, *et seq.*

**77** https://www.affairedreyfus.com/2014/02/book.html
https://www.amazon.ca/Dreyfus-Politics-Emotion-Scandal-Century/dp/0312572980

**78** See: Robert Harris, *AN OFFICER AND A SPY*, [an historical novel: "Author's Note/. . . the true story of the Dreyfus affair, perhaps the greatest miscarriage of justice and political scandal in history, which in the 1890s came to obsess France and the ultimately the entire world . . ." – as a narrative of Colonel Georges Piquart], HUTCHINSON, Random House, London, SW1V 2SA; ISBN 978-0-09-194456-8; 2013.

**79** ". . . I accuse the three handwriting experts. . .". Robert Harris, quoting Émile Zola, "J'ACCUSE . . . !" Paris, *L'Aurore*, 13 January 1898; page 372, et seq.; at page 374.

**80** ". . . I accuse the three handwriting experts, Messrs. Belhomme, Varinard and Couard, of submitting reports that were deceitful and fraudulent, unless a medical examination finds them to be suffering from a condition that impairs their eyesight and judgement. . .".
https://www.marxists.org/archive/zola/1898/jaccuse.htm

**81** https://en.wikipedia.org/wiki/J%27Accuse . . .!
https://en.wikipedia.org/wiki/%C3%89mile_Zola
https://en.wikipedia.org/wiki/Dreyfus_affair

**82** It should be noted that the Dreyfus Affair had a prolonged and profound impact on the Third Republic and indirectly far-reaching influence on the foundation of the state of Israel. And prompted what may be the most famous editorial of-all-time.

**83** *". . . J'accuse les trois experts en écritures, les sieurs Belhomme, Varinard et Couard, d'avoir fait des rapports mensongers et frauduleux, à moins qu'un examen médical ne les déclare atteints d'une maladie de la vue et du jugement . . .,"* Émile Zola, "J'ACCUSE . . .!" Paris, *L'Aurore*, 13 January 1898.
**84** Roger Cohen, *"France Moves to Atone by Elevating Alfred Dreyfus as Antisemitism Spreads* / The National Assembly voted to promote the Jewish army captain to brigadier general more than 130 years after he was falsely convicted of espionage and imprisoned in French Guiana," *New York Times*, 03 June 2025. https://www.nytimes.com/2025/06/03/world/europe/france-antisemitism-dreyfus.html
**85** ". . . *I accuse the three handwriting experts* . . ." Robert Harris, quoting Émile Zola, "J'ACCUSE . . .!" Paris, *L'Aurore*, 13 January 1898; at page 374.
**86** https://www.amazon.ca/AFFAIRE-DREYFUS-L-%C3%89MILE-ZOLA/dp/2760616371
**87** See *C&EN*, 12 August 2002, page 35; and *ibid.*, letters, 21 October 2002, page 8.
**88** See: "The Old Woodstock Typewriter," as Chapter 19, page 355 *et seq.*, in John Chabot Smith, *Alger Hiss/The True Story*, Penguin Books, 1977.
**89** http://algerhiss.com/history/the-hiss-case-the-1940s/the-typewriter/the-serial-number-1978/ http://algerhiss.com/history/the-hiss-case-the-1940s/the-typewriter/forgery-by-typewriter/
https://www.nytimes.com/2008/09/12/nyregion/12tytell.html?mtrref=www.bing.com&gwh=76E5432166D43FD18A67DB48764B0388&gwt=pay
**90** [with one missing]
**91** Victor Mather, *"A Ten Commandments Tablet Goes Up for Auction, and Questions Abound* / Bids may rise to a million or more, Sotheby's estimates, though some experts say that fakes and forgeries abound," *New York Times*, 12 Dec. 2024.
https://www.nytimes.com/2024/12/12/arts/ten-commandments-stone-tablet-sothebys-auction.html
**92** Michael Levenson, "A Ten Commandments Tablet Sells for More Than $5 Million at Auction / Sotheby's called the stone the world's oldest inscribed with the Ten Commandments, but some experts had raised questions about its provenance and authenticity," *New York Times*, 18 December 2024.
https://www.nytimes.com/2024/12/18/us/sothebys-ten-commandments-tablet.html
**93** See below:
**94**
- Brian H. Kaye, *Science and the Detective / Selected Reading in Forensic Science*, VCH Verlagsgesellschaft mbH, Weinheim, Germany, 1995, pages 306 and 307, and references cited therein.
- Mary Virginia Orna, ed., *Archaeological Chemistry/Organic, Inorganic Biochemical Analysis*, ACS Symposium Series 625, American Chemical Society, Washington, D.C., 1996:
- Alan D. Adler, "Updating Recent Studies on the Shroud of Turin," as Chapter 17;
- D.A. Kouznetsov, A.A. Ivanov and P.R. Veletsky, "A Re-evaluation of the Radiocarbon Date of the Shroud of Turin Based on Biofractination of Carbon Isotopes and a Fire-Simulating Model," as Chapter 18;
- A.J.T. Jull, D.J. Donahue and P.E. Damon, "Factors That Affect the Apparent Radiocarbon Age of Textiles," as Chapter 19; and
- references cited therein.
- David Va Biema, with Andrea Dorfman, Greg Burke and Martin Penner, "SCIENCE AND THE SHROUD/The relic was declared a fake a decade ago, but millions are expected to venerate it, inspired by those who say there is truth to back their faith," *TIME*, Canadian edition, RELIGION, 20 April 1998, VOL. 151, NO.15, cover, page 1 and pages 37–44.

**95**
- Ian Wilson, *The BLOOD and the SHROUD/New Evidence That the World's Most Sacred Relic Is Real*, The Free Press, Simon and Schuster, New York City, 10020; 1998; ISBN 0-684-85359-8.
https://www.loot.co.za/product/ian-wilson-the-blood-and-the-shroud/kbmb-476-g610
https://www.loot.co.za/product/ian-wilson-the-blood-and-the-shroud/kbmb-476-g610

**96** – Ian Wilson, *The BLOOD and the SHROUD* . . ., 1998, Part 4/Carbon Dating:/Right or Wrong, page 179, *et seq.* At page 179: [AD 1260 → AD 1390 – or – AD 1?]

**97** – Robert A. Rucker, "Summary of Shroud Research," 10 June 2019. http://www.shroudresearch.net/hproxy.php/One-Page-Summaries.pdf

**98** Some Problematic Content: . . . Rucker, "Summary . . ." [re neutron radiation].

**99** Cited literature accuracy. Introduction 2, Introduction 4.3, Chapter 2.

**100** – Adam Lusher, "628-year-old fake news: Scientists prove Turin Shroud not genuine (again)/Forensic analysis of possible bloodstains suggest marks could only have been made by someone adopting different poses, not dead Messiah lying still in tomb before the resurrection," *Independent*, UK, 16 July 2018.
https://www.independent.co.uk/news/world/europe/turin-shroud-latest-fake-forgery-scientific-blood-pattern-spatter-study-carbon-dating-debunked-a8450101.html

**101** – Myra Adams "The Shroud of Turin, Authenticated Again," *NATIONAL REVIEW*, 16 April, 2016.
https://www.nationalreview.com/2016/04/shroud-turin-jesus-christ-blood-relic-sudarium-oviedo/
https://www.nationalreview.com/

**102** and sometimes there would be a kind-of digressing, mostly as endnotes, but still hopefully helpful for understanding. And, sometimes – not too often – just digressing because it would lead to something interesting to some readers – that others can skip over. (And, perhaps some of this digressing would be also be a bit of author-self-indulgence.)

**103** SI = *Le Système International d'Unités Convention du Mètre*, Paris, 1875.

**104** See: Richard Saferstein, *Criminalistics/An Introduction to Forensic Science*, Seventh Edition, Prentice Hall, Upper Saddle River, New Jersey, 07458; 2001; ISBN 0-13-013827-4.

**105** Saferstein, 2001; page 87→89.

**106** See also: J.A. Glenn, *An Introduction to SI Units for students of science and technology*, Heinemann Educational Books Ltd; London W1X 8AH; 1969, 1970; ISBN: 0-435-68170-2.
https://www.gettextbooks.com/isbn/9780435681708/

**107** See also: https://www.cbc.ca/archives/when-a-metric-mix-up-led-to-the-gimli-glider-emergency-1.4754039
https://en.wikipedia.org/wiki/Gimli_Glider

**108** *Weights and Measures Act* RSC 1985, c. W-6, as am. http://laws-lois.justice.gc.ca/eng/acts/W-6/page-1.html

**109** The general convention as typically used, at first use, at various places in this book, is that a substance would be identified by a common chemical name, followed by its symbols, CAS Registry number, with a structural diagram near-by. Thus, for example: "water $H_2O$, . . .". Subsequent use might simply use the common name or symbols. http://physics.nist.gov/cuu/Units/introduction.html
http://physics.nist.gov/cuu/Units/international.html

**110** https://ulstandards.ul.com/wp-content/uploads/2014/12/ULMetricPolicyManual_Dec2014.pdf

**111** For example: millimetres of mercury – mm Hg – still used to report blood pressure.
https://www.ncbi.nlm.nih.gov/books/NBK279251/
https://en.wikipedia.org/wiki/Millimetre_of_mercury

**112** While others are forgotten – conveniently lost to obscure history. 1 Gillette = energy for a LASER beam to penetrate one Gillette brand razor blade ≈ 1.5 joule ≈ 0.3586 calories.
https://www.wired.com/2011/12/st-3st-lasers/
https://en.wikipedia.org/wiki/Gillette#:~:text=Revolution%20in%202025.-In%20science,can%20burn%20through%20four%20blades

**113** 1 Atmosphere ≈ 760 mm Hg ≈ 33.8985 feet $H_2O$
https://en.wikipedia.org/wiki/Standard_atmosphere_(unit)

**114** See also: barn = $10^{-24}$ $cm^2$
https://www.britannica.com/science/barn-measurement
**115** **Criticisms in a *free and democratic society*:** At several places in this book, the author criticizes some Canadian legislative institutions and government agencies and judicial appreciations of science. Most particularly Health Canada – in relation to the *CD&S Act.* Some explanation of context is appropriate. The criticisms are as citizen offerings, in good faith, in a *free and democratic society*, for improvement. The valid necessity of a national health agency to formulate and execute policies is not being challenged; agency good faith is assumed (presumption of regularity}; and much of its work is valid and valuable. But government is not above legitimate criticism including of administrative structure. Citizens are entitled (if not obligated) to opine about government wrong direction, administrative process, policy, and error; and to offer correction. And, similarly for judicial knowledge of science.
**116** See below, for example:
**117** Isaac Asimov, *Asimov on Chemistry*, Anchor Books, Doubleday, Garden City, New York, 1975; ISBN 0-385-04005-9; (Chapters originally appearing as an essays in *The Magazine of Fantasy and Science Fiction*).
**118** Isaac Asimov, *Asimov's Guide to Halley's Comet*, A DELL TRADE PAPERBACK, New York City, 10017; 1985; ISBN 0-440-50434-1; pages 18–21.
**119** Isaac Asimov, *I. Asimov/A Memoir*, Bantam Books, New York City 10036, 1994; ISBN 0-553-56997-X.
**120** https://en.wikipedia.org/wiki/Isaac_Asimov
**121** George Gamow, *Thirty Years that Shook Physics/The Story of Quantum Theory*, Anchor Books, Doubleday and Company, Inc., Garden City, New York.
**122** George Gamow, *One Two Three. . . Infinity/Facts and Speculations of Science*, Bantam Books, New York City, 10016, April 1967; copyright 1947, 1961.
**123** Richard P. Feynman, *The Meaning of It All/Thoughts of a Citizen Scientist*, Perseus Books Group, Cambridge, Massachusetts, 02412; Basic Books, New York City, 10016-8810; ISBN-13: 978-0-465-02394-3; ISBN-10: 0-465-02394-0; 1963, 1998.
**124** Richard P. Feynman, *SIX EASY PIECES / ESSENTIALS OF PHYSICS. . .*, BASIC BOOKS, New York City, 10016-8810; 1963, 2011; ISBN 978-0-465-02527-5.
**125** Richard P. Feynman, *SIX* NOT SO *EASY PIECES / EINSTEIN'S REALITIVITYLY, SYMMETRY, AND SPACE-TIME*, BASIC BOOKS, New York City, 10016-8810; 1963, 2011; ISBN 978-0-465-02528-2.
**126** Richard P. Feynman, *"Surely You're Joking, Mr. Feynman!" / Adventures of a Curious Character / as told to Ralph Leighton / edited by Edward Hutchings, BANTAM BOOKS*, New York City, 1986. ISBN 0-553-25649-1.
**127** https://www.sciencenews.org/blog/context/top-10-richard-feynman-quotations
https://en.wikiquote.org/wiki/Talk:Richard_Feynman
**128** *Quod volimus credimus libenter* ≈ believe what we want to believe.
**129** *Gedankenexperiment.* Introduction 2.2.
**130** To borrow a phrase from the *Canadian Charter of Rights and Freedoms* (being Schedule B, Part I of the *Canada Act* 1982, including the *Constitution Act*, 1982 (1982 c. 11 (United Kingdom)), s. 1.
**131** (Latin) *Stare decisis* = to stand by things decided.
https://www.americanbar.org/groups/public_education/publications/preview_home/understand-stare-decisis/
**132** Perhaps in analogy, in concept, but not quite exactly, to the "full faith and credit" clause of the US Constitution – Article IV, Section 1

[. . .]

> Full Faith and Credit shall be given in each State to the public Acts, Records, and judicial Proceedings of every other State. And the Congress may by general Laws prescribe the Manner in which such Acts, Records and Proceedings shall be proved, and the Effect thereof.
>
> [. . .]

**133** https://www.constitutionalstudies.ca/2019/07/supremacy-of-parliament/

**134** That is, science that humans have understood so far is explainable – this does not mean to say that there are not aspects of science that are yet to be understood. For purposes of this book, every human-declared understanding of science as declared so far must have an explanation that an informed listener is entitled to. It should be accepted that some explanations persist with controversy.

**135** See: Steven Weinberg, *DREAMS OF A FINAL THEORY*, Vintage Books, New York City, 1992, 1993, 1994. ISBN 978-0-679-74408-5.

**136** See: Lee Smolin, *The Trouble with Physics/The Rise of String Theory, the Fall of a Science, and What Comes Next*, Houghton Mifflin Company, Boston & New York City 10003; 2006; ISBN-10: 0-618-91868-X; Chapter 17, "What Is Science?" at page 301.
https://en.wikipedia.org/wiki/The_Trouble_with_Physics#:~:text=The%20Trouble%20with%20Physics:%20The,Time%20Reborn%20(2013)

**137** http://royalsociety.org/ "The Royal Society's motto 'Nullius in verba' roughly translates as 'take nobody's word for it'. It is an expression of the determination of Fellows to withstand the domination of authority and to verify all statements by an appeal to facts determined by experiment."

**138** Edward Neville da Costa Andrade, *Sir Isaac Newton/His Life and Work*, Doubleday Anchor, Garden City, New York; Science Studies Series, S42; reprinted from Macmillan Company, New York, and W. Collins Sons and Co, Ltd., London, first published 1954; page 60.

**139** Canadian, US and Commonwealth literature and law sometimes use concepts from earlier times expressed in Latin; often concisely convenient, if not poetic. Other-than-English words and phrases are used in this book as though their English meanings are known; and they are easily known because they will be defined or easily searched. Thus lawyers get to mis-use, and mis-pronounce, Latin in general ignorance of that language.

**140** *Italics* are used in this book to indicate a somehow special or unusual word usage – often as coming from another language.

**141** https://en.wikipedia.org/wiki/List_of_Latin_phrases
https://en.wikipedia.org/wiki/List_of_Latin_phrases_ (full)

**142** Latin. Introduction 4.1.

**143** *Nullius in verba* ≈ We do not take anybody's word for it.

**144** See: Hannah Fry, "Experiments on Trial/Has the new era of experimentation remembered the lessons of the old?" *The New Yorker*, BOOKS, 02 March 2020, pages 61–65, at page 62.

**145** *ipse dixit* ≈ he himself said it – an assertion without proof.

**146** http://legal-dictionary.thefreedictionary.com/Ipse+Dixit = ". . , [Latin, He himself said it.] An unsupported statement that rests solely on the authority of the individual who makes it. . .".

**147** https://en.wikipedia.org/wiki/Ipse_dixit = ". . . Ipse dixit (Latin for 'he himself said it') is an assertion without proof; or a dogmatic expression of opinion . . . The fallacy of defending a proposition by baldly asserting that it is "just how it is" distorts the argument by opting out of it entirely: the claimant declares an issue to be intrinsic, and not changeable . . . Cicero (106–43 BC) . . . De Natura Deorum. . .".

**148** Pamela R. Metzger, "Cheating the Constitution," Vanderbilt Law Review, 59 Vand. L. Rev. 475 (2006).
https://scholarship.law.vanderbilt.edu/cgi/viewcontent.cgi?article=1589&context=vlr

**149** Pamela R. Metzger, Robert A. Ainsworth Professor in the Courts, Tulane University Law School, New Orleans, Louisiana, 70118.

**150** See, for example: David Epstein, "The DIY Scientist, the Olympian, and the Mutated Gene/How a woman whose muscles disappeared discovered she shared a disease with a muscle-bound Olympic medalist," *ProPublica* and *This American Life*, January 2016; August 2020.
https://www.propublica.org/article/muscular-dystrophy-patient-olympic-medalist-same-genetic-mutation
https://www.thisamericanlife.org/577/something-only-i-can-see
https://www.amazon.com/The-Sports-Gene-Extraordinary-Performance/dp/161723012X

**151** https://diyscientist.blog/
See: Steven Weinberg, *DREAMS OF A FINAL THEORY*, Vintage Books, New York City, 1992, 1993, 1994. ISBN 978-0-679-74408-5.

**152** See: Weinberg, *DREAMS* . . .

**153** Disclaimers – this book – *caveat emptor* – *caveat lector* – *caveat auditor* (Readers should apply their own critical appraisals):

- Although the author has taken care for accuracy, errors and misstatements happen. Timely corrections would be appreciated.
- Much of this book relies on other literature, but readers should be aware that, no matter how carefully that literature is chosen, it too can be flawed, or misunderstood.
- A few of the cited literature items may be of variant controversial opinion, but are relevant and apparently written by their authors in good faith.
- A few of the cited literature items may be included only because they are relevant, perhaps only as a negative example; they would be noted as for example: "NOT RECOMMENDED HERE: Velikovsky, *Worlds* . . .". Introduction 4.2; Chapter 2.
- A few may appear to contain correct analysis, but with problematic parts; they would be noted as for example: "SOME PROBLEMATIC CONTENT: . . . Rucker, 'Summary . . .' [re neutron radiation]". Introduction 4.2; Chapter 2.
- Hopefully, literature of fraudulent intent has been avoided.
- What is written in this book is hopefully correct and up-to-date according to its context.
- Opinions offered in this book are in a US *First Amendment* context.
- This book – as information or opinion – does not offer or provide legal advice, nor should it be taken as, or considered as, legal advice; nor as any other formal professional advice. Legal advice should be obtained from professional, licensed counsel, explicitly engaged for that purpose, on a lawyer-client basis. In the reality of Canada's everyday world professional legal advice for the poor can be hard to find, although there are public-service legal systems and clinics that can be helpful.
- Biblical quotes are intended in a cultural/literature/historical rather than religious context.
- Mention in this book of commercial products, services, lines, manufacturers, suppliers, *etc.* are by way of example only and do not imply approval, endorsement or evaluation, although the products would appear as reliably reputable. Mention here is for reader convenience only; readers should research and decide for themselves.

**154** *Caveat emptor* = let the buyer beware.
= *Caveat emptor, quia ignorare non debuit quod jus alienum emit* = Let a purchaser beware, for he ought not to be ignorant of the nature of the property which he is buying from another party.
*Caveat lector* = let the reader beware.

https://www.merriam-webster.com/dictionary/caveat%20emptor
https://en.wikipedia.org/wiki/Caveat_emptor https://en.wikipedia.org/wiki/Caveat_emptor#Caveatlector
https://en.wiktionary.org/wiki/caveat_lector
https://dictionary.cambridge.org/dictionary/english/caveat
https://www.investopedia. com/terms/c/caveat.asp
**155** *Caveat auditor* = . . . listener beware https://www.investopedia.com/terms/c/caveat.asp
https://en.wikipedia.org/wiki/Caveat_emptor

# Chapter 1
# Science and law

“. . . criminal courts do not necessarily deal in truth, but they deal in evidence . . .” [1].[1]

“. . . I never for a second thought this could happen – that an innocent person could be found guilty. It was an atrocity, that this could happen . . .” [2].

[. . .]

“But if . . . was the traitor . . .?”
“. . . case is over . . . court . . . verdict . . . that is the end of that”.

[. . .]

[Robert Harris, *AN OFFICER AND A SPY*, [historical novel]] [3, 4].[2]

## 1.1 Forensic science

### 1.1.1 Truth

*Scepticism is the first step towards truth.*

Denis Diderot, *Pensées philosophiques* [1746] [5, 6]

This book deals with the theory, observations, methodology, documentation and reporting – and culture – of the systematic search for truth [7][3,4] about human stories,[5] as assisted by chemical science, in relation to the law (and, sometimes, apparently versus the law). As a particular forensic investigation would progress, theories of what happened would be developed, often starting with a site search,[6] and continuing into a scientific laboratory, and perhaps eventually concluding in a courtroom [8, 9, 10] or before an administrative tribunal, or other official panel [11, 12, 13, 14, 15, 16]. And then, perhaps occasionally, become famous, to be argued about in later eras by historians.

Consider as a fictitious example [17] that someone is accused of forging a signature, on a document, using an eventually identifiable pen and ink.[7] For legal purposes, handwriting, ink and video surveillance of the writing can all come into evidence.

1 Introductory quotes. Introduction 1.2.
2 Captain Alfred Dreyfus. *bordereau, petit bleu*. Colonel Hubert Henry. Introduction. 2.3; Chapter 1.
3 Forensic science. Introduction. 2; Chapters 1, 2 and 5.
4 Legal/scientific truth. Chapters 1, 8 and 11.
5 Analytical chemistry and forensically related human story. Chapters 1 and 8.
6 Site. *What it is*. Chapters 1, 2, 3 and 5.
7 Variations of ink analytical chemistry – as examples – *Gedankenexperiment* – continue throughout this book. Introduction 2.2 and 2.3.

https://doi.org/10.1515/9783111618555-002

But, a statement by the accused – admitting crime – might not come into evidence if there were not a legal caution about the rights to silence and legal advice [18, 19].

Legal systems often deal with such scenarios in a seeking of legal truth – seeking justice. Scientific truth is sometimes sought as an aid – this is forensic science. But, these two truths may not be quite the same. That the disciplines of science and law may be called upon in addressing the same kinds of situation has attracted comment [20].

To continue the example [21]: Even if a forensic chemist could confidently [22] identify the ink on the document as having come from the pen found in possession of the accused, such scientific truth might not be allowed as evidence in court. Perhaps the judge would rule it inadmissible if the search was unlawful [23]. Or perhaps the prosecution would fail if a statement of admission was made after accusation, but before the right-to-remain-silent caution.

To further continue the example [24]: Crown counsel [25], perhaps too sure with scientific ink match evidence, fails to call corroborating handwriting and video identification evidence; an acquittal might result because the judge might find a reasonable doubt[8] – the criminal law test. Or, alternatively, Crown counsel forgets to ask an eyewitness to identify the accused in court. This legal truth of "not guilty" might seem in contradiction with the scientific truth; but because of the policy intent of the criminal law rules of engagement, it runs parallel – but not identical – with the scientific truth.

A variation of the example [26]: Suppose that the ink match is allowed in evidence, but the accused testifies that signing was on behalf of a sometimes quite forgetful (now former) friend, with permission. And, suppose the judge accepts this as allowing for a reasonable doubt for the alleged criminal intention – *mens rea*[9] – and acquits on that basis. In this variation, the scientific and legal truths are not even in apparent contradiction because the science has not dealt with that part of the human story where the *moral focus* would be found.

### 1.1.2 Forensic science

*Forensic*[10,11] is a Latin[12] derivation related to ancient Roman public, governmental, legal, *etc.* processes in the *forum* [27] = market place [28, 29, 30]. The term is now associated with court-related and legal processes. *Forensic science* is the science application thereto.

---

**8** Reasonable doubt. Science explanation. Introduction 4.1; Chapters 1.1.1, 10.7, 10.8 and 11.11.

**9** *mens rea.* = mental element – criminal intent. Chapters 1 and 11.

**10** Forensic science. Introduction 2; Chapters 1, 2 and 5.

**11** Analytical chemistry. Expanded definition – more detail, and example. Introduction. 2; Chapters 1, 2, 3, 5 and 8.

**12** Latin. Introduction 4.1.

*Forensics*, as used in this book [31] can include all science that is within the law:
- Legislation [32]
- Enforcement
- Scientific determination
- Determination of fact, law and legal concepts by courts and tribunals

For example [33] – a sequence of *forensic*-science-related activities:
- Public discussion, politicking, editorializing and lobbying about an ink or art supply,[13] for school use – thought by some to contain toxic components – thought by others to be harmless or reasonably manageable[14] – and still others blind to toxicology issues.
- Parliamentary committee hearings – with witnesses, advocates and paid lobbyists.
- Parliament debating and enacting legislation [34] to restrict that ink or art supply, when for school use; and imposing labelling requirements; or not.
- A government inspector investigating a particular school, a routine inspection or from a citizen complaint.
- A scientist working to identify the ink or art supply found at that school by the inspector – to determine what is in it; and intended for evidence; forensic report to be issued.
- Environmental measurements at the school.
- Analytical chemical medical measurements of the students, *etc.*; with informed consent; without consent.
- A government legal counsel determining to prosecute, or not, according to the legislation.
- A trade union leader, representing the art teachers, determining to proceed, or not, with a workplace health and safety grievance, under the collective agreement.
- And similarly for the housekeeping workers – of a different union and collective agreement [35].
- Court and labour tribunal hearings (with legal counsel from all sides) and judgments on the matters – if it goes that far.
- Students refusing to go to class out of concern about possibly toxic materials; debates at student council meetings.
- A student parents group seeking a court injunction.
- A workers' compensation claim, years later, of a housekeeping worker.
- A standards agency publishing an engineering or public health standard.
- Reference to US CFR for engineering or public health requirements [36].

---

**13** Art supply. Forbes art pigment collection. Chapters 1; 2; 3.

**14** ASTM. Crayola label. Chapters 1, 5 and 9.

**This book is mostly concerned with the scientist's item- or site-specific part of the forensic process – mostly as analytical chemistry.**

The term ***criminalistics*** [37, 38, 39] is often appropriately used because of the frequency of forensic science in criminal law, but is avoided in this book because of the author's approach[15] that science as applied to law includes rather more than only the criminal law and its enforcement. And, sometimes an investigation does not initially reveal whether or not an occurrence was a crime – as in the 1996 TWA Flight 800 tragedy [40, 41] – and which government agency is the most appropriate to lead the investigation – the US National Transportation Safety Board or the Federal Bureau of Investigation.

Forensics draws from very many fields – an open-ended list of science and other pursuits – Thus: *forensic*: *engineering, medicine, accounting, art, optometry* [42], *knot tying* [43, 44], *font* [45, 46, 47],[16] *musicology* [48], *literature criticism, bird watching, cigar culture* [49], tattoos [50], *etc.* Just about any human skill or interest can have *forensic* applications and can become a *forensic science*, perhaps to the surprise of the practitioner – for example, a literature professor asked to help identify a writer by language usage characteristics [51, 52]. This book deals mostly with a narrower range of forensic science possibilities – that mostly involve analytical chemistry.

## 1.2 Who decides

In the search for legal truth, it is the judge [53, 54, 55, 56] who, according to the applicable rules and procedures, would eventually determine or confirm the *moral focus*, in the various versions of the story placed before the court. In doing so, a scientific truth might be found as not sufficiently part of the *moral focus* so that for legal purposes, the validity of the scientific evidence and arguments would not be found sufficiently relevant.

However, if the judge is persuaded that a scientific truth would be part of the *moral focus*, then whether or not the scientific evidence and arguments are correct must be dealt with. This is often obvious and assumed for non-novel [57, 58, 59, 60, 61, 62][17] – well-known and accepted – concepts. For example: That fingerprints, competently observed and evaluated are a super-excellent personal identifier.[18] But, otherwise, what of opposing sides' expert testimonies, if any, should be accepted as the *scientific truth* is still the job of the judge. That it would appear that some Canadian judges

15 Author's biases. Introduction 1.2, 2, 4.1 and 4.2; Chapters 10 and 11.

16 Font. Chapters 1 and 7.

17 Manarin, *Forensic Evidence. . .*, 2019. Chapters 1–3.

18 Fingerprint probability. Chapters 1 and 4.
Fingerprint visualization. Fluorescence and phosphorescence. Chapters 6 and 9.
Fingerprint when. Chapters 7 and 8.

are not comfortable with an appreciation of at least high school level science and mathematics suggests that their gatekeeping might be problematic.[19,20]

Thus, the judge is the determiner of both the legal truth and the scientific truth – the gatekeeper for the scientific thought that is allowed into the legal process.

But, what if the legal truth is determined incorrectly because the scientific "truth" was wrong?

Wrongly decided cases can be reconsidered in an appeal court system. However, since that system is mainly concerned with correcting errors of law (such as for jurisdiction and procedure), rather than errors of fact, a wrongly determined (by a trial court [63]) scientific truth may simply stay wrong within the legal system. A theory behind this is that all sides had their chance to bring out the truth of facts to the trial court, and as long as the proper legal procedure was followed, the appeal courts would be reluctant to interfere.

This means that an opposing side which fails in challenging wrong science, for whatever reason, may be stuck with it, if the court accepts it. This might indicate failure in challenging the credentials of a biased or wrongly directed expert or of an incompetent or fraud. But sometimes, it can be a failure of the legislation that imposed the defective science on the litigation process. And sometimes, a failure of the court or tribunal.

## 1.3 Justice and failed justice

[. . .][21]

. . . "But . . . justice . . ."
. . . ". . . an English trial . . . fair a thing as can be."
". . . finest system . . . can make a mistake," . . . ". . . in the hands of men . . . fallible."

[. . .]

[Agatha Christie, *Ordeal by Innocence*, 1958] [64]

**If a law case would be determined on a wrong determination of scientific truth – is it still justice?**[22]

**Maybe yes.** It might be justice in the sense that all would have had their fair chance [65] to present their version in a fair system of legal procedure. Unlike scientists, whose declarations of scientific truth are always open to doubt and revision again and again over the years, decades and centuries, and whose declarations might include that there are not enough data for there to be a decent decision, judges often must decide and declare their view of legal truth relatively quickly, one way or another, even sometimes in the

19 Author's biases. Introduction. 1.2, 4.1 and 4.3; Chapters 10 and 11.
20 re criticism of Health Canada, Judges et al. Introduction 4.1 and 4.3; Chapters 10 and 11.
21 Introductory quotes. Introduction 1.2.
22 Ethics. Lawyers. Experts, *etc*. Introduction; Chapters 1, 5 and 8.

absence of truly compelling evidence so that all involved can get on with the business of their lives. And, a judge is typically *functus officio* [66, 67][23] after deciding a case. There is a concept of *res judicata* = a thing adjudged; that a legal matter has already been decided – not to be considered again [68, 69, 70]. In this sense, unless it is so very important-enough for a special revisitation, regrets would be left to historians to argue about.[24,25]

**But, then again, maybe not.** It might not be justice if the "scientific truth" as accepted by the judge is too much at odds with generally accepted, and essentially undisputed, views of the scientific community. For example, if the trial judge accepts as true the "junk science" [71][26] of an incompetent or fraudulent "expert", and if the appeal courts do not interfere, then justice may have failed.

This becomes most serious for criminal law, where there may be grave moral implications. Canadian [72, 73, 74, 75, 76, 77, 78, 79, 80, 81, 82, 83, 84, 85, 86, 87, 88, 89, 90, 91, 92][27], British [93, 94, 95, 96, 97, 98, 99, 100, 101], and US [102, 103, 104, 105, 106, 107, 108, 109, 110, 111, 112, 113, 114, 115, 116, 117][28] histories are replete with controversaries about serious injustices at law; for some of which there was actual innocence, for some procedural error, for some facts incorrect and some remaining as for controversial arguments with voluminous literature.

Some of these controversies and injustices may be because of innocent error by scientists, lawyers, judges, police, parliamentarians, *etc*. But some because of sloppiness or a culture of sloppiness; or because of failure to strive for proper procedure. And some because of wilful wrongdoing. Some wilful wrongdoing includes bigotry – both specific and systemic – such as the Dreyfus Affair.[29] And, some in the context of long-standing regional bigotry such as the trial at Scottsboro, Jackson County, Alabama, 1931, *et seq.* [118, 119, 120].

These cases should clearly warn that scientists, prosecutors, judges and police can be wrong, if not oblivious to their limited abilities to determine physical reality [121], and that flaws of the "justice" system itself may impede their abilities to do otherwise.

And defence lawyers can be unhelpful too. And Parliament, and the cabinet, can be at fault too, with unwise – if not frankly ignorant – legislation [122]. In the more extreme of criminal cases, a government that does not proactively try to address its legal system's faults at determining physical reality engages in a kind of political repression.

---

23 *Functus officio* ≈ spent legal authority. Chapters 1 and 9.

24 to be argued about by historians. Chapters 1 and 8.

25 Judges, scientists and calendar-deadlines. Introduction 4.1; Chapter 1.3.

26 Science – good, bad and junk. Introduction 4; Chapters 1, 2, 5, 8, 10 and 11.

27 Morin. Fibre evidence contamination. Chapters 1, 3 and 8.

28 Deaths in Police Custody. Expert witness. Chapters 1.3 and 1.5.2.

29 Captain Alfred Dreyfus. *bordereau, petit bleu.* Colonel Hubert Henry. Introduction 2.3; Chapter 1.

And, in cases where an agency of the state is actually engaged in wrongdoing, the very moral legitimacy to govern may be drawn into question [123, 124].

Such cases of failed justice can be difficult to correct; some may never be corrected and remain a blotch.

In some cases of failed justice, it may be best to seek remedy in the context of a democratic society, which, curiously, shares with science, the concepts of published peer scrutiny (*i.e.* for a democratic society, open, public processes where the peers are all citizens) of state affairs and officials, and for which everything should always be open to question, criticism, correction, revision, refinement or rejection. And so too, in analogy, does democracy have rules of engagement for formal decision making by way of continuing *deliberative* conversation [125].[30]

Thus, when the judicial system, or the government, gets science wrong, and persists in it, the best place for remedy over the longer term may be found in the public processes of democracy – in the media and in politics. Maybe remedy for injustice would eventually be found; maybe not; and maybe too late anyhow. What would remain might only be for historians to argue about.[31]

## 1.4 Legal systems

[. . .][32]

*"Mine is a long and a sad tale!" said the Mouse, turning to Alice, and sighing.*
*"It is a long tail, certainly," said Alice, looking down with wonder at the Mouse's tail; "but why do you call it sad?" And she kept on puzzling about it while the Mouse was speaking, so that her idea of the tale was something like this*: (see Fig. 1.1) [126, 127, 128, 129, 130, 131, 132]

[. . .]

[Alice, Dodo, Mouse, *et al.*]

### 1.4.1 Structure

Forensic chemical issues can arise in many legal contexts [133]. Some aspects of law dealt with here are obviously related to forensic chemistry – as for criminal law. Others are not quite so obvious – as for labour law. Canadian law is primarily referred to in this book [134], but to explain forensics, much is similar, if not the same, for US, British and Commonwealth law.

**30** Democracy. Robert's Rules. Chapter 1.3.

**31** to be argued about by historians. Chapters 1 and 8.

**32** Introductory quotes. Introduction 1.2.

——"Fury said to
a mouse, That
he met in the
house, 'Let
us both go
to law: *I*
will prose-
cute *you.*—
Come, I'll
take no de-
nial: We
must have
the trial;
For really
this morn-
ing I've
nothing
to do.'
Said the
mouse to
the cur,
'Such a
trial, dear
sir, With
no jury
or judge,
would
be wast-
ing our
breath.'
'I'll be
judge,
I'll be
jury,'
said
cun-
ning
old
Fury;
'I'll
try
the
whole
cause,
and
con-
demn
you to
death'."

Fig. 1.1: A long and sad tale.

**A practical working definition of *law*, hopefully sufficient for this book** [135, 136]: **The rules of society – of official status – local, provincial and national – and to some extent international – and the methods and procedures of making, changing, interpreting and enforcing those rules; such rules as for the organization and administration of overlapping things and concepts for most everything that goes on in society.**

Importantly, law, in this way, hopefully involves the peaceful settlement of disputes by negotiations, arbitrations, tribunals and courts. By this process – imperfect in actual human practice – many are not too happy with their results,[33] but accept them, to go on otherwise with business and life, in a more or less orderly community. While some who engage in this process have regrets along the way because it can become so cumbersome – maybe it is not worth the trouble? – it is still generally worthwhile for peaceful activity in a hoped-for just and orderly society.

This process is sometimes seen as containing uncomfortable compromises – perhaps to grumble about and not as for complete victory [137]; and perhaps should sometimes include some aspects of remorse and forgiveness [138]. The participants – as well as their supporters and detractors – might have to deal with one another otherwise, on another day. Perhaps sometimes their lawyers have difficulty in persuading them – and themselves – to look at the other side as adversaries in a civil society rather than as enemies. Curiously, criminal law is often quite cordial in these regards; family law, sometimes, not so much so.

In modern times, all of this is usually within the structures of national states, which may be federal or unitary, and which usually have treaty arrangements amongst them. Generally, in Canada [139], the UK [140], the Commonwealth and the USA [141, 142],[34,35] there is a common law system; in continental Europe and elsewhere, a code law system.

Of particular interest for forensic chemistry are those aspects of legal systems for the administration of industry, commerce, labour, natural resources, public health, the environment and criminal law enforcement and judgment. This book is mostly concerned with aspects of criminal law and labour law as they would appear in courts and tribunals, and in the legislatures, governing cabinets and agencies that would bring the laws into effect; and in enforcement.

33 Not liking the results at law. Chapters 1 and 5.

34 Common Law of England. USA. Maryland. Introduction 1.1; Chapter 1.

35 Advocacy. Introduction 4.1; Chapter 1, 5 and 8.

### 1.4.2 Categories of law mainly considered here

There are many categories of law that can be arranged in different and overlapping ways that can be confusing, if not contradictory. Of main concern, here, are those categories where a prosecutor or other government legal counsel seeks to rely on forensic evidence. For purposes here [143], this is mostly for criminal and regulatory/administrative law (mostly workplace health and safety). But other categories of law can also seek to rely on forensics.

**Criminal law, administrative law:**

#### 1.4.2.1 Criminal law

Criminal law in Canada is determined by the federal Parliament [144, 145], but is generally administered by provincial authorities [146]. Generally, in theoretical concept, criminal law may be seen to derive from, and involve, considerations of serious transgressions against society and its morals [147].

In criminal law, the litigants are the state [148] and accused. By this theory, the state is the potentially aggrieved party in that it, in a sense, represents all of society who suffer wrongs. The actual individual victims are witnesses with personal remedies largely found in other forums, if at all. Criminal law should not be seen as a vehicle for personal vengeance, although this is often the way it is seen by some people; at least the vengeance would be moderated and state-administered. The criminal law does, sometimes, however, offer some aspects of restitution, but that is not its essential intent.

The discussion above is on the historical theory of the victim only as witness in criminal law. That role has, in recent years, and continuing, undergone changes in the USA and Canada, in the sentencing phases after conviction [149].

In Canada, there is a presumption of criminal innocence, unless and until all the elements of a crime, required to be proven by the state, *beyond a reasonable doubt* [150],[36] without the help of the accused – are proven in a procedurally correct court:

- *Actus reus* [151] – the physical facts of what happened
- *Mens rea* [152][37] – the mental element – a criminal intent
- Forbidden by federal statute law – mostly the *Criminal Code* [153]

Criminal law can involve forensic chemistry in several ways, with respect to:

- The unlawful possession and use of explosive substances under the *Criminal Code* [154]

36 *Criminal law. Regulatory/administrative law.* proof *beyond a reasonable doubt* test. Chapter 1.4.2.1 and 1.4.2.2; Chapters 10 and 11.

37 *mens rea.* Chapters 1 and 11.

- The unlawful possession, use and trade, of various chemical substances proscribed under the *Food and Drugs Act* [155] and under the *Controlled Drugs & Substances Act* [156, 157]
- Toxicology and its social and criminal implications
- Analysis of evidence
- Law enforcement and legal representation

#### 1.4.2.2 Workplace health and safety – regulatory

Canadian law has special provisions for what goes on in the workplace [158].[38,39,40] Where there is a union or union-organizing activity, **labour law** regulates certain relationships between workers, employers, unions and the state. **Employment law** regulates in a non-union context – often, for example, vacation pay, maximum hours, bounced employer pay-cheques, statute-governed dismissals, maternity leave, *etc.* – typically for less wealthy workers. Employment law can also involve the common law of *wrongful dismissal* – typically for more wealthy workers. The two terms – **labour law** and **employment law** – are sometimes used somewhat loosely and interchangeably.

Workplace health and safety law [159, 160][41] may be regarded as an area of labour law. Although, as early as 1788, *An Act for the better regulation of chimney sweepers, and their apprentices* [161, 162] regulated exposure to a workplace carcinogen, in England [163], extensive chemical-specific regulation is relatively new. In Ontario, for example, such regulation did not start until 1981, with lead Pb as a *designated substance* [164, 165],[42] under the *Occupational Health & Safety Act* [166]. That *Act* was a major component of a then new era of Canada-wide legislation, adding new dimensions to existing legislation that first appeared as the Ontario *Factories' Act* [167, 168, 169] 1884. New was a shifting emphasis to include not only safety but also health, and for worker/employer participatory functions to supplement, with controversary, the older style theory of government enforcement. Along with new concepts and rules for worker rights, the workplace health emphasis necessarily meant dealing with chronic chemical toxicity; and participation meant that workers and employers were to be involved in workplace evaluation and decision making; not without controversary.

Ontario's present *OH&S Act* can deal with chemicals in several ways that might result in forensic involvement, including:

- Allowing a *Director* at the Ministry of Labour to closely control biological, chemical or physical agents by issuing *orders* [170, 171].

---

**38** Neuro-toxicology of lead Pb. Chapters 1 and 3.

**39** Regulatory law. Environmental & workplace health & safety. Chapters 1, 5, 9 and 11.

**40** Labour relations and employment rights. Chapters 1 and 5.

**41** Workplace health & safety = OH&S. Chapters 1 and 5.

**42** Neuro-toxicology of lead Pb. Chapters 1.4.2.2 and 3.5.

- Allowing for assessment for hazardous materials [172].
- Allowing for regulations to require close control of *designated substances* [173, 174].
- Allowing for regulations [175], which in 1986 resulted in a control of exposure regulation [176] setting ". . . out specific legal limits on exposures to more than 600 different toxic agents" [177].
- Requiring safety data sheets to be available [178].[43]
- Allowing for regulations to adopt, by reference, codes or standards, and require compliance thereto [179].
- The *OH&S Act* was amended in 1987 to provide for chemical *right-to-know*, as Ontario's part of the Canada-wide federal/provincial WHMIS scheme.[44] This is Canada's hazard communication standard.
- There are also general *OH&S Act* applications that can apply to many workplace situations [180].

In Canada, workplace health and safety law can be investigated and enforced by labour inspectors, who have rights to enter premises, issue orders and recommend prosecutions by government legal counsel – Crown counsel. Regulatory prosecution in court is similar to criminal law, but without the concept of *mens rea*; and is usually in the provincial legislative jurisdiction, although also often federal. As with criminal law, the litigants are the state and the accused; the accused are frequently corporations.

And, as with criminal law, the regulatory law standard of proof is as *beyond a reasonable doubt* [181].[45]

There are two alternative regulatory prosecutional possibilities, depending on the nature of the alleged offence [182]:

- Strict liability offences – the *defence of due diligence* replaces *mens rea.*
- *Absolute liability* – neither *mens rea* nor *due diligence* are available for defence – it only matters that the event happened.[46]

### 1.4.3 Some other categories – for mention here

#### 1.4.3.1 Civil litigation – tort – contract

**Tort** [183, 184] is the law whereby a person (either natural or corporate) claiming injury to person, property or business, sues an allegedly offending other person for damages and may seek various judicial orders.

43 SDS. Chapters 1 and 6.

44 Consumer protection. Hazard communication. WHMIS. Chapters 1 and 5.

45 *Criminal law. Regulatory/administrative law.* Proof *beyond a reasonable doubt* test. Chapters 1.4.2.1 and 1.4.2.2; Chapters 10 and 11.

46 Absolute liability offence. Chapters 1 and 8.

**Contracts** [185] allow agreements to be made and enforced between persons (either natural or corporate). Contract law is often related to sale of goods law.[47]

Tort and contract law are distinct from *criminal* and *regulatory* law. For *criminal* and *regulatory*, as indicated above, the litigants are the state and accused. For tort and contract, any aggrieved-claiming individual – and their adversaries – with acceptable standing before the court (which might – or not – include the state or its agencies), can be litigants, and can have the opportunity for remedy – sometimes achieved, but often not. This is called *civil litigation*. Such *civil* suits can involve chemicals [186] and chemists and engineers; and chemical forensic issues may arise – sometimes in a hugely tragic context [187]. Typically – but not always – *civil* and *criminal* lawyers don't mix their practices.

Canadian, UK, Commonwealth and US tort and contract law is based on the system of the *Common Law of England* – with some exceptions where it is code-based (at least historically) – Scotland [188], Louisiana [189] and the *Québec Civil Code* [190]. The common law system is largely a tradition of judicial history – *stare decisis* [191]. The code system – of the rest of the world – is largely by legislation. Both systems may evolve with increasing similarities.

In this civil law, a judge may be called upon to decide, on a *balance of probabilities* (for criminal and regulatory it is the *beyond a reasonable doubt* test), who is liable for causing what damage to whom, to what extent and at what price. A theoretical intention would appear to be to encourage responsible behaviour, so that, for example, someone who is asked to supply a hazardous chemical to a child will be very, very careful [192, 193]. But, by a counter-theory, court decisions are sometimes seen as to not discourage innovation (risk taking). These civil suits rarely involve workplace injuries in Canada because of workers compensation law which bars most civil suits. Related to civil suits is **insurance law** – a commercial device to spread risk, which can often effectively place insurance companies in the role of enforcers of carefulness; and the insurer is often *subrogated* to act for the insured litigant (who perhaps might be surprised with the on-behalf litigation).

#### 1.4.3.2 Consumer protection

**Consumer protection law**[48] is for the public as they buy or rent goods and services, often for personal or family use. There are several branches. At the Canadian **provincial** level, there is legislation to protect consumers in a commercial sense such as the Prince Edward Island *Consumer Protection Act* [194, 195]. Also, at the provincial level are **product liability law**[196] and **personal injury and property damage law** [197] as under **tort law** in the common law system, and under the *Québec Civil Code* [198].

47 Sale of goods. Chapter 1.

48 Consumer protection. Hazard communication. WHMIS. Chapters 1 and 5.

Consumer protection law includes the Canadian federal ***Hazardous Products Act*** [199] and regulations under it, for the health and safety of users of consumer goods. Much of this is chemically related. This *Act* would be enforced by federal inspectors and by prosecutions for which chemical forensic evidence might be called for. This *Act* is also related to hazard communication law – **WHMIS**.

#### 1.4.3.3 Sale of goods

**Sale of goods law** can also be related to consumer protection. Based on common law concepts – it is legislated at the provincial level [200] – for example, the New Brunswick *Sale of Goods Act* [201] – and governs the rights and duties of vendors and purchasers of goods (but not of services). Sale of goods cases can sometimes involve chemical toxicology, as in *Ashington Piggeries Ltd. et al. v Christopher Hill Ltd.* [202, 203, 204] where toxic animal feed was sold (see Fig. 1.2) [205, 206, 207, 208].[49,50,51,52]

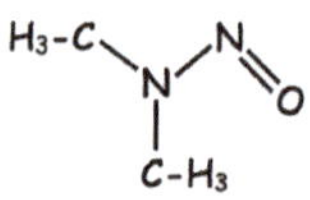


Fig. 1.2: Dimethylnitrosamine, CAS 62-75-9, 74.83 g/mol.

Or, an unintended chemical reaction, as in *Vacwell Engineering v BDH Chemicals* [209] – Judge Rees: (see Fig. 1.3) [210, 211, 212, 213, 214, 215, 216] [53]

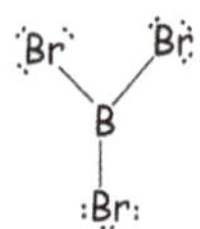


Fig. 1.3: Borontribromide B $Br_3$, CAS 10294-33-4 250.52 g/mol.

$$BBr_3 + 3H_2O \rightarrow B(OH)_3 + 3HBr$$

$$HBr + H_2O \rightarrow H_3O^+ + Br^-$$

[. . .]

. . . The claim arises out of an accident which happened on 26th April 1966 at Vacwell's laboratory premises . . . in the county of Surrey . . . when a Mr. Neale, who was in charge of Vacwell's applied physics department, was working side by side with a Russian physicist, Mr. Strouzhinski . . .

A sketch . . . shows Mr. Neale working at a sink on the left and Mr. Strouzhinski working at a sink on the right. The two sinks were made of a ceramic material and each was about 16 inches × 12 inches × 8 inches or 9 inches in depth. The operation in which the two men were engaged at

49 *N*-Nitrosodimethylamine CAS 62-75-9. Chapter 1.4.3.3

50 Consumer protection. Hazard communication. WHMIS. Chapters 1 and 5.

51 Molecular structure. Chapters 1, 3, 4 and 6.

52 Molecular structural drawing. CAS. Chapters 4.4 and 4.6.

53 B $Br_3$ CAS 10294-33-4. Chapters 1.4.3.3 and Chapter 4.

the relevant time was that of putting into the sinks glass ampoules containing a chemical called boron tribromide. The sinks contained water . . . I find that there were in each sink at the relevant moment of time certainly more than 40 ampoules and there may well have been as many as 100 ampoules in each.

. . . As might be expected, Mr. Neale does not remember what happened immediately after the explosion. He says that there was sudden white flash, . . . There can be no doubt that a violent explosion had taken place as the two photographs exhibited demonstrate. A Mr. Horner, who was an assistant divisional officer of the London Fire Brigade . . . saw that the damage included part of the roof of the laboratory which had been blown off and that the partitioned walls were shattered and indented and there had been a violent explosion. As a result Mr. Strouzhinski was killed and Mr. Neale was injured. Fortunately Mr. Neale's injuries were such that he was only detained in hospital for about a week and was kept at home for only two weeks.

As a result of this accident, Vacwell claim damages from B.D.H., who were the suppliers of the ampoules of boron tribromide, alleging that there are breaches of contract and of duty at common law . . . . The damages claimed by Vacwell in the action include a claim for £74,689 in respect of the damage to the laboratory premises caused directly by the explosion and, in addition, there is a claim for loss of profit in Vacwell's business amounting, I was told at the Bar, to some £300,000. The issues of fact and of law, therefore, have a considerable importance to the parties to this action.

[. . .]

#### 1.4.3.4 Food and drug

**Food and drug law** may be regarded as a specialized area of consumer protection law and of health law, but as well, in Canada it can have aspects of criminal law to it.

US Food and drug law has its historical origins with the Chicago stockyards, the presidency of Theodore Roosevelt,[217] and *The Jungle*.[218, 219, 220, 221, 222, 223]. The developing concepts likely carried into Canada. Curiously, Upton Sinclair's novel would appear as intended for social, economic, civic corruption and labour criticism, but it was widely seen as revealing extensive filth in US food production [224, 225, 226, 227, 228]:[54]

[. . .]

. . . a man could run his hand over these piles of meat and sweep off handfuls of the dried dung of rats. These rats were nuisances, and the packers would put poisoned bread out for them; they would die, and then rats, bread, and meat would go into the hoppers together. This is no fairy story and no joke; the meat would be shoveled into carts, and the man who did the shoveling would not trouble to lift out a rat even when he saw one – there were things that went into the sausage in comparison with which a poisoned rat was a tidbit. There was no place for the men to wash their hands before they ate their dinner, and so they made a practice of washing them in the water that was to be ladled into the sausage . . . .

[. . .]

**54** Prince Otto von Bismarck. Upton Sinclair, *Jungle*. Chapter 1.

As with consumer protection law, Food and drug law seeks the protection of the public, and much of the other law categories above are also applicable. However, there is Canadian federal legislation of particular interest: *Food and Drugs Act* [229] and *Controlled Drugs & Substances Act* [230].[55] These statutes and regulations under them deal with *food purity* and *additives* and with *controlled, restricted* and *narcotic drugs*. Forensic chemistry should have a high profile in such law because of the substances being regulated or prohibited as well as because of the analytical chemistry involved in administration and enforcement. In addition, the regulation of drugs often appears in the courts as part of Canada's criminal law.[56]

At the provincial level, food and drug law can be related to **public health law** [231] and to those aspects of **health care law**,[232, 233, 234] which involve prescribing of medication and analytical chemistry procedures.

#### 1.4.3.5 Environmental protection

**Environmental protection law**[57] is frequently associated with both chemicals and chemists in the media and in the public mind. Although there was nineteenth-century industrial environmental legislation – for example the British *Alkali Act, 1863* [235] – modern Canadian regulation of chemicals to protect the environment dates from the 1970s. Presently, such legislation is the federal *Canadian Environmental Protection Act* [236] and various provincial statutes such as the Saskatchewan *Environmental Assessment Act* [237] and *Environmental Management and Protection Act* [238]. Enforcement of such legislation would require forensic involvement.

#### 1.4.3.6 Hazard communication WHMIS

The **Workplace Hazardous Materials Information System** [239, 240, 241, 242, 243, 244][58,59] is Canada's hazard communication standard. It is workplace chemical **right-to-know law**. Recommended in a 1985 federal report [245], it is Canada-wide federal/provincial-coordinated legislation that came into force on 31 October 1988. Generally, WHMIS seeks to give workers a right-to-know by:
- labels
- safety data sheets [246, 247, 248, 249][60,61]
- worker education

---

55 *CD&S Act*. Chapters 1 and 10.

56 Drug law. Chapters 1 and 10.

57 Regularory law. Environmental & workplace health & safety. Chapters 1, 5, 9 and 11.

58 Toxicology. Hazard communication. Chapters 1 and 11.

59 Consumer protection. Hazard communication. WHMIS. Chapters 1 and 5.

60 SDS = Safety Data Sheet. Previously called MSDS = Material Safety Data Sheet. Chapter 1.

61 Disclaimers. Error. Not legal advice. Commercial products. Introduction 2 and 4.1; Chapters 1, 5, 6 and 9.

The legislative structure of WHMIS has its complexities. Generally, Canadian federal legislation is used to set standards; and for trade secrets (which can legitimately be valuable commercial property) to be withheld; and federal and provincial legislation for WHMIS application in the work place.

As allowed by the US *Constitution* [250], Article I, Section 8 – interstate commerce clause – there is a US Federal Hazard Communication Standard [251, 252, 253].

> § 1910.1200 Hazard communication.
> (a) Purpose.
> (1) The purpose of this section is to ensure that the hazards of all chemicals produced or imported are classified, and that information concerning the classified hazards is transmitted to employers and employees. The requirements of this section are intended to be consistent with the provisions of the United Nations Globally Harmonized System of Classification and Labelling of Chemicals, Revision 3. The transmittal of information is to be accomplished by means of comprehensive hazard communication programs, which are to include container labelling and other forms of warning, safety data sheets and employee training.
>
> (2) This occupational safety and health standard is intended to address comprehensively the issue of classifying the potential hazards of chemicals, and communicating information concerning hazards and appropriate protective measures to employees, and to pre-empt any legislative or regulatory enactments of a state or political subdivision of a state, pertaining to this subject. Classifying the potential hazards of chemicals and communicating information concerning hazards and appropriate protective measures to employees may include, for example, but is not limited to, provisions for: developing and maintaining a written hazard communication program for the workplace, including lists of hazardous chemicals present; labelling of containers of chemicals in the workplace as well as of containers of chemicals being shipped to other workplaces; preparation and distribution of safety data sheets to employees and downstream employers; and development and implementation of employee training programs regarding hazards of chemicals and protective measures. Under Section 18 of the Act, no state or political subdivision of a state may adopt or enforce any requirement relating to the issue addressed by this Federal standard, except pursuant to a Federally approved state plan.
>
> [. . .]

It should be noticed that hazard communication requirements do not necessarily mean that a worker would be in actual jeopardy; but rather the theory would seem to be that there is a reasonable concern, and the worker is entitled to be informed so as to self-decide, often in consultation with union or other worker advocates. For example, to continue the art forgery fictitious example [254] (in Chapter 3)[62]: Suppose that the well-known, but long-dead, artist, whose possible work was being scrutinized, was also historically thought to have suffered copper Cu, arsenic As [255] and mercury Hg [256] poisoning related to the greens and reds of the painting.[63] Hazard communica-

**62** Art forgery analysis. Chapters 1 and 3.
**63** Forbes art pigment collection. Chapters 1, 2 and 3.

tion concepts would imply that the forensic analytical chemist should be so informed, notwithstanding that there might not be an actual danger.[64]

#### 1.4.3.7 Workers' compensation

**Workers compensation law** can be related to forensic chemistry because of compensation for workplace injuries and diseases that are chemically caused.

After observing systems elsewhere, including the German system established by Prince Otto von Bismarck [257],[65] Ontario Chief Justice Sir William Meredith, [258] in his 1912 and 1913 reports, recommended what was to become the Ontario workers compensation system. It may be regarded as historic compromise between labour and capital that resulted in 1914 Ontario legislation [259, 260], and which was followed, and further developed over the years, in all provinces [261]. Workers [262] let go of their right to sue their employers (it became *barred by statute*) in exchange for compulsory no-fault insurance, financed by employers and administered by a workers compensation board (now in Ontario: Workplace Safety and Insurance Board). Although often criticized from both sides, there seems little serious effort by either side to return to **tort law** for legal remedy of workplace accidents or injuries.

Workers' compensation law is related to OH&S law in that workers look to the WSIB for financial remedy for lost wages and for medical and rehabilitation costs after there has been an OH&S failure.

Workers compensation can deal with chemically caused injuries and diseases:

- the same way as for other workers' claims – WSIB determines if the injury or disease arose out of work; and there may be an onus, or onus reversal, on the worker to convince the WSIB of this;
- by a *presumption* that certain listed occupational diseases were caused by certain co-listed processes or chemicals [263].

WSIB evaluation of disease – which is often of a chronic nature – has been criticized by labour's side as inadequate. Those concerns have resulted in a tribunal system in Ontario [264].

#### 1.4.3.8 Profession regulation

As with any **profession**, trade or calling, chemists and engineers are free to operate commercially, subject to legal and licensing requirements [265].[66] But, since many chemically related enterprises are legally dealt with under other professional or trade headings (such as teaching, clinical analysis, pharmacy, medicine and law) or are

---

64 Workplace health & Safety. Hazard communication. WHMIS. Chapters 1, 2 and 4.

65 Prince Otto von Bismarck. Upton Sinclair, *Jungle*. Chapter 1.

66 Profession. Chapters 1 and 5.

under a corporate or government umbrella, chemists per se [266] are left somewhat unregulated. Engineers are regulated under the provincial designation of *Professional Engineer* (*P.Eng.*) [267]. But, particularly as regards as to chemists, Canadian law is rather less extensive.

A person may self-identify as chemist or engineer, either personally or within a corporate structure, and as such may consult, invent, direct a corporation, write, teach, sell, serve the state as a forensic scientist or testify as an expert witness, without the supervision or sanction of a chemical professional body. But not completely so.

What may not be done is claim to be a *Chartered Chemist* (*C.Chem.*) [268], *Professional Chemist* or *Professional Engineer* (*P.Eng.*) [269], without fulfilling the conditions set out in various provincial legislation that go with these titles (or other legislatively reserved titles).

The Québec *Loi sur les chimistes professionnels – Professional Chemists Act* [270] restricts the "practice of professional chemistry" to members of the *Ordre des chimistes du Québec* [271]. For engineers in every province there are analogous legislated restrictions, for example, in the Ontario *Professional Engineers Act* [272]. However, the *Association of the Chemical Profession of Ontario Act, 1984* [273] appears to mostly reserve the terms *Chartered Chemist* (*C.Chem.*) and *Professional Chemist*.

These professional designations would appear to be less consequential than for the health sciences professions: Medical Doctor MD, Registered Nurse RN, Pharmacist, *etc.* – for which there are also legislation-required licensures. And, likewise, for various trades: electrician, plumber, *etc.*

But, importantly, for many kinds of engineering work, a sign-off is required by a *P.Eng.* for the process to be legally valid or allowed. And there can be consequences if something goes wrong [274, 275].

#### 1.4.3.9 Education

**Education law** affects students, teachers and administrators – chemical education included. There is provincial legislation for the organization and financing of schools [276], colleges [277] and universities [278] and for labour relations with their faculties and staff. And for licensing of teachers, pre-KG–12. In addition, the Canadian provincial and federal financing of research at universities and other institutions [279, 280] can involve legal questions as to ownership, rights, duties and liabilities with respect to supplies, equipment, salaries, grants and students; and with respect to intellectual property.

Another aspect of education law – and with it, insurance law – that can be related to forensic chemistry is: **liability for student health and safety**. For example, in *James et al. v River East School Division No. 9, etc.* [281, 282, 283, 284, 285, 286, 287], there was facial scarring and serious eye injury of an 18-year-old student from nitric acid spattering in a school chemistry lab experiment gone wrong, 22 September 1972 [288, 289] (see Fig. 1.4):[67]

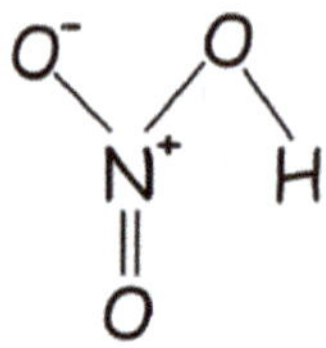

$HNO_3 + H_2O \rightleftharpoons H_3O^+ + NO_3^-$

Fig. 1.4: Nitric acid, CAS 7697-37-2, 63.013 g/mol.

From the MANITOBA QUEEN'S BENCH – Judge Deniset, 1975:

> [. . .]
>
> The facts in the present case . . . are that a Grade XII student was instructed to conduct a laboratory experiment which was dangerous. It was foreseeable that an accident could happen. One did.
>
> After this accident, better instructions were issued.
>
> Goggles were available. None were recommended on this occasion by the teacher . . . His excuse that the students knew about the goggles and that none requested them, is not valid. The Greater Winnipeg Minor Hockey Association does not recommend the use of helmets when playing league games. You put on a helmet, or you don't play.
>
> Students must be told, when necessary "wear goggles".
>
> I find here a failure to instruct properly, a failure to caution and to supervise properly and that an unfortunate and foreseeable accident occurred, and that it could have been avoided if the defendants had not been negligent and if . . . [the teacher] had not omitted to do what he should have done in the circumstances.
>
> [. . .]
>
> After due consideration of all these factors and the whole of the evidence regarding the injuries sustained by this young girl, I allow as general damages the sum of $25,000.

| Plaintiffs will have judgment for: [290] | |
|---|---|
| Special damages | $ 5,921.65 |
| General damages | $25,000.00 |
| | $30,921.65 |

> Plaintiffs will have their costs to be taxed [291].

67 Molecular structural drawing. CAS. Chapters 4.4 and 4.6.

Effective protective eyewear enforcement apparently did not quite reach the Toronto District School Board, as late as 2001, as appeared in a website photograph, Sir Robert L. Borden Business and Technical Institute. Students appeared as obviously doing chemistry lab without protective eyewear [292].

More recently, the (should not have been so) popular Rainbow Demonstration was still being used in US schools as late as 2015. This involved a too-difficult-to-control use of a flammable liquid in a classroom demonstration that involved an open flame. The important spectral principles, known from the nineteenth century, can be safely shown otherwise [293, 294, 295, 296, 297, 298, 299, 300].[68]

Lab safety [301, 302] extends to university labs, where there is now more widespread careful attention, especially after an incident in California: Professor Patrick Harran and UCLA were involved in serious and extensive litigation after the death in 2009 of an employed researcher. There were issues of a pyrophoric organo-lithium reagent, storage of a flammable liquid and worker oversight. This was most newsworthy because it involved issues of criminal responsibility [303, 304, 305, 306, 307, 308] (see Fig. 1.5).[69]

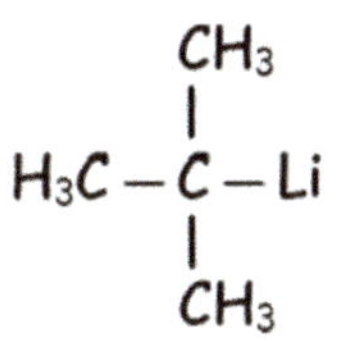


Fig. 1.5: *t*-Butyl lithium, CAS 594-19-4.

And, at Kingston, Ontario, 2018, there was criminal conviction of a Queen's University graduate student for intentionally poisoning a co-worker using chemistry-lab-obtained *N*-nitrosodimethylamine [309, 310, 311, 312][70,71] (see Fig. 1.6).

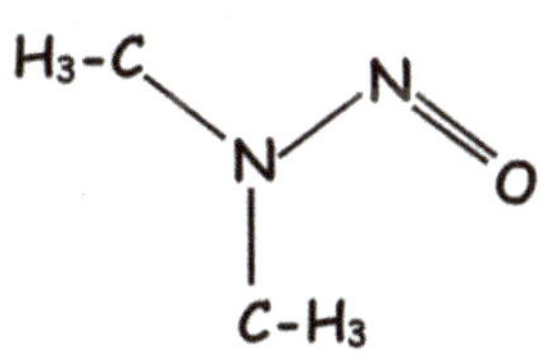


Fig. 1.6: *N*-Nitrosodimethylamine, CAS 62–75-9.

68 Beckman DU. Bunsen, Kirchhoff. cover. Chapters 1 and 6.
69 Molecular structural drawing. CAS. Chapters 4.4 and 4.6.
70 *N*-Nitrosodimethylamine CAS 62-75-9. Chapter 1.
71 in-court-drawn-artwork. Chapters 1.4.3.9 and 6.27.

As news reported [313], there was effectively in-court mention of acute and latent toxicology [314] and of $LD_{50}$.[72] The in-court evaluation of these topics would appear as problematic.[73,74,75] However, more appropriate scientific interpretations would not likely have much benefited the accused because *N*-nitrosodimethylamine is undoubtedly toxic in some way or another, and the central criminal law issue was intent to put the victim in harm's way with what was believed to be an effective poison.

Also reported within this Ontario case were similar incidents in China. But, no reference was made to a toxic apple attributed to Robert Oppenheimer, when he was studying in England, ~1925 [315, 316, 317].[76]

And, at Bethlehem, Pennsylvania, 2020, a Lehigh University chemistry student plead guilty to poisoning a roommate with thallium [318, 319].

#### 1.4.3.10 Patent

**Patents** may be of interest to chemists and chemical engineers for commercial rights for their inventions of devices, processes or molecules [320]. The constitutional basis in the United States is found in Article I, Section 8: "The Congress shall have Power . . . To promote the Progress of Science and useful Arts, by securing for limited Times to Authors and Inventors the exclusive Right to their respective Writings and Discoveries". Canadian Patent law is based on the federal *Patent Act* [321]. A patent is a kind of **intellectual property** right that may be sold or licensed by its owner.[77]

While detailed data of patent-protected inventions are declared in public documents, there is also a process of keeping **trade secrets** [322, 323] – for what is not patentable, or for what the owner does not want to patent for commercial or other reasons, such as when the patent process is too expensive, or when patent-enforcement would be considered as too difficult. Trade secrets may be of interest in trying to protect inventions, processes and formulations by simply not telling. (However, trade secret disclosure may be required under **WHMIS**.)

Patent and trade secret law can overlap with **employment law** [324] when, as often would happen, the chemist or engineer is not inventing independently, but as an employee, so that the employer may be the patent or trade secret owner, perhaps subject to the terms of a contract called a patent agreement or an employment no-competition clause.

Another kind of intellectual property that chemists and engineers can create is **copyright** – articles, books and computer software that they might write are protect-

72 Toxicology. $LD_{50}$. Chapter 1.

73 Toxicology. Hazard communication. Chapters 1 and 11.

74 Consumer protection. Hazard communication. WHMIS. Chapters 1 and 5.

75 SDS. Chapter 1.

76 Hazards of found chemicals. Chapters 1 and 8.

77 Intellectual property. Chapters 1 and 5.

able under the Canada federal *Copyright Act* [325, 326], and which co-ordinates, at least in concept, with international arrangements.

Also related is **trade mark law** – the Canada federal *Trade Marks Act* [327], which deals with the legal reservation of various words and symbols for commercial purposes.

#### 1.4.3.11 International arms control

**International arms control law** becomes a chemical forensics issue for the policies and practices for enforcement of the **Chemical Weapons Convention** [328]. This international treaty, ratified by Canada [329], includes as a method to control chemical weapons with the international inspection of chemical industries to monitor certain chemicals and precursors [330].

## 1.5 The law of evidence

### 1.5.1 Evidence in court

There are many science-based facts that a Canadian court or tribunal would accept as obvious – not questioned – for example: That Toronto weather includes temperatures both above and below 0 °C so that sometimes there is solid ice on the ground, and sometimes liquid water, sometimes both, sometimes none. And, for court-accepted weather facts of a particular day, the judge can take *judicial notice* [331] of an official weather report, or even a news report, offered by legal counsel of one or another of the participants; or perhaps just look out the courtroom window. But many other facts – essential to the litigation – must be proven to the court. This involves the law of *evidence* [332][78] – and that is what is of concern here.

Generally, evidence into court can come by way of:

- Declarations referred to in legislation, for example: *Notices* under the *Canada Evidence Act* [333]. And, for example, police-seized business records [334].
- Documents of some legal status, for example: a *real property deed* and *mortgage*.
- Testimony – *viva voce* [335, 336][79] – of witnesses, called by one side or another. This is often how documents of legal status come before the court. And, usually the only way that other documents, physical items and recollections come before the court. Such a witness is allowed to testify only for what is directly personally known and not heard from someone else – *hearsay* [337, 338] is not allowed. Nor is opinion allowed – that is reserved for an *expert*.

---

**78** Evidence. Chapters 1 and 10.

**79** = by voice.

- Sometimes the litigants can agree – through counsel – about what should be accepted as evidence. This may be done by way of an *agreed statement of facts*.
- However, the content of the law is not included as evidence. For example: That *peace officers* – police – have *Criminal Code* and *Highway Traffic Act* rights to stop and examine motor vehicles and drivers, within the legislated limits, as interpreted in case law, is a matter of law rather than evidential fact. Law is not for a witness to determine; it is for the judge after hearing from the lawyers. However, sometimes what a witness would perceive of the law can be regarded as a fact offered before the court, but that perception can be wrong.

It is important to realize that the evidence presented is not automatically accepted as true or credible – determining that is the job of the judge or tribunal panel – after hearing arguments of legal counsel. That is what the court or tribunal process is for.

The law of evidence has its complexities and complications – that can prompt much litigation – so that there are well-known – and not-so-well-known – exceptions to the generalities above. But the above descriptions are generally applicable for purposes of this book. The exception of concern here is for *experts*, who may give *opinion evidence*.

### 1.5.2 Forensic expert witnesses

This *evidence* exception is that a witness, *qualified* by the court as an *expert*, may give *opinion evidence*, within the bounds of the expertise recognized by the court [339, 340, 341, 342, 343, 344].[80,81,82,83,84]

Qualification is a formal process that would occur just before the anticipated *expert* testimony. Typically, the anticipated *expert* is sworn-in; the *curriculum vitae* and other credentials are presented and legal counsel seeking the expertise asks questions – direct examination – to try to persuade the judge to allow qualification. Other counsel, who would have seen the *CV* beforehand, then cross-examine [345] – perhaps to try to persuade the judge to not allow qualification.

If qualified by a decision of the judge, that witness would be allowed to give a scientific or engineering *opinion* to the court – within the bounds of the qualification – and that testimony would be the next step. If not qualified, evidence would not be heard from that witness.

The now-qualified *expert* would then testify as to the science or engineering of concern to the case being heard. That testimony has a special status.

**80** Expert. Chapters 1 and 9.
**81** DRE. Expert witness. Chapters 10 and 11
**82** Science gate-keeping. Chapter 1.
**83** Ethics. Lawyers. Experts. Introduction; Chapters 1, 5, 8 and 9.
**84** Deaths in Police Custody. Expert witness. Chapters 1.3 and 1.5.2.

Legal counsel who sought the expertise would ask questions to support the eventual facts that would be argued. Opposing counsel would cross-examine – seeking flaws.

Opposing counsel may repeat the courtroom procedure by calling other *expert* testimony to try to counter the previously called *expert.*

This process of countering experts might sometimes look somewhat chaotic, if not ugly. And sometimes maybe it is. European courts are somewhat known to be kinder to experts, who have somewhat different – more independent – and perhaps more respected – functions there than in Canada, the USA and UK. Arguably, the European system would allow for a more orderly credible version of testimony.

Also, expert witnesses can be expensive, perhaps prohibitively so for criminal accused who are poor.

The special status of the *expert opinion evidence* requires the judge to hear it as presentation of fact. But the judge is not required to accept it as correct, and that is why there may be a series of experts before the court, not agreeing with one another on various points. Opposing legal counsel would argue that the judge should be persuaded one way or another. The judge may reserve on ruling to a later time.

In practical terms, the judge may likely be persuaded, or not, largely based on the credentials used to qualify the *expert,* rather than understanding explanations of the actual science or engineering; and legal counsel would not be too likely to argue about the actual explanations because not too many of them (except practitioners of patent law) would have had too much science or engineering education. Like most of us, they would do what they know and avoid what they don't.

Thus, sadly, the evaluation of unfamiliar scientific opinion evidence, based on a Judge's understanding of essential explanations by the expert witness, would not be too common in a Canadian court.[85] Notwithstanding that this would require only a high-school-level appreciation of science and mathematics [346, 347, 348].

Much of the science and engineering expert evidence heard by the courts is well-known, not controversial, with little argument as to its validity. But some is not [349].[86] Science that is "novel" [350, 351, 352, 353, 354], new to the legal system, presents the most difficult problems; judicial unfamiliarity with science does not help – nor lawyers too [355, 356, 357, 358, 359, 360, 361].[87,88]

The effective reliance on witness credentials rather than critical appraisal of science and engineering content is often satisfactory, but it does introduce potential problems when the science and engineering would be challenged – which may not happen too often – but when it does can be significant [362, 363, 364].

---

**85** re criticism of Health Canada, Judges et al. Introduction 4.1 and 4.2; Chapters 1, 10 and 11.

**86** *R v Bingley*. Chapters 1 and 11.

**87** *Daubert*. Judicial unfamiliarity with science. Introduction 4.1; Chapters 1 and 10.

**88** T&F. Dr. Carl Meyer. Copyright acknowledgement. Chapter 1.

However, in terms of fundamental scientific concept, reliance on credentials rather than explanation presents a serious problem.[89] It is less than scientific. But that is how it is often done in Canadian courts.[90]

Arguably, reliance on credentials rather than critically evaluated explanation can lead to acceptance of opinions as correct, when they are not. Perhaps, sometimes tragically. This can be by honest error, but some "experts" may actually be fraudulent, and some perhaps err because they misjudge their own knowledge range [365, 366, 367, 368]. And sometimes, what had been widely accepted can come into controversary as flawed process [369, 370, 371]. See also Chapter 8.[91]

The above description is for courtroom process. Similar scenarios would play out when lawyers and administrators are involved with science and engineering for tribunal hearings, administrative decision-making, and legislative drafting. And also for policymaking.

### 1.5.3 Police as forensic experts in Toronto drug cases

Disclosure from Crown counsel to Defence counsel, in relation to *Controlled Drugs & Substances Act* [372][92] charges involving *trafficking* in criminalized substances, in some jurisdictions – such as Toronto – often includes an opinion of a police officer to be offered at trial as *expert*[93] in determining whether the quantity of drugs, and other indicators, imply *trafficking* or mere *possession* for personal use. The difference is important because of more severe penalties for *trafficking* rather than *possession*.

This chapter[94] addresses questions of qualification of police officers as expert to determine *trafficking* or *possession*. Data the author has sought – in capacity as legal defence counsel – for clients' alleged drug offences – has included questions of qualification in the context of perhaps more traditionally or classically qualified experts.[95]

Here are noted possible requests for further data and questions that would involve further disclosure than Crown counsel would usually include with the NOTICE FOR EXPERT TESTIMONY. These requests and questions are intended to focus on the witness being able to draw conclusions from data of claimed quantities of drug that would indicate trafficking rather than for personal use:

---

**89** Science and explanation. Introduction 4.1; Chapter 1.

**90** re Criticism of Health Canada, Judges et al. Introduction 4.1 and 4.3; Chapters 10 and 11.

**91** Ethics. Lawyers. Experts. Introduction; Chapters 1, 5, 8 and 9.

**92** War on drugs. *CD&A Act*. Mass (grams) measurement. *W&M Act*. Chapters 3, 6, 8 and 10.

**93** Advocacy. Introduction 4.1; Chapters 1, 5 and 8.

**94** **Alert**: This chapter part is rather case-specific.

**95** Author's biases. Introduction 1.2, 2, 4.1 and 4.3; Chapters 10 and 11. (And also to some extent in Chapters 1.5.3 and 6.20).

### Time of NOTICE

- Because the Crown NOTICE for its expert witness is usually not given until quite close to the 30-day *Criminal Code* requirement [373], it would be suggested that an earlier-than-30-days request is consistent with the spirit of the *Criminal Code* provision – "For the purpose of promoting . . . fair, orderly and efficient presentation . . .".

### Relating to qualification-as-expert as to the court

- Are there academic qualifications – degrees, diplomas, certificates, special courses, *etc.* – not mentioned in the *Curriculum Vitae* – for knowledge in possible relevant basis areas? –
  - Economics
  - Commercial structures and transactions
  - Commercial measurement
  - Finance
  - Sociology
  - Criminology
  - Toxicology
  - Molecular science
  - Epidemiology
  - Addiction research and treatment
  - Mathematical statistics for data surveys and analyses. *Gaussian* distributions?
- Bibliography or list of peer-reviewed publications, written by the witness? Or, other publications – of academic or professional journal or book, or governmental or institutional reports or pamphlets – with writings by the witness, on relevant topics that are published and exposed to public or peer comment?
- Any publications by the witness with an assigned ISBN?
- Bibliography or list of publications, written by others, read or relied-on by the witness?
- In forming an opinion about trafficking or not – in regard to drug masses (grams) claimed – Would such opinion be based on data collections from other past cases? Are there formal written accessible data bases? Published? Or, would the opinion be based on undocumented anecdotal experience and recollection? Are there any data bases of grams of drugs and price that the witness used?
- Did the witness evaluate the data of this case in a blind sort-of-way? Did the witness know that those data are part of a criminal case – and specifically this criminal case? Was the witness otherwise involved in this case?
- Does the witness consider that there is an inherent risk of reasonable apprehension of bias of an expert determiner of trafficking or personal – knowing in-advance about a specific criminal case? This case? Not blind? Not as with *Justitia*? [374]

- Does the witness consider that there is an inherent risk of reasonable apprehension of bias of a law enforcement agency evaluating its own crime statistics without external in-put or consultation? Risk of *Quod volimus credimus libenter*? [375].[96,97]

**Relating to – when – if – the Crown-called expert would testify:**

- For example here – Methamphetamine [376] and Methamphetamine hydrochloride [377] have different molar masses (g/mol) [378, 379]. Could there be other ionizations – with other molar mass values? And, also for cocaine [380] and cocaine hydrochloride? [381]. Which molar mass values are used as data references for the indication of trafficking or personal use in this case?
- In making an evaluation of trafficking or personal use in this case, did the witness take into account concentration, proportion, *tare*, hygroscopy or chirality?
- Please present any data bases that may have been used for grams of drugs and price in this case. Were the listings of grams and price of this case compared to a data base? Published?
- For the mass (grams) data – Did the witness evaluate the measuring process used by Police? Is the witness aware of the INSTRUCTION MANUAL for the FX-5000i Precision Balance, A&D Company Ltd., Tokyo, Japan; 2011; or Mettler Toledo XSR1202S ? [382]
- Is the witness aware of Health Canada *Drug Analysis Service Client Manual for Exhibit Submission and Analysis*, ISBN 9780660054179 ?[98]
- Were any of the issues in the above items considered in any law cases that would be referred to in previous cases that would be mentioned in the *Curriculum Vitae*?

## 1.6 Legislative process – policymaking

> *Laws are like sausages. It's better not to see them being made.*
>
> [Likely not a true quote, apocryphal, attributed to Prince Otto von Bismarck, Chancellor, *Deutsches Reich*, 1871–1890] [383, 384, 385]

Forensic science can come into the legislative process.[99] Policymaking leads to proposals put before Parliament and the provincial legislatures, each in their respective jurisdictions, that would result in statutes, and resulting regulations. For example, the

96 *Quod volimus credimus libenter* ≈ believe what we want to believe.

97 Roman goddess *Justitia* – with scales, blindfold and sword.

98 Mass measurement. Chapters 1.5.3 and 6.20.

99 Legislative process – policymaking. Introduction 4.2; Chapters 1, 10 and 11.

criminalization of possession and trade of certain molecules[100] or the outlawing of certain electric stun devices.

Mostly, Parliament seems to cope with the science satisfactorily; but sometimes not. Sometimes Parliament's getting it wrong can lead to bad science in an important context; other times, merely to confusing results [386], that may not be too important. Thus, some science policy seems to need improvement, even without considering climate change issues.[101,102]

## 1.7 Advocacy

### 1.7.1 Lawyers

The role of defence [387] lawyers – styled in Ontario as *barrister & solicitor* – sometimes called *counsel* – is to advocate on behalf of their clients [388, 389].[103] There are some lawyerly *rules of engagement* that might seem strange to much of the rest of society. In a particular case, the lawyer would represent the client notwithstanding disapproval of the client's actions, and disagreement with the client's views. And, often the lawyer would not – if ever – be sure of what "facts" of the case are actually true [390, 391].

In the criminal law context, although there are actually some innocent accused, the lawyer would still represent the client, even notwithstanding that the client may have privately revealed actual guilt [392] and even (hopefully rarely) having also revealed an unremorseful meanness of character. But criminal law rules of engagement require that the state must prove its case *beyond a reasonable doubt*, without the help of the accused, who has a right to legal counsel, with a *presumption of innocence*. Innocent until the judge would say otherwise, giving reasons.

This is based on a theory that everyone has the right to robust representation within the justice system so that it can function fairly – in overall concept. Some lawyers see this as a duty to society [393, 394, 395, 396]. To some extent, this is also reflected in some lawyers doing *pro bono* work [397, 398, 399, 400].[104]

The system includes, within lawyer/client privilege, confidentiality, at the option of the client. But it does not include frank misrepresentation by the lawyer to the court or other counsel. The tensions and quandaries within this system are governed by the lawyers' licensing agency – the Law Society of Ontario (previously styled as the Law Society of Upper Canada): LSO (LSUC) rules of professional con-

**100** Statutes. Regulations. Criminalized molecular offences. Chapters 1 and 10.
**101** Creationism. Climate change denial. Anti-vaccine. Chapters 1, 2 and 8.
**102** Some science controversary remains. Chapters 1, 2, 10 and 11.
**103** Advocacy. Introduction. 4.1; Chapters 1, 5 and 8.
**104** *pro bono publico*, Latin = for the public good. Freebie legal work.

duct [401, 402]. Other jurisdictions have similar lawyers' governing systems. Advocacy within these bounds – within these rules of engagement – especially by criminal defence counsel – can at times be particularly uncomfortable and problematic; but advocacy is an essential requirement of the job [403]. This is also reflected in fictional and cartoon literature [404, 405, 406, 407, 408, 409, 410, 411, 412].[105]

### 1.7.2 Forensic scientists

In contrast to lawyers, forensic scientists, while professionally engaged on a case, should not be advocates for anyone within the process and, as best as practicable, should be blind to the intricacies of the goings and comings of the human story being considered. The forensic scientist's only advocacy should be to remain true to trying to do and report good science in good faith. Well, that's a theory, anyhow – the scientist's reality of employment may impose messy difficulties.[106]

### 1.7.3 Dual role

However, the forensic scientist can possibly have dual roles as for:[107]

- Doing the analytical chemistry or other science and reporting the results to be considered by a court or tribunal – as blind as practicable to other aspects and
- Assisting investigators, including law enforcement, in developing the narrative of the human story [413].

It may be that a single forensic scientist, or team, can fulfil both roles for a particular investigation. But the inherent potential conflict [414, 415] might introduce a tension that should lead to two separate scientists or teams being engaged. The problem here may not be so much an actual conflict, but rather a reasonable apprehension of conflict. A practical problem may be added costs; and again, well, that's the theory – the scientist's reality of employment.[108]

All of this can be in addition to an academic research role which should not be of concern here.

**105** Ethics. Lawyers. Experts. Introduction.; Chapters 1, 5, 8 and 9.
**106** Ethics. Lawyers. Experts. Introduction 2.3; Chapters 1, 5, 8 and 9.
**107** Dual role. Chapter 1.
**108** Ethics. Lawyers. Experts. Introduction 2.3; Chapters 1, 5, 8 and 9.

### 1.7.4 Who talks to whom and when

It should be noted that lawyers, forensic scientists and others involved in a forensic issue have separate roles that should be directed within each of their own frameworks of operation, along a chain of authority and chronology.

Thus, for example, a defence lawyer would appropriately address issues with Crown counsel, and not, at least initially, with the forensic scientist engaged by the Crown. And, similarly for Crown counsel for a defence-engaged expert witness. And, by this concept, neither defence counsel nor Crown should independently approach the judge.

Everyone should work within the context of their own roles; to deviate from this formalism risks complications that may taint the whole process. For, example, a nightmare scenario [416]: A defence lawyer starts a discussion directly with a Crown forensic expert who inadvertently reveals privileged data; the lawyer would have now achieved a conflict and may now have become a witness; and so too the forensic expert, as a witness of a different sort.

## 1.8 US *qui tam*

*Qui tam* [417][109] is a US anti-corruption federal legal process [418, 419, 420, 421, 422, 423] that derives from the Civil War era. Federal legislation [424] provides fines for anyone who does any business with the federal government fraudulently. The fines can be large – perhaps several times the amount of the fraud. There is a whistle-blower/bounty-hunter aspect, in that anyone who reports the fraud to the US Department of Justice can be rewarded with a proportion of achieved fines. Again, this can be large.

The whistle-blower/bounty-hunter might typically be someone with inside information – a disgruntled former – or soon-to-be-former – employee. Universities can also be liable [425].

Perhaps the theory would be that the rewards should be so large because after blowing the whistle [426], the employee or student would be finished in that industry or entire profession or job description. The huge reward would replace becoming unemployable otherwise.

The bounty-hunter aspect is enhanced in that if the Justice Department does not proceed in federal court, the whistle-blower can self-proceed – thereby acting both for self and for the government [427].

Arguably, *qui tam* distorts the employer–employee relationship. Arguably, it should – in the absence of other process routes for employer–employee tension – for

**109** US *Qui tam*. Chapters 1, 8 and 9.

example, if there is no union collective bargaining or there is no effective government inspectorate. Critics have argued that the *Act* was in need of reform [428].

There are US law firms specializing in *qui tam* [429, 430] that would represent would-be whistle-blower/bounty-hunters, likely on a contingency fee basis. Arguably, this would further distort the employer–employee relationship.

There is apparently no *qui tam* equivalent in Canadian law [431]. Canadian government assurances to protect whistle-blowers would sound hollow in view of the US *qui tam* process. However, since many Canadian businesses contract with the US government, and Canadian universities do research financed with US government grants or contracts, the US *qui tam* process would be open to aspiring Canadian whistle-blowers/bounty-hunters proceeding in US federal court, represented by US law firms.

## 1.9 Toxicology

> [. . .]
> "... cyanide ...... instantaneously"
> [. . .]
> [Agatha Christie, *And Then There Were None*, 1939.][110]

*Forensics* can be related to the *toxicology* of chemical substances and to the *toxicology* of physical processes. Often, the law attempts to define, regulate and limit toxic effects on humans and our environment, sometimes with some success; sometimes following a disaster, with controversary. Some fundamental concepts, briefly summarized[111,112,113]:

Chemical toxicology is described here as the study of harmful effects of chemical substances on biological systems: human, animal and plant. Although forensics is often popularly associated with harm to humans, there can also be harm to animals (wildlife, pets and farm) and vegetation (wild: managed forests, wetlands, *etc.*; farm; and garden and lawn).

### 1.9.1 Acute, chronic and latent toxicology

There are categories of toxicity that overlap – often in complicated ways:

- A chemical substance that does quickly noticed damage is said to be *acutely* toxic. The damage can be injurious or lethal or both. For example: the poisoned whis-

110 Agatha Christie. Acute toxicology. Hydrogen cyanide H–C≡N CAS 74-90-8. Potassium cyanide $K^+$ $CN^-$ CAS 151-50-8. Chapters 1 and 3.

111 Pb Toxicology. Old-stock housing. Sample selection. Chapters 1, 2 and 3.

112 Toxicology. Chapters 1, 3 and 11.

113 Toxicology. Hazard communication. Chapters 1 and 11.

key of Agatha Christie's 1939 mystery – quoted above. Or, for example: the WWI chemical warfare gas chlorine $Cl_2$.[114,115]
- Energy phenomena – such as EM radiation, physical trauma, sound and high pressure – can under various conditions, also be toxic.
- *Chronic* toxicity goes over time; sometimes not noticed until much later – sometimes years. For example: The workplace hazards of benzene to the human blood system [432, 433, 434, 435]. Repetitive trauma encounters can also have chronic consequences [436].
- An acute toxic event can cause chronic results. For example: Many WWI veterans who suffered chlorine $Cl_2$ – or other poison gas – attacks had serious subsequent lifetime respiratory consequences. Or, for example a traumatic injury with lifetime consequences [437, 438, 439].
- Toxic results might not be perceived until after a long time. The mechanisms of damage for such a chronic effect might be as from a succession of small harms accumulated over time or from a chemical presence that remains from a time-ago that does not appear to take effect until later. This may be called *latent* toxicology. For example: Perhaps fibres of the workplace hazard asbestos [440, 441] would remain microscopically embedded in workers' body tissue, long before active disease is noticed.
- Toxicology can also include alcohol and drugs of human abuse and addiction; side effects of medically prescribed pharmaceuticals and iatrogenic effects.

Explaining and quantifying the toxicity of various substances can be difficult because many of the mechanisms would be complex and not be fully known, if even slightly known. The actual biological significance of measurable data may remain obscure, or if known, may still be problematic.

And, complications of attribution can arise, as for example, when an insurer would co-attribute a worker's asbestos-caused cancer with tobacco smoking [442]. Generally, chronic and latent toxicity are the most difficult to assess. But so too, the acute; although graphically presented dose-response data[116] can be useful.

### 1.9.2 Evaluation: $LD_{50}$

Toxicity evaluation of substances can be attempted with multiple short-term exposures using experimental lab animals. More meaningful – but more complex and expensive – would be larger population and longer-time animal studies and large popu-

114 Fritz Haber. $Cl_2$. WWI. Chapters 1 and 4.
115 Mustard gas. Chapters 1.9.1 and 2.3.1
116 Dose/response. Chapters 1 and 8.

lation and long-time human epidemiological studies. Commonly encountered is the $LD_{50}$ test = the chemical dose to kill 50% of a population of lab animals within a set number of days [443, 444, 445, 446].[117] $LD_{50}$ results have acute meaning, are at best limited and problematic and are easy to misinterpret[118]; in addition to being unkind to the rats, mice and rabbits.

$LD_{50}$, and other data, are often stated on safety data sheets[119,120,121] in respect to hazard communication requirements.[122]

Explaining and quantifying toxicity must be distinguished from identifying and measuring suspected toxins. The analytical chemistry is easier than the toxicology, but there are still problems, as for example, indicated in Chapter 3, regarding the measurement of lead Pb in humans [447].[123] Blood Pb measurement for a particular person at a particular time may be easily and accurately done, but its toxicological meaning may be rather difficult to reliably interpret.

### 1.9.3 Workplace health and exposure limits

Law as related to toxicology can be found in several areas: workplace, health and safety, workers compensation, environmental, criminal and civil litigation [448].[124] An aspect of this can be a prominent feature of workplace health regulation – the setting of chemical exposure limits – such as:

- Ontario *OH&S Act* control of exposure regulation [449].
- Canada *Criminal Code* limit on ethanol concentration allowed in blood, when operating a motor vehicle [450].[125]
- *Canada Labour Code* [451] occupational health and safety regulation [452], which references the Threshold Limit Values published by the American Conference of Governmental Industrial Hygienists [453, 454]. These TLVs are found referenced in various other regulatory documents, in other jurisdictions, notwithstanding that they are not intended by ACGIH for such use.

Chemical exposure limits are problematic because, in reality, they do not – and in theory, arguably cannot – determine health effects with specific dividing lines. There

117 $LD_{50}$ fragrance CCOHS disclaimer. Chapters 1, 6 and 9.
118 Toxicology. $LD_{50}$. Chapter 1.
119 SDS. Chapter 1.
120 SDS = Safety Data Sheet. Previously called MSDS = Material Safety Data Sheet.
121 Disclaimers. Error. Not legal advice. Commercial products. Introduction 2.4.1, Chapters 1, 5, 6 and 9.
122 Toxicology. Hazard communication. Chapters 1 and 11.
123 Pb toxicology. Old-stock housing. Pb Sample selection. Chapters 1, 2 and 3.
124 *N*-Nitrosodimethylamine CAS 62-75-9. Chapter 1.
125 Toxicology. Driving impaired. Chapters 1 and 11.

would be a health relationship – but not too precise – because the science is not advanced well-enough. And because, humans have differences, one from another, so that distributions within populations must be considered. And, because human exposures have so many different possibilities. And, because – necessarily – there should be, carefully included, some assumed science and arbitrary definitions. And, even if precision could be claimed by some scientists, universal agreement amongst others may, for legitimate reasons, be hard to find.

The limits are, at best, advisory for actual health purposes. The limits may be used for legal requirements, to trigger law enforcement and to trigger inspectorial interest, but this should not be confused with health *per se*. But they frequently are.

To its credit, ACGIH has tried to explain such concepts in its TLV book [455],[126] offering its numbers as guides for professional occupational hygienists and implying that its TLVs are best not used for law enforcement. This apparently did not persuade the Governor-in-Council for the *Canada Labour Code* regulations; or perhaps it was not noticed.

Exposure limits for law enforcement present another problem: measurement. While *Criminal Code* ethanol in blood measurements, and technicians who measure, are defined in law that would effectively override scientific deficiency (if any), some other measurements would be poorly defined, if at all. And, be poorly attempted.

And, exposure limits declared to trigger legal consequences as at "any detectable level" are especially problematic – and arguably ignorant – because modern analytical chemistry can deliver measurements well below levels of toxicological meaning.[127]

## 1.10 Hazards of found chemicals

Sometimes obviously suspicious, things or situations would appear to require determinations about chemical substances in an urgent context.[128,129,130] Law enforcement and other public service and safety personnel must decide what to do quickly, without the more careful chemical analysis that would follow later. These are situations that are potentially dangerous beyond usual workplace and public health and safety concerns – about drugs, toxins or explosives. Concerns would relate to:

- Facilities that make or store
- Sites of release or explosion
- Containment and isolation of the chemicals
- Isolation, care and monitoring of people involved

**126** Exposure limits. ACGIH. Chapter 1.9.3.
**127** Too sensitive. Toxicology. Chapters 1, 3 and 8.
**128** Pb toxicology. Pb Sample selection. Chapters 1, 2 and 3.
**129** Plutonium. Periodic table. Too sensitive? Chapters 1.10, 3.7.2 and 4.2,3.
**130** Dual role. Chapter 1.

- Restriction of other people
- Collection of evidence
- Public notices and statements

When these would be related to publicly known situations – such as a spill at a commercial site – the forensic scientist may be called upon.

But when related to investigating and prosecuting criminal activity, it becomes more difficult. For example: What to do with the found chemicals at a clandestine drug- or bomb-making lab and how to interpret what is found for criminal prosecution; what is to be the role of the forensic scientist? [456] And how to distinguish unusual but innocent activity [457, 458] from the actual criminal? Or, for example Anthrax [459, 460].

For both defence and Crown counsel, this may present difficulties of interpretation in the criminal law process, starting at the bail hearing [461].

## Notes

**1** From a trial lawyer's affidavit, used on appeal (by the appeal lawyer) – *R v K* [1995] 24 OR (3d) 199B204, at 201, Ontario Court of Appeal, 5 June 1995. The affidavit recounted the trial lawyer's advice, and resulting instructions, in a case where the client originally plead guilty for something he denied doing – as part of a plea bargain.

**2** Guy Paul Morin – quoted by Kirk Makin, Justice Reporter, "Stoic Morin breaks down relating terror of prison/On second day, as inmates lined corridors, 'you could hear a pin drop'," *The Globe and Mail*, 28 October 1997, page A12. www.theglobeandmail.com/.

**3** Robert Harris, *An Officer and a Spy* [an historical novel – about the Dreyfus affair, 1890s, France, involving injustice, political wrong-doing, and forged documents – as a narrative of Colonel Georges Picquart], Hutchinson, Random House, London, SW1V 2SA; ISBN 978-0-09-194456-8; 2013.

**4** At page 226.

**5** "Scepticism is the first step towards truth". – Denis Diderot, *Pensées philosophiques* [1746].
https://www.goodreads.com/quotes/573409-scepticism-is-the-first-step-towards-truth
https://www.saveaquote.com/quotes/truth-and-lies/quote-250623
https://en.wikiquote.org/wiki/Denis_Diderot

**6** CBC Radio, Ideas, "All the World's Knowledge in 28 Volumes: Diderot's radical Encyclopédie," 2019.
https://www.cbc.ca/radio/ideas/all-the-world-s-knowledge-in-28-volumes-diderot-s-radical-encyclop-die-1.5168887

**7** See: The Honourable Louise Arbour, "Searching for Truth," CBC Ideas, 17 January 2025; Horace E. Read Memorial Lecture, Dalhousie University, Schulich School of Law, November 2024.
https://www.cbc.ca/radio/ideas/louise-arbour-search-for-truth-1.7433321

**8** Sometimes with judge alone; sometimes judge with jury; sometimes for appellate courts a panel of judges

**9** See also – regarding juries: https://en.wikipedia.org/wiki/Jury#:~:text=A%20jury%20is%20a%20sworn,of%20up%20to%2015%20people

**10** See also – regarding juries: The Rt. Hon. LORD Denning, *Landmarks in the Law*, Butterworths, London, 1984. ISBN 0-406-17614-0. https://www.gettextbooks.com/isbn/9780406176141/

**11** Such as Parliamentary committees, Royal Commissions of Enquiry, coroners courts – inquests. each with procedural rules.
**12** https://en.wikipedia.org/wiki/Royal_Commission_on_the_Health_and_Safety_of_Workers_in_Mines#:~:text=The%20Royal%20Commission%20on%20the,the%20safety%20of%20underground%20mines
**13** https://www.ontario.ca/page/coroners-inquests
https://www.ontario.ca/laws/statute/90c37
**14** https://www.ontario.ca/laws/statute/90s22
**15** And, sometimes a private arbitration panel – as would have been contracted for in advance by the parties. Such private arbitration is sometimes a good idea – but often not when one of the parties is an individual consumer as against a large business corporation.
**16** See: Emily Flitter, "JPMorgan Chase Pushes an Arbitration Policy," *The New York Times*, BUSINESS, 5 June 2019, page B3.
**17** *Gedankenexperiment*. Introduction 2.2.
**18** Canada:
https://en.wikibooks.org/wiki/Canadian_Criminal_Procedure_and_Practice/Arrest_and_Detention/Arrest_Procedure
https://en.wikibooks.org/wiki/Canadian_Criminal_Procedure_and_Practice/Arrest_and_Detention/Right_to_Counsel
**19** USA: https://en.wikipedia.org/wiki/Miranda_warning
https://en.wikipedia.org/wiki/Miranda_v._Arizona
https://www.washingtonpost.com/archive/entertainment/books/1983/08/07/the-miranda-decision-criminal-wrongs-citizen-rights/9955124b-20b8-4ac6-8b82-3652b79a04e8/
https://www.baltimoresun.com/2010/06/02/eroding-miranda/
https://www.baltimoresun.com/opinion/op-ed/bs-ed-op-0302-juvenile-interrogations-20210301-llha2qrmkjhybdpbt4lh7cz56a-story.html
**20** See David T. Case & Jeffrey B. Ritter, *Disconnects Between Science and the Law/The clash between legal and scientific truth seeking takes on a new dimension in the ever-changing world of electronic commerce, C&EN*, 14 February 2000, pages 49–60; and references cited therein.
https://www.researchgate.net/publication/274590274_Disconnects_between_science_and_the_law
**21** *Gedankenexperiment*. Introduction 2.2.
**22** Within the range of the uncertainties of forensic science.
**23** https://bccla.org/privacy-handbook/main-menu/privacy7contents/privacy7-1-5.html.
**24** *Gedankenexperiment*. Introduction 2.2.
**25** = The prosecutor on behalf of the state.
**26** *Gedankenexperiment*. Introduction 2.2.
**27** *Merriam-Webster* ". . . akin to Latin *foris* outside, *fores* door. . ." ". . .a large public place in an ancient Roman city that was used as the center of business. . .".
https://www.merriam-webster.com/dictionary/forum.
**28** Latin forensic terms

[. . .]

> Many Latin terms are used in the field of forensic science because forensics developed alongside the already established legal profession, which extensively uses . . . The word forensic, . . . the Latin word *forensis*, meaning of the forum. It originally applied to the marketplace areas within ancient Rome . . . governmental debates and actions by courts of law, . . . Entering . . .English . . . 1659, the modern meaning . . . now limited to . . . legal and criminal investigations.

[. . .]

https://www.encyclopedia.com/science/encyclopedias-almanacs-transcripts-and-maps/latin-forensic-terms

**29** See:

- Brian H. Kaye, *Science and the Detective / Selected Reading in Forensic Science*, VCH Verlagsgesellschaft mbH, Weinheim, Germany, 1995, page 3, and references cited therein.

*The Oxford English Dictionary/Second Edition*, Vol. VI, Clarendon Press, Oxford, 1989, page 55.

**30** https://www.vocabulary.com/dictionary/forensic
https://www.merriam-webster.com/dictionary/forensic
https://www.merriam-webster.com/dictionary/forum#learn-more.

**31** With concepts and meanings perhaps somewhat expanded by the author.

**32** In the Canadian system, Parliament (= Commons, Senate & Monarch (represented by the Governor General)) enacts statutes. The Governor-in-council – effectively the cabinet – brings regulations into effect. Both statutes and regulations (to specifically implement parts of the statutes) are considered as legislation. The use of the term legislation should be considered here in this context.

**33** *Gedankenexperiment*. Introduction 2.2.

**34** Followed by Royal Assent and proclamation; and enabling regulations.

**35** And, maybe the two unions have contradictory policies for the same occurrences.

**36** 16 US CFR § 1500.14 https://www.law.cornell.edu/cfr/text/16/1500.14
15 US Code § 1277. Labeling of art materials https://www.law.cornell.edu/uscode/text/15/1277
Illinois Compiled Statutes 105 ILCS 135/) Toxic Art Supplies in Schools Act. http://www.ilga.gov/legislation/ilcs/ilcs3.asp?ActID=1022&ChapterID=17

**37** Richard Saferstein, *Criminalistics/An Introduction to Forensic Science*, Seventh Edition, Prentice Hall, Upper Saddle River, New Jersey, 07458; 2001; ISBN 0-13-013827-4.

**38** Saferstein, *Criminalistics* . . ., 2001; page 1, *et seq.*

**39** Brian H. Kaye, *Science and the Detective/Selected Reading in Forensic Science*, VCH Verlagsgesellschaft mbH, Weinheim, Germany, 1995, page 3.

**40** See: associated press, "Investigators reconsider mechanical malfunction as cause of TWA [Flight 800] crash/Possibility gains credence as bomb theory is weakened," *Baltimore Sun*, 24 September 1996, page 3A.
See also: Sam Roberts, *"James Kallstrom, 78, Who Led F.B.I. Review of T.W.A. Crash, Dies," The New York Times*, OBITUARIES, 10 July 2021, page A20.

**41** https://www.ntsb.gov/investigations/AccidentReports/Reports/AAR0003.pdf
https://en.wikipedia.org/wiki/TWA_Flight_800_conspiracy_theories
https://en.wikipedia.org/wiki/TWA_Flight_800.

**42** Kim Honey, Science Reporter, "Optometrist loves to get felons in his sights/University of Waterloo professor [Graham Strong] has had a hand in building cases against killers, just by examining their eyeglasses," *The Globe and Mail*, SCIENCE & RELIGION, 9 February 1999, page A11.

**43** See Isabel Vincent, Crime Reporter, "Home-grown supersleuths in global demand/From knot analysis to geographic profiling, Canadian police officers are pioneering investigative techniques being used around the world. / Canada's police methods go global," *The Globe and Mail*, 29 December 1997, pages A1 and A6.

**44** https://www.youtube.com/watch?v=KmhGWCvxKF8
https://www.youtube.com/watch?v=kxaWqKM5JyQ
https://www.youtube.com/watch?v=M-i9v9VfCrs

**45** Thomas Phinney/The Font Detective https://thefontdetective.com/

**46** AIH CBC, 16 January 2019.

**47** Aidan Macnab, "Font expert exposes phony trust documents," *Canadian Lawyer Magazine*, 17 January 2019.
**48** https://www.theguardian.com/money/2015/jan/20/how-become-forensic-musicologist
https://www.cbc.ca/radio/thecurrent/the-current-for-aug-5-2019-1.5234526/inspiration-vs-plagiarism-how-copyright-lawsuit-of-katy-perry-s-hit-dark-horse-puts-spotlight-on-music-theft-1.5234532
**49** https://www.cigaraficionado.com/
https://www.cbc.ca/listen/live-radio/1-23-ideas/clip/16068835-puro-cubano-the-meaning-tobacco-cuba
**50** Juliet Macur, "*Trump Has Made Claims About Abrego Garcia's Tattoos. Here's a Closer Look.* / Gang experts say the tattoos on Kilmar Abrego Garcia's knuckles are unlikely to signify gang membership. The president says otherwise," *New York Times*, 8 May 2025.
https://www.nytimes.com/2025/05/08/us/kilmar-abrego-garcia-tattoos-ms-13.html
**51** See Kaye, *Science and the Detective* . . ., 1995, pages 323–326, and references cited therein.
**52** See also: Charles Homans, "The Strange, Post-Partisan Popularity of the Unabomber/When Ted Kaczynski's manifesto appeared 30 years ago, the internet was brand-new. Now his dark vision is finding fans who don't remember life before the iPhone," *New York Times*, 22 March 2025.
https://www.nytimes.com/2025/03/22/magazine/unabomber-ted-kaczynski-luigi-mangione.html
**53** – sometimes assisted by a jury. Also, some law cases are determined by non-court tribunals; these administrative law cases are heard by a panel of one or more quasi-judicial officials.
**54** In many administrative contexts the term "judge" is used for the presiding-person. But often other terminology: "justice," "chancellor," "commissioner, "tribunal chair," "president," "judicial panel," . . .
**55** but see:

- https://en.wikipedia.org/wiki/Jury_nullification
  https://en.wikipedia.org/wiki/Jury_Nullification_(book)
  http://www.thecourt.ca/r-v-krieger-jury-nullification-and-the-limits-of-descriptive-mens-rea/
- Nancy Gertner & Joel Cohen, "What Judges Can Do for the Rule of Law Under Trump," Opinion, Guest Essay, *New York Times*, 26 December 2024.
  https://www.nytimes.com/2024/12/26/opinion/judges-trump-rule-of-law.html

**56** See also [re trial, September 1670, on charge of unlawful assembly, the previous August; London; wherefor the jury was fined and imprisoned until paid, for coming to the wrong verdict.] The Rt. Hon. LORD Denning, *Landmarks in the Law*, Butterworths, London, 1984. Part Five, "Freedom of Assembly," "William Penn and William Mead," at pages 148, 149, 150; ISBN 0-406-17614-0.
https://catalogue.nla.gov.au/catalog/1312655
**57** https://www.quakersintheworld.org/ideas-for-educators/78/The-Penn-Mead-Trial
*Bushel's Case*, Court of Common Pleas,1670, 124 ER 1006, Sir John Vaughan, Judge.
https://en.wikipedia.org/wiki/Bushel%27s_Case
Brian Manarin, *Forensic Evidence in Context: Cases, Materials and Commentaries*; Carswell, THOMSON REUTERS, Canada, 2017, June 2019; ISBN: 9780779880683
https://store.thomsonreuters.ca/en-ca/pdp/forensic-evidence-in-context-cases-materials-and-commentaries/30835077
https://static.legalsolutions.thomsonreuters.com/product_files/relateddocs/220441_ 2019288_111518.pdf
**58** See also: Ronald Joseph Delisle, Don Stuart, David M. Tanovich, Lisa Dufraimont; *EVIDENCE / PRINCIPLES AND PROBLEMS*/Twelfth Edition; CARSWELL/THOMSON REUTERS CANADA, Toronto M1T 3V4; 2018. ISBN 978-0-7798-8655-5. https://digitalcommons.osgoode.yorku.ca/faculty_books/362/
**59** Manarin, *Forensic Evidence* . . ., 2019; [novel scientific theory; *Daubert*], page 788, et seq.
**60** Manarin, *Forensic Evidence* . . ., 2019; KF 9674 F68 2017
http://library.lsuc.on.ca/vwebv/search?searchArg=Forensic+Evidence+in+Context%3A+Cases%2C+Materials+and+Commentaries&searchCode=TKEY%5E&setLimit=1&recCount=10&searchType=1&page.search.search.button=Search

**61** https://www.torontopubliclibrary.ca/detail.jsp?Entt=RDM3754081&R=3754081.
**62** Manarin, *Forensic Evidence. . .*, 2019.
https://books.google.ca/books?id=s-pbtAEACAAJ& source=gbs_ViewAPI&redir_esc=y.
**63** as distinct from an appeal court. The trial court is the court of first instance; or analogously, an administrative tribunal.
**64** Agatha Christie, *Ordeal by Innocence*, 1958; WILLIAM MORROW, Harper Collins, 2018, New York City, 10007; ISBN 978-0-06-288473-2; at pages 73 & 74.
https://www.loot.co.za/product/agatha-christie-ordeal-by-innocence-tv-tie-in/nckv-5541-g390.
**65** But sometimes, when a fair chance has been denied to one of the *parties* – and it sometimes happens – justice fails, and the Appellate Courts have been known to sometimes interfere.
**66** MERRIAM-WEBSTER DICTIONARY https://www.merriam-webster.com/legal/functus%20officio
". . . *functus officio*: of no further official authority or legal effect // the warrant . . . had been returned and was *functus officio* – *Cady v. Dombrowski*, 413 U.S. 433 (1973) // once an arbitrator makes an award she is *functus officio* – used especially of an officer who is no longer in office or of an instrument that has fulfilled its purpose . . .".
**67** [By this concept, a Judge's judgment & orders, after being issued, & the court adjourns, stand on their own as valid enforceable documents, independent of the Judge, and the Judge does not get to revisit them; and even the Judge's explanatory commentary would seem inappropriate. There would be specific exceptions.]
**68** https://www.britannica.com/topic/res-judicata
https://en.wikipedia.org/wiki/Res_judicata
**69** See: Robert Harris, *An Officer and a Spy* [an historical novel: "Author's Note / . . . the true story of the Dreyfus affair, perhaps the greatest miscarriage of justice and political scandal in history, which in the 1890s came to obsess France and the ultimately the entire world . . ." – as a narrative of Colonel Georges Picquart], HUTCHINSON, Random House, London, SW1V 2SA; ISBN 978-0-09-194456-8; 2013.
**70** At page 226.
**71** For example, see "Junk science in courts challenged by C[hemical] M[anufacturers] A[ssociation]," government concentrates, *Chemical & Engineering News*, 16 December 1996, page 25.
**72** Ontario Ministry of the Attorney General, *Report of the Kaufman Commission on Proceedings Involving Guy Paul Morin* [1997], Executive Summary.
https://www.attorneygeneral.jus.gov.on.ca/english/about/pubs/morin/morin_esumm.html
https://www.ontario.ca/feedback/contact-us?id=98538&nid=97157#contactForm.
**73** https://www.thecanadianencyclopedia.ca/en/article/morin-inquiry-revelations
https://en.wikipedia.org/wiki/Kaufman_Report
https://archive.macleans.ca/article/1997/10/13/dubious-science
http://netk.net.au/Canada/Morin5.asp.
**74** See:
- Sean Silcoff, "Probe begins into Morin's wrongful conviction," *The Globe and Mail*, 5 September 1996, page A8;
- "Justice for all," a *Globe and Mail* editorial on the same topic, 6 June 1996, page A18.
- Kirk Makin, "Ontario awards Morin compensation/Unexpected move includes apology for wrongful murder conviction and $1.25-million for family's 12-year ordeal," *The Globe and Mail*, 25 January 1997, pages A1 & A10.

Kirk Makin, Justice Reporter, Whitby, Ontario, "The lawyer who has 'a cause for a client'/James Lockyer is famous for clearing the wrongfully convicted. He's not so big on helping the guilty./'It is hard not to live, eat and sleep each case'." *The Globe and Mail*, column one, 12 May 1999, pages A1 and A12.

75 See Kirk Makin, Justice Reporter, Birmingham, "When a guilty finding raises a reasonable doubt/ Ottawa considers British case-review commission as possible model for reviewing inmates' claims of wrongful conviction," *The Globe and Mail*, national news, 19 December 1998, page A15.
76 See Kirk Makin, Justice Reporter, "Conference hears case for legal reform/Morin's wrongful murder conviction should prompt more change, participants say," *The Globe and Mail*, national news, 13 November 1999, page A11.
77 For example: Tu Thanh Ha, "Quebec court clears dead man of slaying/Ruling that overturns Michel Jetté's 1988 conviction is a Canadian first," *The Globe and Mail*, 22 October 1999, page A3.
78 See:

- "Atoning for the three Ms of the Canadian justice system/Canada should follow Britain's lead in setting up a criminal review system," *The Globe and Mail*, Wednesday, November 24, 1999;
- "Tempering mercy with justice for David Milgaard/Why is it taking so long to reform the system for exonerating the wrongfully convicted?" *The Globe and Mail*, Tuesday, November 23, 1999;
- Timothy Appleby and David Roberts, "What delayed justice for 30 long years/The police already had David Milgaard/Did another suspect just complicate matters?" *The Globe and Mail*, Tuesday, November 23, 1999. www.theglobeandmail.com
- Joyce Milgaard & Peter Edwards, *A Mother's Story: My Battle to Free David Milgaard*, Doubleday Canada, Toronto M5B 1Y3; 1999; ISBN 0-385-25807-0.
- https://quillandquire.com/review/a-mother-s-story-my-battle-to-free-david-milgaard/
- https://www.amazon.ca/s?k=9780385258074&i=stripbooks&linkCode=qs
- Colin Perkel, "Canadian Press "Joyce Milgaard, tireless activist mother of wrongfully convicted son, dies / Milgaard, 89, a mother of four children, died in a care home in Winnipeg on Saturday, her son said," *The Globe and Mail*, 22 March 2020.
  https://www.cbc.ca/news/canada/manitoba/joyce-milgaard-obituary-1.5506203

79 See: Ronna Syed, Shanifa Nasser, "Toronto police identify killer in cold case of 9-year-old Christine Jessop / Sources say killer of 9-year-old Ontario girl in 1984 died by suicide in 2015," CBC News, 15 Oct. 2020.
https://www.cbc.ca/news/canada/toronto/christine-jessop-news-conference-1.5763673
https://www.cbc.ca/radio/asithappens/as-it-happens-the-friday-edition-1.5765309.
80 See: Nate Hendley, *The Boy on the Bicycle: A Forgotten Case of Wrongful Conviction in Toronto*, Five Rivers Chapmanry; 2018; ISBN-10: 1988274516; ISBN-13: 978-1988274515.
https://www.amazon.ca/Boy-Bicycle-Forgotten-Wrongful-Conviction/dp/1988274516
https://www.cbc.ca/radio/thecurrent/the-current-for-december-3-2018-1.4929818/arrested-youth-should-not-be-interrogated-alone-says-man-wrongfully-convicted-of-murder-1.4929826
81 See also: Canadian Press, "Samples sought by Truscott may be lost," *The Globe and Mail*, 3 October 1997, page A10: ". . . A review by Ontario's largest crime laboratory suggests it has lost semen samples that might have helped a man prove he was wrongly convicted of murder 38 years ago. No one at the Centre of Forensic Sciences is quite sure where the samples. . . are now, or whether they still exist . . . Lawyers for Mr. Truscott, convicted of strangling 12-year-old Lynne Harper in 1959 and originally sentenced to death, asked the government for help in finding samples from the crime scene that could be submitted to a DNA test. They hope such a test would prove he did not kill the girl. Records at the centre suggest it once had semen samples found on the victim's body . . . and those records indicate that any samples would have been returned to the Crown . . .". See also Kirk Makin, "Many crime samples gone, coroner says / Case-by-case review, retesting called impossible; even some work seen as enormous task," *The Globe and Mail*, 4 October 1997, page A6.
82 Harvey Cashore, Rachel Ward, Mark Kelley, "B.C. appoints special prosecutor to probe Crown conduct in decade-old child murder," case CBC News, Canada, FIFTH ESTATE, January 2020. https://www.cbc.ca/news/canada/bc-special-prosecutor-autopsy-fifth-estate-1.5430387
https://www.cbc.ca/news/canada/british-columbia/cranbrook-babysitter-charged-with-mur der-1.1123313

83 – Margaret Wente, "Scandal of the Century," *The Globe and Mail*, COMMENT/COUNTERPOINT, 6 January 2004, page A15. [re Klassen case.]
– CBC, Feb. 2004, "Klassen family gets apology from Sask. justice minister." https://www.cbc.ca/news/canada/klassen-family-gets-apology-from-sask-justice-minister-1.469822

**84** Peter Worthington and Kyle Bown, *SCAPEGOAT / How the Army Betrayed Kyle Brown*, Seal Books, Toronto M5B 1Y3; 1997; ISBN 0-770-42755-3.
https://www.amazon.ca/Scapegoat-Peter-Worthington/dp/0770427553
https://en.wikipedia.org/wiki/Somalia_affair
https://nationalpost.com/news/canada/it-takes-everything-veteran-went-off-the-grid-after-torturing-teen-in-somalia-left-him-with-ptsd
https://edmontonjournal.com/news/insight/life-after-somalia-kyle-brown-ptsd-and-the-past

**85** Mefloquine CAS 53230-10-7.

**86** = *lariam* = *mefaquin*.

**87** $C_{17} H_{16} F_6 N_2 O$ 378.312 g/mol (see Fig. 1.7).

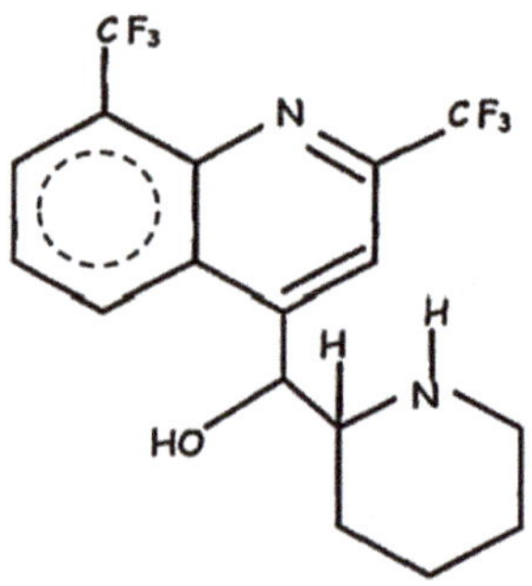


Fig. 1.7: Mefloquine, CAS 53230-10-7.

**88** http://www.cbc.ca/news/health/story/2012/04/10/malaria-drug-mefloquine.html http://www.cbc.ca/gsa/?q=mefloquine.

**89** https://en.wikipedia.org/wiki/Mefloquine Mefloquine CAS 53230-10-7
M.G., "some possible chemical & physical exposures in veterans' health claims," Toronto Workers' Health & Safety *newsletter*, 2012 07. Vol. 20, No. 2, pages 1–4; "*malaria prevention & mefloquine*," at page 4.
2012 07. Vol. 20, No. 2.doc
http://newsletter.workers-safety.ca.s3-website-us-east-1.amazonaws.com/#!/newsletters%2F2012%2007%2F
http://workers-safety.ca/features/newsletters/http://workers-safety.ca/index.html
http://newsletter.workers-safety.ca.s3-website-us-east-1.amazonaws.com/#!/newsletters%2F.

**90** CBC radio Podcasts, *Uncover*, "Satanic Panic," 2020. "In the 1980s, rumours swirled across North America about Satanic cults torturing and terrorizing children. It led to mass hysteria – yet there was no proof of the alleged crimes. So what happened? (Ben Shannon/CBC)"
https://www.cbc.ca/radio/uncover/season-6-satanic-panic-1.5437487

**91** https://www.cbc.ca/news/canada/british-columbia/tammy-marion-bouvette-court-ruling-acquittal-1.7554170

**92** See:

- Kirk Makin, Justice Reporter, Cardiff, "Exoneration comes 46 years too late/Mahmoud Mattan was hanged in 1952 and cleared this year, small . . . [consolation] to his widow and children," *The Globe and Mail,* CRIMINOLOGY & SCIENCE, 18 December 1998, page A17.
- *The Globe and Mail,* CRIMINOLOGY & SCIENCE, 18 December 1998, page A17, "Man acquitted in bombing case," "A convicted IRA bomber who spent 12 years in British prisons was acquitted yesterday in the face of fresh evidence . . . ."

Kirk Makin, Justice Reporter, Birmingham, "When a guilty finding raises a reasonable doubt/Ottawa considers British case-review commission as possible model for reviewing inmates' claims of wrongful conviction, "*The Globe and Mail,* NATIONAL NEWS, 19 December 1998, page A15.

**93** In a non-scientific context, from Charles Dickens, *Bleak House,* Signet Classic, The New American Library, New York City, 1964 and 1980, "In Chancery," Chapter I, page 18:

> [. . .]
>
> Never can there come fog too thick, never can there come mud and mire too deep, to assort with the groping and floundering condition which this High Court of Chancery, the most pestilent of hoary sinners, holds this day in the sight of heaven and earth.
>
> [. . .]

**94** https://www.laphamsquarterly.org/city/never-can-there-come-fog-too-thick.

**95** Selam Gebrekidan, "*British Man Freed From Prison Over Video Misidentification/*Ademola Adedeji was ensnared in a murder conspiracy case with no murder victim. Prosecutors labeled him a gang member and won a conviction and eight-year sentence," *New York Times,* 15 January 2025.
https://www.nytimes.com/2025/01/15/world/europe/uk-conspiracy-gang-video-misidentification-trial-overturned.html

**96** Megan Specia, *"Letby's Lawyers Seek Appeal After Lead Expert' Changed Mind' on Murders/*Lawyers for Lucy Letby, a British nurse found guilty of the murders of seven babies, said that new evidence about the prosecution's lead witness undermined the case against her." *New York Times,* 16 Dec. 2024. ["Lead" here ≠ Pb]
https://www.nytimes.com/2024/12/16/world/europe/letby-murders-appeal.html

**97** Megan Specia, "*Top Doctors Raise Grave Doubts Over Conviction of 'Killer Nurse' Lucy Letby/*After assessing the cases of 17 babies cited at her original trial, a panel of world-renowned specialists said that they had found no evidence that Ms. Letby had murdered anyone," *New York Times,* 4 February 2025.
https://www.nytimes.com/2025/02/04/world/europe/lucy-letby-nurse-uk-appeal-evidence.html

**98** Megan Specia, "*Former Hospital Managers Call for Halt to Inquiry into Lucy Letby Murders/*Former executives at the British hospital where Ms. Letby was a nurse requested that a judge's inquiry into the killings there be suspended amid new evidence," *New York Times,* 18 March 2025.
https://www.nytimes.com/2025/03/18/world/europe/lucy-letby-inquiry-thirlwall.html

**99** Megan Specia, *"One Doctor's Quest for the Truth About Convicted Killer Lucy Letby/*Dr. Shoo Lee's research was used to help convict a British nurse of murdering babies, but he says it should never have been cited," *New York Times,* 21 March 2025.
https://www.nytimes.com/2025/03/21/world/europe/lucy-letby-nurse-murder.html

**100** Michael D. Shear, "*Insulin Test Used to Convict Lucy Letby in Babies' Deaths Was Unreliable, Experts Say/*A new report by neonatal and toxicology experts casts doubt on the validity of the test results that were pivotal to the prosecution's case," *New York Times,* 3 April 2025.
https://www.nytimes.com/2025/04/03/us/politics/uk-lucy-letby-insulin-test-evidence.html

**101** See Alan Berlow, "The Wrong Man," *the atlantic monthly*, Volume 284 No.5, November 1999, pages 66–91.
https://www.theatlantic.com/magazine/archive/1999/11/the-wrong-man/377858/
**102** Michael Ratner, President, Center for Constitutional Rights, New York, 13 October 2002, "The Jogger Convictions," *The New York Times*/letters, 15 October 2002, page A26.
**103** See Francis Russell, "Sacco Guilty, Vanzetti Innocent?" *American Heritage*, June 1962, Volume13, Issue 4.
https://www.americanheritage.com/sacco-guilty-vanzetti-innocent.
https://www.nytimes.com/2025/05/29/arts/design/ben-shahn-jewish-museum.html
https://en.wikipedia.org/wiki/The_Passion_of_Sacco_and_Vanzetti
https://whitney.org/collection/works/1022
**104** Tim Prudente, "Judge exonerates three men in 1983 'Georgetown jacket' school killing," *Baltimore Sun*, 25 November 2019.
https://www.baltimoresun.com/2019/11/25/judge-exonerates-three-men-in-1983-georgetown-jacket-school-killing/
**105** Tom Jackman, "Arrested as teens, three men exonerated after 36 years behind bars for wrongful murder conviction/Baltimore prosecutor's Conviction Integrity Unit finds men are innocent in 1983 slaying of 14-year-old, shot in school hallway for his jacket," *The Washington Post*, True Crime, 25 November 2019.
https://www.washingtonpost.com/crime-law/2019/11/25/arrested-teens-three-men-are-expected-be-exonerated-after-years-behind-bars-wrongful-murder-conviction/
**106** https://www.nytimes.com/2019/11/28/us/conviction-integrity-unit-innocence.html.
**107** Jacey Fortin & Allyson Waller, "[Mississippi] *Death Row Conviction Dismissed Over Doubts in Bite-Mark Evidence* [>¼ century behind bars]," *The New York Times*, 15 January 2021, page A20.
**108** The possibility for error has been an important component in argument for death penalty abolition; but were apparently not mentioned as factors for recent commutations by President Joe Biden.
**109** Aishvarya Kavi, "*Biden Commutes 37 Death Sentences Ahead of Trump's Plan to Resume Federal Executions/*Those affected by the president's action on Monday are still subject to life imprisonment without the possibility of parole. Three men will remain on federal death row," *New York Times*, 23 Dec. 2024.
https://www.nytimes.com/2024/12/23/us/politics/biden-commutes-37-death-sentences.html
**110** Pamela Colloff, "He Dialed 911 to Save His Baby. Then His Children Were Taken Away./The controversial medical diagnosis of shaken baby syndrome can send parents to jail. What if the symptoms are caused by something else?" *New York Times*, 29 December 2024.
https://www.nytimes.com/2024/12/29/magazine/shaken-baby-syndrome-diagnosis.html
https://www.nytimes.com/by/pamela-colloff
**111** Alexandra E. Petri, "*Hawaii Man Wrongfully Convicted of Murder Is Freed After 30 Years/*Gordon Cordeiro, 51, who was serving a life sentence for a 1994 murder on Maui, was released on Friday after DNA evidence was presented in his case," *New York Times*, 24Feb.2025.
**112** Remy Tumin, "*He Served 36 Years for His Wife's Murder and Then Forgave the Man Who Confessed/*Leo Schofield maintained he was innocent all along in his wife's 1987 murder. For "Bone Valley," a podcast about the case, he connected with the man who said he did it," *New York Times*, 9 April 2025.
https://www.nytimes.com/2025/04/09/us/bone-valley-leo-schofield.html
https://www.nytimes.com/2024/04/17/us/leo-schofield-bone-valley-parole.html
**113** Crime Story with Kathleen Goldhar, "Hypnosis, false memory and lies: A Texas Ranger's strange interrogation techniques," CBC, 24 Feb. 2025
https://www.cbc.ca/listen/cbc-podcasts/1399-crime-story/episode/16129931-hypnosis-false-memory-and-lies-a-texas-rangers-strange-interrogation-techniques

**114** See also: https://www.latimes.com/california/story/2024-08-07/crimes-of-the-times-the-george-franklin-repressed-memory-murder-case

**115** Jacey Fortin, *"36 Deaths in Police Custody Should Have Been Called Homicides, Report Finds*/An audit of deaths in Maryland found that in cases of people who died after being restrained by the police, medical examiners often classified the deaths as natural or accidental," New York Times, 15 May 2025.
https://www.nytimes.com/2025/05/15/us/maryland-police-custody-autopsies-audit.html

**116** Hurubie Meko & Jonah E. Bromwich, "*Conviction Reversed in Etan Patz Case That Put Focus on Missing Children*/Pedro Hernandez, a former bodega stock clerk, had confessed years after the 1979 crime to luring Etan Patz, 6, into a basement in SoHo. The case shook New York City," *New York Times*, 21 Juy 2025.
https://www.nytimes.com/2025/07/21/nyregion/etan-patz-conviction-overturned.html

**117** Dan T. Carter, *SCOTTSBORO: A Tragedy of the American South*, revised edition; Louisiana State University Press, Baton Rouge, 70803; 1969, 1979, 1982; ISBN 0-8071-0498-1 https://www.isbns.co.tt/isbn/9780807104989/.

**118** https://en.wikipedia.org/wiki/Scottsboro_Boys.

**119** See also: CBC Radio *Tapestry*, "Champion of the Damned: Bryan Stevenson's fight to fix a broken justice system," 2014.
https://www.cbc.ca/radio/tapestry/champion-of-the-damned-bryan-stevenson-s-fight-tofix-a-broken-justice-system-1.2902489

**120** See: Kirk Makin, Justice Reporter, "Conference hears case for legal reform/Morin's wrongful murder conviction should prompt more change, participants say," *The Globe and Mail*, national news, 13 November 1999, page A11.

**121** Legislation: In the Canadian system, Parliament (= Commons, Senate & Monarch (represented by the Governor General)) enact statutes. The Governor-in-council – effectively the cabinet – brings regulations into effect. Both statutes and regulations (to specifically implement parts of the statutes) are considered as legislation. The use of the term legislation should be considered here in this context.

**122** "In Congress, July 4, 1776/The unanimous Declaration of the thirteen United States of America"
https://www.archives.gov/founding-docs/declaration-transcript
https://www.archives.gov/founding-docs/declaration
https://www.archives.gov/founding-docs/declaration
https://catalog.archives.gov/id/1419123.

**123** [. . .]

> We hold these truths to be self-evident, that all men are created equal, that they are endowed by their Creator with certain unalienable Rights, that among these are Life, Liberty and the pursuit of Happiness.–That to secure these rights, Governments are instituted among Men, deriving their just powers from the consent of the governed, –That whenever any Form of Government becomes destructive of these ends, it is the Right of the People to alter or to abolish it, and to institute new Government, laying its foundation on such principles and organizing its powers in such form, as to them shall seem most likely to effect their Safety and Happiness. Prudence, indeed, will dictate that Governments long established should not be changed for light and transient causes; and accordingly all experience hath shewn, that mankind are more disposed to suffer, while evils are sufferable, than to right themselves by abolishing the forms to which they are accustomed. But when a long train of abuses and usurpations, pursuing invariably the same Object evinces a design to reduce them under absolute Despotism, it is their right, it is their duty, to throw off such Government, and to provide new Guards for their future security.–Such has been the patient sufferance of these Colonies; and such is now the necessity which constrains them to

alter their former Systems of Government. The history of the present King of Great Britain is a history of repeated injuries and usurpations, all having in direct object the establishment of an absolute Tyranny over these States. . . .

[. . .]

**124** For example: Tu Thanh Ha, "Quebec court clears dead man of slaying/Ruling that overturns Michel Jetté's 1988 conviction is a Canadian first," *The Globe and Mail*, 22 October 1999, page A3. http://netk.net.au/Canada/Jette.asp.

**125** See, for example: General Henry M. Robert, *ROBERT'S RULES OF ORDER REVISED*, William Morrow and Company, Inc., New York City, 1971.

**126** – "Fury said to a mouse, That he met in the house, 'Let us both go to law: I will prosecute you.– Come, I'll take no denial; We must have a trial: For really this morning I've nothing to do.' Said the mouse to the cur, 'Such a trial, dear sir, With no jury or judge, would be wasting our breath' 'I'll be judge, I'll be jury,' Said cunning old Fury; 'I'll try the whole cause, and condemn you to death'."

**127** Alice, the Dodo, Mouse, *etc*., . . . in Lewis Carroll, *Alice's Adventures in Wonderland* and *Through the Looking-Glass*, Signet Classics, The New American Library, New York City, 1960 & 1962; from . . . *Wonderland* . . ., "A Caucus-Race and a Long Tale," Chapter III, page 35.

**128** https://www.alice-in-wonderland.net/resources/pictures/alices-adventures-in-wonderland/

**129** [typeset as a mouse tail.]

**130** **[**perhaps similar to a damped trigonometric function.**]** $y = e^{-x} \cos(2\pi x)$
https://en.wikipedia.org/wiki/Damped_sine_wave#:~:text=A%20damped%20sine%20wave%20is,than%20it%20is%20being%20supplied.https://www.graphpad.com/guides/prism/8/curve-fitting/reg_damped_sine_wave.htmhttps://calculushowto.com/damped-sinewave/
https://www.mathsisfun.com/physics/spring.html

**131** https://en.wikipedia.org/wiki/The_Mouse%27s_Tale http://bootless.net/mouse.html

**132** https://upload.wikimedia.org/wikipedia/commons/3/32/Alice_in_Wonderland_Ch.3.jpg
https://upload.wikimedia.org/wikipedia/commons/3/32/Alice_in_Wonderland_Ch.3.jpg
https://en.wikipedia.org/wiki/The_Mouse%27s_Tale

**133** See Gerald L. Gall, *THE CANADIAN LEGAL SYSTEM/Second Edition*, STUDENT EDITION, Carswell, Toronto; 1983; ISBN 0-459-35810-3; page 18, et seq.
https://www.amazon.com/Canadian-Legal-System-Gerald-Gall/dp/0459553984.

**134** (– although this book is published in Germany – not Canada.

**135** *Law*, as used in this book: perhaps somewhat expanded, and perhaps by way of a kind of poetic license by the author.

**136** See also: https://en.wikipedia.org/wiki/Law.

**137** "Complete victory" – perhaps in a sort of vengeful sense. Other phrases sometimes similarly used: "scorched-earth," "winner take all," "slash and burn," "take no prisoners," "total war," "total deadly enemies." Along with such a mind-set might be found concepts of *true believer*, and *sin of certainty*. See: Eric Hoffer, *THE TRUE BELIEVER*, HARPER PERENNIAL MODERN CLASSICS, New York City, 1951, 2010. ISBN 978-0-06-050591-2; and Robert Harris, *Conclave*, Vintage, 2017. ISBN-13: 978-1-101-97290-8
https://www.goodreads.com/book/show/29397486-conclave
https://www.goodreads.com/quotes/8218520-let-me-tell-you-that-the-one-sin-i-have
https://en.wikipedia.org/wiki/The_Sin_of_Certainty
https://www.amazon.ca/Conclave-A-novel/dp/1101972904#:~:text=%E2%80%9CA%20gripping%20read.,final%20twist%20is%20great%20fun.%E2%80%9D

**138** See Giacomo Puccini, *LA FANCIULLA DEL WEST* (*THE GIRL OF THE GOLDEN WEST*), 1910.
https://www.metopera.org/season/2018-19-season/la-fanciulla-del-west/
https://jhiblog.org/2018/12/17/variations-on-a-theme-by-puccini-theologizing-la-fanciulladel-west/
https://archive.org/stream/lafanciulladelwe00pucc/lafanciulladelwe00pucc_djvu.txt
Minnie:

> Ciò vuol dire, ragazzi, che non v'è, al mondo, peccatore cui non s'apra una via di redenzione. Sappia ognuno di voi chiudere in se questa suprema verità d'amore.
>
> Which is to say, boys, that there is no sinner in the world for whom a path to redemption is not open. May each of you learn how to hold this supreme truth of love within you.

Apparently inspired from Biblical Psalms. Chapter 51
https://www.biblestudytools.com/psalms/51.html
https://www.mechon-mamre.org/p/pt/pt2651.htm.

**139** – but not Québec, at the provincial level.
**140** – but not Scotland.
**141** See, for example:

CONSTITUTION OF MARYLAND

[. . .]

DECLARATION OF RIGHTS.

*We, the People of the State of Maryland, grateful to Almighty God for our civil and religious liberty, and taking into our serious consideration the best means of establishing a good Constitution in this State for the sure foundation and more permanent security thereof, declare:*

[. . .] [**emphasis** added]

Art. 5. (a) (1) That the Inhabitants of Maryland are entitled to the **Common Law of England**, and the trial by Jury, according to the course of that Law, and to the benefit of such of the English statutes as existed on the Fourth day of July, seventeen hundred and seventy-six; and which, by experience, have been found applicable to their local and other circumstances, and have been introduced, used and practiced by the Courts of Law or Equity; and also of all Acts of Assembly in force on the first day of June, eighteen hundred and sixty-seven; except such as may have since expired, or may be inconsistent with the provisions of this Constitution; subject, nevertheless, to the revision of, and amendment or repeal by, the Legislature of this State. And the Inhabitants of Maryland are also entitled to all property derived to them from, or under the Charter granted by His Majesty Charles the First to Caecilius Calvert, Baron of Baltimore.

[. . .]

https://msa.maryland.gov/msa/mdmanual/43const/html/00dec.html#:~:text=(1)%20That%20the%20Inhabitants%20of,have%20been%20found%20applicable%20to
https://digitalcommons.law.umaryland.edu/cgi/viewcontent.cgi?referer=&httpsredir=1&article=1025&context=mlh_pubs
https://museumbrawlsandartcrawls.wordpress.com/tag/charles-calvert/.
**142** but Louisiana has a code law basis. https://en.wikipedia.org/wiki/Law_of_Louisiana.
**143** Here = in this book.
**144** *Constitution Act*, 1982 (1982 c. 11 (United Kingdom)). https://laws-lois.justice.gc.ca/pdf/const_e.pdf.
**145** *Constitution Act* s.91 – [paragraph] 27.
**146** *Constitution Act* s.92 – [paragraph] 14.
**147** See: https://www.britannica.com/topic/criminal-law.
**148** "State" as used here refers to the general concept of a governing entity. In a Canadian context "Crown" is often used. In an American context, sometimes as "the people."

**149** See: Jill Lepore, "AMERICAN CHRONICLES/SIRENS IN THE NIGHT/How the victims'-rights revolution has remade American justice," *THE NEW YORKER*, 21 May 2018, pages 48–55.

**150** The Canadian *criminal law* test is proof, by the state, *beyond a reasonable doubt*. Analogously, the same test applies to *regulatory/administrative law*, but perhaps, in practice, not quite as strictly considered. Arguably, the strictness of consideration would be related to the general theoretical proposition that *crime* accusations are for serious moral transitions of individual persons whose personal freedom may be in the balance. In contrast, *regulatory/administrative law* is as for methods of community improvement, encouraged by legal penalties; important in its own right, but dealt with differently.

**151** https://www.law.cornell.edu/wex/actus_reus.

**152** https://www.law.cornell.edu/wex/mens_rea.

**153** Prohibition by criminal statute = *Criminal Code, CD&S Act, etc.*

**154** RSC 1985, c. C-46, as amended, s.79 et seq.
https://laws-lois.justice.gc.ca/eng/acts/c-46/page-12.html?wbdisable=true#:~:text=79%20Every%20one%20who%20has,property%20by%20that%20explosive%20substance
https://laws-lois.justice.gc.ca/eng/acts/c-46/page-1.html.

**155** R.S.C. 1985, as amended, c. F-27. http://canada.justice.gc.ca/stable/EN/Laws/Chap/index.html.

**156** The author, and maybe no one else, sometimes refers to these as *criminalized molecular offences*.

**157** R.S.C. 1985, c. N-1. http://canada.justice.gc.ca/stable/EN/Laws/Chap/index.html Insofar as this is applied as criminal law, there is similar law in the USA – not everyone thinks it is such a good idea – see: Wm. F. Buckley Jr., Ethan A. Nadelmann, Kurt Schmoke, Joseph D. McNamara, Robert W. Sweet, Thomas Szasz and Steven B. Duke, "The War on Drugs is Lost," *National Review*, 12 February 1996, pages 34–48 http://www.nationalreview.com/12feb96/drug.html www.nationalre view.com; and A. Robert Kaufman, "Attack the profit, not the fact, of drugs," OPINION/COMMENTARY, *The Baltimore Sun*, 10 December 1996, page 21A. www.sunspot.net.

**158** See: M.G., *The Law of Occupational Health and Safety in Ontario*, Second Edition, Butterworths, Toronto . . ., 1994; page 2–1, *et seq.*

**159** M.G., *The Law of Occupational Health and Safety in Ontario*, Second Edition, Butterworths, Toronto . . ., 1994.

**160** Workplace health & safery = occupational health & safery = OH&S.

**161** A British statute: 28 George III c. 48, that implied a recognition of soot as an occupational carcinogen (scrotal cancer), and, among other things, imposed (apparently as an occupational hygiene requirement) a weekly bath on the child apprentices who were sent up the chimneys. See *At the Centre*, Canadian Centre for Occupational Health and Safety, October 1987.

**162** See also: *Factories Act*, 1802. M.G., *The Law of Occupational Health and Safety in Ontario*, Butterworths, Toronto, 1994; Chapter 1.3., pages 1.2, *et seq.*

**163** See M.G., "The Chimney Sweep's Disease – Two Views in Late Eighteenth Century England," *At the Centre*, Vol. X, No. 4, pages 8–12, October 1987.

**164** Now: O. Reg. 490/09: DESIGNATED SUBSTANCES https://www.ontario.ca/laws/regulation/090490.

**165** See also: M.G., *The Law of Occupational Health and Safety in Ontario*, Butterworths, Toronto, 1994; Chapter 10.3.d, & Appendix II, pages II-9 and 10.

**166** Now cited as RSO 1990c.O.1, as amended (proclaimed into force 1 October 1979 as SO 1978 c.83). https://www.ontario.ca/laws/statute/90o01.

**167** 47 Victoria c. 39. This early Canadian statute derived from British statutes that started in 1802 as *An Act for the Preservation of the Health and Morals of Apprentices and Others, employed in Cotton and other Mills*, and *Cotton and other Factories* (42 George III c. 73).

**168** *The Ontario Factories' Act, 1884*, 1884 47 Victoria c.39 *Factories' Act*, Proclaimed into force 1 December 1886 – the first of the *Factories' Acts* in Canada – inspired by the British *Acts*, and the ancestor of the present Ontario *Occupational Health & Safety Act*.

**169** Starting with the UK *Factories Act*, 1802 – amended into modern times: *Act for the preservation of the Health and Morals of Apprentices and others employed in Cotton and other Mills, and Cotton and other Factories*, 42 Geo. III. c. 73.
https://en.wikipedia.org/wiki/Health_and_Morals_of_Apprentices_Act_1802#:~:text=The%20Health%20and%20Morals%20of,apprentices%20working%20in%20cotton%20mills
https://www.parliament.uk/about/living-heritage/transformingsociety/livinglearning/19thcentury/overview/earlyfactorylegislation/
**170** Ontario *OH&S Act* RSO 1990, c.O.1, as amended. https://www.ontario.ca/laws/statute/90o01#BK71
**171** s.33.
**172** s.39.
**173** s.70(2) #23, s.35.
**174** See also M.G., *The Law of Occupational Health and Safety in Ontario*, Second Edition, Butterworths, Toronto, 1994, Chapter 10.3.d.
**175** s.70(2)#21. See also M.G., Chapter 10.3.f.
https://www.ontario.ca/laws/statute/90o01#BK128
**176** – now RRO 1990, Reg.833 (previously O.Reg. 654/86, as amended).
https://www.ontario.ca/laws/regulation/900833.
**177** – to quote the government's advertising – see for example *The Globe and Mail*, 10 November 1986, page A10. www.theglobeandmail.com/.
**178** s.38.
**179** s.70(2) #25, #26.
**180** For example, to prevent a toxic happening with routine cleaning materials – or deal with it if it does happen:

| Bleach | + | Ammonia | → water | Sodium hydroxide | + | Chloramine |
|---|---|---|---|---|---|---|
| Na O Cl | + | N–$H_3$ | → $H_2O$ | Na OH | + | $H_2$–N–Cl |
| Sodium hypochlorite | | | | | | Chloramine |
| CAS 7681-52-9<br>CAS 10022-70-5 | | | | | | CAS 10599-90-3 |
| 74.442 g/mol | | | | | | 51.47 g/mol |

https://en.wikipedia.org/wiki/Sodium_hypochlorite https://pubchem.ncbi.nlm.nih.gov/compound/Chloramine

Homar Murillo, "What Happens When You Mix Bleach and Ammonia?" 5 October 2022. ReAgent. https://www.chemicals.co.uk/
https://www.chemicals.co.uk/blog/what-happens-when-you-mix-bleach-and-ammonia?srsltid=AfmBOoqC1-domagKJQ2_kLFfIitAQKKVsNxz0ySfseKjJi348z_qqG3u
https://www.quora.com/What-is-the-chemical-reaction-between-sodium-hypochlorite-and-ammonia
Leigh Krietsch Boerner, "Accidental mix of bleach and acid kills Buffalo Wild Wings employee/Incidents like this, which create chlorine gas, occur more than 2,000 times per year in the US,"
*C&EN*, 13 November, 2019, Volume 97, Issue 45.
https://cen.acs.org/safety/consumer-safety/Accidental-mix-bleach-acid-kills/97/i45

**181** The Canadian *criminal law* test is proof, by the state, *beyond a reasonable doubt*. Analogously, the same test applies to *regulatory/administrative law*, but perhaps, in practice, not quite as strictly considered. Arguably, the strictness of consideration would be related to the general theoretical proposition that *crime* accusations are for serious moral transitions of individual persons whose personal freedom may be in the balance. In contrast, *regulatory/administrative law* is as for methods of community improvement, encouraged by legal penalties; important in its own right, but dealt with differently.
**182** https://www.lawnow.org/whatever-happened-to-r-v-sault-ste-marie-the-due-diligence-defence/
https://scc-csc.lexum.com/scc-csc/scc-csc/en/item/2605/index.do
https://en.wikipedia.org/wiki/R_v_Sault_Ste-Marie_(City_of)
**183** See: Gerald L. Gall, *THE CANADIAN LEGAL SYSTEM/Second Edition*, STUDENT EDITION, Carswell, Toronto; 1983; ISBN 0-459-35810-3; page 18, *et seq.*
https://www.amazon.com/Canadian-Legal-System-Gerald-Gall/dp/0459553984.
**184** John A. Yogis, Q.C., *Canadian Law Dictionary*, Barron's Educational Series, Woodbury, New York, 11797; 1983; ISBN 0-6120-2116-9; at page 212.
https://www.amazon.ca/Canadian-Dictionary-John-Yogis-Q-C/dp/0764125699
**185** John A. Yogis, Q.C., *Canadian Law Dictionary*, Barron's Educational Series, Woodbury, New York, 11797; 1983; ISBN 0-6120-2116-9; at page 50.
https://www.amazon.ca/Canadian-Dictionary-John-Yogis-Q-C/dp/0764125699.
**186** For example, in *Smithson et al v Saskem Chemicals Ltd.* ([1986] 34 CCLT 195).
https://digitalcommons.osgoode.yorku.ca/cgi/viewcontent.cgi?referer=&httpsredir=1&article=1644&context=ohlj
**187** See: Wil Lepkowski, "BHOPAL LITIGATION: Indian Court to review settlement," *C&EN*, 10 July 1989, page 4; and Wil Lepkowski, "Bhopal controversy still active after five years." *C&EN*, 4 December 1989, page 6.
**188** https://en.wikipedia.org/wiki/Scots_law
**189** Louisiana has a code law basis. https://en.wikipedia.org/wiki/Law_of_Louisiana
**190** https://www.legisquebec.gouv.qc.ca/en/document/cs/ccq-1991
**191** http://legisquebec.gouv.qc.ca/en/showdoc/cs/ccq-1991.
**192** https://legal-dictionary.thefreedictionary.com/Stare+decisis
https://en.wikipedia.org/wiki/Precedent.
**193** See *Yachuk and another v Oliver Blais Co. Ltd.* ([1949] AC 386 – on appeal from the Supreme Court of Canada to the Privy Council. In this case, liability was attributed to a gasoline station owner, when, in the summer of 1940, two children ages 7 and 9 years were sold about 1 pint of gasoline for 5¢ in Kirkland Lake, Ontario, and used it for play, with serious burn consequences for the 9 year old.
**194** This was a last case in Canadian history where appeal was to the Privy Council. Now the Supreme Court of Canada is the highest level in the Canadian system. Privy Council appeal possibility remains in a very few other countries.
**195** *Consumer Protection Act*, RSPEI 1988, c C-19
https://www.canlii.org/en/pe/laws/stat/rspei-1988-c-c-19/latest/rspei-1988-c-c-19.html
**196** For example, in a US context see Maureen Rouhi, "Product liability could be the real issue with taggants," *Chemical & Engineering News*, 26 August 1996, page 38.
**197** See: E. P. Belobaba, *Products Liability and Personal Injury Compensation in Canada*: Towards Integration and Rationalization, 1983 (Vol. 1); (Supply and Services Canada Cat. No. RG23-73/1-1983E).
**198** http://legisquebec.gouv.qc.ca/en/pdf/cs/CCQ-1991.pdf
**199** RSC 1985 c. H-3; as amended by S.C. 1987 c. 30, Part 1; and as further amended.
https://laws-lois.justice.gc.ca/eng/acts/h-3/
**200** Canadian sale of goods law derives from English sale of goods law – see P.S. Atiyah, *The Sale of Goods*, Fifth Edition, Pitman Publishing Ltd., London, 1975.

**201** SNB c. S-1.
**202** *Ashington Piggeries Ltd. et al. v Christopher Hill Ltd.* This is an English case, decided by the House of Lords and reported in [1971] 1 All ER 847, that dealt with the toxicity of dimethylnitrosamine to mink.
**203** http://classic.austlii.edu.au/au/journals/SydLawRw/1976/11.pdf
https://en.wikipedia.org/wiki/Ashington_Piggeries_Ltd_v_Christopher_Hill_Ltd.
**204** https://en.wikipedia.org/wiki/N-Nitrosodimethylamine.
**205** *N,N*-Dimethylnitrous amide *N*-nitrosodimethylamine CAS 62-75-9 $C_2 H_6 N_2 O$ 74.83 g/mol.
**206** Dalmeet Singh Chawla, "Graduate student gets prison sentence for poisoning/Zijie Wang gets two 7-year terms to serve concurrently," *C&EN,* 12 December 2018, VOLUME 96, ISSUE 49. https://cen.acs.org/people/Graduate-student-prison-sentence-poisoning/96/i49?utm_ source=Newsletter&utm_medium=Newsletter&utm_campaign=CEN.
**207** https://cen.acs.org/people/Graduate-student-prison-sentence-poisoning/96/i49
Raechel Huizinga and Iain Sherriff-Scott, "Victim testifies in poisoning case, sentencing marked for December/'Like a scene from a horror movie,' victim tells court," *The Journal Queen's University,* 2 November 2018.
https://www.queensjournal.ca/victim-testifies-in-poisoning-case-sentencing-marked-for-december/
**208** https://en.wikipedia.org/wiki/N-Nitrosodimethylamine
https://www.sigmaaldrich.com/catalog/product/sigma/n7756?lang=en&region=US
https://pubchem.ncbi.nlm.nih.gov/compound/n-nitrosodimethylamine
https://pubchem.ncbi.nlm.nih.gov/compound/n-nitrosodimethylamine#datasheet=lcss&section=Top
http://www.molbase.com/en/msds62-75-9-moldata-144856.html
**209** *Vacwell v BDH* [1969] 3 All E.R. 1681 and [1970] 3 All E.R. 553. *All England Law Reports*
https://www.lexisnexis.co.uk/products/all-england-law-reports.html
https://en.wikipedia.org/wiki/All_England_Law_Reports.
**210** Boron tribromide CAS 10294-33-4 B $Br_3$ 250.57 g/mol.
**211** *Vacwell v BDH* [1969] 3 All E.R. 1681 and [1970] 3 All E.R. 553.
**212** See also: M.G., *The Law of Occupational Health and Safety in Ontario, Second Edition,* Butterworths, Toronto and Vancouver, August 1994, ISBN 0-409-90414-7; Chapter 2.13.c, pages 2–28 and 2–29; Appendix II, pages II-6 and II.7.
**213** M.G., *The Law of OH&S . . .,* 2nd Ed., Butterworths, . . . 1994, Chapter 2.13.c.
**214** $\Delta H = -81$ kcal/mol.
H.A. Skinner and N.B. Smith, "The heat of hydrolysis of boron tribromide," *Transactions of the Faraday Society,* Volume 51, 1955.
https://pubs.rsc.org/en/content/articlelanding/1955/tf/tf9555100019
**215** https://en.wikipedia.org/wiki/Boron_tribromide
https://pubchem.ncbi.nlm.nih.gov/compound/Boron-tribromide
https://pubs.rsc.org/-/content/articlelanding/1955/tf/tf9555100019/unauth#!divAbstract
https://cameochemicals.noaa.gov/chemical/253
https://onlinelibrary.wiley.com/doi/abs/10.1002/047084289X.rb244.pub2
https://link.springer.com/article/10.1007/s12209-016-2638-6
https://nj.gov/health/eoh/rtkweb/documents/fs/qr/0244qr.pdf
https://pubchem.ncbi.nlm.nih.gov/compound/Boron-tribromide#section=Refractive-Index
https://www.chemicalbook.com/ChemicalProductProperty_EN_CB5852622.htm
**216** https://www.cdhfinechemical.com/images/product/msds/38_1660199025_BoronTribromide-CASNO-10294-33-4-MSDS.pdf
**217** Ronald Gottesman, "Introduction," pages xxiii, xxiv, xxv, to Upton Sinclair, *The Jungle,* PENGUIN BOOKS; Viking Penguin Inc., New York City, 10010, 1985, 1987; ISBN 0-14-039031-6.
https://inventaire.io/entity/isbn:9780140390315.

**218** Upton Sinclair, *The Jungle*, Doubleday, Jabber & Company, 1906.
**219** https://en.wikipedia.org/wiki/The_Jungle https://www.amazon.ca/s?k=the+jungle+upton+sinclair&hvadid=75110445408153&hvbmt=be&hvdev=c&hvqmt=e&tag=msncahydra-20&ref=pdsl9b4btlkrcce
**220** See also:
Sam Roberts, "*Nick Kotz, 87, Who Won Pulitzer Prize for Exposing Hazards at Meat Plants,*" *The New York Times*, OBITUARY, 20 May 2020, page B10.
https://www.nytimes.com/2020/05/15/business/media/nick-kotz-dead.html#:~:text=He%20was%2087.,him%20as%20it%20rolled%20backward
**221** Jane Mayer, "BACK TO THE JUNGLE/Trump is helping a meat-processing titan leverage the pandemic to strip workers of protections," *The New Yorker*, A Reporter at Large, 20 July 2020, pages 28–39.
**222** – Deborah Blum, *The Poison Squad: One Chemist's Single-Minded Crusade for Food Safety at the Turn of the Twentieth Century*, Penguin, 2018.
https://www.amazon.ca/Poison-Squad-Chemists-Single-Minded-Twentieth/dp/1594205140
**223** – Reviewed as BRIEFLY NOTED, *THE NEW YORKER*, 12 November 2018, page 87.
**224** Upton Sinclair, *The Jungle*, PENGUIN BOOKS; Viking Penguin Inc., New York City, 10010, 1985, 1987; ISBN 0-14-039031-6. [Chapter 14] Page 163, 164. https://inventaire.io/entity/ isbn:9780140390315.
**225** https://en.wikisource.org/wiki/The_Jungle https://en.wikisource.org/wiki/The_Jungle/Chapter_14 https://creativecommons.org/licenses/by-sa/3.0/.
**226** https://strangerthansf.com/reviews/sinclair-jungle.html.
**227** https://www.pagebypagebooks.com/Upton_Sinclair/The_Jungle/Chapter_14_p1.html.
**228** https://web.viu.ca/davies/H321GildedAge/Sinclair.Jungle.ch14.badmeat.htm
**229** R.S.C. 1985 c. F-27, as amended. http://canada.justice.gc.ca/stable/EN/Laws/Chap/index.html.
**230** *Controlled Drugs & Substances Act*, previously *Narcotic Control Act.*
**231** For example, see the Ontario *Health Protection and Promotion Act*, RSO 1990 c.H.7, as amended. www.gov.on.ca/MBS/english/publications/statregs/contents.html.
**232** For example:
**233** Ontario *Regulated Health Professions Act, 1991* https://www.ontario.ca/laws/statute/91r18.
**234** Ontario *Medicine Act, 1991* https://www.ontario.ca/laws/statute/91m30.
**235** 26 and 27 Victoria c. 124. See also articles by Raymond Holland in *Chemistry and Industry*, 3 June 1985, pages 366–371; and by Jeffrey L. Sturchio and Arnold Thackroy in *C&EN*, 9 March 1987, pages 20–29.
**236** *Canadian Environmental Protection Act* SC 1999, c. 33 https://laws-lois.justice.gc.ca/eng/acts/c-15.31/
**237** Statutes of Saskatchewan, *Environmental Assessment Act*, E-10.1
https://publications.saskatchewan.ca/#/products/488
**238** *Environmental Management and Protection Act*, 2010, E-10.22
https://publications.saskatchewan.ca/#/products/31893
**239** https://www.ontario.ca/document/workplace-hazardous-materials-information-system-guide-legislation
**240** M.G., *The Law of Occupational Health and Safety in Ontario*, Second Edition, Butterworths, Toronto . . ., 1994.
**241** Reg Ferguson, Ed.-in-Chief, *WHMIS Compliance Manual*, Carswell, Scarborough, Ontario, 1989 and 1996.
**242** M.G., "WHMIS CAVEAT EMPTOR," *At the Centre*, December 1990.
**243** M.G., "REGULATION OF ONTARIO WORK PLACE CHEMICAL HAZARDS – WHAT THE LAW IS – AND AN OPINION OF WHAT THE LAW OUGHT TO BE," presented at the 73rd Canadian Chemical Conference, Halifax, Nova Scotia, July 1990.
**244** M.G., "WHMIS – A CHARTED OVERVIEW," *Hazardous Materials Management Magazine*, Sept./ Oct. 1989, pages 22 and 23.

**245** *The Final report of the Workplace Hazardous Materials Information System to the Deputy Minister of Labour Canada*, April 1985, Ottawa, (Labour Canada Catalogue No. L-46-1568/85E).
**246** SDS = Safety Data Sheet. Previously called MSDS = Material Safety Data Sheet.
**247** See, for example:
https://www.cdhfinechemical.com/images/product/msds/38_1660199025_BoronTribromide-CASNO-10294-33-4-MSDS.pdf
**248** For example: NIK Test B
http://sds.chemtel.net/webclients/safariland/finished_goods/NIK%20Public%20Safety%20-%20Test%20B%20-%20Nitric%20Acid%20Reagent%20-%20General%20Screening%20Drug%20Test.pdf
**249** "On reading an MSDS," Vol. 17, No. 1, February 2009, pages 7–12.
2009 02. Vol. 17, No. 1. TWH&SLC *newsletter*
http://newsletter.workers-safety.ca.s3-website-us-east-1.amazonaws.com/#!/newsletters%2F2009%2002%2F
**250** https://www.archives.gov/
http://www.archives.gov/exhibits/charters/constitution.html
https://www.archives.gov/founding-docs/constitution-transcript.
**251** US DEPARTMENT OF LABOR, Occupational Safety and Health Administration
https://www.osha.gov/laws-regs/regulations/standardnumber/1910/1910.1200
**252** US § CFR 1910.1200
https://www.osha.gov/dsg/hazcom/ghoshacomparison.html.
**253** 29 CFR § 1910.1200 – Hazard communication.
https://www.law.cornell.edu/cfr/text/29/1910.1200.
**254** *Gedankenexperiment*. Introduction 2.
**255** Historically: Emerald green = Paris green = copper-acetoarsenite + arsenic trioxide – toxic.
https://en.wikipedia.org/wiki/Paris_green
http://www.webexhibits.org/pigments/indiv/overview/emerald.html.
**256** Historically: Cinnabar = mercury sulfide – toxic
https://www.masterpigments.com/cinnabar-pigments/.
**257** https://en.wikipedia.org/wiki/Otto_von_Bismarck
**258** https://www.biographi.ca/en/bio/meredith_william_ralph_15E.html
https://en.wikipedia.org/wiki/William_Ralph_Meredith
https://injuredworkersonline.org/104th-anniversary-of-meredith-report-what-would-the-judge-think-today/
https://collections.ola.org/mon/26005/203796s.pdf
https://collections.ola.org/mon/25005/203796.pdf
**259** *Workmen's Compensation Act* (4 George V, c. 25), proposed by then Ontario Chief Justice Sir William Meredith in his *Report[s] on Laws Relating to the Liability of Employers* (King's Printer, Toronto, 1912, 1913.).
**260** M.G., *The Law of Occupational Health and Safety in Ontario, Second Edition*, Butterworths, Toronto and Vancouver, August 1994, ISBN 0-409-90414-7; page 1–3.
**261** There is also a federal version – linked-in to the provincial.
**262** for most workers.
**263** https://www.ontario.ca/laws/regulation/980175
https://www.ontario.ca/laws/regulation/980175#BK13
https://www.ontario.ca/laws/statute/97w16.
**264** Ontario Workplace Safety and Insurance Appeals Tribunal, Toronto M5G 2P2
https://wsiat.on.ca/en/home/announcements.html https://www.wsib.ca/en
**265** Manarin, *Forensic Evidence . . .*, 2019; Chapter 6 [Professions & Occupations] page 451, *et seq.*
**266** www.vocabulary.com/dictionary/per%20se.

**267** www.peo.on.ca/.
**268** https://www.cheminst.ca/magazine/article/chemists-need-professional-designation-for-the-public-good/
https://www.acpo.on.ca/
https://www.acpo.on.ca/membership/chemists-ring/
**269** www.peo.on.ca/.
**270** RSQ, c. c-15, s.1 and 16. https://www.legisquebec.gouv.qc.ca/en/document/cs/c-15
https://www.legisquebec.gouv.qc.ca/fr/document/lc/C-15
**271** www.ocq.qc.ca/.
**272** RSO 1990 c.P.28, s.12. www.gov.on.ca/MBS/english/publications/statregs/contents.html.
**273** S.O. 1984, c. Pr 10, s.8 and 9. www.gov.on.ca/MBS/english/publications/statregs/contents.html.
**274** See, for example: structural collapse – Algo Centre Mall, Elliot Lake, Ontario:
**275** http://peo.on.ca/index.php?ciid=2289&laid=1
https://www.cbc.ca/news/canada/sudbury/verdict-elliot-lake-mall-collapse-trial-1.3943318
https://www.cbc.ca/news/canada/sudbury/elliot-lake-roof-collapse-class-action-lawsuit-ok-d-by-judge-1.2537034
https://en.wikipedia.org/wiki/Algo_Centre_Mall.
**276** – for example, Alberta *Education Act* https://open.alberta.ca/publications/e00p3
**277** – for example, the Ontario *Colleges Collective Bargaining Act*, R.S.O. 1990, c.C.15.
https://www.ontario.ca/laws/statute/08c15
**278** – for example, the *University of New Brunswick Act*, S.N.B. 1984, c. 40.
https://www.canlii.org/en/nb/laws/astat/snb-1984-c-40/latest/snb-1984-c-40.html
**279** as under the federal *Natural Sciences and Engineering Research Council Act*, R.S.C. 1985, c. N-21. https://laws.justice.gc.ca/eng/acts/N-21/page-1.html.
**280** See also K. Jankowski, "Money for Chemistry Let's Play Chemical Monopoly," in *Canadian Chemical News*, April 1987, p. 7.
**281** E.g., see M.-A. Amour, "Safety in the School Laboratory," *Canadian Chemical News*, Sept. 1987, pages 21|23. See also Claude E. Mays III, Hinsdale, Illinois, "Hazardous spill response," letter to the editor, *Chemical & Engineering News*, 13 May 1996, page 51.
**282** *James et al. v River East School Division No. 9 and Peniuk*; Manitoba Queen's Bench 1975 04 04. 1975 CanLII 1557 (MB QB) [1975] 5 WWR 135
http://canlii.ca/t/gct40
https://www.canlii.org/en/mb/mbqb/doc/1975/1975canlii1557/1975can lii1557.html
**283** Upheld on appeal; Manitoba Court of Appeal 1975 11 03.
64 DLR (3d) 338 [1976] 2 WWR 577
https://www.canlii.org/en/mb/mbca/doc/1975/1975canlii932/1975canlii932.html.
**284** In a US context, see: Kate Shatzkin, "Ex-student hurt in lab sues [Johns] Hopkins [University] seeking $60 million/Man had to quit career as scientist after explosion," *Baltimore Sun*, 21 March 1996, pages 1B &14B.
**285** See also: https://cen.acs.org/safety/lab-safety/Who-pays-when-a-graduate-student-gets-hurt-in-a-British-or-French-lab/99/i16?utm_source
**286** [1998] 29CFR1910.1450 – s. 1910.1450 – Occupational exposure to hazardous chemicals in laboratories.
[2000] 40CFR156.212 – s.156.212 – Personal protective equipment statements. http://www.access.gpo.gov/nara/cfr/index.html
*Regulations for Industrial Establishments*, RRO 1990, Reg.851, s.81. (*Occupational Health and Safety Act*, RSO 1990, c.O.1, as amended).
http://www.e-laws.gov.on.ca/tocRegsE.asp?lang=en
Canadian Standards Association CSA Standard CAN/CSA-Z94.3–92, *Industrial Eye and Face Protectors*.

https://www.csagroup.org/?srsltid=AfmBOoorPLEVikS76olLd4BplU830t8phZnzjfNaX-6BCS7ypkdcqbyH
American National Standards Institute ANSI Standard Z87.1–1989, *Practice for Occupational and Educational Eye and Face Protection.*
**287** American Chemical Society. Kevin MacDermott, "NATIONAL CHEMISTRY WEEK/NCW '99: GREAT CHEMISTRY," *Chemical & Engineering News*, December 20, 1999, Volume 77, Number 51, pages 40–45. http://www.acs.org/ http://pubs.acs.org/cgi-bin/bottomframe.cgi?7751cont http://pubs.acs.org/hotartcl/cenear/991,220/7751ncw.html
*Canada Occupational Safety and Health Regulations*, SOR/86-304, Part XII, s.12.6 (*Canada Labour Code – Part II*) http://laws.justice.gc.ca/en/L-2/SOR-86-304/29634.html
Lab Safety Supply Inc., Janesville, Wisconsin, 53547–1368. www.labsafety.com
www.labsafety.com/store/dept.asp?deptid=623
**288** Nitric acid CAS 7697-37-2 63.013 g/mol.
**289** https://pubchem.ncbi.nlm.nih.gov/compound/Nitric-acid
https://www.britannica.com/science/nitric-acid
https://en.wikipedia.org/wiki/Nitric_acid.
**290** [https://www.law.cornell.edu/wex/special_damages#:~:text=In%20contract%20law%20%2C%20special%20damages,by%20the%20Wex%20Definitions%20Team%20%5D]
**291** [ This refers to which side gets to pay for the other's for court costs.]
**292** The photograph at http://schools.tdsb.on.ca/robertborden/11.html – June 2001 – Sir Robert L. Borden Business and Technical Institute, Scarborough, Ontario M1E 1Z7. The website is no longer functioning and the school no longer exists.
What was apparently shown was a high school chemistry lab session – three at a lab bench, two behind – apparently without any protective eyewear.
This website was searched, at the time the author had other litigation related to the Toronto District School Board – that litigation had nothing to do with chemistry, but this photo came up then incidentally. The school authorities apparently were not too concerned about the eyewear.
**293**

- *The New York Times*, 31 October 2015.
- *Chemical & Engineering News*, 9 November 2015.
- American Chemical Society, "Safety Alert – Stop Using the Rainbow Demonstration," 2014.

https://www.acs.org/content/acs/en/chemical-safety/teach-and-learn/flame-test.html
https://institute.acs.org/acs-center/lab-safety/education-training/safer-experiments/flame-test.html#rainbow
www.csb.gov/videos/after-the-rainbow.
**294** See also: "Flash fire in science lab burns students/Mississauga . . ." Jesse McLean and Jennifer Yang, Staff Reporters, "Flash fire in science lab burns students/Mississauga accident critically injures teen in high school class /Teacher 'saved us all,'" *Toronto Star*, 12 March 2010, pages A1 and A10.
**295** https://www.thestar.com/news/gta/student-in-severe-condition-after-science-class-blast/article_f3f08bb0-8c47-566d-b634-3bf2112238b5.html
**296** [. . .]
Teacher . . . was showing the afternoon chemistry class at St. Marcellinus Secondary School how different chemicals burn when a nearby Bunsen burner ignited the one-litre beaker of methanol in her hand, students said.
She dropped the flask onto the work table, scattering flames onto students in the front row.
[. . .]
**297** https://www.dpcdsb.org/MARCL/
**298** http://www.dpcdsb.org/NR/rdonlyres/DC7923DE-6E02-430F-8296-87422681460 F/58986/Letter_ to_-Parents_Web2.doc [This link no longer works.]

299 [. . .]

. . . Catholic Secondary School

. . .

Mississauga, Ontario . . .

. . .

March 12, 2010

Dear Parents and Guardians:

On Wednesday afternoon, an accident occurred during a routine, teacher-led, Grade 11 Science experiment that, regrettably, resulted in injuries to five students. Four of the students were treated at a local hospital and released, while the fifth student was transferred to Sunnybrook Medical Centre where she is in critical but stable condition.

I commend the actions of the classroom teacher and other staff, whose response quickly brought the situation under control and thank the members of Mississauga's Emergency Services for their very rapid response to our call for aid. As a school community, we continue to pray for the full recovery of everyone who was affected and offer them every support we are able.

We strive to ensure that . . . is a safe, healthy, caring and inclusive place to learn and work. As with all issues that take place within and outside of our school that impact our community, I wanted to bring this matter to your attention.

If you have any questions or concerns, please do not hesitate to contact me.

Please continue to keep . . . in your prayers and especially . . . and . . .

Sincerely,

. . .

Principal

**300** http://newsletter.workers-safety.ca.s3-website-us-east-1.amazonaws.com/#!/newsletters%2F2010%2003%2F

**301** *Safety in Academic Chemistry Laboratories 8TH Edition Best Practices for First- and Second-Year University Students*, a Publication of the American Chemical Society Joint Board – Council Committee on Chemical Safety, ACS, 2017, Washington, DC 20,036; ISBN 978-0-8412-3732-2
https://www.acs.org/content/acs/en/chemical-safety/teach-and-learn/university.html

**302** Benjamin Plackett, "Lab Safety"/Seeking data to make labs safer/Researchers aim to supply evidence to save lives, persevering through lack of recognition and funding, special to *C&EN*, 30 MARCH 2020, pages 26 and 27.

**303** 1,1-Dimethylethyl lithium = *tert*-butyl lithium $(CH_3)$-C-Li 64.055 g/mol CAS 594-19-4

**304** www.commonchemistry.org/ChemicalDetail.aspx?ref=594-19-4
https://en.wikipedia.org/wiki/Tert-Butyllithium
https://en.wikipedia.org/wiki/Patrick_Harran
https://en.wikipedia.org/wiki/Sheri_Sangji_case.

**305** Toronto Workers' Health & Safety Legal Clinic *newsletter* 2013 10. Vol. 21, No. 5, Page 11.

**306** Toronto Workers' Health & Safety Legal Clinic *newsletter* 2013 06. Vol 21, No. 2, 1,1-dimethylethyl Lithium.

**307** Michael Torrice and Jyllian Kemsley, "DEAL REACHED OVER LAB DEATH/PROSECUTION: UCLA professor Patrick Harran to complete community service, pay fine," *C&EN*, 30 June 2014, page 4.

**308** Toronto Workers' Health & Safety Legal Clinic *newsletter* 2012 09. Vol 20. No. 5.

**309** *N,N*-Dimethylnitrous amide, *N*-nitrosodimethylamine CAS 62-75-9 $C_2H_6N_2O$ 74.083 g/mol

**310** Dalmeet Singh Chawla, "Graduate student gets prison sentence for poisoning/Zijie Wang gets two 7-year terms to serve concurrently," *C&EN*, 12 December 2018, VOLUME 96, ISSUE 49. *CHEMICAL & ENGINEERING NEWS*, American Chemical Society, Washington, D.C.
https://cen.acs.org/people/Graduate-student-prison-sentence-poisoning/96/i49?utm_source=Newslet

ter&utm_medium=Newsletter&utm_campaign=CEN
https://cen.acs.org/people/Graduate-student-prison-sentence-poisoning/96/i49
**311** Raechel Huizinga and Iain Sherriff-Scott, "Victim testifies in poisoning case, sentencing marked for December/'Like a scene from a horror movie,' victim tells court," *The Journal Queen's University*, 2 November 2018.
https://www.queensjournal.ca/story/2018-11-02/news/victim-testifies-in-poisoning-case-sentencing-marked-for-december/
**312** https://en.wikipedia.org/wiki/N-Nitrosodimethylamine
https://www.sigmaaldrich.com/catalog/product/sigma/n7756?lang=en&region=US
https://pubchem.ncbi.nlm.nih.gov/compound/n-nitrosodimethylamine
https://pubchem.ncbi.nlm.nih.gov/compound/n-nitrosodimethylamine#datasheet=lcss&section=Top
https://cen.acs.org/people/Graduate-student-prison-sentence-poisoning/96/i49
**313** Raechel Huizinga and Iain Sherriff-Scott; Amelia Rankine, illustrator; "Victim testifies in poisoning case, sentencing marked for December/'Like a scene from a horror movie,' victim tells court," *The Journal Queen's University*, 2 November 2018.
https://www.queensjournal.ca/story/2018-11-02/news/victim-testifies-in-poisoning-case-sentencing-marked-for-december/
**314** – ". . . acute and latent . . ." – in the terminology of this book – not of the news article.
**315** https://en.wikipedia.org/wiki/J._Robert_Oppenheimer
**316** See also: Gregg Herken, *Brotherhood of the Bomb/the tangled lives and loyalties of Robert Oppenheimer, Ernest Lawrence, and Edward Teller*, Henry Holt and Company, New York City, 10,011; 2002; ISBN 0-8050-6588-1.
**317** Katey Rich, "Yes, Robert Oppenheimer Really Did Poison His Professor's Apple/Before he was the world's most famous physicist, he was a depressed college student with a bad revenge plot," ***VANITY FAIR***, 20 July 2023.
**318** https://www.vanityfair.com/hollywood/2023/07/oppenheimer-poison-apple-true-story#:~:text=%E2%80%9CConsumed%20by%20his%20feelings%20of,poison%20might%20not%20have%20been
Dalmeet Singh Chawia, "Former [Lehigh University] chemistry student pleads guilty to poisoning roommate/. . . likely to face 6–12 years in prison," *C&EN*, 30 November 2020, 24 March 2021.
**319** https://cen.acs.org/policy/Former-chemistry-student-pleads-guilty/98/web/2020/11
Dalmeet Singh Chawia, "[Lehigh University] Chemistry student to face trial for allegedly poisoning roommate with thallium," *C&EN*, 18 February 2019.
**320** In a US context see John T. Maynard and Howard M. Peters, *Understanding chemical patents/A guide for the inventor*, Second Edition, American Chemical Society, Washington, D.C., 1991.
**321** RSC 1985, c. P-4, as amended.
https://laws-lois.justice.gc.ca/eng/acts/p-4/
**322** See: Charles Duhigg, "STOP THIEF," *The New Yorker*, 22 October 2018; at page 57.
**323** See also: https://cen.acs.org/materials/polymers/Chemist-convicted-of-stealing-BPA-free-can-liner-trade-secrets-for-a-Chinese-firm/99/web/2021/04?utm_source=Newsletter&utm_medium=Newsletter&utmcampaign=CEN
**324** See, for example: Peter Moon, "Scientist worn down by patent fight/Intellectual Property/An invention that saved his employer millions has cost Joe Dableh hundreds of thousands and his job," *The Globe and Mail*, 13 August 1996, pages A1 and A6. www.theglobeandmail.com/.
**325** R.S.C. 1985, c. C-42 as amended. https://laws-lois.justice.gc.ca/eng/acts/C-42/Index.html
**326** https://canada.justice.gc.ca/eng/terms-avis/index.html
**327** R.S.C. 1985, c. T-13 as amended. https://laws-lois.justice.gc.ca/eng/acts/t-13/
**328** See:
- Stephen Strauss, Science Reporter, "War chemicals not being made in Canada," *The Globe and Mail*, 19 April 1996, page A8 www.theglobeandmail.com/;

- R.G. Sutherland and J.P. Tracey, "The Chemical Weapons Convention: How Will it Affect Canadian Industry?" *Canadian Chemical News*, May 1995, pages 17–20;
- "Protecting Confidentiality: A Critical Task Under the Chemical Weapons Convention," *Canadian Chemical News*, June 1995, pages 18 and 19.
- Madeleine Jacobs, "A Necessary Treaty," editor's page, *Chemical & Engineering News*, 12 August 1996, page 5;
- Lois Ember, "Chemical arms accord debate intensifies," *C&EN*, 5 August 1996, pages 5 and 6;
- Madeleine Jacobs, "Light and Shadow," editor's page, *C&EN*, 29 January 1996, page 5;
- Lois R. Ember, "Failure to Ratify Chemical Arms Pact Would Dampen U.S. Chemicals Trade," *C&EN*, 29 January 1996, pages 19–22;
- Lois R. Ember, "U.S. reveals size of chemical arms arsenal," *C&EN*, 29 January 1996, page 9;
- Lois R. Ember, "Senate committee hears solutions to proliferation of weapons of terror," *C&EN*, 13 November 1995, pages 24 and 25. http://pubs.acs.org/cen/index.html
- M.G., "*The Canadian* Chemical Weapons Convention Implementation Act *Compared to Other Chemical Disclosure Legislation*," a paper presented at the National Meeting of the American Chemical Society, at a session of the Division of Chemistry and the Law, Spring 1997, San Francisco. http://www.acs.org/

http://membership.acs.org/C/CHAL/
https://www.opcw.org/chemical-weapons-convention
https://en.wikipedia.org/wiki/Chemical_Weapons_Convention
**329** *Chemical Weapons Convention Implementation Act*, SC 1994–1995, 42-43-44 Elizabeth II, c.25; Royal Assent 13 July 1995, to come into force on proclamation. The provisions of convention is the schedule to the *Act*.
https://laws-lois.justice.gc.ca/eng/acts/C-27.6/
**330** See also: M.G., *THE CANADIAN* CHEMICAL WEAPONS CONVENTION IMPLEMENTATION ACT *COMPARED TO OTHER CHEMICAL DISCLOSURE LEGISLATION*, paper presented at National Meeting of the American Chemical Society, at session of the Division of Chemistry and the Law, Spring 1997, San Francisco.
**331** https://en.wikipedia.org/wiki/Judicial_notice
**332** Manarin, *Forensic Evidence* . . ., 2019; [evidence], page 193, et seq.
**333** https://www.canlii.org/en/ca/laws/stat/rsc-1985-c-c-5/latest/rsc-1985-c-c-5.html.
**334** s.30.
**335** *Viva Voce* = by voice. From Latin.
**336** https://www.merriam-webster.com/dictionary/viva%20voce
**337** https://en.wikipedia.org/wiki/Hearsay#England_and_Wales
**338** Nancy Shapiro and David Silver, *Evidence 101 – A Primer on Evidence Law*, Koskie Minsky LLP, Toronto.
https://assets.kmlaw.ca/wp-content/uploads/2015/09/NSDS_EVIDENCE101_A_PRIMERONEVIDENCELAW_03jun14.pdf
https://kmlaw.ca/.
**339** Manarin, *Forensic Evidence* . . ., 2019; [expert], page 610, et seq.
**340** Manarin, *Forensic Evidence* . . ., 2019; [novel scientific theory; *Daubert*], page 788, et seq.
**341** Manarin, *Forensic Evidence* . . ., 2019; [*Frye v USA*; *Daubert v Merrell*; *R v Mohan*] page 600, et seq.
**342** Nancy Shapiro and David Silver, *Evidence 101 – A Primer on Evidence Law*, Koskie Minsky LLP, Toronto.
https://assets.kmlaw.ca/wp-content/uploads/2015/09/NSDS_EVIDENCE101_A_PRIMERONEVIDENCELAW_03jun14.pdf
https://kmlaw.ca/

**343** See: Carl Meyer, "Science, Medicine and the U.S Common Law Courts," as Chapter 1 in Carl Meyer, ed., *EXPERT WITNESSING/EXPLAINING AND UNDERSTANDING SCIENCE*, CRC Press LLC, Boca Raton, Florida, 33,431, 1999; ISBN 0-8493-1197-7. (Derived from a symposium of the American Chemical Society, Division of Chemistry and the Law, ACS National Meeting, Las Vegas Nevada, 1997.).
**344** https://www.ppsc-sppc.gc.ca/eng/pub/is-ip/ch6.html
**345** https://en.wikipedia.org/wiki/Cross-examination
**346** – even to understand relativity theory –
**347** See: Albert Einstein, *Relativity/THE SPECIAL AND GENERAL THEORY* [in German 1916]/Translated . . . [English, 1920]; PENGUIN BOOKS, New York City, 10,014; 2006; ISBN 0-014-30982-2. [written, 1916, for an audience ". . .interested in the theory, . . .but . . . not conversant the mathematical apparatus of theoretical physics. . ."].
**348** Einstein, page 7, et seq.
**349** See: Ronald Joseph Delisle, Don Stuart, David M. Tanovich, Lisa Dufraimont; *EVIDENCE/PRINCIPLES AND PROBLEMS*/Twelfth Edition; CARSWELL/THOMSON REUTERS CANADA, Toronto M1T 3V4; 2018. ISBN 978-0-7798-8655-5. At pages 606–614 – *R v Bingley*. https://digitalcommons.osgoode.yorku.ca/faculty_books/362/
**350** Manarin, *Forensic Evidence* . . ., 2019; [novel scientific theory; *Daubert*], page 788, et seq.
**351** Manarin, *Forensic Evidence* . . ., 2019; [*Frye v USA*; *Daubert v Merrell*; *R v Mohan*] page 600, et seq.
**352** Manarin, *Forensic Evidence* . . ., 2019; [expert], page 610, et seq.
**353** See also: Delisle et al.; *EVIDENCE*. . ./12th Edition; CARSWELL. . .At pages 596–603 – [re novel science].
**354** See also: Delisle et al.; *EVIDENCE*. . ./12th Edition; CARSWELL. . .At pages 572, et seq. – *Inman* . . .
**355** See: Carl Meyer, "Science, Medicine and the U.S Common Law Courts," as Chapter 1 in Carl Meyer, ed., *EXPERT WITNESSING/EXPLAINING AND UNDERSTANDING SCIENCE*, CRC Press LLC, Boca Raton, Florida, 33431, 1999; ISBN 0-8493-1197-7. (Derived from a symposium of the American Chemical Society, Division of Chemistry and the Law, ACS National Meeting, Las Vegas Nevada, 1997.) At page 29:
**356**

[. . .]

> The *Daubert* and *Joiner* decisions reflect the fact that the oral, adversary setting of the U.S. common law courts is not conducive for solving interdisciplinary problems, that litigators and judges lack the facilities and the basic scientific training necessary to handle scientific issues that arise in our contemporary society, and that scientists lack the basic training in social sciences necessary to understand the importance of their participation in the legal process.
> Neither *Daubert* nor any other legal rule can substitute for the lack of functional scientific literacy of litigators who have avoided mathematics and science courses since junior high school, but, nevertheless, take it upon themselves to litigate scientific issues. Basic scientific literacy cannot be solved by courses on scientific issues that are currently in litigation. It requires at least rudimentary scientific education for litigators and judges, and an equally basic education in civics for scientists.

[. . .]

**357** From: Carl Meyer, ed., *EXPERT WITNESSING* . . ., CRC Press – as cited in note above – copyright 1999. Reproduced by permission of Taylor & Francis Group.
**358** *William Daubert* et al., *Petitioners v Merrell Dow Pharmaceuticals, Inc.*, 509 US 579 (1993) US Supreme Court. [replaced] *Frye v US*, 293 F. 1013 US Supreme Court (1923). US Federal Rules of Evidence, 1975.
**359** http://en.wikipedia.org/wiki/Daubert_v._Merrell_Dow_Pharmaceuticals
https://www.law.cornell.edu/wex/daubert_standard.

**360** *General Electric Co. v Joiner*, 522 U.S. 136 (1997) US Supreme Court. https://en.wikipedia.org/wiki/General_Electric_Co._v._Joiner https://www.law.cornell.edu/supremecourt/text/522/136.
**361** Meyer, 1999, page 1, notes 1 and 2.
**362** See, for example: "PBS NEWS HOUR/Medical witness [Dr. David Fowler, former Maryland chief medical examiner] for defense in Chauvin trial says Floyd's manner of death 'undetermined'," 14 April 2021. https://www.pbs.org/newshour/show/medical-witness-for-defense-in-chauvin-trial-says-floyds-manner-of-death-undetermined
**363** Meredith Cohn, "CRIMINAL JUSTICE/Dr. David Fowler was a well-respected medical examiner, until he wasn't," *Baltimore Banner*, 15 May 2025.

> [. . .]
> **The work of Maryland's former top forensic pathologist received a critical audit**
> A wide-ranging audit released Thursday of investigations into police-involved deaths in Maryland over many years under the leadership of Dr. David R. Fowler found three dozen cases that should have been classified as homicides but were not.
> [. . .]

https://www.thebaltimorebanner.com/community/criminal-justice/david-fowler-medical-examiner-maryland-HEOSKJIAQBET5DRR7LNIDJL3OU/
**364** Mike Hellgren, "Families sounded alarm years before review of autopsies under former Maryland medical examiner," WJZ CBS Baltimore, 18 May 2025. https://www.cbsnews.com/baltimore/news/maryland-medical-examiner-david-fowler/
**365** Dunning-Kruger Effect https://www.cbc.ca/radio/thesundayedition https://www.cbc.ca/radio/thesundayedition/a-look-at-the-psychological-phenomenon-that-makes-you-cocky-in-your-ignorance-and-how-we-all-have-it-1.5020911
**366** *This American Life* https://www.thisamericanlife.org/585/transcript
**367** https://en.wikipedia.org/wiki/Dunning%E2%80%93Kruger_effect
**368** See also: Canada *Criminal Code* s.657.3

> [. . .]
> PART XX
> **Procedure in Jury Trials and General Provisions (continued)**
> Evidence on Trial (continued)
> **Expert testimony**
> **657.3 (1)** In any proceedings, the evidence of a person as an expert may be given by means of a report accompanied by the affidavit or solemn declaration of the person, setting out, in particular, the qualifications of the person as an expert if
> **(a)** the court recognizes that person as an expert; and
> **(b)** the party intending to produce the report in evidence has, before the proceeding, given to the other party a copy of the affidavit or solemn declaration and the report and reasonable notice of the intention to produce it in evidence.
> **Attendance for examination**
> **(2)** Notwithstanding subsection (1), the court may require the person who appears to have signed an affidavit or solemn declaration referred to in that subsection to appear before it for examination or cross-examination in respect of the issue of proof of any of the statements contained in the affidavit or solemn declaration or report.
> [. . .]

https://laws-lois.justice.gc.ca/eng/acts/c-46/page-160.html#docCont.

**369** Emily Eakin, "Under Scrutiny, A Sanity Study Isn't So Airtight/A book says a psychologist [David Rosenhan] misled to make his landmark 1973 [*On Being Sane In Insane Places*] work stand out." *The New York Times*, [BOOK REVIEW of Susannah Cahalan, *The Great Pretender*], 18 November 2019, page C6.
**370** https://en.m.wikipedia.org/wiki/David_Rosenhan.
**371** https://www.theatlantic.com/health/archive/2013/02/sex-lies-and-separating-science-from-ideology/273169/
https://www.nytimes.com/2001/08/05/world/derek-freeman-who-challenged-margaret-mead-on-samoa-dies-at-84.html
https://en.wikipedia.org/wiki/Margaret_Mead
https://www.colorado.edu/asmagazine/2017/12/01/both-sides-misunderstand-margaret-mead-prof-contends.
**372** *Controlled Drugs & Substances Act* SC 1996, c.19. http://laws-lois.justice.gc.ca/eng/acts/C-38.8/index.html
**373** *Criminal Code* s.657.3(3). https://laws-lois.justice.gc.ca/eng/acts/c-46/page-103.html#h-128722
**374** Roman goddess *Justitia* – with scales, blindfold and sword.
https://en.wikipedia.org/wiki/Lady_Justice#:~:text=Since%20the%2016th%20century%2C
Ilshofen, Baden-Württemberg. https://commons.wikimedia.org/wiki/File:DEU_Ilshofen_COA.svg
**375** *Quod volimus credimus libenter* ≈ believe what we want to believe.
https://en.wikiquote.org/wiki/Julius_Caesar Julius Caesar, *Commentarii de Bello Gallico* [=Commentaries on the Gallic War], Book I, Ch. 14, translated by W.A. McDevitte and W.S. Bohn:
"*Fere libenter homines id quod volunt credunt*" = In most cases men willingly believe what they wish.
**376** See Fig. 1.8.

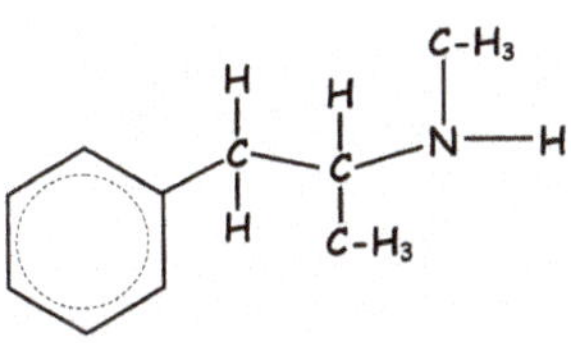


Fig. 1.8: Methamphetamine, CAS 537-46-2, 149.23 g/mol $C_{10}H_{15}N$. mp = 143 °C, bp = 230.33 °C.

https://pubchem.ncbi.nlm.nih.gov/compound/10836 https://commonchemistry.cas.org/detail?ref=537-46-2 https://en.wikipedia.org/wiki/Methamphetamine
**377** See Fig. 1.9.

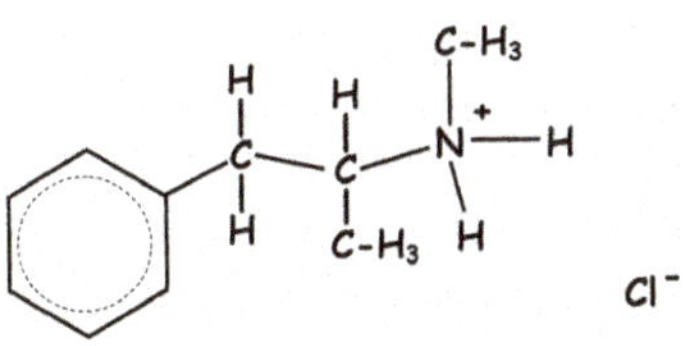


Fig. 1.9: Methamphetamine hydrochloride, CAS 51-57-0 185.69 g/mol, mp=170|175 °C $(C_{10}H_{16}N)^+ Cl^-$, crystal methamphetamine = methamphetamine hydrochloride.

https://en.wikipedia.org/wiki/Methamphetamine https://en.wikipedia.org/wiki/Methamphetamine#Recreational
https://pubchem.ncbi.nlm.nih.gov/compound/Methamphetamine-hydrochloride
**378** See Fig. 1.10.

**379** See Fig. 1.11.

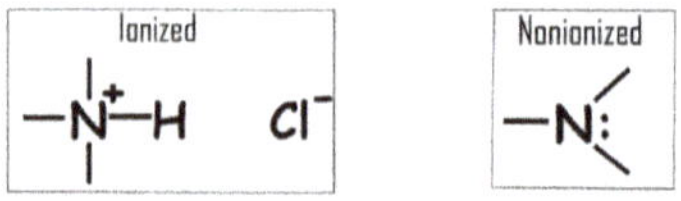


Fig. 1.10: Ionized and nonionized.

**380** See Fig. 1.12.

| | | | | |
|---|---|---|---|---|
| Carbon | **C** | $_6C^{12}$ | 12.011 | g/mol |
| Hydrogen | **H** | $_1H^1$ | 1.008 | |
| Oxygen | **O** | $_8O^{16}$ | 15.999 | |
| Nitrogen | **N** | $_7N^{14}$ | 14.007 | |
| Fluorine | **F** | $_9F^{19}$ | 18.998 | |
| Chlorine | **Cl** | $_{17}Cl^{35}$ | 35.453 | |
| Phosphorus | **P** | $_{15}P^{31}$ | 30.974 | |

Fig. 1.11: g/mol.

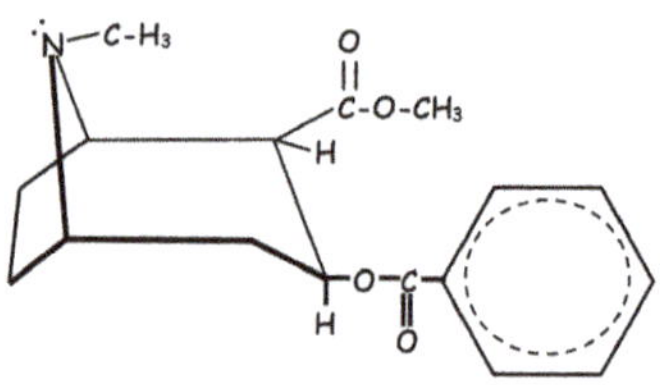

Fig. 1.12: Cocaine, CAS 50-36-2, $C_{17}H_{21}NO_4$, 303.36 g/mol.

**381** See Fig. 1.13.

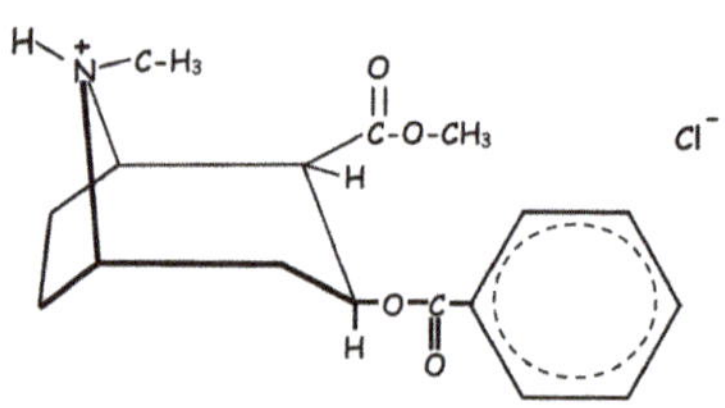


Fig. 1.13: **Cocaine hydrochloride**, CAS 53-21-4 $(C_{17}H_{22}NO_4)^+Cl^-$, 339.82 g/mol.

**382** https://cdn.shopify.com/s/files/1/2006/8997/files/FZ-i_FX-i_Manual.pdf?14788397102087296248
https://pim-resources.coleparmer.com/instruction-manual/11338-65-11339-10-xsr-quick-reference-manual.pdf
https://pim-resources.coleparmer.com/instruction-manual/11338-65-11339-10-xsr-user-manual.pdf

**383** Likely a misquote, apocryphal, attributed to Prince Otto von Bismarck, Chancellor, German Empire, 1871–1890.

**384** https://quoteinvestigator.com/2010/07/08/laws-sausages/
https://www.nytimes.com/2010/12/05/weekinreview/05pear.html
https://professorbuzzkill.com/bismarck-laws-and-sausages/
https://en.wikipedia.org/wiki/Otto_von_Bismarck.
**385** Prince Otto von Bismarck. Upton Sinclair, *Jungle*. Chapter 1.
**386** M.G., "electric stun devices & the workers who use them./– is self-tasering by police to train & demonstrate dangerous – acute, chronic, latent? – / – is it illegal in Ontario? – / – are TASERs a regulated medical device? –" Toronto Workers' Health & Safety Legal Clinic *newsletter*, Vol. 17, No. 1, February 2009, pages 3–6, at page 6.
2009 02. Vol. 17, No.1 TWH&SLC newsletter.
**387** The role of Crown counsel – also lawyers styled as *barrister & solicitor* – in the service of the Crown – is similar to Defence in that ethically keeping to the law is expected, but there might not be such a wide range of problematic client personalities to be represented.
Issues of the constitutional roles of government officials in the USA are argued. See, for example: John Bolton, "Presidents Expect Loyalty. Trump Demands Fealty," New *York Times*, OPINION/GUEST ESSAY, 5 January 2025. https://www.nytimes.com/2025/01/05/opinion/trump-loyalty-cabinet-presidents.html
**388** Manarin, *Forensic Evidence* . . ., 2019; [lawyers], page 74, et seq.
**389** See also: Delisle et al.; *EVIDENCE*. . ./12th Edition; CARSWELL. . .At pages 18–28 – role of lawyers.
**390** – And in a strictly absolute sense the lawyer cannot know for sure what is actually true.
**391** See also: Steven Spielberg, *Bridge of Spies*, 2015. A New York lawyer is kind-of conscripted by the bar association to represent a captured Soviet spy. In an early conversation he explains to his new client why he is not so interested with the facts of the case.
https://en.wikipedia.org/wiki/Bridge_of_Spies_(film)
https://www.imdb.com/title/tt3682448/
**392** – but the lawyer must not have been an actual witness – an inherent serious conflict.
**393** See – but in perhaps a different context:
**394** Francis X. Clines, "AN ISRAELI LAWYER DARES DEFEND AN ACCUSED NAZI," *The New York Times*, 2 March 1987, page 3A.
https://www.nytimes.com/1987/03/02/world/an-israeli-lawyer-dares-defend-an-accused-nazi.html
**395** https://en.wikipedia.org/wiki/John_Demjanjuk
**396** https://www.independent.co.uk/news/world/lawyer-braves-storm-of-abuse-over-ivan-the-ter rible-case-1486946.html
**397** *pro bono publico*, Latin = for the public good. Freebie legal work.
https://www.merriam-webster.com/dictionary/pro%20bono
https://www.merriam-webster.com/dictionary/freebie.
**398** But see: Charles Dickens, *Bleak House*, Signet Classic, The New American Library, New York City, 1964 and 1980, "Attorney and Client," Chapter XXXIX, page 555:
**399**

[. . .]

> The one great principle of the English law is to make business for itself. There is no other principle distinctly, certainly, and consistently maintained through all its narrow turnings. Viewed by this light it becomes a coherent scheme and not the monstrous maze the laity are apt to think it. Let them but once clearly perceive that its grand principle is to make business for itself at their expense, and surely they will cease to grumble.

[. . .]

**400** See also: Scott Cummings, "There's a Darker Reason Trump Is Going After Those Law Firms," *New York Times*, OPINION/GUEST ESSAY, 15 May 2025. https://www.nytimes.com/2025/05/15/opinion/trump-law-firms-attacks.html
**401** https://www.lso.ca/about-lso/legislation-rules/rules-of-professional-conduct

**402** Manarin, *Forensic Evidence . . .*, 2019; [lawyers – rules of professional conduct], page 74, et seq.
**403** See: Marie Henein, *NOTHING BUT THE TRUTH*/A *MEMOIR*; Signal/McLelland & Stewart/Penguin Random House Canada Limited; 2021; Chapter 10 "SORRY, NOT SORRY," pages 171–189. ISBN 9780771039348.
https://www.indigo.ca/en-ca/nothing-but-the-truth-a-memoir/9780771039348.html
**404** Fictional defence lawyer – in Beverley McLachlin, *FULL DISCLOSURE. . .*, 2019.
**405** Beverley McLachlin, *FULL DISCLOSURE/A THRILLER*, Simon and Schuster Canada, Toronto M5A 1J3, 2019; ISBN 978-15011-7279-3; at page 3. [Fictional Vancouver lawyer Jilly Truitt, when first meeting her new client.]
https://www.simonandschuster.ca/books/Full-Disclosure/Beverley-McLachlin/ 9781982116460
https://books.google.ca/books?id=ls9FDwAAQBAJ&dq=BeverleyMcLachlin.com
**406** [A novel by the former Chief Justice of Canada.]
**407** Conversation at the famous lawyer's hospital bed – recovering from Arsenic As poisoning – in Erle Stanley Gardner, *The Case of the Drowsy Mosquito*, 1943.
**408** Erle Stanley Gardner, *The Case of the Drowsy Mosquito*, William Morrow and Company, New York City, 1943. POCKET BOOK EDITION, PERRY MASON/2 in 1, Simon & Schuster of Canada Ltd., 1950, pages 88 and 89.
https://books.google.ca/books?id=N8NCAQAAIAAJ&q=Lieutenant+Tragg&source=gbs_word_cloud_r&cad=3
https://en.wikipedia.org/wiki/Perry_Mason
https://en.wikipedia.org/wiki/Erle_Stanley_Gardner
https://en.wikipedia.org/https://wiki/Erle_Stanley_Gardner_bibliography
http://self.gutenberg.org/
http://articles/eng/perry_mason.
tpl TRL Call number = 813.5 G124.133 https://www.torontopubliclibrary.ca/search.jsp?Ntt=perry+mason+drowsy+mosquito
**409** https://openlibrary.org/books/OL24958393M/The_case_of_the_drowsy_mosquito
https://www.goodreads.com/book/show/1836227.The_Case_of_the_Drowsy_Mosquito
http://vbala99.blogspot.com/
https://www.azquotes.com/quote/696852?ref=stuck-up
https://www.azquotes.com/quote/696852
**410** Fictional conversation between victim's father and Crown Counsel – in Robert Rotenberg. *OLD CITY HALL*, 2009:
**411** Robert Rotenberg, *OLD CITY HALL*, TOUCHSTONE/Simon & Schuster, New York City, Toronto, 2009. page 195. ISBN 978-1-4165-9286-0.
https://robertrotenberg.com/books/oldcityhall https://robertrotenberg.com/
**412** Frank Cotham, *THE NEW YORKER*, 26 August 2019, page 66.
**413** https://www.newyorker.com/cartoon/a20202
Or, perhaps, less often, to assist lawyer advocates for the accused. And sometimes journalists and political advocates.
**414** See: John F. Kelly and Phillip K. Wearne, "Tainting evidence: inside the scandals at the FBI crime Lab," Free Press, New York, 1998. ISBN-10: 0743236416 ISBN-13: 978-0743236416
https://www.amazon.ca/Tainting-Evidence-Inside-Scandals-Crime/dp/0743236416
https://en.wikipedia.org/wiki/FBI_Laboratory
https://en.wikipedia.org/wiki/Federal_Bureau_of_Investigation.
**415** David Burnham, "BUMBLERS IN THE BUREAU," *The Washington Post*, 21 June, 1998
https://www.washingtonpost.com/archive/entertainment/books/1998/06/21/bumblers-in-the-bureau/9b820865-d1f4-448f-b906-384542fcb8c8/
**416** *Gedankenexperiment*. Introduction. 2.

**417** See: Anthony William Deller, of the New York Bar, in *Deller's Walker on Patents*, second edition, the lawyers co-operative publishing co., Rochester, New York 14,603, 1973; Volume Eight, chapter XIX, §590, Pages 3 & 4; describes the Latin derivation, citing several sources: *qui tam pro domino rege, quam pro se ipso, sequitur* = who as well for our lord the king, as for himself, sues.
**418** The description here is for US *qui tam* Federal *False Claims Act*; there are other whistle blower processes, with similar aspects.
See also: https://www.mccaberabin.com/whistleblower-faq/what-does-qui-tam-mean/ https://en.wikipedia.org/wiki/Qui_tam
*qui tam pro domino rege quam pro se ipso in hac parte sequitur* = who sues in this matter for the king as well as for himself.
**419** https://en.wikipedia.org/wiki/Qui_tam.
**420** Annie Karni and Nicholas Fandos, "Second Person Blows Whistle, Legal Team Says/on Trump and Ukrane/'Firsthand knowledge' for Impeachment," *The New York Times*, 7 October 2019, pages A1 and A14.
**421** "Why whistleblowers need protection," BBC Business Daily, October 2019.
https://www.bbc.co.uk/programmes/w3csy7k9.
**422** David Lewis and Wim Vandekerckhove, editors, *SELECTED PAPERS FROM THE INTERNATIONAL WHISTLEBLOWING RESEARCH NETWORK CONFERENCE/OSLO, JUNE 2017.*
https://pdfs.semanticscholar.org/94a0/51bcc1ef155320faf4b63937aba20677d514.pdf.
**423** https://en.wikipedia.org/wiki/Whistleblower https://en.wikipedia.org/wiki/Qui_tam.
**424** US Federal *False Claims Act* 31 USC § 3729–3731
https://www.law.cornell.edu/uscode/text/31/3729
**425**
- *US v Regents of the University of California*, 2001.
- https://casetext.com/case/us-v-regents-of-the-university-of-california-2

*USA, Plaintiff-Appellee v UNIVERSITY OF CALIFORNIA BOARD OF REGENTS*, 2003 https://caselaw.findlaw.com/court/us-9th-circuit/1136605.html
**426** a metaphoric whistle.
**427** See:
- Dan L. Burk, Seton Hall University School of Law, Newark, New Jersey, "'Bounty Hunter' Lawsuits for Research Misconduct under the Federal False Claims Act," CHAL *newsletter*, Fall 1995, pages 11–13, ACS Division of Chemistry and the Law, National Meeting, Chicago.
- Professor William E. Kovacic, George Mason University School of Law, Arlington, Virginia, 22201–4498, "Whistleblower Bounty Lawsuits as Monitoring Devices in Government Contracting," Draft: May 1994.
- William E. Kovacic, "The Civil False Claims Act as a Deterrent to Participation in Government Procurement Markets," *Supreme Court Economic Review*, Vol. 6, pages 201–239, 1998, The University of Chicago Press. https://www.jstor.org/stable/1147106?seq=1

Lucette Lagnado, "Lawyer for Columbia Whistle-Blowers: 'Part Therapist'," *The Wall Street Journal*, MARKETPLACE, 7 January 1999, pages A13 and A15.
**428** Kovacic, 1998, cited above.
**429** https://www.phillipsandcohen.com/whistleblower-practice-areas/qui-tam-false-claims/
https://www.bestlawyers.com/united-states/qui-tam-law.
**430** https://www.phillipsandcohen.com/
https://www.phillipsandcohen.com/what-is-a-qui-tam-case/
**431** But see:

[. . .]

> OSC OFFICE OF THE WHISTLEBLOWER
> ARE YOU A WHISTLEBLOWER?
> Do you have knowledge of, or suspect, a possible violation of Ontario securities law that could harm investors? You may be eligible for a financial award of up to $5 million.

[. . .]

https://safehotline.com/?keyword=whistleblower%20hotline%20providers&matchtype=b&ap=&gad_source=1&gclid=Cj0KCQjwytS-BhCKARIsAMGJyzql1umyhC-it-ys94J_BoFmMV6GFDkIMAtEHLCPUwcM3LkYY0rOnwQaArkNEALw_wcB

**432** https://pubmed.ncbi.nlm.nih.gov/8354177/
https://www.sigmaaldrich.com/CA/en/documents-search?tab=sds&emdredirect=1
http://cepsa.ca/client/documents/benzene.pdf.

**433** See also – re acute and chronic toxicology.

**434** Charles Duhigg, "ANNALS OF EPIDEMIOLOGY/THE PANDEMIC PROTOCOL/the EIS knows what to tell the public in an outbreak," *THE NEW YORKER*, 4 May 2020, page 19.

**435** https://www.newyorker.com/magazine/2020/05/04/seattles-leaders-let-scientists-take-the-lead-new-yorks-did-not

**436** See, for example: Dave Philipps, "*Chronic Brain Trauma Is Extensive in Navy's Elite Speedboat Crews* /The pounding that sailors' brains take from years of high-speed wave-slamming in the Special Boat Teams can cause symptoms that wreck their careers — and their lives," *New York Times*, 12, 14 November 2024.
https://www.nytimes.com/2024/11/12/us/brain-trauma-cte-navy-speedboat.html

**437** Trevor Greene, "Trevor Greene: 'I am still alive. This is my thank-you letter.'/'The right people happened to be at exactly the right place at exactly the right times'," *Maclean's*, 29 April 2016.
https://www.macleans.ca/news/canada/trevor-greene-i-am-still-alive-this-is-my-thank-you-letter/
https://en.wikipedia.org/wiki/Trevor_Greene.

**438** Bryan G Garber, Corneliu Rusu, Mark A Zamorski1, David Boulos, "Occupational outcomes following mild traumatic brain injury in Canadian military personnel deployed in support of the mission in Afghanistan: a retrospective cohort study," BMJ Journals, BMJ Open, Occupational and Environmental Medicine Research, Volume 6, Issue 5, May 2016. https://bmjopen.bmj. com/content/6/5/e010780.

**439** Dave Philipps, "*Soldier's Struggles Began Long Before Las Vegas Blast, Nurse Says*/A Green Beret who had served several combat tours fatally shot himself in a Cybertruck that exploded outside a Trump hotel on New Year's Day," *New York Times*, 4 January 2025.
https://www.nytimes.com/2025/01/04/us/matthew-livelsberger-las-vegas-cybertruck.html

**440** See: Michael A. Kamrin, *TOXICOLOGY/A Primer on Toxicology Principles and Applications/Indoor & Outdoor Air/Drinking Water/Food •Workplace Environment*, Lewis Publishers, Inc., Chelsea, Michigan, 48118; 1988; ISBN 087371-133-5.

**441** Kamrin, *TOXICOLOGY* . . ., Chapter 9, "The Case of Asbestos," page 79, et seq.

**442** Kamrin, *TOXICOLOGY* . . ., page 83, et seq.

**443** Kamrin, *TOXICOLOGY* . . ., page 45, et seq.

**444** Walter E. Harris, *Low Dose Risk Assessment*, Wuerz Publishing Ltd., Winnipeg, 1997; ISBN 0-920063-73-X; page vi.

**445** *Hazardous Products Act*, SCHEDULE/INGREDIENT DISCLOSURE LIST, SOR/2015-17. https://www.canlii.org/en/ca/laws/regu/sor-2015-17/latest/sor-2015-17.html

[. . .]

**PART 1**
**Interpretation**
**Definitions**
**1 (1)** The following definitions apply in these regulations.

[. . .]

**ATE** means an acute toxicity estimate, and includes the $LD_{50}$ and the $LC_{50}$, and the acute toxicity point estimate determined in accordance with the table to Section 8.1.7 (ETA).

**CAS registry number** means the identification number assigned to a chemical by the Chemical Abstracts Service, a division of the American Chemical Society (*numéro d'enregistrement CAS*).

**Chemical name** means a scientific designation of a material or substance that is made in accordance with the rules of nomenclature of either the Chemical Abstracts Service, a division of the American Chemical Society, or the International Union of Pure and Applied Chemistry, or a scientific designation of a material or substance that is internationally recognized and that clearly identifies the material or substance (*dénomination chimique).*

[. . .]

**$LC_{50}$** means the concentration of a mixture or substance in air that causes the death of 50.0% of a group of test animals ($CL_{50}$).

**$LD_{50}$** means the single dose of a mixture or substance that, when administered by a particular exposure route in an animal study, is expected to cause the death of 50.0% of a given animal population ($DL_{50}$).

[. . .]

**446** Chemicals and Materials, "What is a $LD_{50}$ and $LC_{50}$?" Canadian Centre for Occupational Health and Safety, Hamilton, Ontario, 10 May 2024. https://www.ccohs.ca/oshanswers/chemicals/ld50.html

[. . .]

**Disclaimer**
Although every effort is made to ensure the accuracy, currency and completeness of the information, CCOHS does not guarantee, warrant, represent or undertake that the information provided is correct, accurate or current. CCOHS is not liable for any loss, claim or demand arising directly or indirectly from any use or reliance upon the information.

[. . .]

**447** Maggie Astor, "L.A. Firefighters Who Fought Blazes Show Elevated Mercury and Lead Levels/The findings, which compared the firefighters' blood samples against those taken from other firefighters after past fires, suggest unique risks to blazes that burn in populated areas," *New York Times*, 28 March 2025.
https://www.nytimes.com/2025/03/28/well/la-firefighters-mercury-lead-blood-levels.html

**448** *Ashington Piggeries Ltd. et al. v Christopher Hill Ltd.* This is an English case, decided by the House of Lords and reported in [1971] 1 All ER 847, that dealt with the toxicity of dimethylnitrosamine to mink.

**449** – now RRO 1990, Reg.833 (previously O.Reg. 654/86, as amended).
https://www.ontario.ca/laws/regulation/900,833.

**450** *Ethanol.* $CH_3$-$CH_2$-OH CAS 64–17-5 Chapters 1, 4, 11.
**451** *Canada Labour Code*, RSC, 1985, c. L-2. https://laws-lois.justice.gc.ca/eng/acts/L-2/.
**452** *Canada Occupational Health and Safety Regulations* – SOR/86-304, Section 10.19.
https://laws-lois.justice.gc.ca/eng/regulations/SOR-86-304/page-17.html#docCont
**453** *2024/TLVs® and BEIs®/Based on the Documentation of the Threshold Limit Values for Chemical Substances and Physical Agents & Biological Exposure Indices*, ACGIH. ISBN: 978-160-7261-66-7
https://bpsepako.com/wp-content/uploads/2024/08/ACGIH-2024-TLV-Book.pdf
https://www.acgih.org/publications/
*2009/TLVs® and BEIs®/Based on the Documentation of the Threshold Limit Values for Chemical Substances and Physical Agents & Biological Exposure Indices*; ACGIH®; Signature Publications; ISBN: 978-1-882417-95-7; © 2009; 256 pages.
**454** American Conference of Governmental Industrial Hygienists, Cincinnati, Ohio, 45241.
**455** See: ACGIH "POLICY STATEMENT . . ." and "Special Note . . ." in the first pages of its TLV Book.
**456** Donnell R. Christian, *Forensic Investigation of Clandestine Laboratories*, CRC Press, Boca Raton, Florida, 33431, . . .; 2004; ISBN 0-8493-1227-2.
https://www.routledge.com/Forensic-Investigation-of-Clandestine-Laboratories/ChristianJr/p/book/9781032272849?srsltid=AfmBOoqZcCTw2oiEB6DiYUnGEmbomOV85sQGneHrnVk5WUkTIhM7n9K-
**457** Oliver Sacks, *Uncle Tungsten: Memories of a Chemical Boyhood*, Vintage, reprint 2002; ISBN-10: 0375704043; ISBN-13: 978–0375704048.
https://www.amazon.com/Uncle-Tungsten-Memories-Chemical-Boyhood/dp/0375704043
**458** https://en.wikipedia.org/wiki/Uncle_Tungsten
**459** https://www.cdc.gov/anthrax/about/index.html
https://en.wikipedia.org/wiki/Anthrax
https://www.instagram.com/cbcpodcasts/reel/DH9H8BjyO8B/
**460** Anthrax Island, Scotland: https://www.gairlochmuseum.org/gruinard-island
**461** See for example:

- CBC, 5 February 2015.
  http://www.cbc.ca/news/canada/nova-scotia/christopher-phillips-accused-in-chemical-scare-to-stay-in-jail-1.2945901
- *Global NEWS*, 5 February 2015.
  http://globalnews.ca/news/1813427/crown-reviewing-charges-against-man-accused-in-halifax-chemicals-case/
- Toronto Workers' Health & Safety Clinic *Newsletter* April 2010. Vol. 18, No. 2. pages 7–9.

# Chapter 2
# Essential concepts of forensic science

[. . .][1]

*"Please, your Majesty," said the Knave, "I didn't write it, and they can't prove I did: ther's no name signed at the end."*

*"If you didn't sign it," said the King, "that only makes the matter worse. You* must *have meant some mischief, or else you'd have signed your name like an honest man."*

[. . .]

*"That* proves *his guilt, of course," said the Queen: "so, off with . . . [his head]"*

[Alice at a court of justice.] [1]

[. . .]

*". . . forgery . . . different hand . . . not sure . . . heading . . . different ink . . ."*

[Agatha Christie, *Murder at the Vicarage*, 1930.] [2][2]

## 2.1 Essential concepts and propositions

### 2.1.1 Locard's principle

As considered here [3],[3] the essential propositions of *forensic science* are:

- Physical matter – things made of physical matter – can be unique in ways that the forensic scientist tries to discover, categorize, measure and record; and also to determine or estimate the probability of occurrence. These efforts can help in trying to identify things and processes, and to help try to narrate a theory of the related human story.

  For example [4]: Part of a forgery in a written document could be revealed by way of the chemical composition of the ink[4] – describable as to what chemical substances are present and their proportions; and that ink's otherwise random find probability – it is commonly used by writers, or rarely; and also the paper, along with its physical structure.
- Every contact of such matter and things with other matter and things would leave its mark [5] – by exchanges – which the forensic scientist also tries to discover, categorize and measure; again, potentially helpful for the narrative. Also, such transfers can possibly be serially multiple.[5]

---

**1** Introductory quotes, *etc.* Introduction 1.2.

**2** Agatha Christie, *Orient Express*, 1934. *Vicarage*, 1930. Chapters 2, 7 and 8.

**3** Author's biases. Introduction 1.2, 4.1 and 4.3; Chapters 10 and 11.

**4** Variations of Ink analytical chemistry – as examples – *Gedankenexperiment* – continue throughout this book. Introduction 2.3 and 2.2.

**5** DNA and Locard. Chapters 2.1.1 and 6.16.

https://doi.org/10.1515/9783111618555-003

To continue the ink example [6]: If the document were in a coat pocket, it could carry traces of the lint from the garment's fibres. And, traces from the paper fibres could be left in the garment's pocket. And, there might be technology available to measure those exchanged traces; or maybe not (yet?). If the document had fallen on the floor, it could have traces of the building's dust, and maybe left there, traces from paper fibres, or lint acquired from the coat pocket. Or, for another example [7]: In a two-car crash, each would then carry microscopic, if not obviously visible, paint fragments of the other.

- Many of the possible energy interactions involving matter and things may leave their mark, which the forensic scientist also tries to discover, categorize and measure. The energy interaction may have involved changed temperature, pressure, mechanical force, passage of electrical current, oxidation or other chemical reactions, etc.; or collision with fundamental particles of physics or electromagnetic radiation.[6]

  To further continue the ink example [8]: If the document were exposed to a higher temperature than at its creation, or to intense light [9], the ink and paper may have been measurably changed. The intense light may have been exposure to direct sunlight [10, 11, 12, 13]. Or, in the other example, if the cars' velocities could be estimated – perhaps from the skid marks[7] – or perhaps from carried electronic recording devices as part of a modern insurance package – the impact of the crash on the humans in the cars might be estimated [14].

The forensic findings would then be used, along with other-source data, to help construct a theory of the related human story. That theory, even if appearing as obvious, would then be tested as against all indicators that investigators and scientists would be able to imagine, and perhaps theory alternatives would be needed. But perhaps a firm conclusion would be elusive. For example [15]: A confession of the forger who created the document, or a witness report, might be sought to confirm scientific indicators of fraud, but the confession and witness report might be faulty [16, 17].

The above are adapted [18] from *Locard's principle* [19, 20, 21, 22, 23, 24, 25, 26][8,9,10] concerning the mutual exchanges when any two objects come in contact. In the early twentieth century, Dr. Edmond Locard was director of a police scientific laboratory, Lyons, France. Inspired [27, 28] by Sir Arthur Conan Doyle's fictional Sherlock Holmes [29, 30, 31],[11] Locard was a most significant early forensic scientist [32, 33, 34, 35, 36, 37, 38, 39, 40, 41, 42, 43] (see Fig. 2.1).

6 EM. Photons. Chapters 2 and 4.

7 Tire skid. Chapters 2, 3 and 6.

8 Manarin, *Forensic Evidence . . .*, 2019. Chapters 1, 2 and 3.

9 Edmond Locard. Sherlock Holmes. Cocaine. Chapters 2, 3, 4 and 6.

10 Edmond Locard. Sir Arthur Conan Doyle. Forensics. CSI. Chapters 2.1.1 and 5.2.1.

11 Edmond Locard. Sherlock Holmes. Cocaine. Chapters 2, 3, 4 and 6.

Fig. 2.1: "[Sherlock] Holmes working hard over a chemical investigation," *The Naval Treaty*, 1893.

### 2.1.2 Forensic chemistry defined

In the previous chapter, *forensic science* was described as associated with court-related and legal processes.[12] *Forensic chemistry* is further defined here as those aspects of forensic science concerned with the qualitative and quantitative characterization of physical matter, its changes, and of related chemical reactions, kinetics, electromagnetic radiation, thermodynamics and comparisons with other matter, including statistics of occurrence [44]. This largely involves *analytical chemistry* – with which this book is mostly concerned.

12 Forensic science. Introduction 2; Chapters 1, 2, 3 and 5.

### 2.1.3 Analytical chemistry

Analytical chemistry[13,14] = that branch of chemical science involving the chemical determination of substances, as to what they are (qualitative), how much (quantitative) and the uncertainty [45, 46] (see Fig. 2.2) [47, 48, 49]. This definition may be expanded with some more detail so that ***analytical chemistry*** may be employed to chemically determine substances as to what they (see Tab. 2.1).[15,16]

Fig. 2.2: Analytical chemists in the wild west.
[Cartoon by Dr. Nick Kim, Wellington, New Zealand.]

For example [50] (following through ①→⑦ with the above definition): Ink [51] analysed as from writing on a paper [52] document might be determined to contain: [53, 54, 55, 56, 57, 58, 59][17,18,19]

13 Analytical chemistry. Expanded definition – more detail, and example. Introduction; 2; Chapters 1, 2, 3, 5 and 8.

14 Trace metals with Pb. Chapters 2 and 4.

15 Dr. Nick Kim. Cartoons. Copyright acknowledgement. Chapters 2, 6 and 8.

16 Introductory quotes, *etc*. Introduction. top of first page.

17 Most of the ink examples, throughout this book, are for ball point pen inks, but this example is for fountain pen ink. Different chemistries – see endnotes.

18 Qualitative. Quantitative. Error. Chapters 2 and 8.

19 Analytical chemistry. Error. Chapters 2 and 8.

**Tab. 2.1:** ***Analytical chemistry* expanded definition** – to chemically determine what substances –[20]

| | |
|---|---|
| ① are,<br>② were,<br>③ might become;<br>④ or not; | Qualitative |
| ⑤ and how much;<br>⑥ when; | Quantitative |
| ⑦ along with statements of uncertainty. | |

① A ferric ($Fe^{3+}$) tannate[21] – as an ionic oxidation product – as a solid – not water soluble [60, 61, 62, 63];
② that used to be, as its previously not-yet-oxidized water-soluble ferrous ($Fe^{2+}$) tannate complex [64, 65, 66, 67] in water solution, as would be found in an older style fountain pen [68, 69];
③ and that might eventually chemically react with components of the paper, and photochemically react with light;
④ that does not contain non-ionized organic-soluble (not water-soluble) [70, 71] molecules of inks of modern ball point pens [72, 73];
⑤ with perhaps estimates of the density of the ink components found on the paper that might indicate component concentration of the ink that was in the fountain pen [74];
⑥ maybe with estimates of when it was written;[22]
⑦ but without being sure of what the ink actually is – amongst several possibilities. And, maybe a mixture of several substances.

Such analytical chemical findings – of this example [75] – might then hopefully be useful in testing the theories of the human story being investigated. For example, continued so as to try to date the document: A time-ago estimate might be made by finding a fountain pen ink (not here ball point) that might be identified as commercially available not before a certain date, and not too likely used after a later date range. And the extent of the reaction with the paper might allow for another time-ago estimate.[23]

20 Variations of Ink analytical chemistry – as examples – *Gedankenexperiment* – continue throughout this book. Introduction. 2.2 & 2.3.
21 Tannin. Chapter 2.
22 When. Chapters 2, 4 and 7.
23 When. ink. Chapters 2 and 7.

### 2.1.4 Science and chemistry

Even if not explicitly invoked in a forensic investigation, concepts of forensic science and chemistry should be understood in a broader context. Science is described here as the continuing pursuit of and questioning of knowledge of the workings of the physical universe around us.

Observations and measurements are made. Then, theories are constructed to try to explain – account for – those observations and measurements – and to be able to make reliable predictions for the future and other circumstances. And, to explain reports of the past. The simplest theories that work are sought [76] and unnecessary components [77, 78, 79] are avoided [80]. Scientific theories involve mathematical expressions [81]. And, at least from the time of *Principia*, of Isaac Newton, encouraged by Edmond Halley [82], more than three centuries ago, scientists publish and argue about the explanations [83].

Some theories become well established and undoubted – as for classical thermodynamics. Some theories should be thoroughly doubted quickly [84, 85]. But all theories are in some sense, tentative. For example, Newtonian physics might have long been regarded as the final word to describe gravitation and the motion of masses (such as the Earth, Moon and Newton's apple [86]), but that changed with the late-nineteenth and early-twentieth-century observations resulting from the sub-microscopically small, and of the very, very fast. New theories were needed – and quantum mechanics and relativity emerged. Newtonian physics is still quite serviceable for much of science, but not completely for atomic and nuclear structures, nor for near the speed of light [87, 88, 89].[24]

Scientific theories should always be questioned and tested for new observations and measurements [90, 91, 92, 93, 94, 95, 96, 97].[25] And, as already noted above, theories, observations, measurements and mathematical manipulations should be reported and published so that they can be argued about again and again, notwithstanding that for forensics, legislators, judges and lawyers and business investors, would be made uncomfortable. This is also analogously true for the transparency of parliamentary democracy.

Fortunately and conveniently, the forensic science concepts of this book are mostly settled. Mostly, but not entirely.[26,27]

24 Time dilatation of relativity theory: $T = T_0 / \sqrt{(1- (v/ c)^2)}$. Chapters 2 and 7.

25 *Nullius in verba.* Introduction 4.1, Chapters 2 and 5.

26 Some science controversary remains. Introduction 4; Chapters 1, 2, 10 and 11.

27 Science. DRE. Chapters 2, 10 and 11.

### 2.1.5 Science – good, bad and junk

In the Introduction,[28] and in other chapters,[29] reference is made to good and bad science [98].[30,31] Some further discussion is appropriate. The good faith process,[32] argued by Lee Smolin [99], of developing and refining theories, referred to above, is the good science, and it can cope with the wrong as on-the-way towards the correct [100]. In doing good science, the scientist tries to explain and takes care with procedures and data; and is willing to argue when confronted with other opinions. And, for good science, there would always be tests to try to falsify – that is, to carry out experiments designed to reveal if the theory would be wrong [101, 102].

Bad science [103, 104] is to offer a theory too remote from experimental observation – and mathematical expression – and importantly, to fail to do the processes of good science: explanations, care with the procedures and data, and efforts to falsify. Bad science may come along with a non-scientific agenda – as for creationism. This is sometimes called junk science.[105, 106, 107, 108, 109][33] As a former Governor-General [110][34] seemed to try to express a few years ago [111], amid controversy [112, 113], we are confronted with ideas without scientific basis – astrology, for example [114, 115, 116] – on which actual and serious decisions would be made.

Bad science includes that wrong ideas, contrary to Darwin's evolution theory [117],[35] are presented as true alternatives in public education and museums. From the time before H.L. Mencken, reporting in *The Baltimore Evening Sun*, 1925, on *Scopes*, Dayton, Tennessee [118, 119, 120, 121, 122], this remains a problem.

A newer version of bad science against Darwin is creationism [123]. And, similarly in other contexts, bad science is climate change denial and anti-vaccine advocacy [124, 125, 126, 127, 128, 129, 130, 131, 132, 133].[36]

Also, occasionally, a non-scientific opinion is presented in an otherwise scientific context. Sometimes, perhaps, this might be ok; other times, very problematic [134, 135, 136, 137, 138, 139, 140, 141, 142, 143, 144, 145, 146, 147].[37]

---

**28** Science. Introduction 4; Chapter 2.
**29** Chapters 10 and 11.
**30** Science – good, bad and junk. Introduction. 4, Chapters 1, 2, 5, 8, 10 and 11.
**31** Transparency. Opaqueness. Chapters 2, 5 and 10.
**32** Good faith explanation. Chapters 2 and 9.
**33** Cited literature accuracy. Introduction 2; 4.1; Chapter 2.
**34** Governor-General. Chapters 2, 10 and 11.
**35** Creationism. Climate change denial. Anti vaccine. Chapters 1, 2 and 8.
**36** Creationism. Climate change denial. Anti vaccine. Chapters 1, 2 and 8.
**37** Not Recommended Here. Introduction. 4.1; Chapter 2.

## 2.2 Forensics – recent historical

Forensic science applications are of theories constructed, in specific, to account for data typically gleaned from a single recent, human-experienced, incident. That is, experimental work is essentially concerned with data from what is specifically physically found, rather than from repeatable experiments of general concept.

To examine a recent historical incident, forensic science application is, on a very much smaller scale in analogy to other historical sciences – geology, astronomy and archaeology [148].[38] Just as geologists look on and into the ground, astronomers at the heavens [149], and archaeologists at dinosaur remains and at what bygone people left behind (for example, an ancient homestead or place of rituals) [150], for data to construct theories of what happened to form the Earth, the universe and about dinosaurs, and how bygone people lived; so too do forensic scientists look at the remains of a human scene to help construct a theory of what happened there, when and by whom. But while geology, astronomy and archaeology allow for general theories from repeated observations of a seemingly infinite number of similar events, forensic science differs in that it is typically evaluating a uniquely single event, offering perhaps only one chance to collect evidence.

This is to be contrasted with sciences such as physics [151, 152] and chemistry, wherefor experiments are designed, and the theory to be constructed, attempt to explain how things will always continue to happen, everywhere and everywhen in the universe, under certain conditions. These experiments can always be revisited by other scientists, elsewhere and elsewhen, completely re-doing them [153]. In geology, astronomy and archaeology, experimental re-visitation is redone in the sense of taking another look at what natural and history left behind. For forensic science, such re-visitation may not be feasible because the observation method may be destructive, and because such a small and unique part of the universe is being observed.

Forensic science is perhaps most closely related to archaeology – with a similar reliance on chemistry [154]. Perhaps, the differences are the age of the scene, who is interested and why. For forensic science, the scene is typically more recent and of interest to contemporary parties who have some legal/personal/commercial/political concerns. For archaeology, the scene is often ancient and of interest to historians for academic purposes. Overlap is certainly possible – for example, who is actually buried in the grave attributed to President Abraham Lincoln's assassin at Green Mount Cemetery, Baltimore [155, 156], is both a criminal law forensic and an archaeological historical question. And, sometimes forensic scientists might be assisted by their archaeological colleagues [157] and *vice versa* [158].[39]

---

**38** Archeology. Carbon-14. Chapters 2 and 7.

**39** = the other way around. [A statement is also true in reverse.]

## 2.3 The essential propositions expanded

The nature of the discovery, categorization and measurement, as considered above, are considered in this book largely in the context of analytical chemistry. But, in a broader context, the essential propositions may be expanded:

### 2.3.1 What it is

A first task of the essential proposition is often related to determining *what it is* that is found at a search site[40] – such as its physical description, chemical composition and history [159]. Typically, this relates to a physical item – for example: a written document[41] at a fraud scene, a spent ammunition shell [160] at a bank robbery, an art forgery or a chemical residue at an industrial explosion. Perhaps the most modern example is the possibility to analyze for minute amounts of DNA or protein left at a scene [161].[42]

But also, *what it is* may involve an altered condition of something – for example, corrosion of stone statuary [162] at a cathedral; or weathering of a Luxor Obelisk after it was moved to Paris [163].

The search site[43] or scene is typically geographical – for example: a room in a building in a city, or a brownfield.[44] But not always – for example, it could be the bodies of surviving war veterans that carry embedded shrapnel [164].[45]

Or, more subtly, the site could be the bodies of people who suffer illness, with the illness being the *what it is* – for example, veterans exposed to chemical warfare substances [165, 166, 167, 168, 169, 170, 171, 172, 173, 174, 175, 176, 177, 178, 179, 180, 181],[46,47] explosion or artillery-caused concussive injuries [182], workers compensation claimants with work-related chronic soft tissue injury, respiratory tissue of workers with embedded asbestos [183] or inner-city children suffering the neurotoxic effects of lead poisoning from old-stock housing [184, 185, 186, 187, 188].[48]

*What it is* may be natural, such as mud [189] from elsewhere transferred to the scene on shoes, or such as a blood stain. But it also might be human-fabricated [190] with a manufacture history, such as a spent ammunition shell [191] or a discarded ballpoint pen.

**40** Site. *What it is.* Chapters 1, 2, 3 and 5.

**41** Variations of Ink analytical chemistry – as examples – *Gedankenexperiment* – continue throughout this book. Introduction 2.3 and 2.2.

**42** DNA. Electrophoresis. Chapters 2 and 6.

**43** Site. *What it is.* Chapters 1, 2, 3 and 5.

**44** Sample selection. Brownfield. Chapters 2, 3 and 5.

**45** Shrapnel. Chapters 2 and 5.

**46** Fritz Haber, $Cl_2$. Chapters 2 and 4.

**47** Mustard gas. Chapters 1.9.1 and 2.3.1.

**48** Pb toxicology. Pb Sample selection. Chapters 1, 2 and 3.

*What it is* may have several aspects to it. For example, a found written item may be of interest because of – its/the:

- text (original or quoted, and if quoted, from where);
- text's plain language, allegorical [192] or coded [193] meanings; or its social [194] or psychological [195] implications;
- official function (for example: a will, US Federal Reserve Note, Bank of Canada legal tender note, or a movie ticket);
- quirks (for example, written in mirror image, in imitation of Leonardo Da Vinci); [196] – language in relation to its geographic location;
- method of inscription (handwritten with pen, pencil, crayon, spray paint [197], in the air, as revealed by video image, or by machine[49] or a transfer process);
- font [198, 199];
- the medium upon which it is written (paper [200], walls [201], tattoos on people, in a computer memory, *etc.*); or
- ink chemistry.

#### 2.3.1.1 Manufacture history

The manufacture history[50] of *what it is* can be of special forensic interest because, not only are human-fabricated items so common, they can also be rather well documented. The technical specifications, history, production statistics, servicing history [202],[51] and perhaps, reference samples manufactured to be the same, may be kept in the records of – for example:

- The manufacturer (for example: for automotive body paint) [203].
- The manufacture – baking – of bread.[52]
- An industry or professional association (for example: SAE) [204].
- US FBI Standard Reference Files and Collections [205, 206, 207, 208]
- International Ink Library and Digital Ink Library, US Secret Service [209, 210, 211, 212, 213, 214].[53]
- Other organizations' specialized libraries.
- A museum [215] (for example: cellulose nitrate artefacts at the Royal Ontario Museum [216], Toronto).[54]
- The Forbes art pigment collection [217, 218, 219].[55]

49 Woodstock typewriter. Alger Hiss. Introduction 2.3; Chapter 2.
50 Manufacture history. When. Chapters 2 and 7.
51 Manufacture history. Occurrence probability – ink example. Find expectation. Chapters 2, 4 and 9.
52 Bread. Best-before. Chapters 2.3.1.1 and 7.4.1.
53 Ink reference standards. US International Ink Library. ATF. Chapters 2, 3, 4 and 9.
54 Polymers in museum. Chapters 2 and 4.
55 Art supply. Forbes art pigment collection. Chapters 1, 2 and 3.

- Rolex serial number [220].
- Clothing labels, with records of manufacturers or suppliers.[56]
- Private and public collections, for example: a collectors' coin [221, 222, 223].[57]

Although some of such records may be legitimately held confidential for commercial or state secret purposes, they should typically be available to law enforcement agencies, and might be subpoenaed to legal process by others. But, often records are intentionally publicly available – such as:

- Patent documents (which are reported in the *Chemical Abstracts*) [224]
- Standard reference samples from US NIST [225]
- SDSs provided under US hazard communication requirements and WHMIS[58]
- Published engineering standards [226, 227]
- Published scientific and engineering literature
- US *Code of Federal Regulations* (as a specific example: the glass specifications noted on window corners, with a permanent label (US 16 *CFR* § 1201.5) at an Ontario courthouse)[228, 229, 230, 231, 232, 233]
- Soils list [234, 235, 236, 237][59]

Although industrial entities are often reluctant to give out information for trade secret or legal liability reasons – some may take such pride in their product as to publish [238].

It is also possible that a forensic scientist could do original historical research – for example, data for a history of commercial printing inks could be obtained by analysis of samples from old books from a library [239].

Manufactured things can also include intentionally introduced *chemical tags* – for example: in inks [240], or in explosives [241].

#### 2.3.1.2 Qualitative identification

The *qualitative identification* of the chemical component(s) of *what it is* involves *comparative* and/or from *fundamentals* of theory, by way of experimental observations.

***Comparative identification*** here means to identify a single chemical compound, or a mixture, as being the same as some other material, without necessarily having to know what, or exactly what it is. For example: the forensic chemist might have merely determined that the ink of a document is identical to what came from a particular pen, or as on another document or as from a reference material, without delving too much into the theoretical complexities of the molecules that constituted the ink, or the manufacturer's formulation.

---

**56** Crippen murder case. Hyoscine. CAS 51-34-3 Chapters 2 and 7.
**57** 1909 Lincoln Wheat VDB Penny. Cu Sb Zn Chapters 2 and 3.
**58** Workplace Health & Safety. Hazard communication. WHMIS. Chapters 1, 2 and 4.
**59** Soils list. Mud. Chapter 2.

***Fundamental identification*** of a chemical composition – such as for ink – would necessarily require more detailed information. This might involve various chemical and physical analyses to determine, from scratch, molecular structure or the several molecular structures of the compounds of a mixture. By this definition, to fundamentally identify a chemical substance is to observe and measure its properties, as would be expected from the theory of the molecular structure. It might, alternatively, involve use of standard reference samples for a *fundamental* determination by *comparative* method, or further, alternatively, use both *fundamental* and *comparative* methods in a complementary/confirmatory way.

For more on chemical identification, see Chapter 4.[60]

#### 2.3.1.3 Chemical reactions in progress

It should also be noted that *what it is* might involve – and require – knowing more than merely the name of a compound and its molecular structure, or even knowing the several compounds of a mixture. The item found at the scene, in addition to having material existence, might also change over time.[61] Such changes would involve chemical reactions that have reached, or that are still proceeding towards, a conclusion [242].

Such reactions in progress might involve not only the materials of the found item; but also, there may be a loss or gain of material. Also, there might be reactions with the environment of the find scene – such as the oxygen[62] of the air, photons from the sun, moisture, heat, forces of wind or water and biochemistry of bacterial action. Investigating these reactions can be more challenging than mere molecular determinations. Lab handling and storage might cause further and other reactions. Such possibilities strongly suggest that the forensic chemist take special care in the handling and storage of samples. [243]

#### 2.3.1.4 Energy phenomena

Yet another investigatory area: Energy phenomena. These are energy interactions with physical matter that can sometimes be used to determine what a phenomenon was, and how much, for example:

- The application of force for physical breakage (*e.g.*, a broken window) [244, 245]
- Imprint (of a tire in mud) [246, 247, 248, 249]
- Material transfer (a paint chip imbedded in a tree from a car that hit it)
- Transference of heat for an increase or decrease of temperature (a cold or hot object that was placed on a surface)

---

**60** *Comparative/fundamental* identification of molecules. Chapters 2, 3, 4 and 6.

**61** When. Chemical kinetics. Reactions in progress. Chapters 2, 4 and 7.

**62** $CO_2$ CO. Chapters 2, 3, 6 and 8.

- Electromagnetic (visible light of a stated wavelength);[250] and sub-atomic particulate (α [251] and n [252] tracks in a polymer film) radiation;[63,64] and radioactivity (carbon-14 used to date an ancient document [253])
- Electric charge (left from footprints in a carpet [254])
- Sound, vibration, pressure and shock wave phenomena through fluids (such as: air [255, 256] or water) and solids (things of human construction [257], or seismic activity in the Earth [258])
- Voice identification and evaluation [259]
- Magnetic field cartographic orientation of an ancient oven [260, 261]
- Microwave radiation suspected for diplomats' illness; and for police traffic radar operators[65]
- Medical consequences of forces of acceleration and deceleration [262, 263]. For example, on a rocket sled [264, 265, 266] of Major (later Colonel) John Stapp [267]. Sudden start and stop with high velocity causes rapid acceleration and deceleration with great forces. [268, 269, 270]

Much of the above are better dealt with by a physicist, medical researcher or other scientist, or an engineer, rather than a chemist.

#### 2.3.1.5 Corollaries

- Whose it was [271][66,67,68]
- Where it has been before [272, 273, 274, 275, 276][69]
- Related items [277, 278][70,71]

### 2.3.2 How much

For each *qualitative* determination that the forensic chemist might do, a *quantitative* determination might also be called for. For example: mass (kg), absolute concentration (mol/L), relative concentration (mol%), force (nt), kinetic energy (joules), temper-

**63** Subatomic particles. Chapters 2, 4 and 6.
**64** Neutron. Atomic Number. Chapters 2 and 4.
**65** Microwave. Chapters 2 and 6.
**66** DNA. Proteins. Fingerprints. Chapters 2; 6.18.4.
**67** Fingerprint probability. Chapters 1 and 4. Fingerprint visualization. Fluorescence & phosphorescence. Chapters 6 and 9.
**68** Fingerprint when. Chapters 7 and 8.
**69** Soils list. Mud. Chapter 2.
**70** Crippen murder case. Hyoscine. CAS 51-34-3. Chapters 2 and 7.
**71** 1909 Lincoln Wheat VDB Penny. Cu, Sb, Zn, Chapters 2 and 3.

ature (°C), wavelength (light colour; nm), radiation intensity, pressure (Pa), magnetic field strength, cartographic orientation, *etc*.

### 2.3.3 Error

Sometimes forensic results can be quantified with great accuracy, but often not.[72,73] Sometimes, the accuracy is so questionable that the resultant numbers should only be regarded as estimates, and sometimes, the very bases for estimation are questionable. And, sometimes an error can be of a qualitative nature.

Such issues of error are of essential importance to forensic science. These are dealt with in Chapter 8.[74]

### 2.3.4 When

Timing is a special kind of quantification not mentioned above, but implied by the consideration of reactions in progress. Theoretically, all chemical reactions may be characterized by their chemical kinetics – the times taken for various quantities or concentrations of reactants, intermediates and final products to be depleted and created. These times might be measurable from quantity or concentration measures of chemical substances, and the forensic chemist might be able to help relate them to the when of a human story. As with all measurements, error is also of concern.[75]

### 2.3.5 What happened

After considering, as above, what is found at a scene, how much, error possibilities and when, the forensic chemist may be called upon to help explain how this would be related to the human story that left the scene behind.

This might be attempted by developing a chronological theory based on what is found, along with other information. As the theory develops, it might point to further examinations of what was initially found, further searches for other items to examine and theory revision [279].

When a scene is first viewed, a preliminary theory of human stories and scientific explanation may often be implied; otherwise, the forensic scientist might not have much of an idea of what to do. For example, in the forensic ink example, the chemist

72 Qualitative. Quantitative. Error. Chapters 2 and 8.
73 Measurement. Error. Chapters 2, 3 and 8.
74 Error. Chapters 2, 3, 7 and 8.
75 *When*. Chapters 2, 3, 4, 7 and 8.

would assume being called to examine ink on paper, and not an electronic facsimile. The obvious theory would be that a human hand linearly applied ink from a pen onto paper, and that it was not done by a machine all at once by photo reproduction. But, the forensic chemist must bear in mind that the obvious theory could possibly be wrong. This has implications on how to look for things at the scene, and the care to be taken.[76]

Preliminary theory would thus be confirmed or replaced by a more firmly verified theory, or succession of theory refinements, which, hopefully, would eventually be able to uniquely and reliably account for the events and things of the human story.

## Notes

**1** Lewis Carroll, *Alice's Adventures in Wonderland* and *Through the Looking-Glass*, Signet Classics, The New American Library, New York City, 1960 and 1962; from . . . *Adventures* . . ., "Alice's Evidence," Chapter XII, pages 110 and 111.
**2** Agatha Christie, *Murder at the Vicarage*/A Miss Marple Mystery, 1930, Signet, New York City, 10014; 2000; ISBN 0-451-20115-9; at pages 220 and 221.
**3** In the author's terminology.
**4** *Gedankenexperiment*. Introduction 2.2.
**5** Mark here means by leaving evidence – possibly, but not necessarily, directly human-perceivable – sight, feel, odour.
**6** *Gedankenexperiment*. Introduction 2.2.
**7** *Gedankenexperiment*. Introduction 2.2.
**8** *Gedankenexperiment*. Introduction 2.2.
**9** Or not so intense light over time – or to light of different wavelengths.
**10** Perhaps, a faded advertising poster. Or in another example context, a human's sun tan.
**11** Tanith Lee, *Days of Grass*, DAW BOOKS, New York City, 10019; 1985; ISBN 0-88677-094-7. A post-apocalyptic science fiction, wherein teenage Ester sneaks above ground from her small disciplined and secretive community. On return below ground, hours later, she is quickly discovered, because of her suntan – with consequences. The community had been below ground 141 years (at Chapter two page 23); ". . . in the artificially lit dark . . ." (at Chapter One page 9); the leader, Standish, lived in seclusion.
https://www.fantasticfiction.com/l/tanith-lee/days-of-grass.htm
https://www.amazon.ca/Days-Grass-Tanith-Lee/dp/B002M0UOFM/ref=sr_1_1?keywords=Ta nith+Lee+Days+of+Grass&link_code=qs&qid=1585860547&sr=8-1 https://www.pinterest.ca/pin/472244710911308107/
**12** Tanith Lee, *Days of Grass*, 1985; Chapter two, pages 13, 16, 17: ". . . Then they found her out . . . 'Where did you go,' . . . 'to get such a fine tan, young lady?' . . . 'Above Ground. Where else?' . . ."
**13** https://en.wikipedia.org/wiki/Sun_tanning
**14** Randall K. Noon, "Vehicular Accident Reconstruction," in Stuart H. James and Jon J. Nordby, editors, *Forensic Science/An Introduction to Scientific and Investigative Techniques*, CRC Press, Boca Raton, 2003; ISBN 0-8493-1246-9; Chapter 23, pages 433–450. www.crcpress.com
**15** *Gedankenexperiment*. Introduction 2.2.

**76** What to concentrate on. Chapters 2 and 3.

16 See, for example: Piya Chattopadhyay, CBC radio, *Out in the Open*, "If Memory Serves / Piya explores just how much you can really trust your memory"; May 2020.
https://www.cbc.ca/radio/outintheopen/if-memory-serves-1.4945241

[. . .]

> **How police convinced a woman that her bad dream was an actual murder**
> When . . . was a teenager in Reykjavík in the 1970s, she and her boyfriend . . . were detained by police for running an embezzlement scam . . . questioning . . . soon veered toward two totally unrelated murder cases. . . . complex tale of forming false memories under police pressure, and giving testimony . . . Decades later, the case is widely seen as a vast miscarriage of justice.

[. . .]

17 See also: Jan Ramson, "A False Confession Casts Doubt on Bronx Cases/Prosecutor Scrutinizes Detectives' Tactics in 31 Homicides," *The New York Times*, 15 Feb. 2021, pages A1 and A16.

18 By the author; and extended here.

19 See:

- Samuel M. Gerber, ed., *Chemistry and Crime/from Sherlock Holmes to today's courtroom*, American Chemical Society, Washington, D.C., pages 46 and 47, and references cited therein.
- Brian H. Kaye, *Science and the Detective/Selected Reading in Forensic Science*, VCH Verlagsgesellschaft mbH, Weinheim; 1995; ISBN-3-527-29252-7; pages 71 and 72, and references cited therein.
- https://www.samyoung.co.nz/2023/11/locards-exchange-principle.html

20 Brian H. Kaye, *Science and the Detective/Selected Reading in Forensic Science*, VCH Verlagsgesellschaft mbH, Weinheim, Germany; 1995; ISBN-3-527-29252-7.
https://content.e-bookshelf.de/media/reading/L-603243-1e8dab4cde.pdf https://pubs.rsc.org/en/content/articlelanding/1996/an/an996210005n/unauth#! divAbstract

21 Brian Manarin, *Forensic Evidence in Context: Cases, Materials and Commentaries*; Carswell, THOMSON REUTERS, Canada, 2017, June 2019; ISBN: 9780779880683.
https://store.thomsonreuters.ca/en-ca/products/forensic-evidence-in-context-cases-materials-and-commentaries-30835077
https://static.legalsolutions.thomsonreuters.com/product_files/relateddocs/220441_2019288_111518.pdf
https://static.legalsolutions.thomsonreuters.com/product_files/relateddocs/220441_2019288_101542.pdf

22 Manarin, *Forensic Evidence . . .*, 2019; [Locard, *et al.*], page 4, et seq.

23 Manarin, *Forensic Evidence . . .*, 2019; [Sherlock Holmes] page 2, et seq.

24 Manarin, *Forensic Evidence . . .*
https://books.google.ca/books?id=s-pbtAEACAAJ&source=gbs_ViewAPI&redir_esc=y

25 See also: Gil Reavill, *Aftermath, INC. / Cleaning Up After CSI Goes Home*, Gotham Books Penguin, New York City, 1001; 2007. ISBN 978-1-592-40364-6.

26 Included in Gil Reavill, *AFTERMATH. . .* – some commentary about related topics:

- Crime scene.
- CSI TV shows – pages 212→218.
- Forensics.
- Locard's principle, Sir Arthur Conan Doyl, Sherlock Holmes – pages 215→218.
- postmortem interval – pages 50→58.

27 E.J. Wagner, "The French Connection of Sherlock Holmes." [2019, 2020.]
https://ejdissectingroom.wordpress.com/2011/02/25/the-french-connection-of-sherlock-holmes/
https://ejdissectingroom.wordpress.com/

28 Kaye, *Science and the Detective . . .*, 1995, pages 71 and 72.

29 Michael Hardwick, *The Complete Guide to SHERLOCK HOLMES*, Weidenfeld and Nicolson, London SW4 7TA, 1986; ISBN 0 297 78963 5.
30 Hardwick, pages 7 and 13; *et seq.*
31 See also: P.D. James, *Talking About Detective Fiction*, Chapter 2, "The Tenant of 221B Baker Street . . .," pages 27→42; Vintage Books, New York City, 2011, 2009. ISBN 978-0-307-74313-8.
32 Richard Saferstein, *Criminalistics/An Introduction to Forensic Science*, Seventh Edition, Prentice Hall, Upper Saddle River, New Jersey, 07458; 2001; ISBN 0-13-013827-4; pages 4 and 5.
33 https://commons.wikimedia.org/wiki/File:The_Adventure_of_the_Naval_Treaty_01.jpg
34 https://www.euro-book.co.uk/book/isbn/0297789635.html
https://openlibrary.org/works/OL4310622W/The_Complete_Guide_to_Sherlock_Holmes
https://www.ebay.co.uk/itm/Sherlock-Holmes-in-The-Adventure-of-the-Naval-Treaty-drawn-by-Sidney-Paget-/152797082345
https://www.arthur-conan-doyle.com/index.php/The_Adventure_of_the_Naval_Treaty
35 Michael Hardwick, *The Complete Guide to SHERLOCK HOLMES*, Weidenfeld and Nicolson, London SW4 7TA, 1986; ISBN 0 297 78963 5; illustration – between pages 96 and 97.
36 Sir Arthur Conan Doyle, *The Naval Treaty*, 1893. Illustration – Sidney Paget, *Strand*, 1893.
37 https://www.alamy.com/stock-photo-sherlock-holmes-holmes-was-working-hard-over-a-chemical-investigation-33774336.html
38 "SHERLOCK HOLMES – The Naval Treaty – An Appeal to Holmes – By Sir Arthur Conan Doyle" Comic strip EPISODE 1, *The Boston Globe*, 05 December 1930, page 44. Illustrator = Leo O'Mealia.
39 See also: Sir Arthur Conan Doyle, *A Study in Scarlet*, 1887. Illustration – George Hutchinson, 1904.
40 [First meeting of Sherlock Holmes and Dr. Watson.] [A study in Scarlet, 1887.]:
41 https://www.euro-book.co.uk/book/isbn/0297789635.html
https://openlibrary.org/works/OL4310622W/The_Complete_Guide_to_Sherlock_Holmes
https://en.wikisource.org/wiki/A_Study_in_Scarlet
https://www.alamy.com/stock-photo-a-study-in-scarlet-1904-sherlock-holmes-illustrated-by-george-hutchinson-56535569.html
http://www.sherlockholmes-fan.com/a-study-in-scarlet.html
42 Michael Hardwick, *The Complete Guide to SHERLOCK HOLMES*, Weidenfeld and Nicolson, London SW4 7TA, 1986; ISBN 0 297 78963 5; illustration – opposite page 64.
43 "The first meeting of Sherlock Holmes and Dr Watson. Their introduction by Stamford, illustrated by George Hutchinson."
44 Statistics of occurrence = find probability. Also of interest would be proportions of mixture.
45 This is the author's definition adapted and inspired from other sources.
46 See:
- American Chemical Society, "Analytical Chemistry/What Is Analytical Chemistry?" ACS, Washington, DC, 20036; 2016.
  http://www.acs.org/content/acs/en/careers/college-to-career/areas-of-chemistry/analytical-chemistry.html
- John Kenkel, Analytical Chemistry/Refresher Manual, Lewis Publishers, Chelsea, Michigan 48118; 1992; ISBN 0-87371-398-2; Chapter 1.3 Terminology, pages 5 and 6.
- John Keenan Taylor, Quality Assurance of Chemical Measurements, Lewis Publishers, Inc.; CRC Press, Inc., Boca Raton, Florida, 33431; 1987; ISBN 0-87371-097-5; Chapter 22, "Reporting Analytical Data," pages 197–207.
  https://en.wikipedia.org/wiki/Analytical_chemistry

47 Nick Kim
https://en.wikipedia.org/wiki/Nick_D._Kim
48 https://en.wikipedia.org/whttps://www.canada.ca/en/environment-climate-change/services/management-toxic-substances/list-canadian-environmental-protection-act/lead/using-more-lead-free-ammuni

tion/lead-ammunition-executive-summary.html
https://www.handgunsafetycourse.com/handgun/studyGuide/Basic-Components-of-Ammunition/601099_700077856/
https://en.wikipedia.org/wiki/Phoenix_Shot_Tower https://explore.baltimoreheritage.org/items/show/200
https://www.newscientist.com/article/dn13622-gunshot-residue-test-fingers-lead-free-bullets/

**49** See also: Maria Josefi, Writer-editor, US FBI *Handbook of Forensic Science*, February 1995; US Government Printing Office, Superintendent of Documents, Washington, D.C., 20402-9328; ISBN 0-16-045111-6; pages 57 and 58 Lead Analyses.

**50** *Gedankenexperiment*. Introduction 2.2.

**51** Ink https://en.wikipedia.org/wiki/Iron_gall_ink
https://en.wikipedia.org/wiki/Ink
https://sciencing.com/chemical-composition-pen-ink-17194.html
https://www.chemistryworld.com/news/ink-chemistry/3002158.article
https://archive.org/details/jresv15n1p35
https://en.wikipedia.org/wiki/Fountain_pen_ink

**52** See Brian H. Kaye, "Paper, the Forger's Nemesis?" as Chapter 10.2 in *Science and the Detective/Selected Reading in Forensic Science*, VCH Verlagsgesellschaft mbH, Weinheim, Germany, 1995; pages 269–275.

**53** The ink examples in this book are (mostly) for ball point pen ink – not for fountain pen ink. But fountain pen ink is used for the example (tannin-iron sulfate) here. Different chemistries.

**54** The Fe-tannin chemistry here is convenient to help explain the definition of "analytical chemistry": The ink of this example is as would be, or have been, found in older-style fountain pens; sometimes still used. Ink flows from a storage cavity within the pen through a narrow tube, and a point onto the paper. Traditionally, for centuries, variations of that ink have been a water-soluble ferrous ($Fe^{2+}$) complex of tannin. Tannin is class of molecules with polymeric arrangements of a sugar, gallic acid, and polyphenolic molecules, historically derived from oak tree bark, or other plant materials. Generally, when on the paper, the $Fe^{2+}$ complex oxidizes on exposure to air, to not-watersoluble ferric ($Fe^{3+}$) tannate – the insolubility is related to its permanence. (An analogous oxidation occurs with $Fe^{++}$ in haemoglobin in blood → $Fe^{3+}$.) Ink colour can be altered with the addition of dyes, which may fade over years, but the permanent ferric ($Fe^{3+}$) tannate remains. Ball point pen inks are quite differently (than Fe-tannin) formulated – with organic (in chemistry terminology – mostly carbon C and hydrogen H) materials, and solvents. Most pens and their inks are now ball point – and analyses explained in this book are, generally, for ball point, rather than for fountain. There are now other kinds of writing and artists' devices commercially available. Analyses for fountain pen ink could involve many properties – including the tannin composition molecular arrangements

**55** Tannin + iron sulfate $[Fe(H_2O)_6]^{2+}$ $[SO_4]^{2-}$ → ferrous ($Fe^{2+}$) tannate complex and ferrous ($Fe^{2+}$) tannate complex + $O_2$ → ferric ($Fe^{3+}$) tannate

**56** Tannic acid CAS 1401-55-4 $C_{76}H_{52}O_{46}$ – a tannin – a large molecule with a molar mass of 1,701.20 g/mol contains one sugar unit and ten phenolic units (see Figs. 2.3 and 2.4).

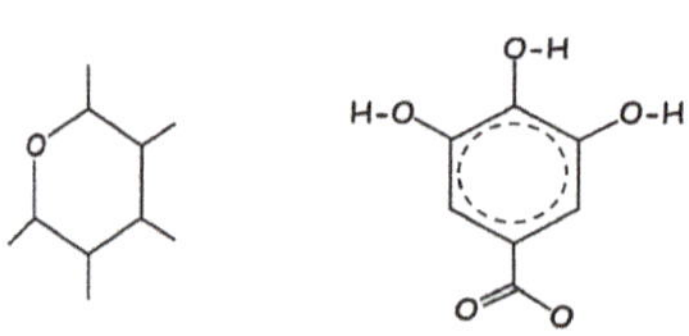


Figs. 2.3 and 2.4: A tannin.

https://www.sigmaaldrich.com/catalog/product/sial/403040?lang=en&region=CA

**57** Tannin: https://en.wikipedia.org/wiki/Tannin
https://en.wikipedia.org/wiki/Gallic_acid
https://www.ncbi.nlm.nih.gov/pubmed/9759559
https://www.britannica.com/science/tannin
https://www.sciencedirect.com/topics/agricultural-and-biological-sciences/tannin
https://www.ncbi.nlm.nih.gov/pmc/articles/PMC3495704/
**58** Ferrous tannate and ferric tannate: https://irongallink.org/ https://pubs.usgs.gov/wsp/1459d/re port.pdf
https://www.researchgate.net/publication/226105873_On_the_Tannic_Acid_Interaction_ with_Metallic_Iron
https://elementsunearthed.com/2014/05/08/making-iron-gall-ink/
https://travelingscriptorium.library.yale.edu/2013/03/21/iron-gall-ink/
**59** Ink https://en.wikipedia.org/wiki/Iron_gall_ink https://en.wikipedia.org/wiki/Ink
https://sciencing.com/chemical-composition-pen-ink-17194.html
https://www.chemistryworld.com/news/ink-chemistry/3002158.article
https://archive.org/details/jresv15n1p35
https://en.wikipedia.org/wiki/Fountain_pen_ink
**60** Fe = iron.
**61** Ferric ($Fe^{3+}$) tannate https://www.lookchem.com/Dictionary/Ferric-Tannate/
https://irongallink.org/ https://irongallink.org/igi_indexedde.html
https://en.wikipedia.org/wiki/Iron_gall_ink
https://en.wikipedia.org/wiki/Iron_gall_ink#cite_note-2
https://en.wikipedia.org/wiki/Iron_gall_ink#U.S._government_%22standard_ink%22_for mula_(1935)
**62** Maria João Melo, Vanessa Otero, Paula Nabais, Natércia Teixeira, Fernando Pina, Conceição Casanova, Sara Fragoso & Sílvia O. Sequeira,"Iron-gall inks: a review of their degradation mechanisms and conservation treatments," *Heritage Science*, volume 10, Article number 145, 16 September 2022. https://www.nature.com/articles/s40494-022-00779-2
**63** Hyunwoo Choi, Hyeong Bin Rheem, Duc Tai Nguyen, Sang Yeong Han, Hojae Lee, Insung S. Choi, "Iron Gall Ink Revisited: Visible Light-Induced, Eosin-Mediated Acceleration of $Fe^{2+}$ Oxidation in Pattern Nanoarchitectonics of $Fe^{3+}$-Tannic Acid Films," *Langmuir*, Vol 40, Issue 43, 17 October 2024. https://pubs.acs.org/doi/10.1021/acs.langmuir.4c03240
**64** Oxygen from atmospheric air (see Fig. 2.5).

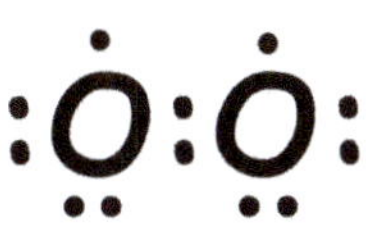

Fig. 2.5: Oxygen, $O_2$.

See: Lejaren A. Hiller, Jr. and Rolfe H. Herber, *Principles of Chemistry*, McGraw-Hill Book Company, Inc., New York, . . ., 1960, page 238.
**65** Oxidation = loss of electrons.
See: Lejaren A. Hiller, Jr. and Rolfe H. Herber, *Principles of Chemistry*, McGraw-Hill Book Company, Inc., New York, . . ., 1960, page 595.
**66** See also: https://www.thoughtco.com/definition-of-oxidation-in-chemistry-605456
**67** Ferrous ($Fe^{2+}$) tannate complex https://en.wikipedia.org/wiki/Tannin
https://en.wikipedia.org/wiki/Iron(II)_sulfate

https://travelingscriptorium.library.yale.edu/2013/03/21/iron-gall-ink/
Figs. 2.3 and 2.4: A tannin.
**68** Older-style fountain pen https://en.wikipedia.org/wiki/Fountain_pen_ink
**69** Most of the ink examples, throughout this book, are for ball point pen inks, but this example – is for fountain pen ink. Different chemistries.
**70** Assumed for this *Gedankenexperiment*. That ball point pen inks are organically, rather than water, soluble is a generality – there are very many formulations so that some such inks would be water soluble.
**71** Generally, in the study of chemistry, it is found that "like dissolves like" – so that organic molecules (generally, composed mostly of carbon and hydrogen atoms) would dissolve in organic solvents and not in aqueous (of water) solvents. And *vice versa* for non-organics. A more careful description of the concept relates to the polarity of the molecules. For purposes here it is sufficient to notice that the older style fountain pen inks would have somewhat different chemistries than many modern ball point inks.
https://pubs.acs.org/doi/abs/10.1021/ed080p447
https://pmc.ncbi.nlm.nih.gov/articles/PMC7880597/
**72** https://travelingscriptorium.com/2013/03/21/iron-gall-ink/
https://travelingscriptorium.com/
https://travelingscriptorium.com/2013/03/21/iron-gall-ink/
https://sciencing.com/ballpoint-pen-ink-made-of-10036405.html
https://en.wikipedia.org/wiki/Ballpoint_pen
https://home.howstuffworks.com/pen4.htm
https://www.ncbi.nlm.nih.gov/pubmed/15568713
**73** Most of the ink examples, throughout this book, are for ball point pen inks, but this example – is for fountain pen ink. Different chemistries.
**74** Dye https://en.wikipedia.org/wiki/Pigment
**75** Details of the analytical chemistry methods for such a determination of Fe and tannin-related inks are not described in this book. They may be different than analyses of ball point pen inks that are described.
**76** Occam's razor. https://en.wikipedia.org/wiki/Occam%27s_razor
**77** Rube Goldberg machines.
https://www.rubegoldberg.com/rube-the-artist/ https://en.wikipe dia.org/wiki/Rube_Goldberg
**78** Hilarie M. Sheets, "One Hand (and Spring and Funnel and Ramp) Washes the Other/Rube Goldberg contraptions to pass a bar of soap compete," *The New York Times*, 11 April 2020, page C6.
https://www.nytimes.com/2020/04/08/arts/design/rube-goldberg-bar-of-soap-challenge.html
**79** CBC Radio as It Happens, "The Rube Goldberg Exercise Bike will fan you and feed you a cookie while you work out," August 2020.
https://www.cbc.ca/radio/asithappens/the-rube-goldberg-exercise-bike-will-fan-you-and-feed-you-a-cookie-while-you-work-out-1.5706434
**80** For example, the concept of the *ether* (*æther*) has long been discarded as the medium of transmission to explain electromagnetic radiation. https://en.wikipedia.org/wiki/Aether_(classical_ element)
**81** See: Jerry Oltion, "Is Math Real?," Science, *Fantasy & Science Fiction*.
https://www.sfsite.com/fsf/2020/jo2011.htm
**82** Sir Isaac Newton, *Philosophiae Naturalis Principia Mathematica* (Mathematical Principles of Natural Philosophy), Royal Society, London, 1687. See Edward Neville da Costa Andrade, *Sir Isaac Newton/ His Life and Work*, DOUBLEDAY ANCHOR, Garden City, New York; SCIENCE STUDIES SERIES, S42; reprinted from Macmillan Company, New York, and W. Collins Sons and Co, Ltd., London, first published 1954; page 70, et seq.

**83** See also: Joshua Rothman, "The Rules of the Game/How does science really work?" *The New Yorker*, Books, pages 67–71 – A review of Michael Stevens, *The Knowledge Machine: How Irrationality Created Modern Science*, Liveright; 2020.
https://www.newyorker.com/magazine/2020/10/05/how-does-science-really-work
https://www.amazon.com/Knowledge-Machine-Irrationality-Created-Science/dp/1631491377
**84** For example, the use of "Perkin's Metallic Tractors" were found to be of no medical benefit. See: Hannah Fry, "Experiments on Trial/Has the new era of experimentation remembered the lessons of the old?" *The New Yorker*, Books, 02 March 2020, pages 61–65, at pages 61 and 62.
**85** And see below: Velikovsky, *Worlds in Collision*
**86** Somewhat apocryphal?
https://en.wikipedia.org/wiki/Isaac_Newton#Apple_incident
https://en.wikipedia.org/wiki/Isaac_Newton
[consider with caution:]
https://www.discovermagazine.com/planet-earth/the-scientific-method-is-a-myth
https://www.amazon.ca/Newtons-Apple-Other-Myths-Science/dp/0674967984
https://www.researchgate.net/publication/260789007_Newton's_Apple_and_other_Myths_about_Science
**87** Relativity theory's speed limit – and the time, space and mass consequences for approaching it – has been in much of science fiction – many, many stories include FTL (faster than light) travel – See, for example:
**88** Paul Anderson, 1970, *Tau Zero*; SF MASTERWORKS, 1988; Orion Publishing Group, London, WG2H 9EA; ISBN 978-1-407-23913-2; pages 54, 55, 56;

$$\tau_0 = \sqrt{(1-(v^2/c^2))}.$$

https://en.wikipedia.org/wiki/Tau_Zero
**89** Richard T. Weidner and Robert L. Sells, *ELEMENTARY MODERN PHYSICS*, ALLYN & BACON, INC., Boston, 1960, 1963; pages 62, 63, 64; *time dilatation*

$$T = T_0/\sqrt{(1-(v/c)^2)}.$$

**90** See: Lee Smolin, *The Trouble with Physics/The Rise of String Theory, the Fall of a Science, and What Comes Next*, Houghton Mifflin Company, Boston and New York City 10003; 2006; ISBN-10: 0-618-91868-X; Chapter 17, "What Is Science," page 289, et seq.
**91** George Gamow, *One Two Three . . . Infinity/Facts and Speculations of Science*, Bantam Books, New York City, 10016, April 1967; copyright 1947, 1961 by the author; page 142, et seq.
**92** Richard P. Feynman, *The Meaning of It All/Thoughts of a Citizen Scientist*, Perseus Books Group, Cambridge, Massachusetts, 02412; Basic Books, New York City, 10016-8810; ISBN-13: 978-0-465-02394-3; ISBN-10: 0-465-02394-0; 1963, 1998; page 4, et seq.
**93** Stephen W. Hawking, *A Brief History of Time/From the Big Bang to Black Holes*, Bantam Books, New York City; 1988; ISBN 0-553-05340-X; page 10, et seq.
**94** Carl Meyer, "Distinguishing Good Science, Bad Science and Junk Science," as Chapter 7 in Carl Meyer, ed., *Expert Witnessing/Explaining and Understanding Science*, CRC Press LLC, Boca Raton, Florida, 33431, 1999; ISBN 0-8493-1197-7. (Derived from a symposium of the American Chemical Society, Division of Chemistry and the Law, ACS National Meeting, Las Vegas Nevada, 1997.)
**95** Edward Neville da Costa Andrade, *Sir Isaac Newton/His Life and Work*, Doubleday Anchor, Garden City, New York; Science Studies Series, S42; reprinted from Macmillan Company, New York, and W. Collins Sons and Co, Ltd., London, first published 1954; page 60.
**96** *Nullius in verba* ≈ We don't take anybody's word for it.

**97** http://royalsociety.org/ "The Royal Society's motto 'Nullius in verba' roughly translates as 'take nobody's word for it'. It is an expression of the determination of Fellows to withstand the domination of authority and to verify all statements by an appeal to facts determined by experiment."
**98** Manarin, *Forensic Evidence . . .*, 2019; [Junk science, . . .], page 686, et seq.
**99** Smolin, page 301, et seq.
**100** See notes above.
**101** https://en.wikipedia.org/wiki/Falsifiability
**102** https://en.wikipedia.org/wiki/Cargo_cult_science
**103** Weinberg diplomatically and politely uses the term ". . . would-be sciences: . . ." – ". . . astrology, precognition, 'channeling,' clairvoyance. telekinesis, creationism, and their kin . . ."
**104** See: Steven Weinberg, *DREAMS OF A FINAL THEORY / The Scientist's Search for the Ultimate Laws of Nature*, Vintage Books, New York City, 1994; ISBN 978-0-679-74408-5; page 48.
**105** Carl Meyer, "Distinguishing Good Science, Bad Science and Junk Science," as Chapter 7 in Carl Meyer, ed., *Expert Witnessing/Explaining and Understanding Science*, CRC Press LLC, Boca Raton, Florida, 33431, 1999; ISBN 0-8493-1197-7. (Derived from a symposium of the American Chemical Society, Division of Chemistry and the Law, ACS National Meeting, Las Vegas Nevada, 1997.)
**106** http://en.wikipedia.org/wiki/Junk_science
**107** NOT RECOMMENDED HERE: Velikovsky, *Worlds . . .*
**108** Immanuel Velikovsky, *Worlds in Collision*, ABACUS, London WC1X 8JL; 1974, 1972, 1950; ISBN 0 349 13573 8.
https://www.abebooks.com/9780349135717/Worlds-Collision-I-Velikovsky-0349135711/plp
**109** [Criticism of Velikovsky, *Worlds . . .*]: Robert Schadewald, "If Continents Can Wander, Why not Planets?," ASIMOV'S SCIENCE FICTION MAGAZINE, 28 September 1981, Vol.5, No.10, pages 84–95.
https://en.wikipedia.org/wiki/Robert_Schadewald
https://www.asimovs.com/
https://en.wikipedia.org/wiki/Asimov%27s_Science_Fiction
**110** https://en.wikipedia.org/wiki/Immanuel_Velikovsky
Linda Solomon Wood, "Science v. religion and the new Governor General under fire," *CANADA'S NATIONAL OBSERVER*, Opinion, Energy, Politics, 06 November 2017.
https://www.nationalobserver.com/2017/11/06/opinion/science-v-religion-and-new-governor-general-under-fire
**111** While then still in office as Governor-General.
**112** The controversy centred on Her Excellency's making public statements unnecessarily controversial to her role as representing the Queen as head of state, with mostly ceremonial functions, but with residual constitutional powers. (This was not the controversary that later resulted in that her resignation.)
**113** https://www.nationalobserver.com/2017/11/06/opinion/science-v-religion-and-new-governor-general-under-fire
https://www.nationalobserver.com/2017/11/04/news/payette-takes-climate-change-deni ers-and-horoscopes-science-conference
https://torontosun.com/opinion/editorials/editorial-julie-payette-speech-oversteps-her-role
https://www.cbc.ca/news/opinion/governor-general-speech-julie-payette-climate-change-1.4384481
https://www.gg.ca/en https://en.wikipedia.org/wiki/Governor_General_of_Canada
**114** Callie Beusman, "Astrology Is Hard, Even if It's Fake / There's an exam, and it involves math. In fact, there are many exams," "Astrology Tries to Leave the Realm of the Absurd," *The New York Times*, 17 Oct. 2018, page D4.
**115** See: Steven Weinberg, *DREAMS OF A FINAL THEORY / The Scientist's Search for the Ultimate Laws of Nature*, Vintage Books, New York City, 1994; ISBN 978-0-679-74408-5; pages 48, 49.

**116** See: David Roney, *About Time / A History of Civilization in Twelve Clocks*, PENGUIN BOOKS, UK, 2021; ISBN 978-0-241-37081-3; pages 39, 40.
**117** https://www.livescience.com/474-controversy-evolution-works.html
https://en.wikipedia.org/wiki/Darwinism
**118** https://www.nationalobserver.com/2017/11/06/opinion/science-v-religion-and-new-governor-general-under-fire
https://www.nationalobserver.com/2017/11/04/news/payette-takes-climate-change-deniers-and-horoscopes-science-conference
**119** Stanley Kramer, *Inherit the Wind*, Spencer Tracy, Fredric March, Gene Kelly, Dick York, Donna Anderson, Harry Morgan; United Artists, 1960.
https://en.wikipedia.org/wiki/Inherit_the_Wind_(1960_film)
https://www.filmsite.org/inhe.html
**120** H.L. Mencken, Baltimore, 1925.
https://archive.org/stream/CoverageOfTheScopesTrialByH.l.Mencken/ScopesTrial Mencken.txt
https://en.wikipedia.org/wiki/H._L._Mencken
https://www.rlhymersjr.com/Online_Sermons/2007/070707PM_MenckenScopesTrial.html
https://www.prattlibrary.org/locations/mencken/
https://en.wikipedia.org/wiki/Enoch_Pratt_Free_Library
**121** https://en.wikipedia.org/wiki/Inherit_the_Wind_(1960_film)
https://www.rottentomatoes.com/m/1010555_inherit_the_wind
https://en.wikipedia.org/wiki/Scopes_trial
**122** Michael Kazin, "This 'Trial of the Century' Is 100. Its Lessons Could Save the Democrats," *New York Times*, OPINION / GUEST ESSAY, 10 July 2025. https://www.nytimes.com/2025/07/10/opinion/scopes-monkey-trial-democrats.html
**123** John Rennie, "Evolution/15 Answers to Creationist Nonsense / Opponents of evolution want to make a place for creationism by tearing down real science, but their arguments don't hold up," *Scientific American*, 01 July 2002. https://www.scientificamerican.com/article/15-answers-to-creationist/
**124** For greater certainty, for the serious discussion of science, the author does not advocate: creationism, climate change denial, anti-vaccine policy; nor Velikovsky's *Worlds in Collision*. And does not recommend their related advocacy literature other than to exemplify what science is not.
**125** Nor: Lysenkoism https://en.wikipedia.org/wiki/Trofim_Lysenko https://en.wikipedia.org/wiki/Lysenkoism
See also: Arthur Koestler, *The Case of the Midwife Toad*; ISBN-10: 1939438454; ISBN-13: 978–1939438454
https://www.amazon.ca/Case-Midwife-Toad-Arthur-Koestler/dp/1939438454
https://en.wikipedia.org/wiki/Paul_Kammerer
**126** See: Peter J. Hotez, M.D., Ph.D., *THE DEADLY RISE OF ANTI-SCIENCE / A Scientific Warning*, Johns Hopkins University Press, Baltimore, 2023, 2025; pages 108→113. ISBN 978-1-4214-5167-1.
**127** See also: David Smith, Washington, "Coronavirus: medical experts denounce Trump's theory of 'disinfectant injection' / Doctors warn US president's musings on disinfectant as a cure for coronavirus could lead to death," *The Guardian*, 24 April 2020. https://www.theguardian.com/ world/2020/apr/23/trump-coronavirus-treatment-disinfectant
**128** See also:
- npr, PUBLIC HEALTH, "Vaccine Controversies Are As Social As They Are Medical," 2014. https://www.npr.org/sections/health-shots/2014/09/30/351242264/vaccine-controversies-are-as-social-as-they-are-medical
- Eula Biss, *On Immunity / An Inoculation*, Farrar Straus and Giroux. https://www.npr.org/books/titles/347663952/on-immunity-an-inoculation

- Paul A. Offit, "Don't Call Kennedy a Vaccine Skeptic. Call Him What He Is: A Cynic," *New York Times*, OPINION / GUEST ESSAY, 13 Jan. 2025. https://www.nytimes.com/2025/01/13/opinion/rfk-jr-is-a-vaccine-cynic.html
- Christina Jewett, "*Kennedy Sought to Stop Covid Vaccinations 6 Months After Rollout* / Robert F. Kennedy Jr. petitioned the F.D.A. to revoke authorization of the shots at a time when they were in high demand and considered life-saving," *New York Times*, 17 January 2025. https://www.nytimes.com/2025/01/17/health/rfk-jr-covid-vaccines.html
- Nicholas Kristof, "Our Health in the Hands of a Man Who'd Make Us Sick," *New York Times*, OPINION, 01 February 2025. https://www.nytimes.com/2025/02/01/opinion/rfk-kennedy-vaccines.html

**129** Apoorva Mandavilli, "*Kennedy Removes All C.D.C. Vaccine Panel Experts* / The U.S. health secretary chose to "retire" members of a committee that makes significant decisions about who receives immunizations, including the vaccines for children," *NewYork Times*, 09 June 2025.
https://www.nytimes.com/2025/06/09/health/rfk-jr-cdc-vaccine-panel.html

**130** See: Hiroko Tabuchi, "*Climate Denial Infuses Reports at U.S. Agency/An Official Under Trump Embeds Climate Denial Language in Scientific Reports*," *The New York Times*, 02 March 2020, pages A1 and A21.

**131** https://www.nytimes.com/2019/09/23/health/anti-vaccination-movement-us.html https://en.wikipedia.org/wiki/Vaccine_hesitancy

**132** See: Chris Turner, *The WAR on Science / Muzzled Scientists and Wilful Blindness in Stephen Harper's Canada*, Greystone Books, Vancouver, 2013. ISBN 978-1-77100-431-2

**133** See: Peter J. Hotez, M.D., Ph.D., *THE DEADLY RISE OF ANTI-SCIENCE / A Scientific Warning*, Johns Hopkins University Press, Baltimore, 2023, 2025. ISBN 978-1-4214-5167-1.

**134** Gordon R. Freeman, Chemistry Department, University of Alberta, Edmonton T6G 2G2, "Kinetics of nonhomogeneous processes in human society: Unethical behaviour and societal chaos," *Canadian Journal of Physics*, 1990, Volume 68 (9), pages 794–798. https://doi.org/10.1139/p90-116
https://www.nrcresearchpress.com/doi/pdf/10.1139/p90-116 ["RETRACTED"]

**135** *Canadian Journal of Physics* advised: "In regard to the publication of this paper, readers are urged to consult the editorials published in the June issue, volume 69 (1991), and the March/ April issue, volume 71 (1993) . . ."

**136** [ABSTRACT] ". . . Families in which both parents have jobs outside the home have relatively low stability, and they do not provide adequate nurture for the children . . . a high probability of developing into unethical or unmotivated adults . . . a relatively high tendency to use drugs and to be sexually irresponsible. Policies are suggested that would reduce societal problems and have large economic benefits."

**137** Stephen Strauss, Science Reporter, "Journal apologizes for article / Chemist had blamed working mothers for offspring's moral weakness," *The Globe and Mail*, Toronto, National News, 31 July 1991, page A7.

**138** For greater certainty, the author does not endorse Dr. Freeman's social policy views, expressed in the *Canadian Journal of Physics* article cited above.

**139** Tomas Hudlicky, " 'Organic synthesis – Where now?' is thirty years old. A reflection on the current state of affairs," as Accepted Article, *Angewandte Chemie*, International Edition; later published, 2020; then removed.
https://ciencianarua.net/wp-content/uploads/2020/06/anie.202006717.pdf
https://sites.krieger.jhu.edu/chemdna/files/2020/06/10.1002anie.202006717.pdf

**140** https://onlinelibrary.wiley.com/doi/full/10.1002/anie.202006717

[. . .]

THIS ARTICLE HAS BEEN WITHDRAWN

[. . .]

Angewandte Withdrawal
Withdrawal: T. Hudlicky, " 'Organic synthesis – Where now?' is thirty years old. A reflection on the current state of affair". Angew. Chem. Int. Ed. 2020, DOI:
10.1002/anie.202006717
The above article, published online on 4 June 2020 in Wiley Online Library (wileyonlinelibrary.com), has been withdrawn by agreement between the journal's Editor in Chief and Wiley-VCH Verlag GmbH & Co. KGaA, Weinheim. The withdrawal has been agreed as the opinions expressed in this essay do not reflect our values of fairness, trustworthiness, and social awareness. It is not only our responsibility to spread trusted knowledge, but to also stand against discrimination, injustices, and inequity. While diversity of opinion and thoughts can spur change and debate, this essay had no place in our journal.

[. . .]

**141** See also: Laura Howes, "Ethics/Essay criticizing efforts to increase diversity in organic synthesis deleted after backlash from chemists / Publishing of a personal essay in Angewandte Chemie points to deeper problems within the community, chemists say," *C&EN*, 08 June 2020.
https://cen.acs.org/research-integrity/ethics/Essay-criticizing-efforts-increase-diversity-inorganic-synthesis-deleted-after-backlash-from-chemists/98/web/2020/06?utm_source=News letter&utm_medium=-Newsletter&utm_campaign=CEN
https://mail.yahoo.com/b/folders/1/messages/AC501nR6ePWDXuJYtwRNkOouQ2w?showI mages=-true&offset=0&action=printMessage
https://www.cbc.ca/news/canada/hamilton/brock-hudlicky-1.5605791
https://brocku.ca/tomas-hudlicky/
https://www.chemistryworld.com/news/angewandte-essay-calling-diversity-in-chemistry-harmful-decried-as-abhorrent-and-egregious/4011926.article
https://www.chemistryworld.com/news/angewandte-essay-calling-diversity-in-chemistry-harmful-decried-as-abhorrent-and-egregious/4011926.article
https://www.chem.iastate.edu/news/angewandte-chemie
https://www.science.org/content/blog-post/not-winning-friends-not-influencing-people
**142** https://www.change.org/p/brock-university-fire-tomas-hudlicky [Petition; August 2020] ". . . Fire Tomas Hudlicky . . ."
**143** Laura Howes, "Ethics/Essay deleted after backlash / Personal essay published in *Angewandte Chemie* points to deeper diversity problems within the community, chemists say," *C&EN*, 15 June 2020, pages 4 and 5. CEN.ACS.ORG
**144** Bibiana Campos Seijo, Editor in Chief, "We need to do better," *C&EN*, Editorial, 15 June 2020, page 3. CEN.ACS.ORG
**145** Also: Jordan Peterson, COMMENT, "The activists are now stalking the hard scientists," *NATIONAL POST*, Toronto, 27 June 2020, pages A6 and A7.
https://nationalpost.com/opinion/jordan-peterson-the-activists-are-now-stalking-the-hard-scientists
**146** Nicholas Tibollo, "Remembering Professor Tomáš Hudlický, and how poorly Brock University treated him,"
*The Niagara Independent*, 18 May, 2022.
https://niagaraindependent.ca/remembering-professor-tomas-hudlicky-and-how-poorly-brock-university-treated-him/
**147** For greater certainty, the author does not endorse Dr. Hudlicky's views on diversity and student unconditional submission to professors, in the withdrawn *Angewandte Chemie* article cited above.
**148** And in some sense biology – but perhaps restricted only to Planet Earth?

**149** and at things from the heavens – such as reported life-suggesting evidence in a meteorite from Mars – See Richard A. Kerr, NEWS/“Ancient Life on Mars?” *Science*, Vol.273, No.5277, 16 August 1996, pages 864–866.

**150** See L.A. Barba, A. Ortiz, K.F. Link, L. López Luján and L. Lazos, “Chemical Analysis of Residues in Floors and the Reconstruction of Ritual Activities at the Templo Mayor, Mexico,” Chapter 12, in Mary Virginia Orna, ed., *Archaeological Chemistry/Organic, Inorganic Biochemical Analysis*, ACS Symposium Series 625, American Chemical Society, Washington, D.C., 1996; pages 139–156.

**151** Although it must be recognized that cosmology, in its fundamental concepts, is an important part of physics.

**152** Neglecting here multiverse speculations.

**153** (The “Big Bang” of cosmological theory doesn’t fit neatly within this description.)

**154** See:
- Pamela S. Zuer, “Archaeological Chemistry/Physical Science helps to unravel human history,” *Chemical & Engineering News*, SPECIAL REPORT, 21 February 1983, pages 26–44.
- Mary Virginia Orna, ed., *Archaeological Chemistry/Organic, Inorganic Biochemical Analysis*, ACS Symposium Series 625, American Chemical Society, Washington, D.C., 1996.

**155** See:
- Robert A. Erlandson, “Request to dig up the remains in Booth grave will get a hearing in court,” *Baltimore Sun*, 15 November 1994, page 12B www.sunspot.net;
- “Cemetary chairman wants full hearing on Booth petition,” *Baltimore Sun*, 28 October 1994, page 6B;
- “Who’s buried in Booth’s grave? Historians, family want to check,” *Baltimore Sun*, 25 October 1994, page 1A;
- Reuters, “New Scrutiny on John Wilkes Booth,” *The New York Times*, 25 October 1994, page A16.

**156** https://en.wikipedia.org/wiki/Green_Mount_Cemetery
https://www.youtube.com/watch?v=UCuqTU6JtJE
https://www.youtube.com/watch?v=ZOPegyKaydI

**157** See “Dig at Stanton site set to begin,” *The Toronto Star*, 14 July 1996, page A12: “An archaeological team begins a forensic dig tomorrow at the site where Julie Stanton’s body was found. . . .”

**158** https://www.scribbr.com/definitions/vice-versa/#:~:text=Vice%20versa%20is%20an%20adverb,if%20the%20order%20is%20reversed

**159** Although it may be trivially obvious, it should not overlooked that *what it is*, and other related information, may be imprinted on it by way of code numbers, if not in plain language. For example, the labelling on the side of a consumer-purchased pen: “. . . PIGMA MICRON 03 / Micro pigment ink for waterproof . . . 0.35 mm line width . . . CONFORMS TO ASTM-D4236 Item No. XSDK03. SAKURA COLOR PRODUCTS CORP. JAPAN.”
See:
https://www.artistsemporium.net/product/00/084511306400/Sakura-Pigma-Micron-03-Black-XSDK0349
https://www.astm.org/Standards/D4236.htm
https://www.astm.org/ABOUT/faqs.html https://www.astm.org/
ASTM International, West Conshohocken, Pennsylvania, 19428–2959.
https://en.wikipedia.org/wiki/List_of_ASTM_International_standards
https://www.hamilton.edu/documents/Art%20Hazard%20Labeling%20Considerations.pdf
https://outlinesart.com/product-info/what-is-astm-d-4236/

**160** See: re use of computer technology and mathematical algorithms to recognize microscopic patterns: Tim Hornyak, “Forensic firm [Forensic Technology Inc., Montreal www.fti-ibis.com] aims to be No.1 with a bullet/Police from New York to Pretoria use identification technology to match spent ammunition with guns,” *The Globe and Mail*, 29 July 1999, page T4.

**161** See:

- Roland A.H. van Oorschot and Maxwell K. Jones, "DNA fingerprints from fingerprints," *Nature* volume 387, page767 (1997) https://www.nature.com/articles/42838; reviewed in ASSOCIATED PRESS, "People shed DNA all over the place, researchers find / Australian scientists trace genetic fingerprints," *Baltimore Sun*, 19 June 1997, page 9A http://www.nature.com/ www.sunspot.net .
- Ian Findlay, Andrew Urquhart, *et al.*, *Nature*, 389, 555 (1997) – and a news report on this – "DNA forensic profiles from single cells," science/technology concentrates, *C&EN*, 13 October 1997, page 17.

**162** For example, see: Nigel J. Bunce, *Environmental Chemistry*, Wuerz Publishing Ltd., Winnipeg, 1991; cover and page iii.

**163** One of two Luxor Obelisk moved from Egypt, 1830s, to *Place de la Concorde*, Paris:
https://en.wikipedia.org/wiki/Luxor_Obelisks
https://archaeology-travel.com/france/luxor-obelisk/

**164** Shrapnel = small pieces of metal that fly through the air when a bomb or similar weapon explodes and are intended to injure people. Cambridge Dictionary, 2019. https://dictionary.cam bridge.org/dictionary/english/shrapnel
Original device invented, 1784, Lieutenant Henry Shrapnel, British Royal Artillery. Retired as Lieutenant General, 1837. Some controversary with attribution – was it the invention of French engineer Bernard Forest de Bélidor, 1760? https://en.m.wikipedia.org/wiki/Henry_ Shrapnel

**165** See Lois Ember, "Probe of troops' exposure to chemical arms faulted," *Chemical & Engineering News*, 23 September 1996, pages 40 and 42. See also: "Pentagon seeks independent study of Gulf War illnesses," *C&EN*, 7 October 1996, page 21. See also: NEW YORK TIMES NEWS SERVICE, "U. S. gulf war panel widens probe into chemical weapons / But Pentagon committee resists turning over inquiry to outsiders," *Baltimore Sun*, 13 Nov. 1996, page 3A.

**166** See Steven Lee Myers, "Chemical Damage Seen in Gulf War Veterans," *The New York Times*, National Report, 01 December 1999, page A15.

**167** Poison gas warfare.

**168** Raffi Khatchadourian, "A Reporter at Large/Operation Delirium/*Decades after a risky Cold War experiment, a scientist lives with secrets*," *The New Yorker*, 17 December 2012, cover panel, page 2, 46–64. www.newyorker.com

**169** Daniel Charles, *Master Mind/The Rise and Fall of Fritz Haber, the Nobel Laureate Who Launched the Age of Chemical Warfare*, Harper-Collins Publishers Inc., New York City, 10,022; 2005; ISBN 0-06-056272-2.

**170** https://www.youtube.com/watch?v=QQkmJI63ykI

**171** Edwin E. Slosson, M.S., Ph.D., Creative Chemistry/Descriptive of Recent Achievements in the *Chemical Industries*, The Century Co., New York City, 1919.

**172** http://openlibrary.org/books/OL7111329M/Creative_chemistry https://archive.org/details/creativechemist02slosgoog/page/n329/mode/2up

**173** Edgewood Arsenal was apparently the main production facility for US gas munitions during WWI. See Chapter XII – Fighting with Fumes – Slosson at page 224:

[. . .]

> . . . A gas plant was started at Edgewood, Maryland, in November, 1917. By March it was filling shell and before the war put a stop to its activities in the fall it was producing 1,300,000 pounds of chlorin, 1,000,000 pounds of chlorpicrin, 1,300,000 pounds of phosgene and 700,000 pounds of mustard gas a month. . . .

[. . .]

at page 229:

> [. . .]
> . . . Of the 925 cases requiring medical attention at the Edgewood Arsenal 674 were due to mustard. During the month of August 3½ per cent. of the mustard plant force were sent to the hospital each day on the average. But the record of the Edgewood Arsenal is a striking demonstration of what can be done in the prevention of industrial accidents by the exercise of scientific prudence. In spite of the fact that from three to eleven thousand men were employed at the plant for the year 1918 and turned out some twenty thousand tons of the most poisonous gases known to man, there were only three fatalities and not a single case of blindness. . . .
> [. . .]

Edgewood Arsenal was apparently the US response to the work of Fritz Haber in Germany.
**174** Mustard gas Sulfur mustard 1-Chloro-2-[(2-chloroethyl)sulfanyl]ethane S $(CH_2\,CH_2\,Cl)_2$
CAS 505-60-2 159.07 g/mol (See Fig. 2.6)

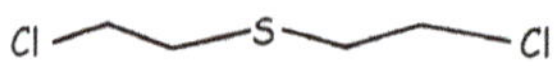

Fig. 2.6: Mustard gas.

https://en.wikipedia.org/wiki/Mustard_gas#:~:text=Mustard%20gas%20or%20sulfur%20mustard,as%20well%20as%20other%20species.
**175** See also: https://www.nytimes.com/2024/12/25/us/syria-chemical-weapons.html
**176** Both Chiam Weitzman and Fritz Haber were chemists. While Weitzman did work for the production of British explosives during WWI (arguably giving political leverage later for the issuing of the Balfour Declaration), Haber developed gas warfare for Germany. Haber's wife Clara's suicide, 1915, is said to be related to her dislike of her husband's war work.
Haber's earlier work, resulting in the Haber-Bosch process, arguably allowed that the Kaiser wouldn't run out of ammunition during WWI. The Haber-Bosch process (catalytic hydrogenation of $N_2$, at raised pressure and temperature, to make $NH_3$) is still important for the production of fertilizers, dyes and explosives.
Haber was a friend of Albert Einstein at the Kaiser Wilhelm Institutes – now the Max Planck Institutes. Although Haber, a Nobel laureate, was a famous and very patriotic German, he died in sad exile from the Nazis in Switzerland, 1934. He was of Jewish origin. Had he lived longer there was a possibility that he could have gone to British Palestine, at the suggestion of Weitzman. Max Planck organized Haber's memorial service in Germany.
**177** See also about Fritz Haber and poison gas in: Thomas Levenson, *EINSTEIN in BERLIN*; BANTAM BOOKS; 2003, 2004; pages 121→124. ISBN 0-553-37844-9. https://thomaslevenson.com/einstein-in-berlin
**178** CBC Podcasts, True Crime, Crime Story with Kathleen Goldhar, "The Gas Man: The unlikely figure behind some of Iran's deadliest crimes," 23 Dec. 2024. https://www.cbc.ca/listen/cbc-podcasts/1399-crime-story/episode/16117557-the-gas-man-the-unlikely-figure-behind-some-of-irans-deadliest-crimes
**179** Toronto Workers Health & Safety Legal Clinic *newsletter* 2013 02. Vol. 21 No.1
**180** John Bryden, *Deadly Allies / Canada's Secret War –1937–1947,* McClelland and Stewart, Inc., Toronto M5G 2E9; 1989; ISBN 0-7710-1724-3.
**181** CBC Podcasts, True Crime, Crime Story with Kathleen Goldhar, "The Gas Man: The unlikely figure behind some of Iran's deadliest crimes," 23 Dec. 2024. https://www.cbc.ca/listen/cbc-podcasts/1399-crime-story/episode/16117557-the-gas-man-the-unlikely-figure-behind-some-of-irans-deadliest-crimes
**182** Ann C. McKeea and Meghan E. Robinsona, "Military-related traumatic brain injury and neurodegeneration," 2014. HHS Author ManuscriptsPMC4255273
Alzheimer's Dement. Author manuscript; available in PMC 2014 Dec 4.

Published in final edited form as: Alzheimer's Dement. 2014 Jun; 10(3 0): S242–S253. doi: 10.1016/j.jalz.2014.04.003
https://www.ncbi.nlm.nih.gov/pmc/articles/PMC4255273/
**183** See: "Homeless exploited to strip asbestos / Soup kitchen recruits got no training or pay, [US federal] prosecutors charge," [from Los Angeles *Times*], *Baltimore Sun*, 25 April 1998, pages 1A and 7A.
**184** See:
- Ernest F. Imhoff, "Poisoned in childhood / Poisoned by lead paint as a child," *Baltimore Sun*, 30 October 1996, pages 1B and 4B;
- Scott Shane, *Sun* Staff, "A child's suffering puts lead on trial; Paint poisoning suit seeks compensation for Baltimore family," *Baltimore Sun*, Monday, 27 September 1999, FINAL Edition, TELEGRAPH Section, page 1A;
- Jane E. Brody, "Lead linked to delinquency / Childhood exposure to metal cited in study of criminality," New York Times Service, *The Globe & Mail*, 8 February 1996, page A12.
- Brian H. Kaye, *Science and the Detective/Selected Reading in Forensic Science*, VCH Verlagsgesellschaft mbH, Weinheim, Germany; 1995; ISBN-3-527-29,252-7; Chapter 9.10, "Poisonous Plates and Sweet Wine," pages 244–250, and references cited therein.

**185** Brian H. Kaye, *Science and the Detective / Selected Reading in Forensic Science*, VCH Verlagsgesellschaft mbH, Weinheim, Germany; 1995; ISBN-3-527-29,252-7.
https://pubs.rsc.org/en/content/articlelanding/1996/an/an996210005n/unauth#! divAbstract
**186** *Baltimore Sun*:
**187** **Funds near release to fight lead poison** House panel reviewing revised state-city plan to clean affected homes (Sep 14 2000)
**Lead's lethal passage: One family's anguish** Poisoned: One woman's struggle to cope with dangerous impulses and a lead-damaged brain stands as a warning to a city racked by street violence and drug abuse. (Sep 10 2000)
**Experts fault enforcement of city lead laws** Dearth of inspectors, prosecutors limits effectiveness of rules; Poorest children poisoned (Sep 8 2000)
**$50 million Pledged to fight lead poisoning;** Glendening, O'Malley outline campaign to protect city children; 'A moral obligation' (Sep 8 2000)
**Links to lead paint poison sites on the Internet** Where to go for more information about Baltimore's paint plague (Sep 8 2000)
**City Council measure would require lead poisoning tests for children** Proposal expected to pass; those under age 7 targeted (May 23 2000)
**Studies suggest link between lead, violence** Experiment on rats indicates exposure hinders brain growth; Analysis tracks lead, crime (May 9 2000)
**Landlords can be liable, appellate court rules** Ignorance of hazards no excuse, judges say (Apr 21 2000)
**Beilenson urges bill requiring lead tests for 1-, 2-year-olds** Similar state measure not as far-reaching, says city health commissioner (Apr 13 2000)
**Lead paint warning renewed** Baltimore City's top health official says threat is statewide (Apr 12 2000)
**Assembly wraps up in a flurry** Anti-lead measure, health insurance bill, teacher raises pass; Big surplus fuels spending; Legislation to halt racial profiling in traffic stops defeated (Apr 11 2000)
**House OKs bill aimed at lead poison** Tests would be required of city infants in first year (Mar 28 2000)
**Lead fight takes wide approach** City, state, federal agencies coordinate anti-poisoning efforts; A $3.1 million boost; Baltimore prosecutes landlords, seeks to raise awareness (Mar 23 2000)
**Plan calls for stricter lead test standards** All children up to age 2 would be examined (Feb 29 2000)
**$50 million pledged to fight lead poisoning** Glendening, O'Malley outline campaign to protect city

children; 'A moral obligation' (Jan 29 2000)
**Lawmakers back bill on lead paint** Rosenberg, Hoffman support proposal to make lawsuits easier; Cleanup to be announced; Manufacturers would share damages based on Md. sales (Jan 28 2000)
**Governor promises city more money to fight lead** Clergy, legislators say action against poisonings is overdue (Jan 22 2000)
**A lord of the slums takes an apprentice** Links: A paper trail through a web of corporations reveals a lucrative merger of real estate and drugs. In poisoned surroundings, a mother and son pray for deliverance. (Dec 19 1999)
**188** **Special report** Jan 20 2000 Lead's lethal legacy engulfs young lives Epidemic: With poison in their blood, thousands of Baltimore's children contribute to unsettled classrooms and violent neighborhoods. Simple fix: conscientious landlords
**189** Richard Saferstein, *Criminalistics/An Introduction to Forensic Science*, Seventh Edition, Prentice Hall, Upper Saddle River, New Jersey, 07458; 2001; ISBN 0-13-013827-4; pages 108→112.
**190** That is, in an industrial factory or work shop.
**191** https://www.youtube.com/watch?v=60ahob95TUw
**192** – which is popular in fiction – for example., the nursery rhyme in Agatha Christie, *And Then There Were None*, 1939, Berkley Books, New York City, 1991, Chapter II, Part VI, at pages 22 and 23.
**193** For example, see Clifford Krauss, New York Times Service, "Police use graffiti to solve crimes / Scrawls on walls have evolved from blobs and doodles into codes to mark off gang turf and advertisements for drugs and guns," *The Globe and Mail*, 7 October 1996, page A11.
See also: Michelle Shephard, Education Reporter, #Greater Toronto, " 'Prime turf' for gangs / Police to principals about school danger: take off 'blinders' and face the problem," *The Toronto Star*, Sunday, 22 November 1998, page A5:
Kim Stezinar, a Toronto Police intelligence unit employee, described the more subtle gang signs. "The gang graffiti is straight to the point. It's not pretty like the colourful tagging you see," she said. She encouraged teachers to examine notebooks or backpacks for gang graffiti and to take a close look at art projects make sure a gang sign is not hidden in a painting .
**194** See: Daniel J. Kevles, "E Pluribus Unabomber," *The New Yorker*, 14 August 1995, pages 2 and 4.
**195** See: David Johnston, "Bomber Is Called Killer Who Is Not on a Political Mission/Investigators start to blur their elaborate portrait of the Unibom suspect," *The New York Times*, 6 November 1995, page B8.
**196** https://www.mos.org/leonardo/activities/mirror-writing
**197** See: Clifford Krauss, 1996, cited above.
**198** Thomas Phinney/The Font Detective https://thefontdetective.com/
**199** AIH CBC, 16 January 2019. https://www.cbc.ca/radio/aih
**200** See Brian H. Kaye, "Paper, the Forger's Nemesis?" as Chapter 10.2 in *Science and the Detective/ Selected Reading in Forensic Science*, VCH Verlagsgesellschaft mbH, Weinheim, Germany, 1995; pages 269–275.
**201** See: Clifford Krauss, 1996, cited above.
**202** For example of a Rolex watch – see, for example: Crime Story with Kathleen Goldhar, "Bonus | Sea of Lies from Uncover," CBC, 22 April 2025.
https://www.cbc.ca/listen/cbc-podcasts/1399-crime-story/episode/16141508-bonus-or-sea-of-lies-from-uncover
**203** See John I. Thornton, in Richard Saferstein, ed., *Forensic Science Handbook* [Volume I], Prentice-Hall, Inc., Englewood Cliffs, New Jersey, 1982, Chapter 10, "Forensic Paint Examination." pages 545–547, and 555–560, and references cited therein.
**204** Formerly the Society of Automotive Engineers, Inc. – now – The Engineering Society for Advancing Mobility Land Sea Air and Space7, 400 Commonwealth Drive, Warrendale, Pennsylvania 15096–0001.

**205** Maria Josefi, Writer-editor, US FBI *Handbook of Forensic Science*, February 1995; US Government Printing Office, Superintendent of Documents, Washington, D.C., 20402-9328; ISBN 0-16-045111-6.
https://archive.org/details/FBI_Handbook_of_Forensic_Science/mode/2up
https://ia800205.us.archive.org/11/items/FBI_Handbook_of_Forensic_Science/FBI_Hand book_of_Forensic_Science.pdf http://www.dianepublishing.net/product_p/0788103857.htm

[page ix] [. . .]

**Standard reference files and collections**

The FBI Laboratory maintains these collections so that evidence may be compared to the following standard files:

Adhesives
Ammunition
Checkwriter Standards
Duct Tape/Electrical Tape
Explosives and Related Items
General Rifling Characteristics
Hairs and Fibers
National Automotive Image File
National Automotive Paint File (Foreign and Domestic)
National Motor Vehicle Certificate of Title
National Motor Vehicle Altered Numbers File
National Vehicle Identification Numbers (VIN) File
Office Equipment Standards (typewriters, copiers, printers, etc.)
Reference Firearms Collection
Safe Insulation
Safety Paper Standards
Shoe Sole Design Standards
Tire Tread Design Standards
Watermark Standards

[. . .]

[page 47] [. . .]

Ink Dating: Standard ink reference files necessary for possible association of a questioned ink with a manufacturer are not in the FBI Laboratory. However, the Secret Service in Washington, D.C., does maintain a standard ink library.

FBI requests should be sent directly to the FBI Laboratory. Other agencies should contact the Secret Service Laboratory directly.

**Limitations on Dating.**

A limited number of cases have been successfully matched to a standard ink not in existence on the date the document was allegedly prepared. The laboratory cannot determine how long writing ink has been on a document.

[. . .]

**206** US FBI *Handbook of Forensic Science*, 1984
https://www.ncjrs.gov/pdffiles1/Digitization/95271NCJRS.pdf

**207** Kim Waggoner, Editor, US *FBI Handbook of Forensic Science*, 2011.
https://www.goodreads.com/book/show/14611023-fbi-handbook-of-forensic-science

**208** US FBI *The Handbook of Forensic Services*, 2019
http://www.fbi.gov/about-us/lab/handbook-of-forensic-services-pdf
https://www.fbi.gov/file-repository/handbook-of-forensic-services-pdf.pdf/view

**209** (Derived from the standard ink library of the Bureau of Alcohol, Tobacco, and Firearms (ATF), US Treasury Department.)

**210** International Ink Library and Digital Ink Library, US Secret Service, Forensic Services Division.
https://www.nist.gov/oles/forensic-database-questioned-documents-table
https://www.secretservice.gov/press/releases/2018/09/secret-service-looks-science-crime-solving-during-forensic-focus-week
https://www.secretservice.gov/press/releases/2019/06/secret-service-dedicates-forensic-services-division-library-honor-former

**211** Gerald M Laporte, Marlo D Arredondo, Tyra S McConnell, Joseph C Stephens, Joseph C Stephens, Antonio A Cantu, Douglas K Shaffer "An Evaluation of Matching Unknown Writing Inks with the United States International Ink Library," June 2006, *Journal of Forensic Sciences* 51(3):689–92.
https://onlinelibrary.wiley.com/doi/abs/10.1111/j.1556-4029.2006.00144.x

**212** Cedric Neumann, Pierre Margot, "Considerations on the ASTM Standards 1789–04 and 1422–05 on the Forensic Examination of Ink," Sep 2010.
https://www.researchgate.net/journal/Journal-of-Forensic-Sciences-1556-4029

**213** https://nowiknow.com/the-trail-ink-left-behind/
https://web.archive.org/web/20120827190722/http://economictimes.indiatimes.com/news/news-by-industry/et-cetera/new-digital-ink-library-to-speed-up-document-investigation/arti cleshow/6275878.cms?
https://www.popularmechanics.com/technology/news/a13875/secret-service-ink-foren sics-library/
http://www.forensicsciencesimplified.org/docs/how.html

**214** See Richard L. Brunelle and A.A. Cantu, in G. Davies, ed., *Forensic Science*, ACS Symposium Series 13, American Chemical Society, Washington, D.C., 1975, Chapter 14, "Ink Analysis – A Weapon Against Crime by Detection of Fraud," particularly at page 137.

**215** See: Stephen C. Stinson, "Chemists Learn To Preserve Historical Polymers While Probing Their Nature," *Chemical & Engineering News*, 9 September 1996, pages 34–37. This article reports on a symposium of the Division of Polymer Chemistry, 25 August 1996, American Chemical Society National Meeting, Orlando, on polymers in museums – see *Book of Abstracts*, Part II, POLY, papers 13–19 and 41–48.

**216** Stinson, at pages 34, 35, 36; and ACS symposium, Julia Fenn, paper 18.

**217** Forbes Collection, Straus Center for Conservation and Technical Studies, Harvard University.

**218** R. Leopoldina Torres, "A Short History of a Pigment Collection (and Art Conservation in the United States)," 2 October 2013.
https://www.harvardartmuseums.org/article/a-short-history-of-a-pigment-collection-and-art-conservation-in-the-united-states
https://www.harvardartmuseums.org/teaching-and-research/research-centers/straus-center-for-conservation-and-technical-studies

**219** Simon Schama, "Onward and Upward with the Arts / Blue as Can Be / A color archive's treasures reveal a history of pigment," *The New Yorker*, 3 September 2018, pages 28–32.
https://www.newyorker.com/magazine/2018/09/03/treasures-from-the-color-archive

**220** See note above.

**221** For example: US 1909 Lincoln Wheat VDB Bronze Composite Penny.
VDB = Victor D. Brenner. https://www.usacoinbook.com/encyclopedia/coin-designers/vic tor-d-brenner/

**222** Diameter = 19 mm
3.11 g.
95% Cu 5% Sb and Zn
27 995 000 Minted, Philadelphia.
USA COIN BOOK [April 2020]
https://www.usacoinbook.com/coins/342/small-cents/lincoln-wheat-cent/1909-P/vdb/
https://www.usacoinbook.com/ https://en.wikipedia.org/wiki/List_of_copper_alloys

**223** See also: Ernest W. Flick, *Paint & Ink Formulations Database*, William Andrew Publishing, Norwich, USA; January 2005; ISBN10 0815515081; ISBN13 9,780,815,515,081. https://www.bookdepository.com/Paint-Ink-Formulations-Database-Ernest-W-Flick/ 9780815515081
**224** See also: John T. Maynard and Howard M. Peters, Understanding Chemical Patents/A Guide for the Inventor, Second Edition, American Chemical Society, Washington, D.C., 1991.
**225** National Institute of Standards and Technology (formerly *N*ational *B*ureau of *S*tandards), US Department of Commerce, Gaithersburg, Maryland, 20899-0001.
**226** Such as those of ASTM International, West Conshohocken, Pennsylvania, 19428–2959. https://www.astm.org/ABOUT/faqs.html https://www.astm.org/
**227** https://en.wikipedia.org/wiki/ASTM_International https://www.astm.org/Standard/standards-and-publications.html
**228** Kitchener, Ontario. January 2014.

Oldcastle®
LAMINATED 35
16 CFR 1201-I,II
ANSI Z97.1–2009
SGCC 4526 HUAD

**229** Oldcastle® LAMINATED 35
https://obe.com/ https://oldcastleproductguide.com/
https://oldcastleproductguide.com/products/search?utf8=%E2%9C%93&product_cate gory=Architectural+Glass&query=
**230** 16 CFR 1201-I,II http://www.law.cornell.edu/cfr/text/16/part-1201
**231** ANSI Z97.1-2009
"Safety Glazing Materials Used In Buildings – Safety Performance Specifications And Methods Of Test (Includes Errata)"
http://webstore.ansi.org/RecordDetail.aspx?sku=ANSI+Z97.1-2009
**232** SGCC 4526 HUAD.
Safety Glazing Certification Council, Sackets Harbor, New York 13685 www.sgcc.org
**233** See also: From inside a passenger waiting shelter GO Brampton, Ontario, 3 June 2019:

*Ever Temp*
SAFETY TEMPERED
VITRE TREMPE
16 CFR 1201–11
ANSI Z97.1–1984
CAN/CGSB-12.1-M

**234** For example: from soil or sand carried on a car's underbody.
See:

- Raymond C. Murray, in Richard Saferstein, ed., *Forensic Science Handbook* [Volume I], Prentice-Hall, Inc., Englewood Cliffs, New Jersey, 1982, Chapter 13, "Forensic Examination of Soil," pages 653–671, and references cited therein.
- John McPhee, Annals of Crime/"The Gravel Page." *The New Yorker*, 29 January 1996, pages 44–69.
- Tom Horton, *Sun* Staff, "Soil type vital to the life on it / Ground: Scientists have decided that Sassafras-type soil, which is found in 18 of Maryland's counties, should be the official state dirt," *Baltimore Sun*, 10 September 1999. https://en.wikipedia.org/wiki/List_of_U.S. _state_soils

**235** https://sis.agr.gc.ca/cansis/nsdb/index.html
**236** Natural Resources Conservation Service, US Department of Agriculture / Soils – Maryland.
https://www.nrcs.usda.gov/conservation-basics/conservation-by-state/maryland/soils-maryland
**237** https://en.wikipedia.org/wiki/List_of_U.S._state_soils

**238** See: Joseph R. Boldt, Jr. and Paul Queneau (Technical Editor), *The Winning of Nickel/Its Geology, Mining, and Extractive Metallurgy*, Longmans Canada Limited, Toronto, 1967.
**239** Not without the librarian's permission, of course.
**240** See R.L. Brunelle, in R. Saferstein, ed., *Forensic Science Handbook* [Volume I], Prentice-Hall, Inc., Englewood Cliffs, New Jersey, 1982, Chapter 14, "Questioned Document Examination," page 716.
**241** Both chemical and microparticle – called *taggants* – see:
- Maureen Rouhi, "Research council to study taggants," *Chemical & Engineering News*, 09 September 1996, pages 10 and 11.

This has been a controversial issue in the USA – see also:
- Maureen Rouhi, "Product liability could be the real issue with taggants." *C&EN*, 26 August 1996, page 38.
- *C&EN*, 5 August 1996, "Isotopic labels form basis of new chemical markers."
- Linda Raber, "Clinton drops demand for explosives taggants," *C&EN*, 5 Aug. 1996, page 7.
- Michael Kramer, "Without a Clue/Why Congress balks at a method for tracing bombs," *Time*, 12 Aug. 1996, page 21.
- Sara Mosle, Comment/"Without a Trace / A timely lesson from a forgotten East Texas tragedy." *The New Yorker*, 15 May 1995, pages 7 and 8.

**242** – That is, at chemical equilibrium; or for reactions where $K \rightarrow \infty$, when the reactants are spent.
**243** For reactions in progress, time contamination is also possible. This might be dealt with by accurate and precise time record keeping, and by cooling to try to stop the reaction – such as by refrigeration, ice (melting point = 0 °C), "dry ice" (solid $CO_2$ – sublimes at -78.5°C) or liquid nitrogen (boiling point = -195.8°C). Since reaction rates slow with decreasing temperature, the colder the better – except that too cold may be related to crystallization damage to complex bio-molecules.
**244** Richard Saferstein, *Criminalistics/An Introduction to Forensic Science*, Seventh Edition, Prentice Hall, Upper Saddle River, New Jersey, 07458; 2001; ISBN 0-13-013827-4; pages 103–107
**245** https://www.icp.org/browse/archive/objects/bullet-through-plexiglas
**246** Richard Saferstein, *Criminalistics/An Introduction to Forensic Science*, Seventh Edition, Prentice Hall, Upper Saddle River, New Jersey, 07458; 2001; ISBN 0-13-013827-4; pages 443–449
**247** Brian H. Kaye, *Science and the Detective/Selected Reading in Forensic Science*, VCH Verlagsgesellschaft mbH, Weinheim; 1995; ISBN-3-527-29252-7; Chapter 3, "Footprints and other Vestages," pages 45–68.
**248** William J. Bodziak, "Forensic Footwear Evidence," in Stuart H. James and Jon J. Nordby, editors, *Forensic Science/An Introduction to Scientific and Investigative Techniques*, CRC Press, Boca Raton, 2003; ISBN 0-8493-1246-9; Chapter 16, pages 297–311. www.crcpress.com
**249** William J. Bodziak, "Forensic Tire Impression and Tire Track Evidence," in Stuart H. James and Jon J. Nordby, editors, *Forensic Science/An Introduction to Scientific and Investigative Techniques*, CRC Press, Boca Raton, 2003; ISBN 0-8493-1246-9; Chapter 17, pages 313–326. www.crcpress.com
**250** See also: Arthur C. Clarke, *TALES OF TEN WORLDS*, "LET THERE BE LIGHT," DELL, New York City, 1962, 1964; pages 98–105.
https://en.wikipedia.org/wiki/Let_There_Be_Light_(Clarke_short_story)
**251** For example, see: Track Etch alpha-particle detectors – see H. Ward Alter and Robert L. Fleishcer, "Passive Integrating Radon Monitor for Environmental Monitoring," *Health Physics*, 40, pages 693–702, May 1981.
**252** For example, see: S. Prête, A. Aroua, M. Boschung, M. Grecescu, J.-F. Valley and Ch. Wernli, "The Fission Track Detector Revisited: Application to Individual Neutron Dosimetry," *Health Physics*, 71(2), pages 128–134, August 1996.
**253** ${}_6C^{14} \rightarrow {}_7N^{14} + {}_{-1}\beta^0$ $t_{1/2}$ = 5730 years.

**254** See Brian H. Kaye, *Science and the Detective/Selected Reading in Forensic Science,* VCH Verlagsgesellschaft mbH, Weinheim, Germany, 1995; pages 48–50, and references cited therein.
**255** See Fig. 2.7.

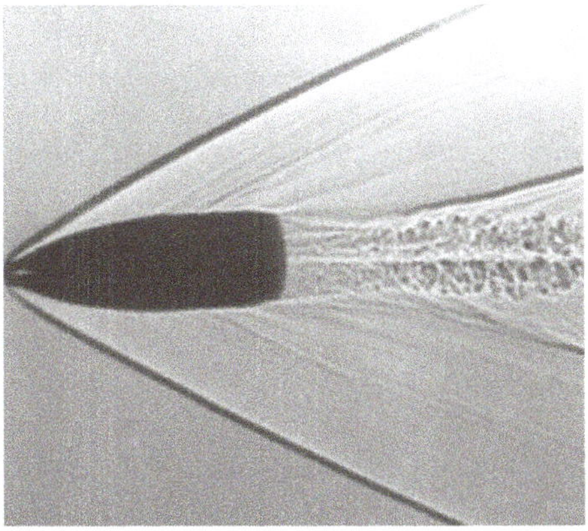

Fig. 2.7: Shockwave from fast-moving object in air. Edgewood Arsenal (copied from a photo inherited from the author's father).

**256** See also: Harold E. Edgerton & James R. Killian, Jr., *flash! / SEEING THE UNSEEN BY ULTRA HIGH-SPEED PHOTOGRAPHY,* Charles T. Branford Company, Boston, 1954; page 45.
**257** See:
- Stephen Strauss, "Heartbeat can reveal stowaways / Scientists have developed a device that can find people hidden in wheeled vehicles by detecting vibrations caused by their hearts," *The Globe and Mail,* 12 September 1996, page A8.
- Carl Young and Robert Yohman, "Rapid, Noninvasive Contraband Detection by Acoustic Response," *Journal of Forensic Sciences,* 40(3), May 1995, pages 470–474.

**258** From earthquake or nuclear weapons tests – see Michael Heylin, "After 40 Years, Treaty To Ban All Nuclear Weapons Tests May Be At Hand," *Chemical & Engineering News,* 20 May 1996, pages 10–18, at page 17.
**259** See:
- Harry Hollien, Laura Geison and James W. Hicks, Jr., "Voice Stress Evaluators and Lie Detection," *Journal of Forensic Sciences,* Vol.32, No.2, March 1987, pages 405–418;
- Frank Horvath, "Detecting Deception: The Promise and the Reality of Voice Stress Analysis"; *Ibid.,* Vol.27, No.2, April 1982, pages 340–351;
- J.W. Gocke and W.A. Oleniewski, "Voiceprint Identification in the Courtroom," *Ibid.,* Vol.18, No.3, July 1973, pages 232–236;
- Jared J. Wolf, "Efficient Acoustic Parameters for Speaker Recognition," *The Journal of the Acoustical Society of America,* Vol.51, No.6 (Part 2), 1972, pages 2044–2056;
- Oscar Tosi, Herbert Oyer, William Lashbrook, Charles Pedrey, Julie Nicol and Ernest Nash, "Experiment on Voice Identification"; *Ibid,* Vol.51, No.6(Part 2), 1972, pages 2030–2043;
- James J. Hennessy and Clarence H.A. Romig, "A Review of the Experiments Involving Voiceprint Identification"; *Journal of Forensic Sciences,* Vol.16, No.2, April 1971, pages 183–198;
- Mark Haggard Stephen Ambler and Mo Callow, "Pitch as a Voicing Cue"; *The Journal of the Acoustical Society of America,* Vol.47, No.2(Part 2), 1970, pages 613–617;
- Richard H. Bolt, Franklin S. Cooper, Edward E. David, Jr., Peter D. Denes, James M. Pickett and Kenneth N. Stevens, "Speaker Identification by Speech Spectrograms: A Scientists = View of its

Reliability for Legal Purposes," *The Journal of the Acoustical Society of America*, Vol.47, No.2 (Part 2), 1970, pages 597–612; https://pubmed.ncbi.nlm.nih.gov/5439660/
- Peter D. Bricker and Sandra Pruzansky, "Effects of Stimulus Content and Duration on Talker Identification," *Ibid*, Vol.40, No.6, 1966, pages 1441–1449;
- I. Pollack, J.M. Pickett and W.H. Sumby, "On the Identification of Speakers by Voice," *Ibid*, Vol.26, No.3, May 1954, pages 403–406;
- K.N. Stevens, E.C. Williams, J.R. Carbonell and Barbara Woods, "Speaker Authentication and Identification: A Comparison of Spectrographic and Auditory Presentations of Speech Material," *The Journal of the Acoustical Society of America*, Vol.44, No.6, 1968, pages 1596–1607; https://pubmed.ncbi.nlm.nih.gov/5705846/
- Jim Krane, New York, "Voice recognition an inexact science, audio analysts say," *The Globe and Mail*, Toronto, 14 November 2002, page A4.

**260** Rae Ellen Bichell, "Iron Age Potters Carefully Recorded Earth's Magnetic Field – By Accident," NPR INTERNATIONAL, February 2017. https://www.npr.org/sections/thetwo-way/2017/02/14/515032512/iron-age-potters-carefully-recorded-earths-magnetic-field-by-accident

**261** C. Manoharan, P. Sutharsan, R. Venkatachalapathy, S. Vasanthi, S. Dhanapandian and K. Veeramuthu, "Spectroscopic and rock magnetic studies on some ancient Indian pottery samples," pages 39–49, *Egyptian Journal of Basic and Applied Sciences*, Volume 2, 2015, Issue 1, 2019
https://www.tandfonline.com/doi/full/10.1016/j.ejbas.2014.11.001

**262** Jeffrey T. Barth, Jason R. Freeman, Donna K. Broshek, and Robert N. Varney, "Acceleration-Deceleration Sport-Related Concussion: The Gravity of It All," *J Athl Train.* 2001 Jul-Sep; 36(3): 253–256. https://www.ncbi.nlm.nih.gov/pmc/articles/PMC155415/

**263** https://www.sciencedirect.com/science/article/abs/pii/0002961057905408
- J.P.Stapp, "Human exposures to linear deceleration. I. Preliminary survey of the aft-facing seated position," Air Force Technical Report No. 5915. June, 1949.
- J.P. Stapp, "Human exposures to linear deceleration. II. The forward facing position and the development of a crash harness," Air Force Technical Report No. 5915. December, 1951.
- J.P.Stapp, "Crash protection in air transports," Aeronautical Engineering Rev. 1953; 12: 71.

**264** See: https://en.wikipedia.org/wiki/John_Stapp

**265** https://commons.wikimedia.org/wiki/File:Stapp_hits_the_brakes,_hard.gif

**266** https://baylorline.com/john-paul-stapp/#:~:text=He%20became%20president%20of%20the,at%20the%20age%20of%2089.

**267** See Fig. 2.8.

Fig. 2.8: Rocket sled (copied from a photo autographed by Maj. Stapp; inherited from the author's father).

**268** d(mv)/dt = -m dv/dt = -m **a** = **-F** (**boldface** here indicates vectors.)

**269** Robert Resnick and David Halliday, *Physics*, Part I, John Wiley and Sons, Inc., New York, 1966; page 191.

**270** See also:
https://commons.wikimedia.org/wiki/File:Stapp_hits_the_brakes,_hard.gif

https://en.wikipedia.org/wiki/John_Stapp
https://commons.wikimedia.org/wiki/File:Rocket_sled_track.jpg
https://airandspace.si.edu/stories/editorial/man-behind-high-speed-safety-standards
http://www.nmspacemuseum.org/halloffame/detail.php?id=46
https://www.nytimes.com/1999/11/16/us/john-paul-stapp-89-is-dead-the-fastest-man-on-earth.html
https://www.af.mil/News/Article-Display/Article/555822/60-years-later-earning-the-title-the-fastest-man-on-earth/
https://en.wikipedia.org/wiki/Murphy%27s_law

**271** Which might be suggested from DNA analysis of deposited blood (see Matt Bai, "Cold Case Confidential / Old murder cases are being re-opened – and solved," *Newsweek*, 12 January 1998, page 70) or other bio-fluid, from fingerprints, or from code numbers or other inscriptions. Also of interest might be who has been carrying what – for example, see A. Maureen Rouhi, "Detecting Illegal Substances / / Sampling Body Heat, Skin Flakes," special report, *C&EN*, 29 September 1997, page 29.

**272** For example: from soil or sand carried on a car's underbody.
See:

- Raymond C. Murray, in Richard Saferstein, ed., *Forensic Science Handbook* [Volume I], Prentice-Hall, Inc., Englewood Cliffs, New Jersey, 1982, Chapter 13, "Forensic Examination of Soil," pages 653–671, and references cited therein.
- John McPhee, Annals of Crime/"The Gravel Page." *The New Yorker*, 29 January 1996, pages 44–69.
- Tom Horton, *Sun* Staff, "Soil type vital to the life on it / Ground: Scientists have decided that Sassafras-type soil, which is found in 18 of Maryland's counties, should be the official state dirt," *Baltimore Sun*, 10 September 1999. https://en.wikipedia.org/wiki/List_of_U.S. _state_soils

**273** And, where it has not been might also be forensically indicated – for example, an item identified as a pork product would suggest that it does not come from a kosher restaurant.

**274** https://sis.agr.gc.ca/cansis/nsdb/index.html

**275** Natural Resources Conservation Service, US Department of Agriculture / Soils – Maryland.
https://www.nrcs.usda.gov/conservation-basics/conservation-by-state/maryland/soils-maryland

**276** https://en.wikipedia.org/wiki/List_of_U.S._state_soils

**277** That is, paired or sets of items – for example, if a right-hand glove is found, it might be of interest to look for the left; or, if several playing cards are found, to look for the rest of a game. Or pyjama top and bottom.

**278** Or if a US 1909 Lincoln Wheat VDB Penny were found, other collectors' coins might be looked for.

**279** Theory revision might include that what was found was irrelevant to the story. It also might allow that the initial theory was wrong, that nothing much of interest happened, or what did happen was not human-caused (it was a natural occurrence or animal-caused – such as the dog did it). It might also allow that an item was so contaminated by forensic or pre-forensic handling, or the environment. that the information that might have been derived from it has become hopelessly garbled or lost.

# Chapter 3
# Analytical approach

[. . .]
*. . . sniffing . . . at the glass . . . very cautiously . . . finger . . . tip of . . . tongue* [1].
*". . . cyanide . . ."*
[. . .][1]

[Agatha Christie, *And Then There Were None*, 1939.] [2, 3, 4, 5, 6, 7, 8, 9][2,3]

*[They go camping; set up their tent under the stars and retire for the night. Later Sherlock Holmes awakens, and then wakes Dr. Watson and asks:*
*"Look up at the stars – what do you deduce?"*

*Dr. Watson: [philosophizing at some length about the wonders of the universe, its immense size, and of the possibility of life on a hypothetical planet going around a distant star; . . .]*

*Sherlock Holmes: "You've missed it – Somebody's stolen our tent!"*] [10, 11, 12][4]

## 3.1 Analytical chemistry

As indicated in the previous chapter, the science of this book is mostly concerned with *analytical chemistry*.[5,6] Forensic chemistry was discussed in the context of what was found at a site and the further forensic laboratory analyses that would follow to help construct a scientific theory to be invoked for a human story. This chapter examines how to deal with the sequence and progress of those analyses.[7]

### 3.1.1 Measurement

But first – when considering analytical chemistry – some comment about ***measurement***[8]:

1 Introductory quotes, *etc.* Introduction 1.2.
2 Agatha Christie. Acute toxicology. Hydrogen Cyanide H-C ≡ N CAS 74-90-8. Potassium cyanide $K^+$ $CN^-$ CAS 151–50-8. Chapters 1 and 3.
3 Agatha Christie – quotes. Chapters 1, 2, 3, 7 and 8.
4 Edmond Locard. Sherlock Holmes. Cocaine. Chapters 2, 3, 4 and 6.
5 Analytical chemistry. Expanded definition – more detail, and example. Introduction 2; Chapters 1, 2, 3, 5 and 8.
6 Forensic science. Introduction; Chapters 1, 2, 3 and 5.
7 Analyses. Credibility. Chapters 3 and 5.
8 War on drugs. *CD& SAct.* Mass (grams) measurement. *W&M Act.* Introduction; Chapters 3, 6, 8 and 10.

https://doi.org/10.1515/9783111618555-004

Although some *qualitative* analytical chemical determinations may be satisfactory for some people, for some purposes; for others, *quantitative* determinations – with numerical measurements – are often necessarily important [13].[9]

Measurements may be of a single characteristic of a single item – for example, the mass [14, 15, 16, 17, 18, 19, 20][10,11] of something found at a site.[12]

Or, maybe a statistical determination composed of measurements of several – often many – of members of a group of things – for example, workplace chemicals measured as concentration levels in blood samples from workers. Further treatment of these concepts is considered in Chapter 8,[13] with discussion of error.

Most measurements are – or should be – in *Le Système International*[21, 22, 23] and its definitions and conventions should be assumed. Likewise, descriptions of chemical substances should be in terms used by the Chemical Abstracts Service, a division of the American Chemical Society. However, if other persistent historical systems are used, they must be relatable to SI and CAS.[14,15]

When measurements are reported, it is also often important to include how the measurements were made, with what instrumentation, within what conditions;[24] the SI units; and, if appropriate, comparison with published values [25].[16] And, without error estimates, many measurements are of limited value, some useless, and some worse than useless.[17]

If numbers are logarithmic or exponential, they would not have SI or any other units. It is also important to specify base 10 or e [26].[18]

### 3.1.2 What the measurement means

When making measurements and considering the resulting data, it may also be important to consider the actual meaning beyond the numbers. This would involve questioning why the measurements were made and assumptions of the concept [27].

For example [28]: A high school student writes a university-education-intended entry examination, as part of the application process. Typically, this would be a US-nationally standardized [29], computer-marked, multiple-choice test (such as SAT [30] or ACT [31] (or have been SSQT [32, 33, 34, 35, 36]) [37, 38, 39] apparently purport-

9 ***Qualitative*** = what it is. ***Quantitative*** = how much. Chapters 3.1.1 and 8.1.

10 Mass measurement (grams). Accuracy and precision too extensive? Chapters 3.1.1, 3.7.3 and 6.20.

11 Mass, Weight. Chapters 3 and 8.

12 Site. *What it is*. Chapters 1, 2, 3 and 5.

13 Measurement. Error. Chapters 2, 3 and 8.

14 SI CAS. Introduction 3.2; Chapters 3 and 4.

15 Chemical names as used in this book. Introduction 3.2; Chapters 3 and 4.

16 1909 Lincoln Wheat VDB Penny. Cu Sb Zn. Chapters 2 and 5; Chapter 3.1.1.

17 Measurement. Error. Chapters 2, 3, 7 and 8.

18 Napier. $Log_{10}$ $Log_e$. Chapters 3 and 6.

ing to determine some kind of fitness for success in a higher education institution – and beyond. A result in the 99th percentile would be good; 30th percentile, not so.

But what is the test actually measuring [40]: Aptitude? Intellectual ability, aside from the subject matter? Knowledge of a body of subject matter? Speed in writing tests? Gaming ability? Financial ability of parents to pay for preparation training? [41] Experience in test-taking? Pre-trained ability to successfully write any test of that format? Not getting too nervously upset? Some occurrence of chance?

Likely not being measured: Appreciation, love and respect of the subject matter. Resolve, in good faith, to try to better learn and understand the subject matter. Intellectual joy. Moral commitment. Personal bias to the concept of an inherent value of the pursuit of knowledge in human civilization, other than as a marketable commodity.

Maybe not being fully measured: Prediction of future success in further education or eventual livelihood pursuit; or future success otherwise.

What the test result would more reliably provide is a convenient ranking number for an administrator [42], or machine, to decide with, when there are too many applicants [43] for too few seats. Perhaps the result would be little beyond a test result congratulatory – or regrets – notice system. Perhaps there is really little of cultural or intellectual importance to ranking students, anyhow [44, 45, 46, 47, 48, 49]. Perhaps a decision process inspired by the biblical Gideon would have been good enough.[19]

(There likely would be possible testing regimes that would actually be more useful for considered evaluations, helpful for academic pursuits, but maybe not as useful for the testing of this example.)

As another example, directly related to forensic chemistry: $CO_2$ may be measured in a workplace atmosphere because that is the equipment available, for a situation when something else – gas or particulate – would actually be important to monitor.[20]

### 3.1.3 Counting measurement

Another kind of measurement: Things of the same kind, that are counted, as either present or not. Similarly for events. For example: A population census [50]. And, a count of disease-positive, or -negative, public health tests [51, 52].[21]

It should be noted that there can be errors in count or variations, if not confusion in definition, and things and events can be missed, or misattributed.

Precise details of definition, and perhaps relevancy, of what is being counted can be problematic [53]. As can the methodology of the counting, which can include com-

---

19 Selection protocol. Gideon. Chapters 3.1.2 and 3.5.2.

20 $CO_2$ CO. Chapters 2, 3, 6 and 8.

21 Semmelweis. Handwashing. Counting measurement. Introduction 4.1; Chapters 3 and 8.

bining counts of different observers who might use differing definitions [54, 55] and methods, with different competencies, and might include counts at different times. Care as to definition and method – and their constancy – should be taken when trying to combine counts of several sources [56].

### 3.1.4 Changed circumstances

Changes in circumstances can happen for counting measurements, and this may affect the evaluations; and should be taken into consideration for error assessment. For example: During the time window of a census of a defined geographic area, some people are born, others die, become too ill to participate, immigrate, emigrate, flee, or avoid or lie to the census taker (perhaps fearing the immigration officer, or fearful about an outstanding arrest warrant). During a public health count, some people may be cured, others die, hide, follow advice that is non-scientific or unwise, or behave in intentionally unhealthy ways; the counting methods and protocols may change; the geographic boundaries may change.

## 3.2 The things of evidential interest

It may be that things thought to be of interest by various investigators at a site[22] – police, labour inspector, insurer – are brought to the forensic chemist. The investigators should be expert-enough to properly choose the items, care for them and document [57] what is being done. Alternatively, they might consult with the forensic chemist or someone specially trained in forensic chemical investigation and collection. They should know to take proper care to protect the scene.

Many kinds of things of evidential interest are typically well known, with extensive, established, explanatory and encyclopaedic literature [58, 59, 60, 61] – such as for ink, bullet and shell casings, paint chips, seized drugs, explosion residue and blood. These are things for which forensic science is advanced well-enough that they may often be considered as obvious and standard. And, it would seem that the standardized economic availability of science equipment, procedure and personnel largely drives what is looked for, and how it is dealt with. But, it must be remembered that as science advances, so do the lists of kinds of things of forensic interest. For example, DNA and protein material, not visible to the human eye, and sometimes not from blood, unthought-of of as evidence only a few decades ago, are now important. Thus, issues of scientific availability and economy to finance new forensic process, become important. Analogously, video recording available as evidence that used to

22 Site. *What it is*. Chapters 1, 2, 3 and 5.

rare is now very common; arguably to now walk down a city street and not be multiply video-recorded – from business premises, motor vehicles, residence entrances, and body-born cameras – may have become rare.

Any search activity has the potential to change the site and found item, and thereby cause loss or destruction of evidence. Taking proper care to protect that evidential history and record, therefore, includes ample photography to record an overview of where everything was, as well as for close-up detail and for what the human eye missed or could not see; and to document the forensic investigative process as it might deconstruct and reconstruct the scene. The least intrusive methods are preferred, but some level of intrusion is unavoidable.

Forensic scene photography[23] should include not only the photographs themselves, but also information about the camera, its location and setup (which might include a labelled sketch or an overview photograph by another camera), and also information about the recording media (which used to be photographic film, but is now some version of digital electronic format). Also, as appropriate, a calibration device to show length and angle should be placed at the scene to appear in the photographs, with a north-pointing arrow [62], a dated and encoded sign identifying the investigating agency and its photographer, and a true-colour chart – all of this might be simply included in standard placards. Copies [63] of all the photographs should be readily available as referenced in the final report (but multiple repeats may wisely be avoided in the report).

What is chosen for forensic chemical analysis and how it is looked for would depend on a succession of progressively refined theories of the chronology of what happened, starting with a preliminary theory that may involve some reasonable speculation.

For example [64], for a highway crash where there is a skid,[24] the investigators should naturally be interested in the road surface condition (wet?), the tire treads (worn?), and the ethanol-in-blood concentration, or drug, of the driver (impaired?). The preliminary theory might turn out to be correct and sufficient: The crash resulted from a drunken driver with bald tires on a wet road. But maybe not. To continue the example alternatively [65]: The driver was sober, the tires were new, the road was dry and the crash was because of an oil slick leaked on the road by a tanker truck, and the forensic chemist would be able to help identify the manufacturer of the slick substance by way of mixture profile analysis;[25] the police officer having taken a sample from the road surface for completeness of investigation, but not knowing its significance at the time.

---

**23** Photography. f/d $\sqrt{2}$. Chapters 3 and 6.

**24** Tire skid. Chapters 2, 3 and 6.

**25** Mixture. Chapters 3 and 4.

## 3.3 What forensic aspects to concentrate on

Once something has been chosen for forensic analysis, the succession of progressively refined theories mentioned above now influences the choice of what the analysis should be. That is, what aspect of the item should the forensic chemist focus attention on, what should be left for later, left to others (such as biologists, physicists and electronic engineers) and what should be left?[26] For example [66]: For the forged document investigation,[27] the forensic chemist would essentially be concerned with the chemical composition/mixture profile of the ink on the paper and in the pen, and on other documents written by the suspect. Otherwise, it is important to protect the examined item to preserve evidence that should be considered by other investigators [67] – for fingerprints, handwriting analysis, pollen or dust particles, bio-residue, paper fibre [68], *etc.*

Thus, each investigation should be as carefully non-destructive as possible. For example: A handwriting expert who actually touches a questioned document may inadvertently put extraneous fingerprints on the paper, exude bio-substances (later of concern for DNA or protein analysis), or add microscopic dust to its surface, thereby damaging those kinds of evidence.

An example of destructive investigation – not realized at the time – comes from archaeology – the nineteenth-century excavations of the site that included archeological layers of ancient Troy [69].

Another influence on the choice of analysis is the availability of lab facilities, equipment and personnel. For example [70]: A labour inspector might use carbon dioxide measurements at the site – a workplace – to determine air quality, knowing that other factors should be taken into account, simply because $CO_2$-measuring equipment is all that is available. In such a situation, the inspector should clearly inform other investigators about the limitations of the measurements,[28] including that the measurements might not have relevant meaning.

## 3.4 Sample handling

As indicated above, there are possibilities that the item of interest could be contaminated by investigative process – the finding and subsequent handling and processing [71, 72].[29] There can also be contamination from the environment in which it was found – increasingly over time and changing conditions – and from the lab environment in which it is stored, awaiting analysis.

---

26 What to concentrate on. Chapters 2 and 3.

27 Variations of Ink analytical chemistry – as examples – *Gedankenexperiment* –continue throughout this book. Introduction 2.3 and 2.2.

28 $CO_2$ CO. Chapters 2, 3, 6 and 8.

29 *Morin.* Fibre evidence contamination. Chapters 1, 3 and 8.

Therefore, it is important that there be formal protocols for the collection, transport, storage and handling [73] of forensic samples. Generally, this means they are brought into contact only with chemically inert and clean tools and containers. Although inertness and cleanliness are relative to the chemistry of the sample, tools and containers of glass, stainless steel and high molar mass polymers are commonly satisfactory. This also generally means that care be taken concerning ambient air, humidity, light and temperature [74]; and time – that the analysis be performed without undue delay. This also necessarily means clear and available documentation of all aspects of collection, transport, storage and handling, including the formal protocols.

The analysis should be as non-destructive of the sample as possible, or at least be minimally destructive, either in the sense that the analytical method is non-destructive *per se* (*e.g.*, a spectrum of visible reflected light [75][30]), or that if it is possibly destructive, only a smallest portion of the total sample be used. Non-destructive analysis leaves the sample for further or other analyses; this may be important for disputed forensic conclusions.

It is important to realize that proper preserving and handling for one kind of analysis may be destructive for another, so that one forensic scientist should be aware of the precautions needed for colleagues' work. For example: As mentioned above, the preservation of fingerprint, dust or pollen evidence before handling a document for ink analysis.

It is also important to appreciate that preservation of the sample may be important for other kinds of analyses that might be thought of only in the future. For example: Preservation of bio-samples stored from long ago, before DNA science had been thought of [76].

Consideration should also anticipate that the scene may be difficult to preserve. For example: Tire marks on a roadway[31] – police cannot hold-up other traffic for too long. Such anticipation should include much photography.[32]

The concern for non-contamination of the sample has important legal significance [77, 78].[33] Proper and careful handling and documentation help protect the integrity and validity of the evidence that the legal authorities would be concerned with and may wish to argue about, and for which a judge may wish to hear argument [79].

There is another kind of evidential integrity and validity that legal authorities are especially concerned about – the *chain of custody* [80][34,35] of an item for forensic analysis. A secure chain means that after the item is initially found, it is kept in the secure custody of identified official personnel (*e.g.*, locked in the personal-access-only locker

30 Colour. Reflection spectroscopy. Chapters 3 and 6.
31 Tire skid. Chapters 2, 3 and 6.
32 Photography. f/d √2. Chapters 3 and 6.
33 *Morin*. Fibre evidence contamination. Chapters 1, 3 and 8.
34 Chain of custody. Chapters 3 and 5.
35 Manarin, *Forensic Evidence*, 2019. Chapters 1, 2 and 3.

of a police officer – only one lock and key – with the key kept personally; or otherwise by formally documented access protocol), and that transfer from one official person to another be receipted (*e.g.*, the police officer personally hands the item to the forensic chemist, who signs for it for secure custody in the chemist's personal-access-only locker). Such a secure chain necessarily includes documentation – by notations in both the police officer's and the chemist's notebooks, on the receipt, and/or the item's tag, label or container, or by a formally documented access protocol. This might also include the possibility of court testimony from the official persons.

A simple break in the chain (*e.g.*, [81] the chemist left it out on the lab bench while at lunch) might cast a slight doubt on the evidence, which in criminal law cases can lead to dismissal of the charges. Defence lawyers would love this; the concept is so obvious to the judge, and so easy to understand, unlike many other aspects of forensic evidence. The judge needs to know nothing of science, nor care, to reject the evidence.

## 3.5 Site and lab work

Once the forensic chemist has custody of the *forensic item* or *sample* of interest from the site, the laboratory analytical sequence will start.[36] This will be guided by the lab's core documents and standard operating procedures, in consideration of published scientific literature.[37]

### 3.5.1 Initial observations

As mentioned above, initial observations of the *forensic sample* are often obvious and closely related to the preliminary theory and speculations. They may eventually confirm the preliminary theory – but further observations might lead otherwise.

To continue the fictitious concocted example above [82]: The forensic chemist, having accepted custody of the suspected forged document contained in a clear polymer envelope[38] might notice:

- a machine-cut sheet of white paper, of standard size;
- onto which has been pre-printed a form in English, with check boxes to be filled in and a place for signature and dating;
- the pre-printed parts were mostly in Times New Roman font, with black printers' ink, with a few words emphasized in red Gothic;

**36** Site. *What it is*. Chapters 1, 2, 3 and 5.

**37** Lab core documents. SOP, scientific literature. Chapters 3, 4, 5 and 8.

**38** Polymer envelopes. Chapters 3, 5 and 9.

- the pre-printed form has an identifier date in small Arial Narrow print at the bottom, along with other apparent coded indicators;
- the alleged – suspected – forgery is near the bottom of the page, in a place for signature, in Roman-looking script, quite unintelligible, in light blue ink that appears to be handwritten with a ballpoint pen.

In another polymer envelope is:
- an ordinary-looking ballpoint pen that writes with light blue ink on blank white paper, which appears to the unaided eye to be the same as the ink on the questioned document;
- and that ink colour appears unusual – a lighter shade of blue – when compared to most blue ball pen inks.

These initial observations suggest continuing with an initial theory: the alleged forgery with light blue ball pen ink could have come from the accompanying pen. Obvious contrary theory possibilities might be immediately considered. For example [83]: If the date written on the document pre-dates ballpoint pens [84, 85], the initial theory would be challenged. And similarly, for example: If the ink writing obviously appears to have come from a fountain pen or an artist's device – with apparently different ink chemistry.

### 3.5.2 Sample selection

The next steps proceed with trying to confirm/deny/test the initial theory. The very next step would be to select and isolate that part of the sample to undergo chemical analysis. In a different context, selection protocols have, perhaps, biblical origin – how does Gideon choose only 300 of his too many soldiers [86, 87, 88]?[39,40,41]

Continuing the example [89]: According to the initial theory, the forensic interest is in the light blue allegedly forged signature and the ink in, and from, the pen – and for the time being, nothing else – other inks can be ignored.

If it is then assumed that all of the ink in the pen, and as written on the paper, is homogeneous (that is, all the same), then the chemist may simply choose any part of the ink of the alleged forgery to be compared with any sampling from the pen. This may well be a good assumption for this example. But more generally, sample selection can pose major problems, as for example for art forgery analysis[42] or brownfield[43]

39 Error. Sampling. Sampling plan. Chapters 3, 4 and 8.
40 Sample selection. Chapters 3, 5 and 9.
41 Selection protocol. Gideon. Chapters 3.1.2 and 3.5.2.
42 Art forgery analysis. Chapters 1 and 3.
43 Sample selection. Brownfield. Chapters 2, 3 and 5.

survey, where homogeneity should not be assumed; or for quantitative determination where grains of one substance are thought to be mixed in with those of another (*e.g.*, a large package of some cocaine in almost-look-alike filler).

Where non-homogeneity is a possibility, sample selection, as a major problem, must be dealt with, by methods of considering statistical randomness and formal selection protocols – sampling plans may become important [90, 91, 92, 93, 94, 95, 96, 97, 98, 99, 100].[44]

Consider another fictitious example [101]: A painting is offered for sale as a newly discovered work by a well-known, but long-dead, artist. But, really [102], it is a high-quality forgery, with modern materials that would have been unknown to that artist.[45] A skeptical art appraiser, consulted by a potential buyer, is baffled by the authentic appearance of style and colour. And indeed, by eye comparison to a colour chart and surface reflection visible spectroscopy,[46] of selected parts of the painting confirm that if it is a forgery, it is forensically smart – except perhaps for some of the yellow areas [103, 104];[47] further forensic work is indicated.

Continuing the example [105]: On the advice of a forensic chemist, micro-samples of the oil paint would be taken from the surface of the painting,[48,49] grouped for colour by eye with comparison to a colour chart or with reflection visible spectroscopy and characterized by mass spectroscopy.[50] These would then be compared to similar samples taken from authenticated genuine paintings of the well-known artist.[51]

But, this would be destructive testing, perhaps at its most frightening. Not only would damage be done to the suspected forgery, but also to genuine works of art that would be used as reference standards. And, even forgeries can be works of art. Such destructiveness in this testing proposal would likely be sufficient cause to not do it; to try another method [106, 107, 108]. Destructiveness of testing methods was impliedly of concern for the scientific work done on the Shroud of Turin.[52,53,54]

Ideally, accepting temporarily, for the sake of argument, the taking of micro-samples, many should be taken at random and by a formal protocol [109]. For example, a *Cartesian* grid [110][55] might be drawn on a photograph of the painting, and surface samples taken from the painting at regular x and y intervals. And since oil painting can involve several layers, there should also be z intervals (but $z \ll x$ & y) [111, 112].

---

**44** Error. Sampling. Sampling plan. Chapters 3, 4 and 8.
**45** Art supply. Forbes art pigment collection. Chapters 1, 2 and 3.
**46** Colour. Reflection spectroscopy. Chapters 3 and 6.
**47** Forbes art pigment collection. Chapters 1, 2 and 3.
**48** **Alert**: this would be destructive testing that likely should not be done – read on.
**49** Destructive and non-destructive testing. Chapters 3, 5 and 8.
**50** Mass spectroscopy. GC MS. Chapters 3 and 6.
**51** **Alert**: this would be destructive testing that likely should not be done – read-on.
**52** Shroud of Turin. Carbon-14. Introduction 2; Chapters 3 and 7.
**53** Art analysis. Colour change. Chapters 3 and 7.
**54** Raman spectroscopy. Paint analysis. Chapters 3 and 6.
**55** *Cartesian* Grid. Chapters 3, 6 and 8.

Such selections might be assumed as random for the painting as a whole. But, there may be areas of the painting that are of special interest. The by-eye or the surface reflection visible spectroscopic results might suggest special attention for the yellows, so that they might be considered by a separate random choice protocol aside from the whole-painting x, y, z grid. What is described here may be expected to be very difficult and time-consuming.

However, there may be a further problem because a *Cartesian* grid might not actually be random for the painting as a whole [113]. If the artist started the painting with a to-be-eventually-painted-over-grid [114, 115, 116, 117, 118], then a bias might be introduced. It might be better, but more difficult, to use another method for random site selection [119].

As another example [120]: Soil [121, 122, 123] samples taken in a brownfield [124, 125] in a three-dimensional *Cartesian* grid, to obtain random samples of suspected, previously left industrial or other wastes or pollutants, at various depths, at various times [126]. For such environmental studies, the randomness of sampling for the resulting data – the selection protocols – sampling plan [127, 128][56] – may be a most important consideration. There are the same problems as with the art forgery example. In analogy to the artist's grid, there could have been prior engineering and landscaping plans involving a grid or actual selective land use [129, 130, 131]. And, the interpretation would be more difficult, because rather unlike – or more-so – than artwork, a soil and debris field may be slowly in motion, have continuing chemical and biological processes, have been worked on with machinery, have more added to it, have been rained on, pooped-on by geese, *etc.*

Selection protocols can be even more difficult for suspected polluted water courses or bodies because of the fluidity, precipitation or settling of some substances as solids or denser liquids [132, 133], freeze and thaw, *etc.* – so that the bottom might become a kind of underwater brown field.

Sometimes, it would be especially important for the randomness protocol to be secret – at least temporarily. For example: For a horse race with financial benefits for winning ticket holders and horse owners. Which substances and horses would be subject to pre-race doping testing, and when?

### 3.5.3 Sample selection – Pb in humans

Sample selection is particularly problematic for measuring and evaluating toxins in biological systems. For example, lead Pb is an anciently known and more recently historically widely used metal [134, 135, 136].[57] It is widely used for water municipal piping;

56 Error. Sampling. Sampling plan. Chapters 3, 4 and 8.
57 Pb toxicology. Pb Sample selection. Chapters 1, 2 and 3.

sheathing for insulated electrical cable; architectural, automotive and other paints; ammunition; electrical solder; jewelry; toys; for dinner- or drink-ware; toothpaste tubes [137]; for burial containers of human remains; a component of glass; *etc.* Continuing legacies from at least some of these historical uses remain [138, 139, 140, 141, 142].[58]

Pb is known to be especially toxic to children [143, 144, 145, 146], whose sources may include architectural paint in old-stock housing, municipal water and toys. The analytical chemistry for Pb measurement is not too difficult. Pb concentration can be measured in the blood system, but the sample selection as to when it is measured can be problematic, including choosing which children would be measured, and from which neighbourhoods [147, 148].

But also problematic is the health significance of blood Pb [149].

Blood Pb measurements may be indirect indicators of health harm. The health damage of concern for Pb in children is largely neurological. Pb exposure can vary over time, as can storage in and exchanges between various tissues, such as in bones. Neurological diagnosis can be difficult – including differentiation from other causes and health issues.

Pb is measured in blood because that is where it can be practicably done, and the blood system is connected to other systems – to give readable numbers. Typically, other human body parts are not directly sampled (at least initially). The conclusions drawn must be in the context of the complexities noted above; such conclusion may be only unclear indicators [150].

### 3.5.4 Screening tests

With the selected sample part, the forensic chemist can now continue with further lab work, the first of which may be *screening tests*. These are quick and economical procedures – in terms of time, effort, resources and cost – that can indicate a probably correct forensic answer or answer range, to confirm a speculation/suspicion of chemical identity, but with limited reliability. And, importantly, a screening test might allow the chemist to determine that the sample is not the speculated/suspected substance. *Screening tests* can allow the forensic chemist to decide which samples should be subjected to less economical but more reliable analysis.

For example, a forensic chemist asked to make a determination of an alleged heroin sample might use the Marquis test, Froehde's reagent or Mecke's test [151, 152, 153]. Such tests are sometimes called *colour* or *spot tests* because a colour change for a micro-amount of sample – perhaps a drop (or few) of a solution on a spotting plate – might occur when a drop (or few) of a particular reagent is added (such as Froehde's reagent for heroin: purple → olive green on standing [154]). Commercially

58 Neuro-toxicology of lead Pb. Chapters 1.4.2.2 and 3.5.

prepared kits for many of these tests are available; often used by law enforcement officers.[59]

As another example, a screening test could be by simple thin layer chromatography[60] run on a small plate, such as might be done for an alleged cocaine [155][61] sample, before mass spectroscopy or infrared spectroscopy would be done, or perhaps, avoided.

For the document forgery example mentioned above [156], a screening test would have already been done when the chemist initially observed the appearance to the eye of the light blue ink. Had the pen produced a darker blue line, or a green line, on the white paper (here a screening test), then the chemist could have quickly concluded [157] that this pen was not used for the alleged lighter blue forgery. Further screening tests might include examining the ink lines for obvious differences under ultraviolet light, or perhaps photography with various other light sources, filters and photo-receiving media, and perhaps also by observing the ink drawn on white paper when subjected to heat, strong light or oxygen.

### 3.5.5 Analytical/instrumental methods

If a screening test is reliable enough for negative results – for example [158]: If the pen drew dark blue, or green – then more expensive analytical work can be avoided. Otherwise, it must be undertaken.

In modern labs, this usually involves procedures with instrumentation (see in Chapter 6)[62,63] to reveal characteristics that are uniquely and theoretically linked to molecular structure (see in Chapter 4).[64] Several methods might be used, in sequence or alternatively, for example: UV–visible spectroscopy, infrared spectroscopy,[65] TLC,[66] mass spectroscopy and gas chromatography-mass spectrometry (GC-MS).[67,68]

Such procedures usually start by isolating and preparing part of the found sample. For example [159]: After being photographed [160] (to best preserve the evidence), there could be a spectroscopic measurement of reflected light[69] from a selected part

**59** Screening tests. Wet chemistry. Chapters 3, 4, 5 and 6.
**60** TLC. Chapters 3 and 6.
**61** TLC. Chapters 3 and 6.
**62** Screening test → instrumentation. Chapters 3 and 6.
**63** Chemical identification. Chapters 3, 4 and 6.
**64** Molecular structure. Chapters 1, 3, 4 and 6.
**65** *Genesis*. light. EM radiation. Photon. UV. IR. Microwave. Chapters 3, 4 and 6.
**66** TLC. Chapters 3 and 6.
**67** **Spoiler alert**: The instrumental methods would be too extensive for ink analysis. TLC alone would be quite sufficient. Chapter 3.
**68** Analytical/instrumental methods. Chapters 3 and 6.
**69** Ink analysis. Reflection spectroscopy. Chapters 3 and 6.

of the light blue ink. This would be non-destructive. More conveniently [161], colour comparison by the human eye could be done.[70]

Procedures could continue by again isolating and preparing part of the found sample for other kinds of analysis: The forensic chemist might scrape off [162, 163][71] with a clean sharp scalpel, a very small portion of the alleged forgery from the document, being careful to avoid any other ink. Then take a similar scraping from an apparently blank spot of the paper, as a control.

The two scrapings would then be put into a solvent – to dissolve [164] the ink – known to be non-reactive with ball pen ink, and transparent to visible and ultraviolet light – for example, spectroscopic-grade cyclohexane: [165, 166, 167, 168, 169][72,73]

After removing any paper fibres [170], the two sample parts would then be put into two UV–visible spectral cells, which would then be observed in a UV–visible spectrophotometer – each run versus a pure cyclohexane blank [171].

A similar solvent extraction for UV–visible examination would also be done for ink lines from the pen – drawn on a small, clean glass surface [172]. Similarly taken ink lines could be taken from another pen, as a comparison. All of the spectra would be compared to each other, and, importantly, to available ink industry reference spectra or standard samples.

These UV–visible spectra would have been run because of the coincident convenience with the solvent extraction. Generally, however, UV–visible spectra are mainly useful as screening tests and not reliable for unambiguous identification of similar molecules.[74] And also, it is most likely that the inks are mixtures of different kinds of molecules, so that the forensic chemist would also want to do chromatographic analysis, with perhaps infrared spectral analyses or mass spectral analyses of the components of the chromatographically separated mixtures.[75]

TLC[76] would be the method of choice for this analysis because of its sensitivity and economy. And because its use for ink analysis has been well-published. TLC would allow separation of the different molecular components of the mixture and their removal for preparation for IR spectroscopy.

If it were found that the sample parts would be too small for TLC to be useful, then gas-phase column chromatography[77] might be tried; however, for GC, preserving the separated components of the mixture for IR analysis may be more difficult [173].

70 Colour comparison. Chapters 3.5.5 and 6.33.2.
71 Destructive and non-destructive testing. Chapters 3, 5 and 8.
72 Too much scientific technical detail here? Introduction 2.1.
73 Screening test → instrumentation. Chapters 3 and 6.
74 UV. Chapters 3, 5 and 6.
75 Too much scientific technical detail here? Introduction 2.1.
76 TLC. Chapters 3 and 6.
77 GC. Chapters 3 and 6.

Another alternative would be to use GC in conjunction with mass spectroscopy.[78] GC-MS would be the best alternative for the most reliable determination; however, the equipment is more complex. For both TLC and GC, the ink samples from the UV–visible analysis can be used, still dissolved in the cyclohexane [174].

The TLC-separated components, as scraped from the TLC plate, could each be prepared for IR[79] by being mixed with Nujol to form a mull or made into a KBr pellet [175], with the cyclohexane evaporated. Then, IR spectroscopy could be done.[80,81]

## 3.6 Forensic conclusions

The results of the lab work hopefully imply logical conclusions.[82] For purposes of this fictitious example [176], assume that the ink of the alleged forged signature matches both the ink in the pen and the ink of an industry standard. Real-life forensic conclusions could perhaps (and wishfully) be similar, but, alternatively, might indicate otherwise, or that other samples might be needed. In real life, there might also be some doubt in the conclusion.

## 3.7 Analysis too extensive? – Too sensitive – or not enough?

### 3.7.1 Too extensive?

In general, the question of how extensive an analytical chemistry investigation is enough for a reliable conclusion can be on-going.[83,84,85] However extensive it may have been, an opposing lawyer with a desperate case and a willing fee-paying client [177, 178, 179] may always want further analytical chemistry, because of obstinate client instructions. In the legal system, this decision, which is eventually the job of the judge or a tribunal panel, may depend as much on judicial policy as on science. It may also be influenced by how confused and impatient the judge or tribunal panel might be for all this.

---

**78** GC-MS. Chapters 3 and 6.

**79** IR. Chapters 3 and 6.

**80** Spoiler alert: The instrumental methods would be too extensive for ink analysis. TLC alone would be quite sufficient. Chapter 3.

**81** Too much scientific technical detail here? Introduction 2.1.

**82** Variations of Ink analytical chemistry – as examples – *Gedankenexperiment* – continue throughout this book. Introductions 2.3 and 2.2.

**83** Analysis – how extensive? Chapters 3 and 6.

**84** Mass measurement (grams) accuracy and precision too extensive? Chapters 3 and 6.

**85** Forensic chemistry choices. Chapters 3 and 4.

Several issues can arise:

- First is the question of what level of chemical complexity is sufficient for chemical identification. As explained in Chapter 4,[86] chemical substances can, in concept, be uniquely defined – quite exactly – including subtle details. For example [180], if there is a workplace regulation of *o*-xylene,[87] and that molecule only, then the analysis must be so specific as to reveal only *o*-xylene and neither the quite similar *m*-xylene nor *p*-xylene. An analysis that cannot distinguish between the three is not sufficient. But, *o*-xylene-specific analysis need not go into still further details, such as for isotope differences.

  If, however, the regulation would be for any xylene (*o*-, *m*-, *p*-), then the less detailed designation would be ok.
- Second, even when a chemical substance with a very specific definition is being looked for, the analytical method need not always be the most exacting, depending on the context in which that substance would be found. Again, for the xylene example, if the analysis revealed xylenes – but not specific as to *o*-, *m*- or *p*- – and an inspector found a bottle with an *o*-xylene label, and none other at the site, for many purposes, this might be sufficient for a legal finding as specific to *o*-xylene.

To continue the above ink example [181],[88] the chemical identities of the components can remain unknown – and likely would actually be unknown, except in a general way. The presence of identifiable TLC patterns would be sufficient, because there would be a sufficiently comprehensively standard sample library for comparison. For this, the analytical chemistry to confirm or determine the detailed molecular structure is not needed for the forensic conclusion.

Arguably, the ink analysis described above,[89] other than TLC, is too extensive, and reliable conclusions could be drawn based on the TLC patterns alone,[90] without resorting to the more time-consuming and expensive spectroscopy [182]. This argument is supported by the observation of Brunelle and Cantu [183, 184][91] that all of the ~3,000 inks then [185] in the ATF library[92] are distinguishable by TLC [186, 187, 188].

For the example here, for those readers who are persuaded that the spectroscopy can and should be avoided, or limited – restricted to only whole inks, this book at least provides information about the theoretically more complete spectral analysis.

---

**86** Chemical substances described. Chapters 3 and 4.

**87** *o*-Xylene. 106.17 g/mol. Chapters 3.7.1, 4.2.1 and 4.5.1; Figs. 4.2a and 4.10.

**88** Variations of Ink analytical chemistry – as examples – *Gedankenexperiment* – continue throughout this book. Introduction 2.3 and 2.2.

**89** Ink analysis. Chapter 3.

**90** See **Spoiler alert**, above: The instrumental methods would be too extensive for ink analysis. TLC alone would be quite sufficient. Chapter 3.

**91** Brunelle and Cantu. Chapters 3 and 4.

**92** Ink reference standards. US International Ink Library. ATF. Chapters 2, 3, 4 and 9.

To further continue the ink example [189]: Based on TLC alone, and in the absence of any other evidence to contradict, the ink may now be identified as coming from the seized pen or from another pen of identical manufacture. Is this an absolute determination? No, but it provides a reasonable *moral certainty*; and it is not worth [190] the extra time, trouble or expense to do IR of each chromatically revealed component.

Would a complete spectral analysis provide an absolute determination? Yes, almost, but for this, some questions of uniqueness must be dealt with – in Chapter 4.[93]

Another example where the analytical method need not be the most exacting, depending on the context in which the substance would be found, and for which there would be a sufficiently comprehensively standard sample library for comparison, is marihuana determination by the *Duquénois-Levine*[191, 192] test and TLC. Bailey [193][94] reported using these methods, of limited specificity by themselves, but in comparison with a large roster of other possible substances. His argument for the sufficiency of *Duquénois-Levine* and TLC is persuasive.

### 3.7.2 Too sensitive – or not enough?

Modern analytical chemistry can deliver results for very small samples and concentrations.[95] Perhaps, sometimes in the range of $<10^{-9}$ [194] moles [195, 196, 197, 198, 199][96] or moles/litre [200].[97,98] And, future analytical chemistry should be expected to get even better [201]. But, sometimes such small amounts – sometimes called trace – are so small as to be below what would be of forensic significance. The small amount may be from an already present background or otherwise of innocent origin. It also may be that the background can be of larger amounts.

Examples:

- Poppy seeds found with usual foods can result in analytical chemistry reports for failed drug tests. Amounts of the forbidden chemical would most likely be below those for actual drug-abuse or health concerns. Injustice can happen when test reports are used for parole applications [202] and child custody decisions [203, 204, 205, 206, 207].[99]
- A comment that some US paper money – now all as Federal Reserve Notes – often bears some trace of cocaine [208] – apparently turns out to have some validity beyond a cinema scene [209]. However, while such forensic findings may be evi-

**93** Fundamental identification of molecules. Chapters 2, 3, 4 and 6.
**94** K. Bailey. Chapters 3 and 4.
**95** Too sensitive. Toxicology. Chapters 1, 3 and 8.
**96** Mole. Avogadro number. Chapters 3, 4 and 7.
**97** Octafluoronaphthalene CAS 313-72-4. Chapter 3.
**98** Mass measurement (grams). Accuracy and precision too extensive? Chapters 3.1.1, 3.7.3 and 6.20.
**99** Poppy. Too sensitive. Chapters 1, 3 and 8.

dence of actual criminal contact – for use or trafficking – it could also indicate innocent trace presence from when other not-so-innocent Notes previously passed through automatic counting machines used by banks. Arguably, even such traces can be discovered by modern analytical chemistry. Cocaine found on Federal Reserve Notes should be forensically evaluated with care [210, 211, 212].

- Exposure limits declared under the Canada *Criminal Code* for blood drug concentration, with consequences as at "Any detectible level,"[213] would appear to ignore that modern analytical chemistry may be able to deliver measurements well-below levels of forensic or toxicological concern [214, 215, 216, 217, 218, 219, 220, 221, 222, 223, 224].[100]
- Consider a contrived, fictitious example [225]: A geological engineer speculatively examines a site as a potential Silver Ag mine. Nothing to do with Uranium U or Thorium Th. Returning to the home office, the engineer is stopped at an international border when a new, improved, supersensitive mass spectrometer reveals shoe-borne U-235 – the weaponizable atomic component, but in quantity $10^{-25}$ moles – far, far below the critical mass. But an "any detectible level" zero tolerance policy leads to detention by law enforcement [226].[101]
- But, sometimes very small amounts are not so small as to be below public health and forensic significance: biological contaminants in blood [227, 228]. And similarly for DNA Samples.[102]

### 3.7.3 Summary

To summarize, generally: chemical analysis, for qualitative identification, notwithstanding a very specific definition of what is being looked for, need not always be the most specific or exacting, depending on the context in which that substance would be found – including confirmation with other not completely specific tests. But this must be done with thoughtful care and is not a licence to avoid more careful analysis needed otherwise.

Similarly, lower detection limits should be reasonably related to actual forensics or toxicology; Canada's criminalization, in concept, of "any detectable level" would cause a serious problem for scientists, though apparently not so much of a concern for law-makers.[103]

---

**100** Toxicology. Chapters 1, 3 and 11.

**101** Plutonium. Periodic Table. Too sensitive? Chapters 1 and 10; Chapters 3.7.2 and 4.2.3.

**102** DNA – too sensitive? Chapters 3.72 and 6.16.

**103** Re criticism of Health Canada, Judges, *et al.* Introduction 4.1 and 4.3; Chapters 10 and 11.

## Notes

**1** [Tasting the evidence – especially at a murder scene – would not now be regarded as appropriate forensic analytical chemistry procedure.]
**2** [Dr. Armstrong determines the fatal poison in Anthony's whiskey and soda as likely cyanide of Potassium.]
**3** Agatha Christie, *And Then There Were None*, 1939, Berkley Books, New York City, 1991, Chapter IV, Part IV → Chapter V, Part I, at pages 46, 49→51.
**4** Hydrogen cyanide, H–C ≡ N (Prussic acid) CAS 74-90-8 27.0253 g/mol.
**5** https://en.wikipedia.org/wiki/Hydrogen_cyanide
**6** Potassium cyanide ($K^+$ $CN^-$) CAS 151-50-8 65.12 g/mol.
**7** https://en.wikipedia.org/wiki/Potassium_cyanide
https://en.wikipedia.org/wiki/Cyanide_poisoning
https://pubchem.ncbi.nlm.nih.gov/compound/Potassium-Cyanide
https://www.health.ny.gov/environmental/emergency/chemical_terrorism/cyanide_general.htm
**8** In *Sleeping Murder*. . ., the drink was brandy, poison not indicated, and not fatal because of prompt attendance by a near-by police surgeon.
**9** Agatha Christie, *Sleeping Murder: Miss Marple's Last Case*, HarperCollins Publishers, London, W6 8JB; 1976, 2002; ISBN 978-0-00-712106-9; pages 282, 283, 299.
**10** A Sherlock Holmes joke, of uncertain origin, found in variations in several sources, editorially summarized for purposes here to illustrate an approach to analyze evidence.
**11** Tim Radford, science editor, "Scientists close in on world's funniest joke," *The Guardian International Edition*, UK news, 20 December 2001. https://www.theguardian.com/uk/2001/dec/20/humanities.research
**12** http://www.hep.caltech.edu/~arakitin/jokes/sherlock.html
https://www.sunnyskyz.com/funny-jokes/20/Sherlock-Holmes-and-Dr-Watson-Go-Camping
https://www.sherlockforum.com/forum/topic/2488-sherlock-holmes-jokes/
https://unijokes.com/joke-4856/
**13** Analytical chemistry is the branch of chemical science for the chemical determination of substances, as to what they are (*qualitative*), how much (*quantitative*), and the uncertainty . . . .
**14** **Gram** is the SI measure of **mass,** but often referred to as **weight.** However, "weight" is so commonly used that it often passes without comment. From Sir Isaac Newton's defining equation:

| Weight | = | mass | × | the acceleration of gravity on the Earth. |
|---|---|---|---|---|
| Weight (nt) | = | mass (kg) | × | 9.81 ($m/s^2$). |

(1 kilogram = 1,000 grams).

**15** In Police documents, **gram** is the SI measure used, but often referred to as **weight**.
**16** John Kenkel, *Analytical Chemistry / Refresher Manual*, Lewis Publishers, Chelsea, Michigan 48118; 1992; ISBN 0-87371-398-2; Chapter 2.2 BALANCES, pages 18→24.
**17** Richard Saferstein, *Criminalistics/An Introduction to Forensic Science*, Seventh Edition, Prentice Hall, Upper Saddle River, New Jersey, 07458; 2001; ISBN 0-13-013827-4; pages 91 and 92.
**18** See: John Kenkel, *Analytical Chemistry / Refresher Manual*, Lewis Publishers, Chelsea, Michigan 48118; 1992; ISBN 0-87371-398-2; Chapter 2.2 BALANCES, pages 18→24.
**19** Kenkel, page 18.
**20** SI = *Le Système International d'Unités Convention du Mètre*, Paris, 1875.
https://www.nist.gov/pml/owm/metric-si

**21** SI is the essential foundation for the Canada federal *Weights & Measures Act.*
**22** *Weights and Measures Act* RSC 1985, c. W-6. http://laws-lois.justice.gc.ca/eng/acts/W-6/page-1. html
**23** Measurement Canada, Ottawa K1A 0C9 http://www.ic.gc.ca/eic/site/mc-mc.nsf/eng/home
**24** Within what conditions – such as temperature, pressure, humidity, spin velocity and voltage.
**25** For example: a US 1909 Lincoln Wheat VDB Penny (VDB = Victor D. Brenner).
https://www.coincollecting.com/lincoln-wheat-pennies-key-dates-and-values#:~:text=A%20few%20coins%20of%20note,the%20edge%20of%20the%20coin.
**26** https://pubs.acs.org/doi/10.1021/ed072p954
**27** See also: Hannah Fry, "WHAT REALLY COUNTS," BOOKS, *THE NEW YORKER*, 29 March 2021, pages 70→83.
**28** *Gedankenexperiment.* Introduction 2.2.
**29** But not government-established, nor likely directly closely government-regulated.
**30** https://satsuite.collegeboard.org/sat
**31** https://www.act.org/content/act/en/products-and-services/the-act.html#:~:text=The%20ACT%20Test%20Overview,not%20affect%20your%20composite%20score
**32** (SSQT = Selective Service Qualification Test. SSS = Selective Service System.)
**33** https://www.ets.org/research/policy_research_reports/publications/report/1983/icfq.html
**34** (1966, during the US Vietnam engagement, SSQT used by SSS to help determine educational deferment draft status.)
**35** (Also see: Patricia M. Shields, "THE BURDEN OF THE DRAFT: THE VIETNAM YEARS," *Journal of Political & Military Sociology*, University of Florida Press, Vol. 9, No. 2 (Fall 1981), pages. 215→228.
https://www.jstor.org/stable/45293572)
**36** https://unodigitalhumanitiesprojects.omeka.net/exhibits/show/vietnam-draft-1964-1973/1966#:~:text=In%20March%201966%2C%20the%20Director,and%2080%20for%20graduate%20students.
**37** But see: David Leonhardt, *"The Misguided War on the SAT* / Colleges have fled standardized tests, on the theory that they hurt diversity. That's not what the research shows," *New York Times*, 07 January 2024.
https://www.nytimes.com/2024/01/07/briefing/the-misguided-war-on-the-sat.html
**38** See also: Orlando Mayorquín,*"Problems With New California Bar Exam Enrage Test Takers and Cloud Their Futures* / Many aspiring lawyers could not access the test and cited delays or other issues. A lawmaker has called for an investigation," *New York Times*, 02 March 2025.
**39** See also: https://kaplan.com/ https://kaplan.com/all-prep
**40** See:
- CTV News, "Trump touts five-word memory test as showing he is 'cognitively there'," 23 July,2020.
- https://www.ctvnews.ca/world/trump-touts-five-word-memory-test-as-showing-he-is-cognitively-there-1.5035687
- Brooklyn Neustaeter, CTV News, "Canadian doctor behind cognitive test says Trump's score is 'normal performance'," 23July 2020. https://www.ctvnews.ca/health/canadian-doctor-be hind-cognitive-test-says-trump-s-score-is-normal-performance-1.5036251

**41** See also:
https://en.wikipedia.org/wiki/Battle_Hymn_of_the_Tiger_Mother
https://en.wikipedia.org/wiki/Helicopter_parent
**42** (Or Draft Board).
**43** (Or too many seeking Draft deferment).
**44** See: Stephen Murdoch, *IQ: A Smart History of a Failed Idea*, John Wiley and Sons, Inc., Hoboken, New Jersey; 2007; ISBN 978-0-471-69977-4. https://www.publishersweekly.com/978-0-471-69977-4
**45** See also: Elizabeth Bruenig, "An I.Q. Score as a Death Sentence / It's illegal to execute intellectually disabled people. Yet it may happen." *The New York Times*, Opinion, 13 January 2021, page A26.

46 Shawn Hubler, "*No SAT or ACT for University of California /. . . in Admissions,*" *The New York Times*, 22 May 2020, pages A1 and A24.
47 See also: Brooklyn Neustaeter, CTVNews.ca Writer, HEALTH | Explainer, "Canadian doctor behind cognitive test says Trump's score is 'normal performance'," 23 July, 2020.
https://www.ctvnews.ca/health/article/canadian-doctor-behind-cognitive-test-says-trumps-score-is-normal-performance/
48 See also: Meredith Broussard, "*When Algorithms Do The Grading / In-person final exams were canceled, so computers stepped in,*" *The New York Times*, OP-ED, 09 September 2020, page A27.
49 See also: "The Campus Tour Has Been Cancelled / How the pandemic has thrown college admissions process into a kind of slow-motion chaos. One of the biggest changes: most colleges have stopped requiring the SAT. For decades, there's been a debate over whether schools should drop the test. What's it mean that it finally happened?"; *This American Life*, PRX The Public Radio Exchange; March 2021.
https://www.thisamericanlife.org/734/the-campus-tour-has-been-cancelled
50 Population census – here determined by an actual count – and not by mathematical estimation formula using chosen indicators.
51 See, for example: Holden Thorp, "I Was Diagnosed With Autism at 53. I Know Why Rates Are Rising," *New York Times*, OPINION / GUEST ESSAY, 20 March 2025.
https://www.nytimes.com/2025/03/20/opinion/autism-diagnosis-rfk-vaccines.html
52 See: Jeffrey S. Rosenthal, *Struck by Lightning / The Curious World of Probabilities*; Harper Perennial / HarperCollins Publishers Limited, Toronto M4W 1A8; 2005, 2006; Chapter 5, pages 63→77, "Murder Most Foul / Measuring Trends." ISBN 13: 978-0-00-639495-2 ISBN 10: 0-00-639495-7
53 See for example: Beth Blauer and Jennifer Nuzzo, "Standards For Covid Testing," *The New York Times*, OP-ED, 24 November 2020, page A23.
54 For example, as for maternity deaths – see: Austin Frakt, "U.S. Could Prevent Two-Thirds of Deaths Related to Pregnancy," *The New York Times*, BUSINESS, The Upshot, 14 July 2020, page B3.
55 CDC NCHS, Centers for Disease Control and Prevention, National Center for Health Statistics, US Department of Health & Human Services, Maternal Mortality
https://www.cdc.gov/nchs/maternal-mortality/index.htm
https://www.cdc.gov/nchs/maternal-mortality/implementation.htm
https://www.cdc.gov/nchs/maternal-mortality/evaluation.htm
56 CBC aih 04 September 2020, virus school count https://www.cbc.ca/radio/asithappens
57 Including in their own note books – see Karen Jakob, *The Complete Guide to Police Writing*, The Carswell Company Limited, Toronto, 1984, "The Memorandum Book," Chapter 8, pages 105→109.
58 See, for example:
59 – Richard Saferstein, *Criminalistics / An Introduction to Forensic Science*, Seventh Edition, Prentice Hall, Upper Saddle River, New Jersey, 07458; 2001; ISBN 0-13-013827-4.
60 – Stuart H. James and Jon J. Nordby, editors, *Forensic Science / An Introduction to Scientific and Investigative Techniques*, CRC Press, Boca Raton, 2003; ISBN 0-8493-1246-9.
https://www.routledge.com/corporate/about-us/crc-press?srsltid=AfmBOorPh3Ze66mryh8nPkQur-PCQ9usoKChoHv4urB4izjFHHG82eZ2
61 – Brian H. Kaye, *Science and the Detective / Selected Reading in Forensic Science*, VCH Verlagsgesellschaft mbH, Weinheim, Germany; 1995; ISBN-3-527-29,252-7.
62 Showing map co-ordinate location, and magnetic declination; if necessary.
63 In modern terms copies of photographs would be as a computer-readable files. From when photographic film was used, this would include contact prints of the negative strips to verify that all photographs have been accounted for by checking the manufacturer's frame sequence numbers.
64 *Gedankenexperiment.* Introduction 2.2.

**65** *Gedankenexperiment.* Introduction 2.2.
**66** *Gedankenexperiment.* Introduction 2.2.
**67** See: the Centre of Forensic Sciences, Public Safety Division, Ministry of the Solicitor General, *Laboratory Aids for the Investigator*, Fourth edition 1984, "Documents Section," pages 69→75, at page 70; Toronto M7A 2G8; Publications Ontario, $6.50.
**68** Although there might also be chemical reaction – ink and fibre. More generally, paper analysis, including microscopic observation, can be an important forensic pursuit – see Brian H. Kaye, *Science and the Detective / Selected Reading in Forensic Science*, VCH Verlagsgesellschaft mbH, Weinheim, Germany, 1995, "Paper, the Forger's Nemesis?," Chapter 10.2, pages 269→275.
**69** https://www.thisamericanlife.org/689/transcript
https://whc.unesco.org/en/list/849/ https://en.wikipedia.org/wiki/Troy
**70** *Gedankenexperiment.* Introduction 2.2.
**71** Ontario Ministry of the Attorney General, *Report of the Kaufman Commission on Proceedings Involving Guy Paul Morin*, [1997], Executive Summary.
https://www.attorneygeneral.jus.gov.on.ca/english/about/pubs/morin/morin_esumm.html
https://www.ontario.ca/feedback/contact-us?id=98538&nid=97157#contactForm
http://netk.net.au/Canada/MorinList.asp
**72** [Chapter II]

[. . .]

- The fibre evidence was contaminated within the Centre of Forensic Sciences. The timing and precise origin of the contamination cannot now be determined. However, it remains possible that this contamination tainted Ms. Nyznyk's earliest findings. No inferences can safely be drawn from any alleged fibre similarities, given the existence of this in-house contamination.

[. . .]

**73** "Handling" here means how something is dealt with or processed, rather than actually touching it with bare or gloved human hands, which very often should be avoided – for protection of both the item and the human.
See also: https://www.merriam-webster.com/dictionary/handling
**74** Especially for biological samples.
**75** Assuming no photochemical reactions.
**76** See, for example: Michael Levenson, "*DNA Profile Identifies Remains as Mother Who 'Just Packed Up Her Stuff and Left'* [~50 years ago]," *The New York Times*, 28 March 2021, page 23.
**77** For example: fibres that were regarded as evidence in *Moran* – but might have been from the clothing of a forensic lab investigator.
**78** As a corollary: non-contamination of the handling personnel by hazardous substances has important occupational health and safety significance, and they should insist on being informed and protected. Chapters 1 and 3.
**79** See: "Standard Practice for Receiving, Documenting, Storing, and Retrieving Evidence in a Forensic Science Laboratory," ASTM E 1492–92; included in John J. Lentini, "ASTM Standards for Forensic Sciences," *Journal of Forensic Sciences*, 40(1), January 1995, pages 146→149.
ASTM E1492 - 11(2017) https://www.astm.org/Standards/E1492.htm
**80** Manarin, *Forensic Evidence* . . ., 2019; [ASSUMPTION OF INTEGRITY chain of evidence], page 7.
**81** *Gedankenexperiment.* Introduction 2.2.
**82** *Gedankenexperiment.* Introduction 2.2.
**83** *Gedankenexperiment.* Introduction 2.2.
**84** Possibly ~1941, Argentina; RAF, UK, during WWII; USA, after WWII.
**85** https://en.wikipedia.org/wiki/Ballpoint_pen

**86** Judges Chapter 7 שופטים https://www.mechon-mamre.org/p/pt/pt0707.htm
https://www.bible.com/bible/114/JDG.7.NKJV
**87** A Hebrew - English Bible According to the Masoretic Text and the JPS 1917 Edition
https://www.mechon-mamre.org/p/pt/pt0101.htm
https://www.mechon-mamre.org/p/pt/pt0.htm
**88** The meaning taken here is in the context of a selection protocol that should be random – or at least sort of random. There is biblical/theological interpretation otherwise not invoked here.
**89** https://www.whatchristianswanttoknow.com/gideon-bible-story-summary-with-lesson/
https://en.wikipedia.org/wiki/Gideon
*Gedankenexperiment.* Introduction 2.2.
**90** Sampling plan = How to choose and document what small part of the item or place of interest is used for measurement, when the entire item is not used, or the entire place is not sampled. This involves statistical considerations. This helps determine what is actually being measured. It should be noted that item or place may be too large – or far too large – to be dealt with other than by way of small statistically chosen samples.
**91** See below:

**92** – Lawrence H. Keith, *Environmental Sampling and Analysis: A Practical Guide*, Lewis Publishers, CRC Press, Inc., Boca Raton, Florida, 33431; 1991; ISBN 0-87371-381-8.

**93** – Neil T. Crosby and Indu Patel, *General Principles of Good Sampling Practice*, published for the Laboratory of the Government Chemist by The Royal Society of Chemistry, Cambridge CB4 4 WF, 1995. ISBN 0-85404-412-4.https://www.isbns.net/author/Indu_Patel

**94** – Lawrence H. Keith, editor, *Principles of Environmental Sampling*, ACS Professional Reference Book; Washington, D.C., 20,036; 1988; fourth printing 1993; ISBN 0-8412-1437-9.

**95** – Gavin D. Armstrong, Ohio Environmental Protection Agency, "Field Screening – Quick & Dirty" Is Rapidly Earning the Reputation of "Efficient & Cost Effective," as Paper 1, *Proceedings/The Eleventh Annual Waste Testing & Quality Assurance Symposium*, ACS & EPA, Washington, D.C., 23→28 July 1995.

**96** – John Keenan Taylor, *Quality Assurance of Chemical Measurements*, LEWIS PUBLISHERS, INC.; CRC Press, Inc., Boca Raton, Florida, 33431; 1987; ISBN 0-87371-097-5.

**97** – Taylor, *Quality Assurance* . . ., Chapter 8, "Principles of Sampling," page 55, *et seq.*

**98** – Lawrence H. Keith, *Environmental Sampling and Analysis: A Practical Guide*, LEWIS PUBLISHERS, INC.; CRC Press, Inc., Boca Raton, Florida, 33431; 1991; ISBN 0-87371-381-8.

**99** – Keith, *Environmental Sampling* . . ., Chapter 1, "Planning and Sampling Protocols," page 7, *et seq.*

**100** – See also: ISO/IEC 17,25:2005 Clause 5.10.2 h

**101** *Gedankenexperiment.* Introduction 2.2.
**102** "Really" only in the context of this *Gedankenexperiment.*
**103** "Yellows" – as part of the concoction for this *gedankenexperiment.*
**104** Yellow artists' paints – historically, containing cadmium Cd
https://news.artnet.com/art-world/7-deadly-art-materials-to-watch-out-for-1081526
https://www.jacksonsart.com/blog/2014/05/07/cadmium-crisis/
https://en.wikipedia.org/wiki/Cadmium_pigments
https://www.independent.co.uk/arts-entertainment/art/features/cadmium-the-rare-paint-pigment-faces-a-europewide-ban-and-artists-are-seeing-red-9756636.html
**105** *Gedankenexperiment.* Introduction 2.2.
**106** Niraj Chokshi, "How Art-Forgery Slueths Learned to Love the Bomb / The period of nuclear testing gave at least a temporary boost to radiocarbon dating," *The New York Times*, SCIENCE, 11 July 2019, page D4.

**107** Sophie Haigney, "Time Robs 'The Scream' of Its Color / Experts study how notable paintings lose pigment," *The New York Times*, Arts, 19 February 2020, pages C1 and C2.
https://www.nytimes.com/2020/02/07/arts/design/the-scream-edvard-munch-science.html

**108** See also: Cibele Bugno Zamboni, Marcelo Miyada Redígolo, Vinicius Takami Miura, Isolda Costa, Maria Luiza Emi Nagai, Pablo Antonio Vásquez Salvador, Dalton Giovanni Nogueira da Silva, "Non-destructive analysis in the study of historical photographs by pXRF and ATR-FTIR spectroscopies," *Journal of Forensic Sciences*, TECHNICAL NOTE, February 2021.
https://doi.org/10.1111/1556-4029.14680
https://onlinelibrary.wiley.com/doi/full/10.1111/1556-4029.14680

**109** That is, at random, using a formal protocol for a randomizing method – otherwise "random" and "formal protocol" would seem to be in contradiction.

**110** Cartesian Graph Paper https://www.printablepaper.net/category/cartesian

**111** Likely difficult to do – but remember this is a *Gedankenexperiment*.

**112** See also:

- Jonathan Feakins, "Alba Álvarez-Martín is making mass spec more accessible in museums / The cultural heritage scientist is adapting a medical imaging technique to analyze paintings in Amsterdam's Rijksmuseum," *C&EN*, ART & ARTIFACTS, 02 January 2025.
- Alba Alvarez-Martin, Jusal Quanico, Teresa Scovacricchi, Ermanno Avranovich Clerici, Geert Baggerman, Koen Janssens, "Chemical Mapping of the Degradation of Geranium Lake in Paint Cross Sections by MALDI-MSI," *Analytical Chemistry*,2023, 95, 49, 18,215–18,223, 23 November 2023. https://doi.org/10.1021/acs.analchem.3c03992

**113** See: Jeffrey S. Rosenthal, *Struck by Lightning / The Curious World of Probabilities*; Harper Perennial / HarperCollins Publishers Limited, Toronto M4W 1A8; 2005, 2006; Chapter 2, at page 7, *What Are the Odds of That? / Coincidence and surprise*. Chapter 16, pages 234→246, "Ignorance Chaos and Quantum Mechanics / Causes of Randomness." ISBN 13: 978-0-00-639495-2 ISBN 10: 0-00-639495-7

**114** Although regular to the grid, they would be random to the painting. If however, if it is suspected that the artist might have used, or forger might have anticipated, random by x,y grid, the x,y coordinates should be chosen by a formal random number generation process.

**115** For example, Graham Sutherland's studies for his portraits of Winston Churchill, Lord Beaverbrook, *et al.*, appear to have a grid pattern.

**116** (The not-well-received Churchill portrait is famous because Lady Churchill had it destroyed, before Sir Winston's death. According to Sutherland: an "act of vandalism".)

**117** https://www.telegraph.co.uk/news/winston-churchill/11730850/Secret-of-Winston-Churchills-unpopular-Sutherland-portrait-revealed.html
https://en.wikipedia.org/wiki/Sutherland%27s_Portrait_of_Winston_Churchill
https://en.wikipedia.org/wiki/Graham_Sutherland
https://www.pinterest.ca/car00672293/graham-sutherland-portraits/
https://www.cbc.ca/news/canada/new-brunswick/beaverbrook-the-crown-1.3897600
https://beaverbrookartgallery.org/

**118** (There is no provenance question of the Sutherland works, Beaverbrook Art Gallery, Fredericton, New Brunswick E3B 1C4. The example here is only to show an artist's use of a grid in a study.)

**119** See: Keith, Lewis, 1991, page 16, *et seq.*

**120** *Gedankenexperiment*. Introduction 2.2.

**121** Soil ≈ a top layer of the Earth's surface – of various kinds – easily crumbled – generally derived from weathering-in-place of previously existing material, mixed with organic matter near the surface. Generally, soil may be considered as what land, in its top layer, is composed of. In the context here potentially containing various pollutants, including dumpings from elsewhere.

**122** https://en.wikipedia.org/wiki/Soil

**123** James Gilluly, A.C. Waters, A.O. Woodford, *PRINCIPLES OF GEOLOGY / Second Edition*, W.H. Freeman and Company, San Francisco and London, 1959; page 43, *et seq.*
**124** ≈ a polluted land area. – at surface, or near-below, and maybe elsewhere.
**125** https://19january2021snapshot.epa.gov/brownfields/overview-epas-brownfields-program_.html
https://en.wikipedia.org/wiki/Brownfield_land
https://www.epa.gov/sites/default/files/2019-11/documents/brownfields-program-overview-ne.pdf
**126** Various times ≈ different eras – including recent – of human-directed development.
**127** Virginia T. McLemore, Kathleen S. Smith, Carol C. Russell, editors, *Sampling and Monitoring for the Mine Life Cycle*, Society for Mining, Metallurgy, and Exploration, 28 May 2014; [sampling] page 63.
https://books.google.ca/books?id = S6ylAwAAQBAJ&pg = PA79&lpg = PA79&dq = All + analytical + measurements + are + wrong:++it%27 s + just + a + question + of + how + large + the + errors + are, + and + whether + they + are + acceptable&source = bl&ots = CEaUXpIpA6&sig = ACfU3U1ElAgWh7C7M9U-kyLtLXn3VMca76A&hl = en&sa = X&ved = 2ahUKEwimw4KQsuPmAhUC7qwKHQQnAl0Q6AEwCnoECAgQAQ#v = onepage&q = All%20analytical%20measurements%20are%20wrong%3A%20%20it's%20just%20a%20question%20of%20how%20large%20the%20errors%20are%2C%20and%20whether%20they%20are%20acceptable&f = false .
https://www.amazon.com/Sampling-Monitoring-Management-Technologies-Influenced/dp/0873353552#
**128** [page 63.]

[. . .]

> 3.4.8 Sampling Theory and Errors
> The final results of an analytical determination will never be any more reliable than the reliability of the sampling. . . . sampling is the most important step . . . . Many modern analytical techniques have errors < 5%, whereas poor sampling can result in errors ≥100%. If a sample was not properly collected or is not representative of the target population, then it doesn't matter how much QA/QC is performed by the laboratory or during other aspects of the investigation.

[. . .]

**129** As an example, see a sampling pattern, reported in 2003, used to examine horticulturally used land in New Zealand:
**130** S.K. Gaw, N. Kim, A.L. Wilkins, "Contaminated horticultural land in the Auckland region: an historic legacy," *Chemistry in New Zealand*, 67(1), 55–59, 2003. At page 56: "*Figure 1.* 'Z' sampling pattern used on cropping areas – each aggregate sample contained 10 soil cores [depth = 7.5 cm] collected from an area # one hectare [=$10^{+4}$ $m^2$]."
https://shado-ns.massey.ac.nz/massey/expertise/profile.cfm?stref = 819040
https://nzic.org.nz/cinz
**131** See also:
- New Zealand Government, Ministry for the Environment, *Contaminated Land Management Guidelines No. 5 / Site Investigation and Analysis of Soils* (Revised 2011), Ministry for the Environment, Wellington, New Zealand, February 2004, Revised October 2011, ISBN: 978-0-478-37260-1, ME number: 1073. www.mfe.govt.nz
- *Land and Soil Monitoring: A guide for SoE and regional council reporting*, Land Monitoring Forum, New Zealand. 2009.

**132** For example: Mercury Hg pollution – Grassy Narrows, Ontario – Minamata disease.
**133** https://www.ontario.ca/page/mercury-contamination-english-and-wabigoon-rivers-near-grassy-narrows-wabaseemoong-independent-nations
https://www.cbc.ca/news2/interactives/children-of-the-poisoned-river-mercury-poisoning-grassy-nar rows-first-nation/

https://www.cbc.ca/news/canada/thunder-bay/grassy-narrows-youth-report-1.4931731
https://en.wikipedia.org/wiki/Ontario_Minamata_disease
**134** Lead, Pb [Xe] $4f^{14}5d^{10}6s^{2}6p^{2}$.
https://www.rsc.org/periodic-table/element/82/lead
https://www.ecowatch.com/coronavirus-plastic-use-2645882816.html
https://www.packagingnews.co.uk/features/history-of-the-world-in-52-packs-6-tooth paste-tubes-03-09-2015
http://castboolits.gunloads.com/archive/index.php/t-11118.html
https://www.sciencedirect.com/science/article/abs/pii/S0002817773620272
**135** https://en.wikipedia.org/wiki/Health_of_Vincent_van_Gogh
http://www.vggallery.com/visitors/summary.pdf
https://en.wikipedia.org/wiki/Vincent_van_Gogh
https://en.wikipedia.org/wiki/Theo_van_Gogh_(art_dealer)
**136** See:

- Katherine Kornei, "*Sure, the Romans Were Smart. But They Could Have Been Smarter* / Exposure to lead from mining probably lowered I.Q. levels in the empire, research has found. It might be the world's first case of widespread industrial pollution," *New York Times*, 06 Jan. 2025. https://www.nytimes.com/2025/01/06/climate/ancient-rome-lead-poisoning.html
- Prachi Patel, "Lead pollution in the air impacted all ancient Romans / Evidence from ice cores shows widespread lead pollution that may have caused IQ decline throughout the empire," *Chemical & Engineering News*, 06 January 2025.
- Joseph R. McConnell, Nathan J. Chellman, Andreas Plach, & Andrew I. Wilson, "Pan-European atmospheric lead pollution, enhanced blood lead levels, and cognitive decline from Roman-era mining and smelting," *Proc. Natl. Acad. Sci.* 2024, DOI: 10.1073/pnas.2419630121. https://doi.org/10.1073/pnas.2419630121

**137** https://www.theguardian.com/us-news/2025/apr/17/toothpaste-lead-heavy-metals
https://tributedentistry.com/lead-in-toothpaste-2025/#: ~:text = %F0%9F%A7%B4%20Did%20Toothpaste%20Tubes%20Used,made%20from%20lead%2Dtin%20alloy
**138** See, for example: Toronto Workers' Health & Safety Legal Clinic *newsletter*, "U of T Engineering Strategies & Practice course," June 2013, Vol.21 No 2, pages 3 & 4.
**139** http://s3.amazonaws.com/newsletter.workers-safety.ca/newsletters/2013%2006/2013%2006.%20Vol.21%20%20%20%20No.2.pdf
**140** "Lead Poisoning Poster Design," Team 142: Edward Liu, Rami Nessim, Xu (Dennis) Wu, Maria Haddad, Yufan (Faith) Jin
2013 04 18. Pb ESP Poster design.pptx
2013 04 18. Pb Disposal Presentation.pptx
**141** Alexander Aciman, "I Went Down a Lead-Testing Rabbit Hole. Here's What You Need to Know About Home Tests," 11 December 2024, *New York Times*, "HEALTH AND FITNESS / Wirecutter."
https://www.nytimes.com/wirecutter/reviews/home-lead-test-review/
https://www.amazon.ca/dp/B0CF6SV33S?ascsubtag = F0401JEVD9SPW03BT2W4KZGNPXB87&linkCode = gs2&tag = thewire0f-20&th = 1
**142** However, within the link to **Lumetallix**, there was apparently no scientific indication of how this process works. No references to:

- Product specifications that would reveal the analytical chemistry.
- Links to technical literature from the manufacturer or supplier.
- Scientific literature.
- Safety data sheets for the chemical contents of the marketed kit, nor the UV light – as might be required by US cfr for hazard communication or Canada WHMIS.
- US or other patent numbers.

**143** https://pmc.ncbi.nlm.nih.gov/articles/PMC4961898/
https://www.atsdr.cdc.gov/
https://www.mayoclinic.org/diseases-conditions/lead-poisoning/diagnosis-treatment/drc-20354723
**144** Jim Haner, "Victims of Lead Unnoticed by Courts/Justice System Fails to Track Poisoning, Despite Its Crime Effects; 'Not Even on the Radar'," *Baltimore Sun*, 08 October 2000.
**145** Julie Bosman and Jamie Kelter Davis, *"Milwaukee's Lead Crisis: Flaky Paint, Closed Schools and a C.D.C. in Retreat* / Some children were exposed to lead and investigators found flaking paint inside aging schools. Two federal experts, expected to help guide the response, have lost their jobs," *New York Times*, 18 April 2025.
https://www.nytimes.com/2025/04/18/us/milwaukee-schools-lead-cdc.html
**146** Erin Rode, "L.A. Officials Test Residents for Lead After Fires / People in and around the burn zone may have been exposed to the toxic chemical through ash from the wildfires," *New York Times*, 05 May 2025.
https://www.nytimes.com/2025/05/05/well/los-angeles-fires-lead.html
**147** Erica L. Green, *"A Legacy of Poisoned Water: 'Damaged Kids' Fill Flint's Schools," The New York Times*, 07 November 2019, pages A1, A14, A15.
**148** Kathleen Gray and Julie Bosman, "[Criminal] *Charges for 9 Officials* [including a former governor] *Give Flint* [Michigan] *Some Relief, but 'Trust Is Gone' / Pain and Anger Endure in the Water Crisis," The New York Times*, 15 January 2021, pages A1, A18.
**149** Ab Latif Wani, Anjum Ara, Jawed Ahmad Usmani, "Lead toxicity: a review," Interdiscip Toxicol. 2015 Jun; 8(2): 55–64. https://www.ncbi.nlm.nih.gov/pmc/articles/PMC4961898/
**150** See also: M.G., "Some possible chemical & physical exposures in veterans' health claims," Toronto Workers' Health & Safety *newsletter*, 2012 07. Vol.20 No.2, pages 1→4; *"lead* $82Pb^{207}$," at page 3: and references cited therein.
2012, 07. Vol. 20, No. 2. Doc.
http://newsletter.workers-safety.ca.s3-website-us-east-1.amazonaws.com/#!/newsletters%2F2012%2007%2F
**151** See Jay A. Siegel, "Forensic Identification of Controlled Substances" as Chapter 3 in Richard Saferstein, ed., *Forensic Science Handbook Volume II*, Prentice Hall, Englewood Cliffs, New Jersey, 1988; page 100; and that Chapter's appendices and references.
**152** https://en.wikipedia.org/wiki/Froehde_reagent
**153** https://dancesafe.org/product/mecke-reagent-testing-kit/
https://en.wikipedia.org/wiki/Marquis_reagent
**154** Siegel, Saferstein – see note above.
**155** Siegel, Saferstein – page 95, for suggested TLC systems.
**156** *Gedankenexperiment.* Introduction 2.2.
**157** – assuming here no fading or bleaching effects.
**158** *Gedankenexperiment.* Introduction 2.2.
**159** *Gedankenexperiment.* Introduction 2.2.
**160** Such photography might also include photo-microscopy to record the mechanical tracks left by the pen point; and photographs of the fluorescence images of fingerprints found on the paper. Chapter 6.
**161** Convenience of procedure may be dependent on availability of various lab setups.
**162** The method described here is necessarily partially destructive of other aspects of the evidence. A method recommended by Brunelle (page 713) uses the blunted point of a hypodermic needle to punch out several small plugs of the paper bearing the ink writing; this method is also likely more convenient; but care should be taken not to capture any ink on the other side of the page.

**163** R. L. Brunelle, in Richard Saferstein, ed., *Forensic Science Handbook* [Volume I], Prentice-Hall, Inc., Englewood Cliffs, New Jersey, 1982, Chapter 14, "Questioned Document Examination," particularly at pages 680, 7096711, 712→717, and references cited therein.
**164** It is important that the ink of the sample dissolve completely, to assure that its components have not gone into solution selectively. See in note below.
**165** Cyclohexane is chosen as the solvent because of its transparency in the UV–visible spectral region, and because as an organic solvent, the inks (also organic) are likely to dissolve in it. But cyclohexane is a non-polar organic, so that the solubility might not be too extensive. A better solvent to capture the inks in solution would be pyridine (recommended by R. L. Brunelle, in Richard Saferstein, ed., *Forensic Science Handbook* [Volume I], Prentice-Hall, Inc., Englewood Cliffs, New Jersey, 1982, Chapter 14, "Questioned Document Examination"; at page 713; but this polar organic solvent absorbs in the UV region. Pyridine UV absorption does not necessarily prevent the observation of a UV spectrum with pyridine as the solvent, but it makes it more difficult. The solvent absorption can be coped with by the match of the sample and blank cells. The difficulty with such a procedure is that if the match is off, the spectrum will be too; and keeping the match accurate becomes more difficult with increasing absorbance. A solvent/transparency compromise might be appropriate by using a pyridine/cyclohexane mixture (UV visible. Chapter 6) (Brunelle. Chapter 5)
**166** Cyclohexane. CAS 110-82-7 $C_8$ $H_{12}$ 84.162 g/mol.
https://en.wikipedia.org/wiki/Cyclohexane#: ~:text = Cyclohexane%20is%20a%20cycloalkane%20with, which%20it%20is%20sometimes%20used).
**167** https://webbook.nist.gov/cgi/cbook.cgi?ID = C110827&Mask = 400.
**168** Pyridine CAS 110-86-1 $C_5$ $H_5$ N 79.102 g/mol. https://en.wikipedia.org/wiki/Pyridine
**169** https://webbook.nist.gov/cgi/cbook.cgi?ID = C110861&Mask = 400
**170** Perhaps by simply letting them settle to the bottom of the flask and removing the liquid above with a syringe.
**171** See the above note re solvent choice.
**172** The use of a glass surface would allow for easy recovery of the ink, but also requires an assumption that there is no reaction with paper fibre, or selective retention by the fibre. It is also necessary that there be a waiting time to allow the ink's solvent(s) to evaporate.
**173** depending on whether or not the GC is set up for sample collection.
**174** See note above.
**175** See in Chapter 6.
**176** *Gedankenexperiment.* Introduction 2.2.
**177** For this lawyer, it would be practical that the "willing-fee-paying client" should actually be a "willing-wealthy-fee-paying client who has paid up-front," having committed in writing to understanding that the legal advice was that the case was desperate with a loser defence; with no legal fee refunds.
**178** Otherwise, up-front payment has important practical importance: If the lawyer goes on-record as defence counsel before the court without being paid, that lawyer is still counsel-of-record until and unless the judge says that removal would be allowed; and the judge may require a formal written application for that purpose; which application might not be allowed. The lawyer might be stuck because it is the judge who finally determines this.
**179** According to the first two rules of a fictional defence lawyer, in Beverley McLachlin, *PROOF*, Simon and Schuster Canada, Toronto M5A 1J3, 2024; ISBN 978-1-6680-0361-9; at page 63: – up-front payment and keeping a personal distance.
**180** *Gedankenexperiment.* Introduction 2.2.
**181** *Gedankenexperiment.* Introduction 2.2.
**182** And arguably, not only do the UV-visible spectra only serve as screening tests, but they also invite possible solubility problems – see notes above.

**183** Richard L. Brunelle and A.A. Cantu; in Geoffrey Davies, ed., *Forensic Science*, ACS Symposium Series 13, American Chemical Society, Washington, D.C., 1975, Chapter 14, "Ink Analysis – A Weapon Against Crime by Detection of Fraud," pages 134→141, and references cited therein; page 136.]

**184** Richard L. Brunelle and Robert Reed, *Forensic Examination of Ink and Paper*, Hardcover (June 1984) Charles C Thomas Pub Ltd; ISBN: 0398049351
https://www.amazon.ca/Forensic-Examination-Paper-Richard-Brunelle/dp/0398060398

**185** Then in the ATF library – now the US International Ink Library – with the many more.

**186** https://www.nist.gov/oles/forensic-database-questioned-documents-table
https://www.popularmechanics.com/technology/news/a13875/secret-service-ink-foren sics-library/
https://onlinelibrary.wiley.com/doi/abs/10.1111/j.1556-4029.2006.00144.x
https://www.ojp.gov/ncjrs/virtual-library/abstracts/forensic-analysis-blue-ballpoint-pen-inks-using-cap illary

**187** Hal Harris, "Forensic Chemistry of Inks," University of Missouri St. Louis, 2007. (Reference: Michael Quigley and Hongyan Qi, *J. Chem. Educ.* 1991 68(7) 597 and JCE Editorial Staff, JCE 2000 77(2) 176A&B)
https://www.umsl.edu/ ~ chickosj/chem11/Labmat/ForensicChemistryOfInks.pdf

**188** Michael N. Quigley and Hongyan Qi, "A chemistry whodunit: Forensic examination of pen inks," *J. Chem. Educ.*, 1991, 68 (7), p 596, DOI: 10.1021/ed068p596. https://pubs.acs.org/doi/abs/10.1021/ed068p596

**189** *Gedankenexperiment.* Introduction 2.2.

**190** for now. Perhaps in the future, instrumentation would become available to do this economically – perhaps GC coupled with IR. See in Chapter 6; and see Brian C. Smith, *Fundamentals of FourierTransform Infrared Spectroscopy*, CRC Press, Boca Raton, Florida, . . ., 1996, pages 167→172.

**191** J.A. Siegel, "Forensic Identification of Controlled Substances," as Chapter 3 in R. Saferstein, ed., *Forensic Science Handbook* Volume II, Prentice Hall, Englewood Cliffs, New Jersey, 1988, pages 87→92.

**192** *Duquénois-Levine*, Siegel, Saferstein, page 124.

**193** See also: K. Bailey, "The value of the Duquénois Test for Cannabis – a Survey," *Journal of Forensic Sciences*, 1979, 24(4), 817→841.

**194** $<10^{-9}$ = less than $10^{-9}$ = less than 0.000000001

**195** 1 mole = mass /molar mass

**196** See: Isaac Asimov, *ASIMOV ON CHEMISTRY*, Anchor Books, Doubleday, Garden City, New York, 1975; ISBN 0-385-04005-9; Chapter 6 "To Tell a Chemist," page 89, *et seq.* (Originally appearing as an essay in *The Magazine of Fantasy and Science Fiction*, May 1965.)

**197** Avogadro number = $N_0$ = 6 . 022 140 76 × $10^{+23}$ molecules/mol

**198** 1 mole = 6 . 022 140 76 × $10^{+23}$ particles.

**199** https://www.britannica.com/science/Avogadros-number
https://en.wikipedia.org/wiki/Avogadro_constant

**200** Mass measurement (grams). Accuracy and precision too extensive? Chapters 3.1.1, 3.7.3; 6.20.

**201** Walter E. Harris, *Low Dose Risk Assessment*, Wuerz Publishing Ltd.,Winnipeg,1997; ISBN 0-920063-73-X; page 9, *et seq.*

**202** Even when poppy-seed-containing muffins were served at an approved half-way house – a situation reported to legal counsel who consulted the author. Parole was lost.

**203** http://time.com/5360817/poppy-seed-bagel-drug-test/
https://thetakeout.com/can-poppy-seeds-really-make-me-fail-a-drug-test-1832874359

**204** US *Federal Register*, Vol. 73, No. 228, 25 November 2008, Notices, page 71,858, *et seq.*; Department of Health and Human Services, Substance Abuse and Mental Health Services Administration, Mandatory Guidelines for Federal Workplace Drug Testing Programs.
https://www.govinfo.gov/content/pkg/FR-2008-11-25/pdf/E8-26726.pdf

**205** at page 71,900:

> [. . .]
>
> [Section 13.4] (d) When an HHS-certified laboratory reports a positive result for opiates on the primary (Bottle A) urine specimen, the MRO must determine that there is clinical evidence in addition to the urine test result of illegal use of any opium, opiate, or opium derivative (*e.g.*, morphine/codeine) listed in Schedule I or II of the Controlled Substances Act. However, this requirement does not apply if the laboratory confirms the presence of 6-acetylmorphine (*i.e.*, the presence of this metabolite is proof of heroin use) or the morphine or codeine concentration is equal to or greater than 15,000 ng/mL and the donor does not present a legitimate medical explanation for the presence of morphine or codeine at or above this concentration. Consumption of food products must not be considered a legitimate medical explanation for the donor having morphine or codeine at or above this concentration.
>
> [. . .]

**206** A.D. Fraser, J. Zamecnik, J. Keravel, L. McGrath, J. Wells, "Experience with urine drug testing by the Correctional Service of Canada," *Forensic Science International*, Volume 121, Issues 1→2, 15 September 2001, Pages 16→22. https://www.sciencedirect.com/science/article/abs/pii/S0379073801004479
**207** See also: Jan Ransom, "Inmates Received Harsh Punishments Over False Positives for Drugs," *The New York Times*, 21 November 2019, page A22.
**208** And fecal matter traces.
**209** https://www.thrillist.com/entertainment/nation/legend-of-cocaine-island-netflix-review
https://www.rottentomatoes.com/m/the_legend_of_cocaine_island
https://www.gq.com/story/the-great-cocaine-treasure-hunt
**210** Jonathan Oyler, William D. Darwin, and Edward J. Cone, Addiction Research Center, National Institute on Drug Abuse, National Institutes of Health, Baltimore, Maryland 21224, "Cocaine Contamination of United States Paper Currency," *Journal of Analytical Toxicology*, Vol. 20, July/ August 1996.
https://academic.oup.com/jat/article/20/4/213/838490
**211** https://www.cnn.com/2009/HEALTH/08/14/cocaine.traces.money/
https://www.marketwatch.com/story/this-is-exactly-how-often-cocaine-and-feces-show-up-on-your-dollar-bills-2017-07-11
https://en.wikipedia.org/wiki/Contaminated_currency https://www.theguardian.com/world/2009/aug/17/cocaine-dollar-bills-currency-us
https://www.snopes.com/fact-check/cocaine-on-money/
**212** https://www.cnn.com/2009/HEALTH/08/14/cocaine.traces.money/
https://en.wikipedia.org/wiki/Contaminated_currency
**213** Blood Drug Concentration Regulations: SOR/2018-148
http://gazette.gc.ca/rp-pr/p2/2018/2018-07-11/html/sor-dors148-eng.html

> [. . .]
>
> **Blood Alcohol Concentration and Blood Drug Concentration**
>
> **Summary offence**
>
> 1 For the purpose of paragraph 253(3)(b) of the *Criminal Code*, the prescribed blood drug concentration for tetrahydrocannabinol (THC) is 2 ng of THC per mL of blood.
>
> **Hybrid offence – drugs**
>
> 2 For the purpose of paragraph 253(3)(a) of the *Criminal Code*, the prescribed blood drug concentration for each drug set out in column 1 of Tab. 3.1 to this section is set out in column 2.

Tab. 3.1: Blood drug concentration regulations.

| Item | Column 1 drug | Column 2 concentration |
|---|---|---|
| 1 | Tetrahydrocannabinol (THC) | 5 ng/mL of blood |
| 2 | Lysergic acid diethylamide (LSD) | Any detectable level |
| 3 | Psilocybin | Any detectable level |
| 4 | Psilocin | Any detectable level |
| 5 | Phencyclidine (PCP) | Any detectable level |
| 6 | 6-Monoacetylmorphine | Any detectable level |
| 7 | Ketamine | Any detectable level |
| 8 | Cocaine | Any detectable level |
| 9 | Gamma hydroxybutyrate (GHB) | 5 mg/L of blood |
| 10 | Methamphetamine | Any detectable level |

**214** LSD CAS 50-37-3 https://en.wikipedia.org/wiki/Lysergic_acid_diethylamide (see Fig. 3.1).

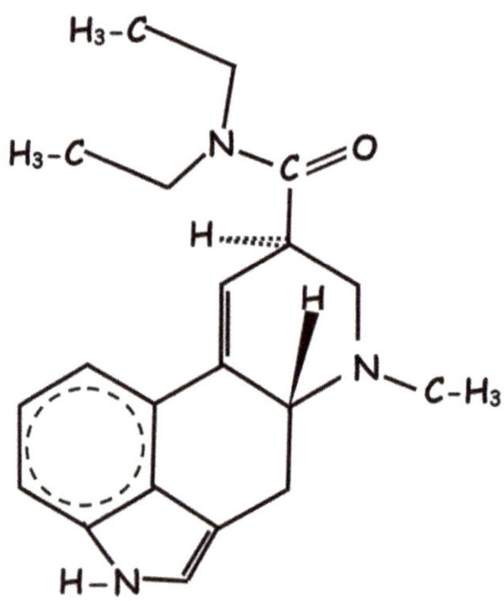


Fig. 3.1: LSD CAS 50–37-3.

Lysergic_acid_diethylamide *LSD* CAS 50-37-3 323.44 g/mol. $C_{20}H_{25}N_3O$.
**215** Methamphetamine CAS 537-46-2 149.233 g/mol (see Fig. 3.2a).
**216**

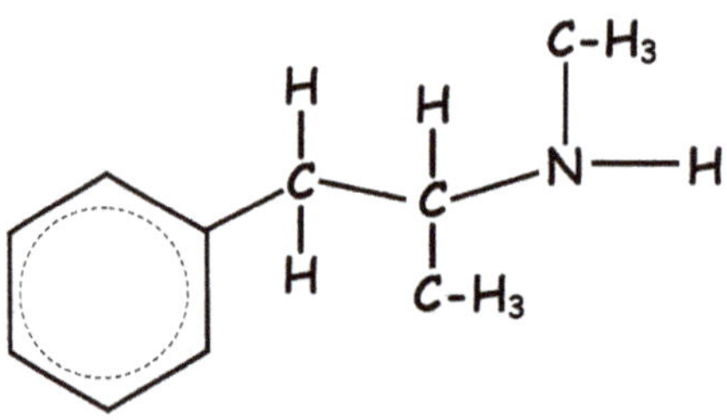


Fig. 3.2a: CAS 537-46-2.

| | | | | |
|---|---|---|---|---|
| $C_{10}$ | 10 × 12.011 | = | 120.11 | |
| $H_{15}$ | 15 × 1.008 | = | 15.12 | |
| N | 1 × 14.007 | = | 14.01 | |
| | Σ | = | 149.24 | g/mol. |

**217** Methamphetamine hydrochloride CAS 51-57-0 185.694 g/mol. mp = 171–175 °C (see Fig. 3.2b).

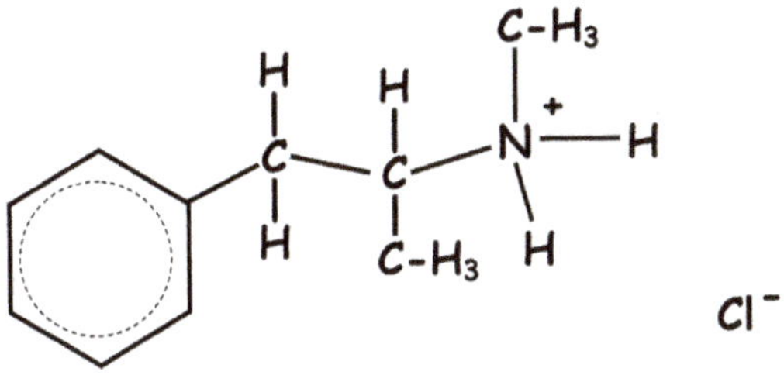

Fig. 3.2b: CAS CAS 51-57-0.

**218** γ-Hydroxybutyrate GHB ($C_4H_8O_3$) 104.10 g/mol CAS 591-81-1 (see Fig. 3.3) [γ = Gamma]

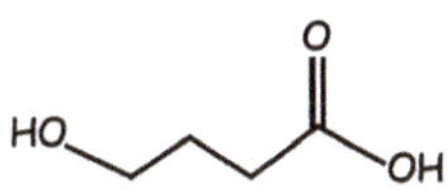

Fig. 3.3: GHB.

https://www.drugs.com/illicit/ghb.html https://en.wikipedia.org/wiki/Gamma-Hydroxybutyric_acid
**219** See Daniel M. Perrine, *The Chemistry of Mind-Altering Drugs / History, Pharmacology, and Cultural Context*, American Chemical Society, Washington, DC, 1996; ISBN 0-8412-3253-9.
**220** Perrine, page 193, *et seq.* – at pages 194 and 196 4–28
**221** http://en.wikipedia.org/wiki/Methamphetamine
**222** US National Institute of Standards and Technology, Chemical Science and Technology Laboratory http://webbook.nist.gov/cgi/cbook.cgi?ID = 537-46-2
**223** https://www.cas.org/cas-data/cas-registry
**224** https://jpdb.nihs.go.jp/jp14e/14data/Part-I/Methamphetamine_Hydrochlori.pdf
**225** *Gedankenexperiment*. Introduction 2.2.
**226** But, this is not to argue that very stringent control of nuclear materials is not appropriately important; government surveillance authorities should be overly cautious. See, for example:

- *Guardian* [Australia],"Sydney 'science nerd' may face jail for importing plutonium in bid to collect all elements of periodic table / Emmanuel Lidden, 24, to learn fate after breaching nuclear non-proliferation laws by shipping samples of radioactive material to parents' suburban home," Australian Associated Press, 21 Mar 2025.
  https://www.theguardian.com/australia-news/2025/mar/21/emmanuel-lidden-sydney-science-nerd-importing-plutonium-ntwnfb
- David Southwell, Australia, "His package delivery locked down his street as special agents in hazmat suits swooped. Now he will be the first person ever to be sentenced under a decades-old law . . ." *Daily Mail*, 14 December 2024 .

https://www.dailymail.co.uk/news/article-14193785/Emmanuel-Steven-Lidden-plutonium-arncliffe.html

- https://www.luciteria.com/element-cubes/plutonium-for-sale

[. . .]

> It may seem like a joke at first but, yes, this is actual plutonium for sale . . . .
> with a big asterisk (read on)

[. . .]

> *This item is now discontinued.*

[. . .]

https://www.luciteria.com/?srsltid = AfmBOoo4C1cJDBHpLlAUmlPuA3awtBFR2xhMkCtiMHdFZB5tB5_1jnOs

https://www.luciteria.com/elements-for-sale

- https://en.wikipedia.org/wiki/Plutonium

In addition to being of national security concern, and radioactive, Plutonium Pu, Uranium U and Thorium Th are quite toxic otherwise.

**227** André Picard, *The Gift of Death/Confronting Canada's Tainted Blood Tragedy*, HarperCollins, Toronto M5R 3L2, 1995; ISBN: 0-00-255415-1.

https://www.harpercollins.ca/9780006385752/the-gift-of-death/

**228** *Final report. Commission of Inquiry on the Blood System in Canada*, 1997.

http://epe.lac-bac.gc.ca/100/200/301/hcan-scan/commission_blood_final_rep-e/

https://en.wikipedia.org/wiki/Royal_Commission_of_Inquiry_on_the_Blood_System_in_Canada

# Chapter 4
# Uniqueness of matter

[. . .][1]

"Now, if you'll only attend, Kitty, and not talk so much, I'll tell you all my ideas about Looking-Glass [1][2] House. First, there's the room you can see through the glass – that's just the same as our drawing room, only the things go the other way . . . . Well then, the books are something like our books, only the words go the wrong way . . ."

"How would you like to live in Looking-Glass House, Kitty? I wonder if they'd give you milk in there? Perhaps Looking-Glass milk isn't good to drink . . ."

[. . .]

[Alice & Kitty & molecular chirality.] [2, 3, 4][3]

. . . *"Let there be light." And there was light.* יְהִי אוֹר; וַיְהִי-אוֹר . . .

[*Genesis* בְּרֵאשִׁית : electromagnetic waves, photons and spectroscopy.] [5, 6, 7][4]

*ANYONE WHO IS NOT SHOCKED BY QUANTUM THEORY HAS NOT UNDERSTOOD IT.*

[Niels Bohr] [8, 9, 10, 11]

*I don't like it, and I'm sorry I ever had anything to do with it.*

[Erwin Schrödinger] [12, 13, 14, 15][5]

*I think I can safely say that nobody understands quantum mechanics.*

[Richard Feynman, American physicist, 1967.] [16]

*Nothing is real.*

[John Lennon] [17, 18, 19, 20, 21, 22, 23][6]

*You believe in a God who plays dice, and I in complete law and order.*

[Albert Einstein, Letter to Max Born] [1944] [24, 25, 26]

. . . *Pepper spray is not a chemical irritant. It's not chemical. . . .*

[US Attorney General William Barr, 2020] [27, 28, 29][7] [see Fig. 4.1]

---

**1** Introductory quotes. Introduction 1.2.

**2** = Mirror.

**3** Alice's Kitty's Looking-Glass milk. Chirality. Chapter 4 intro. quote. Chapters 4 and 8.

**4** *Genesis*. light. EM radiation. Photon. UV. IR. Microwave. Chapters 3, 4 and 6.

**5** Schrödinger's cat. Introduction 2; Chapter 4.

**6** Beatles, Strawberry Fields. quantum mechanics. Chapter 4 intro. quote.

**7** molecular structural drawing. CAS. Chapters 4.4 and 4.6.

https://doi.org/10.1515/9783111618555-005

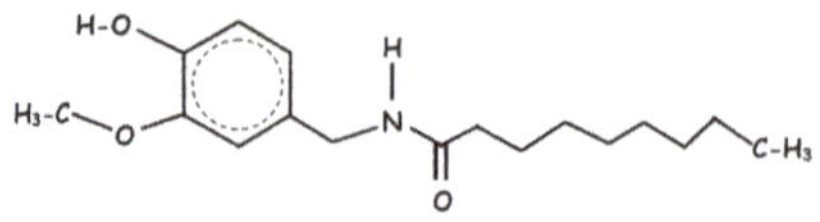


Fig. 4.1: CAS 2444-46-4.

# 4.1 Uniqueness of matter and chemical science

## 4.1.1 Questions about the uniqueness of matter

Questions in forensic science about the nature and uniqueness of matter essentially ask how to describe and define what a substance is, and if one substance is the same as another and what variations are allowed to still be considered same.[8,9,10]

Also, importantly, if necessary, forensic science is called on to characterize a sample on the basis of mixture proportions of its components.[11]

## 4.1.2 Chemical science

Chemical science is the study of the nature and identity of matter [30], the reactions of various kinds of matter with other kinds, interactions of energy with matter, related electromagnetic radiation, particle thermal collisions, related energy and entropy changes. That study, when tried to be taken to the fundamentals of modern physics, can lead to some very strange concepts for those of us who persist in our everyday world, [31, 32, 33, 34, 35, 36, 37, 38, 39, 40, 41, 42, 43, 44][12,13,14] including that all physical matter of personal human experience is mostly empty space [45].

Fortunately, for this book, it is sufficient to consider that matter is composed of a hierarchy that starts with a few kinds (neglecting here, others of modern physics) of very small (submicroscopic) particles [46, 47, 48][15,16,17] that are the "building blocks"[49]

8 ". . . what variations are allowed to still be considered same." – Yes, this is self-contradictory in a strictly literal sense. But please read on.

9 What a substance is. Mixture proportion of components. Chapter 4.

10 Chemical substances described. Chapters 3 and 4.

11 What a substance is. Mixture proportion of components. Chapter 4.

12 See introductory quotes.

13 Schrödinger's cat. Introduction 2; Chapter 4.

14 Astrology & occult. crystal ball. Quantum mechanics. Governor-General. Prime Minister Mackenzie King. Harry Houdini. Chapters 4 and 10; Chapter 6.26.

15 Fundamental particles. Chapters 4 and 7.

16 Submicroscopic particles: $p^+$, electrons $e^-$ and neutrons n. Chapters 4 and 7.

17 Subtle variations. Chapter 4.

for not quite-as-small things, which are, in turn, further larger "building blocks," and so on, until there are items of human size experience. And further on, to items of planetary and cosmic dimensions [50].

## 4.2 Atoms, molecules, etc.

### 4.2.1 Atoms, molecules, etc. – hierarchy

Further, fortunately, for both the author and the reader, forensic chemistry is usually concerned only with some parts of this hierarchy[18]:

- ***Atoms***, which are assemblies of very small particles, are the "building blocks" for –
- ***Molecules*** and *crystal structures*, which are assemblies of atoms.

Large numbers of *atoms* and *molecules* can, in bulk, be in various *phases*[19]:

- *Gases* – that can be in containers of variable shape and volume, and able to flow.
- *Liquids* – that can be in containers of variable shape, but of definite volume within the container, and able to flow.
- *Solids* – of definite shape and volume, and generally not able to flow.
- Other phases, under various conditions of pressure and temperature that need not be considered here [51].

Within these parts of the hierarchy, there are also various possibilities, so that:

- *Atoms* can have variation of *isotope*, some radioactive, some not.[20]
- *Molecules* can have variation of *isotope* for the atoms within their structures; and can have *isomeric* variations of structure that may be neglected for some purposes, but not others.
- *Solids* are assemblies of molecules, atoms, or *ions*, into groupings of definite bulk volume and shape, and which can have different internal arrangements (such as for different crystal structures) and linkages (hydrogen bonding and polymerization). Some atoms and smaller molecules can be linked in long lines (and sometimes planes). These *polymers*, often mostly of carbon and hydrogen, can be very large, with molar masses (g/mol) in the thousands. (In common language, many of these might be called plastics – although in more precise scientific language, there is an important distinction between thermoplastic polymers and thermoset polymers).[21] There are very many naturally occurring biopolymers, but also very many polymers of forensic interest are synthetic, often petroleum-derived.

---

**18** Too much scientific technical detail here? Introduction 2.1.

**19** Phases. Melting points. Chapters 4 and 6.

**20** Isotope. Radioactive. Carbon-14. Chapters 4, 6, 7.

**21** Polymers. Chapter 4.

- *Atoms and molecules* can be in *ionic* form – having additional or fewer electrons so that they carry an electrical charge. A molecule might also be called radical [52], if it temporarily has an extra electron that is part of a chemical reaction mechanism [53].
- *Solutions* are volumes of liquids that contain *atoms, molecules, ions, etc.* as *solutes* at various concentrations, in various *solvents*. For example, common edible table salt [54] = $Na^+Cl^-$ is a solid ionic crystal matrix of sodium $Na^+$ and $Cl^-$. *Brine* = as soluble in water $H_2O$, as $Na^+$ and $Cl^-$, is an ionic *solution*. An example of an organic nonionic solution: *o*-xylene in *n*-heptane[22,23] (see Fig. 4.2a and b) [55].[24]

Solution concentrations are typically expressed as *moles/litre* or *grams/litre*. Much of practical and experimental chemistry is performed and observed as for solutes in solution – often quite dilute. One theoretical view of solutes, in dilute solution, is to consider them analogous to gas particles in a vacuum (the solvent assumed a kind of nonexistent for this hypothetical purpose).

A)

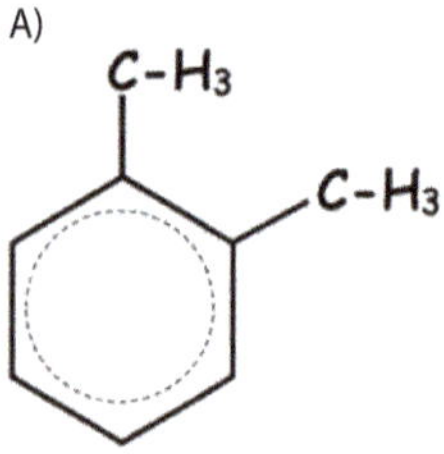


B)

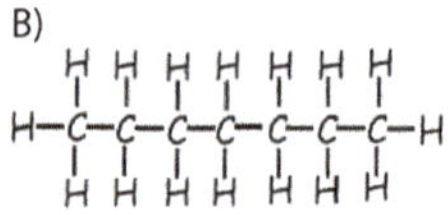

Fig. 4.2: (a) *o*-Xylene. (b) *n*-Heptane.

## 4.2.2 Energy relationships

The atoms, molecules and other assemblies of very small particles can interact with electromagnetic radiation; EM is often considered as photons [56, 57, 58, 59, 60, 61, 62,

22 Molecular structural drawings. Chapter 4.

23 Chemical names as used in this book. Introduction 3.2; Chapters 3 and 4.

24 *o*-Xylene. 106.17 g/mol. Chapters 3.7.1, 4.2.1 and 4.5.1; Figs. 4.2a and 2.10.

63, 64, 65, 66, 67, 68, 69, 70, 71, 72, 73, 74, 75, 76].[25] This involves energy relationships. EM absorption and emission is observable and measurable with spectroscopic equipment. Spectroscopy is of essential [77] importance in most every branch of chemistry – further described in Chapter 6.[26,27] The EM continuum is usually described according to approximate categories [78] (see Tab. 4.1).

Tab. 4.1: Electromagnetic spectral regions.

| | | nm | 1 nm = $10^{-9}$ m |
|---|---|---|---|
| Gamma ray | γ | ~0.001 | |
| X-ray | | ~0.1 | |
| Ultraviolet | = UV | ~200 | Electronic |
| Visible | | ~400 | Electronic |
| Infrared | = IR | ~800 | Vibrational |
| Microwave | | ~10,000,000 | Rotational |
| Radio | | ~1,000,000 000,000 | |

Energy interactions can also involve thermal collisions and vibrations of atoms and molecules.

Thermodynamics is also of essential importance in every branch of chemistry, and every chemical reaction, solvation, change and interaction can be thought of as a thermodynamic process.[28] However, although an important theoretical underpinning, it is often not of immediate practical concern to forensic science practice – except for temperature.

### 4.2.3 Periodic table

Although it may not too often be explicitly referred to for every-day practical use in relation to much of forensic science, mention of the *Periodic Table of the Elements* is appropriately necessary.[29,30] Each type of *atom* is called an *element*. All elements are to be found on the modern periodic table [79, 80, 81, 82, 83, 84, 85, 86, 87, 88, 89, 90, 91, 92][31] (see Fig. 4.3).

This is "periodic" because its arrangement has trends of properties that sometimes can be simply shown. The present published version has 118 elements [93], the

25 Andlmu, "EM-Wave," Wikimedia Commons. Copyright acknowledgement; Chapter 4.
26 *Genesis*. light. EM radiation. Photon. UV. IR. Microwave. Chapters 3, 4 and 6.
27 EM. Photons. Chapters 2 and 4.
28 Energy. Thermodynamics. Entropy. "Time's Arrow." Chapters 4; 6; 7.
29 Too much scientific technical detail here? Introduction 2.1.
30 Plutonium. Periodic Table. Too sensitive? Chapters 1 and 10; Chapters 3.7.2 and 4.2.3.
31 Periodic table. Chart of nuclides. Chapters 4 and 7.

| | | | | | | | | | | | | | | | | | |
|---|---|---|---|---|---|---|---|---|---|---|---|---|---|---|---|---|---|
| 1 H | | | | | | | | | | | | | | | | | 2 He |
| 3 Li | 4 Be | | | | | | | | | | | 5 B | 6 C | 7 N | 8 O | 9 F | 10 Ne |
| 11 Na | 12 Mg | | | | | | | | | | | 13 Al | 14 Si | 15 P | 16 S | 17 Cl | 18 Ar |
| 19 K | 20 Ca | 21 Sc | 22 Ti | 23 V | 24 Cr | 25 Mn | 26 Fe | 27 Co | 28 Ni | 29 Cu | 30 Zn | 31 Ga | 32 Ge | 33 As | 34 Se | 35 Br | 36 Kr |
| 37 Rb | 38 Sr | 39 Y | 40 Zr | 41 Nb | 42 Mo | 43 Tc | 44 Ru | 45 Rh | 46 Pd | 47 Ag | 48 Cd | 49 In | 50 Sn | 51 Sb | 52 Te | 53 I | 54 Xe |
| 55 Cs | 56 Ba | 57* La | 72 Hf | 73 Ta | 74 W | 75 Re | 76 Os | 77 Ir | 78 Pt | 79 Au | 80 Hg | 81 Tl | 82 Pb | 83 Bi | 84 Po | 85 At | 86 Rn |
| 87 Fr | 88 Ra | 89** Ac | 104 Rf | 105 Db | 106 Sg | 107 Bh | 108 Hs | 109 Mt | 110 Ds | 111 Rg | 112 Cn | 113 Nh | 114 Fl | 115 Mc | 116 Lv | 117 Ts | 118 Og |

| | | | | | | | | | | | | | | |
|---|---|---|---|---|---|---|---|---|---|---|---|---|---|---|
| * | 58 Ce | 59 Pr | 60 Nd | 61 Pm | 62 Sm | 63 Eu | 64 Gd | 65 Tb | 66 Dy | 67 Ho | 68 Er | 69 Tm | 70 Yb | 71 Lu |
| ** | 90 Th | 91 Pa | 92 U | 93 Np | 94 Pu | 95 Am | 96 Cm | 97 Bk | 98 Cf | 99 Es | 100 Fm | 101 Md | 102 No | 103 Lr |

Fig. 4.3: Modern periodic table.

heaviest of which, newly discovered, created in modern nuclear reactors, are unstable, with very short lifetimes. Most other elements are stable, or with longer lifetimes.

Every aspect of the study of chemistry involves *atoms* – defined by their number of *protons* = number of *electrons* = *atomic number*. The number of *neutrons* defines the various *isotopes* of an atom.[32]

The arrangement of atoms of the periodic table may be theoretically explained by starting with Hydrogen H, the simplest (top left on the table) – with only one proton, one electron and no neutrons; atomic number = 1. All other elements are describable by adding (going to the right, and down the rows), in theoretical concept, one by one, protons with equal numbers of electrons. There are various numbers of neutrons. The three-dimensional geometric probabilities of electronic structures, outside the proton- and neutron-containing nucleus, are predictable (at least in approximation) from quantum mechanical considerations, and can be used to try to account for the table's periodic arrangement. The arrangements of electrons are of essential interest to chemists. Much of the modern understanding of chemistry is related to these concepts.

The periodic table is a most important accomplishment of science – and of human civilization. It was substantially advanced with the nineteenth century work of Dimitri Mendeléev [94, 95], well-before the discovery of the neutron, the modern concept of atomic number and quantum mechanics. Mendeléev, was able to predict the existence of then-yet-to-be-discovered elements. Perhaps a modern – too whimsical – speculation is that the periodic table is so universal that humans and intergalactic aliens (if ever existent, if ever would meet) could recognize each other's periodic tables [96].

There is also a *Chart of Nuclides* that lists, with data, the known isotopes [97, 98].[33]

## 4.3 Molecules

Although there must be no absolute exclusion for other parts of the hierarchy, it is probably safe to say that forensic chemists are largely concerned with identification of molecules in their various forms, including ionic structures, and with mixtures of different molecules and ions. It is therefore appropriate to consider the theoretical characterization of molecules in some more (but not too much) detail.

**32** Neutron. Atomic Number. Chapters 2, 4 and 7.

**33** Periodic table. Chart of nuclides. Chapters 4 and 7.

## 4.4 Molecular structure

A chemical substance with a molecular structure may be said to be uniquely defined as a particular arrangement of its atoms and chemical bonds in three-dimensional space [99, 100].[34,35]

For a chemical compound, its component atoms are said to be connected by chemical bonds of various lengths and strengths [101]. Such bonding may be regarded as the various sharings of electrons by atoms, with the atoms being held in relative place ranges by physical forces. This can happen in several different ways that are not of immediate concern for purposes here.

**A chemical's uniqueness is defined and described with its molecular structural drawing** [102]. [36] A chemist tries to use this to explain the various properties of molecules.

A simple example:

*Methane* [103, 104], CAS *Registry Number* 74-82-8,[37] is a central *carbon* C atom *single bonded* to four *hydrogen* H atoms – equiangular = 109½° bonds [105] – the tetrahedral bond angle [106]. The C–H single bond is a representation of a sharing of an electron pair, C:H, between the two atoms [107]. A molecular model may be used to artificially visualize *methane* (see Fig. 4.4) [108, 109].

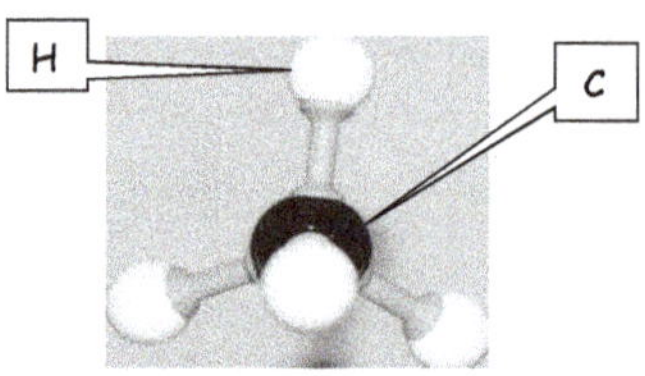


Fig. 4.4: Methane molecular model.

For convenience, a chemist might reduce this three-dimensional representation (see Fig. 4.5) [110, 111, 112, 113, 114][38,39,40] to symbols on the two-dimensional plane of a printed page, simplified as shown in Fig. 4.6.

And, a still further simplification, as shown in Fig. 4.7, makes printing on the plane of a page – as here – reasonably convenient.

34 Too much scientific technical detail here? Introduction 2.1.

35 Molecular structure. Chapters 1, 3, 4 and 6.

36 molecular structural drawing. CAS. Chapters 4.4 and 4.6.

37 SI CAS. Introduction 3.2; Chapters 3 and 4.

38 The exaggerated wedge-shaped lines here are to try to give an image of receding from, or coming toward, the viewer who would be facing the plane of the page.

39 Where the dashed (exaggerated wedge-shaped) lines here represent the hydrogen as below the plane of the page, with the carbon in the plane of the page.

40 Where the solid (exaggerated wedge-shaped) lines here represent the hydrogen as above the plane of the page, with the carbon in the plane of the page.

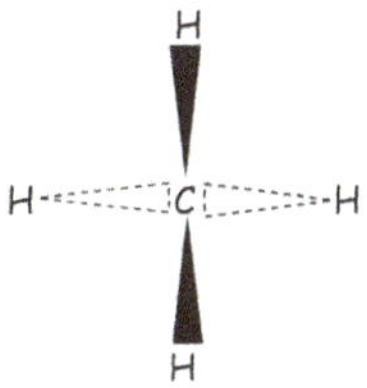

Fig. 4.5: Three-dimensional drawing of methane.

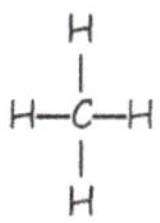

Fig. 4.6: Two-dimensional drawing of methane.

$CH_4$

Fig. 4.7: Simplified two-dimensional drawing of methane.

But it must be remembered that all these are short-hand drawings for three-dimensional assemblies. They are schematic representations, rather than actual pictures of a "true" reality.[41]

Molecules are submicroscopically small. For example, the C–H *single-bond* shown above (called a σ – *sigma* bond)

= 1.091 Ångström units [115, 116, 117, 118, 119]
= 0.00000001091 m

And there are very many of them. For example, 16 g of *methane* = 1 *mol* [120, 121][42] contains

= $6.02214076 \times 10^{23}$ molecules = Avogadro's number molecules/mol [122, 123, 124, 125][43]
= 602,214,076,000,000,000,000,000.

For purposes here, the reader need not be too concerned about the intricacies of molecular structures[44]; rather, it would be sufficient to realize that there are molecular structures and that their even subtle differences result in unique CAS *Registry Numbers*[45] that can appear on legal lists.

However, benzene[126, 127, 128, 129] CAS 71-43-2, and *benzenoid* compounds (also called *phenyl*) are so commonly found and important, that it is worthwhile to show the symbols here. *Benzene* is considered as a hexagonal [130] planar ring of *carbon* atoms with attached *hydrogens* (see Fig. 4.8).

**41** Molecular structural drawings. Chapter 4.
**42** Mole. Avogadro's number. Chapters 3, 4 and 7.
**43** Avogadro's number. Chapters 3, 4 and 7.
**44** Too much scientific technical detail here? Introduction 2.1.
**45** SI CAS. Introduction 3.2; Chapters 3 and 4.

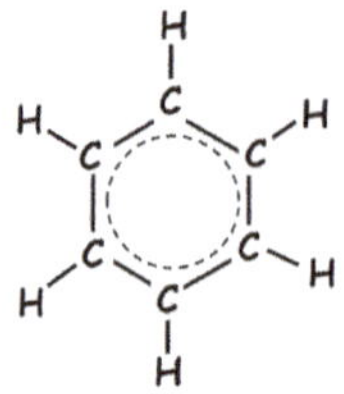

Fig. 4.8: Benzene.

The circular dotted line shown here inside and alongside the -C-C- single (σ – *sigma*) [131] bonds represents another kind of chemical bonding (π – *pi*) [132, 133, 134]. Benzene is often simply symbolized as in Fig. 4.9 [135, 136, 137, 138, 139, 140, 141, 142].[46,47]

Fig. 4.9: Simplified drawing of benzene.

## 4.5 Subtle variations

### 4.5.1 Isomers, congeners, *etc.*

Arrangements of atoms into molecules can have infinite variations,[48,49] and with the same or similar number of atoms, some molecules can share very similar properties. Some molecules can have the same number of the same atoms, but arranged differently – isomers, for example (see Figs. 4.10, 4.11 and 4.12) [143, 144, 145, 146].[50,51]

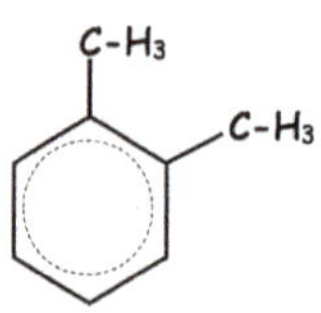

Fig. 4.10: *o*-Xylene, CAS 95-47-6.

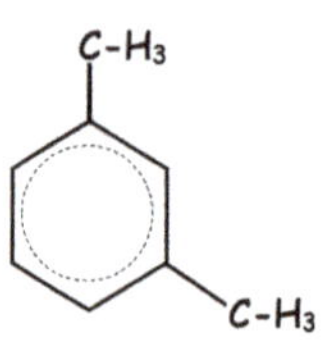

Fig. 4.11: *m*-Xylene, CAS 108-38-3.

46 Naphthalene. Double bond? CAS 91-20-3. Chapters 4 and 8.
47 molecular structural drawing. CAS. Chapters 4.4 and 4.6.
48 Too much scientific technical detail here? Introduction 2.1.
49 Identification of molecules. *CD&S Act.* Chapters 4 and 10.
50 *o*- = *ortho*-, *m*- = *meta*-, *p*- = *para*-Xylene. 106.17 g/mol. Chapters 3 and 4.
51 *o*-Xylene. 106.17 g/mol. Chapters 3.7.1, 4.2.1 and 4.5.1; Figs. 4.2a and 2.10.

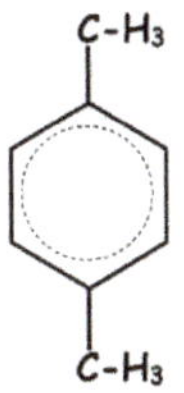


Fig. 4.12: *p*-Xylene, CAS 106-42-3.

Other molecules are related, so that they contain similar numbers of atoms with slightly different arrangements – ***congeners*** [147, 148], for example[52,53] see Fig. 4.13a and b [149, 150, 151, 152, 153, 154, 155, 156].[54]

A)

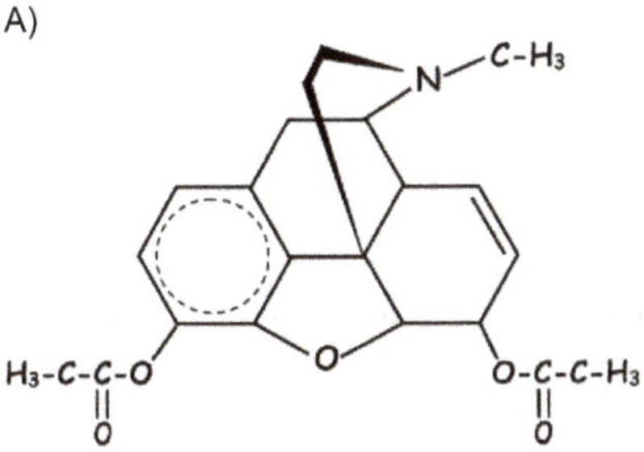


B)

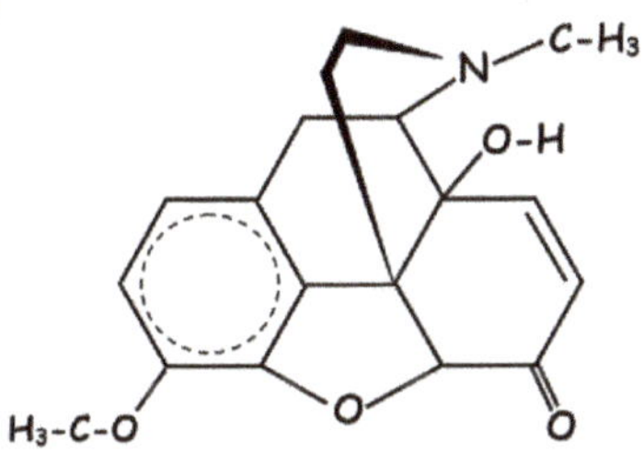


Fig. 4.13: (a) Heroin, CAS 61-27-3. (b) Oxycodone, CAS 76-42-6.

And, for example see Fig. 4.14a [157, 158] and 4.14b [159, 160, 161, 162].

Another example is two allotropes of the element phosphorus. An *allotrope* is something like an isomer or congener but refers to the arrangement of atoms, all of the same element [163]. For example: white and red phosphorus have different properties and very different toxicities. The less toxic red was eventually used in the manufacture of matches – preventing the development of "phossy jaw" in workers. General William Booth, founder of the Salvation Army, pioneered the use of red, rather than white, in its factory. Matches made with white phosphorus were specifically

52 *Heroin*. Bail. CAS 1502-95-0 CAS 561-27-3. Chapters 4 and 5.

53 *Oxycodone*. CAS 76-42-6. Chapter 4.

54 Molecular structural drawing. CAS. Chapters 4.4 and 4.6.

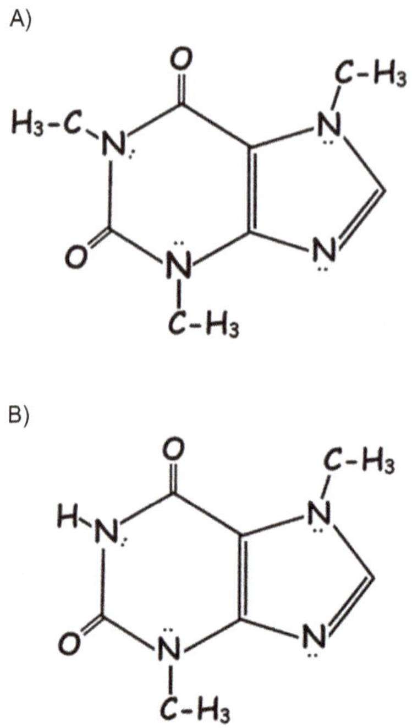


Fig. 4.14: (a) Caffeine, CAS 58-08-2, $C_8H_{10}N_4O_2$, 194.194 g/mol. (b) Chocolate = *Theobromine* $C_7H_8N_4O_2$, CAS 83-67-0, 180.167g/mol.

taxed in many countries to eliminate its use in favour of red [164, 165, 166, 167, 168, 169, 170, 171, 172, 173, 174, 175, 176, 177].

### 4.5.2 Chirality

For some molecules, there can also be subtle variation of molecular three-dimensional structure (see Fig. 4.15) [178, 179, 180].[55]

The concept of *chirality*: The nonsuperimposable mirror image of a molecule is a *different* molecule, albeit with a nearly identical repertoire of properties [181, 182, 183, 184, 185, 186, 187, 188, 189, 190, 191, 192, 193, 194]. This is usually not of much concern for forensic analytical chemistry, but is an important concept when considering biological systems, including for pharmaceuticals and drugs of abuse and recreation – and Alice's Kitty's Looking-Glass milk.[56]

This difference, for those systems where biological chemical reactions depend on chiral matching of reactants, can be profound. A working pharmaceutical or drug might be inert, or otherwise toxic, in its chiral version(s). There is no mention of

55 Introductory quotes. Introduction, top of the first page.

56 Alice's Kitty's Looking-Glass milk. Chirality. Chapters 4 and 8.

Fig. 4.15: Alice and Tweedledum and Tweedledee.

this in the legislated lists of criminally forbidden molecules of the *Controlled Drugs & Substances Act*; many Canadian judges might not notice.[57]

To explain chirality in more detail, consider the example of *glyceraldehyde*, the smallest of the sugar molecules, which has the structure (represented for three dimensions [195, 196, 197, 198, 199])[58,59,60] (see Fig. 4.16) [200, 201, 202].

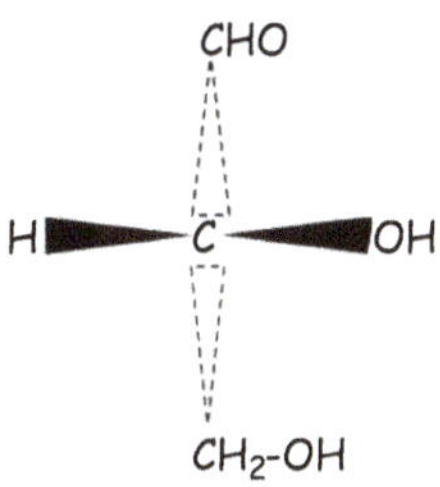


Fig. 4.16: Glyceraldehyde.

This three-dimensional arrangement in space defines this molecule uniquely [203], and the Chemical Abstracts Service of the American Chemical Society uses for this

57 Re criticism of Health Canada, Judges, et al. Introduction 4.1 and 4.3; Chapters 10 and 11.

58 The exaggerated wedge-shaped lines here are to try to give an image of receding from, or coming toward, the viewer who would be facing the plane of the page.

59 Where the dashed (exaggerated wedge-shaped) lines here represent the hydrogen as below the plane of the page, with the carbon in the plane of the page.

60 Where the solid (exaggerated wedge-shaped) lines here represent the hydrogen as above the plane of the page, with the carbon in the plane of the page.

structure, the *CAS Index Name*: *(R)-2,3,-dihydroxypropanl*, and has assigned *CAS Registry Number*: 453-17-8. This molecule is commonly called ***D-glyceraldehyde*** and is conveniently represented on a two-dimensional page (Fischer projection [204, 205]) as in Fig. 4.17.[61,62]

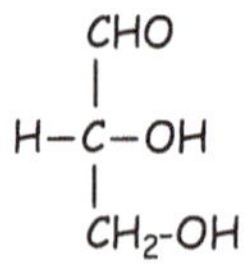


Fig. 4.17: Fischer projection.

But, this molecule was chosen as the example here, because it is a classic [206] of the two possibilities of *chirality* – shown on a two-dimensional page in a pseudo-kind of way (See Fig. 4.18):

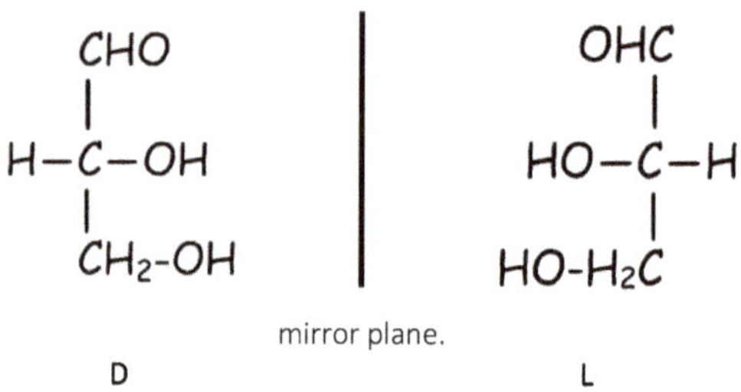


Fig. 4.18: D and L mirror plane [207].

***D-Glyceraldehyde*** is shown on the side of the mirror plane labelled **D.** The structure shown on the other side of the mirror plane, labelled **L,** is called *(S)-2,3-dihydroxypropanal* or ***L-glyceraldehyde***, CAS 497-09-6 [208]. The **L**-structure may be seen to be different from the **D** by constructing three-dimensional models of both *D-* and *L-glyceraldehyde* and trying to superimpose one on the other – it can't be done – similar to left and right hands [209] (see Fig. 4.19) [210, 211, 212].[63]

Because of the similarities of these *two* molecules, a ***racemic mixture*** (both D and L) [213, 214] (±)-*2,3-dihydroxypropanal* – would historically have often been used for reasons of economy [215]. This has been assigned CAS 56-82-6.

---

**61** Molecular structural drawings. Chapter 4.

**62** Chemical names as used in this book. Introduction 3.2; Chapters 3 and 4.

**63** Furfur, "Hand drawing," Wikimedia Commons. Copyright acknowledgement; Chapter 4.

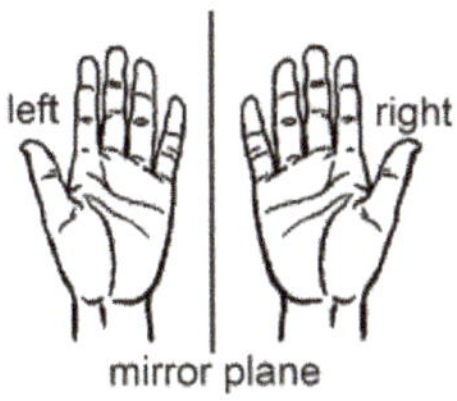


Fig. 4.19: Left and right hands.

The central carbon atom in these chiral versions is called an *asymmetric carbon*, sometimes signified as C* (see Fig. 4.20) [216].[64]

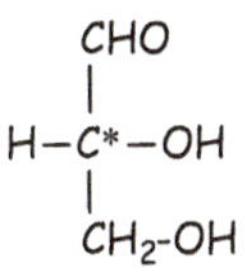


Fig. 4.20: Asymmetric carbon.

The property to notice is that for the visualization from the asymmetric C* – looking in the four tetrahedral directions – something different is seen in each direction.

### 4.5.3 Isotopes

There are other possibilities of subtle difference – such as for a molecule containing variations of *isotope*.[65]

All atoms of each element are the same – as to the number of protons $p^+$ and electrons $e^-$. And the number of protons always = the number of electrons for a neutral (not ionized) atom. But atoms may also have varying numbers of neutrons n for what are called *isotopes*.

Molecules can contain atoms of isotopic variation. Because it is the electron configuration that determines chemical properties, the slight isotopic mass differences are usually related to small, often imperceptible, effects.

Some – not all – isotopes are radioactive – such as carbon-14; so that for example, a radioactive isomer of D-*glyceraldehyde* might be synthesized for specialized research purposes [217, 218] (see Fig. 4.21).

64 But beware about C* symbolism – it has quite different definition/meaning in two unrelated contexts:

C* = an asymmetric carbon – indicating chirality.

C* = radioactivity of an atom – here, carbon.

65 Submicroscopic particles: protons $p^+$, electrons $e^-$, neutrons n. Chapters 4 and 7.

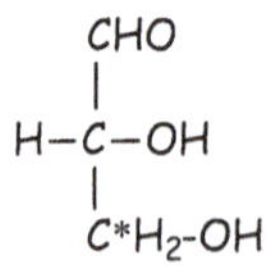

Fig. 4.21: With carbon-14.

C* in Fig. 4.21 denotes carbon-14, a beta minus ($\beta^-$) emitter [219, 220].[66,67]

### 4.5.4 Polymers

Another possibility of subtle difference is for ***polymers***[68] – molecules composed of strings of repeats of smaller molecular units [221].

Ordinary-size *monomer* (monomer = ×1) molecules, *dimer* (= ×2), . . ., appear as assembled into very large molecules – *polymer* (= ×n) [222], and can have molecular masses in the thousands, and beyond (n can be very large). *Polymers* are characterized [223] not only by the molecular structure of their monomers and their linkages, but also by the statistical spread of the large molecular masses, by inter-linkages and interactions of these large molecules, and by additives (as chemically part of the large molecules; or as other things mixed in, but not chemically combined). These characteristics can vary according to the *formulations* of production. Although there are many polymeric natural products, it is synthetic polymers [224] that are often of forensic interest. These include all commercially produced plastics and thermosetting resins [225].

While there can be very great differences – both chemical and physical – amongst polymers – within any particular polymer group, subtle differences would always be expected. Often, this has to do with the length of the polymer chain, as characterized by the statistics of their properties.

### 4.5.5 Hydration of the chemical substance

Some chemical substances combine with water $H_2O$ in a molecule-like arrangement, so that there are two possibilities. For example: for hexavalent chromium – Sodium

66 Isotope. Radioactive. Carbon-14. Chapters 4 and 6.

67 But beware about C* symbolism – it has quite different definition/meaning in two unrelated contexts:

C* = an asymmetric carbon – indicating chirality.

C* = radioactivity of an atom – here carbon.

68 Polymers. Chapter 4.

dichromate [226, 227, 228] shown here[69,70] as ***anhydrous*** – without water molecules – $\mathbf{Na_2Cr_2O_7}$ – 262 g/mol CAS 10588-01-9 (See Fig. 4.22):

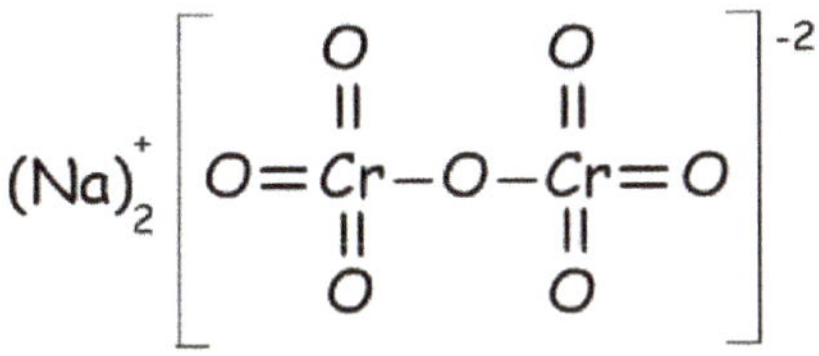

Fig. 4.22: Sodium dichromate.

But, it is usually as dihydrate – with water molecules $\mathbf{Na_2Cr_2O_7 \cdot 2H_2O}$ – 298 g/mol, CAS 7789-12-0. The detailed structures of the hydration are often not well known.

### 4.5.6 Moisture content of the whole package: $CH_3–CH_2–OH$

Bulk amounts of some chemical substances also contain water [229].[71] For many, but not all, purposes this does not matter much. For example: When ethanol – grain or drinking alcohol – $\mathbf{CH_3–CH_2–OH}$ in pure form – no water $\mathbf{H_2O}$ – is exposed to the moisture of the atmosphere it takes on ~5% water. It is said to be ***hygroscopic***. Thus, when a container of absolute ethanol ≈100% is unsealed, it would soon diminish to ~95% [230]. For beverages – largely, water, anyhow this does not matter, but for ethanol to be used as a solvent in a spectral or chemical process, it might matter much.

### 4.5.7 Halide ions that balance charge

Many chemical substances are commonly available in both an ionized and nonionized [231] form – often this is for pharmaceutical purposes, and therefore, also for drugs of recreation and abuse. See for example, cocaine, below, for some data details.

The chemical difference is that the bridge nitrogen in the ionized form (shown in Fig. 4.23c, with detail in box, see Fig. 4.23b) has taken on an additional hydrogen atom to give a positive charge at the nitrogen $N^+$. This is then balanced by the presence of a chloride, $Cl^-$. As ionized, the melting point would increase substantially; vaporization, even if possible without decomposition, would require still higher temperatures.

The cocaine–nonionized (shown in Fig. 4.23a) [232] with a lower melting point would, therefore, be more easily vaporized for inhalation to go into the blood system

69 Molecular structural drawings. Chapter 4.

70 Chemical names as used in this book. Introduction 3.2; Chapters 3 and 4.

71 *Ethanol.* $CH_3–CH_2–OH$, CAS 64-17-5. Chapters 1, 4 and 11.

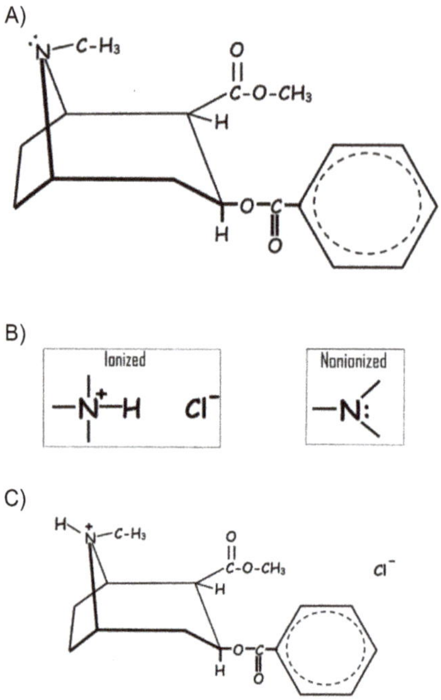


Fig. 4.23: (a) Cocaine. (b) Ionized and nonionized. (c) Cocaine hydrochloride.

directly from the air/blood interface in the lungs [233]. The ionized form would have a different mechanism of entry [234].

### 4.5.8 Virus mutations

Medical identification of a virus may include mutations of slightly different chemical structure [235, 236] that may change some of its properties, but might, (hopefully) likely, still allow the same vaccine to be used.

### 4.5.9 Microparticles

Micropaticles [237, 238, 239, 240, 241] are solids – often resulting from manufactured polymers. Sizes are in the $10^{-6}$ m [242] range, with properties related to their large surface-to-volume ratio. They can have environmental and biological effects – including in the human food chain; there are also medical uses [243]. Micropaticles are of serious toxicological concern [244, 245].

### 4.5.10 Sampling plan

The importance of sampling processes for the analytical chemistry of forensic evidence, described elsewhere in this book,[72] apply as well, to the subtle variations, *etc.*, described in this chapter.

### 4.5.11 Forensic chemistry choices

Whether or not to consider such subtleties – as *chirality, isomeric, isotopic, polymeric* differences, hydration, moisture content or halide ions – as different enough to attract forensic attention is a judgment call of the forensic chemist, in the context of the investigation.[73] This may also be influenced by anticipated concerns of the judge and the opposing lawyers, even if those concerns would not make scientific sense. In Canadian *Controlled Drugs & Substance Act* cases, judicial jurisdiction for discretion in molecular decision making is effectively removed by s.51; many judges may be oblivious to this.[74,75]

What is a subtle difference that may be ignored in one context may be important in another. For example [246] an isotopic analysis for carbon-14 in *glyceraldehyde* would usually not [247, 248] be necessary; but if the forensic investigation relates, as a fictitious example, to a theft from a research lab that had *glyceraldehyde-carbon-14*, then it might constitute the most important piece of evidence.

## 4.6 Chemical names – CAS

The system, in concept, to uniquely define chemical substances by structure is described above. But, there are millions known [249, 250],[76] some with more than one name [251]. More are reported in world scientific literature daily. Systematic scientific chemical naming [252], therefore, becomes critically important in keeping track of what is what. The Chemical Abstracts Service, a division of the American Chemical Society, provides the basis of a system used worldwide [253, 254, 255, 256, 257].[77,78] Although systemic chemical names do not originate with CAS, the worldwide use of its *Registry Numbers* does.

72 Error. Sampling. Sampling plan. Chapters 3, 4 and 8.
73 Forensic chemistry choices. Chapters 3 and 4.
74 Re criticism of Health Canada, Judges, *et al.* Introduction 4.1 and 4.3.
75 Identification of molecules. *CD&&S Act*. Chapters 4 and 10.
76 molecular structural drawing. CAS. Chapters 4.4 and 4.6.
77 SI CAS. Introduction 3.2; Chapters 3 and 4.
78 Chemical names as used in this book. Introduction 3.2; Chapters 3 and 4.

Since 1907 [258], from Columbus, Ohio, the American Chemical Society has been involved with the *Chemical Abstracts* [259]. It has contained written abstracts and data of historic and recent chemical scientific articles in world chemical literature, and other sources.[79] Until 2010, it was published regularly and frequently as print-on-paper; data are now available continually by-way-of online systems [260].

A *Registry Number* is assigned sequentially – as new substances enter the CAS system – daily by the thousands. Started in 1965, the CAS *Chemical Registry System*, today, continues to assign unique identifier numbers to newly reported substances. These are the CAS *Registry Numbers*; each new entry also would have a systematic chemical name.

Although there can be complications, the simple description above is mostly sufficient for purposes here, to understand the listing of distinctly different chemicals in legal lists. For example, from the (former) WHMIS ingredient disclosure list [261, 262][80,81] (see Tab. 4.2 for extracts.):

Tab. 4.2: Ingredient disclosure list – extracts.

| **Item** | | **Column I** | | **Column II** |
|---|---|---|---|---|
| | | **ingredient** | | |
| | | **Chemical identity** | **CAS* registry number** | **Concentration (%-weight/weight)** |
| | | [. . .] | | |
| 204. | *(67)* | Boric acid | 10043-35-3 | 1 |
| 205. | (228) | Boric anhydride | 1303-86-2 | 1 |
| 206. | *(1636)* | Boron tribromide | 10294-33-4 | 1 |
| 207. | *(1654)* | Boron trichloride | 10294-34-5 | 1 |
| 208. | *(1674)* | Boron trifluoride | 7637-07-2 | 1 |
| 209. | *(1675)* | Boron trifluoride diethyletherate | 109-63-7 | 1 |
| 210. | *(1676)* | Boron trifluoride dihydrate | 13319-75-0 | 1 |
| | | [. . .] | | |
| 216. | *(324)* | Bromoacetone | 598-31-2 | 1 |
| 217. | *(331)* | Bromoacetyl bromide | 598-21-0 | 1 |

Even if another name is more popularly used – and the formal designations are often not used in Canadian legal proceedings – it is important to realize that a particular chemical substance may be uniquely described by its systematic chemical name, CAS *Registry Number* and its three-dimensional spatial structural diagram. From a scientific perspective, these formal designations should appear at least somewhere in the forensic documentation and report; otherwise detracts from credibility.

**79** Scientific literature. Chapters 4 and 5.

**80** B $Br_3$ CAS 10294-33-4. Chapters 1 and 4.

**81** Workplace health & Safety. Hazard communication. WHMIS. Chapters 1, 2 and 4.

While the systematic name can tell a chemist about what the structural diagram would look like [263], the CAS *Registry Number* contains within it no structural or other chemical data, *per se*.

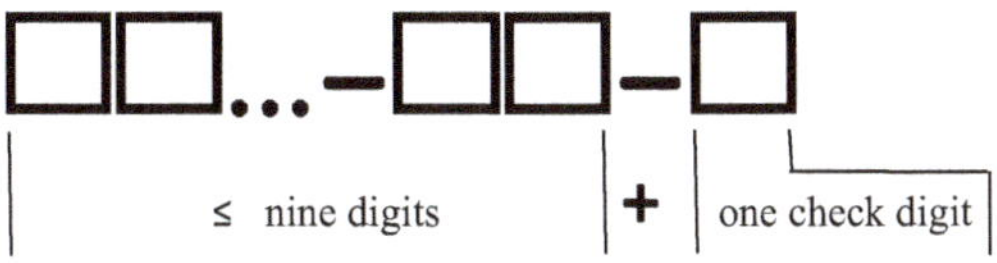


Fig. 4.24: CAS numbering.

A CAS *Registry Number* [264] has the general form described in Fig. 4.24: [265] – ten digits or less, in three hyphenated parts. The last digit (on the right) is a *check digit* computed by formula from the previous digits (to its left) that allows for a 90% quick verification that [][] . . . -[][]-[] could really be in the CAS system. (A similar process is used to verify Canadian Social Insurance Numbers – so that if you try to make up a fake SIN, or get it wrong innocently, it would likely be noticed.)

To see how good a lead-in they are to scientific information, readers might simply try to enter any of the CAS *Registry Numbers* shown in this book as examples into most-any online search engine.

Further and more detailed information may be found at [266]:

https://www.cas.org/cas-data/cas-registry

https://www.cas.org/about/faqs

Much of recent US, EU, and Canadian legislation include CAS *Registry Numbers*; some remaining Canadian legislation is not so updated [267].

Thus, the *Controlled Drugs & Substances Act* [268] still neither uses CAS *Registry Numbers* nor molecular structural diagrams. For example, from its *SCHEDULE I* of forbidden things:[82]

> [. . .]
>
> 2. Coca (Erythroxylon), its preparations, derivatives, alkaloids and salts, including:
>
> (1) Coca leaves
>
> (2) Cocaine (benzoylmethylecgonine)
>
> (3) Ecgonine (3-hydroxy-2-tropane carboxylic acid)
>
> [. . .]

Also, Canadian courts adopt some "street" terminology; and most judges and lawyers would likely be somewhat oblivious to the actual science – and scientific identity – of the criminally proscribed molecules they are litigating.[83]

**82** Alice's Kitty's Looking-Glass milk. Chirality. Chapters 4 and 8.

**83** re criticism of Health Canada, Judges, *et al.* Introduction 4.1 and 4.3.

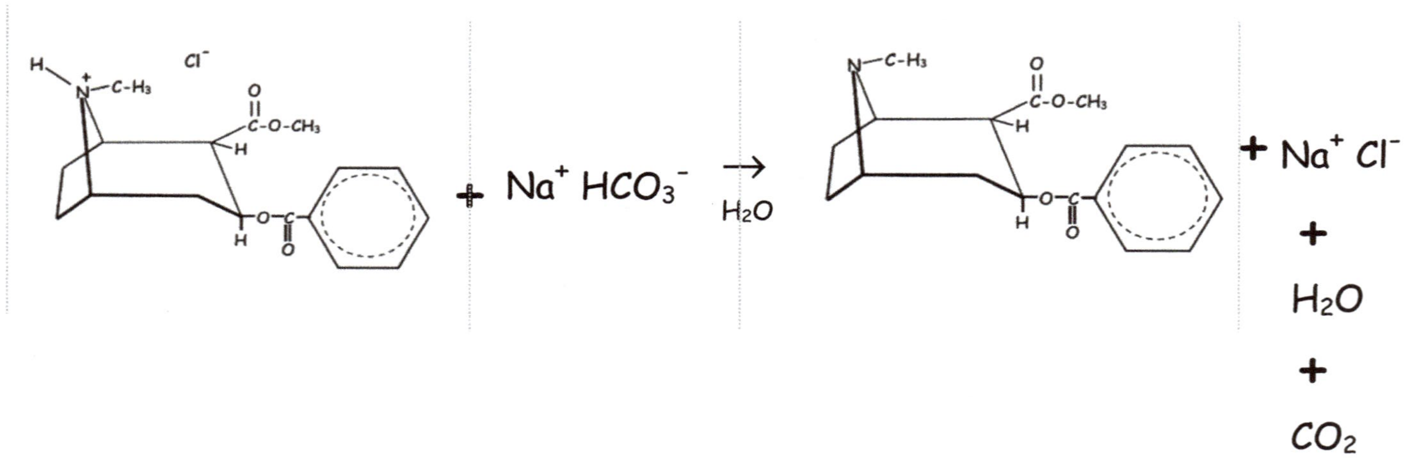


| 53-21-4 **cocaine hydrochloride** 339.82 g/mol. mp= 197°C. | sodium hydrogen carbonate | 50-36-2 **cocaine** 303.36 g/mol. mp= 98°C. | sodium chloride (salt) + water + carbon dioxide. |
|---|---|---|---|
| = "cocaine" (in Canadian court terminology.) | = baking soda. | ="crack cocaine" & "crack" (in Canadian court terminology.) | |

Fig. 4.25: Cocaine conversion reaction.

As further examples of CAS *Registry Numbers* and molecular structural diagrams, a simple conversion reaction [269, 270, 271, 272, 273, 274, 275, 276, 277, 278, 279, 280, 281, 282, 283][84,85] [284] (see Fig. 4.25) [285, 286].

While all known chemicals should have systematic chemical names and CAS *registry numbers,* and be describable by three-dimensional structural drawings, at least one has a Limerick [287, 288, 289, 290]:

*A mosquito was heard to complain*
*That a chemist had poisoned his brain*
*The cause of his sorrow*
*Was paradichloro*
*Diphenyltrichloroethane.*

Using this now as an example: several of the names are shown for this famous [291, 292] chemical pesticide commonly called *DDT* (see Fig. 4.26) [293, 294, 295, 296, 297, 298, 299, 300].[86]

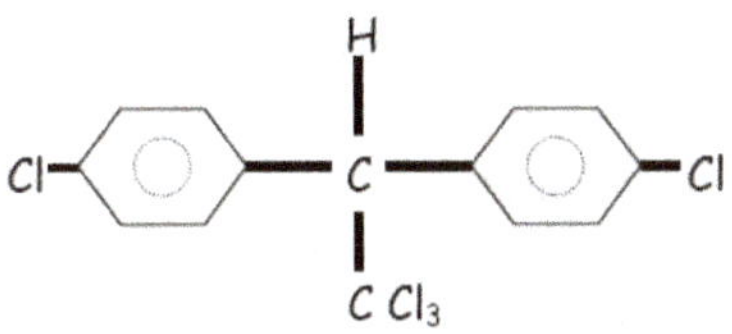


CAS 50-29-3 DDT p,p'-DDT
Dichloro-diphenyl trichloroethane
1,1,1-trichloro-2,2 bis(p-chlorophenyl)ethane

Fig. 4.26: DDT.

## 4.7 Identification of a molecule by its characteristics

Although, as indicated earlier, a molecule is defined by its structure, this definition has little practical value without methods to identify the actual structure.

Since molecular structures are too small to be seen by the human eye, even with powerful optical microscopes, these methods are related to expressions of physical effects by large numbers of the molecule.

84 Edmond Locard. Sherlock Holmes. Cocaine. Chapters 2, 3, 4 and 6.
85 Cocaine. CAS 53-21-4 CAS 50-36-2. Chapters 4, 6 and 10.
86 DDT CAS 50-29-3. Chapter 4.

A ***fundamental identification***, as mentioned in Chapter 2,[87] of a molecule may be made by examining a sufficient number of characteristics to make a unique theory-based determination.[88] For organic molecules, a common method is *infrared spectroscopy*, where the appearance of *absorption peaks* at certain *wave numbers* is uniquely associated with the *masses* of the atoms and the strengths and geometries of their *bonds*; and the absence of any extraneous peaks. IR is described in more detail in Chapter 6,[89] along with a discussion of its limitations (for *chirality* and instrumental sensitivity limits for very subtle differences).

Another theory-based molecular structural determination method is mass spectroscopy, often linked with gas chromatography. GC-MS is dealt with in Chapter 6.[90]

However, such theory-based methods are usually not directly used for forensic determinations. Rather, a *fundamental identification* is made using a *standard reference sample* [301] or *reference literature*.[91] With such references, a forensic chemist can let others do the theoretical determination of structure (often difficult, requiring special efforts, resources and skills; and maybe extra patience), while merely confirming an identification, by comparison with the same method – a fundamental determination by comparative method. Again, IR is a common technique, as is GC-MS – see Chapter 6.[92]

*References* can also be used for comparison for other physical characteristics expressed by a molecule – such as chromatographic retention time or distance, UV-visible spectroscopic absorption,[93] melting point, solubility and refractive index – see Chapter 6. But, these methods are more appropriate as *screens*[94] than for fundamental determination [302].[95,96]

## 4.8 Mixtures – mixture profile

Many, if not most, forensic samples are mixtures of molecules of different identities, so that even after the forensic chemist has decided on the level of molecular subtlety

87 Fundamental identification of molecules. Chapters 2, 3, 4 and 6.
88 Chemical identification. Chapters 3, 4 and 6.
89 Molecular identification by IR. Chapters 4 and 6.
90 GC-MS. Chapters 4 and 6.
91 Lab core documents. SOP. Scientific literature. Chapters 3, 4, 5 and 8.
92 GC-MS. Chapters 4 and 6.
93 UV. Chapters 3 and 6.
94 Screening tests. Wet chemistry. Chapters 3, 4, 5 and 6.
95 K. Bailey. Chapters 3 and 4.
96 Brunelle & Cantu. Chapters 3 and4.

to be concerned with, there are then questions of what different molecules are present and of the significance, if any, of their proportions.[97,98,99]

Identification of a forensic sample may depend on comparison of a mixture profile of its components with other evidential samples and reference standards. For example, a probability of finding trace metals in otherwise pure lead Pb, may help identify a bullet [303].[100]

### 4.8.1 Mixture profile – glycol example

Consider, as a fictitious example [304],[101] a forensic sample that would appear to contain what was intended to have been a single substance. A police detective has seized a bottle of *methanol* from the home of an arson suspect, and the forensic chemist is asked to compare this *methanol* with some unconsumed *methanol* – thought to have been intended as an accelerant – recovered from the fire scene. From the discussion above, the reader would know that all *methanol* is the same [305] (see Fig. 4.27) – a liquid.

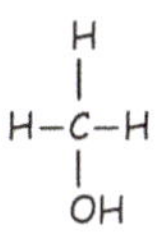

Fig. 4.27: Methanol, CAS 67-6-1.

All methanol molecules in the container would be identical. No chirality; isotopic variations negligible.

What might be different, and what the forensic chemist would investigate here, are other non-*methanol* substances – additives or impurities – that are in with the *methanol*, for one reason or another. Since most commercially available *methanol* likely has at least a little of other things in it, the forensic chemist would also look for an identifiable pattern and then try to assess its uniqueness.

To continue the fictitious example:[306] The alleged arsonist acquired the *methanol* from a bulk supplier whose source was very pure. But that bottle had been used previously for an *ethylene*/ *±-propylene glycol* mixture (20/80 by mol) for use as antifreeze (see Fig. 4.28a and b) [307, 308].[102]

That former antifreeze bottle had not been cleaned. An analysis of the contents of that seized bottle now containing methanol showed (see Tab. 4.3):

97 What a substance is. Mixture proportion of components. Chapter 4.

98 Too much scientific technical detail here? Introduction 2.1.

99 Mixture. Chapters 3 and 4.

100 Trace metals with Chapters 2 and 4.

101 Occurrence probability – ink example. Find expectation. Chapters 2, 4 and 9.

102 Fictional literature. Ethylene glycol. CAS 107-21-1. Introduction 2.2; Chapters 4 and 7.

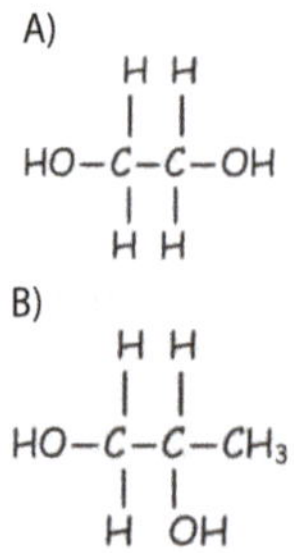

Fig. 4.28: (a) Ethylene glycol, CAS 107-21-1. (b) ±-Propylene glycol, CAS 57-55-6.

Tab. 4.3: Methanol in seized bottle.

| | mol% = |
|---|---:|
| *Methanol* | 97 |
| *Ethylene glycol* | 0.6 |
| *±-Propylene glycol* | 2.4 |

These results appear consistent with the suspected history of the bottle. Analysis of unconsumed fluid from the fire scene showed (see Tab. 4.4):

Tab. 4.4: Methanol unconsumed at fire scene.

| | mol% = |
|---|---:|
| *Methanol* | 95.5 |
| *Ethylene glycol* | 0.9 |
| *±-Propylene glycol* | 3.6 |

– which might be consistent with that mixture losing some of its *methanol* concentration by evaporation.

The question the forensic chemist should now consider is the uniqueness of the *methanol/ethylene glycol/propylene glycol* = 97/0.6/2.4 mixture, and the significance of:

$$0.6/2.4_{\text{seized bottle}} = 0.9/3.6_{\text{fire scene}} = 20/80_{\text{original anti-freeze}}$$

To answer this, the forensic chemist must do more than just analyse in the lab. Perhaps with the help of the detective, they should try to determine the occurrence [309] of the 97/0.6/2.4 mixture in the bottle to other finds in the area [310] of the fire.

In one variation of this example, if it is found that the supplier routinely provided and used empty antifreeze bottles in which to sell other solvents, then the alleged arsonist's bottle of impure *methanol* would hardly be expected to be unique, and the forensic evidence would be of diminished prosecutorial value, particularly if there is no other evidence. The alleged arsonist's lawyer would like this.

But, in a contrary variation of the example, if the alleged arsonist already had an emptied antifreeze bottle, to be presented to and filled by the supplier, then the 97/0.6/2.4 occurrence might be quite unique. And, if the detective were to also seize a similar full bottle of antifreeze from the alleged arsonist's home, that the forensic chemist determined was *ethylene glycol/propylene glycol* = 20/80, consistent with its own label, the alleged arsonist's lawyer [311] might wish to advise of the advantages of plea bargaining.

In the above example, the occurrence of *methanol/ethylene/propylene glycol* = 97/0.6/2.4 might be called a ***mixture profile,*** and, as indicated above, its uniqueness is a question of a probability of occurrence that must necessarily take the forensic chemist beyond the sample analysis at hand.

### 4.8.2 Occurrence probability – DNA, fingerprints

For example, in another context, for DNA [312] matching of human tissue, the frequency of the forensic results in the human population of concern should be considered. Assuming here, no arguments on the accuracy of the analysis, perhaps $<10^{-6}$ chance [313] of duplicate match would be considered good enough [314] for Crown counsel to feel confident that, aside from other factors, a jury could convict, based on corroborating evidence. And, for $<10^{-9}$, corroborating evidence might not be needed.

Similar reasonable speculative estimates would apply for fingerprints – consistent with no duplicate human fingerprints having ever been found [315, 316, 317].[103,104]

### 4.8.3 Occurrence probability – ink example

In the ink example[105] [318, 319], it may be necessary to consider the local popularity of the pen ink colour found in possession of the accused. If it is assumed here that the TLC patterns are a sufficient indicator to establish the identity of the ink [320], then it might be appropriate to survey the population of ballpoint pens that the accused could have had access to. That survey might be considered as from a sampling random grab [321, 322] of pens from the regional population [323].

The ease or difficulty of doing such a survey may be related to the nature of the regional ink industry (*e.g.*, if there is only one pen and ink company that produces

**103** Occurrence probability – ink example. Find expectation. Chapters 2, 4 and 9.

**104** Fingerprint probability. Chapters 1 and 4. Fingerprint visualization. Fluorescence and phosphorescence. Chapters 6 and 9. Fingerprint when. Chapters 7 and 8.

**105** Occurrence probability – ink example. Find expectation. Chapters 2, 4 and 9.

only a dozen different inks), and the writing habits of the region's citizenry (*e.g.*, if a large plurality use only dark blue ink). And, perhaps other factors, such as if the ink colour of interest is found in pens sold by a street vendor just outside where the forgery was alleged – very rare overall, but very common in the immediate neighbourhood of interest. Or, if that kind of pen can only be purchased abroad, so that kind of ink is found only in pens of a few recently returned travellers.

An alternative – and easier method – to estimate the local find expectation of the pen ink colour found in possession of the accused could be to examine already existing sales statistics.

To continue the ink example, for the alternative mentioned above, the regional banking industry (anti-fraud division) provides data [324] (see Tab. 4.5).

Tab. 4.5: Ink inventory.

| **Colours of the entire inventory of ballpoint pen inks [325]** | **(Totals of pen sales for most recent 5 years) × ($10^{-6}$) =** | **[326, 327]** |
|---|---|---|
| Royal blue | 4 | |
| Governor-General's vice-regal blue | 3.1 | |
| Bank examiner's blue-black | 1.1 | |
| Citizens' dark blue | 14 | |
| Copy editor's blue [a light blue ink] | 0.1 | [106] |
| Rose red | 3.2 | |
| Poets' green | 1 | |
| Orange | 0.1 | |
| Traffic sign luminescent orange | 1.2 | |
| Leaf green | 2 | |
| Censor's black | 5 | |
| Σ = | 37.8 | |

The forensic chemist does TLC analysis for samples of all of these inks, exactly repeating the method and specifications used for the case evidence (see above, Chapter 5 references cited therein). There are, thus, 14 TLC patterns to compare, one with another – the 12 from the data chart above plus one from the pen and one from the questioned document. The TLC patterns may be considered as sufficient identifiers of each ink type.

Suppose the chemist finds that all of the dozen industry-provided samples are different one from another, but that one of them – light blue – has a TLC pattern identical to that of the case evidence. An approximate inference might then be that, of a five-year local distribution of 37.8 million ballpoint pens, the light blue ink of copy editor's blue has a general find expectation of 0.1/37.8 ≈ 0.27% – that might safely be

**106** Spoiler alert: the ink of interest in this *Gedankenexperiment*.

called as <½%. But logically, this number would still have only an approximate and tenuous relationship to an actual occurrence probability expectation for the pen – and even within this *Gedankenexperiment* should be regarded with care.

Although the forensic chemist is entitled to conclude from this that light blue ballpoint ink is indeed rare, determining the significance of this to the law case would eventually be the job of the judge. The significance to the law case in this fictitious concocted example [328] would be that the pen of concern is unlikely to be in the possession of nearly everyone, so that the alleged forger is severely implicated. (The defence lawyer might want to privately review, again, to the accused, about the advantages of a plea bargain.)

As indicated in the Introduction, fictitious examples as above are concocted [329] with simplicity in mind, so as to try to more clearly explain concept – here occurrence frequency. The oversimplification of this particular example must be emphasized. In particular, the small, finite and totally known number of ink colours should not be expected in real-life forensic work. More generally [330], such determinations could be very difficult, and completely satisfactory conclusions elusive.

Also, the above example only deals with the *presence* aspect of a *mixture profile*, because TLC, by appearance alone, does not so easily give *quantity* measurements (other than in a gross way). For ballpoint pen ink identification this is likely sufficient.[107] For a *mixture profile* with both *presence* and *quantity* measurements, GC[108] would be an appropriate method – for example, to try to characterize retail-purchased gasoline samples [331].

## 4.9 Reactions in progress

*What is* found at the forensic scene may include things that are chemically changing in time – chemical reactions in progress.[109,110,111] These problems in chemical kinetics are typically more challenging to the forensic chemist than merely identifying molecular compounds and mixtures. Experimental reality may be that such analyses may be very difficult to do, with little reference reported in chemical literature about the detailed chemical kinetic mechanisms.

**107** Ink reference standards. US International Ink Library. ATF. Chapters 2, 3, 4 and 9.
**108** GC. Chapters 4 and 6.
**109** When. Chemical kinetics. Reactions in progress. Chapters 2, 4 and 7.
**110** Chemical kinetics. Reactions in progress. Chapters 2, 4 and 7.
**111** When. Chapters 2, 4 and 7.

## Notes

1 Looking-Glass = mirror https://dictionary.cambridge.org/dictionary/english/looking-glass
2 Alice to the black kitten in Lewis Carroll, *Alice's Adventures in Wonderland* and *Through the Looking-Glass*, Signet Classics, The New American Library, New York City, 1960 & 1962; from . . . *Looking-Glass* . . ., "Looking-Glass House," Chapter I, page 129. In modern biochemical terms, Kitty should have some very serious concerns about the chirality of the looking-glass milk's components.
3 And at page 127:
4 https://www.alice-in-wonderland.net/wp-content/uploads/2book2.jpg
https://www.alice-in-wonderland.net/wp-content/uploads/2book3.jpg
https://www.alice-in-wonderland.net/wp-content/uploads/2book4.jpg
https://www.alice-in-wonderland.net/resources/pictures/tweedledum-tweedledee/
https://www.alice-in-wonderland.net/resources/pictures/through-the-looking-glass/
https://www.alice-in-wonderland.net/resources/faq/#copyright
5 *Genesis* Chapter 1 1→6 בְּרֵאשִׁית
6 A Hebrew–English Bible According to the Masoretic Text and the JPS 1917 Edition
https://mechon-mamre.org/p/pt/pt0101.htm
https://mechon-mamre.org/p/pt/pt0.htm
7 Also: https://en.wikipedia.org/wiki/Let_there_be_light
8 John Gribbin, *In Search of Schrödinger's Cat/Quantum Physics and Reality*, Bantam Books, New York City, 10036; 1984; ISBN 978-0-553-34253-6.
9 John Gribbin, . . . *CAT* . . ., quoting Niels Bohr.
10 Gribbin, on back cover.
11 Several variations of this quote:
https://en.wikiquote.org/wiki/Niels_Bohr#:~:text=have%20understood%20it.-,Those%20who%20are%20not%20shocked%20when%20they%20first%20come%20across,not%20understood%20a%20single%20word
12 John Gribbin, . . . *CAT* . . ., quoting Erwin Schrödinger.
13 John Gribbin, *In Search of Schrödinger's Cat / Quantum physics and reality*, Bantam Books, New York City, 10036; 1984; ISBN 978-0-553-34253-6.
14 Gribbin, . . . *CAT* . . ., on introductory page v.
15 Also: https://www.goodreads.com/quotes/393059-i-don-t-like-it-and-i-m-sorry-i-ever-had
https://www.quotetab.com/quote/by-erwin-schrodinger/i-dont-like-it-and-im-sorry-i-ever-had-anything-to-do-with-it
https://medium.com/starts-with-a-bang/einstein-schr%C3%B6dinger-and-the-story-you-never-heard-7588341288b9
16 https://www.sciencenews.org/blog/context/top-10-richard-feynman-quotations
17 https://mitpress.mit.edu/9780262060165/the-character-of-physical-law/
John Gribbin, *In Search of Schrödinger's Cat/Quantum Physics and Reality*, Bantam Books, New York City, 10036; 1984; ISBN 978-0-553-34253-6.
18 Introductory page v – quoting John Lennon – "Nothing is real."
19 Although words in the Beatles song might be stretched for reference to quantum concepts – it would appear unlikely that was intended.
20 https://en.wikipedia.org/wiki/Strawberry_Fields_Forever
https://www.amazon.com/Nothing-Real-UK-BEATLES/dp/B00EV9416K
21 The Beatles,"Strawberry Fields Forever,"1967:
https://www.azlyrics.com/lyrics/beatles/strawberryfieldsforever.html
22 https://jonfogelberg.substack.com/p/the-quantum-journey-of-strawberry

**23** "Strawberry Fields Forever - The Beatles (Acoustic cover by Emily Linge)" [2023] :
https://www.youtube.com/watch?v=12Q-lfj9KEU
**24** See: Ian Sewart, *Does God Play Dice ? / The Mathematics of Chaos*, PENGUIN BOOKS, London W8 5TZ; 1989, 1990; "Prologue / Clockwork or Chaos ?," "[Chapter ] 1 / Chaos from Order," at page 1. ISBN 0-14-012501-9.
**25** https://www.azquotes.com/quote/565846#google_vignette
https://en.wikipedia.org/wiki/Religious_and_philosophical_views_of_Albert_Einstein
**26** [Albert Einstein, Princeton, 1944. Letter copy with variation of quote wording]:
https://winners-auctions.com/en/items/letter-from-einstein-to-max-born-with-extensive-discussion-of-einsteins-famous-remark-about-quantum-physics-g-d-doesnt-play-dice/

[. . .]

> In our scientific expectations we have developed into antipodes. You believe in the dice throwing God and I in a complete order in a world of objective existence which I am trying to catch in a wildly speculative way. I believe firmly, but I hope that somebody will find a more realistic way or a more solid basis than I have given. The great initial success of the Quantum Theory cannot persuade me to believe in the fundamental dice-playing, although I know that the younger colleagues interpret this as a result of sclerosis. Once it will become manifest which instinctive attitude had been right.

[. . .]

**27** Michael Tomasky, "Why Does Trump Lie? He has nothing but contempt for the institutions that exist to keep presidents in check," *The New York Times*, Opinion, 11 June 2020.
https://www.nytimes.com/2020/06/11/opinion/trump-lies.html

> The lies and obfuscations pile up. No, it wasn't tear gas used to clear Lafayette Park for President Trump's Bible-waving photo-op last Monday night, Attorney General William Barr said on CBS's "Face the Nation" on Sunday. Rather it was "pepper balls," he said. "Pepper spray is not a chemical irritant. It's not chemical." Wrong, according to The Washington Post; pepper balls are very much a chemical irritant. The paper awarded the nation's top law enforcement officer four Pinocchios for his claim.

**28** Michael Tomasky, "Why Does Trump Lie? He has nothing but contempt for institutions that could expose him," *The New York Times*, OP-ED, 12 June 2020, page A25.
**29** Pelargonic acid vanillylamide PAVA
*N*-[(4-Hydroxy-3-methoxyphenyl)methyl]nonanamide
Pelargonyl vanillyl amide CAS 2444-46-4 293.407 g/mol
https://en.wikipedia.org/wiki/Pepper-spray_projectile
https://en.wikipedia.org/wiki/Nonivamide
**30** Of concern here is the *scientific* nature and identity of matter. Other natures and identities – of serious and important concern in their own contexts – are not of concern here. For example: That an item is kosher כשר; manufactured in a union shop; not manufactured with child labour; ethically mined? However, forensic science may help in the determination/verification of such non-*scientific* natures and identities.

(But see: Richard P. Feynman, *"Surely You're Joking, Mr. Feynman!" / Adventures of a Curious Character / as told to Ralph Leighton / edited by Edward Hutchings, BANTAM BOOKS*, New York City, 1986. ISBN 0-553-25649-1. "Is Electricity Fire?," page 253→261. The famous American physicist had a conversation with rabbinical students about Talmudic interpretation for pushing elevator buttons on Shabbos.)

**31** See John Gribbin, *In Search of the Big Bang/Quantum Physics and Cosmology*, Bantam Books, New York City, 10103; 1986; ISBN 0-553-34258-4.
**32** Gribbin, . . . *BANG* . . . page xiii, et seq.; page 214 et seq.
**33** See Lincoln Barnett, *The Universe and Dr. Einstein*, BANTAM BOOKS, New York City, 10019; . . . 1948 . . . Revised . . . 1979 . . .; ISBN 0-553-13060-9.
**34** Barnett, . . . *Universe* . . ., page 23 et seq.
**35** Newtonian – classical – physics is sometimes seen as having a kind of theoretical determinism to it, in that, in concept, every particle in the universe would be either at rest or calculable as moving from somewhere/when in time and space of the past to a predictable somewhere/when in the future. This is understandable as a kind of imaginable physical mechanical clockwork model – that is given mathematical expression.

Not so much so for modern physics: quantum mechanics deals in hypothetical probabilities of finding a particle anywhere in the universe, with a mathematical accounting that eludes a unique clockwork model that in some aspects does not seem to even need particles. And, similar for the mathematical expressions of general relativity, with its concepts of gravity-bending of space-time.

And, there are also mathematical issues – See Ian Sewart, *Does God Play Dice ? / The Mathematics of Chaos*, PENGUIN BOOKS, London W8 5TZ; 1989, 1990; "Prologue / Clockwork or Chaos ?," "[Chapter ] 1 / Chaos from Order," page 1→22. ISBN 0-14-012501-9. And see also: Jeffrey S. Rosenthal, *Struck by Lightning / The Curious World of Probabilities*; Harper Perennial / HarperCollins Publishers Limited, Toronto M4W 1A8; 2005, 2006; Chapter 16, pages 234→246, "Ignorance Chaos and Quantum Mechanics / Causes of Randomness." ISBN 13: 978-0-00-639495-2 ISBN 10: 0-00-639495-7

In these contexts, for those of us who are used to trying to persist in our everyday world, to make it to the next day without too many complications of contemplating what is "real," this can become disconcertingly confusing.
**36** See also: Steven Weinberg, *DREAMS OF A FINAL THEORY / The Scientist's Search for the Ultimate Laws of Nature*, Vintage Books, New York City, 1994; ISBN 978-0-679-74408-5.
**37** See also: Weinberg, *DREAMS* . . ., Chapter VI, "BEAUTIFUL THEORIES," page 132, *et seq.*
**38** See: David Eliot Brody and Arnold R. Brody, *The Science Class You Wish You Had . . ./ The Seven Greatest Scientific Discoveries in History and the People Who Made Them*, A Perigee Book, Berkley Publishing Group, New York City, 10016; 1997; ISBN 0-399-52313-8; Part Two. The Structure of the Atom, page 51, *et seq.*
**39** Ursula K. Le Guin, "Schrödinger's Cat," *THE COMPASS ROSE*, Short Stories, BANTAM PAPERBACK, Toronto, . . ., 1983; ISBN 0-553-23512-5; pages 41→49; from *Universe 5*, 1974.
**40** Arthur Koestler, "The Perversity of Physics," as Chapter 2 in *THE ROOTS OF COINCIDENCE*, Pan Books, Picador, London SW10 9PG; 1974; ISBN 0-330-24167-2; pages 50→81.
**41** CBC Radio, 2019, *Ideas*. "Finding meaning in the universe with astrophysicist/Hubert Reeves, Part 1/ The Origins of Us, Part 1 of a 2-part series."
https://www.cbc.ca/radio/ideas/finding-meaning-in-the-universe-with-astrophysicist-hubert-reeves-part-1-1.5143324
**42** CBC Radio, April 2019, *Ideas*, "'Finding wonder in the face of existential dread': Grandeur of the universe gives comfort to physicist / 'Nothing is permanent. / Nothing lasts over sufficiently long timescales,' Brian Greene says." https://www.cbc.ca/radio/ideas
https://www.cbc.ca/radio/ideas/finding-wonder-in-the-face-of-existential-dread-grandeur-ofthe-universe-gives-comfort-to-physicist-1.5532008
**43** See: CBC Radio Ideas, "The Relativity Revolution: Albert Einstein and the making of the modern world / From the cosmos to cellphones, Einstein's ideas have revolutionized how we live our lives today," December 2019.
https://www.cbc.ca/radio/ideas/the-relativity-revolution-albert-einstein-and-the-making-of-the-modern-world-1.5382204

**44** Frederic Brown, "It Didn't Happen," in *Transit of Earth*, PLAYBOY SCIENCE FICTION, Playboy Press, Chicago, 60611; 1971; at pages 95→110.
https://www.librarything.com/work/704545 https://www.librarything.com/author/brownfredric
*Playboy* Magazine, October 1963. https://www.librarything.com/work/5537527
**45** https://education.jlab.org/qa/atomicstructure_09.html
https://education.jlab.org/qa/how-much-of-an-atom-is-empty-space.html
https://chem.libretexts.org/Courses/Indiana_Tech/EWC%3A_CHEM_1020_-_General_Chemistry_I_(Budhi)/02%3A_Atoms_and_Elements/2.05%3A_2.5_The_Structure_of_The_Atom#:~:text=The%20diameter%20of%20an%20atom,m%E2%80%94about%20100%2C000%20times%20smaller.
https://www.youtube.com/watch?v=cwkZ–b0kCE
**46** A universe of protons $p^+$, electrons $e^-$, neutrons n and photons is sufficient to explain much of chemistry, and many chemists, in practice, can be satisfied with only these. For this book this is satisfactory. But some of chemistry and most of physics require rather more that readers of this book should be aware of, at least casually. Modern particle physics has a universe of many more kinds of fundamental particles, and there are substantial attempts to relate them to theories of radiation & wave mechanics, space and time, and to unify those theories.
**47** Generally, the notation used here is that Roman or Greek letters symbolize a particle, or an assembly of particles into an atom or an ion, and are written with a subscript and a superscript. An additional + or – superscript $^{+-}$ is for the electrical charge. (Ions are atoms with too few or too many electrons for the atom to be of neutral charge; there are positive and negative ions; for example, $Na^+$ $Cl^-$.)

The subscript is the atomic number = the number of protons in an atom = the number of electrons. This is largely related to chemical properties, which are determined by electronic structure.

The superscript is the atomic mass number = a rounded number for the total of protons + neutrons in an atom or ion.

For example for the isotope of an element called cobalt-60

$_{27}\mathbf{Co}^{60}$
Co = Cobalt.
$[Ar]3d^7 4s^2$ = electron configuration.
60 = atomic mass number
58.93 = atomic mass (g/mol).
27 = atomic number.

This notation might be over-extended for fundamental particles themselves

proton $(_1p^1)^+$
electron $(_1e^0)^-$
neutron $(_0n^1)$
electromagnetic radiation considered as photon $(_0\gamma^0)$

**48** https://en.wikipedia.org/wiki/Photon
**49** "Building blocks" = fundamental component units ≠ literally blocks.
**50** https://earthsky.org/space/what-is-the-electromagnetic-spectrum
https://imagine.gsfc.nasa.gov/science/toolbox/emspectrum1.html
https://www.britannica.com/science/electromagnetic-spectrum
**51** For example: phases of ice. https://en.wikipedia.org/wiki/Ice
https://web.archive.org/web/20090318135129/http://blogs.static.mentalfloss.com/blogs/archives/20311.html
**52** here a scientific – not a political – term.

53 https://en.wikipedia.org/wiki/Radical_(chemistry)
54 Mark Kurlansky, Salt/A World History, Alfred A. Knopf Canada, Toronto, 2002; ISBN 0-676-97268-3 www.randomhouse.ca
55 n-Heptane CAS 142-82-5 https://pubchem.ncbi.nlm.nih.gov/compound/Heptane
56 Protons p+, electrons e–, neutrons n, and photons hγ: See notes above.
57 https://en.wikipedia.org/wiki/Photon
58 (See Fig. 4.29)

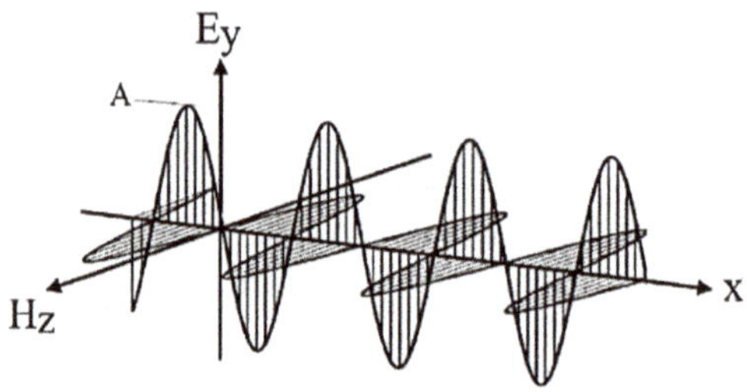


Fig. 4.29: EM wave shown in three dimensions.

"From Wikimedia Commons, the free media repository," author = And1mu. 2016. "EM-Wave," According to Commons terms; altered here. https://commons.wikimedia.org/wiki/File:EM-Wave.gif
59 $E_y$ = electric field strength vector (see Fig. 4.30).
$H_z$ = magnetic field strength vector.
x = geometric direction of propagation.
λ = wavelength
A = amplitude

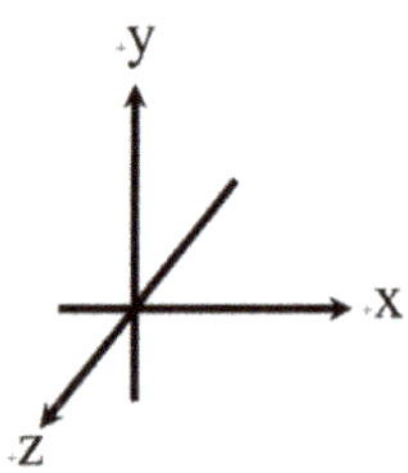


Fig. 4.30: xyz.

60 And 1mu, "EM-Wave," Wikimedia Commons, 2016. Copyright acknowledgement; Chapter 4.
61 https://www.mathsisfun.com/algebra/trig-sin-cos-tan-graphs.html (see Fig. 4.31).

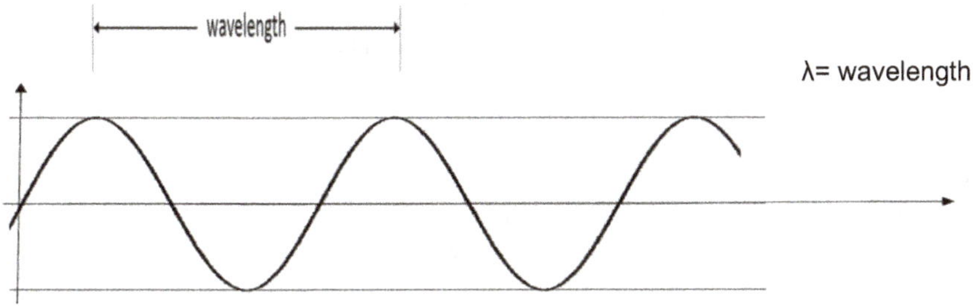


Fig. 4.31: Wave shown in two dimensions.

**62** The fundamental physics of light can be puzzling and confusing. Light can be described as wave or particle phenonium that can be explained alternatively – for different – experimental observations – called wave/particle duality – complementarity – Copenhagen. The wave theory regards light as represented by propagation in space with electric field vectors $E_y$ as a sine function, alternating at right angles as magnetic field vectors $H_z$ – as presented by Maxwell. Often, only the electric is shown. The propagation in space is in the x direction – shown as geometric distance, or as time.

The wave version is experimentally observed with diffraction gratings. The particle version is experimentally observed by assuming photons can impart momentum to electrons – as presented by Einstein.
https://www.nobelprize.org/prizes/physics/1921/summary/
https://www.aps.org/apsnews/2005/01/einstein-photoelectric-effect

Engineers involved with microwave and radio wave, *etc.* phenomena tend to refer to wave concepts – chemists and physicists involved with ultraviolet & visible light, infrared, *etc.* tend to refer to both photons and waves. Although necessarily important as underpinning theory, these quantum mechanical concepts would not be of too much practical concern for applying forensic chemistry.

**63** See also: Jack G. Calvert and James N. Pitts, Jr., *Photochemistry*, John Wiley and Sons, Inc., New York, 1966; page 4.

**64** "Spectroscopy/Molecular energy levels"; Wikiversity; April 2020. https://en.wikiversity.org/wiki/Wikiversity:Main_Page
https://en.wikiversity.org/wiki/Spectroscopy/Molecular_energy_levels
https://en.wikiversity.org/wiki/Spectroscopy/Rotational_spectroscopy
https://en.wikiversity.org/wiki/Wikiversity:Welcome

**65** See John Gribbin, *In Search of the Big Bang/Quantum Physics and Cosmology*, BANTAM BOOKS, New York City, 10103; 1986; ISBN 0-553-34258-4.

**66** Gribbin, . . . *BANG* . . ., page 216, et seq.

**67** John Gribbin, *In Search of Schrödinger's Cat / Quantum physics and reality*, BANTAM BOOKS, New York City, 10036; 1984; ISBN 978-0-553-34253-6.

**68** John Gribbin, . . . *CAT* . . ., page 159, et seq. "The Copenhagen Interpretation"

**69** Werner Heisenberg, *Physics & Philosophy / The Revolution in Modern Science*, Harper Perenial, New York City, 10022, 1962, 2007; ISBN 978-0-06-120919-2.

**70** Heisenberg, *Physics & Philosohy* . . ., page 18, *et seq.* ". . . Copenhagen Interpretation . . ."

**71** "molecular structures" here = molecules, ions, atoms, their intermediary formations, *etc.*

**72** EM spectrum. Magnitudes of things:
https://earthsky.org/space/what-is-the-electromagnetic-spectrum
https://imagine.gsfc.nasa.gov/science/toolbox/emspectrum1.html
https://www.britannica.com/science/electromagnetic-spectrum

73 See also: C.A. Parker, *Photoluminescence of Solutions/With Applications to Photochemistry and Analytical Chemistry*, Elsevier, Amsterdam . . ., 1968; at page 3 – Table 1 – "APPROXIMATE SIZES OF QUANTA."
74 https://calculator-converter.com/nanometers-to-meters.htm
75 "Spectroscopy/Molecular energy levels"; Wikiversity; April 2020.
https://en.wikiversity.org/wiki/Wikiversity:Main_Page
https://en.wikiversity.org/wiki/Spectroscopy/Molecular_energy_levels
https://en.wikiversity.org/wiki/Spectroscopy/Rotational_spectroscopy
https://en.wikiversity.org/wiki/Wikiversity:Welcome
76 https://en.wikipedia.org/wiki/Microwave_spectroscopy
https://chem.libretexts.org/Bookshelves/Physical_and_Theoretical_Chemistry_Textbook_Maps/Supplemental_Modules_(Physical_and_Theoretical_Chemistry)%2FSpectroscopy%2FRotational_Spectroscopy%2FMicrowave_Rotational_Spectroscopy
https://www.sciencedirect.com/topics/chemistry/microwave-spectroscopy
77 and arguably, metaphorically, with biblical reference.
78 EM spectrum. Magnitudes of things:
https://earthsky.org/space/what-is-the-electromagnetic-spectrum
https://imagine.gsfc.nasa.gov/science/toolbox/emspectrum1.html
https://www.britannica.com/science/electromagnetic-spectrum
79 American Chemical Society – Periodic Table http://acswebcontent.acs.org/games/pt.html
www.acs.org/content/acs/en/education/whatischemistry/periodictable.html
80 https://www.store.acs.org/eweb/ACSTemplatePage.aspx?site=ACSACS_Store&WebCode=storeCatList&catKey=944d7d91-5947-42f3-aa4f-a5e0d897a8b3
81 http://www.rsc.org/periodic-table/
82 https://www.acs.org/content/acs/en/education/whatischemistry/periodictable.html
83 CBC Radio, 11 Mar 2019, Paul Kennedy, *Ideas / The Music of Matter: 150 years of the Periodic Table*
https://www.cbc.ca/radio/ideas
https://www.cbc.ca/radio/ideas/the-music-of-matter-150-years-of-the-periodic-table-1. 5051355
https://en.wikipedia.org/wiki/Dmitri_Mendeleev
84 See also: Siobhan Roberts, "Reconsidering the Periodic Table / The chart of elements has served chemistry well for 150 years. But it's not the only one," *The New York Times*, 27 August 2019, page D3.
85 See also: Karoliina Pulkkinen, "Values in the Development of Early Periodic Tables," Published online15 April 2020, pages 174→198.
https://www.tandfonline.com/doi/full/10.1080/00026980.2020.1747325?src=recsys
86 American Chemical Society – Periodic Table
http://acswebcontent.acs.org/games/pt.html www.acs.org/content/acs/en/education/whatischemistry/periodictable.html
http://www.rsc.org/periodic-table/
87 Brookhaven National Laboratory – Chart of Nuclides http://www.nndc.bnl.gov/chart/
88 See also: Oliver Sacks, *Uncle Tungsten: Memories of a Chemical Boyhood*, Vintage, reprint 2002; ISBN-10: 0375704043; ISBN-13: 978–0375704048.
https://www.amazon.com/Uncle-Tungsten-Memories-Chemical-Boyhood/dp/0375704043
89 https://en.wikipedia.org/wiki/Uncle_Tungsten
90 See also: Marianne Taylor, *SCIENCE IN BITE-SIZED CHUNKS*, Michael O'Mare Books Limited, London, SW47NC; 2019; pages 58→62. ISBN 978-1-78929-151-3. www.mombooks.com
91 https://www.youtube.com/watch?v=NgJfFCz2Nr8
https://www.youtube.com/watch?v=SqYPrA7upiE
https://www.youtube.com/watch?v=uTc3U2Cqen4
92 Tom Lehrer https://www.youtube.com/watch?v=9p5kY4UjGdk&list=RD9p5kY4UjGdk&start_radio=1

**93** But see: Felicity Nelson, "JAPAN RACES TO STAKE ITS CLAIM ON ELEMENT 119 / Riken is in pole position after US-Russia relations break down. And a US lab chases element 120," *C&EN*, 20/27 January 2025, cover & pages 28 → 32.
https://cendigitalmagazine.acs.org/wp-content/uploads/sites/523/2025/01/10302-PDF-2025-01-27.pdf?utm_email=F464F3842364E3E1934223D642
**94** J.P. McEvoy and Oscar Zarate, *→Introducing Quantum Theory / A Graphic Guide*, Icon Books Ltd.; 2013; ISBN 978-184046-850-2; re Dimitri Mendeléev – Periodic Table, page 100 *et seq.* [Although Mendeléev would not have known of neutrons, nor of the modern concept of atomic number.]
**95** Isaac Asimov, *Asimov on Chemistry*, Anchor Books, Doubleday, Garden City, New York, 1975; ISBN 0-385-04005-9; pages 100 and 101.
**96** See: https://www.ch.cam.ac.uk/news/would-alien-have-periodic-table-yes
https://www.bbc.com/news/science-environment-47008289
**97** Brookhaven National Laboratory - Chart of Nuclides http://www.nndc.bnl.gov/chart/
**98** https://en.wikipedia.org/wiki/Carbon-14

$$1/10^{12} = 1/10^{10}\%$$

**99** Text. Here it is adapted from previous writing by the author.
**100** M.G., "the regulation of named chemicals – What are *CAS Numbers* – why are they used?" Toronto Workers' Health & Safety Clinic *newsletter*, Vol. 16, No. 3, July 2008; pages 4→9.
**101** This is an idealized generalization of chemical bonding. There are various other – sometimes apparently competing – more refined and detailed theories. What is presented here – as simplified – is thought by the author as sufficient for here. While some modern theoretical physicists would be looking for an elegant one-big-theory to explain everything of the universe, chemists – and other scientists and engineers – doing less-universal practical work, seem to be able to cope with an overlapping patchwork of lesser theories. But, the physicists shouldn't be discouraged.
**102** The molecular structural drawings here may vary somewhat from those found in CAS – in terms of spacing, symbols, and the showing of π–bonding – but should be clearly identifiable as for the same molecule.
**103** *Methane* $CH_4$ CAS *Registry Number* 74-82-8 molar mass = 16.04g/mol = (12.01 + 4 × 1.01) g/mol
https://en.wikipedia.org/wiki/Methane https://en.wikipedia.org/wiki/Carbon https://en.wikipedia.org/wiki/Hydrogen
**104** See William H. Brown, *INTRODUCTION TO ORGANIC CHEMISTRY / SECOND EDITION*, Willard Grant Press, Boston, 1978; Chapter 1 "The Covalent Bond and the Geometry of Molecules," 1.9 "The Tetrahedral Carbon Atom," page 9, et seq.
**105** Brown, at page 10 – $sp^3$ hybridization.
**106** The central atom thought of as at the centre of a regular tetrahedron – with bonds pointing toward the four corners.
**107** In the molecule of this example the bonds shown are single bonds – sharing a pair of electrons :
Other molecules might contain double bonds = sharing two pairs of electrons ::
And triple bonds ≡ sharing three pairs of electrons :::
**108** *The Allyn and Bacon Molecular Model Set for Organic Chemistry*, 1984.
**109** See also: John. W. Lehman, *Instruction Booklet for the Allyn and Bacon Molecular Model Set for Organic Chemistry*, Allyn and Bacon, Inc., Boston, 1984. https://www.gettextbooks.com/isbn/9780132019866/
**110** The exaggerated wedge-shaped lines here are to try to give an image of receding from, or coming toward, the viewer who would be facing the plane of the page.
**111** Where the dashed (exaggerated wedge-shaped) lines here represent the hydrogen as below the plane of the page, with the carbon in the plane of the page.

**112** Where the solid (exaggerated wedge-shaped) lines here represent the hydrogen as above the plane of the page, with the carbon in the plane of the page.
**113** See also:
https://www.chem.ucalgary.ca/courses/351/mechanistic_etext/Ch01/ch1-4-2.html
https://chem.libretexts.org/Bookshelves/Organic_Chemistry/Supplemental_Modules_(Organic_Chemistry)/Fundamentals/Structure_of_Organic_Molecules
**114** See:

- William H. Brown, *Introduction to Organic Chemistry*, 2nd ed., Willard Grant Press, Boston, 1978, page 302 et seq.
- Isaac Asimov, *Life and Energy*, Avon [1962]/Discus, New York City, 1972; Chapter 18, "The Workings of Enzymes," at pages 251→254.

**115** See Robert C. Weast, editor, *CRC Handbook of Chemistry and Physics*, 56th edition, CRC Press, Cleveland, Ohio, 1975; "BOND LENGTHS BETWEEN CARBON AND OTHER ELEMENTS," page F-211.
**116**

$$1 \text{ Ångström unit} = 1 \times 10^{-8} \text{ metre}$$

$$1 \text{ nm} = 1 \times 10^{-9} \text{ metre}$$

**117** https://en.wikipedia.org/wiki/Angstrom
**118** Isaac Asimov, *Asimov's Biographical Encyclopedia of Science and Technology / Second Revised Edition*, Doubleday and Company, Inc., Garden City, New York, 1982; ISBN 0-385-17771-2: Anders Jonas Ångström [585]; pages 385 and 386.
**119** Anders Jonas Ångström, 1814→1874, Swedish scientist.
https://en.wikipedia.org/wiki/Anders_Jonas_%C3%85ngstr%C3%B6m
**120** 1mole = mass / molar mass
**121** See: Isaac Asimov, *ASIMOV ON CHEMISTRY*, Anchor Books, Doubleday, Garden City, New York, 1975; ISBN 0-385-04005-9; Chapter 6 "To Tell a Chemist," page 89, *et seq.* (Originally appearing as an essay in *The Magazine of Fantasy and Science Fiction*, May 1965.)
**122** https://www.britannica.com/science/Avogadros-number
**123** Asimov, . . . *On Chemistry*, pages 11, 93, 96 – Avogadro's number.
**124** https://www.mathway.com/popular-problems/Algebra/226346
**125** Lorenzo Romano Amedeo Carlo Avogadro, Count of Quaregna and Cerreto, 1776 →1856, Italian scientist.
https://en.wikipedia.org/wiki/Amedeo_Avogadro https://physics.nist.gov/cgi-bin/cuu/Value?na
**126** See: William H. Brown, *Introduction to Organic Chemistry*, 2nd ed., Willard Grant Press, Boston, 1978, page 302 *et seq.*; Chapter 6 "Benzene and the Concept of Aromaticity," 6.4 "The Structure of Benzene – The Molecular Orbital Model," page 142, et seq.
**127** Benzene. CAS 71-43-2. Chapter 4.
**128** Benzene. CAS 71-43-2 $C_6H_6$ 78.11 g/mol
https://en.wikipedia.org/wiki/Benzene
https://www.cdc.gov/chemical-emergencies/chemical-fact-sheets/benzene.html
**129** https://www.youtube.com/watch?v=NmzMoE3LQHM
**130** (An equilateral and equiangular hexagon )
**131** Brown, at pages 142 and 143.
**132** $sp^2$ hybridization.
**133** Benzene ($C_6H_6$) with $sp^2$ hybridization and $\pi$ bonding should not be confused with cyclohexane $C_6H_{12}$ with $sp^3$ hybridization and $\sigma$ bonding.
**134** Cyclohexane. Chair and boat. CAS 110-82-7, $C_6H_{12}$. Chapters 4 and 6.

**135** An alternative way often used to show benzene is as resonance structure, wherein the hexagonal ring would show double bonds, with two possibilities; the electronic structure (see Fig. 4.32) should actually considered as a kind of averaging thereof, with no actual double bonds:

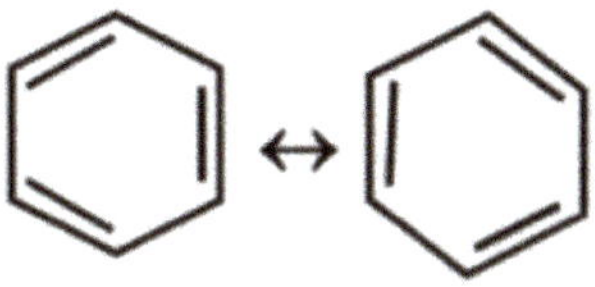

Fig. 4.32: Benzene resonance structures.

**136** See: "A general chemistry Libretexts Textmap organized around the textbook / Chemistry: The Central Science / by Brown, LeMay, Busten, Murphy, and Woodward," 2020, 2019.
https://chem.libretexts.org/Bookshelves/General_Chemistry/Map%3A_Chemistry_-_The_Central_Science_(Brown_et_al.)
https://libretexts.org/
Resonance Structures
https://chem.libretexts.org/Bookshelves/General_Chemistry/Map%3A_Chemistry_-_The_Central_Science_(Brown_et_al.)/08%3A_Basic_Concepts_of_Chemical_Bonding/8.06%3A_Resonance_Structures#:~:text=Resonance%20structures%20are%20a%20set,fractional%20bonds%20and%20fractional%20charges.
Molecular Orbital Theory
https://chem.libretexts.org/Bookshelves/General_Chemistry/Chem1_(Lower)/09%3A_Chemical_Bonding_and_Molecular_Structure/9.08%3A_Molecular_Orbital_Theory
Two Theories of Bonding
https://chem.libretexts.org/Courses/College_of_the_Canyons/Chem_201%3A_General_Chemistry_I_OER/11%3A_Theories_of_Chemical_Bonding
https://pubs.acs.org/doi/10.1021/acs.jchemed.1c00919#:~:text=with%20two%20general%20theories%2C%20valence,VB)%20theory%20and%20MO%20theory
**137** Naphthalene CAS 91-20-3 $C_{10}$ $H_8$ 128.174 g/mol (See Figs. 4.33a & 4.33b.)
https://en.wikipedia.org/wiki/Naphthalene#:~:text%3D8-,.,fused%20pair%20of%20benzene%20rings

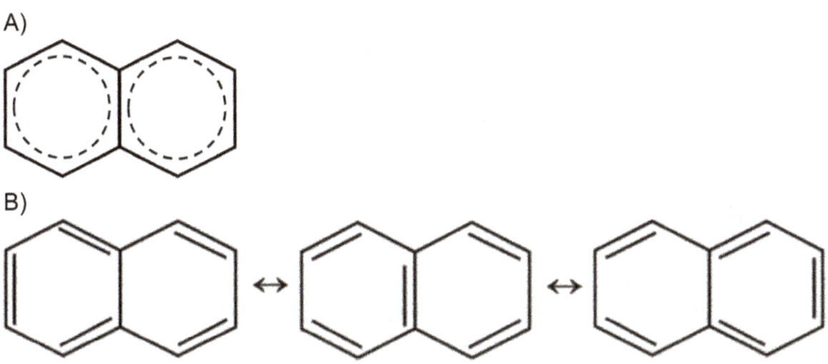


Fig. 4.33: (a) Naphthalene pi bonding structure. (b) Naphthalene resonance structures.

**138** See also: An older view of describing molecular structure: R. B. Sandin & T. H. Evans, "The Structure of Some Naphthalene, Hydrindene and Tetralin Derivatives," *Journal of the American Chemical Society*, Vol.61, 1939, pages 2916→2919; Department of Chemistry, University of Alberta (see Fig. 4.34). https://pubs.acs.org/doi/pdf/10.1021/ja01265a097

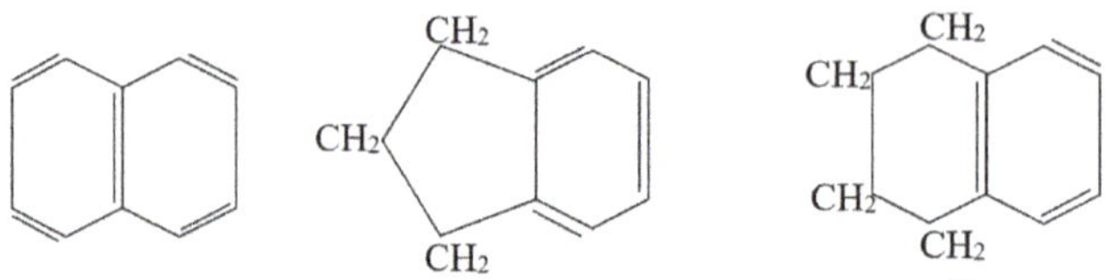


Fig. 4.34: *JACS*, 1939.

**139** *JACS*, 1939, at page 2916.

> . . . There is much chemical evidence in favor of the idea that the double bonds of naphthalene are not free to migrate and that the Erlenmeyer formula (I) is to be preferred . . . .

**140** *JACS*, 1939, at page 2917.

> . . . An examination of Table I definitely points to the Erlenmeyer formula for naphthalene possessing a double bond common to both rings . . . .

**141** Trying to interpret the molecular structures shown, *JACS*, 1939, into more modern terminology:
I naphthalene
II indane
III tetralin

– all of which are better described as for $\pi$ bonding, or represented with resonance structures understood to not be with actual double bonds at fixed locations.
**142** Linus Pauling's *The Nature of the Chemical Bond, and the Structure of Molecules and Crystals* was published 1939. See also: https://www.britannica.com/biography/Linus-Pauling
http://scarc.library.oregonstate.edu/coll/pauling/bond/
https://historyofinformation.com/detail.php?id=3956
https://www.academia.edu/26073847/Pauling_L_The_nature_of_the_chemical_bond_Cornell_Univ_1960_
https://www.abebooks.com/first-edition/Nature-Chemical-Bond-DUST-JACKET-PAULING/4881690120/bd
**143** http://www.chemgapedia.de/vsengine/vlu/vsc/en/ch/12/oc/vlu_organik/aromaten/aro maten/nomenklatur.vlu/Page/vsc/en/ch/12/oc/aromaten/aromaten/nomenklatur2/nomenkla tur2.vscml.html
https://en.wikipedia.org/wiki/Arene_substitution_pattern
**144** CAS 95-47-6 *o*-xylene 106.17 g/mol.
https://en.wikipedia.org/wiki/Xylene https://en.wikipedia.org/wiki/O-Xylene
**145** CAS 108-38-3 *m*-xylene 106.17 g/mol.
https://en.wikipedia.org/wiki/Xylene https://en.wikipedia.org/wiki/M-Xylene
**146** CAS 106-42-3 *p*-xylene 106.17 g/mol.
https://en.wikipedia.org/wiki/Xylene https://en.wikipedia.org/wiki/P-Xylene
**147** https://www.merriam-webster.com/dictionary/congener
**148** https://en.wikipedia.org/wiki/Congener_(chemistry)#:~:text=In%20chemistry%2C%20congeners%20are%20chemical,example%20of%20a%20congener%20grouping.

**149** Daniel M. Perrine, *The Chemistry of Mind-Altering Drugs / History, Pharmacology, and Cultural Context*, American Chemical Society, Washington, DC, 1996; ISBN 0-8412-3253-9; page 63, et seq. – at page 63 2–29
**150** http://en.wikipedia.org/wiki/Diamorphine
**151** http://www.emcdda.europa.eu/publications/drug-profiles/heroin#chemistry
**152** Perrine, Chapter 2.
**153** *heroin-hydrochloride*, CAS 1502-95-0 $C_{21}H_{23}NO_5 \cdot HCl$, 405.87 g/mol.
**154** Paige Williams, "*A Deadly Mistake / Addicts who share a lethal dose of drugs are being prosecuted as killers,*" *The New Yorker*, 10 February 2020, pages 28→39, at page 30.
**155** https://chem.nlm.nih.gov/chemidplus/rn/76-42-6
**156** Paige Williams, "*A DEADLY MISTAKE / Addicts who share a lethal dose of drugs are being prosecuted as killers,*" *The New Yorker*, 10 February 2020, pages 28→39, at page 30.
**157** https://en.wikipedia.org/wiki/Caffeine
**158** https://www.sigmaaldrich.com/CA/en/sds/sial/c0750?userType=undefined
**159** ACS CHEMISTRY AND THE LAW *newsletter* Vol. 20 No.2 Fall 2004 page 11
**160** www,acs.org Molecule of the Week – Archive — 23 December 2002
https://www.acs.org/molecule-of-the-week/archive.html?archive=All
https://www.acs.org/molecule-of-the-week/archive/c/caffeine.html
**161** https://en.wikipedia.org/wiki/Theobromine
**162** https://pubchem.ncbi.nlm.nih.gov/compound/Theobromine#section=2D-Structure
**163** https://en.wikipedia.org/wiki/Allotropy https://en.wikipedia.org/wiki/Allotropes_of_phosphorus
**164** white phosphorus $P_4$ (tetrahedral) CAS 12185–10-3 – very toxic.
**165** red phosphorus $(P_4)_n$ (P amorphous, polymer) CAS 7723-14-0 – not very toxic.
**166** M.G., *The Law of Occupational Health and Safety in Ontario, Second Edition*, Butterworths, Toronto and Vancouver, August 1994, ISBN 0-409-90414-7; Chapter 2.7.b.
**167** Herbert Ellern, in *Kirk-Othmer Encyclopaedia of Chemical Technology*, 3rd ed., vol.15, page 1, et seq., Wiley-Interscience, New York, 1981.
**168** White Phosphorous Match Act of 1912: *An Act to Provide for a Tax upon White Phosphorus, and for Other Purposes*, Washington D.C., 09 April 1912; and REGULATIONS UNDER . . ., Washington D.C., 10 May 1913. http://www.cdc.gov/niosh/enews/enewsv9n12.html http://archive.org/stream/regulation sconc00revegoog#page/n6/mode/2up
**169** M.L. Myers & J.D. McGlothlin, "Matchmakers' "phossy jaw" eradicated," *Am Ind Hyg Assoc J.*, 1996 Apr;57(4):330–2. http://www.ncbi.nlm.nih.gov/pubmed/8901233
**170** http://en.wikipedia.org/wiki/Phosphorus#White_phosphorus_and_related_molecular_forms
http://en.wikipedia.org/wiki/White_phosphorus#References
http://en.wikipedia.org/wiki/Red_phosphorus#Red_phosphorus
http://www.britannica.com/EBchecked/topic/457568/phosphorus-P/280563/Properties-and-reactions#ref146121
**171** http://en.wikipedia.org/wiki/Berne_Convention_%281906%29
**172** General William Booth, Salvation Army http://www.britainunlimited.com/Biogs/Booth.htm
http://en.wikipedia.org/wiki/William_Booth
http://en.wikipedia.org/wiki/Phossy_jaw
http://en.wikipedia.org/wiki/London_matchgirls_strike_of_1888
http://en.wikipedia.org/ wiki/Salvation_Army
**173** Robert C. Weast, editor, *CRC Handbook of Chemistry and Physics*, 56th edition, CRC Press, Cleveland, Ohio, 1975; "The Elements," page B-26 & B27.
**174** L. Bretherick, editor, *HAZARDS IN THE CHEMICAL LABORATORY*, The Royal Society of Chemistry, London, 1981; ISBN 0 85186 419 8; page 442.

**175** https://byjus.com/chemistry/red-phosphorus/
https://www.compoundchem.com/2014/11/20/matches/
https://en.wikipedia.org/wiki/Match
**176** https://www.livescience.com/28932-phosphorus.html
**177** https://www.livescience.com/28932-phosphorus.html
**178** Alice and Tweedledum and Tweedledee, Lewis Carroll, *Alice's Adventures in Wonderland* and *Through the Looking-Glass*, [1865], Signet Classics, The New American Library, New York City, 1960 & 1962; from . . . *Looking-Glass* . . ., "Tweedledum and Tweedledee," Chapter IV, page 157; & cover.
**179** https://www.alice-in-wonderland.net/resources/pictures/tweedledum-tweedledee/
https://www.alice-in-wonderland.net/resources/faq/#copyright
**180** (Tweedledum and Tweedledee may not actually be chiral here – but they suggest images that look almost alike.)
**181** A significant difference is found in the molecules'*optical rotary power*, also called *optical activity* – see Howard A. Strobel, *Chemical Instrumentation / A Systematic Approach to Instrumental Analysis*, Addison-Wesley Publishing Company, Inc., Reading, Massachusetts & London, England, 1960; Chapters 4–13, Optical rotation, pages 100→ 102; and Chapter 10, POLARIMETRY, pages 252→264.
**182** Another significant difference may be found in chemical catalysis in biological systems – where the three dimensional shapes of enzymes acting on their substrates might be bioactive for only one of the chiral forms – so that – from the introductory quote for this chapter – "Perhaps Looking-Glass milk isn't good [for Alice's Kitty] to drink . . ."
**183** To recognize fragrances is reasonably inferred from biological process:
**184** See: John Lanchester, "Scents and Sensibility / *What the nose knows,*" *The New Yorker*, Books, 10 March 2008 Issue (See Fig. 4.35.).
https://www.newyorker.com/magazine/2008/03/10/scents-and-sensibility
**185** In his *New Yorker* article, John Lanchester notes, for example, that the different spearmint and caraway smells come alternatively from the mirror images of otherwise the same molecule.
**186**

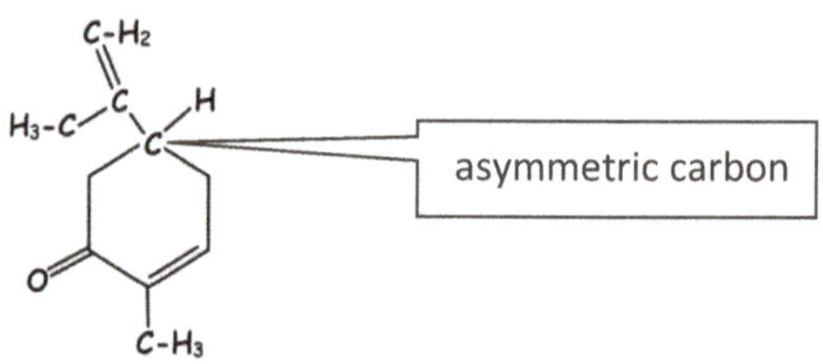


Fig. 4.35: Spearmint caraway.

**187** Carvone https://en.wikipedia.org/wiki/Carvone
https://pubchem.ncbi.nlm.nih.gov/compound/_-_-Carvone#section=Top
**188** [Carvone] [Spearmint] CAS 6485-40-1 (R) L
http://www.commonchemistry.org/ChemicalDetail.aspx?ref=6485-40-1
https://www.sigmaaldrich.com/catalog/product/aldrich/124931?lang=en&region=CA
**189** [Carvone] [Caraway] CAS 2244-16-8 (S) D
https://www.sigmaaldrich.com/catalog/product/aldrich/435759?lang=en&region=CA
**190** [Carvone] [racemic mixture] CAS 99-49-0 (R/S) L D
http://www.commonchemistry.org/ChemicalDetail.aspx?ref=99-49-0

**191** McMaster University - Chem2OB3 Lab Manual; "23dec99; jp."
https://www.chemistry.mcmaster.ca/~chem2ob3/nhw_temp/old_old_labmanual/expt1/2ob3-exp1b.html
**192** https://www.chem.ucla.edu/~harding/IGOC/C/chiral.html
https://en.wikipedia.org/wiki/Chirality
**193** https://www.acsh.org/news/2018/08/01/carvone-one-molecule-two-different-scents-and-flavors-13255
**194** See also: Carl Zimmer, *"A 'Second Tree of Life' Could Wreak Havoc, Scientists Warn* / Research on so-called mirror cells, which defy fundamental properties of living organisms, should be prohibited as too dangerous, biologists said," *New York Times*, 12 Dec. 2024.
https://www.nytimes.com/2024/12/12/science/mirror-life-microbes-research.html
**195** The exaggerated wedge-shaped lines here are to try to give an image of receding from, or coming toward, the viewer who would be facing the plane of the page.
**196** Where the dashed (exaggerated wedge-shaped) lines here represent the hydrogen as below the plane of the page, with the carbon in the plane of the page.
**197** Where the solid (exaggerated wedge-shaped) lines here represent the hydrogen as above the plane of the page, with the carbon in the plane of the page.
**198** See also:
https://www.chem.ucalgary.ca/courses/351/mechanistic_etext/Ch01/ch1-4-2.html
https://chem.libretexts.org/Bookshelves/Organic_Chemistry/Supplemental_Modules_(Organic_Chemistry)/Fundamentals/Structure_of_Organic_Molecules
**199** See:
- William H. Brown, *Introduction to Organic Chemistry*, 2nd ed., Willard Grant Press, Boston, 1978, page 302 et seq.
- Isaac Asimov, *Life and Energy*, Avon [1962]/Discus, New York City, 1972; Chapter 18, "The Workings of Enzymes," at pages 251→254.

**200** See: Brown, at page 302, 303.
**201** (When an organic molecule such as this is considered, the naturally found atomic isotopes are usually assumed – so that it is carbon-12, oxygen-16 and hydrogen-1.)
**202** ${}_6C^{12}$ ${}_8O^{16}$ ${}_1H^{1}$ ${}_{\text{atomic number}}[\ ]^{\text{atomic mass}}$ https://en.wikipedia.org/wiki/Mass_number
**203** Uniquely – but for chirality – read on.
**204** For more detailed Fischer projection rules see: Brown, pages 202, 303.
**205** https://chem.libretexts.org/Courses/Brevard_College/CHE_301_Biochemistry/02%3A_Carbohydrates/2.03%3A_Fischer_Projections
**206** Brown, page 303.
**207** The mirror plane here is for a two-dimensional representation; a three dimensional model must be consulted. In the two-dimensional short-hand drawing here both molecules are identical – no matter how the atoms are placed in the two dimensional drawings to give an illusion of mirror reflection. Their chiral difference is revealed only by consultation with a three-dimensional model. See for example the *methane* model pictured above.
**208** The older **D L** nomenclature, still often used, and used here, has been replaced by an **R S** nomenclature. See Brown at pages 302, 303; and page 94, *et seq.*
**209** https://chemistry.stackexchange.com/questions/66024/are-our-hands-really-chiral
https://i.stack.imgur.com/fol1k.png
https://www.chadsprep.com/chads-organic-chemistry-videos/stereoisomers/
**210** From Wikimedia Commons, the free media repository," author = Furfur. 2017. According to Commons terms; altered here. https://commons.wikimedia.org/wiki/File:Hand_drawing.png
**211** The drawing here is altered for a person holding both left & right hands in front of face – viewing the palms.
**212** Furfur, "Hand drawing," Wikimedia Commons, 2017, Copyright acknowledgement; Chapter 4.

**213** https://en.wikipedia.org/wiki/Racemic_mixture
https://en.wikipedia.org/wiki/Glyceraldehyde
http://www.chim.lu/ech0898.php
https://pubchem.ncbi.nlm.nih.gov/compound/L-glyceraldehyde
http://www.chemspider.com/Chemical-Structure.71347.html
**214** See also: Brown, page 98.
**215** In general, it is the D forms of sugars that are found in natural biological systems, and their chirality is appropriate to the enzymes of metabolism (and similarly, the L forms of amino acids are found in natural proteins). Syntheses by non-biological routes, that do not pay special attention to chirality, result in racemic mixtures – DL. Special methods are used to synthesize L sugars (and D amino acids). Historically, the relative ease of producing these different versions could be reflected in the pricing – for example, for *glyceraldehyde* (calculated for 5 g. from *1993–1994 Janssen Chimica Catalog Handbook of Fine Chemicals*, Janssen Pharmaceutica N.V., Geel, Belgium; page 739): D US$ 329.50, L US$ 630, DL US$ 51.50.
**216** Glyceraldehyde can also be drawn as shown in Fig. 4.36.

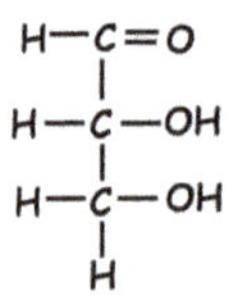

Fig. 4.36: Glyceraldehyde.

**217** *Gedankenexperiment*. Introduction 2.2.
**218** Although this is a fictitious example concocted for illustrative purposes here, it is possible that this molecule could actually exist.
**219** ${}_6C^{14} \rightarrow {}_7N^{14} + {}_{-1}\beta^0 \quad t_{½} = 5{,}730$ years.
**220** The *isotope* of this example is radioactive; many *isotopes* are not.
**221** For example, paraformaldehyde: (See Fig. 4.37.)
https://en.wikipedia.org/wiki/Paraformaldehyde
https://en.wikipedia.org/wiki/Polymerization
https://web.archive.org/web/20080415161207/http://msds.chem.ox.ac.uk/PA/paraformaldehyde.html
**222** n here refers to a counting number; not a neutron.
**223** See Lawrence E. Nielsen, *Mechanical Properties of Polymers and Composites*, Volume 1, Marcel Dekker, Inc., New York, 1974; and I.M. Ward, *Mechanical Properties of Solid Polymers*, Wiley-Interscience, London, . . ., 1971.

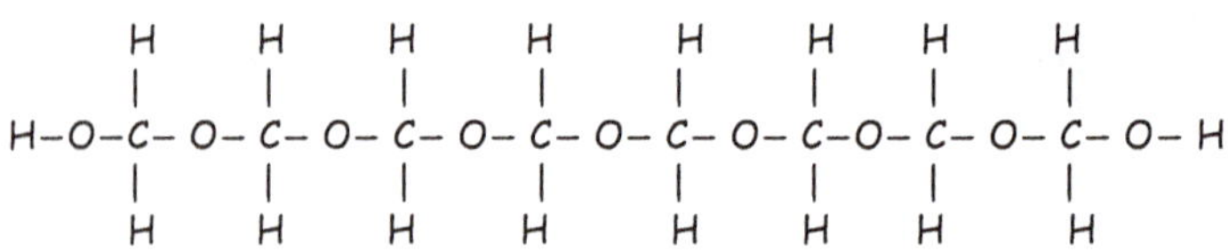

Fig. 4.37: Paraformaldehyde.

**224** Polymers in museum. Chapters 2 and 4.
**225** Both plastics (thermoplastics) and thermosetting resins are often simply referred to as “plastics”
**226** Anhydrous – without water molecules $Na_2Cr_2O_7$ 262 g/mol CAS 10588–01-9
Dihydrate – with water molecules $Na_2Cr_2O_7 \cdot 2H_2O$ 298 g/mol CAS 7789–12-0

**227** $Na_2Cr_2O_7$ Glenn Hess, "CHEMICAL HAZARDS [Oregon] Jury awards $85 million to soldiers exposed to hexavalent chromium in Iraq," *C&EN*, 12 November 2012, page 8.
**228** Toronto Workers Health &Safety Clinic *newsletter*, 2012 12, Vol. 20, No.6.
**229** *Ethanol.* $CH_3$-$CH_2$-OH CAS 64-17-5 46.07 g/mol
**230** https://en.wikipedia.org/wiki/Ethanol#Absolute_alcohol
https://chemistry.stackexchange.com/questions/14392/ethanol-and-water-hygroscopic-equilibrium-concentration
https://www.scienceforums.net/topic/40271-difference-between-pure-ethanol-and-95/
https://www.cs.mcgill.ca/~rwest/wikispeedia/wpcd/wp/e/Ethanol.htm
**231** See: Isaac Asimov, *ASIMOV ON CHEMISTRY*, Anchor Books, Doubleday, Garden City, New York, 1975; ISBN 0-385-04005-9; Chapter 6 "To Tell a Chemist," page 89, et seq. (Originally appearing as an essay in *The Magazine of Fantasy and Science Fiction*, May 1965.)
**232** = an unshared pair of electrons. See Brown, at pages 5, 8, 11 and 12
**233** This is the author's theory.
**234** Cocaine. CAS 5321-4, CAS 50-36-2. Chapters 4, 6 and 10.
**235** See, for example: James Glanz, "*Tests Show Genetic Signature of Virus That May Have Infected President Trump,*" *The New York Times*, 01 November 2020.
https://www.nytimes.com/2020/11/01/science/trump-covid-19-genome.html
**236** See also: Jonathan Corum & Carl Zimmer, "New Mutations, New Variants," *The New York Times*, Science Times, 16 February 2021, page D2.
**237** https://pubs.acs.org/doi/10.1021/acs.est.9b02540
**238** https://www.jtcvs.org/article/S0022-5223(14)01225-2/pdf
**239** https://pmc.ncbi.nlm.nih.gov/articles/PMC10151227/#:~:text=Microplastics%20have%20been%20detected%20in,or%20cause%20local%20particle%20toxicity
**240** https://en.wikipedia.org/wiki/Microparticle
**241** https://pmc.ncbi.nlm.nih.gov/articles/PMC10151227/#:~:text=Microplastics%20have%20been%20detected%20in,or%20cause%20local%20particle%20toxicity
**242** $1 \times 10^{-6}$ metre = 1 μm = 1 micron = 0.001 mm.
https://en.wikipedia.org/wiki/Micrometre#:~:text=The%20micrometre%20(Commonwealth%20English%20as,%2D%22%20%3D%2010%E2%88%926)%3B
https://www.infinitioptics.com/glossary/micron-um#:~:text=Glossary%20Definition,%C2%B5m%20is%20equal%20to%201mm.
**243** https://www.sciencedirect.com/topics/pharmacology-toxicology-and-pharmaceutical-science/microparticle
**244** See, for example: Nina Agrawal & Theodore Tae, "Microplastics Are in Our Brains. What Does That Mean?" New York Times, 08 April 2025.
https://www.nytimes.com/video/well/100000010085694/microplastics-are-in-our-brains-what-does-that-mean.html
**245** David Wallace-Wells, "You Are Contaminated," *New York Times*, 04 August 2025.
https://www.nytimes.com/2025/08/04/opinion/contamination-exposome.html
**246** *Gedankenexperiment.* Introduction 2.2.
**247** Hardly ever, if ever, in the author's view. This example is over-concocted and over-fictionalized for illustrative purposes.
**248** But see: G. Antony, L.W. White, B.R. Landau, "Metabolism of D- and L-glyceraldehyde in adipose tissue: a stereochemical probe for glycerokinase activity," *J. Lipid. Res.* 1969 Sep;10 (5):521–7.
https://pubmed.ncbi.nlm.nih.gov/5808825/
**249** TWH&SLC *newsletter*, Vol.16 No.2 – at page 6.
http://workers-safety.ca/features/newsletters/ http://workers-safety.ca/index.html
http://newsletter.workers-safety.ca.s3-website-us-east-1.amazonaws.com/#!/newsletters%2F

**250** See *CAS Registry Handbook*. According to the 1985 ACS Annual Report (*C&EN*, 21 April 1986, p. 44), the registry contained information on 7 454 721 substances and nearly 11.4 million names. 237 https://www.tandfonline.com/doi/abs/10.1080/00026980.2020.1718303
**251** https://www.tandfonline.com/doi/abs/10.1080/00026980.2020.1718303
**252** https://vdocuments.site/systematic-nomenclature-of-organic-chemistry-.html
**253** Text. Here it is adapted from previous writing by the author.
**254** M.G., "the regulation of named chemicals – What are *CAS Numbers* – why are they used?" Toronto Workers' Health & Safety Clinic *newsletter*, Vol.16, No.3, July 2008; pages 4→7.
http://workers-safety.ca/features/newsletters/
http://workers-safety.ca/index.html http://newsletter.workers-safety.ca.s3-website-us-east-1.amazonaws.com/#!/newsletters%2F
**255** What are *CAS Numbers* . . . This is an instalment in a continuing series on the regulation of named chemicals, started in the April 2008 issue – Vol.16 No.2 – Toronto Workers' Health & Safety Clinic *newsletter*, with the then intention to provide critical appraisals in a collection that might serve as reference material. The table of topics is found in that issue at page 6.
**256** The general convention as typically used, at first use, at various places in this book, is that a substance would be identified by a common chemical name, followed by its symbols, CAS Registry number, with a structural diagram near-by. Thus, for example: "water $H_2O$, . . ." Subsequent use might simply use the common name or symbols.
**257** See below.
**258** https://en.wikipedia.org/wiki/Chemical_Abstracts_Service
**259** http://www.cas.org/
**260** Chemical Abstracts Service "CAS® / A DIVISION OF THE AMERICAN CHEMICAL SOCIETY" http://www.cas.org/index.html
"CAS REGISTRY$^{SM}$ . . . CAS Registry Number® . . ." https://www.cas.org/cas-data/cas-registry
**261** *Hazardous Products Act*, SCHEDULE / INGREDIENT DISCLOSURE LIST, SOR/88-64 (in force 22 Mar 2006. 10 Feb 2015) – now repealed and replaced by SOR/2015-17.
https://www.canlii.org/en/ca/laws/regu/sor-2015-17/latest/sor-2015-17.html
https://www.canlii.org/en/ca/laws/regu/sor-88-64/latest/sor-88-64.html#fn_1-ID0EBAABAABA
**262** *Hazardous Products Act*, SCHEDULE / INGREDIENT DISCLOSURE LIST, SOR/2015-17.
https://www.canlii.org/en/ca/laws/regu/sor-2015-17/latest/sor-2015-17.html

**PART 1**

[. . .]

**Interpretation**

**Definitions**

**1 (1)** The following definitions apply in these Regulations.

[. . .]

**ATE** means an acute toxicity estimate, and includes the $LD_{50}$ and the $LC_{50}$, and the acute toxicity point estimate determined in accordance with the table to section 8.1.7. (*ETA*)

**CAS registry number** means the identification number assigned to a chemical by the Chemical Abstracts Service, a division of the American Chemical Society. (*numéro d'enregistrement CAS*)

**chemical name** means a scientific designation of a material or substance that is made in accordance with the rules of nomenclature of either the Chemical Abstracts Service, a division of the American Chemical Society, or the International Union of Pure and Applied Chemistry, or a scientific designation of a material or substance that is internationally recognized and that clearly identifies the material or substance. (*dénomination chimique)*

[. . .]

**$LC_{50}$** means the concentration of a mixture or substance in air that causes the death of 50.0% of a group of test animals. (*$CL_{50}$*)

**$LD_{50}$** means the single dose of a mixture or substance that, when administered by a particular exposure route in an animal study, is expected to cause the death of 50.0% of a given animal population. (*$DL_{50}$*)

[. . .]

**263** As a practical matter – as molecules that are considered become larger and larger, nomenclature becomes more complex and difficult, and for the very larger less specific (such as for polymers). And similar for the structural diagrams – which become too difficult to do by hand – requiring computer help.

**264** It is entirely numeric – not alphameric.

**265** https://www.cas.org/cas-data/cas-registry

**266** https://www.cas.org/training/documentation/chemical-substances/cas-rn-verified-partner-program

[. . .]

**CAS Registry Number Verified Partner Program**

A CAS Registry Number license is required anytime an organization will "publish" CAS Registry Numbers to the public or use them to support features of a platform that is publicly or commercially available (i.e., as a "search by CAS RN" feature), even if the CAS Registry Number is not displayed as part of the answer.

[. . .]

https://www.cas.org/about/faqs

**267** TWH&SLC *newsletter*, Vol.16 No.2, April, 2008, at page 7.

http://workers-safety.ca/features/newsletters/ http://workers-safety.ca/index.html

http://newsletter.workers-safety.ca.s3-website-us-east-1.amazonaws.com/#!/newsletters%2F

**268** *Controlled Drugs and Substances Act* – Statutes of Canada 1996 Chapter 19, Royal Assent 20 June 1996 – schedule I s.2(2).

https://laws-lois.justice.gc.ca/eng/acts/c-38.8/page-9.html

**269** Daniel M. Perrine, *The Chemistry of Mind-Altering Drugs / History, Pharmacology, and Cultural Context*; American Chemical Society, Washington, D.C.; 1996; ISBN 0-8412-3253-9; page 188, et seq.

http://www.inchem.org/ http://www.inchem.org/documents/pims/pharm/pim139e.htm

**270** Yuji Nakahara and Akiko Ishigami, National Institute of Hygienic Sciences, Tokyo 158, Japan "Inhalation Efficiency of Free-Base Cocaine by Pyrolysis of 'Crack' and Cocaine Hydrochloride," Journal of Analytical Toxicology, Vol. 15, May/June 1991; at page 105:

**Preparation of crack**

[. . .]

To three 5-mL aliquots of 34-mg/mL cocaine hydrochloride aqueous solution were added 1.5, 3, and 4.5 g of sodium bicarbonate. The mixtures in beakers were warmed on a hot plate for 1 h to remove water and prepare three types of white masses of crack (cocaine-sodium bicarbonate 1:10, 1:20, and 1:30).

[. . .]

https://academic.oup.com/jat/article-abstract/15/3/105/757651?redirectedFrom=fulltex

**271** A chart of the proscribed substances of the *CD&S Act* Schedules (showing resonance structures): Steve Chapman, *The Controlled Drugs and Substances Act / Schedules & Structures*, Isomer Design, Toronto M5B 2R3, 2005, 2011; ISBN 0-9697682-7-4.
http://isomerdesign.com/Home/index.php
http://isomerdesign.com/Cdsa/
**272** http://www.inchem.org/ [50–36-2] [53–21-4] http://www.inchem.org/documents/pims/ pharm/pim139e.htm
**273** Farid J. Muhtadi and Abdullah A. Al-Badr, **"COCAINE HYDROCHLORIDE,"** pages 151–231 in Klaus Florey, ed., ***Analytical Profiles of Drug Substances*** / *Volume 15* / Compiled under the auspices of the Pharmaceutical Analysis and Control Section / Academy of Pharmaceutical Sciences; **Academic Press**, 1986, New York.
**274** Terry Mills III & J. Conrad Robertson, ***Instrumental Data for Drug Analysis*** / Second Edition / Volume 1, **Elsevier**, New York, 1987; "Cocaine," pages 524 and 525.
**275** Ray H. Liu & Daniel E. Gadzala, *Handbook of Drug Analysis / Applications in Forensic and Clinical Laboratories*, American Chemical Society, Washington, D.C., 1997; ISBN 0-8412-3448-5; [several index entries].
**276** Daniel M. Perrine, *The Chemistry of Mind-Altering Drugs/History, Pharmacology, and Cultural Context*; American Chemical Society, Washington, D.C.; 1996; ISBN 0-8412-3253-9; page 188, et seq.
**277** http://isomerdesign.com/Cdsa/
**278** *CD&S Act* SC 1996, c. 19. s.4 Schedule I, item 2.
http://laws-lois.justice.gc.ca/eng/acts/C-38.8/index.html
**279** http://www.inchem.org/ [50–36-2] [53–21-4] http://www.inchem.org/documents/pims/pharm/pim139e.htm
**280** Michael Hardwick, *The Complete Guide to SHERLOCK HOLMES*, Weidenfeld and Nicolson, London SW4 7TA, 1986; ISBN 0 297 78963 5.
**281** Sherlock Holmes was a cocaine user – Hardwick, pages 39 and 80.
**282** Hardwick, page 39: [*THE SIGN OF THE FOUR*]

> [. . .]
> [Watson:] 'Which is it to-day,' I asked, 'morphine or cocaine?'
> He raised his eyes languidly from the old black-letter volume which he had opened.
> 'It is cocaine,' he said, 'a seven-percent solution. Would you care to try it?'
> [. . .]

**283** Hardwick, page 80: [*The Yellow Face, Strand,* 1893]

> [. . .]
> [Watson:] . . . Save for the occasional use of cocaine, he had no vices, and he only turned to the drug as a protest against the monotony of existence when cases were scanty and the papers uninteresting.
> [. . .]

**284** **See:** Daniel M. Perrine, *The Chemistry of Mind-Altering Drugs / History, Pharmacology, and Cultural Context*, American Chemical Society, Washington, DC, 1996; ISBN 0-8412-3253-9; pages 184 & 185.
https://pubs.acs.org/doi/10.1021/jm970450a
**285** [*molecular mass*] g/mol [mp = *melting point*] °C.
**286** (An abundance of baking soda may cause additional carbon dioxide to be formed when the cocaine is heated for smoking – giving a crackling sound – hence the name – see: Perrine, *The Chemistry of Mind-Altering Drugs* . . .; page 188. It might be theorized that the creation of the non-ionic form

would allow lower temperature melting and vaporization, without decomposition, and thus greater and quicker absorption into the blood, in the lungs, of the user.)

**287** According to http://www.cs.rice.edu/~ssiyer/minstrels/poems/801.html : – by Dr. D.D. Perrin, late 1950s, Department of Medical Chemistry, John Curtin School of Medical Research, Australian National University, Canberra. This Internet link no-longer works. Attribution of this Limerick remains obscure and unsure as to who wrote it and when.

**288** See also: https://wonderingminstrels.blogspot.com/2001/06/mosquito-was-heard-to-complain-dr-d-d.html
https://allpoetry.com/poem/6654445-Dressed-to-Kill-by-Meter-Maid
http://wonderingminstrels.blogspot.com/2001/06/mosquito-was-heard-to-complain-dr-d-d.html
http://www.atsdr.cdc.gov/CHEM/DDT-st.gif http://www.atsdr.cdc.gov/tfacts35.html
http://wonderingminstrels.blogspot.com/2001/06/mosquito-was-heard-to-complain-dr-d-d.html

**289** https://www.reddit.com/r/chemistry/comments/2ot9dk/chemistry_limericks/

**290** And ironically: Donal O'Regan, "Chemicals secretly buried in County Limerick dug up and then left 'a field away'," *LIMERICK LEADER*, 20 Jan 2019.
https://www.limerickleader.ie/news/home/358723/chemicals-secretly-buried-in-county-limerickdug-up-and-then-left-a-field-away.html
https://www.limerickleader.ie/

**291** See: Rachel Carson, *Silent Spring*, 1962, Fawcett Publications, Inc., Greenwich, Connecticut, 11th printing, April 1970.

**292** See: Rudy M. Baum, Editor-in-chief, *Chemical & Engineering News*, From the Editor, "Rachel Carson," 04 June 2007, Vol.85, No.23, page 5. http://pubs.acs.org/cen/editor/85/8523editor. html
– and cited therein: Naomi Lubick, "DDT's Resurrection," *Environmental Science & Technology*, 2007, *41*, 6323. http://pubs.acs.org/subscribe/journals/esthag/41/i18/html/091507per spective.html

**293** *Silent Spring ACS Matters*, American Chemical Society Weekly Newsletter, 31 October 2012: "50 Years Later, Landmark Status for Rachel Carson's Silent Spring."

**294** www.acs.org/landmarks

**295** Rachel Carson, *Silent Spring*, 1962, Fawcett Publications, Inc., Greenwich, Connecticut, 11th printing, April 1970.

**296** See also: Rudy M. Baum, Editor-in-chief, *Chemical & Engineering News*, from the editor, "Rachel Carson," 04 June 2007, Vol.85, No.23, page 5. http://pubs.acs.org/cen/editor/85/ 8523editor.html
– and cited therein: Naomi Lubick, "DDT's Resurrection," *Environmental Science & Technology*, 2007, *41*, 6323.
https://pubs.acs.org/doi/10.1021/es072616v

**297** See also: Linda Wang, "landmark for [Rachel Carson's 1962 book] 'SILENT SPRING' / Influential book catalyzed the modern environmental movement," *C&EN*, 26 Nov. 2012, page 38.

**298** TWH&SLC newsletter, 2008 07. Vol.16, No.3, page 3.
http://newsletter.workers-safety.ca.s3-website-us-east-1.amazonaws.com/#!/newsletters%2F

**299** TWH&SLC *newsletter*, 2008 07. Vol.16, No.3, page 3.
https://workers-safety.ca/ 2012 12. Vol. 20, No. 6

**300** Richard Sandomir, *"Frank Graham Jr., Nature Writer Who Updated 'Silent Spring,"Dies at 100* / He worked for the Brooklyn Dodgers and wrote about sports but mostly focused on conservation, publishing a sequel to Rachel Carson's exposé on the dangers of pesticides," *New York Times*, 04 June 2025.
https://www.nytimes.com/2025/06/04/science/frank-graham-jr-dead.html

**301** These *standards* are produced and distributed by various organizations, such as:
- the US National Institute of Standards and Technology;
- the United States Pharmacopeial Convention, Inc.;
- the Sigma Chemical Company.

See also Byron Kratochvil and Alan Walton, "The Role of Reference Materials in Quality Assurance for Chemical Measurement," *Canadian Chemical News*, May 1995, Vol.47 (5), p.40, and references cited therein. See in Chapter 5.

**302** Although, arguably, an analysis with a sufficient number of screens can approach reliability of fundamental determination – see K. Bailey, "The value of the Duquénois Test for Cannabis – a Survey," *Journal of Forensic Sciences*, 1979, 24(4), 8176841. See also Brunelle and Cantu cited in Chapter 3.

**303** See: Maria Josefi, Writer-editor, US FBI *Handbook of Forensic Science*, February 1995; US Government Printing Office, Superintendent of Documents, Washington, D,C., 20,402–9328; ISBN 0-16-045111-6; pages 57 & 58 Lead Analyses.

**304** *Gedankenexperiment.* Introduction 2.2.

**305** Methanol. CAS 67-56-1 https://en.wikipedia.org/wiki/Methanol

**306** *Gedankenexperiment.* Introduction 2.2.

**307** Ethylene glycol. CAS 107-21-1 https://en.wikipedia.org/wiki/Ethylene_glycol

**308** ±-Propylene glycol. α-propylene glycol. CAS 57-55-6
https://en.wikipedia.org/wiki/Propylene_glycol
http://www.commonchemistry.org/ChemicalDetail.aspx?ref=57-55-6

**309** Richard Saferstein, *CRIMINALISTICS / An Introduction to Forensic Science*, Seventh Edition, Prentice Hall, Upper Saddle River, New Jersey, 07458; 2001; ISBN 0-13-013827-4; page 64.

**310** That area might be definable simply in geographic terms (such as, a certain neighbourhood of a municipal or rural region), but it might also be definable in social (the people the accused associates with) or economic (where the accused shops) terms.

**311** After looking, unsuccessfully, for search warrant flaws.

**312** Deoxyribonucleic acid. For a survey of this topic see Mark A. Farley & James J. Harrington, eds., *Forensic DNA Technology*, Lewis Publishers, Inc., Chelsea, Michigan, 1991. See also *C&EN* 20 Nov. 1989, pp. 18–30; *special report DNA Profiling*. See also other DNA references cited in Chapter 6.16.

**313** That is, one chance in a million = $1/10^{+6} = 1/1{,}000{,}000 = 1 \times 10^{-6}$

**314** – assuming here no quality assurance problems. See Chapters 5 and 8.

**315** See:

- D.A. Stoney & J.I. Thornton, "A Critical Analysis of Quantitative Fingerprint Individuality Models,"*Journal of Forensic Sciences*, Vol. 31, No. 4, Oct. 1986, pages 1187→1216.
- https://doi.org/10.1520/JFS11901J.
- https://www.astm.org/DIGITAL_LIBRARY/JOURNALS/FORENSIC/PAGES/JFS11901J.htm

**316** – Joseph Polski, Ron Smith, Robert Garrett, *et al.*, "The Report of the International Association for Identification, Standardization II Committee," Document No.: 233,980, 2011, Award Number: 2006-DN-BX-K249; a research report to US Department of Justice.
https://www.ncjrs.gov/pdffiles1/nij/grants/233980.pdf
https://www.academia.edu/30690748/On_the_Individuality_of_Fingerprints

**317** – Richard Saferstein, *CRIMINALISTICS / An Introduction to Forensic Science*, Seventh Edition, Prentice Hall, Upper Saddle River, New Jersey, 07458; 2001; ISBN 0-13-013827-4; "FINGERPRINTS," Chapter 14, pages 395→420. At page 397 – Galton [$<64 \times 10^{-9}$].

**318** *Gedankenexperiment.* Introduction 2.2.

**319** Being followed through in this chapter.

**320** A good assumption.

**321** (Metaphorically).

**322** See: Jeffrey S. Rosenthal, *Struck by Lightning / The Curious World of Probabilities*; Harper Perennial / HarperCollins Publishers Limited, Toronto M4W 1A8; 2005, 2006; Chapter 2, at page 7, "*What Are* *the Odds of That? / Coincidence and surprise.* Chapter 16, pages 234→246, "Ignorance Chaos and Quantum Mechanics / Causes of Randomness." ISBN 13: 978-0-00-639495-2 ISBN 10: 0-00-639495-7

**323** – Richard Saferstein, *Criminalistics / An Introduction to Forensic Science*, Seventh Edition, Prentice Hall, Upper Saddle River, New Jersey, 07458; 2001; ISBN 0-13-013827-4; page 64.

**324** *Gedankenexperiment.* Introduction 2.2.

**325** *Gedankenexperiment.* Introduction 2.2.

**326** Different ways of presenting large numbers in a table. The first shown (at left in Table 4.6), sometimes used by scientists, can be confusing to others.

**327**

Tab. 4.6: Presenting number multiples.

| Total of pen sales for most recent 5 years × $10^{-6}$ | Total of pen sales for most recent 5 years | Total of pen sales for most recent 5 years | Total of pen sales for most recent 5 years (millions) | Total of pen sales for most recent 5 years |
|---|---|---|---|---|
| 4 | 4 × ($10^6$) | 4 million | 4 | 4,000,000 |
| 3.1 | 3.1 × ($10^6$) | 3.1 million | 3.1 | 3,100,000 |
| 1.1 | 1.1 × ($10^6$) | 1.1 million | 14 | 14,000,000 |
| 0.1 | 0.1 × ($10^6$) | 0.1 million | 0.1 | 100,000 |
| 3 | 3 × ($10^6$) | 3 million | 3 | 3,000,000 |
| 3 | 3.2 × ($10^6$) | 3.2 million | 3.2 | 3,200,000 |
| 1 | 1 × ($10^6$) | 1 million | 1 | 1,000,000 |
| 0.1 | 0.1 × ($10^6$) | 0.1 million | 0.1 | 100,000 |
| 1.2 | 1.2 × ($10^6$) | 1.2 million | 1.2 | 1,200,000 |
| 2 | 2 × ($10^6$) | 2 million | 2 | 2,000,000 |
| 5 | 5 × ($10^6$) | 5 million | 5 | 5,000,000 |
| Σ = 37.8 | Σ = 37.8 × ($10^6$) | Σ = 37.8 million | Σ = 37.8 | Σ = 37,800,000 |

1 million = 1,000,000 = $10^{+6}$ = one thousand thousand.
1 billion = 1,000,000,000 = $10^{+9}$ = one thousand million (short scale – now for both UK and USA use).
1 billion = 1,000,000,000,000 = $10^{+12}$ = one million million (long scale – no longer for UK use).
1 trillion = 1,000,000,000,000 = $10^{+12}$ = one million million (long scale).
https://en.wikipedia.org/wiki/Long_and_short_scales
https://en.wikipedia.org/wiki/Billion
https://en.wikipedia.org/wiki/Long_and_short_scales#Current_usage
https://en.wikipedia.org/wiki/Names_of_large_numbers

**328** *Gedankenexperiment.* Introduction 2.2.

**329** *Gedankenexperiment.* Introduction 2.2.

**330** See also:

- H.R. Lukens and D.E. Bryan, "The DS Method for Evidence Characterization," *Journal of Forensic Sciences*, Vol. 19, No. 4, October 1974, pages 855 → 864. –
  Neil T. Crosby and Indu Patel, *General Principles of Good Sampling Practice*, published for the Laboratory of the Government Chemist by the Royal Society of Chemistry, Cambridge CB4 4WF, 1995;
- Lawrence H. Keith, *Environmental Sampling and Analysis: A practical Guide*, Lewis Publishers, CRC Press, Inc., Boca Raton, Florida, 1991.

- B. Budowle, A.M. Giusti, J.S. Waye, F.S. Baechtel, R.M. Fourney, Dwigth E. Adams, Lawrence A. Presley, Harold A. Deadman, and K.L. Monson, "Fixed-Bin Analysis for Statistical Evaluation of Continuous Distributions of Allelic Data for VNTR Loci, for Use in Forensic Comparisons," *American Journal of Human Genetics*, 48, pages 841→855.
- Bruce Budowle and Alan M. Giusti, "Fixed Bin Frequency Distributions for the VNTR Locus D5S110 in General United States Reference Databases," *Journal of Forensic Sciences*, Vol. 40, No.2, March 1995, pages 236→238.
- Major Nu En Huang and Bruce Budowle, "Fixed Bin Population Data for the VNTR Loci D1S7, D2S44, D4S139, D5S110, and D17S79 in Chinese from Tiwan," *ibid.*, pages 287→290.

**331** Gasolines are mixtures of $C_4$→$C_{10}$ hydrocarbons (~boiling point range 40→180°C), which may include additives and impurities. A GC *mixture profile* of all observable components, or even of a few selected component types, might be used to characterize each gasoline, or at least to distinguish one from another. Variations of ASTM D 5580-94 (*1995 Annual Book of ASTM Standards . . . Volume 05.03 . . .*, pages 744→751) *Standard Test Method for Determination of Benzene, Toluene, Ethylbenzene,* p/m-*Xylene,* o-*Xylene, $C_9$ and Heavier Aromatics, and Total Aromatics in Finished Gasoline by Gas Chromatography* might be usable for such purpose.

# Chapter 5
# Site and lab procedure, and documentation

[. . .][1]

*"I see you're admiring my little box," the Knight said in a friendly tone. "It's my own invention – to keep clothes and sandwiches in. You see I carry it upside-down, so that the rain can't get in." "But the things can get out," Alice gently remarked. "Do you know the lid's open?"*
*"I didn't know it," the Knight said, a shade of vexation passing over his face. "Then all the things must have fallen out! And the box is no use without them." . . .*

[. . .]

[Alice and the White Knight] [1]

## 5.1 To try to assure credibility – lab practices

In Chapter 3,[2] the forensic chemist's lab work was referred to – as for the progress of the analyses. But other considerations are needed for the credibility of the result. That credibility is important for everyone who would scrutinize the process, and who might question the result, especially some of the law participants [2][3] who – or whose clients – might not like the results.[4] These other considerations, which in a sense might seem somewhat removed from the actual science, are important.

As indicated in the Introduction, although this book is directed towards forensic science[5] – with specialization in analytical chemistry – much is shared with science more generally, so that chemistry and other science concepts are sometimes referred to – with implications towards forensic analytical chemistry.[6,7]

**The other considerations are discussed further.**

1 Introductory quotes. Introduction 1.2.

2 Analyses. Credibility. Chapters 3 and 5.

3 Advocacy. Introduction 4.1, Chapters 1, 5 and 8.

4 Not liking the results at law. Chapters 1 and 5.

5 Forensic science. Introduction 2; Chapters 1, 2, 3 and 5.

6 Analytical chemistry. Expanded definition – more detail and example. Introduction 2; Chapters 1, 2, 3, 5 and 8.

7 Forensic analytical chemistry; chemistry more generally. Otherwise than forensic – useful for understanding; and perhaps a few digressions. Introduction 2.4; Chapter 5.1.

https://doi.org/10.1515/9783111618555-006

## 5.2 Site, lab and samples

### 5.2.1 Site examination

Parts of an investigation that would later merge into forensic chemistry laboratory work would typically start at a site[8] where an incident would have happened. This would be a known physical location – for example[9]: The *Wagon Lit* compartment, First Class Berth No. 2, *Orient Express* [3], where a stabbed dead traveler [4] was discovered, with subsequent investigation by the famous Belgian detective [5, 6, 7, 8, 9, 10, 11, 12, 13, 14, 15, 16, 17, 18].[10,11,12]

Or, for example [19]: An outside factory storage yard where there was an explosion. Sometimes, the location may be unknown as to extent – for example: A chemical spill into the aquifer – ground water – how deep? – how widespread? – migrating in which direction(s)? – how fast? And sometimes, the location may be unknown – as for missing hazardous chemicals – known from inventory control where they came from, that they are missing, but not where they went [20, 21, 22].

There could be other kinds of sites – for example, within the bodies of workers harmed by exposure to hazardous chemicals; or shrapnel.[13] Such exposures might be contemporary or from long ago, and could be on-going.

Procedure for examining and collecting evidence for a site with a known location and geometry – for example, the factory explosion – is perhaps easiest to visualize and describe. There would be an apparent debris area or field to photograph and measure, and from which, to collect items after documenting their locations (with care taken to avoid changing or contaminating the area). The area might actually be a volume – for example, an enclosed space – an inside storage room – wherein, the air might contain particulates or unusual gases and debris on walls, floor and ceiling. The area or volume examination would also have a time scale.[14]

There is reference literature [23, 24, 25, 26, 27, 28] [29, 30][15,16] for site examination. Fictional literature too (see Fig. 5.1) [31, 32, 33, 34, 35]:[17,18]

---

8 Site. *What it is.* Chapters 1, 2, 3 and 5.

9 Fictional literature. Christie. Poe. Introduction 2.2; Chapters 5 and 8.

10 *Saturday Evening Post,* Agatha Christie, *Murder in the Calais Coach* [. . . *Orient Express*], 1933. Chapter 5.2.1.

11 *Saturday Evening Post. Murder on the Orient Express.* Chapter 5.

12 Introductory quotes. Introduction top of first page.

13 Shrapnel. Chapters 2 and 5.

14 Time scale. Chapters 5 and 7.

15 Edmond Locard. Sherlock Holmes. Cocaine. Chapters 2, 3, 4 and 6.

16 Edmond Locard. Sir Arthur Conan Doyle. Forensics. CSI. Chapters 2.1.1 and 5.2.1

17 Fictional literature. Christie. Poe. Introduction 2.2; Chapters 5 and 8.

18 Introductory quotes. Introduction top of first page.

Fig. 5.1: Purloined Letter.
[Edgar Allan Poe, THE PURLOINED LETTER, 1845, 1910].

*Nil sapientiae odiosius acumine nimio.* — Seneca [36, 37, 38]

[. . .]

". . . we searched every where . . . . the entire building, room by room; devoting the nights of a whole week to each . . . the furniture of each apartment . . . . opened every possible drawer; and I presume you know that, to a properly trained police agent, such a thing as a secret drawer is impossible . . . . After the cabinets we took the chairs. The cushions we probed with the fine, long needles you have seen me employ. From the tables, we removed the tops."

[. . .]

"Sometimes the top of a table, or other similarly arranged piece of furniture, is removed by the person wishing to conceal an article; then the leg is excavated, the article deposited within the cavity, and the top replaced. The bottoms and tops of bedposts are employed in the same way."

[. . .]

". . . we examined the rungs of every chair in the hotel, and, indeed the jointings of every description of furniture, by the aid of a most powerful microscope. Had there been any traces of recent disturbance we should not have failed to detect it instantly. A single grain of gimletdust, for example, would have been as obvious as an apple. Any disorder in the glueing – any unusual gaping in the joints – would have sufficed to insure detection."

. . . looked to the mirrors, between the boards and the plates, and . . . the beds and the bedclothes, as well as the curtains and carpets . . . .

". . . then we examined the house itself. We divided its entire surface into compartments, which we numbered, so that none might be missed; then we scrutinized each individual square inch throughout the premises, including the two houses immediately adjoining, with the microscope, as before."

[. . .]

"All the grounds are paved with brick. They gave us comparatively little trouble. We examined the moss between the bricks, and found it undisturbed."

. . . papers, of course, and into the books of the library . . .

> ". . . we opened every package and parcel; we not only opened every book, but we turned over every leaf in each volume, not contenting ourselves with a mere shake, according to the fashion of some of our police officers. We also measured the thickness of every book-cover, with the most accurate admeasurement, and applied to each the most jealous scrutiny of the microscope. Had any of the bindings been recently meddled with, it would have been utterly impossible that the fact should have escaped observation. Some five or six volumes, just from the hands of the binder, we carefully probed, longitudinally, with the needles."
>
> [. . .]
>
> . . . We removed every carpet, and examined the boards with the microscope. . . .
>
> . . . And the paper on the walls . . .
>
> [. . .]
>
> . . . into the cellars . . .
>
> [. . .]
>
> ". . . what would you advise me to do?"
>
> "To make a thorough re-search of the premises."
>
> "That is absolutely needless," replied G – . "I am not more sure that I breathe than I am that the letter is not at the Hotel."
>
> [. . .]

But, as in Poe's famous story [39], something important really can be missed:

> [. . .] [40]
>
> . . . Mr. [B] . . . hid the tapes in the ceiling above the bathroom in the house . . . where he and Ms. [H] . . . were living, and police missed them during a 71-day search of the home. [41]
>
> [His lawyer] Mr. [M] . . . and Ms. [MacD] . . . acting on instructions received in a phone call from Mr. [B] . . ., entered the house . . ., and left with the six tapes in a shoe box . . . . [42, 43][19]
>
> [. . .]

Although site examination is quite important, often with dedicated professionals, the focus of this book is on the analytical chemistry of some of what is found [44].

If there would be on-site chemical analysis, it should usually be for screening tests[20] – to go later for laboratory confirmation (positive or negative) [45]. For example [46], on arrest, an accused's seized bag of suspicious-looking substance might screen-test positive for heroin, only to be revealed later, as not heroin[21] – perhaps, police misapplied the test kit [47] (unfortunately for the accused, at the bail hearing, only the screen results would have been available, but not yet the negative lab result, so that bail might be denied) [48].

Typically, a properly contained sample of something of apparent interest is brought from the site to the forensic analytical chemist at a laboratory.

---

**19** Ethics. Lawyers. Experts. *Etc.* Introduction 2.3; Chapters 1, 5, 8 and 9.

**20** Screening tests. Wet chemistry. Chapters 3, 4, 5 and 6.

**21** *Heroin.* Bail. CAS 1502-95-0 CAS 561-27-3. Chapters 4 and 5.

### 5.2.2 General lab set-up

Mostly, a forensic chemistry lab would be expected to look like any other chemistry lab, with much of the same standard furnishings and equipment – such as bench tops, water lines (of municipal source), distilled water (from other source), sinks (with ordinary and special plumbing), oxygen and gas connections, electrical outlets, electrical ground (earth) connections ⏚ [49], balances, glassware, fume hoods, hazard warning signs, safety and first aid equipment, *etc.*; and various instrumentation.[22] And, as with many labs, it may be the nature of the instrumentation that determines some (or most) of the lay-out. Lay-out would also be influenced by the volume and routineness of the lab's work. There is reference literature on chemistry lab procedures [50].

### 5.2.3 Integrity of sample and analysis process

An important feature of a forensic chemistry lab relates to security – for the tracking of what and who goes and comes, and the barring of unauthorized what and whom. This would also include tracking within the lab. And, all of this should generate a document trail.

This security is concerned with the integrity of the forensic sample and the analysis process – to make sure that it has not been interfered with, by accident or intent; and to assure that sample handling and analytical method are appropriate, and do not interfere with other aspects of the investigation. Part of this integrity is simply the sample chain of custody;[23] other parts are found in various documentation.

There would be documentation as provided for in the lab's core documents, including its standard operating procedures:

## 5.3 Documentation

The credibility, reliability and integrity of the lab's eventual report depend not only on doing good science,[24] but also necessarily include documenting what is done and how.[25,26]

---

22 Forensic analytical chemistry; chemistry more generally. Otherwise, than forensic – useful for understanding; and perhaps a few digressions. Introduction 2.4; Chapter 5.2.2.

23 Chain of custody. Chapters 3 and 5.

24 Science – good, bad and junk. Introduction 4; Chapters 1, 2, 8, 10 and 11.

25 Forensic analytical chemistry; chemistry more generally. Otherwise than forensic – useful for understanding; and perhaps a few digressions. Introduction 2.4.

26 Lab documentation. Introduction 4.1; Chapter 5.

The document trail [51] that would be relevant to the lab's written report allows for an important concept: *Nullius in verba* [52, 53, 54, 55].[27,28] Scientific results must be explainable and reproducible by other scientists elsewhere – on the basis of reports, as supported by documentation. For forensic science, a scarce availability of the samples – unpolluted – may make reproducing the work more difficult – perhaps, prohibitively.

Missing parts of the document trail should raise serious questions as to the credibility of the reported results.[29] As should a failure or refusal to explain the document trail.[30]

### 5.3.1 Lab's core documents and standard operating procedures

**A forensic laboratory should have an array of documents related to the control and recording of what the lab is supposed to do, and does do** [56].[31] This array would essentially function as the lab's own library and would be catalogued and organized as such. These core documents' various titles might include: *Operations Manual, Rules of Engagement, Directives, Organizational Manual, Laboratory Manual* [57], *Standard Operating Procedures, etc.* [58, 59, 60, 61].[32] These might function as a kind of lab-like constitutional and legislative documents – to borrow legal terms for analogy. For convenience, here, the acronym, **SOP** = standard operating procedure – is sometimes used, kind-of generally, for any of these, without a more careful definition. This acronym, here, may refer to both the core documents array and to particular procedures, as the context would require. (Readers are cautioned that this use here may be at variance with others' use elsewhere – here, its use includes a more general sense.)

As indicated above, there should be core documents or SOPs – or SOP-related references – for

- security.

And, otherwise for most every aspect of lab operation – such as[33]:

- Mission statement
- Chain of authority organizational chart within the lab
  - where that lab fits within its controlling agency, and
  - where that agency fits within its larger administrative environment;

27 *Nullius in verba* ≈ We do not take anybody's word for it.

28 *Nullius in verba*. Introduction 4.1; Chapters 2 and 5.

29 Author's biases. Introduction 1.2, 4.1 and 4.3; Chapters 10 and 11.

30 Document trail. Documentation. Chapter 5.3 and Chapter 10.

31 Core documents. SOPs. Chapters 5 and 9.

32 Health Canada Guide. Chapters 5 and 9.

33 Lab core documents. SOP, scientific literature. Chapters 3, 4, 5 and 8.

- Memoranda of understanding [62][34]
  - or contracts with other agencies;
- Various forensic analysis methods [63] that the lab does
  - as to specific procedures that might be used within those analyses;
  - as to general lab procedures [64];
- Method validations;[35,36]
- Handling procedures [65];
- Quality assurance – test samples, proficiency monitoring, peer review, *etc.*;
- Accreditation and other credentials;[37]
- Processing of reports of work done – production, distribution – public and confidential;
- Instrumentation list and library of manufacturer and supplier literature;
- Instrument operation and maintenance;
- Equipment, standard samples, reagents – purchases and rentals;
- Sample acceptance, storage, retention and return
  - environmental-friendly disposal;
- Lab personnel and public health and safety;
- Budgetary directives and documents;
- Employee loyalty declarations and security clearances;
- Workers' collective agreements;
- SOP format;
- Copies of, or references to, legislation and case law – for any of the above – if appropriate;
- *Etc.*

Some of these might effectively be set out in legislation – such as for measuring airborne concentrations under the Ontario *Occupational Health & Safety Act* [66]; and for the measuring of ethanol in breath (and therefore blood) for drivers of motor vehicles under the Canada *Criminal Code* [67, 68].[38] Or, an SOP might effectively be set out in an engineering-type standard [69].

SOPs might be required by agency or government policy for various specific reasons – such as, simply for economy (the same reagents can be purchased in larger lots at cheaper unit costs), or to forestall allegations of perceived unfairness (one accused's sample was analyzed differently – with a different legal result).

Generally, the core documents and SOPs serve to contain the lab's processes within defined bounds with accessible instructions and useful record-keeping. The

**34** RCMP mou. Chapters 5 and 10.

**35** Method validation. Chapters 5 and 8.

**36** Reasonable doubt. Science explanation. Introduction 4.1; Chapters 1.1.1, 5.3.1, 10.7, 10.8 and 11.11.

**37** Accreditation. QA ISO. ANAB. Chapters 5, 8, 10 and 11.

**38** *Criminal Code* ethanol in blood. SOP. Chapters 5 and 11.

document trail of what the forensic lab does, and is supposed to do, facilitates good lab work.

While many possible variations are expected in core document and SOP format, content, and organization, and in procedure and performance, the concepts implied by the above list should apply to all forensic labs. When making a critical appraisal of a lab's work, these concepts should be looked for. For example, if it would be noticed that there is no mention of quality assurance documentation – by whatever description – in whatever equivalent terminology – wherever in the lab's systems – the lab's eventual report should be regarded as having seriously damaged credibility, for that reason alone, although Canadian courts might not notice.[39]

Likewise, significant deviation from what is prescribed in core documents and SOPs, without considered and presented reasons, should be seen as damaging the lab's credibility and the value of its reports.

More about documentation – generally.

### 5.3.2 The scientific literature

Generally, the method, theory and discussions about the science work done on something of forensic interest will have been reported somewhere in the scientific literature – contained in various textbooks or periodicals, or other formal publications.[40,41] Such a reference is necessary for the documentation assembled for the science to eventually support the theory of the related human story. Sometimes reports in popular literature and news items can be included, especially for an overview. All literature should be open to critical appraisal.

Thus, for the ink analysis example [70], the forensic chemist might variously cite [71][42]:

- Richard L. Brunelle and Robert Reed [72, 73];
- Richard L. Brunelle and Kenneth R. Crawford [74];
- R. L. Brunelle [75];[43]
- Richard L. Brunelle and A.A. Cantu [76];
- R.L. Kuranz [77];
- Joel Harris [78];
- Renna A. Merrill and Edward Bartick [79];
- Antonio A. Cantu [80];

39 re criticism of Health Canada, Judges, *et al.* Introduction 4.1 and 4.3; Chapters 10 and 11.

40 Lab core documents. SOP. Scientific literature. Chapters 3, 4, 5 and 8.

41 Impairment and science literature. Chapters 5 and 11.

42 Variations of Ink analytical chemistry – as examples – *Gedankenexperiment* – continue throughout this book. Introduction 2.3 & 2.2.

43 Destructive and non-destructive testing. Chapters 3, 5 and 8.

- Antonio A. Cantu (US Senate Hastings Impeachment hearings) [81, 82, 83, 84, 85, 86];[44]
- Jane A. Lewis [87];
- Frank H. Norwitch and Howard Seiden [88, 89, 90];
- Paul L. Freese [91];
- Roger W Jones and John F McClelland [92];
- Mark Maremont [93].

If it happens that, in attending to a particular case, the forensic chemist makes a significant modification of reported methods, develops an essentially new method, or builds or modifies specialized equipment not otherwise available, then that would have to be set out in detail in the final forensic report.[45] And, if the new method or equipment is significant enough, it should be published in the scientific literature.[46]

It should be emphasized that without reference to the chemical or other scientific literature or a reporting of a new method, chemical forensic conclusions should not be seen as worth much. Notwithstanding that such a reference can be very brief [94] and might look perfunctory or *pro forma* [95],[47] it is still necessary [96]. This is related to method validation.[48] And, this is related to the concept that science is explainable.[49]

### 5.3.3 Tags and receipts

#### 5.3.3.1 Tags and receipts – on paper

As mentioned above,[50] the chain of the forensic sample custody is important. That such a chain is unbroken would be evidenced by a system of tags and receipts, and notebook entries.

To continue the ink example [97], the actual physical acceptance by the forensic chemist of the two clear polymer envelopes [98] would follow a procedure – perhaps:

- A police officer (badge [][][])[99] enters the lab (after having presented credentials and signing in at the security desk).
- And hands the two envelopes to the forensic chemist (employee No. [][][]).
- The next entry box on the attached property tag is filled-in and they both initial it.

44 US Senate Hastings Impeachment hearings, 1989. Chapters 5 and 9.
45 Forensic report. Chapters 5 and 9.
46 Patent. Chapters 1 and 5.
47 = Made or carried out in a perfunctory manner or as a formality.
48 Method validation. Chapters 5, 8 and 11.
49 Explanation. Method validation. Introduction 4.1; Chapters 5.3.2 and 10.3.
50 Forensic sample chain of custody. Chapters 3 and 5.

- The police officer leaves the lab (signing out at security) and makes police notebook [100, 101] notations.
- The forensic chemist makes a lab notebook entry – see below – and secures the samples in storage.

In this procedure, the tag would always travel with the item – and can also be photocopied for receipts. When (if) the chemist returns the item to the police, the next box is filled in, and so on, for all such transfers.

#### 5.3.3.2 Tags and receipts – electronic

As with the lab notebook, dealt with just below, the system of tags and receipts may often include modern computer data-keeping and management. Similar concerns and cautions are appropriate.

### 5.3.4 Lab notebook

#### 5.3.4.1 Lab notebook – on paper

A lab notebook[51,52,53,54] is a chemist's on-going record of work done; it is the essential and central reference for the reports to be written, and it is where to look to try to solve problems that might later arise, including from criticisms of the reports – even years later. According to H.M. Kanare [102, 103, 104], "[t]he guiding principle for note keeping is to write with enough detail and clarity that another scientist could pick up the notebook at some time in the future, repeat the work based on the written descriptions and make the same observations that were originally recorded."

For disputed forensic conclusions, Kanare's "another scientist" might not be so hypothetical – and might be called by an opposing lawyer to try to persuade a court, or public opinion, otherwise. And, historians and journalists might also be looking at the data, years later; see, for example: Nicola Sacco and Bartolomeo Vanzetti, Boston, 1920 [105, 106, 107, 108];[55] and Steven Truscott, Ontario, 1959 [109, 110].

Generally, a lab notebook should have permanently bound and numbered pages, and be of substantial construction. It should also have identification/title pages, which include declarations of the lab's policy concerning notebooks and a clear indication or declaration of its importance as a valuable original document. While loose-leaf

51 **Alert**: See both: Lab notebook – paper. Electronic alternative. Chapters 5.3.4.1 and 5.3.4.2.

52 Forensic analytical chemistry; chemistry more generally. Otherwise than forensic – useful for understanding; and perhaps a few digressions. Introduction 2.4.

53 Lab core documents. SOP. Scientific literature. Chapters 3, 4, 5 and 8.

54 Lab notebook. Chapters 5 and 9.

55 Sacco, Vanzetti. Chapters 1 and 5.

binders of lab notes, various kinds of file folders and releasable binders, and computer storage, may sometimes be convenient or even appropriate, the older-style bound book has an advantage that separate pages cannot get mixed-up or lost by simple accident, and intentional tampering may be more difficult to cover-up. And, even in situations where unbound notes are used, the forensic chemist might still be well-advised to keep a chronological registry of sample numbers, *etc.*, in a bound diary calendar book, with cross-references to the other paperwork – as a kind of mini-lab notebook.

The contents and style of lab notebook entries would vary, depending on the work to be done, its repetitiveness, volume and routineness. Generally, the advice of Kanare [111] is worthwhile to consider, even if some of it might seem too fussy. If a forensic chemist is following a procedure from reference documents, such as a procedures manual [112, 113] or an SOP, a proper reference to such a document may be sufficient, provided it and the appropriate data are clearly enough identified (for the " . . . other scientist").

For large batches of routine forensic analyses, **check-box** and **fill-in-the-blank** kinds of **forms** or worksheets may be used, and may even be computerized. But, that convenience should not obscure from the forensic chemist, the problems of unbound notes indicated above; and although such forms might be contained within the lab's report, it should not be confused with that report itself.

An example, concocted fictionally for illustrative purposes here, would show – possibly – what such a form might include: [114, 115][56]

- Descriptive title – for example: LAB NOTES PROCESS SHEET FOR MARIHUANA DETERMINATION [116, 117, 118][57]
- Location of the lab
- Case process number for the lab [][][];[119] for the Police [][][]
- Various tag numbers [][][] [][][] [][][] [][][] [][][]
- This PROCESS SHEET's control number [][][]
- page numbers [][] → [][]
- Identification of where and when the sample for testing is from – for example: Cst. [][][][], [][][][] Police, delivered, [][] [][][][] 20[][], [][]:[][] []m [120]
- Analyzed using procedures of SOP [][][] (which might follow methods and concepts presented by J.A. Siegel [121] and by K. Bailey [122])
- Date, times and location of the analysis
- Identification of the analyst – [][][][]; employee identification [][][]
- General appearance of the sample to be identified and/or sketch or photo:
- Mass = [][][].[][]g [123]

**56** Report – Short Form. Chapters 5 and 9.

**57** Tetrahydrocannabinol THC
CAS 1972-08-3
Chapters 5 and 9.

- Dimensions = [][][].[][]cm × [][][].[][]cm × [][][].[][]cm. Geometry: [][][][] [124]
- [][].[][]°C [][][].[][] % humidity
- Microscopy: plant material present? Cystolithic [125, 126] hairs observed? Photo? Drawing?
- Duquénois-Levine?
- TLC: Solvent = [][][][]?[127] Solvent mixture? Photo? Drawing?
- IR = infrared spectroscopy (usually not necessary for this kind of analysis – see both Siegel and Bailey, cited above)?
- Conclusion: the analyzed sample – ✓
  - ☐ is ☐ is not marihuana – plant material.
  - ☐ does ☐ does not contain THC.
- Lab Accreditation = [][][][][128] Accrediting agency: [][][][][58]
- ISO: [][][][]
- Analyst credentialed? [][][][]
- Lab's exchequer or budgetary declaration?
- Interagency memorandum of understanding – legal citation?
- QA reports? Publicly available?
- Workplace health and safety; worker's compensation?
- This PROCESS SHEET's freedom of information status?[59]
- Confidentiality status? Until when?
- Legal declaration? Disclaimer?
- Relevant legislation citation: [][][][]
- Signed, dated and witnessed attestation? Anti-corruption declaration?
- Next administrative step?
- What to do with this form now? To whom does it go next? Does not go?
- Distribution?
- See notebook pages: [][][] → [][][]

Another notebook-keeping practice would be to firmly attach experimental spectra, photographs, print-outs and so on to the pages of the bound book; but if there are too many, this would be too cumbersome, and they should be archived elsewhere. At least, identification numbers and location instructions for these items should be recorded in the notebook. Leaving spectra and other data stored in operating or unsecured computer memories is asking for trouble.

A lab notebook would contain the kinds of data listed in the above form, but in the bound format mentioned above; the form might be attached within the notebook; perhaps, pre-printed on its pages. But, the lab notebook would also contain all of the

58 Accreditation. QA ISO. ANAB. Chapters 5, 8, 10 and 11.
59 Freedom of info. legislation. Chapters 5, 8 and 10.

other work that the analyst would do in chronological sequence and might note other steps not included in the form. If a form is used, that would also be noted in the notebook, if not attached within it.

Notwithstanding the above commentary, attaching other papers to lab notebook pages is obviously impracticable when there are many analyses – perhaps many thousands, as for labs serving in the war on drugs.[60] But the lab notebook should still record adequate references to the documents of analysis within a logical system of location codes or directions; this would be practicable because the lab notebook would still record the work of the individual scientists during their workdays.

#### 5.3.4.2 Lab notebook – electronic

With the advent of modern computer[61] data-keeping and management, much of the commentary about lab notebooks above may be seen as becoming obsolete, if not already so. This is ok; especially for the very large numbers of analyses of large labs. However, the forensic scientist must be aware that the precautions inherent in the paper-based systems – readily available, reliable, tamper-resistant, easily readable, showable to the jury, *etc.* – are not lost with electronic data keeping – or are not brought into doubt – or are not unintentionally given over to the control of others not so concerned with the intricacies and special responsibilities of forensic science, and are not exposed to hackers. There is a kind of analogy with citizen voting – paper ballots remain the best device against fraud, mix-up, hacking and error.

The forensic scientist, ultimately responsible for the reported conclusions, is also to be held responsible for the integrity of the data keeping and processing. The advantage of the lab notebook is/was that many gaffs could quickly look obvious; the scientist is directly responsible for ensuring that the electronic record-keeping keeps to that standard. The forensic scientist should be careful not to lose – to the programmer or controller of the computer system, hackers [129, 130],[62] or biases within outside-sourced programmes [131] – the tight control that a physical lab notebook would provide. If the forensic scientist does not address this and is not completely confident of a computer equivalent of that tight control remaining, then there would be a serious credibility loss.[63] And, if there were to be found a glitch, or a hack into the computing process, or a bias within outside-sourced programmes, even indirectly, such an occurrence must appear in the notes and report, and is relevant to be considered in any legal process.[64]

**60** War on drugs – Canada. Chapters 5.3.4.1 and 10.1.

**61** Alert: See both: Lab notebook – paper. Electronic alternative. Chapters 5.3.4.1 and 5.3.4.2.

**62** LifeLabs. Chapters 5 and 9.

**63** Notes. Signature. electronic. Chapters 5 and 9.

**64** Glitch in computer process. Chapters 5 and 9.

### 5.3.5 Reference documents

Many documents can be conveniently shortened – and made more manageable – by referencing to other documents.[65] Even very specialized forensic work will include routine procedures, with routinely used chemicals and equipment. And, sometimes, an entire kind of forensic work will be routine. Document referencing is a common practice in law, where, for example, the *Occupational Safety and Health Regulation* [132], under the *Canada Labour Code* [133], contains building standards provisions, simply by referencing the *National Building Code* [134]. It is also a common engineering practice, where, for example, specifications for certain equipment are sufficiently described by reference [135] to CSA [136] standards. There are several kinds of reference documents, including:

- A lab's self-developed core documents and SOPs, perhaps compiled in a manual. For example: K Br pellet preparation for IR spectroscopy,[66] might simply be noted in the lab notebook and later in the report as SOP No. [][][] [137]. This is satisfactory as long as the SOP reference is clear, and the SOP, as it was at the time the work was done, is readily available [138].
- Documentation for standard reference materials – for which, if used, the documentation that accompanies them is essential. These standards are produced and distributed by various organizations, such as the US National Institute of Standards and Technology [139], the United States Pharmacopeial Convention, Inc. [140], the American Petroleum Institute [141]. Often, commercially available materials from reputable suppliers may serve as standards, for example, the Sigma Chemical Company [142, 143][67] Reference standards might also be in-lab produced.
- The scientific literature.[68]
- Standards published by various organizations – a frequent engineering practice. These are published documents, with reference numbers, that address issues of definitions, engineering, compatibility, quality, health and safety, *etc.* For example – American National Standards Institute; American Society for Testing and Materials [144, 145, 146, 147, 148];[69,70] International Organization for Standardization (ISO) [149, 150].

---

**65** Lab core documents. SOP, scientific literature. Chapters 3, 4, 5 and 8.

**66** K Br. Nujol. IR. Chapters 5 and 6.

**67** Disclaimers. Error . . ., Not legal advice . . . Commercial products, . . . Introduction 2 and 4.1.

**68** Lab core documents. SOP. Scientific literature. Chapters 3, 4, 5 and 8.

**69** Disclaimers. Error . . ., Not legal advice . . . Commercial products, . . . Introduction 2 and 4.1.

**70** ASTM. Crayola label. Chapters 1, 5 and 9.

### 5.3.6 Spectra and other instrumental print-outs

Very much of modern chemical analysis involves instrumentation and machine display and recording of data [151] – so that in the ink example, there might be IR spectra. As already noted above, an instrument's print-outs might be attached to the appropriate notebook page or clearly cross-referenced there. The same procedure should be followed for photographs and drawings, or recorded in the electronic data-keeping equivalent.

### 5.3.7 Report

The forensic report – the statement of findings and conclusions – and a most important kind of document to be issued by the forensic lab – is dealt with as Chapter 9.[71]

### 5.3.8 Quality

Throughout this chapter, it would be assumed that the lab is well run – its samples and equipment are handled in a careful and orderly way, with clean handling and good housekeeping, with good document control, and with a conscious, serious effort to produce a quality product: Reliably credible and clearly reported and explained analytical chemistry results. All of this should be addressed by way of a quality assurance programme – see in this chapter and in Chapter 8.[72]

### 5.3.9 Publication or other public access

There should be public accountability for forensics – of its reports and documents that lead to the reports. This is consistent with the usual practice – in principle, at least – of legal matters being open to public access.[73,74,75] The public nature of the court process is seen to be essential for a *free and democratic society* [152].[76] Depending on circumstances, this sometimes leads to actual public knowledge or even publication; notwithstanding that it would often seem that hardly anyone would be interested. The mechanisms of becoming public include that courts are public forums

71 Proper scientific report. *CD&S* Act. Health Canada. Introduction 4.3; Chapters 5, 8, 9, 10 and 11.

72 Accreditation. QA ISO. ANAB. Chapters 5, 8, 10 and 11.

73 Public. Introduction 4; Chapters 5 and 8.

74 **But see:** "However . . . confidentialities . . ." below.

75 Public process. Introduction 4.1; Chapters 5, 8 and 10.

76 *Free & democratic society*. Canadian constitution. *Charter*. Introduction 4.1; Chapter 5.

[153], the proceedings of which are recorded for transcription for anyone who pays the official court reporters' fees [154]. Also, anyone who pays the photocopying fees is normally entitled to copies of case documents.

Aside from court process, everyone has, with limitations, rights of access to government documents under the Canada federal *Access to Information Act* [155], and its various provincial counterparts [156]; this can be disappointing.[77] The US federal equivalent is the *Freedom of Information Act* [157].

Also, although the Canada federal *Auditor General Act* [158], and its various provincial counterparts [159], would not give individuals rights of access to specific source information [160], published reports of these independent [161] agencies can be helpful.

However, within this public nature of the legal process, there can – and should – be considerable confidentialities. There is a necessary tension between the public and the confidential [162]. This tension is further discussed below – in the more general context of transparency and opaqueness of government and other institutional process – secrecy.

### 5.3.10 Record and sample retention

The importance of documentation and the document trail is frequently indicated in this chapter. This is related to a concept that the documentation should have an independence from its issuer; and be available for some considerable time for the scrutiny of others, still independent of its issuer. This means that a forensic lab should have a record retention policy that declares explicit protocols for listing what records exist, for how long [163], and what their public/confidentiality status is [164].

There should also be protocols for the records' ultimate archival status or disposal. And, if there is to be disposal, there should be protocols of method and documentation of execution.

Record retention should include data observed for the scientific measurements – in raw forms and, as appropriate, at various stages of processing. Much of these would not appear in the final report, but should be referenced in that report as archived and accessible.[78]

The kinds of records contemplated here would include not only those of the scientific work, but also as related to lab administration and finance, and the tracking of samples, chemicals, equipment, personnel and records.

In the author's view, the record retention time should be as close to forever as practicable, unless there is a specific reason for earlier disposal (*e.g.*, a legal require-

77 Freedom of info. legislation. Chapters 5, 8 and 10.
78 Data archive. Chapters 5 and 9.

ment to dispose – as distinct from a legal minimum retention time). Forestalling legitimate critics' scrutiny is not an acceptable reason. The reason for such long-term retention is so that others – often adversarial others – would have sufficient opportunity to research, even decades later, for either a specific legal or political purpose, or for more general historical research [165, 166, 167].[79] But this should be considered in a context of the personal "right to be forgotten" [168].

In the author's view, the long-term availability of forensic lab records for others' scrutiny is important for forensic credibility [169], and is a responsibility of lab leadership personnel discussed below.

The same concepts that apply to record retention should also apply to forensic samples – for the same reasons, although keeping physical property in storage may be difficult, perhaps prohibitively so.

### 5.3.11 Security

As mentioned above, security may be of special concern in a forensic lab. And, while it may be part of everyone's job, it may also be necessary that larger labs have security-dedicated personnel who are to act by formal, documented protocols of their own, and who are answerable directly to senior management, and perhaps to external official agencies. Such a security regime should generate its own documentation.

## 5.4 Lab organization

### 5.4.1 Legal/budgetary

As with any other institution, a forensic chemistry lab has a legal and physical existence. It may be part of a personal- or corporate-owned business or institution, or it may be operated by some level of government (a Ministry, board, or Crown corporation). It would have various ownership or lease rights to facilities, equipment, inventory, client lists and intellectual property;[80] and it would have financial accounting and employees (perhaps with a trade union).

While most of these matters would not be of great concern in evaluating a lab's forensic work, there should at least be some documentary appearance to identify them. Such identification can be important because it can carry the lab's reputation – good or bad. If the reputation is good, then, for a private or corporate lab, its trade name, marks and brand may become valuable intellectual property.

**79** Sacco, Vanzetti. Chapters 1 and 5.
**80** Intellectual property. Chapters 1 and 5.

A lab's legal existence is also related to contract and tort law – in the ability to sue (*e.g.*, to collect its analysis fees) and to be sued (for wrong results that caused client damage, or correct results that were damaging by being late). A lab's legal existence may also allow it [170] to be prosecuted by the Crown for regulatory infractions.

### 5.4.2 Organization and leadership responsibility

All forensic labs (even including – although perhaps to a lesser extent – a one- or a few-person operation) should have a formal, documented organizational/management structure – with a system of core documents described above. This would usually be influenced by the legal structure. There should be a clear chain of authority, from operating personnel (chemists at the lab bench) to leadership personnel (a senior supervisor or lab director; a Minister of the Crown and deputies) (see Fig. 5.2) [171, 172].[81,82]

Fig. 5.2: Lab core documents (Sidney Harris Chemistry Cartoon).

81 Introductory quotes. Introduction top of first page.

82 Sidney Harris Chemistry Cartoon. "www.CartoonStock.com". Copyright acknowledgement. Chapter 5.

In the author's view, the responsibility that everyone in the lab bears for the production of reliably credible analytical results bears more heavily on the leadership personnel because they have control over the organizational functions, the pace of work, employee payment practices [173] and policies. As leaders, they should be identifiable and answerable for everything that happens on their watch. This is important when something goes wrong – and proper answers are called for[83] – where the buck stops [174].

### 5.4.3 Personnel – scientific qualification

Although most anyone can be trained for specific forensic science tasks – and do them well – the usual qualification[84] for forensic science personnel would include a diploma or degree in chemistry or related science, engineering or technology – that, importantly, should include at least some exposure to humanities. These may be within college and university programmes that are forensic-science-directed. Such formal education should give the scientist an understanding, beyond mere technical skills, that may be necessary when the unexpected appears, when a philosophical overview would be needed, or if something goes wrong, for example, if a moral quandary arises – to identify and address it. Participation in professional organizations [175] and professional designations [176][85] is also important.

With modern automated instrumentation, it is now possible, if not usual, for a forensic worker to be able to produce more, quicker and better work than was possible in previous eras; but possibly with less understanding, or even with hardly any understanding at all.[86] [177, 178]

Care must be taken to avoid confusing uneducated presenting of results with an actual understanding of scientific concepts and meaning. Simple procedural and mechanical training for a modern automated scientific instrument is not sufficient, and should not be confused with thoughtfully and professionally evaluated and explained evidence.

As an example [179], in another context, consider a person who wants to write a will – to leave a fortune to good and loyal friends, and not to family. Quite impressive-looking forms might be purchased cheaply at stationery stores [180]; but this is no proper substitute for professional legal advice from a licensed and insured estate lawyer, who is responsible to navigate the sometimes subtle intricacies of a law area, wherein people have been known to viciously feud [181]. The Bold and Gothic lettering of the store-bought form may eventually disappoint would-be heirs.

---

**83** Author's biases. Introduction 1.2; 4.1; 4.3; Chapters 10 and 11.

**84** Accreditation. QA ISO. ANAB. Chapters 5, 8, 10 and 11.

**85** Profession. Chapters 1 and 5.

**86** Personnel – scientific qualification. Automated instrumentation. Chapters 5, 6 and 8.

Perhaps, another analogy, from yet another context – In Disney's *Fantasia*, 1940, the sorcerer's apprentice [182] does not appear to have yet got an adequate knowledge of the magic, needed when something goes wrong [183].

## 5.5 Ethics

A hallmark of a professional is to be held to act within proper ethical and professional bounds, and this relates to the practice of forensic science. Aspects of ethical issues are dealt with in Chapter 8.[87]

## 5.6 Quality assurance

Modern commercial, industrial and laboratory practices should have a strong focus on producing quality products and services, and for assuring consumers of that quality. This not only looks at the technology of the products and services, but also the organizational functions to produce them. Various aspects of quality assurance are dealt with in the context of how to try to limit and cope with error, in Chapter 8.[88,89]

## 5.7 Other responsibilities

In a hopefully by-gone era, when the chemistry lab work was finished and reported, so too were the remains of the experimental work, no longer of concern. Nor were other aspects – both during the lab work and afterwards. No longer – hopefully – in our modern era.

While these other aspects – of importance on their own – are not of direct concern when considering forensic results, failures in these regards may enter into legal proceedings with the issue of credibility. And, even if lab personnel credibility survives exposure of such failures, it could lead to other legal activity and public policy criticisms in the context of these aspects.

---

**87** Ethics. Lawyers. Experts. *Etc.* Introduction 2.3; Chapters 1, 5, 8 and 9.

**88** Accreditation. QA. ISO. ANAB. Chapters 5, 8, 10 and 11.

**89** QA. ISO. Chapters 5.6; 9.1.1; and 8.

### 5.7.1 Environmental: waste management

In that, hopefully, by-gone era, the liquid waste might simply have been poured down the drain, and the solid put out with the municipal trash.[90] Such pollution, environmental [184] and waste management [185] matters are now not only regulated, but also properly seen as a lab responsibility – from the chemist at the lab bench to the leadership in the management and executive offices.

### 5.7.2 Workplace health and safety

Similarly, lab and other [186] forensic worker health and safety [187, 188], and workers' compensation, have become relevant concerns that must be addressed.

For example[91]: Failure of chemistry lab ventilation, Nova Scotia, 2008, resulted in a worker fatality – lung failure after exposure to trimethylsilyldiazomethane (see Fig. 5.3) [189, 190].

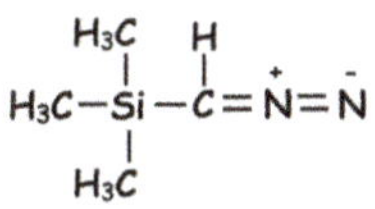

Fig. 5.3: TMSD, CAS 18107-18-1, 114.22 g/mol.

The lab operator plead guilty – under provincial regulatory legislation – to a charge of failing to provide proper workplace ventilation, resulting in a fine = [US $47,000] [191, 192, 193, 194, 195, 196, 197, 198, 199, 200, 201, 202, 203, 204, 205, 206].

### 5.7.3 Hazard communication: WHMIS

Related to worker health and safety and workers' compensation, in Canada, the Work-place Hazardous Materials Information System, has become relevant to concerns that must be addressed in a chemistry lab. And, similarly in the USA, with the Federal Hazard Communication Standard.[92]

---

**90** Regularory law. Environmental and workplace health and safety. Chapters 1, 5, 9 and 11.

**91** Forensic analytical chemistry; chemistry more generally. Otherwise than forensic – useful for understanding; and perhaps a few digressions. Introduction 2.4.

**92** Consumer protection. Hazard communication. WHMIS. Chapters 1 and 5.

### 5.7.4 Labour relations and employee rights

And, similarly, labour relations and employment law have become relevant concerns that may arise in a chemistry lab.[93]

### 5.7.5 Consumer protection

And, since commercially operated forensic labs produce a marketable product, consumer protection issues may also arise.[94] These labs should be held responsible, at law, to not inflict a defective product on their customers. This could be addressed by both government-inspected regulatory law and by way of civil litigation in tort or contract. However, consumers might not be aware that the product would be defective.

## 5.8 Transparency and opaqueness of government and other institutional process – secrecy

### 5.8.1 Confidential versus public

As indicated above, for publication or other public access to forensic reports and other documents, there is a necessary tension between the public and the confidential. This extends into the more general context of transparency and opaqueness of government and other institutional process.[95,96]

In Chapter 2,[97] it is argued that science progresses with the developing and refining of theories. In Introduction 4, it is argued that this is a public process; so that proposed explanations are available to potential critics.[98] Also argued in Introduction 4 is that there should be an accountability to the citizenry – the workings of the government should be clearly visible – transparency. Thus, the doing of science should be a public process, from both scientific and democratic perspectives.

However, forensic science, intertwined as it is with legal, institutional and personal processes, must necessarily include significant confidentialities. At least for some specific investigations, at least, temporarily. Transparency might only extend to particular participants, and in a required sequence. For example, in a criminal investigation, police would confide to Crown counsel; Crown, in-turn, to defence counsel;

93 Labour relations and employment rights. Chapters 2 and 5.

94 Consumer protection. Hazard communication. WHMIS. Chapters 1 and 5.

95 Confidential *verses* public. Introduction 4; Chapters 5.8; and 8.

96 Health Canada. DAS. Ethics. Introduction 4.3; Chapters 5 and 10.

97 Transparency. Opaqueness. Chapters 2, 5 and 10.

98 Public. Introduction 4; Chapters 5 and 8.

in-turn to the court; *et seq.* And, that chronological order of confiding must be maintained. The forensic lab personnel would be the beginning part of all of this confiding.

But, these necessary confidentialities should not become an excuse for not explaining the science to those entitled to argue about it, when their time would come to do so. And, in due course, confidential, with explained exceptions, should usually come into the public process.

In Chapter 10,[99] it is opined that the Canadian government has been a pioneer in opaqueness. In the absence of any national security or privacy issues, there has been a failure of scientific disclosure and explanation, without justifiable reason. In this, Parliament has enacted anti-science legislation that allows bad science, and the government has invoked it to do bad science. It is bad science, at least for that reason alone, that the scientific disclosure and explanation are missing.[100]

### 5.8.2 Temporary allowance

Confidentiality should temporarily apply for a specific forensic investigation that would involve a court or tribunal process, until those involved can consult, as needed, and so that the court or tribunal can decide.

Thus, there are some kinds of investigations for which temporary secrecy might be necessary – for selection protocols,[101] for sample taking. Examples:

- Doping testing for horse racing [207]. Which horses to be tested for what chemical, and when, should not be known by anyone but the testing personnel (and preferably, not even them), so that those who might profit from racing industry investment and bets cannot interfere.
- Brownfield testing for environmental regulation. What tested for, where, when and how deep.[102]
- Workplace drug testing. What tested for, where, when and what job categories, at random, or for what reasons.
- Air purity testing at workplaces [208]. When and where.
- Food inspection, including storage temperature, at restaurants.
- Purity and process inspection at food processing plants.
- Purity and process inspection at pharmaceutical processing plants [209].
- Viscosity and evaporation characteristics testing for petroleum and other chemicals carried in highway and railway tankers [210].

---

**99** Transparency. Opaqueness. Chapters 2, 5 and 10.

**100** Re criticism of Health Canada, Judges, et al. Introduction 4.1 and 4.3; Chapters 10 and 11.

**101** Sample selection. Chapters 3, 5 and 9.

**102** Sample selection. Brownfield. Chapters 2, 3 and 5.

Preferably, for all of these, what is tested for, where, when, in what ways and under what circumstances, should be known only to the testing personnel. And, as far as practicable, not even to them – using selection protocols involving randomness that helps keep the tester blind.[103]

### 5.8.3 Manhattan Engineer District

> . . . *fear that* . . . *would do it first*. . . .
>
> [Thomas Powers, *HEISENBERG'S WAR* . . ., 1993.] [211, 212][104]

A historical example where scientific confidentiality from the public, and many others [213], was needed is found in the work of the Manhattan Engineer District [214].[105] The U-235 critical mass for a workable atomic bomb would be a very real secret; well-known, years after WWII, to be only a few kilograms [215, 216, 217, 218, 219, 220, 221, 222, 223, 224, 225, 226].

But, in Albert Einstein's famous letter to President Franklin Roosevelt, 1939 [227, 228, 229, 230, 231, 232, 233, 234, 235] Einstein, Leo Szilard, Enrico Fermi and Jean Frédérick Joliot-Currie, apparently thought it might be tonnes. Atomic research might provide for dangerous new weaponry, but Little Boy and Fat Man [236] in aircraft over Japan (not then yet even thought of) would not be possible [237].

But later, Manhattan-related work would proceed with the few kilograms fact [238, 239, 240].

Although Alsos [241] later discovered insufficient atomic progress, what Third Reich scientists knew, thought and intended, remains a puzzling controversy [242, 243].

By one version, Werner Heisenberg, Otto Hann, Ernest von Weizsäcker, Paul Harteck [244, 245, 246], Kurt Diebner, *et al.*, of the Third Reich's atomic project, apparently did not realize kilograms, rather than tonnes, until after Hiroshima [247, 248], when they heard a news report when they were interned at Farm Hall, with the British listening-in [249]. They apparently thought that a deliverable atomic bomb would not have been possible; that not enough U-235 could have been separated from U-238 [250]; and too heavy for then-available aircraft.

But, notwithstanding the controversy about what Reich scientists knew, thought and intended, the Allied secrecy about the U-235 critical mass seems to have been well justified (see Fig. 5.4) [251].[106,107]

---

**103** Justitia. Blind. Introduction 4.1; Chapters 5 and 8.

**104** Introductory quotes. Introduction top of first page.

**105** ${}_{92}U^{235}$ U-235 Einstein. FDR. Critical mass. Chapters 5 and 6.

**106** Introductory quotes. Introduction top of first page.

**107** WWII poster (used by US government).

Fig. 5.4: WWII poster (used by US government).

## Notes

**1** Lewis Carroll, *Alice's Adventures in Wonderland* and *Through the Looking-Glass*, Signet Classics, The New American Library, New York City, 1960 and 1962; from . . . *Glass*, "It's my own invention", Chapter VIII, pages 205 and 206.
**2** Such as the author, when as legal counsel.
**3** Fictionally delayed on its north-westward journey, in a snow drift, likely in what is now Croatia; ~~1933.
**4** = Samuel Ratchett. https://agathachristie.fandom.com/wiki/Cassetti
**5** Agatha Christie, *Murder on the Orient Express*, 1934, William Morrow/Harper Collins, New York City, 10022; 2011; ISBN 978-0-06-207349-5; page 38.
https://www.fantasticfiction.com/c/agatha-christie/murder-on-orient-express.htm
**6** Agatha Christie, *Murder on the Orient Express*, 1934, Pocket Books, New York City, 1960.
**7** Nancy Blue Wynne, *An Agatha Christie Chronology*, Ace Books, New York City, 10036; 1976; pages 42→45; ISBN 0441 10445 2. https://www.goodreads.com/book/show/140378.An_Agatha_Christie_Chronology
**8** https://en.wikipedia.org/wiki/Murder_on_the_Orient_Express
**9** https://www.agathachristie.com/film-and-tv/murder-on-the-orient-express/facts-about-murder-on-the-orient-express
https://www.biblio.com/book/murder-calais-coach-murder-orient-express/d/877715751
https://en.wikipedia.org/wiki/Murder_on_the_Orient_Express
https://www.saturdayevening post.com/
**10** https://bodiesfromthelibrary.com/2017/03/13/murder-on-the-orient-express-v-stamboul-train/
**11** Agatha Christie, *Murder on the Orient Express*, first published as "Murder in the Calais Coach," *Saturday Evening Post*, 6 installments, 1933.

**12** Illustrated by William C. Hoople; *Saturday Evening Post*, 7 October 1933, page 20 [Hercule Poirot and Dr. Constantine at the crime scene – a private compartment on the *Orient Express*].

**13** *Saturday Evening Post*, 7 October 1933, page 20: "A Woman's Handkerchief," Said the Doctor. "Our Friend the *Chef De Train*, Was Right. There is a Woman Concerned in This"

**14** *Saturday Evening Post*, 7 October 1933, page 20 ≈ [. . . *Orient* . . . Harper Collins . . ., at page 62.]

**15** https://www.saturdayeveningpost.com/2020/03/christie/

**16** Jeffersen, *The Paperback Palette*, "The Cover Art of Agatha Christie's MURDER ON THE ORIENT EXPRESS," 2019. https://paperbackpalette.blogspot.com/2019/06/the-cover-art-of-agatha-christies.html

**17** *Saturday Evening Post*, 7 October 1933, *Wagon Lit* diagram, page 48.

**18** See also: Agatha Christie, *Murder on the Orient Express*, 1934, William Morrow/Harper Collins, New York City, 10022; 2011; ISBN 978-0-06-207349-5; *Wagon Lit* diagram, page 77.
https://www.fantasticfiction.com/c/agatha-christie/murder-on-orient-express.htm

**19** *Gedankenexperiment*. Introduction 2.2.

**20** Not only hazardous chemicals – also other things:

**21** – Daniel Leblanc, "Missing gold never left the Mint," Parliamentary Reporter, *The Globe and Mail*, 27 Nov. 2009. https://www.theglobeandmail.com/news/politics/missing-gold-never-left-the-mint/article1347654/

**22** – "Mounties called in as $110 K in gold missing from mint / Mint says employee terminated after inventory reveals missing gold," CBC News, 26 April 2018.
https://www.cbc.ca/news/canada/ottawa/ottawa-mint-gold-theft-missing-1.4637799

**23** See below:

**24** Richard Saferstein, *Criminalistics/An Introduction to Forensic Science*, Seventh Edition, Prentice Hall, Upper Saddle River, New Jersey, 07458; 2001; ISBN 0-13-013827-4; page 33, et seq.

**25** Detective Peter Code, Toronto Police Service, "Forensic Crime Scene Investigation," as Chapter 3, in David Rose and Lisa Goos, *DNA / A Practical Guide*, Thomson Reuters Canada Limited / Carswell, Toronto M1T 3V4; 2004, 2009; KF 9666.5 A6R67 2004; ISBN 978-0-7798-1685-9.

**26** Marilyn T. Miller, "Crime Scene Investigation," in Stuart H. James and Jon J. Nordby, editors, *Forensic Science/An Introduction to Scientific and Investigative Techniques*, CRC Press, Boca Raton, 2003; ISBN 0-8493-1246-9; Chapter 8, pages 115→135. www.crcpress.com

**27** Maria Josefi, Writer-editor, US FBI *Handbook of Forensic Science*, February 1995; US Government Printing Office, Superintendent of Documents, Washington, D.C., 20402-9328; ISBN 0-16-045111-6; "The Crime Scene," page 14; "Practical Suggestions Regarding Crime Scene Administration and Management," pages 15→19.
https://archive.org/details/FBI_Handbook_of_Forensic_Science/mode/2up
https://ia601209.us.archive.org/15/items/FBI_Handbook_of_Forensic_Science/FBI_Handbook_of_Forensic_Science.pdf

**28** Michael Karagiozis and Richard Sgaglio, *Forensic Investigation Handbook: An Introduction to the Collection, Preservation, Analysis and Presentation of Evidence*; Charles C Thomas Publisher Ltd., Springfield, Illinois, 62704-4730; 2005.
https://www.ccthomas.com/details.cfm?P_ISBN13=9780398075798

**29** And some literature for what might be done with the site afterwards. See Gil Reavill, *AFTERMATH, INC. / CLEANING UP AFTER CSI GOES HOME*, GOTHAM BOOKS Penguin, New York City, 1001; 2007. ISBN 978-1-592-40364-6.

**30** Included in Gil Reavill, *AFTERMATH* . . . – some commentary about related topics:
- Crime scene.
- CSI TV shows – pages 212→218.

- Forensics.
- Locard's principle, Sir Arthur Conan Doyl, Sherlock Holmes – pages 215→218.
- postmortem interval – pages 50→58.

**31** Edgar Allan Poe, "The Purloined Letter," 1845, 1910.
https://en.wikipedia.org/wiki/The_Purloined_Letter
https://en.wikipedia.org/wiki/C._Auguste_Dupin#/media/File:The_Purloined_Letter.jpg
https://en.wikipedia.org/wiki/C._Auguste_Dupin

**32** The Works of Edgar Allan Poe *by Edgar Allan Poe* The **Purloined Letter**
https://etc.usf.edu/lit2go/147/th-works-of-edgar-allan-poe/5357/the-purloined-letter/
https://poestories.com/read/purloined https://etc.usf.edu/lit2go/147/the-works-of-edgar-allan-poe/
https://en.wikipedia.org/wiki/The_Purloined_Letter
https://www.gutenberg.org/files/2148/2148-h/2148-h.htm
https://www.gutenberg.org/files/2148/2148-h/2148-h.htm#linkpurloin
https://en.wikipedia.org/wiki/Edgar_Allan_Poe
https://en.wikipedia.org/wiki/C._Auguste_Dupin
https://www.britannica.com/topic/The-Purloined-Letter
https://www.amazon.com/The-Purloined-Letter/dp/B001PU0V2G

**33** Edgar Allan Poe, "The Purloined Letter," 1845, 1910.
https://etc.usf.edu/lit2go/147/the-works-of-edgar-allan-poe/5357/the-purloined-letter/
[Paris, 33, Rue Dunôt, Faubourg St. Germain, after dark, autumn 18-, the story's narrator reports of a meeting with his friend, Le Chevalier C. Auguste Dupin (Poe's fictional amateur detective), interrupted by Monsieur G–, the Prefect of the Parisian police].

**34** https://en.wikisource.org/wiki/Tales_(Poe)/The_Purloined_Letter
https://en.wikisource.org/wiki/Tales_(Poe)
https://en.wikisource.org/wiki/Author:Edgar_Allan_Poe https://www.eapoe.org/

**35** See also: P.D. James, *Talking About Detective Fiction*, Chapter 2, "The Tenant of 221B Baker Street . . .," pages 33, 34; Vintage Books, New York City, 2011, 2009. ISBN 978-0-307-74313-8.

**36** *Nil sapientiae odiosius acumine nimio.* = Nothing is more hateful to the sense than too much cunning.
https://groups.google.com/forum/#!topic/alt.language.latin/eFyn9kttoo4

**37** *Nil sapientiae odiosius acumine nimio* = Nothing is more hateful to wisdom than too much cleverness
https://www.proz.com/kudoz/latin-to-english/other/141379-nil-sapientiae.html

**38** But likely not Seneca. Stavros Theodorakis, "The Motto in Edgar Allan Poe's 'The Purloined Letter'," 07 Aug 2010.
https://www.tandfonline.com/doi/abs/10.3200/ANQQ.22.1.25-27?journalCode=vanq20

**39** Wherein there was an interference with the search process.

**40** Henry Hess and Michael Grange, "Bernardo's ex-lawyers charged / Two removed damning tapes from house the day before Homolka struck plea bargain," *The Globe & Mail*, 24 January 1997.

**41** Emphasis added here.

**42** For which the lawyer got into big and protracted trouble, and about which the Law Society of Ontario amended the professional rules of conduct. See the full *Globe and Mail* article.

**43** Mr. B . . . was eventually convicted of some of the most serious crimes in Ontario's history. His lawyer, Mr. M . . ., although unsuccessfully criminally prosecuted for his handling of the tapes, was subsequently the centre of focus for very significant changes in the law of lawyers' handling physical evidence.

**44** Although Poe's story of *The Purloined Letter* did not involve analytical chemistry.

**45** (But there was no outside lab work on the *Orient Express*. Hercule Poirot, Agatha Christie's, Belgian detective, did his own forensic science procedure – a thermal process – on-site – to reveal an important clue; Harper Collins, 2011, at page 66).

**46** *Gedankenexperiment*. Introduction 2.2.

**47** This *Gedankenexperiment* is inspired by an actual case from the author's law practice – where it appeared that police may have misapplied a commercially available test kit.

**48** See also Jan Ransom, "Inmates Received Harsh Punishments Over False Positives for Drugs," *The New York Times*, 21 November 2019, page A22.

**49** https://energyeducation.ca/encyclopedia/Grounding#:~:text=Electrical%20grounding%2C%20otherwise%20known%20as,with%20a%20dedicated%20grounding%20prong.

**50** Such as:

- Writing Team for the Chemical Technician Curriculum Project, Robert L. Pecsok, Kenneth Chapman and Wade H. Ponder; eds., *Chemical Technology Handbook / Guidebook for Industrial Chemical Technologists and Technicians*, American Chemical Society, Washington, D.C., 1975.
- Julian Tyson, *Analysis / What Analytical Chemists Do*, Royal Society of Chemistry Paperbacks, 1994.

**51** In the present era of electronic data processing, what used to be called a *paper trail* might now best be called a *document trail*; for purposes here, the concept is the same.

**52** *Nullius in verba* ≈ We do not take anybody's word for it.

**53** "The Royal Society's motto 'Nullius in verba' roughly translates as 'take nobody's word for it'. It is an expression of the determination of Fellows to withstand the domination of authority and to verify all statements by an appeal to facts determined by experiment." https://royalsociety.org/
https://royalsociety.org/about-us/who-we-are/history/

**54** Edward Neville da Costa Andrade, *Sir Isaac Newton/His Life and Work*, Doubleday Anchor, Garden City, New York; Science Studies Series, S42; reprinted from Macmillan Company, New York, and W. Collins Sons and Co, Ltd., London, first published 1954; page 60.

**55** The Latin motto of the Royal Society, London, freely translated by E.N. da C. Andrade in *Sir Isaac Newton, His Life and Work*, Doubleday Anchor (reprinting a book first published 1954 by Macmillan, New York; and Collins, London), Garden City, New York; page 60, as: **we do not take anybody's word for it.** And, at page 61, Sir Isaac is quoted as replying to an (incorrect) report of astronomical observations contrary to Newtonian physics: "It may be so, there is no arguing against facts and experiments."

**56** supposed to do, and does do – past and present – and as best as practicable, future.

**57** US Food and Drug Administration, *ORA Laboratory Manual*.
https://www.fda.gov/science-research/field-science-and-laboratories/field-science-laboratory-manual

**58** See below For example:

**59** *Guide to the Drugs Directorate Laboratory Activities Quality Assurance Program*; Minister of National Health and Welfare Canada, Minister of Supply and Services Canada; February 1991; Catalogue No. H42-2/26-1990; ISBN 0-662-5750-5.

**60** Canada Health Protection Branch, *Regional Laboratory Quality Assurance Manual*, Document Number 038, 1 October 1997.

**61** Health and Welfare Canada, Field Operations Directorate, Health Protection Branch, *Laboratory Quality Assurance Manual/For Use in Field Operations Directorate Laboratories*, Ottawa, 1977.

**62** For example, "MEMORANDUM OF UNDERSTANDING Between the DEPARTMENT OF NATIONAL HEALTH & WELFARE . . . as represented by the Minister . . . and the ROYAL CANDIAN MOUNTED POLICE . . . as represented by the Solicitor General . . . CONCERNING THE PROVISION OF AN ANALYSIS SERVICE OF SEIZED DRUGS AND DRUG PARAPHERNALIA," Ottawa, 1984, 1985.

**63** For example:
- analysis for organics in factory air,
- analysis for ethanol in blood,
- analysis for pharmaceuticals in race horses,
- DNA profiling of human fluids,
- analysis for narcotic drugs,
- ink identification,
- analysis for organics in the city water and sewage systems,
- forensic sample retention terms of reference,
- analysis for pesticides in market-vended vegetables,
- paint and pigment analysis.

**64** For example:
- archival procedures for lab-generated documents,
- public access terms of reference for lab-generated documents,
- forensic sample retention terms of reference,
- procedures to run IR spectra,
- procedures to prepare K Br pellets for IR spectra,
- procedures to run visible–UV spectra,
- procedures to run TLC plates, including solvent mixes,
- in-lab photography,
- equipment maintenance.

**65** Such as:
- "Standard Practice for Receiving, Documenting, Storing, and Retrieving Evidence in a Forensic Science Laboratory," ASTM E 1492–92; included in John J. Lentini, "ASTM Standards for Forensic Sciences," *Journal of Forensic Sciences*, 40(1), January 1995, pages 146→149. http://www.astm.org/
- "Recommended Standards and Procedures of the Canadian Society of Forensic Science Alcohol Test Committee," *Canadian Society of Forensic Science Journal*, Vol. 28, No. 1, March 1995, pages 1→53.
- https://www.tandfonline.com/doi/abs/10.1080/00085030.1995.10757469

**66** O. Reg. 490/09: DESIGNATED SUBSTANCES, under Ontario *Occupational Health and Safety Act*, RSO 1990, c. O.1 https://www.ontario.ca/laws/regulation/090490#BK7

[. . .]

***Measuring airborne concentrations***

**24.** (1) An employer shall ensure that procedures for monitoring, sampling and determining airborne concentrations of a designated substance and worker exposure to airborne concentrations of a designated substance,

(a) comply with,

(i) a standard method for workplace air sampling and analysis or another method recognized in industrial hygiene practice, and

(ii) the rules set out in Part I of Schedule 1 and

(b) are performed by, or under the direction of, a person who is qualified because of knowledge, training and experience in industrial hygiene practice.

(2) If a direct-reading instrument is used to determine airborne concentrations of a designated substance, the employer shall ensure that the instrument is used, calibrated and maintained in accordance with the manufacturer's instructions.

***Posting of monitoring results***

**25.** Whenever results become available under a control program that relate to the monitoring of airborne concentrations of a designated substance and worker exposure to airborne concentrations of a designated substance, the employer shall,

(a) promptly post the results in a conspicuous place or places where they are most likely to come to the attention of workers who would be affected by them and leave them posted for no less than 14 days;

(b) provide a copy of the results to the joint health and safety committee; and

(c) keep the results for no less than five years.

[. . .]

SCHEDULE 1

**PART I**

**AIRBORNE MEASUREMENT AND CALCULATION OF EXPOSURE**

1. Airborne concentrations of a designated substance are expressed as,
   (a) parts of the agent per million parts of air by volume (ppm);
   (b) milligrams of the agent per cubic metre of air ($mg/m^3$); or
   (c) fibres per cubic centimetre of air (f/cc).
2. In determining exposure to airborne concentrations of the designated substance, no regard shall be had to the wearing or use of respirators.
3. Air sampling of the airborne concentrations of the designated substance is not required for the full period of a work day or a work week if the air sampling is representative of airborne concentrations of the substance likely to be present during the full period.
4. The average concentrations of a designated substance to which a worker is exposed shall be determined from analysis of air samples taken as being representative of the exposure of the worker to the designated substance during work operations.
5. Subject to section 6 of this Schedule, the time-weighted average exposure to an airborne designated substance in a work day or work week shall be calculated as follows:
   1. The cumulative daily or weekly exposure shall be calculated using the following formula:
      C
      $_1T_1 + C_2T_2 . . . + C_nT_n$
      where,
      $C_1$ is the concentration found in an air sample, and
      $T_1$ is the total time in hours to which the worker is taken to be exposed to concentration $C_1$ in a work day or a work week.
   2. The time-weighted average exposure shall be calculated by dividing the cumulative daily exposure by eight and the cumulative weekly exposure by 40 respectively.
6. A determination of time-weighted average exposure for extended work shifts may be calculated using the methodology set out in the Guide for the Adjustment of Permissible Exposure Values for Unusual Work Schedules (March 2015), published by Quebec's Institut de recherche Robert-Sauvé en santé et en sécurité du travail (IRSST), using the applicable TWA set out in Table 1 of this Regulation.
7. Short-term exposures to the designated substance in any 15-min period are determined from a single sample or from a time-weighted average of sequential samples taken during that period.

[. . .]

67 Where there is, in effect, an SOP by the listing of approved instruments – RSC 1985, c. C-46, as am., s. 254, et seq. and regulations.

**68** See also:
- "Canadian Society of Forensic Science Alcohol Test Committee Recommended Operational Procedures," December 2018. https://www.csfs.ca/wp-content/uploads/2018/12/2018-12-18-Operational-Procedures.pdf
- K.L. Blake, T.C. Cherlet, A. Dion, P.M. Harding, Robert Langille, Teri Martin, D.J. Mayers, V.M. Mendes, B.K. Wong, H. Pruden, T.C. Cherlet, "Canadian Society of Forensic Science Alcohol Test Committee Equipment Standards and Evaluation Procedures," *Journal of the Canadian Society of Forensic Science*, 47(4):200–213, October 2014 DOI: 10.1080/00085030.2014.959750
- https://www.researchgate.net/publication/280164034_Canadian_Society_of_Forensic_Science_Alcohol_Test_Committee_Equipment_Standards_and_Evaluation_Procedures
- "Recommended Standards and Procedures of the Canadian Society of Forensic Science Alcohol Test Committee," *Canadian Society of Forensic Science Journal*, Vol. 28, No. 1, March 1995, pages 1→53.

**69** See John J. Lentini, "ASTM Standards for Forensic Sciences," *Journal of Forensic Sciences*, 40(1), January 1995, pages 146→149.

**70** *Gedankenexperiment.* Introduction 2.2.

**71** Most of the ink examples, throughout this book, are for ball point pen inks (rather than for fountain pen ink as for one example in Chapter 2 – different chemistries).

**72** – Richard L. Brunelle and Robert Reed, *Forensic Examination of Ink and Paper*, Hardcover (June 1984) Charles C Thomas Pub Ltd; ISBN: 0398049351 $82.95
- https://www.amazon.com/Forensic-Examination-Paper-Richard-Brunelle/dp/0398049351/ref=sr_1_1?keywords=0398049351&qid=1565568517&s=gateway&sr=8-1

**73** Richard L. Brunelle, Robert W. Reed, *Forensic Examination of Ink and Paper*, Charles C Thomas Publisher · Ltd., Springfield, Illinois, 62704-4730; 1984.
https://www.ccthomas.com/details.cfm?P_ISBN13=9780398060398

**74** – Richard L. Brunelle and Kenneth R. Crawford, *Advances in the Forensic Analysis and Dating of Writing Ink*; Charles C Thomas Publisher · Ltd., Springfield, Illinois, 62704-4730; 2003. https://www.ccthomas.com/details.cfm?P_ISBN13=9780398073473

**75** – R. L. Brunelle, in Richard Saferstein, ed., *Forensic Science Handbook* [Volume I], Prentice-Hall, Inc., Englewood Cliffs, New Jersey, 1982, Chapter 14, "Questioned Document Examination," particularly at pages 680, 709→711, 712→717, and references cited therein.

**76** – Richard L. Brunelle and A.A. Cantu; in Geoffrey Davies, ed., *Forensic Science*, ACS Symposium Series 13, American Chemical Society, Washington, D.C., 1975, Chapter 14, "Ink Analysis – A Weapon Against Crime by Detection of Fraud," pages 134→141, and references cited therein; page 136.

**77** – R.L. Kuranz, *Journal of Forensic Sciences*, Vol. 19, No. 4, October 1974, pages 852→ 854. "Technique for the Separation of Ink Dyestuffs with Similar *RF* Values"
https://www.astm.org/DIGITAL_LIBRARY/JOURNALS/FORENSIC/PAGES/JFS10475J.htm

**78** – Joel Harris, "A Preliminary Report on the Nondestructive Examination of Ballpoint Pen Ink on Questioned Documents by FT-IR Spectroscopy," *Canadian Society of Forensic Science Journal*, Vol. 24, No. 1 (1991), pages 5→21, and references cited therein. https://www.tandfonline.com/doi/abs/10.1080/00085030.1991.10756979

**79** – "Analysis of Ballpoint Pen Inks by Diffuse Reflectance Infrared Spectrometry," *Journal of Forensic Sciences*, Vol. 37, No. 2, March 1992, pages 528→541.
- Cantu, A., "The Paper Mate® Ink on the Howard Hughes "Mormon Will"," *Journal of Forensic Sciences*, Vol. 31, No. 1, 1986, pp. 360→364, Jan. 1986, https://doi.org/10.1520/JFS11895J. ISSN 0022–1198

**80** https://www.astm.org/DIGITAL_LIBRARY/JOURNALS/FORENSIC/PAGES/JFS11895J.htm

**81** A. Cantu, in *REPORT OF THE* [US] *SENATE IMPEACHMENT TRIAL COMMITTEE . . . AGAINST JUDGE ALCEE L. HASTINGS*, "HEARINGS / SENATE IMPEACHMENT TRIAL COMMITTEE . . ." US Government Printing Office, Washington, DC, 1989; pages 691 and 693.
Google books https://books.google.com
https://www.google.ca/books/edition/Report_of_the_Senate_Impeachment_Trial_C/6JSbAAAAMAAJ?hl=en&gbpv=1&dq=%5Bpage+691%5D+%5BCurriculum+Vitae+%E2%80%93+Antonio+A.+Cantu+Page+8%5D&pg=PA691&printsec=frontcover
https://guides.loc.gov/federal-impeachment/alcee-hastings
**82** https://www.senate.gov/artandhistory/history/common/briefing/Impeachment_Hastings.htm
https://heinonline.org/HOL/LandingPage?handle=hein.cbhear/imptalhb0001&div=1&src=home
https://www.govinfo.gov/content/pkg/GPO-CRECB-1989-pt18/pdf/GPO-CRECB-1989-pt18-4.pdf
https://catalog.loc.gov/vwebv/search?searchCode=LCCN&searchArg=90600189&searchType=1&permalink=y
**83** Google books https://books.google.com
https://www.google.ca/books/edition/Report_of_the_Senate_Impeachment_Trial_C/6JSbAAAAMAAJ?hl=en&gbpv=1&dq=%5Bpage+691%5D+%5BCurriculum+Vitae+%E2%80%93+Antonio+A.+Cantu+Page+8%5D&pg=PA691&printsec=frontcover

[. . .]

> [page 691] [*Curriculum Vitae* – Antonio A. Cantu Page 8]
>
> R.L. Brunelle and A.A. Cantu, "Ink Analysis – A Weapon Against Crime by Detection of Fraud," a chapter in the American Chemical Society Symposium Series No. 13., "Forensic Science", Geoffery Davis – Ed (1975).
>
> Jerry D. Kelly and Antonio A. Cantu, "Proposed Methods for Comparative Evaluation for the Establishment of a Standard System for Ink Identification," Journal of the AOAC, 58 122 (1975).
>
> D.A. Crown, R.L. Brunelle and A.A. Cantu, "Parameters of Ballpen Ink Examination," J. For. Sci. 21, 917 (1976).
>
> R.L. Brunelle, J.F. Negri, A.A. Cantu and A.H. Lyter III, "Comparison of Typewriter Ribbon Inks by Thin Layer Chromatography," J. For. Sci. 22, 807 (1977).
>
> Stepen Cain, A.A. Cantu, A.H. Lyter III, and R.L. Brunelle, "A Scientific Study of Pencil Lead Components," J. For. Sci. 23, 643 (1978).
>
> Carl Sensi and A.A. Cantu, "Infrared Luminescence: Is it a Valid Method to Differentiate Among Inks?," J. For. Sci. 27, 196 (1982).
>
> A.A. Cantu, "On the Relative Aging of Ink," (in preparation).

[. . .]

**84** Google books (see Fig. 5.5) [. . .] [page 693].

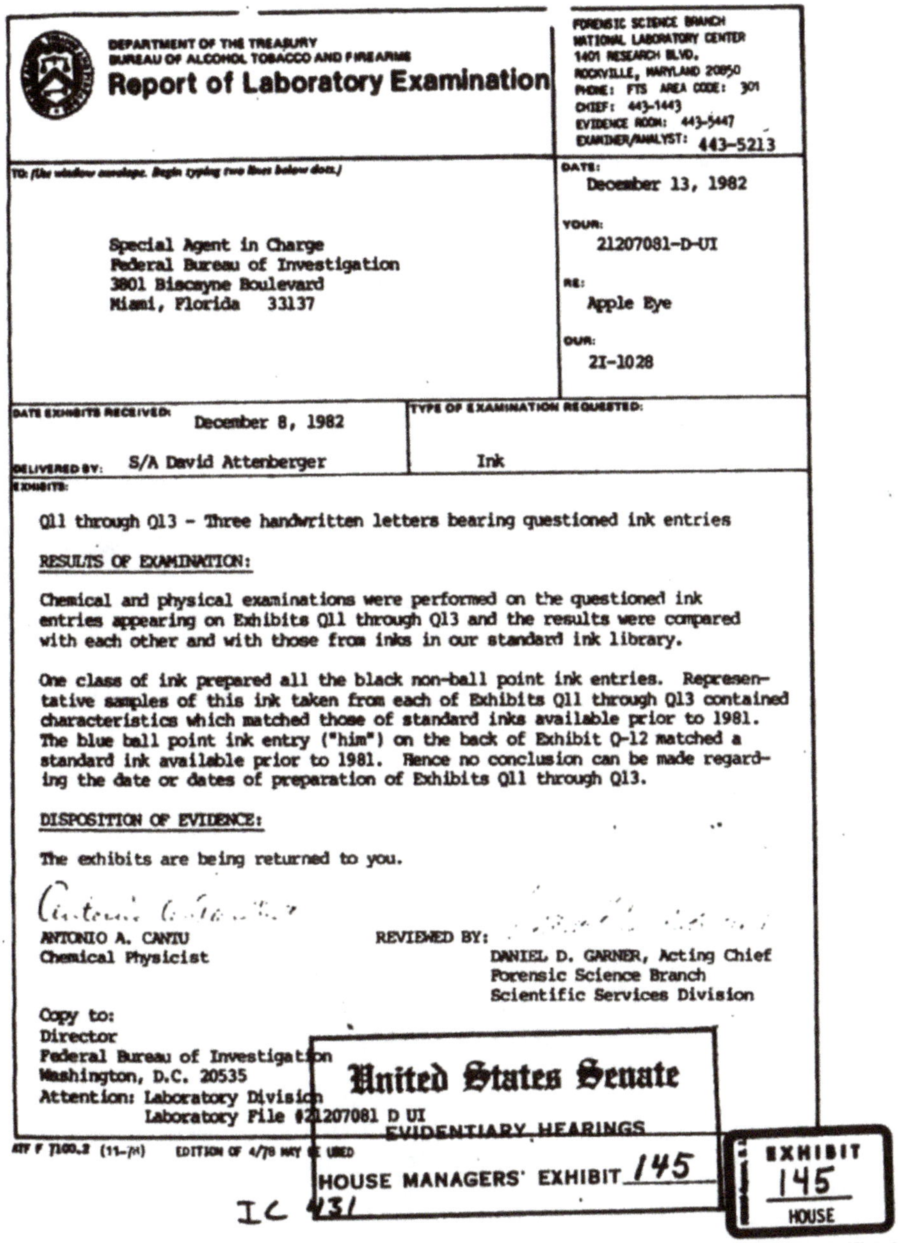

DEPARTMENT OF THE TREASURY
BUREAU OF ALCOHOL TOBACCO AND FIREARMS
**Report of Laboratory Examination**

FORENSIC SCIENCE BRANCH
NATIONAL LABORATORY CENTER
1401 RESEARCH BLVD.
ROCKVILLE, MARYLAND 20850
PHONE: FTS AREA CODE: 301
CHIEF: 443-1443
EVIDENCE ROOM: 443-5447
EXAMINER/ANALYST: 443-5213

TO: (Use window envelope. Begin typing two lines below dots.)

Special Agent in Charge
Federal Bureau of Investigation
3801 Biscayne Boulevard
Miami, Florida 33137

DATE: December 13, 1982

YOUR: 21207081-D-UI

RE: Apple Eye

OUR: 21-1028

DATE EXHIBITS RECEIVED: December 8, 1982

DELIVERED BY: S/A David Attenberger

TYPE OF EXAMINATION REQUESTED: Ink

EXHIBITS:

Q11 through Q13 - Three handwritten letters bearing questioned ink entries

RESULTS OF EXAMINATION:

Chemical and physical examinations were performed on the questioned ink entries appearing on Exhibits Q11 through Q13 and the results were compared with each other and with those from inks in our standard ink library.

One class of ink prepared all the black non-ball point ink entries. Representative samples of this ink taken from each of Exhibits Q11 through Q13 contained characteristics which matched those of standard inks available prior to 1981. The blue ball point ink entry ("him") on the back of Exhibit Q-12 matched a standard ink available prior to 1981. Hence no conclusion can be made regarding the date or dates of preparation of Exhibits Q11 through Q13.

DISPOSITION OF EVIDENCE:

The exhibits are being returned to you.

ANTONIO A. CANTU
Chemical Physicist

REVIEWED BY:
DANIEL D. GARNER, Acting Chief
Forensic Science Branch
Scientific Services Division

Copy to:
Director
Federal Bureau of Investigation
Washington, D.C. 20535
Attention: Laboratory Division
Laboratory File #21207081 D UI

ATF F 7100.2 (11-78) EDITION OF 4/78 MAY BE USED

United States Senate
EVIDENTIARY HEARINGS
HOUSE MANAGERS' EXHIBIT 145

EXHIBIT
145
HOUSE

IC 431

Fig. 5.5: US Senate Hastings evidence forensic report (US Treasury & US Senate evidence document).

**85** REPORT 2d Session, HOUSE OF REPRESENTATIVES 100–810 IMPEACHMENT OF JUDGE ALCEE L. HASTINGS REPORT OF THE COMMITTEE ON THE JUDICIARY TO ACCOMPANY H. Res. 499.
https://www.senate.gov/about/powers-procedures/impeachment/impeachment-hastings.htm
https://www.c-span.org/video/?8478-1/judge-hastings-impeachment-trial

**86** Pierre Thomas, "FBI ROLE IN IMPEACHMENT PROBED, *Washington Post*, 26February 1997.
https://www.washingtonpost.com/archive/politics/1997/02/26/fbi-role-in-impeachment-probed/16e98fab-4f65-4e88-9c3e-109ee47400a5/

**87** – "Thin-Layer Chromatography of Writing Inks – Quality Control Considerations," *Journal of Forensic Sciences*, 41(5), September 1996, pages 874→877.
https://www.ncjrs.gov/App/Publications/abstract.aspx?ID=165765
– https://doi.org/10.1520/JFS14015J ISSN 0022–1198

**88** Frank H. Norwitch and Howard Seiden, "Questioned Documents," as Chapter 19, in Stuart H. James and Jon J. Nordby, editors, *Forensic Science / An Introduction to Scientific and Investigative Techniques*, CRC Press, Boca Raton; 2003; ISBN 0-8493-1246-9; pages 357→373.www. crcpress.com

**89** Norwitch and Seiden, ". . . Ink Examination," pages 364→366.

**90** Norwitch and Seiden, references cited therein, at page 373.

**91** – Paul L. Freese, 1986. "Howard Hughes and Melvin Dummar: Forensic Science Fact Versus Film Fiction." *Journal of Forensic Sciences* 31 (January).
– Ink on the Howard Hughes 'Morman Will'," *Journal of Forensic Sciences*, Vol. 31, No. 1, Jan. 1986, pages 360→364.
https://law.jrank.org/pages/11317/Will-Howard-Hughes-Mormon-Will.html

**92** Mass Spectrometry, Roger W Jones and John F McClelland, *Forensic science international* 231 (1–3):73–81, September 2013.

**93** – Mark Maremont, Staff Reporter, "In Corporate Crimes, Paper Trail Often Leads to Ink Analysts'Door / Mightier Than Broker's Word, 2 Ballpoints Could Help Land Martha Stewart in the Pen," THE WALL STREET JOURNAL, updated 01 July 2003 12:01 am ET.
https://www.wsj.com/articles/SB105701075148372400

**94** Perhaps simply as a literature citation.

**95** = Made or carried out in a perfunctory manner or as a formality
https://www.merriam-webster.com/dictionary/pro%20forma

**96** Although it is arguable that some common methods are well known enough amongst chemists that only mentioning the name of the method is sufficient.

**97** *Gedankenexperiment*. Introduction 2.2.

**98** Polymer envelopes. Chapters 3, 5 and 9.

**99** In the symbolism here: [][][] = an identification number.

**100** See Karen Jakob, *The Complete Guide to Police Writing*, The Carswell Company Limited, Toronto, 1984, "The Memorandum Book," Chapter 8, pages 105–109.

**101** See Michael Souliere, "The ABCs of police note-taking," BLUE LINE, 13 May 2015.
https://www.blueline.ca/the_abcs_of_police_note-taking-3288/ https://www.blueline.ca/

**102** Howard M. Kanare, *Writing the Laboratory Notebook*, American Chemical Society, Washington, D.C.; 1985; ISBN-0-8412-0933-2, ISBN-0-8412-0906-5; at page 1.

**103** https://files.eric.ed.gov/fulltext/ED344734.pdf
https://che.k-state.edu/docs/people/emeritus/aikens/writing/laboratory_notebook_rules.pdf
https://eric.ed.gov/?id=ED344734
https://pubs.acs.org/doi/10.1021/ed066pA74.2

**104** See also:
– Kevin M. Cox, Ocala, Florida; and Lloyd Taylor, Lexington, Mass; two letters to the editor, *Chemical & Engineering News*, 26 August 1996, page 4;
– and Dave English, Orange, California, letter to the editor, *C&EN*, 23 September 1996, page 12.

- Robert L. Pecsok, Kenneth Chapman and Wade H. Ponder; eds., *Chemical Technology Handbook / Guidebook for Industrial Chemical Technologists and Technicians*, American Chemical Society, Washington, D.C., 1975, Chapter 10, "Laboratory Notebooks," pages 81→87.
- Analogous to the chemist's lab notebook, and of possible relevance in forensic cases, is the investigator's notebook – see Karen Jakob, *The Complete Guide to Police Writing*, The Carswell Company Limited, Toronto, 1984, "The Memorandum Book," Chapter 8, pages 105→109.

**105** See Francis Russell, "Sacco Guilty, Vanzetti Innocent?" *American Heritage*, Vol. XIII, No. 4, June 1962, pages 4→9 and 107→111. In 1927, Nicola Sacco and Bartolomeo Vanzetti, were executed for two murders near Boston in 1920. The court proceedings were in an American anti-anarchist and anti-Italian-immigrant era. The case drew world-wide attention. In his article, Mr. Russell reports on his 1961 re-examination and re-evaluation of ballistic and other evidence.

**106** Walter F. Rowe, "Firearms Identification," as Chapter 8 in R. Saferstein, ed., *Forensic Science Handbook* Volume II, Prentice Hall, Englewood Cliffs, New Jersey; 1988; ISBN 0-13-326877-2; pages 414→416; and references cited therein.

**107** https://en.wikipedia.org/wiki/Sacco_and_Vanzetti#Later_evidence_and_investigations

**108** https://en.wikipedia.org/wiki/Sacco_and_Vanzetti

**109** See Canadian Press, "Samples sought by Truscott may be lost," *The Globe and Mail*, 3 October 1997, page A10: ". . . A review by Ontario's largest crime laboratory suggests it has lost semen samples that might have helped a man prove he was wrongly convicted of murder 38 years ago. No one at the Centre of Forensic Sciences is quite sure where the samples . . . are now, or whether they still exist . . . . . . . Lawyers for Mr. Truscott, convicted of strangling 12-year-old Lynne Harper in 1959 and originally sentenced to death, asked the government for help in finding samples from the crime scene that could be submitted to a DNA test. They hope such a test would prove he did not kill the girl. Records at the centre suggest it once had semen samples found on the victim's body . . . and those records indicate that any samples would have been returned to the Crown . . . ." See also Kirk Makin, "Many crime samples gone, coroner says / Case-by-case review, retesting called impossible; even some work seen as enormous task." *The Globe and Mail*, 4 October 1997, page A6.

**110** https://www.cbc.ca/news2/background/truscott/
https://www.thestar.com/news/canada/2007/08/28/court_acquits_truscott.html

**111** See the above note – the whole book.

**112** For example, see:

- US FDA *Field Science – Laboratory Manual* [2021].
  https://www.fda.gov/science-research/field-science-and-laboratories/field-science-laboratory-manual

**113** – AOAC International Official Methods of Analysis, 22nd Edition (2023)
https://www.aoac.org/official-methods-of-analysis/ https://www.aoac.org/

**114** *Gedankenexperiment*. Introduction 2.2.

**115** Readers are cautioned that this is a creation of the author, for illustration purposes, that has not been reality or lab-tested. Some of it might be regarded as too fanciful.

**116** Tetrahydrocannabinol, THC
CAS 1972-08-3
314.469 g/mol.
$\Delta^9$-THC (see Fig. 5.6).

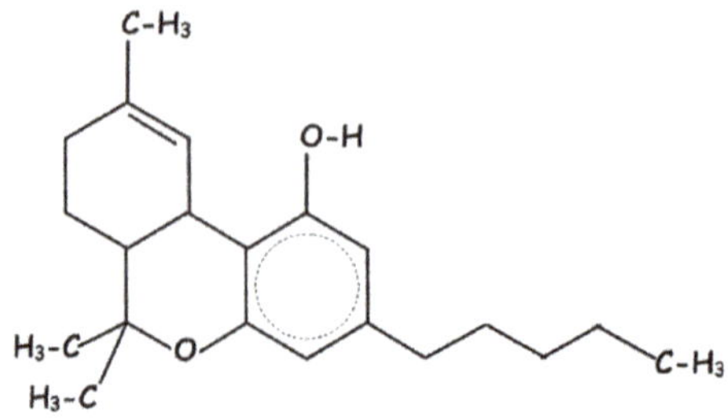


Fig. 5.6: CAS 1972-08-3.

**117** – J.A. Siegel, "Forensic Identification of Controlled Substances," as Chapter 3 in R. Saferstein, ed., *Forensic Science Handbook* Volume II, Prentice Hall, Englewood Cliffs, New Jersey, 1988, pages 87→92.
– Duquénois-Levine, Siegel, Saferstein, page 124.
– K. Bailey, "The Value of the Duquénois Test for Cannabis – A Survey," *Journal of Forensic Sciences*, 1979, 24(4), pages 817–841. (Keith Bailey, M.A., D.Phil., Chief, Drug Identification Division, Drug Research Laboratories, Health Protection Branch, Health & Welfare Canada, Tunney's Pasture, Ottawa K1A 0L2.)
– Daniel M. Perrine, *The Chemistry of Mind-Altering Drugs / History, Pharmacology, and Cultural Context*; American Chemical Society, Washington, D.C.; 1996; ISBN 0-8412-3253-9; page 347, et seq. – at page 347, 7–10.

**118** http://en.wikipedia.org/wiki/Tetrahydrocannabinol
http://www.commonchemistry.org/ChemicalDetail.aspx?ref=1972-08-3&terms=THC
https://pubchem.ncbi.nlm.nih.gov/#tab/sidsrcname=ChemIDplus&query=1972-08-3&input_type=text

**119** In the symbolism here: [][][] = an identification number.

**120** In the symbolism here: [][] [][][][] 20[][], [][]:[][] []m. = a date: | day | month | year | time – am or pm. |

**121** J.A. Siegel, "Forensic Identification of Controlled Substances," as Chapter 3 in R. Saferstein, ed., *Forensic Science Handbook* Volume II, Prentice Hall, Englewood Cliffs, New Jersey; 1988; ISBN 0-13-326877-2; pages 87→92; and references cited therein.

**122** K. Bailey, "The value of the Duquénois Test for Cannabis – a Survey," *Journal of Forensic Sciences*, 1979, 24(4), 817→841; and references cited therein.

**123** In the symbolism here: [][].[][] = an SI quantity.

**124** In the symbolism here: [][][][] = a brief description of shape.

**125** See:
– Gerald T.Mitosinka, John I. Thornton, Thomas L. Hayes, "The Examination of Cystolithic Hairs ot Cannabis and other Plants by means of the Scanning Electron Microscope," *Journal of the Forensic Science Society*, Volume 12, Issue 3, July 1972, pages 521→529
https://www.sciencedirect.com/science/article/abs/pii/S0015736872707173
– Justin J. McShane, "The Myth of Specific Identification of Marijuana in Criminal Court. Part 3: What Is Microscopic Morphological Examination? Is It a "Good" Test?" [2012].
https://thetruthaboutforensicscience.com/the-myth-of-specific-identification-of-marijuana-in-criminal-court-part-3-what-is-microscopic-morphological-examination-is-it-a-good-test/

**126** Suzanne Bell, *A Dictionary of Forensic Science*, Oxford University Press, 2012, Print ISBN-13: 9780199594009 Online, 2013; 2024.
https://www.oxfordreference.com/display/10.1093/acref/9780199594009.001.0001/acref-9780199594009-e-0291#:~:text=A%20microscopic%20feature%20found%20on,CaCO . . . %20 . . .

cystolithic hair

A microscopic feature found on the upper surface of the leaves of . . . marijuana . . .

**127** In the symbolism here: [][][][] = chemical names of solvents and % proportions.

**128** In the symbolism here: [][][] = an identification number.

**129** "LifeLabs paid ransom to cyber attackers who may have compromised personal information of 15 million Canadians / Customer information could include name, address, email, login, passwords, date of birth, health card number and lab test results," *Financial Post*, Toronto. M4W3L4; 18 Dec. 2019. https://financialpost.com/technology/lifelabs-hack-may-have-compromised-personal-info-of-15-mln-canadians-2

**130** https://www.lifelabs.com/

**131** See, for example, in another context: CBC Radio, Ideas, "How algorithms create a 'digital underclass' / Princeton sociologist Ruha Benjamin argues bias is encoded in new tech," September 2019. https://www.cbc.ca/radio/ideas/how-algorithms-create-a-digital-underclass-1.5269959 https://www.cbc.ca/radio/ideas

**132** SOR/86-304, as am., s.2.2(1). https://laws-lois.justice.gc.ca/eng/regulations/SOR-86-304/page-2.html

[. . .]

**DIVISION I**

Buildings Standards

**2.2 (1)** The design and construction of every building, the construction of which begins on or after the day of the coming into force of this subsection, shall meet the requirements of the National Building Code.

**(2)** Every building, the construction of which begins before the day of the coming into force of this subsection, shall, if feasible, meet the requirements of the National Building Code.

**(3)** The renovation of any building or part of a building shall, if feasible, meet the requirements of the National Building Code.

**(4)** If it is not feasible for an employer to comply with the requirements of subsection (3), the employer shall, before the proposed renovations start, notify the work place committee or the health and safety representative.

[. . .]

**133** *Canada Labour Code* RSC, 1985, c. L-2 https://laws.justice.gc.ca/eng/acts/L-2/

**134** ". . . *The National Building Code of Canada 2015* . . ., published by . . . [National Research Council Canada] and developed by the Canadian Commission on Building and Fire Codes, sets out technical provisions for the design and construction of new buildings . . . ."
https://nrc.canada.ca/index.php/en/certifications-evaluations-standards/codes-canada/codes-canada-publications/national-building-code-canada-2015
https://nrc.canada.ca/index.php/en/certifications-evaluations-standards/codes-canada/codes-canada-publications

**135** O. Reg. 213/91: CONSTRUCTION PROJECTS, under Ontario *Occupational Health and Safety Act*, RSO 1990, c. O.1 s.26.1 https://www.ontario.ca/laws/regulation/910213%20 https://www.ontario.ca/laws/statute/90o01

[. . .]

(3) The components of any system listed in subsection (2) shall be designed by a professional engineer in accordance with good engineering practice, and shall meet the requirements of any of the following National Standards of Canada standards that are applicable:

1. CAN/CSA-Z259.1-05: Body Belts and Saddles for Work Positioning and Travel Restraint.

2. CAN/CSA-Z259.2.5-12: Fall Arresters and Vertical Lifelines.
3. CAN/CSA-Z259.2.2-98 (R2004): Self-Retracting Devices for Personal Fall-Arrest Systems.
4. CAN/CSA-Z259.2.3-99 (R2004): Descent Control Devices.
5. CAN/CSA-Z259.10-06: Full Body Harnesses.
6. CAN/CSA-Z259.11-05: Energy Absorbers and Lanyards.
7. CAN/CSA-Z259.12-01 (R2006): Connecting Components for Personal Fall Arrest Systems (PFAS).
8. CAN/CSA-Z259.14-01 (R2007): Fall Restrict Equipment for Wood Pole Climbing. O. Reg. 85/04, s. 5 (2); O. Reg. 443/09, s. 1; O. Reg. 345/15, s. 5 (2).

[. . .]

**136** Canadian Standards Association CSA Group, Toronto M9W 1R3 https://www.csagroup.org/ http://www.companylisting.ca/Canadian_Standards_Association1/default.aspx

**137** In the symbolism here: [][][] = an identification number.

**138** That is available in the edition as it was at the date of the report – subsequent revisions are not satisfactory for this purpose.

**139** US National Institute of Standards and Technology (formerly National Bureau of Standards), US Department of Commerce, Gaithersburg, Maryland, 20899-0001.
Standard Reference materials https://www.nist.gov/srm

**140** United States Pharmacopeial Convention, Inc., 12601 Twinbrook Parkway, Rockville, Maryland, 20852-1790. US Pharmacopeia http://www.usp.org/ https://www.usp.org/reference-standards
USP Reference Standards https://www.usp.org/reference-standards

**141** API, Washington, DC 20001-5571.
API Standards https://www.api.org/products-and-services/standards/purchase

**142** Sigma Chemical Company, St. Louis, Missouri, 63178.

**143** See also Byron Kratochvil and Alan Walton, "The Role of Reference Materials in Quality Assurance for Chemical Measurement," *Canadian Chemical News*, May 1995, pages 40–42, and references cited therein. See also Chapter 4.

**144** ASTM, ASTM International, ASTM Headquarters, West Conshohocken, Pennsylvania, 19428-2959.
http://www.astm.org/
https://www.astm.org/Standard/standards-and-publications.html

**145** American National Standards Institute, ANSI Headquarters, 11 W 42nd Street, New York City. http://www.ansi.org/

**146** For example, ASTM D4236 – 94(2016)e1 / ASTM D4236 / "Standard Practice for Labeling Art Materials for Chronic Health Hazards" https://www.astm.org/ https://www.astm.org/Standards/D4236.htm

**147** For example: Crayola crayons, as art materials, are sold in retail packaging with notice of ASTM conformity (as to labeling). ". . . Conforms to . . . ASTM D 4236 . . . ALL CRAYOLA ART MATERIALS ARE NONTOXIC . . ." CRAYOLA LLC, Easton, Pennsylvania, 18044.
https://www.crayola.com/
https://www.crayola.com/support/product-safety.aspx
https://www.greenamerica.org/green-living/are-art-supplies-toxic

**148** ACMI = The Art and Creative Materials Institute, Inc. https://acmiart.org/

**149** ISO Central Secretariat address: International Organization for Standardization (ISO), 1, rue de Varembé, Case postale 56, CH-1211 Genève 20, Switzerland.

**150** See Norm A. Hagan, *Symposium/A Guide to Accessing Standards Information in Canada*, SCC, Ottawa, ISBN 0-920360-40-8.

**151** Not too many years ago this involved rolls of chart paper and an ink pen (or other writing system) that followed an electronic signal from a detector. Often problematic, because when, for example, a chromatographic signal would appear the pen would suddenly move faster and not work. And, when it did work, too many data-containing rolls of paper to unroll for reading, and then to store. Modern computer-received and recorded data systems avoid much of this.

**152** To borrow a phrase from the *Canadian Charter of Rights and Freedoms* (being Schedule B, Part I of the *Canada Act* 1982, including the *Constitution Act*, 1982 (1982 c. 11 (United Kingdom)), s. 1.

**153** With some limitations – for example under the *Young Offenders Act*, RSC 1985, as am., c.Y-1, s.38 and 39; and the Canada *Criminal Code*, RSC 1985, c.C-46, as am., s.542(2) and 276.3. See also, Timothy Appleby, "[Hamilton] *Spectator* under fire for printing teen's [criminal] record / Story about fugitive stirs police complaint," *The Globe and Mail*, NATIONAL NEWS, 13 November 1999, page A11.

**154** Typically several $/page for transcription; less for subsequent copies thereof. Consequently, lawyers might quietly compete with one another to not be first to ask.

**155** RSC 1985c. A-1.

**156** For example, the Ontario *Freedom of Information and Protection of Privacy Act*, RSO 1990, c.F.31.

**157** See, for example, on the Food and Drug Administration's Internet site http://www.fda.gov/: "FDA's Electronic Freedom of Information Reading Room" http://www.fda.gov/foi/foia2.htm; and "A Handbook for Requesting Information and Records from FDA," http://www.fda.gov/ opacom/background ers/foiahand.html

**158** RSC 1985 c.A-17. https://laws-lois.justice.gc.ca/eng/acts/a-17/

**159** For example, the Ontario *Auditor General Act*, RSO 1990, c. A.35.
https://www.ontario.ca/laws/statute/90a35
https://www.gov.on.ca/MBS/english/publications/statregs/contents.html

**160** Although audit officials might exercise a discretion to investigate suggestions from the public.

**161** They report directly to Parliament and the provincial Legislatures, rather than to the governments – for example: the federal *Auditor . . . Act*, s.3, 7 and 8; and the Ontario *Auditor General Act*, s.6 and 12. http://legis.acjnet.org/

**162** See, for example: Sarah Maslin Nir, "In Rush to Release Kennedy Files, Personal Information Went Public, Too / Social Security numbers and other personal details were included in the 64,000 pages of documents that the Trump administration declassified this week," *New York Times*, 20 March 2025.
https://www.nytimes.com/2025/03/20/us/jfk-assassination-files-personal-information.html

**163** See, for example, by analogy:ONTARIO REGULATION 490/09, *DESIGNATED SUBSTANCES*, *Occupational Health & Safety Act*. https://www.ontario.ca/laws/regulation/R09490 – retention of exposure medical records – s.30 and 31. https://www.ontario.ca/laws/regulation/090490#BK40

**164** s.25, s.29(3)(b), (6), (7).
https://www.ontario.ca/laws/regulation/090490#BK28
https://www.ontario.ca/laws/regulation/090490#BK40

**165** See, for example:
- Francis Russell, "Sacco Guilty, Vanzetti Innocent?." *American Heritage*, June 1962, Volume 13, Issue 4. https://www.americanheritage.com/sacco-guilty-vanzetti-innocent

**166** See also:
- Canadian Press, "Samples sought by Truscott may be lost," *The Globe and Mail*, 3 October 1997, page A10: ". . . A review by Ontario's largest crime laboratory suggests it has lost semen samples that might have helped a man prove he was wrongly convicted of murder 38 years ago. No one at the Centre of Forensic Sciences is quite sure where the samples . . . are now, or whether they still exist . . . . . . . Lawyers for Mr. Truscott, convicted of strangling 12-year-old Lynne Harper in 1959 and originally sentenced to death, asked the government for help in finding samples from the crime scene that could be submitted to a DNA test. They hope such a test would prove he did not kill the girl.Records at the centre suggest it once had semen samples found on the victim's body . . . and those records indicate that any samples would have been returned to the Crown . . . . "

- Kirk Makin, "Many crime samples gone, coroner says / Case-by-case review, retesting called impossible; even some work seen as enormous task," *The Globe and Mail*, 4 October 1997, page A6.

**167** See:

- Robert A. Erlandson, "Request to dig up the remains in Booth grave will get a hearing in court," *Baltimore Sun*, 15 November 1994, page 12B; "Cemetery chairman wants full hearing on Booth petition," *ibid.*, 28 October 1994, page 6B; "Who's buried in Booth's grave? Historians, family want to check," 25 October 1994, page 1A.
- Reuters, "New Scrutiny on John Wilkes Booth," *The New York Times*, 25 October 1994, page A16.

**168** https://gdpr.eu/right-to-be-forgotten/
https://en.wikipedia.org/wiki/Right_to_be_forgotten

**169** And willful misdirection or tampering is very serious misconduct – e.g. see Paul Koring and Jeff Sallot, "Top brass fostered secrecy, probe told / Papers renamed, colonel testifies / Access law scorned by senior officials, inquiry told." *The Globe and Mail*, 30 August 1996, pages A1 and A4: "OTTAWA – Colonel Geof Haswell, the only officer charged to date in connection with document tampering, told the Somalia inquiry yesterday that the top brass and senior bureaucrats at National Defence Headquarters were obsessed with managing and controlling the flow of information. . . . The documents, sought by a CBC reporter under the access law, were altered to make them conform to a previous batch that were released to the journalist informally . . ."
See also Luke Fisher, "Somalia sideshow / An inquiry fixes on documents, not lives lost," *MacLeans*, 02 September 1996, page 15.

**170** As a *corporate person* – a corporate entity – as distinct from its shareholder owners, directors and employees – who might also be prosecuted in their own *natural person* capacities. The identity of the *corporate person* is sometimes hard to figure out – perhaps sometimes intentionally – it may be either the lab itself as an incorporated entity, or another incorporated entity that the lab is a part of.

**171** https://www.cartoonstock.com/cartoonists/sidneyharris?srsltid=AfmBOoolZd78H4WG0b8xkso-_AkIYmDw3SXBGnlaJffX0JZvo16QPl4l # S. Harris Chemistry Cartoons.

**172** Sidney Harris Chemistry Cartoon. Copyright acknowledgement. Chapter 5.

**173** For example: That the chemist would be paid for piece work, or be salaried, may relate to quality issues when there a fast pace of work. An analogy from the mining industry: Underground workers who would drill for and install long steel bolts to control rock blast, when paid per installed bolt (piece work) might be tempted to use shorter-than-required so as to be able to apparently do more. A difficult-to-catch safety hazard.
https://en.wikipedia.org/wiki/Rock_bolt
http://miningst.com/rock-bolting/material-for-installing-bolts/installation-and-procedures/
https://www.sciencedirect.com/science/article/pii/S1674775517300239
https://www.e-mj.com/features/rockbolting-technology-keeps-mines-safe-and-secure/
https://www.contractortalk.com/threads/piece-work-good-or-bad.30889/

**174** President Harry Truman – desk sign. https://www.trumanlibrary.gov/education/trivia/buck-stops-here-sign
https://featherfoster.wordpress.com/2017/08/21/harry-truman-poker-and-the-buck-stops-here/
https://en.wikipedia.org/wiki/Buck_passing#%22The_buck_stops_here%22

**175** Such as the American Chemical Society. https://www.acs.org

**176** Such as *C.Chem.* or *P.Eng.*

**177** See also Ursula Franklin, 1989 CBC Massey Lecture # 4.
https://www.cbc.ca/radio/ideas/the-1989-cbc-massey-lectures-the-real-world-of-technology-1.2946845

**178** CBC radio, ideas, Massey Lectures, "The humane world of Ursula Franklin, a scientist who wanted us to question technology," 1989 Massey Lectures, December 2020.
https://www.cbc.ca/radio/ideas/the-humane-world-of-ursula-franklin-a-scientist-who-wanted-us-to-question-technology-1.5825485

**179** *Gedankenexperiment.* Introduction 2.2.
**180** https://seniors.lovetoknow.com/Will_Kits
**181** For example, in another context – See *This American Life*, "668: / The Long Fuse / Act Two / Babies Got Bank." "On his deathbed [1926], a wealthy man in Toronto decides to make some trouble. Hundreds of babies are involved. Stephanie Foo tells the story. (25 minutes)."
https://www.thisamericanlife.org/668/the-long-fuse/act-two-41
https://www.thisamericanlife.org/668/transcript
https://discover.hubpages.com/education/The-Toronto-Stork-Derby
**182** Played by a famous mouse.
**183** https://disney.fandom.com/wiki/The_Sorcerer%27s_Apprentice
https://disney.fandom.com/wiki/Fantasia
**184** For example: *Environment Management Act*, Statutes of British Columbia 2003 c.53. https:// www.canlii.org/en/bc/laws/stat/sbc-2003-c-53/latest/sbc-2003-c-53.html
**185** See ACS Task Force on Laboratory Waste Management, *Laboratory Waste Management: A Guidebook*, American Chemical Society, Washington, D.C., 1996.
**186** See F.C.G. Souter, C. Van Netten, and R. Brands, "Morbidity in Policemen occupationally exposed to fingerprint powders," *Int. J. Environ. Health Res.*, 1992, 2(2), pages 114→119 (abstracted in *CA Selects / Forensic Chemistry*, ACS, CAS, Issue 7, 5 April 1993, page 6, 118: 108770 v, Columbus 43210-0012.
**187** In a US context, see:
- Jay A. Young,Warren K. Kingsley and George H. Wahl, Jr., *Developing a Chemical Hygiene Plan*, American Chemical Society, Washington, D.C., 1990, and other ACS publications cited therein;
- §29 CFR1910.1450, "*Occupational exposure to hazardous chemicals in laboratories*" (reprinted from the *Federal Register* by Young, *et al.*, Appendix D, pages 31–48).
  https://www.osha.gov/laws-regs/regulations/standardnumber/1910/1910.1450
- D.G. Schmidt, "The Role of the Chemical Hygiene Officer," presented at a symposium of the ACS Division of Chemistry and the Law, ACS National Meeting, Orlando, 25 August 1996 (abstract published in the CHAL *newsletter*, Spring 1996, page 8; and in the *Book of Abstracts / Part I / . . .*, CHAL, 006).

**188** See also:
- M.-A. Amour, "Safety in the School Laboratory" in *Canadian Chemical News*, Sept. 1987, pages 21→23.
- Keith Furr, ed., *CRC Handbook of Laboratory Safety, 3rd Edition*, CRC Press, Boca Raton, Florida, . . ., 1990.

**189** *Trimethylsilyldiazomethane*, TMSD CAS 18107-18-1 114.22 g/mol.
**190** http://en.wikipedia.org/wiki/Trimethylsilyldiazomethane
**191** Toronto Workers' Health & Safety *newsletter* 2011 10. Vol. 19, No. 2.
**192** In *Chemical & Engineering News*, Washington, DC: "FIRM FINED FOR CHEMIST'S DEATH / SAFETY: Sepracor Canada admits lack of lab ventilation in worker fatality case." "DRUGMAKER SEPRACOR CANADA pleaded guilty in . . . court [in Nova Scotia] on May 2 to one charge of failing to provide proper workplace ventilation . . . U.S.$47,000 fine . . . [Roland] Daigle died on Oct. 8, 2008, from lung failure after exposure to trimethylsilyldiazomethane . . . when lab fume hoods were not operating because of roof work . . . ."
**193** TMSD Jyllian Kemsley, "FIRM FINED FOR CHEMIST'S DEATH / SAFETY: Sepracor Canada admits lack of lab ventilation in worker fatality case," *C&EN*, American Chemical Society, Washington, DC, 20036, 09 MAY 2011, page 15.
**194** http://pubs.acs.org/cen/news/89/i19/8919notw8.html
**195** *C&EN* refers to *Clin. Toxicol.*, DOI: 10.1080/15563650903076924).
**196** https://www.cbc.ca/news/canada/nova-scotia/guilty-plea-in-n-s-drug-lab-death-1.1094083
**197** http://www.chemicalbook.com/ChemicalProductProperty_EN_CB9432840.htm

**198** http://en.wikipedia.org/wiki/Sepracor
**199** http://www.cbc.ca/news/canada/nova-scotia/story/2010/09/15/ns-sepracor-daigle-court.html
**200** http://www.dcnonl.com/article/id38708
**201** http://pubs.acs.org/cen/news/89/i19/8919notw8.html
**202** http://www.osha.gov/pls/imis/establishment.inspection_detail?id=311266522
**203** http://www.cbc.ca/news/canada/nova-scotia/story/2010/05/25/ns-lab-death-charges-windsor.html
**204** See also: "Mercury Poisoning Kills Lab Chemist," *SCIENCE*, June 1997.
https://www.sciencemag.org/news/1997/06/mercury-poisoning-kills-lab-chemist#:~:text=In%20a%20tragic%20end%20to,She%20was%2048
**205** Dimethylmercury ($H_3C$-Hg-$CH_3$) CAS 593-74-8 230.66 g/mol https://en.wikipedia.org/wiki/Dimethylmercury#:~:text=A%20highly%20volatile%2C%20reactive%2C%20flammable,easily%20absorbed%20through%20the%20skin
**206** See also: Neil Vigdor, "Missouri Man Gets 12 Years in Prison for Trying to Buy a Chemical Weapon," *The New York Times*, 8 April 2021, page A19.
**207** https://en.wikipedia.org/wiki/Equine_drug_testing https://en.wikipedia.org/wiki/Doping_in_sport
**208** O. Reg. 490/09: DESIGNATED SUBSTANCES, under Ontario *Occupational Health and Safety Act*, R.S.O. 1990, c. O.1 https://www.ontario.ca/laws/regulation/090490
**209** Sometimes by US authorities at Canadian facilities.
**210** https://en.wikipedia.org/wiki/Lac-M%C3%A9gantic_rail_disaster
**211** Thomas Powers, *HEISENBERG'S WAR / The Secret History of the German Bomb*, BACK BAY BOOKS, Little, Brown and Company, Boston, . . ., 1993; ISBN 0-316-71623-5.
**212** Powers, Introduction page vii.
**213** Including Missouri Senator and later Vice President Harry Truman.
https://www.atomicheritage.org/profile/harry-truman
**214** https://en.wikipedia.org/wiki/Manhattan_Project
https://www.osti.gov/opennet/manhattan-project-history/Events/1945-present/med_45-46.htm http://www.aasc.ucla.edu/cab/index.html
**215** Powers, page 447.
**216** Richard Rhodes, *The Making of the Atomic Bomb/25th Anniversary Edition*, Simon and Schuster Paperbacks, New York City, 10020; 2012; ISBN 978-1-4516-7761-4; page 296, et seq.
https://www.thriftbooks.com/w/the-making-of-the-atomic-bomb_richard-rhodes/253126/item/2235028/?utm_source=google&utm_medium=cpc&utm_campaign=pmax_canada_high_17770447165&utm_adgroup=&utm_term=&utm_content=&gad_source=1&gclid=CjwKCAiAt4C-BhBcEiwA8Kp0CRyjZ5mbP9Jw6cSP7Cx6a3REEAjohvqiBrFmQoJKBsruLKrccHnnhBoCH7cQAvD_BwE#idiq=2235028&edition=1879081
**217** Jim Baggott, Atomic/The First War of Physics/and the Secret History of the Atom Bomb: *1939–1949*, Icon Books, LONDON, M7 9DP; 2009; ISBN 978-184831082-7; Chapter 3, page 57, et seq., at pages 60 and 61.
**218** Michael Frayn, *Copenhagen*; presented in Toronto by David and Ed Mirvish https://www.mirvish.com The Winter Garden Theatre, 189 Yonge Street, January 2004; starring Michael Ball, Martha Henry and Jim Mezon; directed by Diana Leblanc. January 2–22, 2004. National Arts Centre/Neptune Theatre co-production. Reviewed by M.G. in American Chemical Society DIVISION OF CHEMISTRY AND THE LAW *newsletter*, Vol. 20, No. 1, Spring 2004, pages 18 and 19.
http://acs.org/ http://chemistryandthelaw.org/
**219** John Cornwell, *Hitler's Scientists / Science, War and the Devil's Pact*, Viking, New York City, 10014; 2003; ISBN 0-670-03075-9; page 222, et seq., re Paul Harteck pages 224 and 417.
**220** Michael Frayn, *Copenhagen*, Anchor Books, New York, August 2000; ISBN 0-385-72079-3. https://www.amazon.ca/Copenhagen-Michael-Frayn/dp/0385720793

**221** Nicholas Dawidoff, *The Catcher Was a Spy: The Mysterious Life of Mo Berg*, Random House, Pantheon Books, New York, 1994; ISBN 0-679-41566-1.
**222** Werner Heisenberg, *Physics & Philosophy / The Revolution in Modern Science*, Harper Perennial, New York, London, . . . / Harper Collins, 2007, New York City, 10022; 2007; ISBN 978-0-06-120919-2.
**223** Gitta Sereny, *Albert Speer/His Battle with Truth*, Vintage Books, New York City; 1995; ISBN 0-679-76812-2; page 317, et seq.
**224** Ruth Lewin Sime, *Lise Meitner/A Life in Physics*, University of California Press, 1996, 1997; ISBN 0-520-20860-9; page 354, et seq.
**225** William J. Broad, "An Atomic Spy / . . . at Los Alamos," *The New York Times*, Science Times, 28 January 2020, pages D1 and D6.
**226** https://www.youtube.com/watch?v=NgJfFCz2Nr8
https://www.youtube.com/watch?v=SqYPrA7upiE
**227** 2 August 1939.
**228** "Albert Einstein's Letter to President Franklin D. Roosevelt about the Atomic Bomb (1939)"; in *The US Constitution and Other Key American Writings*, Word Cloud Press, San Diego, California, 92121; 2015; ISBN 978-1-62686-393-4 ISBN 978-1-62686-393-4; at pages 259 and 260.
**229** Powers, pages 63 and 64.
**230** https://en.wikipedia.org/wiki/Einstein%E2%80%93Szil%C3%A1rd_letter
**231** https://en.wikisource.org/wiki/Albert_Einstein_to_Franklin_D._Roosevelt_-_August_2,_1939
**232** https://www.atomicheritage.org/history/einstein-szilard-letter-1939
https://www.albuqhistsoc.org/source-documents/fdrs-response-einstein-letter/
**233** See also: Laura Fermi, *ATOMS IN THE FAMILY / My Life with Enrico Fermi*, University of Chicago Press; 1954, 1961; page 198, et seq. ISBN 0-226-24367-2.
https://www.bibliovault.org/BV.book.epl?ISBN=9780226243672
**234** https://www.youtube.com/watch?v=FSvhQVYksIc
**235** The actual historical effect of this letter with FDR had its complications, and may have been less than often presented. See: https://www.youtube.com/watch?v=W3ba1Y8L_yg&t=34s
**236** Little Boy – Hiroshima 6 August 1945 Uranium-235.
https://en.wikipedia.org/wiki/Little_Boy Fat Man – Nagasaki 9 August 1945 Plutonium https://en.wikipedia.org/wiki/Fat_Man
**237** From the Einstein letter, 1939:

> [. . .]
>
> This new phenomenon would also lead to the construction of bombs, and it is conceivable – though much less certain – that extremely powerful bombs of a new type may thus be constructed. A single bomb of this type, carried by boat and exploded in a port, might very well destroy the whole port together with some of the surrounding territory. However, such bombs might very well prove to be too heavy for transportation by air.
>
> [. . .]

**238** https://www.youtube.com/watch?v=LduH7613QXw
**239** https://www.youtube.com/watch?v=6VSrGDOrWXc
**240** https://www.youtube.com/watch?v=W3ba1Y8L_yg&t=34s
**241** Alsos = the US military project, closely following (both in time and geography) the Allied invasion, to investigate Third Reich atomic progress. Powers, pages 369 and 370.
**242** Powers and other writers discuss this considerable controversy.
**243** See also: Dr. Kathy Joseph, Kathy Loves Physics, YouTube, "Heisenberg and Bohr's 1941 Copenhagen Meeting: What Happened?"

https://www.youtube.com/watch?v=zIxh3RSdoVM
https://www.amazon.com/stores/author/B0B57TQDHD/about

**244** John Cornwell, *Hitler's Scientists/Science, War and the Devil's Pact*, viking, New York City, 10014; 2003; ISBN 0-670-03075-9; page 222, *et seq.*, re Paul Harteck pages 224 and 417.

**245** https://en.wikipedia.org/wiki/Paul_Harteck
https://www.rpi.edu/ https://guides.archives.rpi.edu/repositories/2/resources/87

**246** https://www.youtube.com/watch?v=6zIJTwQ2blU

**247** Little Boy – Hiroshima 6 August 1945 Uranium-235 https://en.wikipedia.org/wiki/Little_Boy
Fat Man – Nagasaki 9 August 1945 Plutonium https://en.wikipedia.org/wiki/Fat_Man

**248** See also: John Hersey, *Hiroshima*, Vintage Books, New York, 1946. 1973, 1985. ISBN 0-679-72103-7.

**249** Powers, page 441, *et seq.*

**250** Powers, page 445.

**251** https://en.wikipedia.org/wiki/Loose_lips_sink_ships https://military-history.fandom.com/wiki/Loose_lips_sink_ships
https://time.com/4591841/loose-lips-sink-ships-posters/

# Chapter 6
# Analysis, instrumentation and methods

[. . .][1]

אֶבֶן שְׁלֵמָה וָצֶדֶק יִהְיֶה-לָּךְ, אֵיפָה שְׁלֵמָה וָצֶדֶק יִהְיֶה-לָּךְ . . .

*A perfect and just weight shalt thou have; a perfect and just measure shalt thou have . . .*

*Deuteronomy* דְּבָרִים[1, 2, 3][2]

[. . .]

(see Fig. 6.1) [4, 5, 6, 7][3]

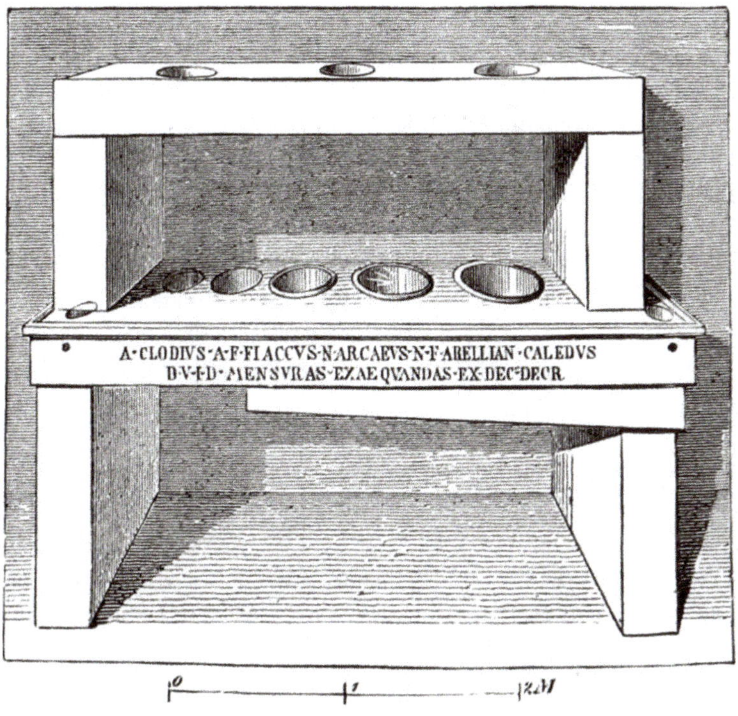


Fig. 6.1: (*Mensa Ponderaria* = table of standard measures – Pompeii).
[. . . German Archaeological Institute . . . *Pompe II Its Life and Art*, . . . MacMillan .., New York, 1902 – see endnote].

**1** Introductory quotes. Introduction 1.2.

**2** *Leviticus. Deuteronomy.* Chapters 6 and 8.

**3** A·CLODIVS·A·F·FIACCVS·N·ARCAEVS·NXF·ARELLIAN·CALEDVS D·V·I·D·MENSVRAS·EXAEQVANDA-S·EX·DECR. "'Aulus Clodius Flaccus, the son of Aulus, and Numerius Arcaeus Arellianus Caledus, the son of Numerius, duumvirs with judiciary authority, in accordance with a decree of the city council, caused the measures to be made equal' to the Roman measures."

https://doi.org/10.1515/9783111618555-007

# 6.1 Introduction

## 6.1.1 Fundamentals – essential concepts

As indicated in the Introduction, this book is neither intended as an encyclopaedic reference nor a handbook as to specific methods.[4] However, as also indicated there and again in this Chapter,[5] important scientific methods are listed, along with some forensic problems they can be applied to, together with literature references. Other methods of forensic chemical relevance are also listed.

But, yet, a few other scientific methods – of lesser general application – and with less direct – perhaps minimal – relevance to forensic analytical chemistry are also listed; these may be of interest to some readers.[6] These methods are described only in bare outline, or merely with literature references – to lead to further information, if the reader is interested. The remote relevance of some of these would simply be that they involve electromagnetic radiation concepts, and EM is so important otherwise for forensic analytical chemistry.

As indicated in Chapter 3,[7] various methods can be used for chemical identification. These could be for unique, theory-based molecular fundamental identification – directly, followed by confirmation by comparison to standard reference samples or reference literature. But, in practice, the theory-based molecular fundamental identification is often not needed, and is avoided – as for ink analysis.[8] Sometimes, identification efficiency can be increased by educated guess and screening tests – then, confirmed – or rejected – with more sophisticated methods.

In Chapter 4, the importance of the energy relationships of atoms and molecules with EM was mentioned[9], and this may be observable and measurable with spectroscopic equipment. This is dealt with in many sections in the chapter. The concepts of EM spectroscopy are of fundamental importance for understanding forensic chemistry.[10,11]

Also of fundamental importance, for understanding forensic chemistry, are the concepts of chromatography and mass spectroscopy [8]. These are also dealt with immediately below.

---

**4** An explain book. Not encyclopaedic. References. Introduction 2 and 3.1; Chapter 6.

**5** Concept. Scientific methods. Introduction 2.1; Chapter 6.

**6** As a kind of law practice notebook. Author's biases. Introduction 1.2; Chapter 6.

**7** Chemical identification. Chapters 3, 4 and 6.

**8** Variations of Ink analytical chemistry – as examples – *Gedankenexperiment* – continue throughout this book. Introduction 2.3 and 2.2.

**9** Energy. Thermodynamics. Entropy. "Time's Arrow." Chapters 4, 6 and 7.

**10** Beckman DU. Bunsen, Kirchhoff. cover. Chapters 1 and 6.

**11** *Genesis*. light. EM radiation. Photon. UV. IR. Microwave. Chapters 3, 4 and 6.

**The fundamentals – in the next sections – the most important and commonly used methods – comprise the essential concepts attempted to be explained in this chapter – to try to fulfil the intention of this book.** The remainder of this chapter, while hopefully interesting and useful, and important in their own contexts, are not here the central focus.[12,13]

**Modern analytical chemical methods mostly involve instrumentation**[14] – often with sophisticated built-in automation and computerized record keeping. And, sometimes including computerized libraries of reference data to identify molecules. But, in understanding the processes, it is important to separate, in thought, the fundamental principles of physics and chemistry from the automation. These fundamentals are what are of concern here.

Screening tests – many of which are of an older style wet chemistry – are not dealt with extensively here. They are, however, still often useful.[15]

Generally, the descriptions below flow from considering the analytical and instrumental chemical concepts, rather than from the forensic problem. The method is introduced with a summary of its basis – theoretical physical and chemical principles – in more detail for the most important and most commonly used – less so, otherwise – although otherwise here includes methods important in their own context.[16] First, the most important ones are addressed; then, the others follow in alphabetical order.

The list here is certainly not exhaustive – reflecting what the author is aware of and considers most important, or is of some interest. Readers might consider this as always to be continued.

### 6.1.2 Most important and commonly used methods

The most important and commonly used scientific methods are listed next below (Other methods follow further below.)

12 An exception to this is DNA analysis – very important in modern forensic science – but not dealt with as extensively in this book. (But more extensively in this second edition, than in the first edition.)

13 DNA. Importance in forensic science. Electrophoresis. Introduction 2.3; Chapter 6.

14 Screening test → instrumentation. Chapters 3 and 6.

15 Screening tests. Wet chemistry. Chapters 3, 4, 5 and 6.

16 Most importantly: DNA and electrophoresis.

## 6.2 UV = ultraviolet-visible spectroscopy

### 6.2.1 Theory

Atoms and molecules can interact with light[17,18] – that is, with electromagnetic radiation – often considered as photons – so as to absorb in the ultraviolet (UV, ~200 → ~400 nm) and visible (~400 → ~800 nm) wavelength regions of the spectrum – as described in Chapter 4.[19]

For UV-visible spectroscopy, absorptions may be explained in terms of energy *transitions*, upward within the molecules from one *electronic* energy level to another. This may be shown diagrammatically in an over-simplified [9] way (see Fig. 6.2).

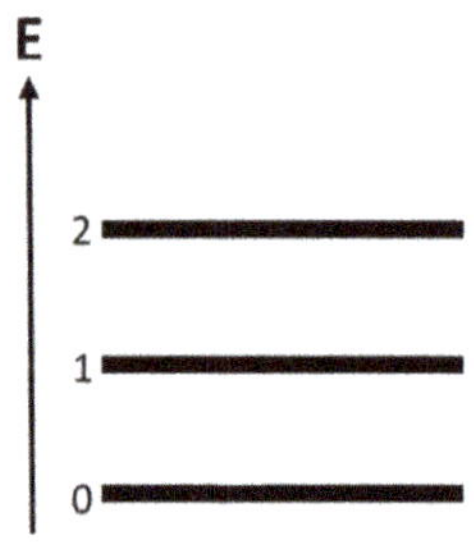


Fig. 6.2: Energy levels.

In Fig. 6.2, the 0 or *ground electronic* level is where nearly all the molecules of the light-absorbing substance are found at thermal equilibrium, at room temperature. The absorption process may, for example, involve the transition of a molecule from electronic energy level $E_0 \rightarrow E_1$ [10].

Generally, the various energy levels are associated with a probability of where a rapidly moving electron may be found in three-dimensional space, surrounding the atomic nucleus. A particular geometry would be associated with an energy level.

Electronic transitions may be illustrated in the simplest way by considering the *hydrogen* atom, and by the over-simplification of the *Bohr* model of the atom [11, 12, 13, 14, 15, 16, 17, 18].

A negatively charged *electron* ⊖ orbits in one of several possible concentric rings [19] around a nucleus (containing one positively charged *proton* ⊕); not in-between rings nor anywhere else unless freed entirely from the atom when that atom would be ionized. That Newtonian physics does not work here was a major innovation of Niels Bohr in 1913, it was a starting lead into the development of quantum mechanics in the years following (see Fig. 6.3).

---

**17** UV. Chapters 3 and 6.

**18** Too much scientific technical detail here? Introduction 2.1.

**19** *Genesis*. light. EM radiation. Photon. UV. IR. Microwave. Chapters 3, 4 and 6.

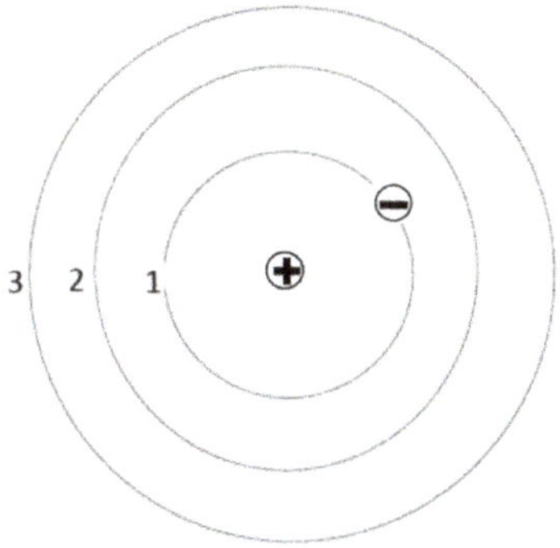


Fig. 6.3: Bohr atom – electron in most inner ring.

This is not to scale; including that the ring diameters would be thousands of times the size of the nucleus.

If the atom absorbs light of the correct energy magnitude, the electron jumps from one orbit to another – for example here, from the first → the second orbit (see Fig. 6.4).

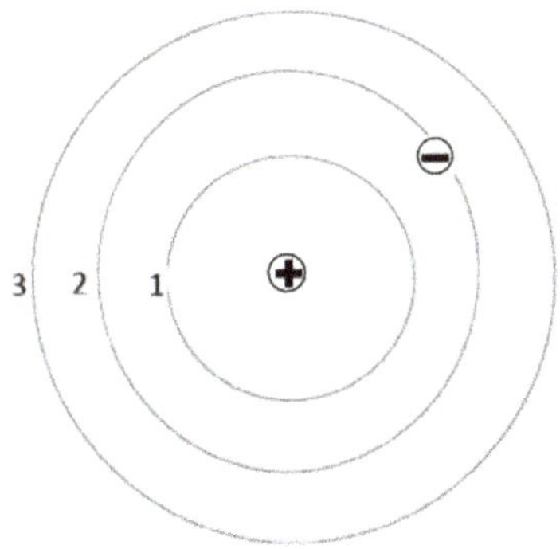


Fig. 6.4: Bohr atom – electron in next higher ring.

No account is made for what might go on in-between during the transition. On the Bohr model's *energy level diagram*, this may be represented (see Fig. 6.5) [20, 21].

The *energy* of this *transition* [22], can be related to the wavelength of the absorbed light (see Tab. 6.1).[20]

If the electron falls down from the second → the first orbit, the atom emits light of the same energy.[21,22]

20 Too much scientific technical detail here? Introduction 2.1.

21 UV absorption, fluorescence and phosphorescence spectroscopy. Chapters 6.2 and 6.18.

22 UV absorption spectroscopy is dealt with in Chapter 6.2. The related fluorescence and phosphorescence spectroscopy is dealt with in Chapter 6.18. Some of the cited literature references apply for both Chapters 6.2 and 6.18. Some of the figures are relevant to both Chapters 6.2 and 6.18.

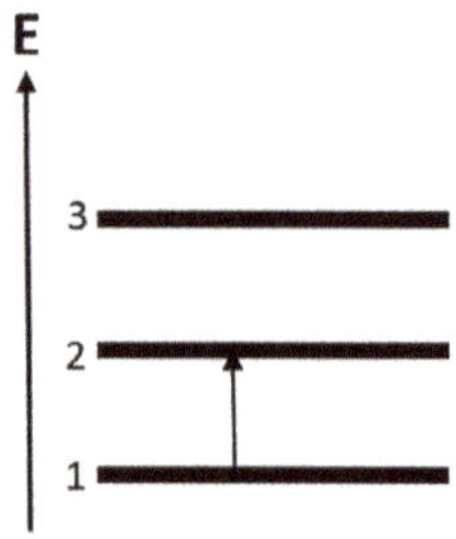


Fig. 6.5: Bohr atom energy transition upward.

Tab. 6.1: E = hν.

| | | | $\Delta E = h\nu = hc/\lambda = hc\bar{\upsilon}$ [23] | |
|---|---|---|---|---|
| Where | $\Delta E$ | = | $E_2 - E_1$ = *Energy* difference between levels [24] | (J) |
| | $\nu$ | = | *Frequency* of the absorbed light [25] | (1/s) |
| | $\lambda$ | = | *Wavelength* [26] | (nm) |
| | $\bar{\upsilon}$ | = | *Wavenumber* = $1/\lambda$ | (1/cm) |
| | h | = | $6.626 \times 10^{-34}$ = Planck's constant [27] | (J. s) |
| | c | = | $2.998 \times 10^{8}$ = speed of light in a vacuum [28] | (m/s) |

Modern instrumentation often refers to $\upsilon$ or $\bar{\upsilon}$, rather than $\lambda$; both $\upsilon$ and $\bar{\upsilon}$ are conveniently directly proportional to the energies of transition, while $\lambda$ is inversely proportional.

Systems that are more complex than a simple one-proton and one-electron atom, such as the organic molecules that forensic chemists are often called upon to identify, also have *electronic* energy levels, although not so easily theorized as the Bohr model, and are explained with more complex geometries. This involves values of $\bar{\upsilon}$, and includes that the theoretical probability of transition is related to the strength of absorption and the value of $\varepsilon$, defined below [29].

The Bohr model works well-enough for the hydrogen atom – one proton and one electron – but not for experimental accountings and predictions of energy levels and the wavelengths of larger atomic and molecular systems. Satisfactory experimental correlations come from modern quantum mechanics, with the Psi functions $\psi$ of the Schrödinger wave equation [30]. But, the Bohr model sufficiently informs of the concepts of quantized energy levels, *etc.* without the more complex mathematics of solving for $\psi$ that is not needed for the forensics here.

When UV or visible light passes through an experimental sample in a transparent (to UV) container – a quartz-walled three-dimensional rectangular cell[23] – as dissolved in a liquid solvent – a liquid solution – measurements can be made (see Fig. 6.6).

23 Quartz spectral cell. Chapter 6.

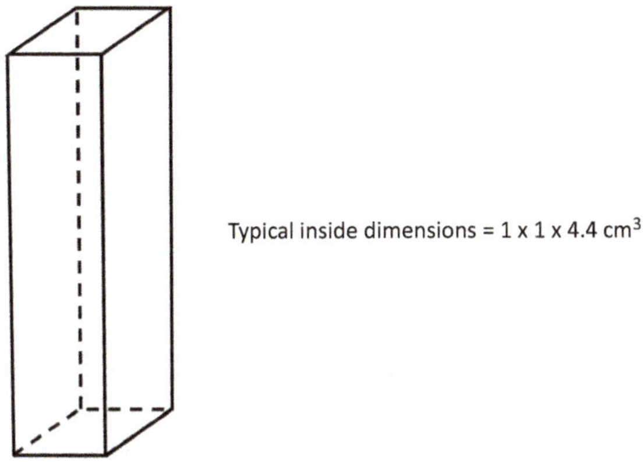


Fig. 6.6: Spectral cell.

Some of the entering light having intensity $I_0$ at wavelength λ is absorbed, so that its intensity is diminished to I by the time it leaves.

Fig. 6.7: Filled spectral cell $I_0$ and I.

I is measured, shown (diagrammatically top view) above, by a photoreception device (not shown here) [31] that transfers the photon signal into an electrical signal, which is measured and recorded. Many measurements are made within multiple ranges of λ. Historically, with Beckman instrumentation, this was a laborious process (plotting of data by hand, point by point, perhaps over several days), but now with automated instrumentation, it takes minutes or less [32, 33, 34, 35, 36, 37, 38, 39].[24,25]

$I_0$ is measured as the entering light passes through an identical reference cell – containing only the solvent – not absorbed, as shown below. This is observed at the same time, for each λ, when I is measured. Again, now automated (see Figs. 6.7 and 6.8, and Tab. 6.2).[26]

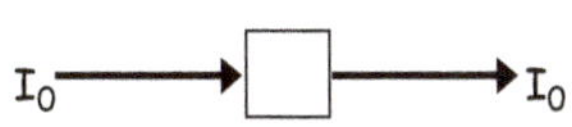


Fig. 6.8: Blank spectral cell $I_0$.

24 Beckman. Bunsen and Kirchhoff. cover. Chapters 1 and 6.

25 Cover illustration of this book. Chapter 6. See endnote.

26 Too much scientific technical detail here? Introduction 2.1.

Tab. 6.2: $T = e^{-A}$ Beer-Lambert equation.

| | | | | |
|---|---|---|---|---|
| Then: | | | $T = I/I_0 = e^{-A} = e^{-\varepsilon cl}$ at each λ of measurement | [40, 41] |
| Where | T | = | *Transmittance* | |
| | $I_0$ | = | Incident light intensity – before absorption | |
| | I | = | Light intensity after absorption | |
| | λ | = | *Wavelength* | (nm) |
| | e | = | Natural or Napierian log base = 2.71828 [42, 43, 44][27,28] | |
| And | A | = | ε c l = *Absorbance* [45][29] | |
| Where | ε | = | *Extinction coefficient* = *absorptivity* [46] | (L × cm/mol) |
| | c | = | Concentration of molecule in the solvent. | (mol/L) |
| | l | = | Cell path length, usually = 1 cm | (cm) |
| So that | A | = | $\log_e I_0/I = (\log_e I_0 - \log_e I)$ [47][30] | |
| | ε | = | $(\log_e I_0 - \log_e I)/c\,l = A/c\,l$ | |
| | c | = | $(\log_e I_0 - \log_e I)/\varepsilon\,l = A/\varepsilon\,l$ | |

ε is related to the probability of an electronic transition – of theoretical spectral importance for understanding the molecule – but usually not needed for forensic identity considerations here.

*Absorptivity* ε becomes the spectral y-axis. Thus, a spectral display may be as for ε or *Absorbance* A, and may be expressed as $\log_{10}$ or $\log_e$ = ln. For qualitative examination, the log base does not matter. But for quantitative, it does, when there is a comparison with literature-reported reference standards – and the forensic scientist would need to check on this.

The spectral x-axis is given as λ, ῡ or υ (see Fig. 6.9).

With ε known, from literature reference or independent experiment, UV-visible spectroscopy can be conveniently used to measure the concentration c, with reasonable accuracy. These measurements are made at selected wavelengths λ at convenient locations on the spectrum – usually, at the top of an absorption peak.

Real molecules present spectra that can include several electronic transitions with different probabilities of transition that are related to different ε at various wavelengths. And within the *electronic* energy levels are *vibrational* energy levels, related to modes of vibration of the various parts of the molecule (see IR spectroscopy

27 UV absorption spectroscopy is dealt with in Chapter 6.2. The related fluorescence and phosphorescence spectroscopy is dealt with in Chapter 6.18. Some of the cited literature references apply for both Chapters 6.2 and 6.18. Some of the figures are relevant to both Chapters 6.2 and 6.18.

**28** Napier. $\log_{10}$ $\log_e$. Chapters 3 and 6.

**29** *Extinction coefficient* = *absorptivity*. ε Chapters 6 and 8.

**30** Napier. $\log_{10}$ $\log_e$. Chapters 3 and 6.

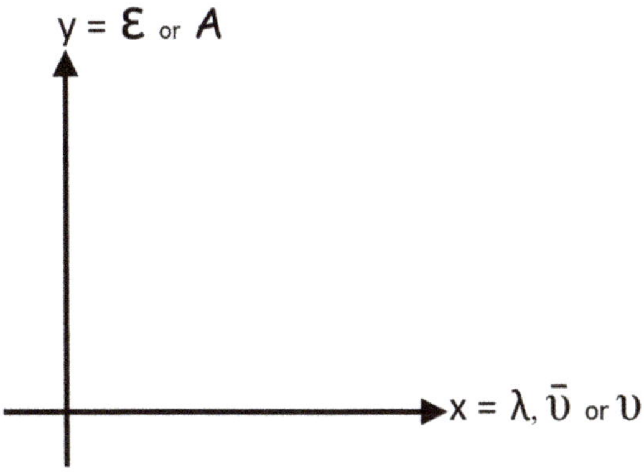


Fig. 6.9: $y = \varepsilon$ $x = \lambda$.

Fig. 6.10: Electronic, vibrational and rotational energy levels.

below). And within the *vibrational* energy levels are *rotational* energy levels, related to modes of rotation of the molecule, represented, oversimplified, in Fig. 6.10 [48][31].

UV absorption spectra result from transitions upward, usually between the ground electronic and vibrational level to various vibrational levels in the next higher electronic level. Represented, oversimplified, in Fig. 6.11 [49].

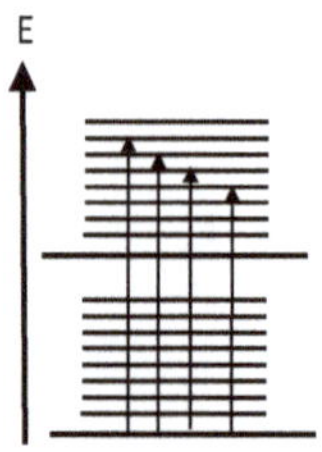


Fig. 6.11: Transitions for UV.

A UV spectral example: *ortho*-fused benzocyclooctene [50, 51, 52, 53, 54, 55, 56, 57, 58, 59, 60, 61, 62] (see Fig. 6.12).

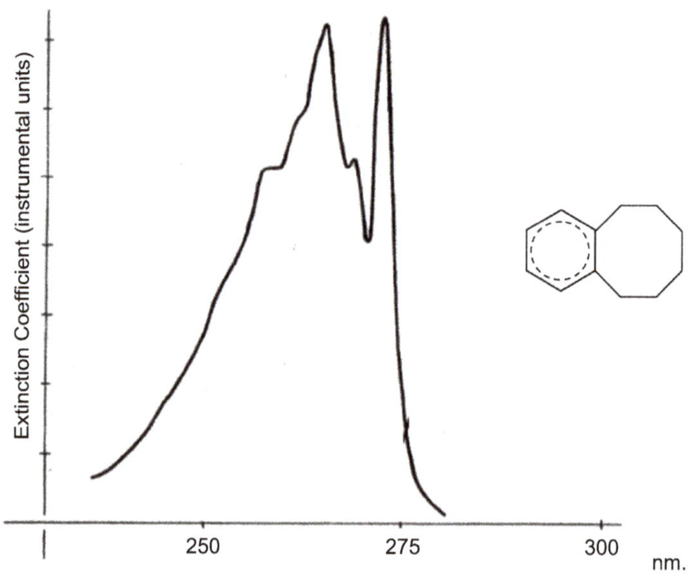


Fig. 6.12: Ortho-fused benzocyclooctene UV spectrum.
[from author's work, 1972 – see endnotes].

31 Too much scientific technical detail here? Introduction 2.1.

For another UV spectral example: *1-phenylnaphthalene* [63, 64, 65, 66][32] (see Fig. 6.13).

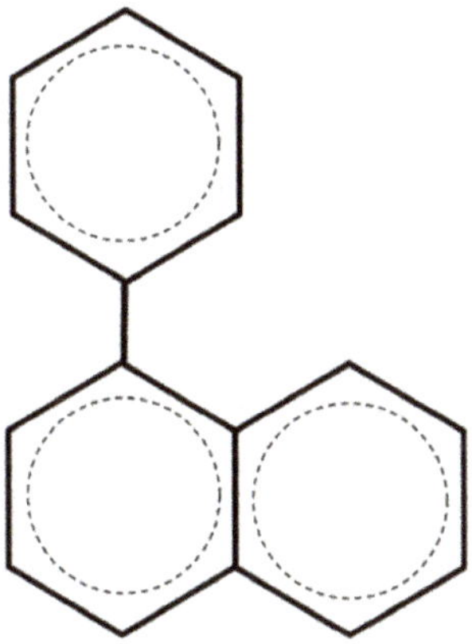

Fig. 6.13: Phenylnaphthalene. CAS 605-02-7.
See: US NIST standard reference data (SRD).

See also: *UV ATLAS of Organic Compounds*, 1966, from which data for ε may be obtained [67, 68].

### 6.2.2 To identify molecules – limited

Since the appearance of shape and magnitude of a spectrum are related to the physical aspects of the molecule that generated it, such a spectrum may be used to identify that molecule.

The theory is that, since atoms can be characterized by their atomic masses and since the energy-interaction-related three-dimensional geometric assembly of those atoms into a molecule consequently produces a unique [69] energy level diagram, the resultant spectrum would also be unique for that molecule, if the spectrum could be examined in sufficient detail.

But, for larger molecules, the UV-visible spectra typically do not show sufficient detail to view vibrational structure for a unique determination. This is related to spectral broadening with increasing molecular mass, and that higher molecular mass molecules are usually examined in solution, wherein there are solvent effects.

What this means for forensic chemistry is that while UV-visible spectroscopy might be useful to say that a substance is of a molecular category and is not of another certain molecule, it can usually not be used alone for unambiguous identification – either fundamental or by comparison. For more unambiguous spectral identification, readers should look to other methods – such as IR – see below.

32 molecular structural drawing. Chapter 4.4.

### 6.2.3 Instrumentation

The UV-visible spectrum (such as shown above) for molecules may be recorded by a *spectrometer* [70, 71]. In principle, in such an instrument, *monochromatic* [72] light of various wavelengths, within the range of wavelengths of the UV-visible spectrum, is directed through a sample of that substance. The part of the instrument that allows wavelength choice is called a *monochromator.*

Ideally, the light source produces *white* light – of all the wavelengths in the UV and visible range [73]. That source might be a high-pressure Hg–Xe arc lamp [74], there are many other light source possibilities, including other kinds of lamps; and LASERS may be used to provide specific wavelengths (not *white*). The *white* beam is then *diffracted* from a *grating* [75], or in other, often older, instruments, *refracted* by being passed through a *quartz prism* [76] – to allow for wavelength choice [77]. A mechanical and geometric arrangement allows the beam to continue only as the chosen wavelength. It passes through the sample and the blank cells to a detector. An electrical signal from the detector can then be processed to record the spectrum. There are very many commercially available *spectrometer* versions, with many different arrangements and with modern electronic controls. A simplified typical diagram, with prism, to explain the concept is given below – see Fig. 6.14 [78, 79, 80, 81, 82, 83, 84].[33,34]

In practice, the light source may not be uniform – there may be an emission spectrum, with various intensities; and the detector may have wavelength-dependent sensitivities; so that some instrumentation might have corrections to electrical signal leaving the detector.

As indicated, for a *spectrometer*, as shown here, it is assumed that the monochromatic light intensity, after it passes through the *blank cell,* is of intensity $I_0$, same as that would have entered the *sample cell.* Some of the $I_0$ is absorbed by the sample so that the intensity, on leaving, is I. Typically, the experimental arrangement is that the two cells are a matched pair [85], the *blank cell* contains only the transparent solvent and the *sample cell* contains the sample dissolved in that solvent [86]. This is assumed equivalent to the sample being contained in an imaginary solvent and cell of total transparency (at the wavelength of interest).

Also, as indicated above, although ε is of theoretical importance [87] for interpreting spectral intensity, it is A or T that is often displayed as the y-axis of the spectrum. The forensic chemist can usually be content with this to determine the shape of the spectral peaks and to, thereby, determine their wavelengths. The peak wavelengths are used to identify the molecule [88]. The x-axis is the wavelength scale.

---

**33** Beckman DU. Bunsen, Kirchhoff. cover. Chapters 1 and 6.

**34** Too much scientific technical detail here? Introduction 2.1.

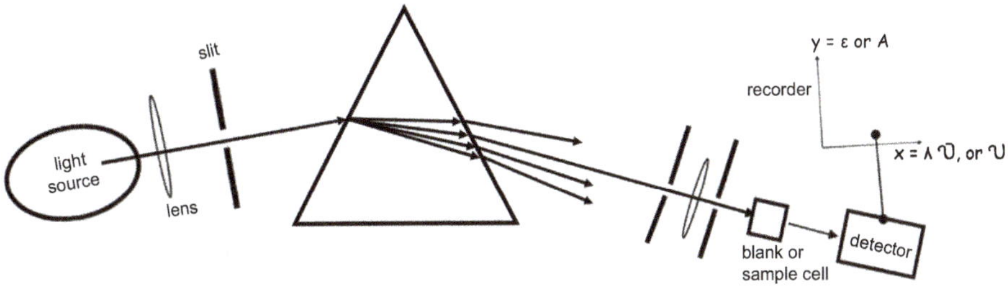


Fig. 6.14: UV-visible *spectrometer*.

The above treatment assumes that the substance of interest is alone in the solvent and cell, or at least is not sharing space with other substances with which it might react, or which are not transparent in the appropriate wavelength range; also, usually in dilute solution. It is, therefore, necessary to ascertain purity, which can often be done conveniently by chromatography (see below), or sometimes even by critically examining the UV-visible or IR spectra. If other substances are found, it may be possible to correct the issue (by arithmetically correcting A; or perhaps by a physical correction, by adding the proper amount of the impurity to the blank cell; or by purification, usually, by chromatography) [89].

In spectral analyses, the possibility of absorbance by impurities is an ever-present problem that the forensic chemist must always look for and try to cope with. Often, the problem can be managed, but sometimes it is so severe that techniques other than spectral analysis should be used. A forensic spectral analysis that has not addressed potential impurity problems should not be used in evidence.

### 6.2.4 Concentration measurements

Although UV-visible spectral analysis should usually not be relied upon for unique molecular identification of a sample, it can be used to determine a concentration in solution[35], if there has been a unique identification, otherwise. In contrast, it is often inconvenient to rely on IR for such quantification.

See above for how to determine concentration – with the equations[36]:

$$T = I/I_0 = e^{-A} = e^{-\varepsilon c l}$$

$$c = 2.303(\log_{10} I_0 - \log_{10} I)/\varepsilon l = A/\varepsilon l \ (\text{mol/L})$$

**35** Quartz spectral cell. Chapter 6.

**36** Too much scientific technical detail here? Introduction 2.1.

The choice of wavelength, λ, at which the concentration would be measured, is important for better accuracy. For example, to measure *1-phenylnaphthalene* dissolved in cyclohexane [90, 91][37], choose a wavelength that would appear to optimize absorption intensity and flatness of curve, such as at λ = 289 nm in the spectrum referred to above (see Fig. 6.13).

In general, such $\varepsilon_\lambda$, may be known from the chemical literature [92] (see above), or better yet, if possible, from measurements made from a standard reference sample with the same spectrometer [93] as used in the experiment that measures the sample, at nearly the same time.

The spectral cell path length = l = 1 cm [94][38] is often sufficient for accurate results. Thus, if at λ = 289 nm, an unknown *1-phenylnaphthalene* sample is measured, then c can be calculated. (From the spectrum referred to above, it would appear that $\varepsilon_{\lambda=289\ nm}$ ≈ **9,000** (L/mol cm) (expressed as for $\log_{10}$)).[39]

The error in such measurement is influenced by the errors in determining A and l; and by the accuracy of the literature reported $\varepsilon_\lambda$ at the narrow range of wavelength that is called λ, and the flatness of the spectral curve at λ. Error [95] of ±a few % is not untypical for such measurements.

An example of concentration measurement based on these concepts is found in the earlier versions of ethanol-in-breath measuring devices used to evidence impaired driving. Chemical reactions that involve ethanol and potassium dichromate result in changes in the transmission of visible light (~420 nm) through the sample, which under standardized conditions can indicate concentration [96, 97].[40]

## 6.3 IR = infrared spectroscopy

Importantly, molecules' modes of energy involvement are also related to the *vibrational* energies of the various force interactions – called *chemical bonds* – of the atoms of the molecule. The *vibrational* energy levels fit within the *electronic* levels shown above; this is related to infrared spectroscopy (IR) spectroscopy[41,42,43].

And, see below for yet another mode of energy involvement – related to the *rotation* of the molecule in space [98] – for microwave spectra.[44]

37 Cyclohexane. Chair and boat. CAS 110-82-7, $C_6H_{12}$. Chapters 4 and 6.

38 For the notation here: path length = l = 1 cm.
Roman letter lowercase L Arabic number one.
(Unfortunately the symbolism here looks confusingly the same.)

39 *Extinction coefficient* = *absorptivity*. ε Chapters 6 and 8.

40 Concentration measurement – IR. Chapters 6 and 11.

41 Too much scientific technical detail here? Introduction 2.1.

42 *Genesis*. light. EM radiation. Photon. UV. IR. Microwave. Chapters 3, 4 and 6.

43 IR. Chapters 3 and 6.

44 Microwave. Chapters 2 and 6.

### 6.3.1 Theory

The same theory [99], described earlier for UV-visible, can be applied to IR spectroscopy, except that the energy *transitions* of interest are within the *vibrational* levels of the molecule's (usually) *ground electronic* level; shown diagrammatically above.

Such transitions produce spectra in the infrared region, which is of particular interest here:

$\sim 800 \to \sim 100,000\,\text{nm}.$[45]

### 6.3.2 To identify molecules – almost uniquely

That same theory, as described above to use UV-visible spectroscopy to identify molecules, can be applied to infrared spectroscopy, except that **IR** can often be used, in practice, for unique molecular determination [100] (well, almost unique).[46]

In practice, over the range of interest, there are often enough sharp-appearing IR peaks for a theory-based[47] molecular structural determination. Such a determination would take into account that various modes of *molecular vibration* will show up as peaks, in predictable wavelength regions, to indicate parts of a structure, and that the presence or absence of various peaks can indicate what that structure could and could not be.

Sometimes, if the molecule is simple enough, it is possible to uniquely assign it a molecular structure based on IR theory alone. For larger and more complex molecules, the possibilities for subtle variations of structure (see Chapter 4) to go unnoticed, increase (and mirror images would go unnoticed [101]).

But, as mentioned in Chapter 4, most forensic molecular identifications are not theory-based directly, but are made using a *standard reference sample* or *reference literature*. Thus, if there are a sufficient number of IR peaks, if the wavelengths of all them of the unknown sample match all those of the *standard* or the *literature*, exactly, if there are no extraneous peaks, if there is no argument about the veracity of the *standard* or the *literature*, and if all instrumentations are of high-enough resolution, then a unique absolute identification will have been made (except for some subtleties such as mirror image). The theory behind this is that even very slight changes in isotopic atomic mass – slightest of variations of molecular structure – would cause slight, observable, IR peak wavelength shifts.[48]

Since, in real-life practice, unfortunately, the above qualifications may not be met and impurities can exist, the forensic chemist should always keep aware of the possi-

---

**45** EM radiation. Photon. UV. IR. Chapters 3, 4 and 6.

**46** Molecular structure. Chapters 1, 3 and 4.

**47** Molecular identification by IR. Chapters 4 and 6.

**48** Too much scientific technical detail here? Introduction 2.1.

bilities of mis-identification of molecular subtleties by IR. But, with care, IR is an excellently reliable method.

For an example, IR spectra – of molecular simplicity:
H-Br, *hydrogen bromide*, CAS 1003510-6 [102].
See: US NIST Standard Reference Data (SRD) [103, 104, 105].

And, for an IR spectrum of greater complexity, such as likely to be encountered by the forensic chemist: *Cocaine*, CAS 50-36-2 [106, 107, 108, 109][49,50,51] (see Fig. 6.15).

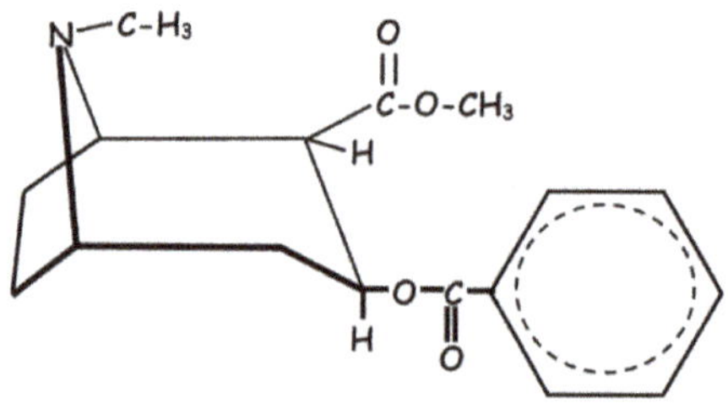


Fig. 6.15: Cocaine, CAS 50–36-2.
See: US NIST Standard Reference Data (SRD).

### 6.3.3 Instrumentation

Similar concepts of instrumentation, as described above for UV-visible, are applicable to IR with the light source – an electrically heated element – as the first major difference [110]. A second major difference is that the very many modern versions of IR spectrophotometers are often non-dispersive Fourier-transform IR[111] that use wavelength interference rather than diffraction grating [112,113]. Many literature reports of IR-reference spectra are now of FTIR [114].

A third major difference between IR and UV-visible spectrophotometers is in the sample containers. As mentioned above, UV-visible samples are contained in quartz cells, typically of 1 cm path length. IR samples are often contained in **K Br pellets** [115] or in a mineral oil – **Nujol** [116,117][52]; both with typically shorter path lengths (~ <2 mm) [118].

To make the pellets, the IR-transparent K Br powder is thoroughly mixed with the sample, and then pressed. Ideally, the resulting pellet appears to the human eye as a small transparent disc. A blank pellet can also be made.

49 Cocaine. CAS 53-21-4, CAS 50-36-2. Chapters 4, 6 and 10.
50 molecular structural drawing. Chapter 4.4.
51 Edmond Locard. Sherlock Holmes. Cocaine. Chapters 2, 3, 4 and 6.
52 K Br. Nujol. IR. Chapters 5 and 6.

To use **Nujol**, the sample is thoroughly mixed with the oil so that an evenly distributed viscous suspension results. This is then sandwiched between two smooth discs of IR-transparent NaCl [119]. A blank Nujol assembly can also be made.

### 6.3.4 Concentration measurements

In theory, the same method to determine concentration, as described for UV-visible can be used for IR, although, typically, not as conveniently, for practical reasons.

For both the KBr pellet and the Nujol, the sample preparation and concentration would be a function of sample preparation, rather than of the forensic problem; and maintaining and monitoring the sample path length would require extra attention or additional measurements [120], and the short path lengths, along with the sharpness of the samples' IR peaks [121], can introduce error problems.

However, reliable concentration measurements are made with an ethanol-in-breath measuring device, for which an IR beam passes through ethanol vapour as it moves through a length of tube [122, 123, 124].[53]

## 6.4 TLC = thin layer chromatography

### 6.4.1 Theory

The several components of a mixture of chemical compounds can often be separated and identified *chromatographically*. Such separation depends on each of the components entering the *mobile phase* of the chromatographic system and being carried in that phase over (or through, and in contact with) a *stationary phase*. The components' differing affinities to the *stationary phase* allow for their separation from each other as the *mobile phase* moves along because they travel at different speeds.

The mobile phase is typically a fluid flow of a gas or liquid. The stationary phase is, typically, a solid coated with a tightly attached liquid, into which the analysed compound dissolves, and then evolves. There are many variations.

Perhaps, the most easily recognizable example of chromatography to the non-chemist is the spreading and separation of colours [125] from an ink drop [126] placed on a paper napkin. The paper structure, sub-microscopically coated with water, is the stationary phase; the capillary action of the water solvent of the ink is the mobile.

Paper chromatography – with a more sophisticated set-up than a napkin – can be used for chemical analyses with commercially available papers for the purpose.

53 Concentration measurement – IR. Chapters 6 and 11.

**Thin layer chromatography**[127][54,55] would be the next simplest, most convenient and most economical scientific chromatographic set-up. It is widely recognized as a very good method, within its limitations. A common TLC version uses silica or alumina spread for the stationary phase as a thin layer on a glass or polymer plate [128][56]. The mobile phase is a mixture of liquid organic solvents that are carried upward along the plate by capillary action.

As the mixture of components of the sample being examined – spotted at the end of the plate – are carried in solution in the mobile phase, they *adsorb* onto the surface of the stationary phase with different strengths, and, therefore, dissolve into the mobile phase at different rates. The result is that sample's components travel at different rates up the plate, so that, after a time, they have travelled different distances.

### 6.4.2 To identify molecules – limited

These distances – called retention distances, or *retardation factors*, $R_f$ – are characteristic of the components and, therefore, can be used as an aid to identify them.

But, since $R_f$ is essentially only related to the strength of molecular attractions (often in non-standard experimental set-ups), it is too simple a measure for reliable unique identification of molecules, in comparison to reference materials. Since other corroborations are usually required, TLC[57] is often used as a screen test.

### 6.4.3 Set-up

To accomplish a TLC separation, a solution of the sample [129] is concentrated by evaporation and is then *spotted* (that is, deposited in drops) near (but not at) the very end of the plate.

That end of the plate is then placed downward into (but not as far in as the *spot*) a pool of the solvent mix that is used as the mobile phase. As the plate leans against its container, the mobile phase moves up the plate by capillary action. When the solvent front appears to be almost at the top of the plate, the plate is removed and the solvents are allowed to evaporate [130].[58]

**54** Too much scientific technical detail here? Introduction 2.1.

**55** TLC. Chapters 6 and 9.

**56** Disclaimers. Error . . ., Not legal advice . . . Commercial products, . . . Introduction 2; 4.1; Chapters 1, 5, 6 and 9.

**57** TLC. DNA. Chapter 6.

**58** SDS. Chapters 1 and 6.

Often, the travelled components – now separated as elongated spots – can be seen with the unaided human eye, such as for textile dyes or the components of ink (see Fig. 6.16) [131].

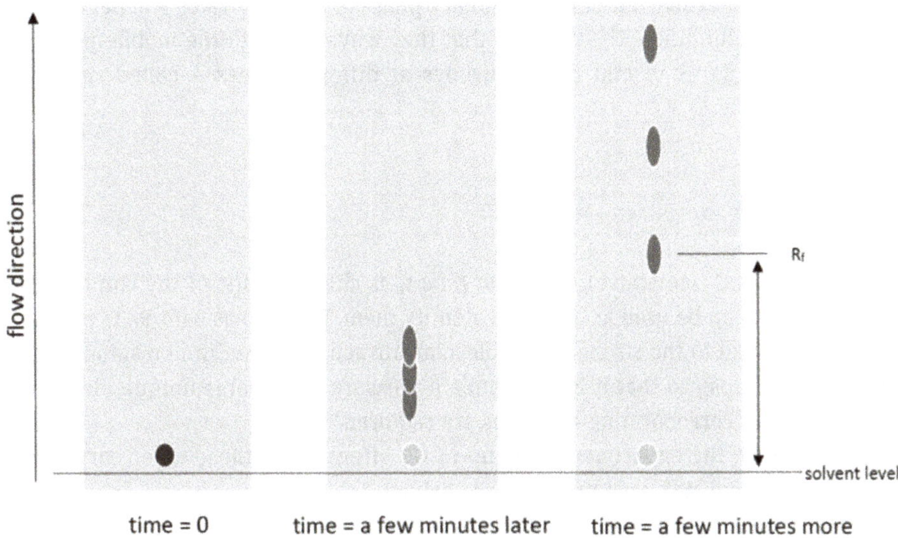


Fig. 6.16: TLC.

But, also often, they must be seen by a visualization process that might include a developing/colouring material being sprayed onto the plate or by viewing the plate under ultraviolet light to see shadows of the component spots against the luminescing background of a luminescent material mixed into the silica, and sometimes as luminescence directly from the spots.

## 6.5 GC = gas chromatography

### 6.5.1 Theory

As with TLC (above), **gas chromatography** (GC)[132][59,60] involves both a stationary and a mobile phase for the separation of a mixture of chemical compounds. The GC-stationary phase is often a high molar mass liquid [133] adsorbed onto inert granular solid surfaces contained in a length of tubing. The mobile phase is usually a carrier

59 Too much scientific technical detail here? Introduction 2.1.
60 Gas chromatography. GC. Chapters 3, 6 and 8.

gas of helium or nitrogen, flowing under pressure through the liquid-coated solid grains.

The sample mixture being examined is first vaporized (perhaps at an elevated temperature). As its components are then carried by the gas of the mobile phase, they dissolve into and evolve out of the stationary phase at different rates – depending on their relative solubilities. The result is that they travel through the mobile phase at different rates so as to end their journeys at different times – called *retention times*, $t_R$.

### 6.5.2 To identify molecules – limited

As with $R_f$ of TLC (see above), *retention time* $t_R$ is characteristic of the components and, therefore, can be used as an aid to identify them. But, also as with $R_f$, $t_R$ is essentially only related to the strength of molecular attractions (often in non-standard experimental set-ups), so that it is too simple a measure for reliable unique identification [134]. Other corroborating measures are required.

Fortunately, the experimental set-up of GC often allows those other corroborations to be conveniently accomplished (see below).

### 6.5.3 Set-up

To accomplish a GC separation, the liquid stationary phase, adsorbed onto its granulated solid support, is packed into a long (ranging from a few centimetres to several metres) narrow (a few millimetres or less) column (typically of coiled stainless steel or glass tubing) [135]. The stationary phase typically has a dry-sand-type look. The column is then fitted into instrumentation of commercially available manufacture – there are many variations – called a *gas chromatograph* – that allows the carrier gas to flow from a heated injection port through the column to a detector, recorder, and/or a sample collector. The GC includes gas flow controls and programmable temperature controls for the column. There are several different kinds of detectors to convert to an electrical signal [136]. The sample injection is usually by a syringe. The data readout from the recorder is called a *chromatogram* [137] – shown here, sketched diagrammatically (see Fig. 6.17) [138].

Ideally [139], each of the *peaks* shown above represents a component separated from the injected mixture. As indicated above, $t_R$ can aid in the identification of the component.

As indicated above, $t_R$ is considered to be a reasonably good indicator. Variations of instrumental design and conditions – length of column, nature of the packing, pressure, temperature, carrier gas, *etc.* – make use of reference literature less sure for identification. More sure identification can be achieved by injecting and comparing

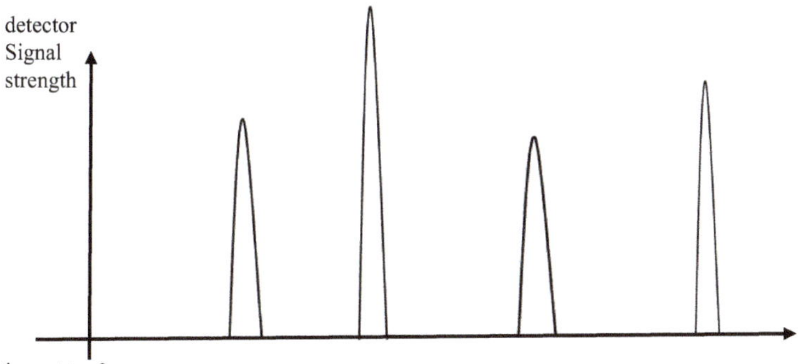


Fig. 6.17: Chromatogram.

known reference standards that might be available. And further sure identification might be achieved by introducing a deuterated known of a suspected component into the sample being analysed; this slightly altered molecule should produce a peak very near the suspected component.

### 6.5.4 Quantity measurement

The vertical axis represents the strength of the detector signal so that, ideally [140], the area under the peaks shown above is proportional to the number of molecules detected for each peak. By comparing several peak areas and by calibration of the GC system with injections of known quantity, it may be possible to quantify the components of the mixture, within error ranges appropriate to the GC system. GC is often used for quantitative determinations, in conjunction with other more qualitatively reliable methods.

## 6.6 MS = mass spectroscopy

### 6.6.1 Theory

Perhaps, an earliest version of mass spectroscopy was developed by Ernest Lawrence – merging into the Manhattan Engineer District – that was used to separate ${}_{92}U^{235}$ from ${}_{92}U^{238}$ to accumulate weapon-grade uranium [141, 142, 143].[61] Modern versions of MS are now used extensively – for rather different intention and purpose –

**61** ${}_{92}U^{235}$ U-235 Einstein. FDR. Critical mass. Chapters 5 and 6.

to separate ion fragments of organic molecules as part of the process of identifying those molecules – and hence its importance for forensic analytical chemistry.[62]

For a simple version of mass spectroscopy [144], a small (often heated) sample of a single substance, in vapour phase, is fragmented and ionized [145] by collision with electrons of an electron beam and accelerated (now, as a number of ions) in an electric field (maintained at a voltage difference) through a strong magnetic field, at an angle [146]. The ions are affected by the magnetic field so that their lines-of-flight curve according to their mass-to-charge ratios, m/z [147,148] – thus ending at different locations at a target; there, detected and recorded. A *mass spectrum* can thus be generated for the originally introduced sample – shown here by a schematic diagram [149, 150] (see Fig. 6.18).

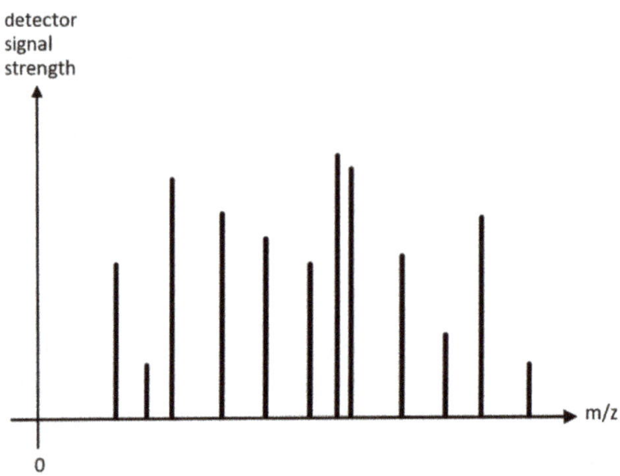


Fig. 6.18: MS.

The many fragments can be reviewed to try to theoretically reconstruct the starting molecule that would be identified; perhaps, something like solving a jigsaw puzzle with molecular structural diagrams.

See Fig. 6.19 for a simple example (not yet quite the jigsaw puzzle) – H-Br *hydrogen bromide* – CAS 10035-10-6, 80.912 g/mol – a sketch of what a mass spectrum would look like – approximately to scale [151, 152, 153, 154, 155].

Each vertical line represents an ionized fragment of the molecule (assume, usually z = +1). The several fragments shown in this sketched MS might be identified as the two bromine isotopes in their natural abundances [156, 157].

62 Too much scientific technical detail here? Introduction 2.1.

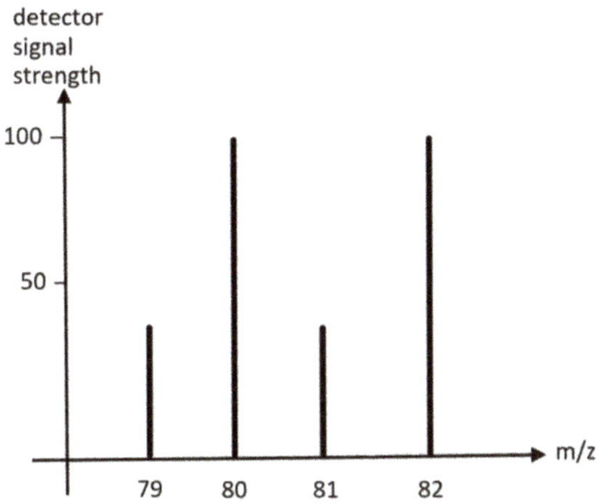


Fig. 6.19: H Br MS, CAS 10035-10-6.

For another example: *tetrahydrocannabinol*, $\Delta^9$-THC, 314.45 g/mol. CAS 1972-08-3 (see Fig. 6.20) [158].[63]

See: US NIST Standard Reference Data (SRD) [159, 160, 161].

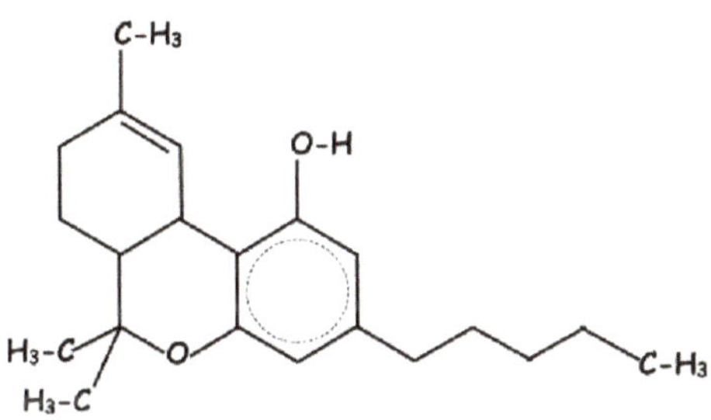


Fig. 6.20: THC, CAS 1972-08-3.

Perhaps, in this example, two fragments could be speculated on and tentatively confirmed (see Fig. 6.21).

With enough fragments (more than the two shown here) identified in this way, a very sure determination of *tetrahydrocannabinol* can be achieved.

Another example: A large *cocaine* [162] fragment, with z = +1, might be theorized as shown – see Fig. 6.22a–c [163, 164, 165, 166, 167, 168].

Electron impact typically creates many ionic fragments – for a full ion scan MS analysis. But there are other, perhaps described as milder, fragmentation methods – QqQ multiple reaction monitoring (MRM), and selected reaction monitoring (SRM)

63 molecular structural drawing. Chapter 4.4.

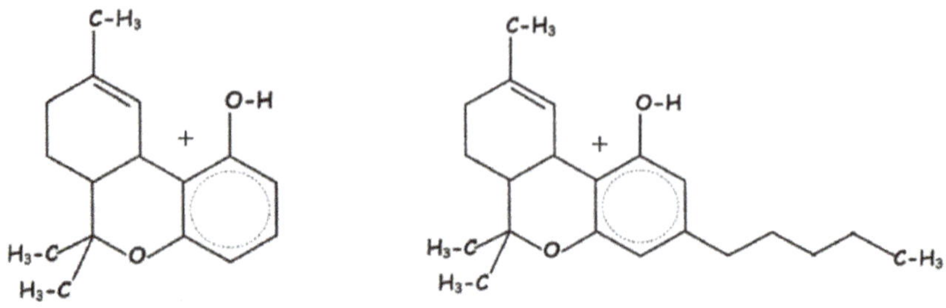


Fig. 6.21: THC fragments.

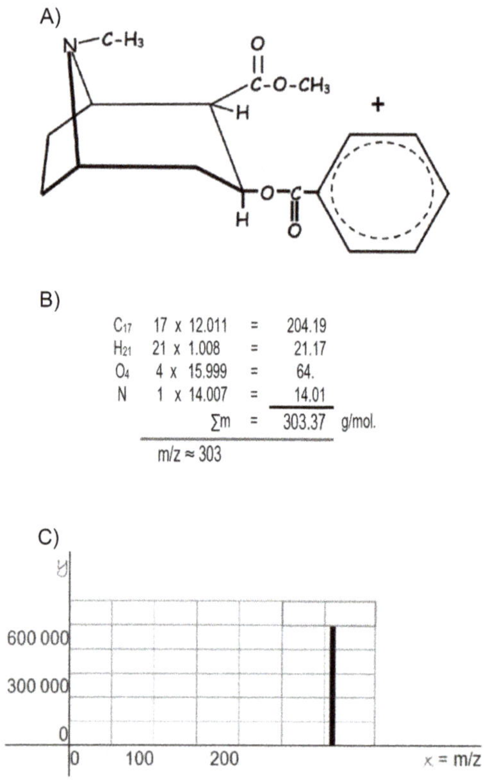


Fig. 6.22: (a) MS fragment structure, (b) fragments in g/mol and (c) sketched chart.

[169, 170, 171, 172][64] that result in fewer fragments – perhaps, only two or three. The fewer and selected fragments would be identified using an automated software library along with other data, such as chromatograph retention times [173].

64 GC-MS. Chapters 6 and 11.

### 6.6.2 To identify molecules – almost uniquely

From the displayed molecular mass data of the fragments and their relative abundances, a theory of molecular structural possibilities for the introduced sample can be constructed for a full ion scan MS analysis. The whole (or nearly whole) molecular ion gives a good identity indicator because it essentially reveals the molar mass (or nearly so). It may be that the data – whole molecular ion, along with many fragments, is sufficient to assign a molecular structure uniquely – questions of subtleties such as chirality may have to be resolved with additional information. As indicated in Chapter 4, to assign molecular structure is to identify a substance.

MRM and SRM, resulting in fewer fragments, provide a less sure identification. But when coupled with other data, such as chromatographic retention times, and compared with data in a large library, they can provide reasonably reliable identification.[65]

### 6.6.3 Instrumentation

Instrumental set-ups for a number of modern versions of MS are described by Kenkel [174] and by Saferstein [175]. These employ the principles described above. There are many and varied commercially available systems.

(An historical observation: MS instrumentation, of years-ago, would typically be contained in a separate (sometimes-large) climate-controlled room, require daily attention even when not in use, and have data readouts on large scrolls that weren't quite-so-user-friendly for handling and preserving. Dedicated personnel may have been employed. And the interpretation was manual. Modern MS instrumentation comes in table-top units with convenient electronic data evaluation and recording. A few minutes observation now might have historically taken a day.)

A practical problem with forensic MS identifications is the questions related to sample purity and how to deal with a mixture of substances. A practical solution is to use GC to purify the sample or to separate the mixture – to be fed into MS. Instrumentation is now available to automatically do this [176]. If MS can be used for an almost unique molecular identification, then GC-MS – combining MS analysis with GC *retention time* – effectively allows for a unique identification.

## 6.7 GC-MS = gas chromatography-mass spectroscopy

As indicated above, MS combined with GC, when used with a large library of molecular data, can provide an almost unique molecular identification. Variations of auto-

65 Analysis – how extensive? Chapters 3 and 6.

mated gas chromatography-mass spectroscopy (GC-MS) instrumentation, with internal supplier-installed libraries are commercially available; or can be equipped with libraries of the analytical lab [177].[66,67,68]

These instruments allow a small sample into an injection port to quickly result in graphical and tabular data. Typically, GC data is shown with *retention times*, $t_R$, with MS data shown at each $t_R$. The $t_R$ *vs* detector signal strength might be shown on a two-dimensional x, y Cartesian grid; and m/z at right angles on the third z Cartesian dimension [178].[69] Often, acceptably, this is as for MRM, rather than electron impact (see Figs. 6.23 and 6.24) [179, 180, 181, 182, 183].[70]

Such well-automated computer-driven instrumentation can also print out the molecular identities. But care is needed by the human chemist – not a robot – who would be signing-off on the report – to critically evaluate the results; automated equipment can get it wrong. Automated systems do not replace human-reasoned evaluations and explanations, and it is a serious error-in-principle for a court or tribunal to consider otherwise.[71]

When properly applied and interpreted, GC-MS is amongst the best of the analytical chemical methods; and, with modern automated equipment, amongst the most convenient.

## 6.8 HPLC = high-performance liquid chromatography

High-performance liquid chromatography (HPLC) is in concept similar to GC (see above), but the mobile phase is a liquid, pressurized by a mechanical pump. HPLC can be used for qualitative identification and quantitative determination [184].

## 6.9 Other methods (alphabetically)

Some important and commonly used scientific methods are listed above; next below are other methods – also important – but more briefly dealt with here.

**66** Too much scientific technical detail here? Introduction 2.1.

**67** Mass spectroscopy. GC MS. Chapters 3 and 6.

**68** GC-MS. Chapters 3, 4 and 6.

**69** *Cartesian* grid. Chapters 3, 6 and 8.

**70** Dr. Nick Kim. Cartoons. Copyright acknowledgment. Chapters 2, 6 and 8.

**71** Personnel – scientific qualification. Automated instrumentation. Chapters 5, 6 and 8.

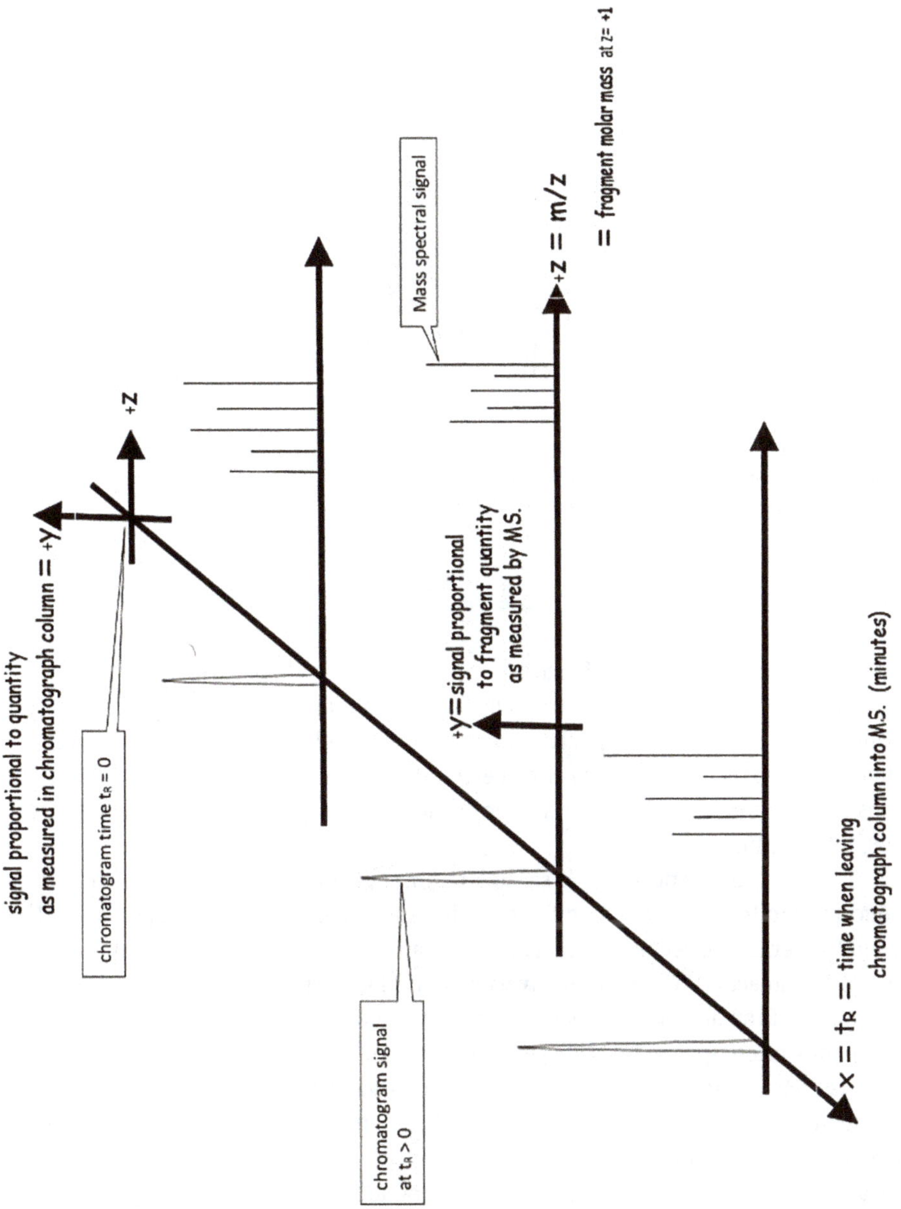


Fig. 6.23: GC-MS sketch.

Fig. 6.24: Lunch.
[cartoon by Dr. Nick Kim, Wellington, New Zealand].
*"Okay – who put my lunch through the mass spectrometer . . . ?"*

## 6.10 Atomic absorption and emission spectroscopy

Spectral observation to identify atoms.

Ultraviolet and visible spectroscopy (see above), and emission spectroscopy (see below), are described as applicable for examining molecules, although the explanation above starts by explaining the oversimplified Bohr model of the simplest atom. This explanation is also appropriate for examining the spectra of atoms [185, 186, 187, 188, 189, 190, 191].

Typically, the experimental set-ups to observe atomic spectra seem simpler in concept than for molecules, and the resulting spectra are simpler in appearance. This would generally be related to atoms being smaller and seemingly simpler in construction than molecules. There are many variations of experimental set-up.

Many of the atoms of the periodic table are solids when humans usually encounter them – often, spoken of as at room temperature. Thus, atomic absorption spectroscopy is typically done at elevated temperatures – vaporized in a flame. The atoms are excited by incoming light, from outside the flame, to a higher electronic energy level; the incident light absorption is measured as an intensity decrease, after passing through the flame. There is a system of slits and lenses. A record of this process, on a photosensitive surface, is the AA spectrum [192].

Similarly, for atomic emission spectroscopy, a minute amount of an element is introduced into the flame. At high temperatures, the atom is sent to a higher energy level; when it falls back down to a lower level, light is emitted, and it passes through a system of slits and lenses, to be recorded as an emission spectrum on a photosensitive surface [193].

Atomic emission spectroscopy derives from the nineteenth-century work of Robert Wilhelm Bunsen and Gustav Robert Kirchhoff, Heidelberg.[72]

These AA and emission spectra, often, may be observed for atoms even when they are associated with other atoms as ±ion pairs. Thus, for example, salt $Na^+Cl^-$ would give its characteristic sodium yellow emitted light.

Atomic spectra may be useful to identify and quantify metals, such as found in bullets from a crime scene and gunshot residue [194].

## 6.11 Boiling and melting point determination: temperature measurement to assist in identifying molecules

Substances exist in various *phases* [195]. When a *solid* is subjected to higher *temperature*, it turns into a *liquid* – *melting*.[73] Further increase – to *boil* – turns it into a gas – vapour. *Melting* and *boiling points* are listed in chemical literature references [196]. *Melting point* can be observed in a sample for a limited qualitative identification when the substance is pure. Alternatively, if the substance has been reliably identified otherwise, melting point may be a good indicator of purity [197].

Observation of a sample, melting in a small glass tube, can be simply done visually [198], often aided with a magnifier lens system (if not an actual microscope system), with a slow-enough temperature increase, with an electric or other heating arrangement.

Temperature is associated with atomic and molecular collisions and molecular vibrations. A high enough temperature can result in overcoming the bonding – attraction forces – that keep a substance solid. Temperature measurement (°C) [199] can be done by visual observation with a mercury (Hg) thermometer in thermal contact with the substance, or an electronic display from a thermocouple device. A simple indication of the accuracy of the Hg thermometer or thermocouple can be by measuring the ambient temperature with another thermometer [200, 201]; and with a stirred ice and water mixture at 0 °C.

Melting point data assumes a transition from solid to liquid. But complications might arise because, for some substances, that transition does not happen – for example, sublimation from solid to gas – often determinable from a published *phase diagram* [202]. Phase changes are also related to ambient pressure. And for some, when the temperature is too high, the substance decomposes.

Melting point use in analytical chemistry is less popular with the advent of convenient modern instrumentation for other methods. But, for drama: *The French Connection*, 1971 [203, 204, 205, 206] – in a scene confirming drug purity with melting temperature.

---

**72** Beckman DU. Bunsen, Kirchhoff. Cover. Chapters 1 and 6.

**73** Phases. Melting points. Chapters 4 and 6.

*Boiling point* would not find too much use in modern forensic analytical chemistry [207]. Historically, water boiling point was used to determine altitude, by-way-of barometric pressure measurement [208, 209]. Also, for drama: *Mountains of the Moon*, 1990 [210, 211].

## 6.12 Centrifugal force measurement

Canadian [212] law criminalizes some automatically, or semi-automatically, openable knives – particularly including by "... centrifugal force ..." [213, 214][74] This should raise some issues of physics – both of theory and measurement. The author, as legal counsel, has tried to raise such scientific evidential issues – apparently, not too-well appreciated by Crown counsel. Typically, in the author's limited experience, these criminal charges are of lesser importance than other charges of the same incident and fall-in – and may fall-away – with more global plea bargaining [215, 216, 217, 218, 219, 220, 221, 222, 223, 224].

## 6.13 Carbon dioxide ($CO_2$) measurement [225, 226, 227, 228, 229, 230, 231, 232]

$CO_2$ CO $O_2$ Chapter 2, 3, 6, 8 and 10.

## 6.14 Dogs

Molecules of gases and molecules that evolve from solids and liquids can be recognized in very low concentration by dogs. [233, 234, 235, 236, 237, 238, 239, 240, 241, 242, 243][75] Asta helped Nick Charles solve the crime in *The Thin Man*, 1934 [244, 245, 246, 247, 248, 249][76]. Laika warned of an earthquake in "DOG STAR." [250, 251] In real-life, dogs can assist with source location and identification of various substances by odour signatures of single kinds of molecules and mixtures. This has forensic practical use regarding: drugs, explosives, food [252, 253] people, pollutants, human illness, human remains, various physical things, *etc.* And, rats too – for sensing odours of explosives from landmines [254, 255]. And humans too can recognize some odours [256, 257, 258].[77]

Also, but not chemically related, a help-dog, assisted a complainant-witness at a criminal trial, Dutchess County Courthouse, Poughkeepsie, New York, 2011; raising a de-

74 As a kind of law practice notebook Author's biases. Introduction 1.2; Chapter 6.

75 Disclaimers. Error . . ., Not legal advice . . . Commercial products, . . . Introduction 4; Chapters 1, 5, 6 and 9.

76 Skippy as Asta; William Powell & Myrna Loy as Nick & Nora Charles.

77 $LD_{50}$ fragrance CCOHS Disclaimer. Chapters 1, 6 and 9.

fence appeal issue. News-reported: ". . . comforts traumatized children and aided a teenager on the stand in a rape trial, . . ." ". . . Both prosecutors and defence lawyers have described [Rosie, a golden retriever therapy dog] . . . as adorable . . ." ". . . Cute as the dog was, the defence said, Rosie's presence 'infected the [jury] trial with such unfairness' that it constituted a violation of their client's constitutional rights. . . ." [259].

## 6.15 Electron microscopy

Electron microscopy allows for observation of various materials at magnitudes smaller than would be possible, with the wavelengths of light-dependent optical microscopes [260, 261, 262, 263].

## 6.16 Electrophoresis: DNA analysis

*Deoxyribonucleic acid*[78] is a large biomolecule, sometimes thought of as a kind of "blueprint" [264, 265] for all life on Earth. It is not an actual single specific molecule, but rather it is a specific kind of molecular structure. The DNA "central dogma" and its relationship to genetic expression has been known for more than five decades [266]. An important DNA property is that it is of two oppositely matching strands that can self-copy. Eventually, important aspects of *evolution* theory are understood to involve chance DNA changes and copying errors.

DNA works as a source for a template for the creation of other kinds of large molecules – *proteins* – that have two general functions: as the structural material of biological systems and as the substance of catalytic molecules called *enzymes* that typically govern the kinetic chemical processes of biology and botany. In a sense, *enzyme* activity is biological and botanical process. *Enzymes* function with fittings of their molecular three-dimensional shapes [267] – wherefore chirality is important. Those enzymatic molecular three-dimensional shapes can form by the complicated folding [268] of linearly-linked amino acids according to a DNA-coded sequence.

DNA [269] is a long twin-polymer chain composed variously of *adenine*, *thymine*, *guanine* and *cytosine* [270], attached to a backbone of linked *glucose* [271] molecules. As a simplified description, A, T, G and C [272] form a kind of alphabet for the *genetic code*. The various sequencings of A, T, G and C, thousands of units long, in *chromosomes*, serve as the "blueprint" for the cascades of interactions of these and other molecules, eventually resulting in controlling, building and maintaining life's structures and functions.

---

**78** DNA. Importance in forensic science. Electrophoresis. Introduction 2.3; Chapter 6.16.

The entire DNA sequence in detail – the total *genome* – could, in theory, tell much (with limitation), but certainly not all, about the entity that contains it. That entity could be a human person.

Technology to reveal a total *genome* in detail is available, but other analytical chemical determinations that would not actually reveal that detail are easier and cheaper to do, and these are used for DNA as a unique identifier. Thus, identification matches can be made with crime-scene-obtained DNA – a boon for forensic law enforcement [273, 274, 275, 276, 277, 278, 279, 280, 281, 282, 283].

In recent years and continuing, the cost of knowing an entire *genome* has been dropping towards a personally affordable range [284, 285, 286, 287, 288, 289] so that a retained bio-sample can be used to reveal much about a person – beyond merely as an identifier. This might allow police to know their suspects, insurers their applicants, and employers their workers, better than ever before.

Such personal *genomic* data raise human rights issues. Medical coverage and employment might be denied or allowed with discrimination, and invaded privacy.

DNA – not of total genome can also raise some issues [290, 291]. Data of some blood relatives – using familial DNA concepts – could become known, because they would biologically share some data [292, 293, 294]. Use of this familial DNA has been discussed in Canada [295]. Familial DNA is now used to help identify suspected criminals [296] – or persons-of-interest [297].

Commercially available kits are available for forensic methods of DNA identification, which, are far short of revealing the entire genome. These convenient methods – which are well-automated are what are presently used for forensic purposes, rather than total genome sequencing. These methods involve short lengths of DNA that don't code for genetics, but can still provide for identifications with very high probabilities.

Electrophoresis has similarities to TLC,[79] but with paper or gel with a voltage difference between its ends, and with sample migration involving electrical charge [298, 299]. For kits designed especially for forensics, the electrophoresis is with narrow capillary tubes [300].

DNA identifications can be very excellent and reliable, but it should be noted that the usual precautions and commentary for quality assurance and error control still apply.[80] And, if not applied carefully enough can lead to wrong determinations.

An aspect of DNA forensics to be especially careful about is its sensitivity.[81] Observed samples found in body cavities, and on surfaces, can be of very slight quantity. DNA forensics is workable because its process amplifies these slight quantities (by replication) – so as to be large-enough for analysis. This means that slight samples that might have been innocently transferred to where they are found can cause con-

79 TLC. DNA. Chapter 6.

80 DNA QA and error. Chapter 6.16; Chapter 8.

81 DNA – too sensitive? Chapters 3.72 and 6.16.

fusing results. And, such transfers can possibly be serially multiple.[82] Care must be taken to realize and avoid such possible confusion. Very clean, careful and explained procedures are necessary [301].

Related to DNA as criminal law evidence is the Vitullo Kit [302] – Rape Kit and also identification of human remains [303, 304].

## 6.17 Electron spin resonance

Electron spin resonance, also called electron paramagnetic resonance, is a spectroscopic method that involves matter with unpaired electrons, observed in a magnetic field with microwave radiation, and in circumstances at which energy-level splitting is detected [305, 306]. ESR, of lesser interest for forensic analytical chemistry, has been used in archaeological dating studies (relating to long-ago temperature treatment of flint items) [307, 308, 309].[83]

## 6.18 Fluorescence and phosphorescence spectroscopy

### 6.18.1 Theory

*Fluorescence* is the re-emission of light by molecules, after having absorbed electromagnetic radiation.[84] Re-emission is at longer wavelength = lower energy. With reference to the energy level diagram shown above for UV – Chapter 6.2[85,86] – fluorescence, for an organic molecule, may be shown; over-simplified as in Fig. 6.25 [310, 311].

The transitions here are from a ground level molecule with electrons as spin paired ↑↓ here – called a *singlet* state – to a first above level *singlet* – absorption – $S_0 \rightarrow S_1$. This is related to a $\pi \rightarrow \pi^*$ transition, associated with the $\pi$ bonding of the benzene ring.[87] Then, the re-emission is the fluorescence – $S_1 \rightarrow S_0$. $\pi^* \rightarrow \pi$.

For example, for *ortho*-fused benzocyclooctene[313][88] (see Fig. 6.26) [314, 315].

*Fluorescence* is defined as $S_1 \rightarrow S_0$, a radiative transition, with no change in spin multiplicity and characterized by very short lifetimes ~$1 \times 10^{-9}$ second → ~$1 \times 10^{-8}$ sec-

82 DNA & Locard. Chapters 2.1.1 and 6.16.

83 ESR. Chapters; Chapters 6 and 7.

84 Too much scientific technical detail here? Introduction 2.1.

85 UV absorption, fluorescence and phosphorescence spectroscopy. Chapters 6.2 and 6.18.

86 UV absorption spectroscopy is dealt with in Chapter 6.2. The related fluorescence and phosphorescence spectroscopy is dealt with in Chapter 6.18. Some of the cited literature references apply for both Chapters 6.2 and 6.18. Some of the figures are relevant to both Chapters 6.2 and 6.18.

87 Too much scientific technical detail here? Introduction 2.1.

88 *Orthofused-benzocycloalkene*. -hexene. -pentene -alkene. Chapters 6.2.1 and 6.15.1.

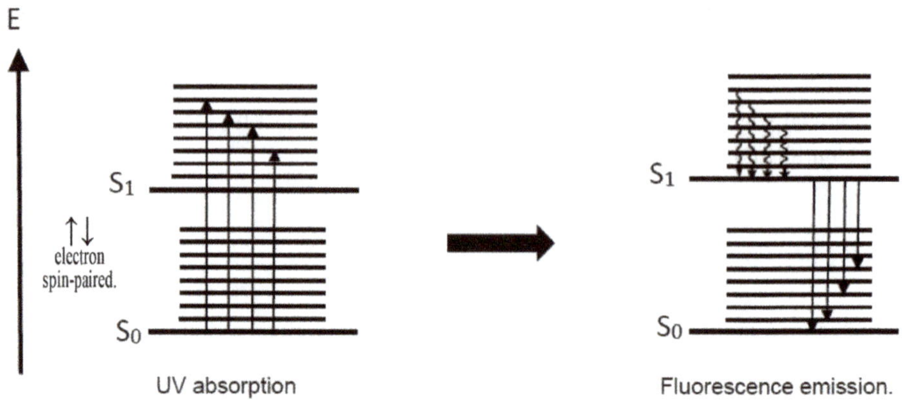


Fig. 6.25: UV absorption and emission fluorescence energy levels [312].

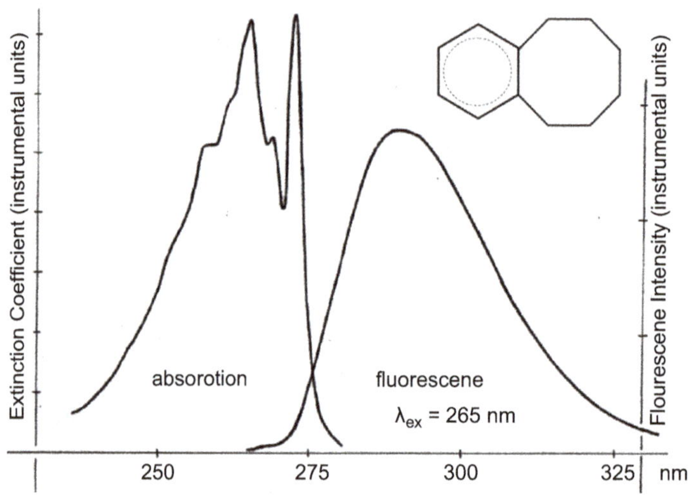


Fig. 6.26: Fluorescence spectrum ortho-fused benzocyclooctene.
[from author's work, 1972 – see endnotes].

ond [316]. *Fluorescence* lifetime can be used as an identifying characteristic of the molecule (historically, the lifetime was a defining characteristic).

***Phosphorescence*** may be shown diagrammatically; over-simplified:

The transition is from a ground level molecule with electrons as spin-paired ↑↓ – $S_0 \rightarrow S_1$, then, to an above ground level as spin-parallel ↑↑ – called a *triplet* state – $S_1 \rightarrow T_1$, and, then, re-emission – *phosphorescence* – $T_1 \rightarrow S_0$ – ↑↑ → ↑↓.

*Phosphorescence* is defined as $T_1 \rightarrow S_0$, a radiative transition, with a change of spin multiplicity and is characterized by longer lifetimes ~$1 \times 10^{-4}$ second → >1 second

[317]. *Phosphorescence* lifetime can also be used as an identifying characteristic of the molecule (historically, it was a defining characteristic).

For example, for *ortho*-fused benzocyclooctene [318, 319] (see Figs. 6.27 and 6.28).

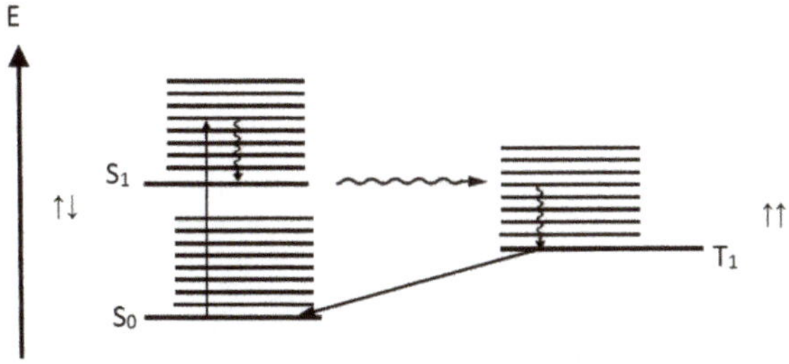


Fig. 6.27: Phosphorescence energy levels.

*Fluorescence* and *phosphorescence* are both more generally described as ***luminescence***. For heavier atoms and molecules, the theoretical distinctions between fluorescence and phosphorescence may break down and the term *luminescence* would be better used [320].

### 6.18.2 To identify molecules – limited

As with UV and visible spectroscopy,[89] the use of fluorescence and phosphorescence spectroscopy to identify molecules is of limited use, for similar reasons.

### 6.18.3 Instrumentation

Fluorescence may be measured by observing the re-emitted light as coming from the centre of the quartz sample cell. Light of a chosen wavelength, to excite fluorescence, is directed into the cell containing a solution of the molecules that are expected to fluoresce – observed at right angles to the direction of the exciting light.

Because fluorescence is so sensitive to impurities, including atmospheric oxygen, the sample cell must have been air-evacuated and sealed. This somewhat cumbersome degassing process is not further described here. The particular problem with oxygen is that its ground-state triplet electronic configuration causes fluorescence quenching. This too is not further described here. When preparing the fluorescence sample cells, care must also be taken that the solution is dilute enough to prevent self-re-absorption that would hide the fluorescence signal. This too is not further de-

**89** UV & visible spectroscopy. Chapter 6.2.

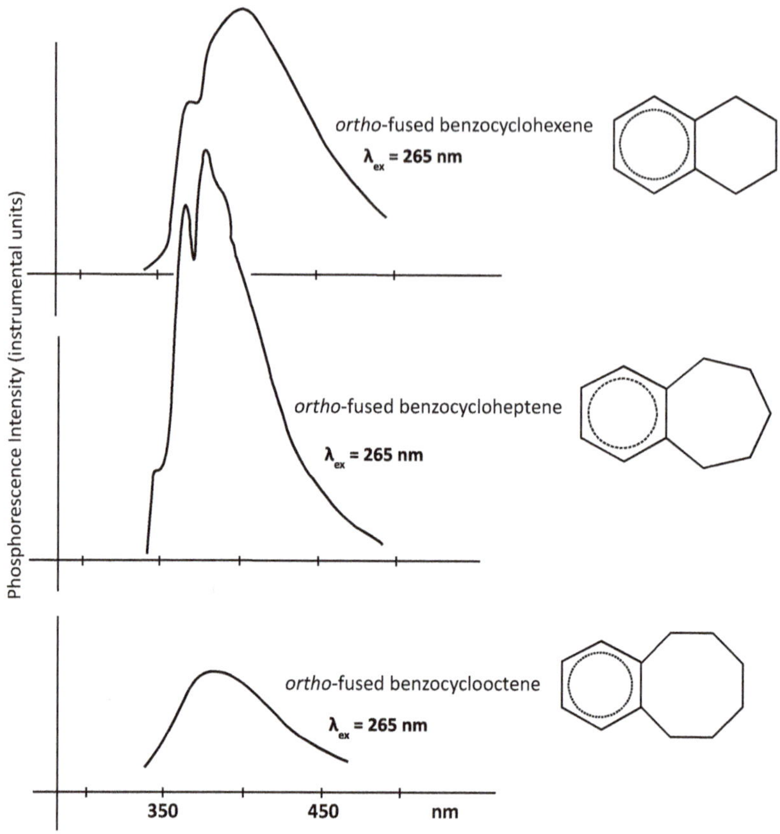


Fig. 6.28: Phosphorescence spectra.
[from author's work, 1972 – see endnotes].

scribed here. Fluorescence is observed and measured for several chosen exciting wavelengths. The fluorescence spectrum is observed with instrumentation that is similar to the spectrometer described above. However, the fluorescent light is scanned using the diffraction grating or prism for each of the chosen exciting wavelengths.

As for UV-visible spectrometers, there are very many commercially available versions of fluorescence-measuring instruments.

A simplified typical diagram to show the concept of observing fluorescence [321, 322] (see Fig.6.29).

An example of a fluorescence spectrum is shown in Fig. 6.26, along with the related absorption spectrum. A kind of fluorescence efficiency can be measured as a *quantum yield* that can be used as an identifying characteristic of the molecule [323, 324, 325]:

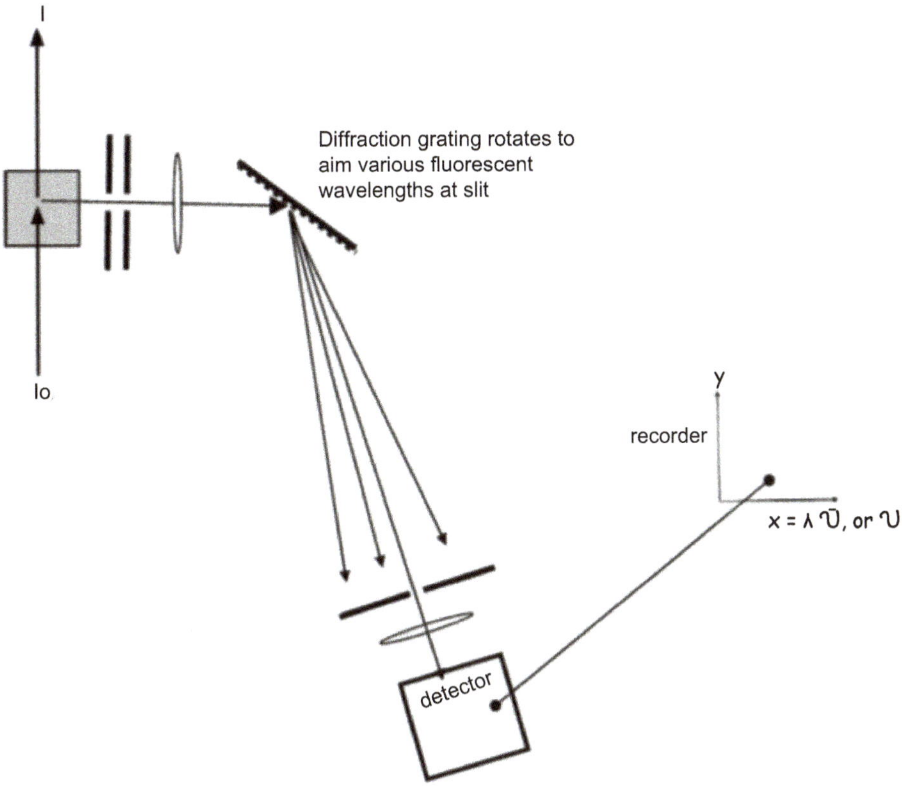


Fig. 6.29: Fluorescence, phosphorescence spectrometer.

$$\Phi_f = (\text{quanta of light fluoresced})/(\text{quanta of light absorbed})$$

Fluorescence observation and measurement instrumentation might be modified to observe reflected light.[90]

### 6.18.4 Application to forensic science

A frequent method to visualize, by eye, traces of substances left on a surface is to try to observe luminescence from ultraviolet light – for those substances that luminesce. Or, substances that do not luminesce, as shadow, on a luminescent background. This might include luminescence from bio-substances exuded from human skin and left as latent fingerprints; and luminescence from other substances, used to develop latent

90 Colour. Reflection spectroscopy. Chapters 3 and 6.

fingerprints [326, 327, 328, 329, 330, 331, 332, 333, 334, 335, 336, 337, 338, 339][91,92,93]. Although luminescence may allow some identification of substance, its main use is often for visualization by eye and for photography. Similarly, luminescence can be used to observe TLC plates, and often the luminescence material would have been added to the thin layer so that the substance on the plate would be observed as shadow.

## 6.19 Length measurement – tire skid

To estimate car speed (km/h).

The length (metres) of a tire skid mark visible on a roadway can be used to calculate the speed of the vehicle just as the skid started when the brakes were applied [340, 341, 342, 343, 344, 345, 346].[94,95,96]

The length can be measured with the modern equivalent of a surveyor's Gunter's chain, [347] and a frictional drag constant would also be measured at the site [348, 349]:

$$\text{Car speed} = 15.946 \times \sqrt{(\text{(drag constant)} \times \text{(skid distance (metres))})}\ \text{(km/h)}$$

Modern computer monitoring and data-keeping, perhaps as part of a car insurance package, would make this method obsolete [350].

## 6.20 Mass measurement

Mass (SI unit = gram) is a physical property found in Sir Isaac Newton's famous equation[351, 352, 353]. Many scientific studies involve measurement of mass. But also, there are many other human-engaged mass measurements that are not usually thought of by many people as scientific study: For postal service, buying and selling groceries, athletic competition, health evaluation, airline baggage, buying and selling tomatoes or anthracite, *etc.* These processes can be so routine with such readily available and reliable equipment, that their reports would most usually not include scientific technical details or explanation; and perhaps hardly anyone would think to ask about the science, or care. So too, perhaps, for police reports for the *Controlled Drugs and Substances Act* [354]; but not quite. Addressed here are questions of Crown-

91 DNA. Proteins. Fingerprints. Chapter 2; Chapter 6.18.4.

92 Fingerprint probability. Chapters 1 and 4. Fingerprint visualization. Fluorescence and phosphorescence. Chapters 6 and 9.

93 Fingerprint when. Chapters 7 and 8.

94 Too much scientific technical detail here? Introduction 2.1.

95 Tire skid. Chapters 2, 3 and 6.

96 As a kind of law practice notebook Author's biases. Introduction 1.2; Chapter 6.

disclosed mass (gram) data that the author [355] has sought for alleged drug offences.[97,98,99]

Disclosure as initially received, in relation to *CD&S Act* charges involving criminalized substances, typically includes Police-reported mass (grams) quantity, but has neither scientific technical detail nor explanation of the mass measurement process. Requests for such as pre-trial forensic disclosure have sometimes in the past yielded some data. But more recently, such requests are simply not replied to in a scientifically meaningful way – and this would appear to be Crown policy [356].

Defence argument [357] – anticipated for trial litigation – would perhaps include Police-reported forensic measurement of drug quantities. This Chapter part is to try to address forensic issues that would arise for this very case-specific measurement.[100]

Dealt with here are forensic data, and related reports, of Police evidence collection and processing, regarding measurement of quantity – reported by Police as mass (grams),[101] of substances – samples of which were transferred to the Health Canada Drug Analysis Service. Questions and issues that would arise – and comment [358]:

- Identification of the device used to measure mass (grams). This device would have an actual physical existence, and would be in possession of Police, who would have direct-contact – hands-on – used it. Toronto police have used an FX-5000i Precision Balance, A&D Company Ltd., Tokyo, Japan [359].
- That Toronto Police [360, 361, 362, 363] (and Peel Region Police[364]) have used scientifically sophisticated precision balances with extensive operating manuals raises a particularly case-specific question [365][102]: Why would such expensive precision balances be used, when inexpensive and simpler-to-use balances or scales would be quite appropriate for drug law enforcement? And the simpler-to-use balances or scales would not likely have such detailed and extensive operating manuals that Defence counsel would cross-examine about.[103]
- References or links to the published operating manual of the measuring device, or other literature. The operating manual would be essential for the use of the device, and therefore a main source for cross-examination of personnel who made and reported the measurements – as to: proper installation, proper use, cautions about incorrect use, scientific measurement error, calibration, servicing, electronic data storage and on-paper-print-out, *etc.*

---

97 Author's biases. Introduction 1.2, 2, 4.1 and 4.3; Chapters 10 and 11. (And also to some extent in Chapters 1.5.3 and 6.20.)

98 War on drugs. *CD&A Act.* Mass (grams) measurement. *W&M Act.* Chapters 3, 6, 8 and 10.

99 Mass measurement (grams). Accuracy and precision too extensive? Chapter 3; Chapter 6.20.

100 **Alert**: This chapter part is very case-specific – and perhaps tediously argued. Some readers would find it convenient to merely scan-over or skip.

101 Mass measurement (grams). Accuracy and precision too extensive? Chapters 3.1.1, 3.7.3 and 6.20.

102 Mass measurement (grams). Accuracy and precision too extensive? Chapters 3.1.1, 3.7.3 and 6.20.

103 Mass measurement (grams). Accuracy and precision too extensive? Chapters 3.1.1, 3.7.3 and 6.20.

- If an A&D FX-5000i – Was it installed [366] levelled [367, 368, 369], and set to zero [370], as according to its Instruction Manual, 2011 [371]? https://cdn.shopify.com/s/files/1/2006/8997/files/FZ-i_FX-i_Manual.pdf?1478839710 2087296248
- Error statement for the mass measuring process used (other than only the accuracy capability of the measuring device indicated in its manual).
- Standard Operating Procedures for personnel who made and reported the measurements. SOP documents [372]. Training documentation of those personnel. Accreditation. *Curriculum Vitae.*
- For the reported mass measurements – Was there any measure of proportion or concentration of criminalized substance contained within the bulk of what was found? Or consideration of *tare*? [373] Or consideration of selection protocol for what was sent to DAS? Or consideration of contamination at the find site; or selection protocol, there?
- Revelation of the Health Canada *Drug Analysis Service Client Manual for Exhibit Submission and Analysis*, ISBN 9780660054179, is requested as from the Police agency which would have that document and would be following its directions. (Health Canada policy and practice would likely prevent disclosure of this document – a copy is <u>not</u> sought as from DAS.) [374, 375, 376, 377].[104]
- Confirmation is requested that Police were following ISBN 9780660054179 directions; which might possibly affect the validity and relevance of the declarations of the DAS *Certificates of Analyst.*
- If it exists, a copy of the federal inspection certificate sticker as under *Weights & Measures Act* [378, 379], s. 2 & 8 is requested. Measurements for *CD&S Act* trafficking would appear related to "trade" as under the *W&M Act*, so that the device should have the sticker. The absence of a *W&M Act* sticker would allow for argument that the measurements are not reliable-enough for proof beyond a reasonable doubt.[380] Analogously, many thousands of inspection certificate stickers are routinely displayed in food stores [381, 382, 383], and at gasoline self-serves [384, 385, 386, 387, 388, 389].
- Copies of calibration certificate stickers from private contractors – if there were any – are requested. If there were – from Clear Tech, Mississauga, Ontario L4Z 2H6? And, the related *Certificate of Calibration*?
- Did the mass measuring device have computer compatibility to generate an electronic record of the measurements? Are there such electronic records? Were there on-paper-print-outs [390]?
- Hygroscopy [391]? Humidity and temperature data for where and when the substance was seized. And, humidity and temperature versus time records, and for security access, for the location at the Police facility where the seized substance is stored.

104 Mass measurement. Chapters 1.5.3 and 6.20.

### 6.20.1 Some further comment to try to explain about the above-requested data

These data and reports would be from a Police process, at a Police facility, by Police personnel, and not part of the works and reports of a Health Canada *Analyst. CD&S Act* s.51 [392, 393] effectively allows the Federal Crown and the Health Canada Drug Analysis Service to not disclose a proper scientific report, and historically they do not so disclose. (In this regard, as Federal Crown counsel informed in 2018 [394, 395, 396, 397, 398], CAN-P-1578 [399] supersedes reporting requirements of ISO/IEC 17025 [400, 401, 402, 403, 404, 405, 406, 407]). (Effectively, CAN-P-1578 would appear as a Canadian "Catch-22"[408] for ISO/IEC 17025.)

However, the Police data sought here are not under the restrictive jurisdiction of s.51. Those Police data would be from before when DAS became involved. The normal law of evidence would apply to those Police data, which would therefore be disclosable.

Also, importantly, Police data sought here are essentially of quantity measurement; DAS data are usually qualitative[105] only. It is anticipated that the litigation in these cases would be concerned – in quite some detail – with questioning the reliability of Police-reported mass [409] measurements. Qualitative declarations of substance identity, of DAS *Certificates of Analyst*, are not questioned here; they are recognized here, by operation of law, as conclusive evidence of what they declare. The limited data the *Certificates* impose are to be regarded as the legal truth before the court, regardless of the actual science, and regardless of the actual scientific knowledge of DAS personnel. And, notwithstanding the quality assurance calamity that led to the Griffiths [410, 411] report, 2001 (Griffiths, although mentioning QA, does not actually delve into it.). No DAS forensic data are sought here; only Police forensic data are sought.

The Police forensic mass determination data are of significant importance for evidence here because the quantity of illegal drug – determined from Police process, and not by DAS – is relevant to severity of sentence. The quantity would also relate to argument about possession for the purpose of trafficking or personal use.

The Police forensic mass determination data have additional significant importance: To persuade a presiding Ontario Court of Justice or Superior Court judge that an accused was criminally in possession of a drug substance, the Crown would have to prove actual quantity beyond so small that the court would not consider it as zero – *de minimis non curat lex* [412, 413]. To prove this requires reliable Police-source forensic data; this is not a Health Canada DAS matter. Disclosed DAS *Certificates* usually do not indicate quantity, or proportion of quantity. Nor hydrochloride, chirality or hygroscopy [414, 415, 416, 417]. The *Certificates* indicate only the presence of a criminally forbidden molecule. In this context, it should be observed that GC-MS instru-

105 **Qualitative** = what the chemical substance is. **Quantitative** = how much substance is found.

mentation that might be used by DAS – perhaps, for example, Agilent instrumentation [418, 419] – might have a sensitivity in the range of $10^{-9}$ mol/L [420, 421, 422, 423, 424, 425, 426, 427].[106]

There would be evidence of some criminalized substance seized by Police. But, if not presented as a forensic scientific examination of mass, proportion and selection, the court should not accept it as reliable. There would not be proof beyond a reasonable doubt of any quantity significantly above zero. And, even if reports of sample bulk mass were to be accepted, there would still be no indication of proportion of criminalized drug within that bulk.

It is emphasized that while measuring mass (grams) can be easily-done, it is not a trivial process, but involves sophisticated scientific performance and explanation (even for simpler-to-use balances or scales – but in a less detailed way.). It is rather more than simply copying-over numbers appearing on a display screen – presented without evidential meaning, expert or otherwise.

Also involved is interpretation of *W&M Act* provisions. It is curious that *W&M Act* inspection certificate stickers are thought of as needed to sell and buy tomatoes, but not when citizens are sent to prison for crime.

It is also curious that data sought here could be so easily provided – even aside from legal disclosure requirements – *Stinchcombe* [428]. For example, the mass measurement device must exist, and there must have been direct-contact hands-on operation by Police personnel; and it would have a manufacturer's or supplier's label [429]. And, it would obviously either have – or not have – a *W&M Act* inspection certificate sticker.

## 6.21 Metal detection

Metallic elements of the periodic table have the property to conduct electricity – often explained with the concept of an "electron sea" [430]. The outer electrons of these elements are thought of as flowing free of their individual atoms. Metals may be detectable at a distance by observing interaction with an applied radiofrequency electromagnetic signal [431]. There are very many commercially available electronic – often hand-held – devices used as metal detectors, finding forensic use in security and item location. A related property of metals is the response to various manifestations of magnetism – related to metals' spin pairings, or not, of outer electrons [432]. A simplest detection device may be an ordinary compass. Or, a hand-held magnet that can be used to distinguish between Canadian-minted silver Ag coins (before 1967) [433], and those of nickel Ni.

---

**106** SI. Mass measurement (grams). Accuracy and precision too extensive? Chapters 3.1.1, 3.7.3 and 6.20.

## 6.22 Microscopy – optical and photo-microscopy

Optical microscopy allows for observation to identify, match, measure and record very small things using visible light wavelengths [434, 435, 436, 437, 438, 439, 440, 441, 442].

## 6.23 Microwave spectroscopy

Energy transitions between molecular rotational energy levels are in the microwave spectral region [443, 444, 445][107,108]. This EM radiation is the same as otherwise described in this book – which is why it is mentioned here – but at wavelengths longer than UV, visible or IR, as used for analytical chemistry. Microwave spectroscopy does not find much use for forensic analytical chemistry. It may be a forensic topic because of health physics concerns [446, 447, 448, 449, 450].

## 6.24 Neutron activation analysis

Neutron bombardment can cause a change in the atomic nucleus of atoms of a sample and the resulting $\gamma$ radiation can indicate the identity of those atoms [451, 452, 453, 454]. A disadvantage of the method is that the neutron source may be from a nuclear reactor, at a large facility. Neutron activation analysis can be used for qualitative identification of various materials, such as gunshot residue and metallic poisons [455, 456, 457].

## 6.25 NMR = nuclear magnetic resonance

Nuclear magnetic resonance (NMR) involves interaction of radiofrequency radiation with energy transitions that are related to the spin of atomic nuclei in an external magnetic field [458, 459, 460]. It can be used for qualitative identification of organic molecules. NMR appears to not be commonly used for forensic applications; however, NMR-based imagining methods are widely used for medical diagnoses [461, 462].

**107** *Genesis*. light. EM radiation. Photon. UV. IR. Microwave. Chapters 3, 4 and 6.
**108** Microwave. Chapters 2 and 6.

## 6.26 Occult and astrology

(see Fig. 6.30) [463]
Caution: Neither occult nor astrology are recommended as forensic methods.

Fig. 6.30: Crystal ball.

*. . . I do not believe that ghosts or spirits exist. . . .*

Bess Houdini, at a séance, Hollywood, California, 1936. [464, 465][109]

Curiously, during the earlier years of the twentieth century, when quantum mechanics and relativity theory were developing, the occult[110] was also popular [466]. While some of the developing concepts of modern physics might seem no less bazar than some of the occult, as we persist in our everyday world (well-enough-explained by Newtonian physics), modern physics has the advantage of being tested as against experimental observation and measurement, in the context of published scientific enquiry and mathematical expression.

**Neither occult nor astrology are recommended as forensic methods** [467, 468, 469, 470, 471, 472, 473, 474, 475, 476].[111,112]

However, occult and astrology have persistent recurring popularity [477], so that it would be reasonable to speculate that their practitioners might have encountered some true facts of human activity from consultations with their clients [478, 479]. But, occult and astrology practitioners would be considered as ordinary witnesses. Also possible could be that a person being investigated, and known to be an occult or astrology adherent, might have some predictable behaviour.

**109** Astrology and occult. Crystal ball. Quantum mechanics. Governor-General. Prime Minister Mackenzie King. Harry Houdini. Chapters 4 and 10; Chapter 6.26.

**110** **Caution**: Neither occult nor astrology are recommended as forensic methods.

**111** **Caution**: Neither astrology nor occult are recommended as forensic methods.

**112** Astrology & occult. crystal ball. Quantum mechanics. Governor-General. Prime Minister Mackenzie King. Harry Houdini. Chapters 4 and 10; Chapter 6.26.

## 6.27 Photography

It may be assumed that hand-drawn representations of evidence, including for human face identification and depictions of past events, would have had an historic role in forensic science. That role has almost entirely been taken over by photography[113,114] [480], which now has a very prominent place in forensic science (although, in-court-drawn-artwork still appears in news reports) [481].[115] Photography is now essential to record a scene being investigated and to get a better than the human eye image of an object. And, real-time video recording of events being investigated is now routinely sought.

This has gone from mechanical devices dependent on photosensitive wet chemistry – labour-intensive [482] – to well-automated electronics accomplished with minimal human labour – or sometimes, even human thought.

Photography started in the nineteenth century [483,484] and by the early twentieth, mechanical cameras with photochemically producible images fixed on a physical surface were widely available. A latent image on the physical surface was then developed and fixed by wet chemical processes (with then-under-appreciated chemical health hazards). Glass plates (fragile) with photosensitive coating were early-used surfaces, then, nitrate-based film (not fragile, but very flammable), and later, acetate (not flammable "safety film"). By mid-twentieth century, most films were acetate-based (breakage and flammability became forgotten problems). By the late twentieth and into the twenty-first century, electronic recording, processing and storage came to replace the physical and chemical (becoming – if not already – forgotten, by many people).

Essential photographic concepts remain from the early twentieth century and there are very many commercially available versions of photographic cameras, with many different arrangements. A simplified typical diagram to schematically show some concepts (see Fig. 6.31a) [485, 486, 487, 488, 489, 490, 491, 492, 493, 494].

Light – from very many possible sources – reflected from the surfaces of a three-dimensional [495] real object, projects onto a surface within the camera as a two-dimensional virtual image.

The image-bearing light comes in through the front of the camera housing – through *lens* [496], *aperture* and *shutter*; then onto a photosensitive recording surface.

The *aperture* controls the amount of light coming into the camera – as light quantity/area [497]. The control can be by way of thin metal leaves, adjusted to reduce the circular entry area. This can be represented by focal length / diameter = f/d [498, 499]. This is often called an "f/number,"[500] "f/stop,"[501] or "f-stop" or "f stop." "f/number" = f/d. *Aperture* front view [502] (see Fig. 6.31b) [503].

---

**113** Too much scientific technical detail here? Introduction 2.1.

**114** Photography. f/d √2. Chapters 3 and 6.

**115** In-court-drawn-artwork. Chapters 1.4.3.9 and 6.27.

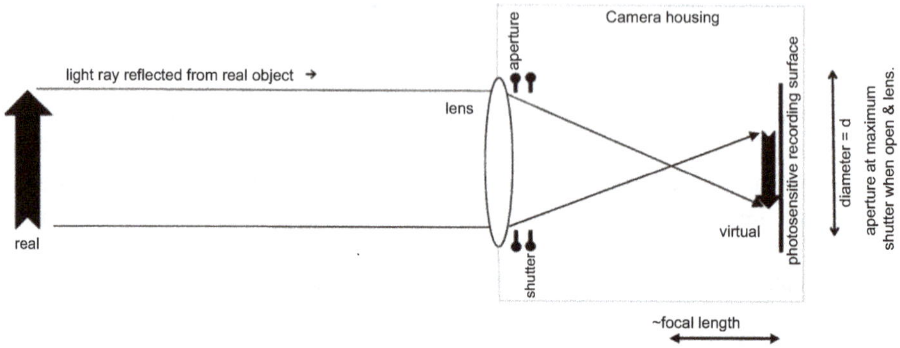


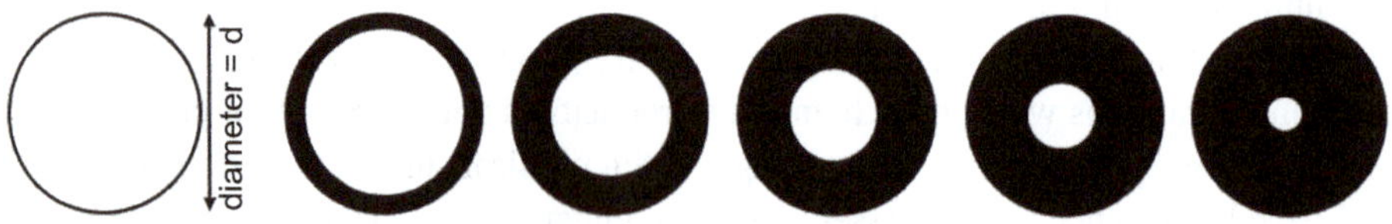


Fig. 6.31: (a) Photo camera. (b) Aperture.

For each "stop" setting, the light quantity / area decreases from a maximum "f/ number" = f/d = 0.5. After this largest possible *aperture*, the subsequent smaller *apertures* are usually designated as "f/number" = 0.7, 1.0, 1.4, 2.0, 2.8, 4.0, 5.6, 8.0, 11.0, *etc*. For each of these, "f/number" = previous $\times\sqrt{2}$ [504]. And, each has ½ the area of the previous to allow in half the amount of light [505].[116]

The *shutter* can be open or closed to allow in the image for an amount of time, from a small fraction of a second to many seconds. This is the *exposure time*, usually in fractions of seconds [506, 507]; it is also a control of the amount of light coming in. The *shutter* may be of metal leaf design or otherwise, such as a kind-of fast-moving curtain (see Fig. 6.32).

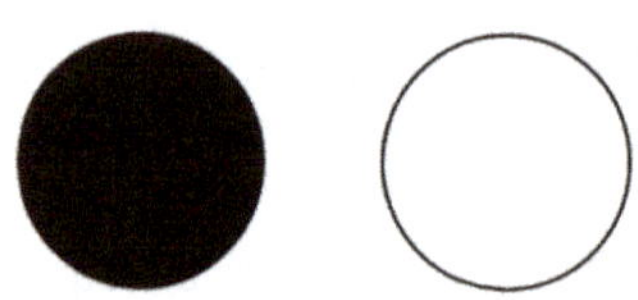


Fig. 6.32: Camera shutter.

After passing through the *lens*, *aperture* and *shutter*; the image then goes onto a photosensitive recording surface (see Fig. 6.33).

116 Too much scientific technical detail here? Introduction 2.1.

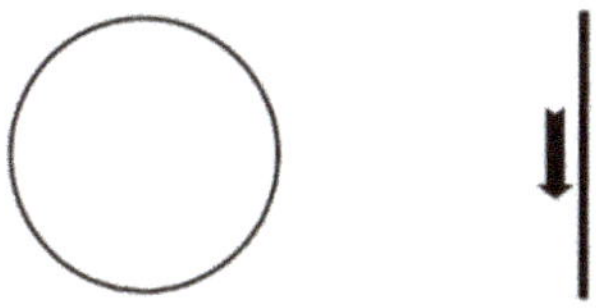

Fig. 6.33: Image on photo-receiving surface.

The "*lens* speed" [508] in this system is characterized by f/d = "f stop".

The sensitivity of the photo-receiving surface, if chemical photographic, has a "film speed" [509]. The photography process can also be related to the chemical-in-water-solution development process, by controlling and monitoring temperature, concentration and process times.

The image-receiving film has a grain size; larger for faster films, but also resulting in poorer resolution.

To summarize, the interacting effective variations and controls of the light quantity coming into the camera, as appearing on the film, are:
- *Lens* speed – aperture.
- *Shutter* speed.
- Film speed – related to grain size and resolution.
- Film processing – choice of chemicals, temperature, concentration, time.

Before ~mid-twentieth century, most photography was only of black and white. Then, colour came to prevail.

Towards the end of the twentieth century and into this – the twenty-first, most of the machinery, control, processing and record-keeping of photographic imagery became computer electronic. Film use has disappeared, replaced by computer receptors and memory. This extensive automation of photography means that photographers of now can get better pictures, while getting away with being unaware of the above-described technical matters; but essential concepts remain.

There has long been some possibility of manipulation of photo-images. An earliest and most famous was a published altered photograph of Theodor Hertzl [510, 511, 512, 513] shown standing near Kaiser Wilhelm II on horseback, near Jaffa, Ottoman Palestine, 1898. Hertzl, a well-known Austrian journalist, famous for his reporting and commentary of the Dreyfus affair and founder of modern Zionism, was interested in trying to persuade the German Emperor to suggest to the Ottoman Sultan to allow possible Jewish settlement in that region of the Sultan's empire. Hertzl was actually close-by and did meet the Kaiser, but not as in the altered photograph – apparently, by modern standards, not too hard to notice as altered [514, 515, 516, 517].

But, with modern computerization, hard-to-notice fake photomanipulation becomes a most prominent possibility and concern. Modern photo-fakery is rapidly becoming more and more doable, with more sophisticated software becoming more easily available [518, 519]. This raises serious concerns about verifying and controlling the use of photographs as reliable forensic evidence.

## 6.28 Radar location of things

"Radio detection and ranging", better known by its WWII-era acronym RADAR [520], was extensively used in the defence of England, with electromagnetic radiation detected as reflected back from incoming *Luftwaffe* [521] attackers [522]. At radio-frequency wavelengths, electronic equipment operation is engineered in EM, rather than in photon terminology; it is the same as otherwise described in this book – but at much longer wavelengths than for usual analytical chemistry application.[117]

RADAR continues for many kinds of application in various legally defined and assigned wavelength regions. RADAR can be used to ground-penetrate so that the reflected EM radiation can be used to locate buried objects, including for archaeological study of burial sites [523].

RADAR is also well-known to monitor highway vehicle speed [524].

## 6.29 Radioactivity

As indicated in Chapters 4 and 7, many atomic isotopes are radioactive and many are not. Radioactivity, as considered here, involves processes producing [525][118,119]:

- $\alpha$ = alpha particles
- $\beta^-$ = beta particles
- $\gamma$ = gamma rays

These processes occur variously from elements of the periodic Table. The end results in the time scheme of the universe are isotopes of lead [526].

For example, of forensic interest, are occurrences of radon gas and the subsequent radon daughters, which are of workplace and environmental concern [527, 528, 529, 530, 531].

Another example of forensic interest and of workplace concern, historically, was the use of radium in the painting of the glow-in-the-dark dials of watches and instrumentation – resulting in worker – ". . . radium girls . . ." – health harm and death [532, 533, 534, 535, 536, 537, 538, 539, 540, 541, 542, 543, 544, 545].

Also, for example, carbon-14 dating – see Chapter 7.

Radioactivity can be measured in several different ways with very many commercially available instruments – some hand-held.

Radioactive materials are also used for medical diagnosis [546] and treatment [547].

---

117 *Genesis.* light. EM radiation. Photon. UV. IR. Microwave. Chapters 3, 4 and 6.

118 Isotope. Radioactive. Carbon-14. Chapters 4, 6 and 7.

119 Subatomic particles. Chapters 2 and 6.

## 6.30 Radiofrequency tags

Much of modern equipment incorporates radiofrequency-activated devices for operation and identification. Also, radiofrequency tags are used to locate things [548]. These devices are often very small and energized to by the radiofrequency signal that would activate them.

## 6.31 Raman spectroscopy

Raman spectroscopy involves light scattering from a molecule with a change from the incident wavelength when scattered. In forensic science, it has been used for qualitative identification of paints and pigments, with the advantage that observation can be at a distance – an inherently non-destructive method [549, 550, 551, 552, 553].

## 6.32 Refractive index

In empty space – vacuum – a light beam travels in a straight line [554, 555] at constant velocity[120]

$$c = 2.997\,924\,58 \times 10^{+8}\ \mathrm{m/s}$$

– which is also the maximum velocity possible according to relativity theory, which also states that that the velocity of light is the same as measured by all observers, regardless of their own movements. This description is for both wave and particle theory of the physics of light [556].

When a light beam passes from empty space into a transparent medium [557] – such as water or glass – its velocity is reduced, with an associated change of angle [558, 559, 560] (see Fig. 6.34).

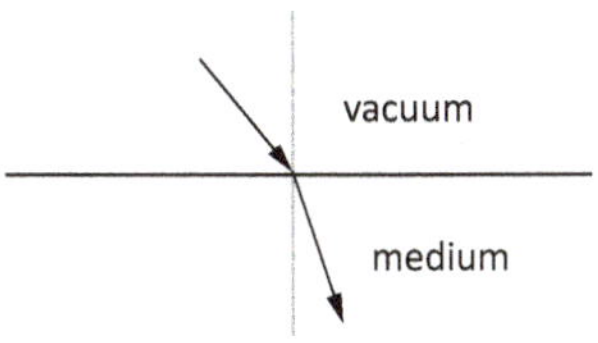


Fig. 6.34: Refractive index.

120 Astrology and occult. Quantum mechanics. Governor General. Chapters 4, 6 and 10.

This is measured, at particular wavelengths, as *index of refraction* [561]:

$$= n = (\sin(\text{angle of incidence})/\sin(\text{angle of refraction}))$$

– a characteristic of the substance that is the medium and may be used for screening confirmation for qualitative identification, or as a purity check [562, 563].

The concept of light refraction is important for optical-use materials of various kinds of silica glass and quartz in the construction of scientific instrumentation: prisms for spectrometers, lenses for hand-held magnifiers, lenses for telescopes and microscopes, lenses for photographic cameras, *etc.*

The physics essentials of lenses are often described, as simplification, for thin lenses and simple schematic representations may be drawn as such. However, in various instrumentations and cameras, the lens would actually be a system that is a composite of several carefully engineered optical parts; and there are very many different arrangements that are commercially available.

## 6.33 Reflectance and colour

### 6.33.1 Reflectance spectroscopy

As indicated above, fluorescence spectral observation and measurement instrumentation might be modified to observe reflected light.[121] A fictionalized example [564] in Chapter 3 suggested that the colour of a surface could be examined for forensic identification purposes. A spectrum of reflected light could be compared to reflected light from other surfaces, including of standards that might include reference to colour charts [565, 566].[122,123,124]

When UV-visible *white light* [567] – or other incident light of selected wavelengths – is directed onto a surface, some (assuming that the surface is not a totally reflecting mirror) of it would be absorbed, involving atomic and molecular energy processes

Such absorption might eventually result in photochemical reactions, or be re-emitted as fluorescence or phosphorescence, or somehow result in producing heat.

Some of the light might be *Raman*-scattered. But some – often much – of incident light would be reflected. That reflective process would be directed by the shape and texture of the surface.

Of interest here is the reflection of the source light at various wavelengths and intensities.

---

**121** Fluorescence. Reflection spectroscopy. Chapter 6.

**122** Too much scientific technical detail here? Introduction 2.1.

**123** Colour. Reflection spectroscopy. Colour. Reflection spectroscopy. Chapters 3 and 6.

**124** Ink analysis. Reflection spectroscopy. Chapters 3 and 6.

For instrumental spectral observation, a simplified typical diagram to show a concept of observing surface reflection spectra – indicated above as having similarities with fluorescence spectral observation and measurement instrumentation[125] (see Fig. 6.35).

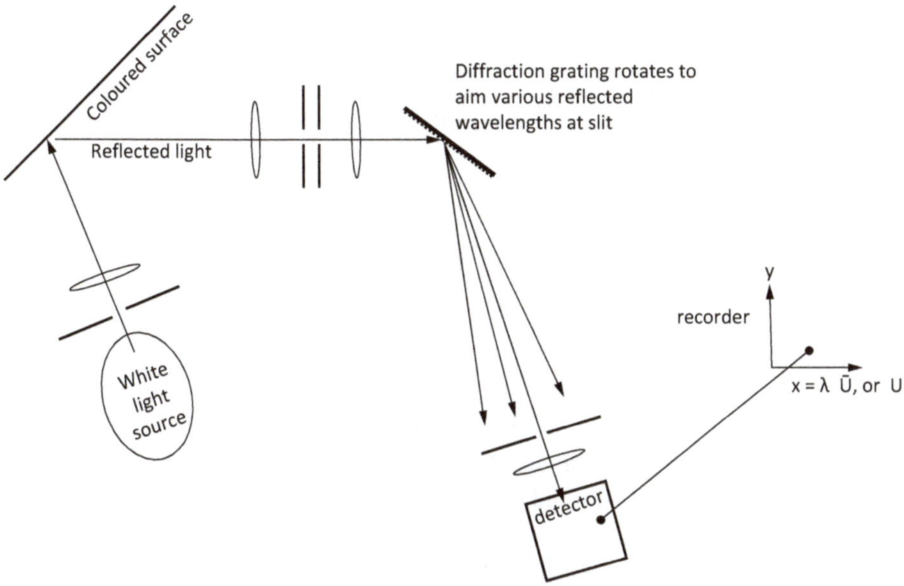


Fig. 6.35: Reflectance spectrometer.

Unlike absorption, fluorescence and phosphorescence spectra, for which the incident light beam passes through the sample and interacts with molecular orbital electrons, this reflection spectroscopy would not be so easily-related to a theory of the chemistry of the sample. Although, reflection would be related to surface processes, observation, to a great extent, would be related to the light source. Reflection would be components of the light source minus what had been absorbed at the reflection surface [568, 569]. Reflection spectroscopy can also be used to measure thickness of thin film coatings [570].

### 6.33.2 Visible light colour

Instrumentation as suggested above, which might be assembled from parts of other spectral equipment would not be so commonly found or so much used in forensic sci-

125 fluorescence. Reflection spectroscopy. Chapter 6.

ence; it might, in practice, not be much better than human visual comparisons (perhaps, aided with colour photography) to a colour chart [571].[126]

For the reflected light, a distinction, in concept, is made between instrumental spectral observation and what the human eye would see – as visualized within the complexities of the human photobiochemical-biophysical-optical system [572].

This is the concept of human-perceived colour [573, 574, 575].

Colour may be observed as light as perceived by the human eye –

- from a producing source – for example, a light bulb, the sun, a chemical reaction, such as the flame of a candle, or high-temperature-radiation from molten glass [576];
- from combined sources;
- having been affected by a substance as it passes through it – for example, white light entering a filter and seen coming out on the other side as green [577, 578, 579]; or
- having been affected by a substance after reflection from its surface from a light-producing source.

What the human eye would see as related to these processes would not be the same as instrumental-spectral observation. Photobiochemical-biophysical-psychological systems of human colour vision would observe a conglomeration of wavelengths and intensities with those systems' own specialized terminology and measurement, while spectral instrumentation would observe a spectrum simply as wavelength *versus* signal intensity. While it might often appear to be the same as at a specific wavelength of the EM spectrum, a colour observed by the human eye is, often, rather a complex mixture of wavelengths and intensities.

However, colour comparisons of samples and references by human eye observation when viewed side-by-side, with instrumentation designed for that purpose, can be accurately done and might help with forensic identification using human observation and colour photography.

## 6.34 Thermoluminescence analysis

When the crystal structures of some substances, at high temperatures, are exposed to ionizing radiation in their environments, some electrons can be energized and remain trapped in crystal defects after cooling. If reheated at a later time, *thermoluminescent* light may be emitted and measured to calculate the time interval – for archaeological dating. The substances could include ceramics at the time of firing, or lava, or bone

**126** Colour comparison. Chapters 3.5.5 and 6.33.2.

material. *Thermoluminescence* would be of minimal use for forensic analysis – it might be a method to identify fraudulently-claimed ancient pottery [580, 581, 582].

## 6.35 Viscosity measurement

Viscosity of chemical substances [583, 584, 585] can be measured for slow laminar flow of fluids. This includes organic chemical liquids. That viscosity is temperature-dependent can have forensic meaning, at least in fictional literature [586, 587].

## 6.36 Wet chemistry – screening tests

As mentioned above [588], screening tests[127] are not dealt with extensively here [589, 590, 591, 592, 593]. In analytical chemistry, these usually spot chemical reaction tests – often done as colour tests – often, conveniently used commercially available kits [594, 595, 596, 597][128] – are usually not now used for reliable final determination. Many kits are marketed for law enforcement, environmental, and industrial use; but also some for personal or home use [598, 599].[129]

In fictional literature, Sherlock Holmes has been illustrated as doing such a colour test (he had neither GC-MS, nor commercially available kits, nor used protective eyewear) [600, 601, 602].[130]

These – now screening – tests used to be the mainstay of analytical chemistry – sometimes called wet chemistry [603] because the colour producing chemical reactions were typically in water solution – or at least, in liquid solution. Screening, however, is still useful to the analytical chemist to avoid the more expensive analyses that would have been screened out.

An example: Used by police for quick in-the-field chemical confirmation colour tests of heroin – perhaps, not always quite correctly as according to kit instructions [604, 605, 606, 607, 608, 609, 610, 611, 612, 613, 614].

## 6.37 X-ray crystallography

X-ray diffraction from crystal structures can be used in identification of substances [615, 616, 617, 618, 619, 620].

**127** Screening tests. Wet chemistry. Chapters 3, 4, 5 and 6.
**128** Disclaimers. Error . . ., Not legal advice . . . Commercial products, . . . Introduction 4.
**129** Disclaimers. Error . . ., Not legal advice . . . Commercial products, . . . Introduction 4.
**130** Edmond Locard. Sherlock Holmes. Cocaine. Chapters 2, 3, 4 and 6.

## Notes

1 *Deuteronomy* Chapter 25 דְּבָרִים
2 A Hebrew – English Bible According to the Masoretic Text and the JPS 1917 Edition https://www.mechon-mamre.org/p/pt/pt0525.htm https://www.mechon-mamre.org/p/pt/pt0.htm
3 *Deuteronomy*, 25. 15 (13→15); *The Torah . . . A New Translation . . .*, The Jewish Publication Society of America, Philadelphia, 1962, page 370.
4 August Mau, German Archaeological Institute In Rome, *Pompe II Its Life and Art*, Translated into English by Francis W. Kelsey, New Edition, Revised and Corrected, The MacMillan Company, New York, 1902.
5 August Mau, Francis Kelsey, Translator; *Pompeii, Its Life and Art*, 2013, The Project Gutenberg EBook, CHAPTER XI, "THE BUILDINGS AT THE NORTHWEST CORNER OF THE FORUM, AND THE TABLE OF STANDARD MEASURES," Fig. 34. – Table of standard measures, *mensa ponderaria*; Page 94.
https://www.gutenberg.org/files/42715/42715-h/42715-h.htm
https://www.gutenberg.org/files/42715/42715-h/images/fig034fs.jpg
6 ACLODIVSAFFIACCVSNARCAEVSNFARELLIANCALEDVS
DVIDMENSVRASEXAEQVANDASEXDECR.
"'Aulus Clodius Flaccus, the son of Aulus, and Numerius Arcaeus Arellianus Caledus, the son of Numerius, duumvirs with judiciary authority, in accordance with a decree of the city council, caused the measures to be made equal' to the Roman measures."
7 http://www.pompeiiinpictures.com/pompeiiinpictures/R7/7%2007%2031.htm
8 The "spectroscopy" of "mass spectroscopy" is a bit of confusing terminology – and does not refer to EM.
9 However, this oversimplification is good-enough to explain the concepts needed here.
10 The convention here for the Bohr model is that the ground level is called $E_1$. For other treatments it may be called $E_0$. This results from source literature, and what it is called should not matter for purposes of this book.
11 See:

- Isaac Asimov, Asimov's Biographical Encyclopedia of Science and Technology / Second Revised *Edition*, Doubleday & Company, Inc., Garden City, New York, 1982, pages 700→702 [1101].
- Gordon M. Barrow, *Introduction to Molecular Spectroscopy*, McGraw-Hill Book Company, Inc., New York, . . ., 1962, pages 8→11.
- Gerhard Hertzberg, *Atomic Spectra and Atomic Structure*, Dover Publications, New York City, 2nd edition, first published by Dover 1945 (translation by J.W.T Spinks of *Atomspektren und Atomstruktur*, Prentice-Hall, Inc.,1937), page 13 et seq. (see also Asimov, *ibid.*, page 803 [1286]).

12 Jack G. Calvert and James N. Pitts, Jr., *Photochemistry*, John Wiley and Sons, Inc., New York, 1966; page 30, et seq.
13 Calvert and Pitts, page 31.
14 $\bar{\upsilon} = 1/\lambda = \Delta E/hc = (1/hc)\,(E_1 - E_2) = (2\pi^2\, m_e\, e^4/ch^3)\, Z^2\, [(1/n_2^2) - (1/n_1^2)]$.
15 The Bohr model could not adequately account for spectra beyond the very simple atomic system of hydrogen. The more successful – and complex – theory is based on the wave mechanics theory of Erwin Schrödinger published in 1926:

- See Asimov, pages 710 and 711 [1117].
- Walter J. Moore, *Physical Chemistry*, Prentice-Hall, Inc., Englewood Cliffs, New Jersey, 1962 and 1963, page 581, et seq.

16 – Werner Heisenberg, *PHYSICS AND PHILOSOPHY / THE REVOLUTION IN MODERN SCIENCE*, HARPERPERENNIAL, 1958, 2007, New York City, 10022; ISBN 978-0-06-120919-2; page 8, et seq.

- Niels Bohr, *Essays 1958–1962 on Atomic Physics and Human Knowledge*, Vintage Books, New York, 1963; page 39.

**17** https://en.wikipedia.org/wiki/Rydberg_formula
https://en.wikipedia.org/wiki/Rydberg_constant
http://scienceworld.wolfram.com/physics/RydbergConstant.html
https://scienceworld.wolfram.com/physics/BohrModel.html

**18** Dr. Kathy Joseph, Kathy Loves Physics, YouTube, "Bohr Model: A Delightful History [CC]"
https://www.youtube.com/watch?v=8IIg4Qt_qv4
https://www.amazon.com/stores/author/B0B57TQDHD/about

**19** The rings are of a path location – not an actual physical-material thing.

**20** The convention here for the Bohr model is that the ground level is called $E_1$. For other treatments it may be called $E_0$. This results from source literature, and what it is called should not matter for purposes of this book.

**21** The representation here is schematic – not from actual calculations.

**22** After a molecule would go to a higher-than-ground energy level, it might return to ground level by a non radiative vibrational process, or be involved in other physical processes, or proceed to react chemically. The spectral absorption process as theorized here is only concerned with promotion from ground to higher energy level.

**23** See:

- Walter J. Moore, *Physical Chemistry*, Prentice-Hall, Inc., Englewood Cliffs, New Jersey, 1962 and 1963, pages 467 and 468.
- Isaac Asimov, *Asimov's Biographical Encyclopedia of Science and Technology / Second Revised Edition*, Doubleday & Company, Inc., Garden City, New York, 1982, pages 571→573 [887].

**24** The convention here for the Bohr model is that the ground level is called $E_1$. For other treatments it may be called $E_0$. This results from source literature, and what it is called should not matter for purposes of this book.

**25** The Greek letter nu υ – not the Roman v.
https://en.wikipedia.org/wiki/Nu_(letter) https://en.wikipedia.org/wiki/Planck_constant

**26** 1 Ångström unit = $1 \times 10^{-8}$ m. https://en.wikipedia.org/wiki/Angstrom#:~:text=The%20angstrom%20(%2F%CB%88%C3%A6%C5%8B,0.1%20nanometre%2C%20or%20100%20picometres
1 nm = $1 \times 10^{-9}$ m.

**27** h = $6.62607015 \times 10^{-34}$ J ×s.

**28** c = $2.99792458 \times 10^{+8}$ m/s.

**29** M.G., A Survey of the photochemistry of the *ortho*-fused bebzocycloalkenes, UNB thesis, 1972. page 46. *ProQuest Dissertations and Theses;* 1972; ProQuest Dissertations and Theses Global UMI Number: DC54154

**30** – There is also the matrix algebra of Werner Heisenberg.

**31** The historically used phototube might now be replaced with other light-measuring devices.

**32** Beckman DU 1941→1976. https://en.wikipedia.org/wiki/DU_spectrophotometer

**33** Although not now of the most modern of forensic and analytical chemistry, the Beckman DU, Fullerton, California, 1954, was initially from 1940 (when Beckman was called National Technical Laboratories) – more recent by almost a century after Kirchhoff and Bunsen. The Beckman instrumentation, in its era, was then thought of as automated instrumentation, although not by present-day standards. The Beckman DU was used in many branches of chemistry – most likely including forensics (see Fig. 6.36).

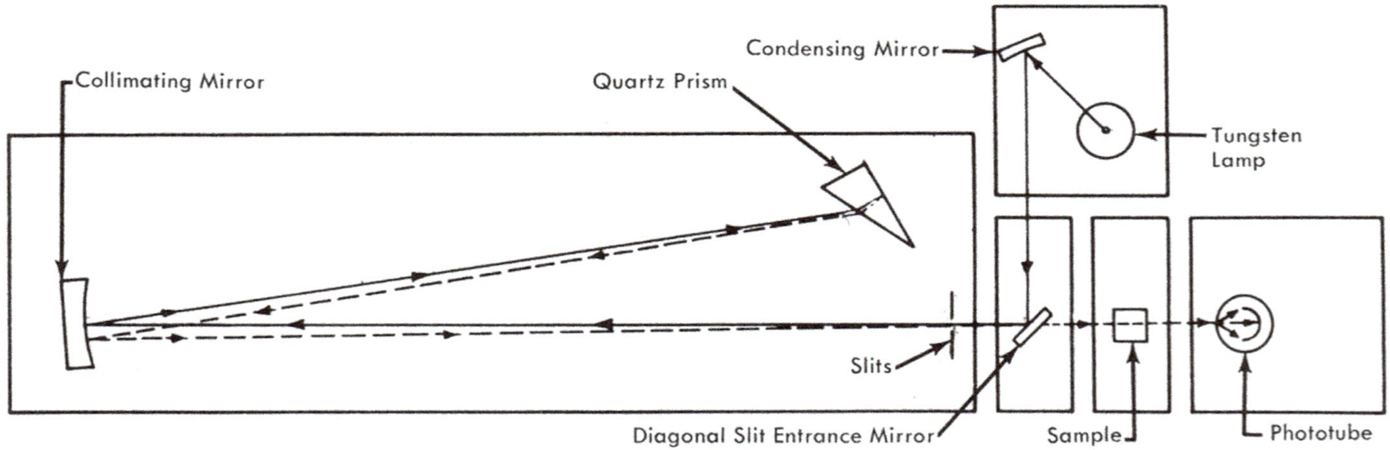


Fig. 6.36: Beckman DU, 1954, Fullerton, California.
https://en.wikipedia.org/wiki/File:Beckman_Manual_1954_Figure_2._Diagram_of_Model_DU_Optical_System_2012-002_b104f197_008_crop.jpg
https://digital.sciencehistory.org/works/r207tp471
https://digital.sciencehistory.org/works/r207tp471/viewer/r781wg113.

**34** (The author often used a Beckman DU, 1966. M.G., "IODINE-TETROHYDROFURAN AND IODINE-FURAN CHARGE TRANSFER COMPLEX FORMATION," B.S. thesis, Rensselaer Polytechnic Institute, Troy, New York, May 1966.)

**35** The **front cover illustration** of both editions of this book, although neither of forensic nor of modern analytical chemistry, was chosen because it gives a recognizable chemistry theme. Spectroscopy is of great importance to the analytical chemistry essential for forensic science. This historically-based cover art (from *gettyimages* ilbusca) is similar to the description in: Gustav Kirchhoff & Robert Bunsen, "Chemical Analysis by Observation of Spectra," *Annalen der Physik und der Chemie* (Poggendorff), Vol. 110, pages 161→189, Heidelberg, 1860 (see Fig. 6.37).

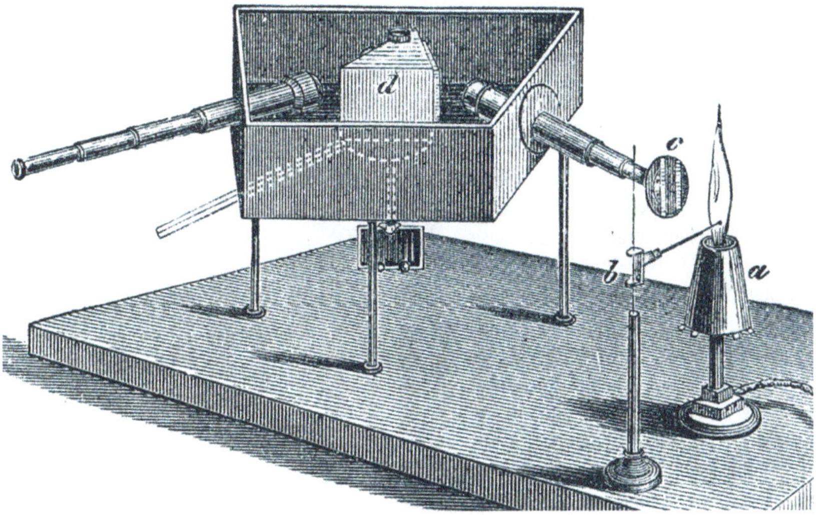


Fig. 6.37: Spectroscopic instrumentation, Heidelberg, 1860.
https://www.chemteam.info/Chem-History/Kirchhoff-Bunsen-1860.html
https://babel.hathitrust.org/cgi/pt?id=umn.31951d00326549e&view=thumb&seq=177.

**36** Rainbow. Workers' Health & Safety Legal Clinic *newsletter*, 2016 08. Vol.24 No.2, pages 3 and 4.
http://newsletter.workers-safety.ca.s3-website-us-east-1.amazonaws.com/#!/2016 08.Vol.24No.2.pdf
http://s3.amazonaws.com/newsletter.workers-safety.ca/newsletters/2016%2008/2016%2008.%20Vol.24%20%20No.2%20.pdf

**37** [...]

> By 1859, Robert Wilhelm Bunsen and Gustav Robert Kirchhoff, both of Heidelberg, had developed a spectroscope that was used to analyse light produced by atoms thermally excited in the flame of a gas-air burner (famously named after Bunsen). By 1860 and 1861 they had discovered two new elements Cs and Rb. Their instrument was also soon used for astronomical observations. Bunsen and Kirchhoff's spectroscope was developed from the well-known phenomenon that minerals put in a flame produced characteristic colours.
>
> [...]

**38** https://en.wikipedia.org/wiki/History_of_spectroscopy

**39** https://www.sciencephoto.com/media/945827/view/kirchhoff-bunsen-spectroscope-19th-century

**40** Jack G. Calvert & James N. Pitts, Jr., *PHOTOCHEMISTRY*, John Wiley and Sons, Inc., New York, 1966; page 21.

**41** Beer Lambert equation – https://chem.libretexts.org/Bookshelves/Physical_and_Theoretical_Chemistry_Textbook_Maps/Supplemental_Modules_(Physical_and_Theoretical_Chemistry)/Spectroscopy/Electronic_Spectroscopy/Electronic_Spectroscopy_Basics/The_Beer-Lambert_Law
https://en.wikipedia.org/wiki/Beer%E2%80%93Lambert_law
Johann Heinrich Lambert https://en.wikipedia.org/wiki/Johann_Heinrich_Lambert
August Beer https://en.wikipedia.org/wiki/August_Beer
**42** https://www.mathsisfun.com/numbers/e-eulers-number.html
https://www.mathsisfun.com/algebra/logarithms.html
**43** https://www.amazon.ca/i-Story-Number-ebook/dp/B003VIWZ8I/ref%2009%3D%2009pd_sbsd_14_2/143-3268445-8463047?_encoding%2009%3D%2009UTF8%26pd_rd_i%2009%3D%2009B003VIWZ8I%26pd_rd_r%2009%3D%200949d09250-f8bf-49d4-b2f9-46fd671c420a%26pd_rd_w%2009%3D%2009WucJv%26pd_rd_wg%2009%3D%2009tygY4%26pf_rd_p%2009%3D%2009f30c29c2-eb2e-4217-8f50-c1711db4eaba%26pf_rd_r%2009%3D%2009MVGQZ7JP78XHX7ZKH6TV%26psc%2009%3D%20091%26refRID%2009%3D%2009MVGQZ7JP78XHX7ZKH6TV
**44** = 2.71828 . . . – see William K. Morrill, *Calculus*, D.Van Nostrand Company, Inc., Princeton, New Jersey . . ., 1956, page 125 et seq.
**45** C.A. Parker, *Photoluminescence of Solutions / With Applications to Photochemistry and Analytical Chemistry*, ELSEVIER, Amsterdam . . ., 1968; page 16, et seq.
**46** Here as $\log_e$ – elsewhere often as $\log_{10.}$ – base e – base 10.
**47** $\ln x = \log_e x = 2.302\,585\,093 \times \log_{10} x$
**48** "Spectroscopy/Molecular energy levels"; Wikiversity; April 2020.
https://en.wikiversity.org/wiki/Wikiversity:Main_Page
https://en.wikiversity.org/wiki/Spectroscopy/Molecular_energy_levels
https://en.wikiversity.org/wiki/Spectroscopy/Rotational_spectroscopy
https://en.wikiversity.org/wiki/Wikiversity:Welcome
**49** M.G., "A Survey of the photochemistry of the *ortho*-fused bebzocycloalkenes," UNB thesis, 1972. page 17.
**50** *Ortho*-fused benzoclclooctene https://en.wikipedia.org/wiki/Tetralin
**51** *Orthofused-benzocycloalkene*. M.G., "A Survey of the photochemistry of the *ortho*-fused bebzocycloalkenes," UNB thesis, 1972. *ProQuest Dissertations and Theses*; 1972; ProQuest Dissertations and Theses Global UMI Number DC54154
**52** M.G., UNB, 1972, pages ii and 3. Abstract and molecular structures.
**53** M.G., UNB, 1972, page 94. UV spectrum. Fluorescence spectrum.
**54** M.G., UNB, 1972, pages 103 and 104. Phosphorescence spectra.
**55** M.G., UNB, 1972, page 26. Quantum yield.
**56** *Orthofused-benzocycloalkene*. M. Grossman, R. Kubela, G.P. Semeluk, and I. Unger, "Spectrofluorometric Studies. XI. Some Aspects of the Photochemistry of Ortho-fused Benzocycloalkenes," Department of Chemistry, University of New Brunswick, Fredericton, New Brunswick; *Canadian Journal of Chemistry*, 50, 3298 (1972), pages 3298→3303. https://www.nrcresearchpress.com/doi/pdf/10.1139/v72-528
**57** *Ortho*-fused benzoclclohexene *tetrahydroanaphthalene tetralin* CAS 119-64-2 132.2 g/mol
https://en.wikipedia.org/wiki/Tetralin http://encyclopedia.thefreedictionary.com/Tetralin
**58** *Ortho*-fused benzoclclopentene *indane* CAS 496-11-7 118.176 g/mol. https://en.wikipedia.org/wiki/Indane
**59** UV absorption, fluorescence and phosphorescence spectroscopy. Chapters 6.2 and 6.18.
**60** UV absorption spectroscopy is dealt with in Chapter 6.2. The related fluorescence and phosphorescence spectroscopy is dealt with in Chapter 6.18. Some of the cited literature references apply for both Chapters 6.2 and 6.18. Some of the figures are relevant to both Chapters 6.2 and 6.18.
**61** *Orthofused-benzocycloalkene*. –Alkene, -pentene, -hexene, -octene. Chapters 6.2 and 6.18 (see Fig. 6.38).

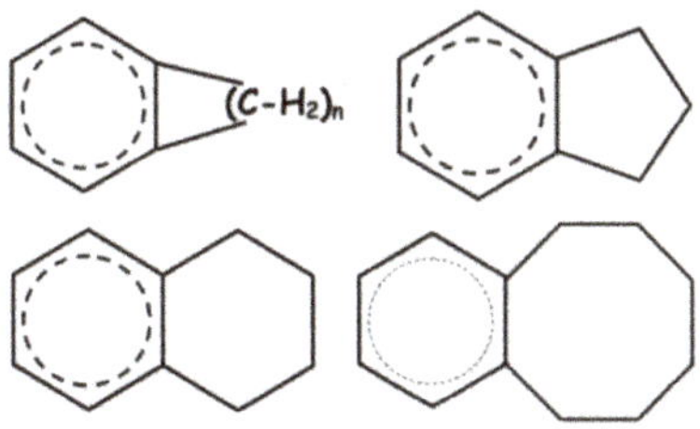


Fig. 6.38: Benzocycloalkenes.

62 See also: C. M. Humby, G. P. Semeluk and R. D. S. Stevens (1970) The Vapour Phase U.V. Spectra of Ortho Fused Benzocycloalkenes, *Spectroscopy Letters*, 3:4–5, 99–114, DOI: 10.1080/ 0038 7017008076300
https://www.tandfonline.com/doi/abs/10.1080/00387017008076300

63 https://pubchem.ncbi.nlm.nih.gov/compound/1-phenylnaphthalene
https://www.alfa.com/en/catalog/B21953/

64 *1-Phenylnaphthalene*, 204.272 g/mol. CAS 605-02-7.

65 NIST UV-visible spectrum https://webbook.nist.gov/chemistry/uv-vis/
https://webbook.nist.gov/cgi/cbook.cgi?ID=C605027&Units=SI&Mask=400#UV-Vis-Spec
UV absorption spectroscopy is dealt with in Chapter 6.2. The related fluorescence and phosphorescence spectroscopy is dealt with in Chapter 6.18. Some of the cited literature references apply for both Chapters 6.2 and 6.18. Some of the figures are relevant to both Chapters 6.2 and 6.18.

66 NIST Chemistry WebBook, SRD 69, National Institute of Standards and Technology, US Department of Commerce, Gaithersburg, Maryland, 20899
https://www.nist.gov/disclaimer
https://www.nist.gov/open/copyright-fair-use-and-licensing-statements-srd-data-software-and-technical-series-publications

67 DMS / *UV ATLAS of Organic Compounds* / Vol. II; Verlag Chemie, Weinheim, and Butterworths, London; 1966; E1/2.
$\lambda = 289\,\text{nm}.\qquad \varepsilon_{\lambda = 289\text{nm}} .\mathbf{9{,}000}(\text{litres/ mol cm})\qquad [\log_{10}$

68 $\Delta E = h\upsilon = hc/\lambda = hc\overline{\upsilon}$
$\overline{\upsilon} = (1/\lambda)$
$\overline{\upsilon} = (1/289)\,\text{nm}^{-1} = (1/.000289)\,\text{cm}^{-1} = 34{,}602.0761\,\text{cm}^{-1}.\mathbf{34{,}600}\text{cm}^{-1}$

69 Well, almost always: the subtle mirror image difference mentioned in Chapter 4.5 would theoretically produce an identical *energy level diagrams*.

70 https://microbenotes.com/uv-spectroscopy-principle-instrumentation-applications/
https://microbenotes.com/uv-vis-spectroscopy/

71 https://www.slideshare.net/mariomS7/uvvis-spectroscopy

72 That is, light **at a single wavelength**. Real practice approximates this with a narrow band of wavelengths. Typically, naturally observed light is *polychromatic* – *i.e.* light of many wavelengths, mixed together – often called *white* light. *Monochromatic* light is achieved with spectral instrumentation – such as the diffraction grating mentioned here.

73 Definition as used here *white light* = light of all wavelengths in the region of interest – here: UV and visible.

74 https://en.wikipedia.org/wiki/Xenon_arc_lamp

75 https://www.merriam-webster.com/dictionary/refract
https://www.yourdictionary.com/diffraction
https://www.ncnr.nist.gov/programs/CHRNS/pdf/Diffraction_of_Light.pdf
https://www.physicsclassroom.com/class/waves/Lesson-3/Reflection,-Refraction,-and-Diffraction

76 https://www.spectroscopyonline.com/view/prisms

77 See: Hazel Rossotti, *COLOUR / Why The World Isn't Grey*, Princeton University Press, Princeton, New Jersey, 08540; 1983; ISBN 0-691-02386-7; at page 61.
https://press.princeton.edu/books/paperback/9780691023861/colour
**78** This diagram is intended to describe concept; it is schematic; it is not to scale, nor in proportion; it is not as an engineering plan or design; none of its parts are presented here as an engineering plan or design.
**79** The *white light* source, shown schematically, may include parabolic & other reflectors, lenses and slits that would help produce parallel *white light* rays narrowly directed towards the lens and slit system of the spectrometer, in front of the prism.
**80** The lens and slit system of the *spectrometer*, just before the *white light* enters the prism, shown schematically, is also intended to produce parallel rays narrowly directed into the prism. In the prism, it is refracted; various wavelengths at slightly different angles (see Fig. 6.39).

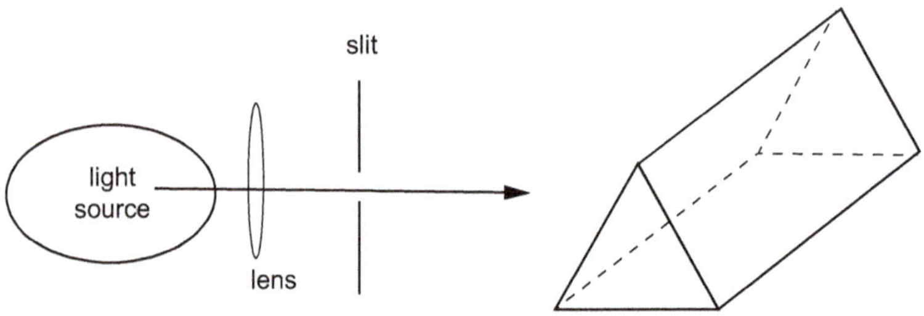


Fig. 6.39: Prism.

**81** The *white light* source, if a high-pressure Hg–Xe electric arc lamp would have a broad emission spectrum, that would likely be reported by the manufacturer and could be verified by the *spectrometer* as without the sample to be examined. If of low pressure, only the Hg emission lines would appear. Other light source systems are possible, including LASER systems.
**82** See: Jack G. Calvert and James N. Pitts, Jr., *PHOTOCHEMISTRY*, John Wiley and Sons, Inc., New York, 1966; page 6; page 725.
**83** C.A. Parker, *Photoluminescence of Solutions / With Applications to Photochemistry and Analytical Chemistry*, ELSEVIER, Amsterdam . . ., 1968; page 131, et seq.
See also: Rainwater, Zim, Perlman, cited below, pages 3, 43; & Rossotti, cited below, page 23.
**84** https://www.knowledgeboat.com/question/a-narrow-beam-xy-of-white-light-is-passing-through-a-glass–808811086487879000
**85** Since in real life the two cells may be less than totally transparent (and there may be light reflection and scattering), the fact that they are matched allows for a reasonable working assumption that they are completely transparent.
**86** Since in real life the solvent may be less than totally transparent (and there may be light reflection and scattering), the same solvent in each of the two cells allows for an assumption of complete transparency (analogous to the matched cells). If the solvent absorbs significantly, then maintaining accurate intensity measurements can, in practice, become very difficult, or impossible.
**87** See Jack G. Calvert and James N. Pitts, Jr., *Photochemistry*, John Wiley and Sons, Inc., New York, . . ., 1966, pages 170→173, particularly Eq.3–41.
**88** Although the value of $\varepsilon$ can also play a role in identification.

89 For a general reference, see also: Howard A. Strobel, *Chemical Instrumentation / A Systematic Approach to Instrumental Analysis*, Addison-Wesley Publishing Company, Inc., Reading, Massachusetts and London, England, 1960; Section 6-2, Beer's law, pages 150→155.
90 Cyclohexane, CAS 110-82-7, $C_6H_{12}$, 84.162 g/mol chair and boat configurations (see Fig. 6.40).

Fig. 6.40: Cyclohexane chair boat.
https://en.wikipedia.org/wiki/Cyclohexane#: ~:text = Cyclohexane%20is%20a%20cycloalkane %20with, which%20it%20is%20sometimes%20used).

91 *1-Phenylnaphthalene*; solvent = "Light petroleum b.p. 100–120°".
92 See note above for *1-phenylnaphthalene*. Care should be taken to determine if the literature reports ε as for $\log_e$ or $\log_{10}$. It does not matter which, except that it must be consistent as for this experiment.
93 For which there need not be concern about $\log_e$ or $\log_{10}$ (see note above), except if ε is to be reported for use elsewhere.
94 For the notation here: path length = l = 1 cm.
Roman letter lower case L Arabic number one.
(Unfortunately the symbolism here looks confusingly the same.)
95 See: Howard A. Strobel, *Chemical Instrumentation / A Systematic Approach to Instrumental Analysis*, Addison-Wesley Publishing Company, Inc., Reading, Massachusetts and London, England, 1960; Chapter 6–4, Sources of error, pages 159→167.
96 See: https://www.mcgill.ca/oss/article/did-you-know-history/breathalyzer-there-was-drunkometer
97 Richard Saferstein, *CRIMINALISTICS / An Introduction to Forensic Science*, Seventh Edition, Prentice Hall, Upper Saddle River, New Jersey, 07458; 2001; ISBN 0-13-013827-4; Chapter 10, FORENSIC TOXICOLOGY, page 265 et seq.; Chapter 5, ORGANIC ANALYSIS, page 136 et seq.
98 See: Walter J. Moore, *Physical Chemistry*, Prentice-Hall, Inc., Englewood Cliffs, New Jersey, 1962 and 1963, pages 595 and 596.
99 Still an oversimplification that is good-enough to explain the concepts needed here.
100 Karen Feinstein, *Guide to Spectroscopic Identification of Organic Compounds*, CRC Press, Inc., Boca Raton, Florida, 33431; 1995; ISBN 0-8493-9448-1.
https://www.amazon.com/Guide-Spectroscopic-Identification-Organic-Compounds/dp/0849394481
101 But see: Prasad L. Polavarapu & Jiangtao He, Vanderbilt University, "Chiral Analysis Using Mid IR Vibrational CD Spectroscopy," *Analytical Chemistry*, 01 February 2004, ACS.
https://pubs.acs.org/doi/pdf/10.1021/ac0415096#:~:text=Vibrational%20circular%20dichroism%20(VCD)%20spectroscopy,enantiomeric%20purity%20of%20chiral%20species
102 Hydrogen bromide, H Br, CAS 10035-10-6.
103 NIST Chemistry WebBook, SRD 69, National Institute of Standards and Technology, US Department of Commerce, Gaithersburg, Maryland, 20899
https://www.nist.gov/disclaimer
https://www.nist.gov/topics/data/public-access-nist-research/copyright-fair-use-and-licensing-statements-srd-data-and
104 https://webbook.nist.gov/cgi/cbook.cgi?ID=C10035106&Mask=80

**105** See also: Gordon M. Barrow, *Introduction to Molecular Spectroscopy*, McGraw-Hill Book Company, Inc., New York, . . ., 1962, fig.7–3, page 137 – showing transition from 0 → 1 vibrational levels, for various rotational levels of H Br gas – related to the molecular linear stretching vibration.
**106** This IR spectrum is of *cocaine* – non-ionized form.
**107** NIST IR spectrum https://webbook.nist.gov/cgi/cbook.cgi?ID=C50362&Mask=80
https://webbook.nist.gov/cgi/cbook.cgi?Spec=C50362&Index=0&Type=IR&Large=on
**108** NIST Chemistry WebBook, SRD 69, National Institute of Standards and Technology, US Department of Commerce, Gaithersburg, Maryland, 20899
https://www.nist.gov/disclaimer
https://www.nist.gov/topics/data/public-access-nist-research/copyright-fair-use-and-licensing-statements-srd-data-and
**109** See also: Terry Mills III and J. Conrad Roberson, Division of Forensic Sciences, Georgia State Crime Laboratory, Atlanta, *Instrumental Data for Drug Analysis*, Second Edition, Volume 1; Elsevier, New York, 1987, pages 524 and 525. For another example – *tetrahydrocannabinol* – see Volume 3, page 2195.
**110** See: Brian C. Smith, *Fundamentals of Fourier Transform Infrared Spectroscopy*, CRC Press, Boca Raton, Florida, . . ., 1996, pages 42 and 43.
**111** For a general treatment of FTIR see Smith, *Fundamentals.*
**112** Smith, pages 8 and 9.
**113** https://chem.libretexts.org/Bookshelves/Physical_and_Theoretical_Chemistry_Textbook_Maps/Supplemental_Modules_(Physical_and_Theoretical_Chemistry)/Spectroscopy/Vibrational_Spectroscopy/Infrared_Spectroscopy/How_an_FTIR_Spectrometer_Operates
**114** See: Mills and Roberson. See Smith, *Fundamentals* . . ., Introduction.
**115** See: Smith, *Fundamentals* . . ., pages 88→93.
**116** See: Smith, *Fundamentals* . . ., pages 93→97.
**117** Plough Inc. CAS 8012-95-1
https://en.wikipedia.org/wiki/Nujol https://en.wikipedia.org/wiki/Nujol
https://homepages.bluffton.edu/~bergerd/classes/CEM222/Handouts/nujol.pdf
https://www.sigmaaldrich.com/catalog/search?term=Nujol&interface=All&N=0&mode=partialmax&lang=en&region=CA&focus=product
**118** https://chem.libretexts.org/Bookshelves/Analytical_Chemistry/Book%3A_Physical_Methods_in_Chemistry_and_Nano_Science_(Barron)/04%3A_Chemical_Speciation/4.02%3A_IR_Spectroscopy
https://www.specac.com/en/documents/user-manuals/2i-03950-2-user-care-guide-for-7mm-pellet-die-asse
**119** – being careful that it is dry – no water $H_2O$ – so as to not dissolve and damage the NaCl disks.
**120** See Brian C. Smith, *Fundamentals of Fourier Transform Infrared Spectroscopy*, CRC Press, Boca Raton, Florida, . . ., 1996, Chapter 5, "Quantitative Analysis," particularly pages 145→149.
**121** For comparison with the literature.
**122** See, for example, for electrochemical and infrared analyses: *INTOX EC/IR II BREATH TEST OPERATOR INSTRUCTIONAL MANUAL*, Virginia Department of Forensic Science, Breath Alcohol Section, July 2008; Richmond, Virginia, 23219.
https://dfs.virginia.gov/laboratory-forensic-services/breath-alcohol/breath-alcohol-manuals/
https://www.dfs.virginia.gov/wp-content/uploads/2013/07/InstructionalManualIntoxECIRII.pdf
**123** See: . . . *Manual*, Virginia . . . 2008; at pages 23–25; Figures 9 and 10.
**124** https://webbook.nist.gov/cgi/cbook.cgi?ID=C64175&Type=IR-SPEC&Index=2
**125** "*Chrom*atography" is related to "colour" – see *The Oxford English Dictionary / Second Edition*, Vol. III, Clarendon Press, Oxford, 1989, page 185.
**126** From an older-style fountain pen.
**127** For a fuller treatment of TLC, see:

- John Kenkel, *Analytical chemistry refresher manual,* Lewis Publishers, Chelsea, Michigan, 1992, pages 215→218; and pages 214→286 concerning chromatography more generally.
- Brian H. Kaye, *Science and the Detective / Selected Reading in Forensic Science,* VCH Verlagsgesellschaft mbH, Weinheim, Germany, 1995, pages 114→118.
- Julian Tyson, *ANALYSIS / What Analytical Chemists Do,* Royal Society of Chemistry Paperbacks, Cambridge CB4 4WF; 1988, 1994; ISBN 0-85186-463-5; page 143, et seq.

**128** Although they can be easily made in the lab – the silica gel or alumina coating is spread onto the glass in water suspension and then dried in an oven – these plates are commonly purchased in quantity ready to use – for example: See *MANDEL SCIENTIFIC COMPANY 1985* catalogue, Rockwood, Ontario N0B 2K0, for different kinds of plates of several manufacturers, pages 195→199. See also Kuranz – cited in Chapter 5.
**129** Perhaps in a solvent of the mix that is to be the mobile phase.
**130** The handling and evaporation of these solvent require appropriate work place health and safety precautions – and therefore reference to the SDSs.
**131** This diagram is sketched to show concept – it is not of a real experiment.
**132** For a fuller treatment see John Kenkel, *Analytical chemistry refresher manual,* Lewis Publishers, Chelsea, Michigan, 1992, Chapter 9 (pages 225–259).
**133** Kenkel, page 235. This version of GC is also called gas liquid chromatography to indicate that the stationary phase is liquid.
**134** This may be seen by examining *retention times* tabulated for various substances in Gunter Zweig and Joseph Sherma, Eds.-in-Chief, *CRC Handbook Series in Chromatography,* CRC Press, Inc., Boca Raton, Florida, 1972 – e.g. Section A, Volume I, Table GC 21, page 26.
**135** Although they can be made in the lab, these columns are now commonly purchased ready for use – *e.g.* see *MANDEL SCIENTIFIC COMPANY 1985* catalogue, Rockwood, Ontario N0B 2K0, pages 223 and 224.
**136** Detector Signal strength is proportional to the quantity under the peak detected
Detector Signal strength = moles detected × k, where k is a constant for the GC set-up, determined by independent calibration experiment.
**137** See, for example: Diane L. Phillips, Ian R. Tebbett, Roger L. Bertholf, "Comparison of HPLC and GC-MS for Measurement of Cocaine and Metabolites in Human Urine," *Journal of Analytical Toxicology,* Vol. 20, September 1996, pages 305→308, at page 306, Figure 2.
**138** Sketched here in an idealized and simplified way (including that the peaked curves would not actually be so exactly symmetrical).
**139** The treatment here is idealized for purposes of simple explanation – in practice other factors and complications can arise – see Kenkel – cited above in Chapter 6.5.1 – for a fuller treatment.
**140** The treatment here is idealized for purposes of simple explanation – in practice other factors and complications can arise – see Kenkel – cited above in Chapter 6.5.1 – for a fuller treatment.
**141** Gregg Herken, *Brotherhood of the Bomb / the tangled lives and loyalties of Robert Oppenheimer, Ernest Lawrence, and Edward Teller,* Henry Holt and Company, New York City, 10011; 2002; ISBN 0-8050-6588-1; page 45.
**142** Thomas Powers, *HEISENBERG'S WAR / The Secret History of the German Bomb,* BACK BAY BOOKS, Little, Brown and Company, Boston, . . ., 1993; ISBN 0-316-71623-5.
**143** Mention about Third Reich mass spectral possibilities. Powers, page 445.
**144** The treatment here is idealized for purposes of simple explanation – in practice other factors and complications can arise – for a fuller treatment. See:

- John Kenkel, *Analytical chemistry refresher manual,* Lewis Publishers, Chelsea, Michigan, 1992, Chapter 6.7, pages 162→165.

- Richard Saferstein, "Forensic Applications of Mass Spectrometry," as Chapter 3 in Richard Saferstein, ed., *Forensic Science Handbook* [Volume I], Prentice-Hall, Inc., Englewood Cliffs, New Jersey, 1982; pages 92→138.

**145** That is, made into charged particles.

**146** The lines of force visualized for that magnetic field are at right angles to the ionic initial lines-of-flight.

**147** This ratio is the molar mass of the fragment divided by its charge. Since these are positive ions, z is > 0 in multiples of the unit charge of an electron $e^-$ – so that z is a small positive integer: usually +1 or +2; most usually +1. When z = +1, the ratio is then simply the molar mass; when z = +2, it is half the molar mass.

**148** Mike Sargent (Ed.), *Guide to achieving reliable quantitative LC-MS measurements*, RSC Analytical Methods Committee, United Kingdom National Measurement System; 2013; ISBN 978-0-948926-27-3.
https://www.studocu.com/my/document/universiti-putra-malaysia/chemistry-i/amc-lcms-guide-tcm18-240030/44098229

**149** This sketched diagram is to describe general concept; not to scale; not drawn at to represent any particular real molecule.

**150** The heights are proportional to quantity.

**151** Relative Intensity v m/z = mass to charge ratio

**152** https://webbook.nist.gov/cgi/cbook.cgi?ID=C10035106&Mask=200#Mass-Spec
https://webbook.nist.gov/cgi/cbook.cgi?Spec=C10035106&Index=0&Type=Mass&Large=on

**153** NIST Chemistry WebBook, SRD 69, National Institute Of Standards And Technology, US Department of Commerce, Gaithersburg, Maryland, 20899
NIST Chemistry WebBook https://webbook.nist.gov/chemistry
https://www.nist.gov/disclaimer
https://www.nist.gov/open/copyright-fair-use-and-licensing-statements-srd-data-software-and-technical-series-publications

**154** Brookhaven National Laboratory – Chart of Nuclides http://www.nndc.bnl.gov/chart/

**155** See: Chart of Nuclides:
$_{35}Br^{71}$ $_{35}Br^{72}$ $_{35}Br^{73}$ $_{35}Br^{74}$ $_{35}Br^{75}$ $_{35}Br^{76}$ $_{35}Br^{77}$ $_{35}Br^{78}$ $_{35}Br^{79}$ $_{35}Br^{80}$ $_{35}Br^{81}$ $_{35}Br^{82}$ $_{35}Br^{83}$ $_{35}Br^{84}$ $_{35}Br^{85}$ $_{35}Br^{86}$ $_{35}Br^{87}$
https://www.nndc.bnl.gov/nudat2/reCenter.jsp?z=35&n=44

**156** https://education.jlab.org/itselemental/iso035.html

[. . .]

**Isotopes with a Known Natural Abundance**
**Mass Number Natural Abundance Half-life**
79 50.69% STABLE
81 49.31% STABLE

[. . .]

**157** $_{35}Br^{79}$ m = 79
$_1H^1 - {}_{35}Br^{79}$ m = 80
$_{35}Br^{81}$ m = 81
$_1H^1 - {}_{35}Br^{81}$ m = 82

**158** Tetrahydrocannabinol THC CAS 1972-08-3 Chapter 6; 9. Fig. 6.20, 9.1

**159** https://webbook.nist.gov/cgi/cbook.cgi?ID=1972-08-3+&Units=SI
https://webbook.nist.gov/cgi/cbook.cgi?ID=C1972083&Units=SI&Mask=200#Mass-Spec
https://webbook.nist.gov/cgi/cbook.cgi?Spec=C1972083&Index=0&Type=Mass&Large=on

**160** National Institute of Standards and Technology, US Department of Commerce, Gaithersburg, Maryland, 20899. http://nist.gov/ http://webbook.nist.gov/

**161** https://webbook.nist.gov/cgi/cbook.cgi?Spec=C1972083&Index=0&Type=Mass&Large=on

**162** Cocaine CAS 50-36-2 303.36 g/mol
**163** m/z = mass / charge.
**164** The x-axis of the mass spectrum print-out, such as sketched here, only for one fragment, would be calibrated as for m/z. It would be most conveniently read as for z = + 1.
**165** The y-axis would indicate relative abundances of various fragments; not used in this calculation, except for the existence of the fragment. The y values shown here are contrived for convenience of example.
**166** See Fig. 6.41.

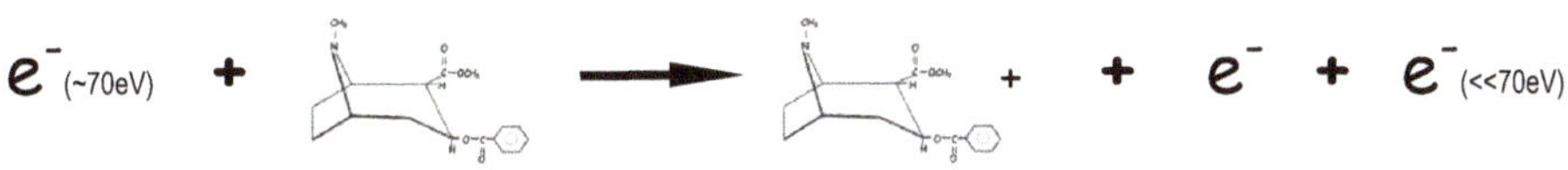


Fig. 6.41: Electron impact.

**167** 1 eV = 1 electron volt
= energy acquired by an electron when accelerated through 1 V
= $1.60 \times 10^{-19}$ J
= $0.3824 \times 10^{-19}$ calorie.
**168** This is a sketch to where the m/z MS peak would look like; not an actual spectrum.
**169** Sargent (Ed.), *Guide* . . . pages 10, 60.
**170** ". . . The traditional approach to mass spectrometry involves scanning a wide mass range to obtain a mass spectrum. However, for LC-MS applications QqQ instruments are typically operated in selected reaction monitoring (SRM) mode which is also called multiple reaction monitoring (MRM) by some suppliers. This involves continuously monitoring a small number of selected transitions (*i.e.* mass numbers) for each analyte, typically one precursor ion to a couple of product ions. . . ."
**171** This involves the separated components from the sample mixture, delivered from the GC or LC line, being fragmented at MS entry by a method milder than electron impact, so that there only perhaps one or two fragment ions, which are characterized by m/z. The molecules are thus characterized by these m/z and the chromatograph retention times, and subsequently identified in a large computer-stored library. This is a reasonably good and reliable method.
**172** MRM = multiple reaction monitoring. SRM = selected reaction monitoring.
**173** See also:
- Jonathan Feakins, "Alba Álvarez-Martín is making mass spec more accessible in museums / The cultural heritage scientist is adapting a medical imaging technique to analyze paintings in Amsterdam's Rijksmuseum," *C&EN*, ART & ARTIFACTS, 02 January 2025.
- Alba Alvarez-Martin, Jusal Quanico, Teresa Scovacricchi, Ermanno Avranovich Clerici, Geert Baggerman, Koen Janssens, "Chemical Mapping of the Degradation of Geranium Lake in Paint Cross Sections by MALDI-MSI," *Analytical Chemistry*,2023, 95, 49, 18215–18223, 23 November 2023.

https://doi.org/10.1021/acs.analchem.3c03992
**174** Kenkel, cited above in Section 6.5.1.
**175** Richard Saferstein, *CRIMINALISTICS/An Introduction to Forensic Science*, Seventh Edition, Prentice Hall, Upper Saddle River, New Jersey, 07458; 2001; ISBN 0-13-013827-4; page 140, et seq.
**176** See John Kenkel, *Analytical chemistry refresher manual*, Lewis Publishers, Chelsea, Michigan, 1992, pages 165, 247 and 248.
**177** https://en.wikipedia.org/wiki/Gas_chromatography%E2%80%93mass_spectrometry
https://www.innovatechlabs.com/newsroom/1841/how-to-interpret-gas-chromatography-mass-spectrometry-

results/
https://www.gmu.edu/depts/SRIF/tutorial/gcd/gc-ms2.htm
**178** Two different traditional uses of z here, not to be confused:
m/z = mass/charge
z = Cartesian third dimension.
**179** Mike Sargent (Ed.), *Guide to achieving reliable quantitative LC-MS measurements*, RSC Analytical Methods Committee, United Kingdom National Measurement System; 2013; ISBN 978-0-948926-27-3. http://www.lgcgroup.com/LGCGroup/media/PDFs/Our%20science/NMI%20landing%20page/Publications%20and%20resources/Guides/AMC_LCMS_Guide.pdf
**180** Sargent (Ed.), . . ., *Guide* . . ., page 1.
**181** *Gas Chromatography-Mass Spectroscopy Background*, The Shared Research Instrumentation Facility, George Mason University, Manassas Virginia, 20110; 1998.
https://www.gmu.edu/depts/SRIF/tutorial/gcd/gc-ms2.htm
http://www.gmu.edu/depts/SRIF/
**182** Cartoon by Dr. Nick Kim, Wellington, New Zealand. With permission.
**183** Dr. Nick Kim. Copyright acknowledgement. Chapters 2; 6; 8.
**184** See:
- John Kenkel, *Analytical Chemistry Refresher Manual*, Lewis Publishers, Chelsea, Michigan, 1992, "High Performance Liquid Chromatography," Chapter 10, page 261 et seq.
- https://en.wikipedia.org/wiki/High-performance_liquid_chromatography
- Richard Saferstein, *Criminalistics /An Introduction to Forensic Science*, Seventh Edition, Prentice Hall, Upper Saddle River, New Jersey, 07458; 2001; ISBN 0-13-013827-4; Chapter 5, Organic Analysis, High-Performance Liquid Chromatography (HPLC), page 127, 128.

**185** Brian H. Kaye, *Science and the Detective / Selected Reading in Forensic Science*, VCH Verlagsgesellschaft mbH, Weinheim, Germany; 1995; ISBN-3-527-29252-7; Chapter 5.1, "Gunshot Residues," at page 114.
**186** Richard Saferstein, *CRIMINALISTICS / An Introduction to Forensic Science*, Seventh Edition, Prentice Hall, Upper Saddle River, New Jersey, 07458; 2001; ISBN 0-13-013827-4; pages 150→157.
**187** John Joseph Fenton, "Forensic Toxicology," in Stuart H. James and Jon J. Nordby, editors, *Forensic Science / An Introduction to Scientific and Investigative Techniques*, CRC Press, Boca Raton, 2003; ISBN 0-8493-1246-9; Chapter 4. "Atomic Absorption Spectrometry," at pages 56 and 57. www.crcpress.com
**188** John Kenkel, *ANALYTICAL CHEMISTRY REFRESHER MANUAL*, Lewis Publishers, Chelsea, Michigan, 1992, "Atomic Spectroscopy," Chapter 7, pages 167→194.
**189** Walter F. Rowe, "FIREARMS IDENTIFICATION," as Chapter 8 in R. Saferstein, ed., *Forensic Science Handbook* Volume II, Prentice Hall, Englewood Cliffs, New Jersey, 1988, page 391, et seq.; "ELEMENTAL ANALYSIS OF BULLETS AND SHOTGUN PELLETS," at page 447.
**190** http://people.whitman.edu/~dunnivfm/FAASICPMS_Ebook/Downloads/CH2_FINAL.pdf
**191** https://www.spectroscopyonline.com/view/micronutrient-analysis-soil-food-determination-icp-oes
http://www.philadelphia.edu.jo/academics/ajaber/uploads/CH%20 %208%20and%2010%20-Basic%20Principles%20of%20Atomic%20 Absorption%20and%20Atomic%20Emission%20Spectroscopy.pdf
**192** https://chem.libretexts.org/Bookshelves/Analytical_Chemistry/Analytical_Chemistry_2.1_(Harvey)/10%3A_Spectroscopic_Methods/10.04%3A_Atomic_Absorption_Spectroscopy
**193** https://chem.libretexts.org/Bookshelves/Analytical_Chemistry/Analytical_Chemistry_2.1_(Harvey)/10%3A_Spectroscopic_Methods/10.07%3A_Atomic_Emission_Spectroscopy
**194** See: W.D. Kinard and D.R. Lundy, "A Comparison of Neutron Activation Analysis and Atomic Absorption Spectroscopy on Gunshot Residue" as Chapter 11 in Geoffrey Davies, ed., *Forensic Science*, ACS Symposium Series 13, Washington, D.C., 1975; pages 976107.
**195** Walter J. Moore, *Physical Chemistry*, Prentice-Hall, Inc., Englewood Cliffs, New Jersey, 1962 and 1963, page 95, et seq.

196 Melting and boiling points are listed in chemical reference literature for confirmation, or not, of qualitative identification and purity of substances – for example see John A. Dean, ed., *Lang's Handbook of Chemistry*, Twelfth Edition, McGraw-Hill Book Company, New York, . . ., 1979, Tables 7.3 Melting points of organic compounds and 7.4 Physical constants of organic compounds, pages 7.48→7.393.
197 See also: Frederick G. Bordwell, *Organic Chemistry*, The Macmillian Company, New York . . ., 1963, §1.6 Criteria for determining the purity of organic compounds, pages 16→18.
198 Bordwell, page 18.
199 https://www.mathsisfun.com/temperature-conversion.html
200 (And by listening to a near-by cricket – perhaps more fun than actually useful.)
°C = (((chirps /25 seconds) / 3) + 4)
201 Catherine Boeckmann, "PREDICT THE TEMPERATURE WITH CRICKET CHIRPS / LEARN TO USE A CRICKET AS A THERMOMETER!" 24June 2019.
https://www.almanac.com/content/predict-temperature-cricket-chirps
202 Walter J. Moore, *Physical Chemistry*, Prentice-Hall, Inc., Englewood Cliffs, New Jersey, 1962 and 1963, page 102.
203 *The French Connection*, 1971– melting point used to determine purity.
204 Opening scene, *The French Connection*, 1971, where illicit heroin purity was determined by melting point.
205 William Friedkin, *The French Connection*, Gene Hackman, Fernando Rey, Roy Scheider, Tony Lo Bianco, Marcel Bozzuffi; twentieth Century Fox, 1971.
https://en.wikipedia.org/wiki/The_French_Connection_%28film%29
http://www.bluelight.org/vb/archive/index.php/t-461789.html
http://www.chem.ucalgary.ca/courses/351/laboratory/meltingpoint.pdf
206 https://www.chemistry.mcmaster.ca/~chem2o6/labmanual/expt1/exp1b-i.html
207 Other than for water, $H_2O$.
208 https://en.wikipedia.org/wiki/Atmospheric_pressure
209 http://www.csgnetwork.com/h2oboilcalc.html
210 *Mountains of the Moon*, Carolco Pictures, 1990, Directed by Bob Rafelson, with Patrick Bergin and Iain Glen – about the nineteenth-century exploratory travels in Africa of Richard Burton and John Speke, and the discovery of the source of the Nile River, and Lake Victoria; and their personal conflict. In the movie, the boiling point of water is used to determine altitude.
211 https://en.wikipedia.org/wiki/Mountains_of_the_Moon_(film)
212 See also: "Gravity Knife' Led to Thousands of Questionable Arrests. Now It's Legal," *The New-York Times*, 31May 2019.
https://www.nytimes.com/2019/05/31/nyregion/ny-gravity-knife-law.html
213 Some scientific methods – of lesser general application – and with less direct involvement with analytical chemistry – are presented (perhaps somewhat self-indulgently) because they arose from the author's law practice; or other interest.
214 *Criminal Code* RSC., 1985, c. C-46; as amended.
https://laws-lois.justice.gc.ca/eng/acts/C-46/section-84.html

[. . .]

**Definitions**

**84 (1)** In this Part,

[. . .]

*prohibited* weapon means

**(a)** a knife that has a blade that opens automatically by gravity or centrifugal force or by hand pressure applied to a button, spring or other device in or attached to the handle of the knife, or

**(b)** any weapon, other than a firearm, that is prescribed to be a prohibited weapon;

[. . .]

**215** Text from a disclosure request letter – while the letter itself is a good candidate for simply being successfully ignored – a plea bargain beneficial to the accused might still result:

[. . .]

**216** . . . Crown,

[. . .]

Toronto. . . .

RE: *R v* . . .

[. . .]

butterfly knife – centrifugal force?

**217** [. . .]

Our discussions at the judicial pre-trial, . . . July, included several issues, including one that we agreed was of lesser importance. In that context, here is my position on that issue:

For disclosure and evidence, the Crown should provide a scientific or engineering description of how the "butterfly knife," allegedly associated with my client,[1] would be ". . . a knife that has a blade that opens automatically by gravity or centrifugal force or by hand pressure applied to a button, spring or other device in or attached to the handle of the knife . . ." [2]

In doing so, I would suggest that the Crown would have to explain how the "butterfly knife" would be opened by a mechanism or process of *gravitational* or *centrifugal force*, rather than *impulse force*. To do this, I would suggest that the Crown would have to present a proper forensic report, by a properly qualified person, cross-examinable, who had studied physics or engineering, and the relevant English language terminology.

- According to the *OED*[3] *centrifugal* force is a force ". . . with which a body moving round a centre tends to fly off from the centre . . ."
- According to Resnick & Halliday,[4] and Tippler,[5] [6] *centrifugal* force is a *pseudo* or *fictitious* force, from the point of view an observer in a non-inertial frame; and there is a centripetal acceleration vector a directed towards the centre.[7] [8] [9] [10]
- As a general description, consider a mass m, attached to a strong massless string of radius r, rotating about a centre point, with a tangential velocity v (see Fig. 6.42). Thus:

**218**

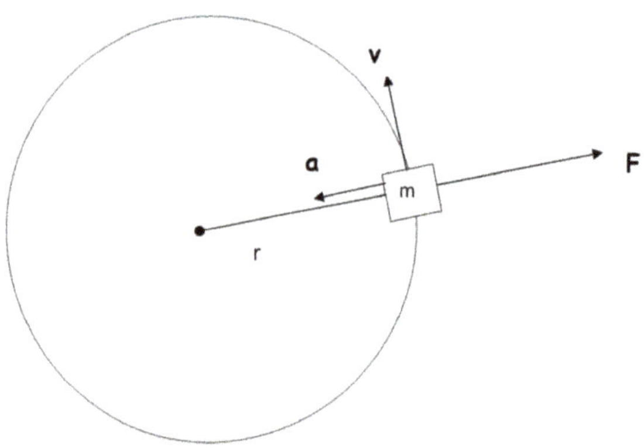


Fig. 6.42: Centrifugal force.

**219** *Centrifugal force* = mass × velocity$^2$ / radius

$F = mv^2/r$ (nt) = (kg)x(m/sec)$^2$/(m)

A *gravitational force*[11 12] would be experienced by a mass as an attraction between its centre of gravity towards that of the earth: [13]

**220** See Fig. 6.43.

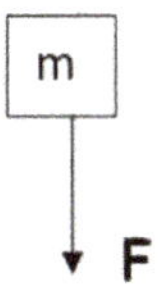


Fig. 6.43: m and F.

**221** $\mathbf{F} = m\,\mathbf{g} = m \times 9.81$ (nt) = (kg) × (m/s$^2$).

- An *impulse force* would be experienced on a moving mass of momentum = $\mathbf{p} = m\,\mathbf{v}$, where there is a change in momentum.[14]

  An *impulse force* would result from a mass moving along a straight line and being stopped- [15] in a short time t:

**222** See Fig. 6.44.

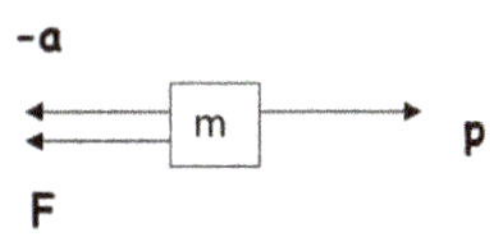


$\mathbf{p} = m\,\mathbf{v}$

$(d\mathbf{p}/dt) = m \times (d\mathbf{v}/dt) = m \times \mathbf{a} = \mathbf{F}$ (kg) x (m/sec)/(sec) = (nt).

Fig. 6.44: P = mv.

**223** – Other possible forces could be those of a *coil spring* [16 17 18] or of an *expanding gas* [19 20 21] in a piston, but these would be most unlikely to be of consideration here.

The Crown seems to have suggested that a police officer opening a knife with a flick would be sufficient to demonstrate *centrifugal* force. I would not agree. And that a flick would not necessarily be *centrifugal force* would be at issue before the court.

The defence position is that if the Crown is to eventually persuade the court that the "butterfly knife" is a "prohibited weapon," within the meaning of the *Criminal Code*, it must do so with a consideration of the actual physics and engineering meaning of the English language words "centrifugal," "pseudo" and "fictitious." And it should do so with a witness who would have done a properly documented forensic study in a quality assured environment such as for ISO/IEC 17025: 2005.[22, 23]

I look forward to your comments and corrections to what I have suggested above. Thank You.

M . . . G . . .,

Barrister & Solicitor.
Toronto,
. . . July 2009.

[. . .]

**224** 1 From the *information* copied in the disclosure: ". . . did have in . . . possession a prohibited weapon . . . butterfly knife . . . *Criminal Code*." . . .

2 *Criminal Code*, PART III, FIREARMS AND OTHER WEAPONS, Interpretation, Definitions, s.84. (1) (a) "prohibited weapon."
http://laws.justice.gc.ca/en/index.html http://laws.justice.gc.ca/en/C-46

3 *The Oxford English Dictionary*, 2nd edition, Clarendon Press, Oxford, 1989, Vol. II, page 1038.

4 Robert Resnick and David Halliday, *Physics*, Part I, John Wiley & Sons, Inc., New York, 1966; page 121; and with reference to pages 64→71, 120.

5 Paul A. Tipler, *Physics*, 2nd edition, Worth Publishers, Inc., New York, 1982; pages 147 & 148; and with reference to pages 67→70.

6 Paul A. Tipler, *Physics*, Worth Publishers, Inc., New York, 1976; page 159, *et seq.*; and with reference to page 65, *et seq.*

7 Resnick & Halliday, 1966, page 66.

8 Tipler, 1982, page 148.

9 $\mathbf{a} = \mathbf{v}^2/r$ Tipler, 1976, page 66.

10 The convention of the physics texts is used here that lighter typeface is used for scalar quantities – such as mass m – and heavier typeface is used for vectors – such as velocity **v**. The arrow → shows the direction of the vector.

11 Resnick & Halliday, 1966, page 93.

12 Tipler, 1976, page 144, *et seq.*

13 **g** = the acceleration of gravity, to a good approximation, on earth. ( $m/sec^2$)

14 Resnick and Halliday, 1966, page 191, *et seq.* and page 212, *et seq.*

15 – here it is –**a** for deceleration. Alternatively, an impulse force could result from a mass at rest accelerating for + **a**.

16 Paul A. Tipler, *Physics*, Worth Publishers, Inc., New York, 1976; pages 154, *et seq.*

17 $\mathbf{F} = k \times x\ \Delta x\ (nt) = (nt/m) \times (m)$;
where k is Hooke's law constant and Δx is the spring's stretch.

18 Resnick and Halliday, 1966, pages 138 and 151.

19 $P = \mathbf{F}/A = (nRT/V)$

20 Tipler, 1976, page 423, *et seq.*

21 Resnick and Halliday, 1966, page 572, *et seq.*

22 http://www.standardsstore.ca/eSpecs/index.jsp

23 http://www.standardsstore.ca/eSpecs/SearchFormAction.do?page=1&searchValue=17025&searchKey=DOCNO&organizations=ISO

**225** $CO_2$ *carbon dioxide* CAS 124-38-9 **O** = **C** = **O** 44.01 g/mol (see Fig. 6.45).

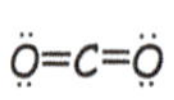

Fig. 6.45: $CO_2$.

https://en.wikipedia.org/wiki/Carbon_dioxide
https://imgbin.com/png/c6zWV7VV/lewis-structure-carbon-dioxide-resonance-diagram-electron-png.

**226** Toronto Workers' Health & Safety Legal Clinic *newsletter* 2012 04. Vol.20 No.1 $CO_2$ Filsingers page 1.

**227** Toronto Workers' Health & Safety Legal Clinic *newsletter* 2012 09. Vol20 No5 $CO_2$ Filsingers page 6.

**228** Toronto Workers' Health & Safety Legal Clinic *newsletter*
2007 01. Vol.15 No.1 page 3. 2007 11. Vol.15 No.4 The BC Sullivan Mine disaster.

**229** Toronto Workers' Health & Safety Legal Clinic *newsletter* 2012 07. Vol20 No3 $CO_2$
**230** Toronto Workers' Health & Safety Legal Clinic *newsletter* 2012 08. Vol20 No4
**231** https://en.wikipedia.org/wiki/Carbon_dioxide_sensor
https://en.wikipedia.org/wiki/Carbon_monoxide_detector
https://www.vaisala.com/sites/default/files/documents/VIM-G-How-to-measure-CO2-Application-Note-B211228EN.pdf
https://www.co2meter.com/
https://ohsonline.com/articles/2016/04/01/carbon-dioxide-detection-and-indoor-air-quality-control.aspx
**232** **CO,** *carbon monoxide*, CAS 630–08-0 28.01 g/mol (see Fig. 6.46).
https://en.wikipedia.org/wiki/Carbon_monoxide

:C≡O: Fig. 6.46: CO.

**233** See: Giuseppe Lippi and Liam M. Heaney, "The 'olfactory fingerprint': can diagnostics be improved by combining canine and digital noses?" Clin Chem Lab Med 2020; 58(6): 958–967; Opinion Paper; DE GRUYTER
https://www.degruyter.com/view/journals/cclm/58/6/article-p958.xml
**234** See: A. Maureen Rouhi, "Detecting Illegal Substances / Unlocking the Secrets of Supersniffing Dogs," special report, *C&EN*, 29 September 1997, pages 24→27.
**235** But see: http://www.mapinc.org/drugnews/v97/n135/a02.html?6580
https://www.sfgate.com/news/article/suit-filed-over-drug-searches-school-officials-3506035.php
**236** See also: DRUG DETECTION DOGS, 255 North El Cielo Road, Suite 205, Palm Springs, California, 92262; 800 627 2364 (NARC DOG) http://www.pe.net/~narcdog/ http://search.yahoo.com/bin/search?p=drug+dogs
**237** Mark Derr, "With Dog Detectives, Mistakes Can Happen," *The New York Times* / Science Times, 24 December 2002, pages D1 and D4.
**238** See: Patricia Marx, "Sit, Stay, Fight Cybercrime / *A yellow lab named Hannah belongs to a new group of police dogs trained to catch child pornographers by sniffing out electronics* / Man's Best Friend / SNIFF PATROL," *The New Yorker*, 11 November 2019, pages 25 and 26. https://www.newyorker.com/magazine/2019/11/11/sit-stay-fight-cybercrime
**239** See: Adam Iscoe, "MAN'S BEST FRIEND / *THE SMELL TEST*," *The New Yorker*, 01 March 2021, pages 16 and 17.
https://www.newyorker.com/magazine/2019/11/11/sit-stay-fight-cybercrime
**240** Cat Warren, "Sniffing Out New (Old) Digs / Recent research highlights how the power of the canine nose allows cadaver dogs to uncover buried remains from ancient human history," *The New York Times*, Science Times, 19 May 2020, pages D1 and D5.
**241** John M. Sweeten, "Odor Measurement and Control For the Swine Industry: Recent Developments," *Journal of Environmental Health*, Vol. 50, No. 5, March/April 1988, pages 282 →286. https://www.jstor.org/stable/44541195
**242** John E. Amoore, *MOLECULAR BASIS OF ODOR*, Charles C. Thomas, Publisher, Springfield, Illinois; 1970. https://pubs.acs.org/doi/pdf/10.1021/ed048pA728.2?rand=3kr671oi
**243** R. H. Wright, *The Science of Smell*, Basic Books, Inc., New York, 1964. https://pubs.acs.org/doi/pdf/10.1021/ed042p693.4
**244** *The Thin Man*, Metro-Goldwyn-Mayer, 1934; Skippy as Asta; and William Powell & Myrna Loy as Nick & Nora Charles. https://en.wikipedia.org/wiki/The_Thin_Man_(film)
https://web.archive.org/web/20170212221218/ http://www.thethinman1934.com/thethinmanposters.html
**245** *The Thin Man* Script [Nick Charles to Asta]
http://www.script-o-rama.com/movie_scripts/t/thin-man-script-transcript-nick.html

**246** *The Thin Man*, TV series,1957→1959, Peter Lawford and Phyllis Kirk. https://en.wikipedia.org/wiki/The_Thin_Man_(TV_series)
**247** Dashiell Hammett, *The Thin Man*, detective novel, 1934. https://en.wikipedia.org/wiki/The_Thin_Man
**248** See also: P.D. James, *Talking About Detective Fiction*, Chapter 4, "Soft-centred and Hard-boiled," pages 82→86; Vintage Books, New York City, 2011, 2009. ISBN 978-0-307-74313-8.
**249** https://en.wikipedia.org/wiki/Skippy_(dog)
**250** Arthur C. Clarke, *TALES OF TEN WORLDS*, "DOG STAR," DELL, New York City, 1962, 1964; pages 161→167.
https://en.wikipedia.org/wiki/Dog_Star_(short_story)
"MOON DOG," *GALAXY MAGAZINE*, Vol. 20, No. 4, New York City, April 1962, pages 188→194.
https://archive.org/details/Galaxy_v20n04_1962-04/page/n95/mode/1up?view=theater
**251** Perhaps earth seismic activity also releases odours.
**252** Aimee Ortiz,*"Man Who Kicked Customs Dog Is Turned Away at D.C. Airport, U.S. Says* / The man, who was traveling on a tourist visa, kicked the dog, a beagle named Freddie who had detected food in his luggage, so hard that he was lifted off the ground, court records say," *New York Times*, 27 June 2025. https://www.nytimes.com/2025/06/27/us/dulles-airport-customs-dog-kicked.html
**253** Linda Qiu, "*Meet the Canine Officers Guarding American Agriculture* / With international travel returning to prepandemic levels, these floppy-eared hounds are sniffing out deadly diseases and pests at border crossings one mango or sausage at a time," *New York Times*, 23 July 2022. https://www.nytimes.com/2022/07/23/us/politics/beagles-airport-security.html
**254** https://www.sciencedirect.com/science/article/abs/pii/S0956566307002618
**255** Nicholas Kristof, "The Giant Rats That Save Lives." *New York Times*, 19 Apr. 2015
**256** Health and Safety Programs, "Scent-Free Policy for the Workplace" Canadian Centre for Occupational Health and Safety, Hamilton, Ontario, 29 August 2024.
https://www.ccohs.ca/oshanswers/hsprograms/scent_free.html
https://www.ccohs.ca/products/posters/scentfree

[. . .]

> **Disclaimer**
> Although every effort is made to ensure the accuracy, currency and completeness of the information, CCOHS does not guarantee, warrant, represent or undertake that the information provided is correct, accurate or current. CCOHS is not liable for any loss, claim, or demand arising directly or indirectly from any use or reliance upon the information.

[. . .]

**257** CBC Ideas https://www.cbc.ca/listen/live-radio/1-23-ideas/clip/16072263-smell-why-invisible-superpower-deserves-attention
**258** And, also fictional literature for human detection of odour. Carolyn Keene, *THE NANCY DREW FILES* ™ / *CASE 54* / *COLD AS ICE*, POCKET BOOKS, New York City, . . ., 1990; ISBN 0-671-70031-6; at page 80.
**259** William Glaberson, "By Helping a Girl Testify at a Rape Trial, a Dog Ignites a Legal Debate," *The New York Times*, 8 August 2011.
https://www.nytimes.com/2011/08/09/nyregion/dog-helps-rape-victim-15-testify.html
**260** Richard Saferstein, *CRIMINALISTICS / An Introduction to Forensic Science*, Seventh Edition, Prentice Hall, Upper Saddle River, New Jersey, 07458; 2001; ISBN 0-13-013827-4; "The Microscope," Chapter 7, [electron microscope] pages 174→178.
**261** Thomas A. Kubic and Nicholas Petraco, "Microanalysis and Examination of Trace Evidence," in Stuart H. James and Jon J. Nordby, editors, *Forensic Science / An Introduction to Scientific and Investigative Techniques*, CRC Press, Boca Raton, 2003; ISBN 0-8493-1246-9; Chapter 14, [electron microscope] at pages 259 and 260. www.crcpress.com

**262** https://www.umassmed.edu/cemf/whatisem/#:~:text=Electron%20microscopy%20(EM)%20is%20a, cells%2C%20organelles%20and%20macromolecular%20complexes
https://www.sciencedirect.com/topics/materials-science/electron-microscopy
https://en.wikipedia.org/wiki/Electron_microscope
**263** See John Sabo, Gary Judd and Stark Ferriss, "Examples of SEM Analyses in Forensic Evidence Applications," as Chapter 8 in Geoffrey Davies, ed., *Forensic Science*, ACS Symposium Series 13, Washington, D.C., 1975; pages 75→82.
**264** with reference, by analogy, to engineering plans printed by a now-obsolete method. Not actually a two-dimensional drawing, nor blue (although a bygone process did give white line copies on blue background.)
**265** https://www.merriam-webster.com/dictionary/blueprint
https://en.wikipedia.org/wiki/Blueprint
**266** See: James D. Watson, *Molecular Biology of the Gene, Second Edition*, W.A. Benjamin, Inc., New York City 10016, 1970; [ISBN] 8053-9603-9; page 330.
**267** and sometimes with other aspects – such as charge location.
**268** See: Jeffrey L. Cleland, EDITOR, *Protein Folding / In Vivo and In Vitro*, ACS SYMPOSIUN SERIES 526, 1993. ISSN 0-8412-2640-7.
**269** Watson, *Second Edition*, pages 634 and 86.
**270** Watson, *Second Edition*, pages 86 and 89.
**271** Watson, *Second Edition*, pages 86 and 91.
**272** Watson, *Second Edition*, pages 629 and 633.
**273** See below:
**274** Suzanne Bell & John M. Butler, ***Understanding Forensic DNA***, Cambridge University Press, Cambridge, CB2 8EA; 2022. ISBN 978-1-009-04401-1.
https://www.cambridge.org/core/books/abs/understanding-forensic-dna/copyright-page/90C34E001E6CAA37F3A38AD8DDEA4CD9
https://www.cambridge.org/
https://www.lehmanns.de/shop/mathematik-informatik/59190146-9781009044011-understanding-forensic-dna
**275** See also: Jeffrey S. Rosenthal, *Struck by Lightning / The Curious World of Probabilities*; Harper Perennial / HarperCollins Publishers Limited, Toronto M4W 1A8; 2005, 2006; Chapter 2, at page 7, "*What Are the Odds of That? / Coincidence and surprise*. Chapter 8, at pages 130, 131, 132, "Ain't Gonna Happen / Very Low Probabilities." ISBN 13: 978-0-00-639495-2 ISBN 10: 0-00-639495-7 https://probability.ca/sbl/
**276** AmpFℓSTR™ Identifiler™ Plus PCR / Amplification Kit / USER GUIDE, Catalog Number 4427368, Publication Number 4440211, Revision H, Forensic or Paternity. Thermo Fisher Scientific, Warrington, Cheshire, WA1 4SR, United Kingdom. 19 September 2018.
https://assets.thermofisher.com/TFS-Assets/LSG/manuals/4440211_AmpFlSTR_IdentifilerPlus_UG.pdf
**277** Susan Herrero, "Legal Issues in Forensic DNA," in Stuart H. James and Jon J. Nordby, editors, *Forensic Science / An Introduction to Scientific and Investigative Techniques*, CRC Press, Boca Raton, 2003; ISBN 0-8493-1246-9; Chapter 32; pages 581→595. www.crcpress.com
**278** "DNA Advisory Board (DAB) Guidelines," in James and Nordby, *Forensic Science . . .*; 2003; Appendix A; pages 597→607.
**279** Richard Saferstein, *CRIMINALISTICS / An Introduction to Forensic Science*, Seventh Edition, Prentice Hall, Upper Saddle River, New Jersey, 07458; 2001; ISBN 0-13-013827-4; "DNA: A New Forensic Science Tool"; Chapter 13, pages 353→394.
**280** Deoxyribonucleic acid. See also:

- Michael L. Baird, "Analysis of Forensic DNA Samples by Single Locus VNTR Probes" as Chapter 3 in Mark A. Farley and James J. Harrington, eds., *Forensic DNA Technology*, Lewis Publishers, Inc., Chelsea, Michigan, 1991, pages 39→49.

See also:
- *C&EN* 20 Nov. 1989, pages 18630; *special report DNA Profiling.*

See also:
- Eric S. Lander, "Invited Editorial: Research on DNA Typing Catching Up with Courtroom Application," *American Journal of Human Genetics*, 48, pages 8196823, 1991;
- Alec J. Jeffreys, Michelle Turner, and Paul Debenham, "The Efficiency of Multilocus DNA Fingerprint Probes for Individualization and Establishment of Family Relationships, Determined from Extensive Casework," *ibid.*, 8246840;
- B. Budowle, A.M. Giusti, J.S. Waye, F.S. Baechtel, R.M. Fourney, Dwigth E. Adams, Lawrence A. Presley, Harold A. Deadman and K.L. Monson, "Fixed-Bin Analysis for Statistical Evaluation of Continuous Distributions of Allelic Data for VNTR Loci, for Use in Forensic Comparisons," *ibid.*, pages 841→855.
- Bruce Budowle and Alan M. Giusti, "Fixed Bin Frequency Distributions for the VNTR Locus D5S110 in General United States Reference Databases," *Journal of Forensic Sciences*, Vol.40, No.2, March 1995, pages236→238;
- Major Nu En Huang and Bruce Budowle, "Fixed Bin Population Data for the VNTR Loci D1S7, D2S44, D4S139, D5S110, and D17S79 in Chinese from Taiwan," *ibid.*, pages 287→290.

**281** See, for example:
- Gina Kolata, New York Times Service, "How will FBI scientists analyse [Monica] Lewinsky's [blue GAP] dress? / *DNA analysis can often be 'better proof than most of us ever see'*," *The Globe and Mail*, 10 Aug. 1998, page A10.
- Jane Mayer, "ASK MR. SCIENCE / *How would a crime laboratory test a garment* [Monica Lewinsky's dress] *for traces of DNA?*", *The New Yorker*, 10 August 1998, page 27.

**282** See also: M.G., "DNA data collection and Canadian labour law . . ., " TORONTO WORKERS' HEALTH & SAFETY LEGAL CLINIC *newsletter*, Vol. 18, No.3, July 2010, pages 10→13.
http://newsletter.workers-safety.ca.s3-website-us-east-1.amazonaws.com/#!/newsletters%2F2010%2007%2F

**283** Deoxyribonucleic acid. For a (dated) survey of this topic see Mark A. Farley and James J. Harrington, eds., *Forensic DNA Technology*, Lewis Publishers, Inc., Chelsea, Michigan, 1991. See also *C&EN* 20 Nov. 1989, pages 18→30; *special report DNA Profiling.*

**284** Nicholas Wade, "Technology Lowers Cost of Decoding a Genome to $50,000 / A $1,000 genome might be possible in two or three years . . ., " *The New York Times*, 11 August 2009, page D3.

**285** "ILLUMINA OFFERS GENOME SERVIVE," *Chemical & Engineering News*, 20 July 2009, page 48; American Chemical Society, Washington, D.C.: "Illumina has launched a personal genome sequencing service for consumers at a price of $48,000 . . ."

**286** Celia Henry Arnaud (Senior Editor)], "CLOSING IN ON PERSONAL GENOME SEQUENCING," *C&EN*, 17 August 2009, page 38; ACS, Washington, D.C.

**287** See also: "Personalized Genomics: Hope or Hype," 2009 Michael Smith Memorial Nobel Forum, CBC *Ideas*, Thursday 24 September 2009. http://www.cbc.ca/ideas/

**288** See also: Steven Greenhouse, "Law Seeks to Ban Misuse of Genetic Testing / Family histories of fatal diseases have been used against workers . . ., " *The New York Times*, 16 November 2009, page B5.

**289** See also: [a commercial full-page newspaper advertisement; cited here as an example of purported availability, not an endorsement], "NEW SCREENING TEST FOR CANCER / Get your DNA tested. /

Genetic screening tests for over 100 diseases / Take control of your health at the DNA Testing Centres of Canada," *Toronto Star*, 07 February 2010, page A14. www.dnatestingcanada.com
**290** See, for example: Andrew J. Campa, "23andMe bankruptcy leaves control of genetic data from millions in limbo," *Los Angeles Times*. 30 March 2025.
https://www.latimes.com/california/newsletter/2025-03-30/23andme-bankruptcy-leaves-control-of-genetic-data-from-millions-up-to-highest-bidder-essential-california
**291** https://www.nytimes.com/2025/06/10/business/23andme-data-lawsuit.html
**292** An historical example of the use of DNA identification by bio-sample of a relative is in the probable identification of the remains of Czar Nicholas II and family. That identification was aided by comparison with DNA data of a grandnephew of the Czar's wife, Alexandra Feodorovna, Empress of Russia. The grandnephew: Prince Philip, Duke of Edinburgh.
**293** https://en.wikipedia.org/wiki/House_of_Romanov#:~:text=Nicholas%20II%2C%20the%20last%20Emperor,members%20of%20the%20imperial%20house
**294** https://www.smithsonianmag.com/smart-news/dna-analysis-confirms-authenticity-remains-attributed-romanovs-180969674/
**295** Privacy and Access to Genetic Research Data in the Biobanking Context / Report on workshop held at the Faculty of Law University of Toronto, Canada / March 13, 2009
http://www.law.utoronto.ca/healthlaw/docs/PrivacyWorkshopReport09_Final.pdf
http://www.law.utoronto.ca/healthlaw/
**296** See, for example: Isabelle Taft, ***"A DNA Technique Is Finding Women Who Left Their Babies for Dead /***
Genetic genealogy is identifying the mothers of deceased newborns found abandoned, shedding light on crimes that went unsolved for years. Women now may face lengthy prison sentences for decades-old chapters of their pasts," *New York Times*, 03June, 2025. https://www.nytimes.com/2025/06/03/us/forensic-genetic-geneology-dna-babies.html
**297** Perhaps, a more diplomatic phrasing – that allows for innocent witnesses.
**298** See: John Kenkel, *Analytical chemistry refresher manual*, Lewis Publishers, Chelsea, Michigan, 1992, Chapter 8.7, ELECTROPHORESIS, pages 220 →224.
**299** Richard Saferstein, *CRIMINALISTICS / An Introduction to Forensic Science*, Seventh Edition, Prentice Hall, Upper Saddle River, New Jersey, 07458; 2001; ISBN 0-13-013827-4; Chapter 13, at page 372.
**300** For example see: AmpFℓSTR™ . . . cited above.
**301** See: Bell & Butler, *Understanding Forensic DNA*, . . ., "The Curse of Sensitivity," as Chapter 6, pages 93→108.
https://www.cambridge.org/core/books/abs/understanding-forensic-dna/copyright-page/90C34E001E6CAA37F3A38AD8DDEA4CD9
https://www.cambridge.org/
https://www.lehmanns.de/shop/mathematik-informatik/59190146-9781009044011-understanding-forensic-dna
**302** https://www.si.edu/object/vitullo-evidence-collection-kit-sexual-assault-examination%3Anmah_2009657
https://www.cbc.ca/listen/live-radio/1-57-the-sunday-magazine/clip/16148315-the-sunday-magazine-may-25-2025
**303** See, for example: Amelia Nierenberg, "*Dig for Children's Remains Begins at Irish Home for Unwed Mothers* / Experts are searching for the remains of hundreds of children who died at the institution run by Catholic nuns until 1961, including bodies disposed of in septic tanks," *New York Times*, 14 July 2025.
https://www.nytimes.com/2025/07/14/world/europe/ireland-excavation-children-tuam.html

**304** Damien Cave, "*How New DNA Science Could Help More Families of the Missing* / Emerging methods are improving the ability to identify even highly degraded human remains," *New York Times*, 4 July 2025.
https://www.nytimes.com/2025/07/04/world/asia/dna-sequencing-identification.html
**305** See: Peter Atkins, *Concepts in Physical Chemistry*, W.H. Freeman and Company, New York City 10010; 1995; pages 107→109; ISBN 0-7167-2928-8.
**306** https://en.wikipedia.org/wiki/Electron_paramagnetic_resonance https://en.wikipedia.org/wiki/Unpaired_electron
**307** See: Anne F. Skinner and Mark N. Rudolph, "Dating Flint Artifacts With Electron Spin Resonance: Problems and Prospects," as Chapter 4; page 37, et seq.; in Mary Virginia Orna, ed., *Archaeological Chemistry / Organic, Inorganic Biochemical Analysis*, ACS Symposium Series 625, American Chemical Society, Washington, D.C., 1996; and references cited therein.
**308** https://en.wikipedia.org/wiki/Electron_spin_resonance_dating
**309** https://geology.com/rocks/flint.shtml
**310** "Fluorescence and Phosphorescence," LibreTexts, 05 June 2019 https://chem.libretexts.org/
https://chem.libretexts.org/Bookshelves/Physical_and_Theoretical_Chemistry_Textbook_Maps/Supplemental_Modules_(Physical_and_Theoretical_Chemistry)/Spectroscopy/Electronic_Spectroscopy/Fluorescence_and_Phosphorescence
**311** Stephen Gregory Schulman, *Fluorescence and Phosphorescence Spectroscopy:* ***Physicochemical Principles and Practice***, Elsevier Science & Technology, 1977, 288 pages. Page 183.
https://books.google.ca/books/about/Fluorescence_and_Phosphorescence_Spectro.html?id=a5ZY_ilwSkcC&redir_esc=y
**312** M.G., A Survey of the photochemistry of the *ortho*-fused bebzocycloalkenes, UNB thesis, 1972. page 17.
*ProQuest Dissertations and Theses;* 1972; ProQuest Dissertations and Theses Global UMI Number: DC54154
**313** C.A. Parker, *Photoluminescence of Solutions/With Applications to Photochemistry and Analytical Chemistry*, ELSEVIER, Amsterdam . . ., 1968; page 43.
**314** M.G., A Survey of the photochemistry of the *ortho*-fused bebzocycloalkenes, UNB thesis, 1972. page 29.
**315** C.A. Parker, Photoluminescence of Solutions/With Applications to Photochemistry and Analytical *Chemistry*, ELSEVIER, Amsterdam . . ., 1968; page 37.
**316** C.A. Parker, *Photoluminescence of Solutions/With Applications to Photochemistry and Analytical Chemistry*, ELSEVIER, Amsterdam . . ., 1968; page 38.
**317** C.A. Parker, *Photoluminescence of Solutions/With Applications to Photochemistry and Analytical Chemistry*, ELSEVIER, Amsterdam . . ., 1968; page 38.
**318** *Orthofused-benzocycloalkene*, -hexene, -pentene, -alkene Chapters 6.2.1, 6.15.1
**319** C.A. Parker, *Photoluminescence of Solutions/With Applications to Photochemistry and Analytical Chemistry*, ELSEVIER, Amsterdam . . ., 1968; page 43.
**320** See also: https://www.sciencedirect.com/topics/physics-and-astronomy/luminescence
**321** – showing here a diffraction grating rather than a prism.
**322** See: Hazel Rossotti, *COLOUR/Why The World Isn't Grey*, Princeton University Press, Princeton, New Jersey, 08540; 1983; ISBN 0-691-02386-7; at page 61.
https://press.princeton.edu/books/paperback/9780691023861/colour
**323** See: Isadore B. Berlman, *Handbook of Fluorescence Spectra of Aromatic Molecules*, Second Edition, Elsevier Inc.; 1971; ISBN 978-0-12-092656-5.
https://www.sciencedirect.com/book/9780120926565/handbook-of-fluorescence-spectra-of-aromatic-molecules#book-description

**324** M.G., A survey of the photochemistry of the *ortho*-fused bebzocycloalkenes, UNB thesis, 1972.
**325** Page 26.
**326** Of interest here is luminescence from exuded bio-substances, and from other substances used to develop them; but most developers would be for visualization of colour change, or by blackening, to contrast with background.
**327** Richard Saferstein, *CRIMINALISTICS / An Introduction to Forensic Science*, Seventh Edition, Prentice Hall, Upper Saddle River, New Jersey, 07458; 2001; ISBN 0-13-013827-4; pages 395→420.
**328** Robert E. Gaensslen and Kimberly R. Young, "Fingerprints," in Stuart H. James and Jon J. Nordby, editors, *Forensic Science / An Introduction to Scientific and Investigative Techniques*, CRC Press, Boca Raton, 2003; ISBN 0-8493-1246-9; Chapter 15, pages 277→296. www.crcpress.com
**329** Brian H. Kaye, *Science and the Detective / Selected Reading in Forensic Science*, VCH Verlagsgesellschaft mbH, Weinheim, Germany; 1995; ISBN-3-527-29252-7; Chapter 2, "Fingerprints," pages 13→44.
https://pubs.rsc.org/en/content/articlelanding/1996/an/an996210005n/unauth#!divAbstract
**330** See: Peter F. Jones, "New Applications of Photoluminescence Techniques for Forensic Science," in Geoffrey Davies, ed., *Forensic Science*, ACS Symposium Series 13, American Chemical Society, Washington, D.C., 1975, Chapter 19, pages 1836196, and references cited therein.
**331** 1,2-*Indanedione* CAS 16214-27-0 146.15 g/mol. $C_9H_6O_2$
**332** See Fig. 6.47.

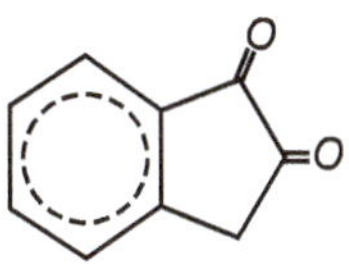

Fig. 6.47: CAS 16214-27-0.

**333** Joseph Sloop and Jonathan Yi, "Derivatives of 1,2-Indanedione: Synthesis and Latent Fingerprint Development Efficacy Studies," April 2013, DOI: 10.13140/RG.2.1.3947.3768
GGC Undergraduate Research Symposium, Georgia Gwinnett College, Lawrenceville, Georgia.
https://www.researchgate.net/publication/277329456_Derivatives_of_12-Indanedione_Synthesis_and_Latent_Fingerprint_Development_Efficacy_Studies
**334** https://www.sciencedirect.com/topics/chemistry/ninhydrin#:~:text=Ninhydrin%20is%20a%20chemical%20reagent,product%20known%20as%20Ruhemann's%20purple
**335** J. Brent Friesen, "Forensic Chemistry: The Revelation of Latent Fingerprints," *J. Chem. Educ.* 2015, 92, 3, 497→504. https://pubs.acs.org/doi/10.1021/ed400597u
**336** http://www.cbdiai.org/Reagents/indane.html
https://www.fbi.gov/about-us/lab/forensic-science-communications/fsc/oct2002/kasper.htm
http://www.bvda.com/EN/prdctinf/en_ind_1.html
http://www.bvda.com/EN/popup/inds_str.html
http://www.chemblink.com/MSDS/16214-27-0_MSDS.htm
http://www.chemblink.com/MSDS/MSDSFiles/16214-27-0_Sigma-Aldrich.pdf https://en.wikipedia.org/wiki/1,3-Indandione
**337** See: F.C.G Souter, C. Van Netten, and R. Brands, "Morbidity in Policemen occupationally exposed to fingerprint powders," *Int.J.Environ.Health Res.*, 1992, 2(2), pages 114-119 (abstracted in *CA Selects/Forensic Chemistry*, ACS, CAS, Issue 7, 05 April 1993, page 6, 118: 108770 v, Columbus 43210-0012.
**338** https://en.wikipedia.org/wiki/Ninhydrin
**339** http://www.pioneerforensics.com/products-1.html
http://nebula.wsimg.com/324667d85cf9a612309bc8f15c71a84c?AccessKeyId = 5FB98090FD4236E549F3&disposition = 0&alloworigin = 1
https://en.wikipedia.org/wiki/1,3-Indandione

**340** The physics of the forces on a decelerating automobile is introduced by Robert Resnick and David Halliday, *Physics / PART I*, John Wiley and Sons, Inc., New York, . . ., 1966; pages 115 and 116 (Example 2).
**341** M.G., "Puzzled by the Law?" AMERICAN CHEMICAL SOCIETY Division of Chemistry and the Law *newsletter*, Volume 21, No. 1, Spring 2005, pages 16 and 17.
**342** https://www.chemistryandthelaw.org/
https://images.magnetmail.net/images/clients/ACS1_D/attach/Fall_2018_NL.pdf
**343** To derive the equation: car speed = 15.946 × (drag constant × skid distance)$^{1/2}$ (km/h)
This would be usable to calculate a car's speed from its visible tire skid marks, and an experimentally found drag constant:

When the wheels lock and the skid starts – leaving a mark on the horizontal road surface – the frictional drag of the tires on the road will cause the car to decelerate in a straight line.
The constant frictional drag force = – f = m × a = (w/g) × a [equation ①]
where
f = frictional drag force
m = car mass
a = acceleration (the car's deceleration = –a)
w = car weight
g = acceleration of gravity = 9.81 m/s$^2$

Therefore: the car's deceleration = – a = – f × g/w [equation ②] * **

In general, an accelerating body, here the car, may be described by: $v_t = v_0 + (a \times t)$ [equation ③]
Where
$v_0$ = initial velocity
$v_t$ = velocity at time t
t = time

When the wheels stop rotating, and the car just starts into the skid, its velocity, which is sought to be calculated here = $v_0$.

The velocity when the car skids to a stop = velocity at time t = $v_t$ = 0. Thus, $0 = v_0 + (-a \times t)$ [equation ④]

Combining equations ② and ④: $v_0 = f\,g\,t / w$ [equation ⑤]

During the time it takes to stop, the car will have an average velocity
= ½ ($v_0 + v_t$) = ½ ($v_0$ + 0) = $v_t$ / 2 [equation ⑥]

In general: average velocity = distance (metres) / time (seconds) = length of skid / t [equation ⑦]

Combining equations ⑥ and ⑦: length of skid / t = $v_0$ / 2
∴ t = 2 × length of skid / $v_0$ [equation ⑧]

Combining equations ⑤ and ⑧: $v_0$ = (f ×g × 2 × length of skid / w)$^{1/2}$
$v_0$ = (f × 9.81 × 2 ×length of skid / w)$^{1/2}$ [equation ⑨]

The frictional drag force results from the friction between the tires and the road, and would be expected to be a constant, here called the drag constant = f / w [equation ⑩]

This is a property between two material surfaces pressed against each other and may be measured by separate experiment. Thus combining equations ⑨ and ⑩: the velocity of the car when it started into the skid

= $(2 \times 9.81 \times \text{drag constant} \times \text{length of skid})^{1/2}$ (metres/second) [equation ⑪]

= $(2 \times 9.81 \times \text{drag constant} \times \text{length of skid})^{1/2} \times (60 \times 60/1000)$ (km/hr) [equation ⑫]

= $15.946\,(\text{drag constant} \times \text{length of skid})^{1/2}$ (km/hr) [equation ⑬] ***

**344** * See Robert Resnick and David Halliday, *Physics*/Part I, John Wiley and Sons, Inc., New York, 1966,

Chapter 16–5, page 395.

**345** ** *Ibid.*, Chapter 6–2, page 115.

**346** *** Some general references:

- http://www.harristechnical.com/skid11.htm
- http://www.frborg-gymhf.dk/eh/opg/fys/kollision.htm
- http://www.accidentreconstruction.com/index.html
- Randall K. Noon, "Vehicular Accident Reconstruction" in Stuart H. James and Jon J. Nordby, editors,
  *Forensic Science/An Introduction to Scientific and Investigative Techniques*, CRC Press, Boca Raton, 2003; ISBN 0-8493-1246-9; page 436, et seq. www.crcpress.com

**347** https://en.wikipedia.org/wiki/Gunter%27s_chain

**348** Car speed = $15.946 \times (\text{drag constant} \times \text{skid distance})^{1/2}$ (km/h).

**349** See also:

https://www.drivesmartbc.ca/police/speed-skidmarks

http://accidentexpert.com/wp-content/uploads/2016/01/Skid-Article-PDF.pdf

https://goodcalculators.com/skid-speed-calculator/

https://itstillruns.com/calculate-skid-speed-5842879.html

**350** However, it might remain useful as a back-up check on computer malfunction, error or fraud.

**351** $\mathbf{F} = m \times \mathbf{a}$ [**Bold face** here indicates vector quantities.]

**352** In Earth's gravitational field mass is human-experienced as weight: $\mathbf{w} = m \times \mathbf{g}$.

**353** See: Robert Resnick and David Halliday, *Physics*, Part I, John Wiley & Sons, Inc., New York, 1966; page 86.

**354** *Controlled Drugs & Substances Act* SC 1996, c.19. http://laws-lois.justice.gc.ca/eng/acts/C-38.8/index.html

**355** Data the author has sought – in capacity as legal defence counsel – for clients' alleged drug offences.

**356** For *CD&S Act* – federal legislation – this is the federal Crown – Ontario Regional Office, Public Prosecution Service of Canada, Toronto.

**357** The author would offer – in capacity as legal defence counsel – for clients' alleged drug offences. The author knows of no other Defence counsel who do this.

**358** No DAS forensic data are sought here; only Police forensic data are sought.
**359** See notes below.
**360** (The device actually used by Police in this case would likely, very generally, appear as described by Dr. Andy Connelly, *The laboratory balance: a practical guide*, 2017. https://andyjconnelly.wordpress.com/2017/02/12/the-laboratory-balance-a-practical-guide/) (see Fig. 6.48).
**361** Creative Commons Attribution-ShareAlike 4.0 International License
**362**

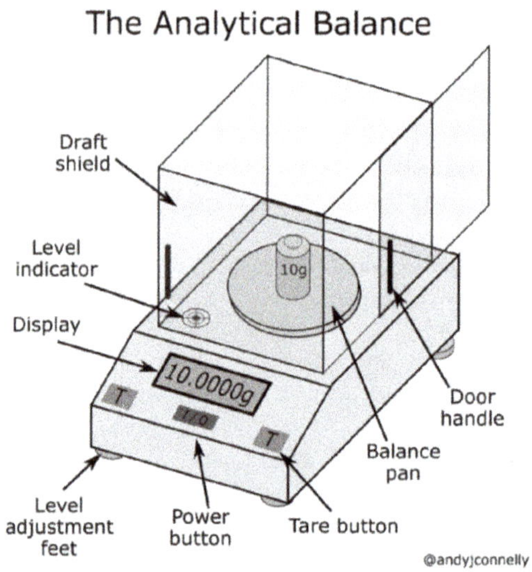


Fig. 6.48: Analytical balance.

From a previous case, of the author, as Defence legal counsel, involving Toronto Police, the measuring device, and related literature, would appear to have been:
Standard-Level Precision Balance, FZ-i/FX-I Series – manufactured by A&D Company, Limited; apparently provided by Cambridge Environmental Products Inc.; and apparently serviced by CLEARTECH:

**A&D FX-5000i – 5200g × 0.01g Precision Balance** [$1290]

Although that balance would have satisfied US FCC and European requirements for radiofrequency and microwave interference, it apparently did not have MEASUREMENT CANADA inspection to verify accuracy for determining mass, nor have displayed on it a certification sticker under the Canada *Weights & Measures Act*.
A&D **Instruction Manual**. References to that MANUAL, & other literature and data:–

- A&D 4D – FX-5000i
  Cambridge Environmental Products Inc., 225 Enterprise Drive, Komoka, Ontario. N0L 1R0
  https://cdn.shopify.com/s/files/1/2006/8997/files/FZiFXi-Brochure.pdf?v=1619118448
  https://cambridgeenviro.com/ https://cambridgeenviro.com/products/balances-scales-fx-5000i
- **FZ-i /FX-i SERIES**
  FZ-120i / FZ-200i / FZ-300i / FZ-500i FZ-1200i /FZ-2000i / FZ-3000i / FZ-5000i FX-120i / FX-200i / FX-300i / FX-500i FX-1200i /FX-2000i / FX-3000i / FX-5000i
  **Precision Balance**

**INSTRUCTION MANUAL 1WMPD4001863B [78 pages]**
A&D Company Ltd., Tokyo, Japan; 2011.
https://www.aandd.jp/ https://www.aandd.jp/support/support.html
https://cdn.shopify.com/s/files/1/2006/8997/files/FZ-i_FX-i_Manual.pdf?14788397102087296248
". . .© 2011 A&D Company Ltd. All rights reserved. No part of this publication may be reproduced, transmitted, transcribed, or translated into any language in any form by any means without the written permission of A&D Company Ltd. . . ."

- CLEAR TECH, Mississauga, Ontario. L4Z 2H6
https://www.cleartech.ca/ https://www.cleartech.ca/Locations

**363** See also A&D MANUAL, page 7.
https://cdn.shopify.com/s/files/1/2006/8997/files/FZ-i_FX-i_Manual.pdf?14788397102087296248

**364** From another, previous and unrelated case, of the author, as Defence legal counsel, involving Peel Police, the measuring device, and related literature, would appear to have been:
**Mettler Toledo XSR1202S [$5925]**

https://www.coleparmer.ca/i/mettler-toledo-xsr1202s-excellence-xsr-precision-toploader-balance-1-210g-x-0-01g-smartpan-weighing-pan-touchscreen/1133887?PubID=UZ&persist=true&ip=no&gclid=EAIaIQobChMIgsmW152J9AIVeWxvBB30SwVEEAYYASABEgJzePD_BwE

https://pim-resources.coleparmer.com/data-sheet/11338-77-11339-10-xsr-data-sheet-small.pdf

https://pim-resources.coleparmer.com/instruction-manual/11338-65-11339-10-xsr-quick-reference-manual.pdf

https://pim-resources.coleparmer.com/instruction-manual/11338-65-11339-10-xsr-user-manual.pdf

This measuring device would appear to not have a federal *Weights & Measures Act*, Measurement Canada inspection certificate sticker.

There would appear to be a FCSA (Frequency Coordination System Association – https://fcsa.ca/) certification sticker ". . . for US and Canada for use of electrical equipment for measurement, control and laboratory use . . ." This would likely relate to microwave and radiofrequency interference rules for electrical equipment, and would likely not be of concern here.

**365** SI Mass measurement (grams). Accuracy and precision too extensive? Chapters 3.1.1, 3.7.3; 6.20.

**366** Installation – See A&D MANUAL, pages 8 & 9.

**367** Levelling – The A&D Precision Balance uses a spirit level – See A&D MANUAL, page 9; "Good" or "No good." https://cdn.shopify.com/s/files/1/2006/8997/files/FZ-i_FX-i_Manual.pdf?14788397102087296248

**368** Sketched (not to scale) appearances for a two-dimensional bubble bull's eye spirit level:
(see Fig. 6.49a and b).

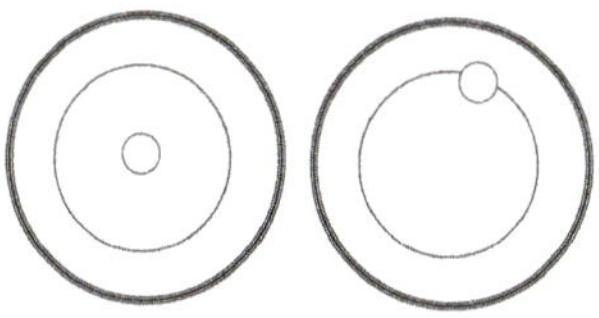

Fig. 6.49: **(a)** Level. **(b)** Not level.

**369** https://en.wikipedia.org/wiki/Spirit_level#:~:text=A

**370** ZERO – See A&D MANUAL, page 10.

**371** A&D MANUAL. https://cdn.shopify.com/s/files/1/2006/8997/files/FZ-i_FX-i_Manual.pdf?14788397102087296248

**372** (Including, if applicable, references to documents derived from Hazard Communication legislation – Canada *WHMIS* & 29 US CFR §1910.1200.)

**373** *Tare* = "... a deduction from the gross weight of a substance and its container made in allowance for the weight of the container ... the weight of the container ... to adjust (a scale on which an empty container has been placed) so as to reduce the displayed weight to zero. Balances are taken for granted now, and many *tare* themselves—among other features. Yet it's still important to remember the basics for accurate use of this most fundamental lab instrument. – *Chemistry* ..." https://www.merriam-webster.com/dictionary/tare
**374** (This is apparently a DAS document intended for Police who would provide samples for exhibits of evidence. DAS would analyze those samples for reports that would be *Certificates of Analyst* of *CD&S Act* s.51. "... [DAS] Client Manual for Exhibit Submission ..." would apparently be a document with some-kind-of secrecy status – see notes immediately below.)
**375** https://publications.gc.ca/site/eng/9.817248/publication.html "... **Drug Analysis Service: client manual. ... Publication information** ... This internal document has been removed at the request of the department (2019-09-20). ... [Ottawa]: Health Canada, 2016. ... 43 p.: ill **ISBN** 9780660054179 **Catalogue number** H129-2/2-2016E-PDF ..."
**376** See Fig. 6.50.

[...]

# Drug Analysis Service : client manual.

"The Drug Identification Service (DIS) was established in the 1960's and originally involved only identification of narcotic, controlled, and restricted drugs. Over the last several years, DAS has met regularly with its clients to obtain their input in regards to service needs and satisfaction and has made adjustments accordingly. The service has since expanded to include other services, and is reflected in its name, the Drug Analysis Service (DAS)"--Background, p. 4.

Permanent link to this Catalogue record:
publications.gc.ca/pub?id=9.817248&sl=0

MARC XML format MARC HTML format

**Publication information**

| | |
|---|---|
| **Department/Agency** | Canada. Health Canada. Regions and Programs Bureau. |
| **Title** | Drug Analysis Service : client manual. |
| **Publication type** | Monograph |
| **Language** | [English] |
| **Other language editions** | [French] |
| **Format** | Electronic |
| **Note(s)** | This internal document has been removed at the request of the department (2019-09-20).<br>Cover title.<br>Issued also in French under title: Service d'analyse des drogues : manuel du client.<br>"September 2016." |
| **Publishing information** | [Ottawa] : Health Canada, 2016. |
| **Description** | 43 p. : ill. |
| **ISBN** | 9780660054179 |
| **Catalogue number** | H129-2/2-2016E-PDF |
| **Departmental catalogue number** | Pub.: 160042 |
| **Subject terms** | Narcotic control |

[...]

Fig. 6.50: DAS.

377 (The apparent secrecy of this document would seem to be indicated on the government website: ". . .**This internal document has been removed at the request of the department** . . ." [emphasis added here]. In view of criminal law disclosure principles & practices, what would be so special about this document is puzzling; but secrecy is apparently a government prerogative. That a government document with some-kind-of secrecy status would have an ISBN seems to be a curiosity.)
**378** *Weights and Measures Act* RSC 1985, c. W-6. http://laws-lois.justice.gc.ca/eng/acts/W-6/page-1.html
**379** http://laws-lois.justice.gc.ca/eng/acts/W-6/page-1.html
**380** (Generally, the process and procedure appear to be that federal government (or government-approved) inspectors inspect all devices involved with the measurement of mass (grams) of traded commodities; then certified by a commonly-used sticker. Defence argument here is not that the absence of a sticker automatically cancels the measurement, but rather that such absence introduces a criminal-law-test reasonable doubt.)
**381** https://www.ic.gc.ca/eic/site/mc-mc.nsf/eng/lm04710.html (see Fig. 6.51).

[. . .]

A Measurement Canada inspection sticker is placed on measuring devices that have been inspected and found to be measuring accurately.

[. . .]

[. . .] 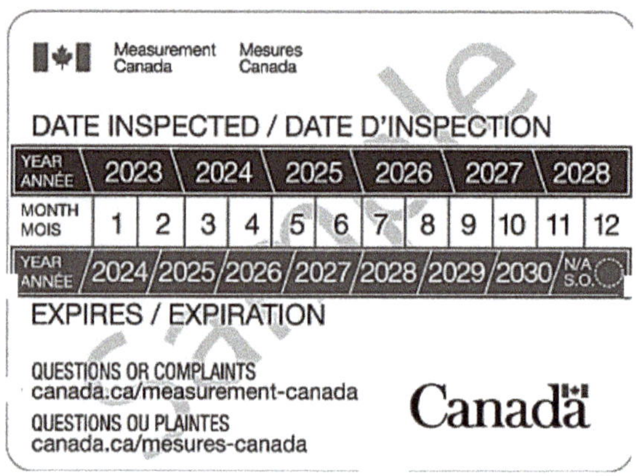 [. . .]



Fig. 6.51: W&M Act.

[. . .]

**382** https://www.ic.gc.ca/eic/site/mc-mc.nsf/eng/h_lm00013.html
**383** https://ised-isde.canada.ca/site/measurement-canada/en/inspections
**384** https://www.ic.gc.ca/eic/site/mc-mc.nsf/eng/lm04920.html
**385** (Historically, attention to weights & measures is perhaps somewhat reflected from *Le Système International d'Unités, Convention du Mètre*, Paris, 1875; *Magna Carta*, 1215, clause 35; *Mensa Ponderaria*, Table of standard measures, Pompei, before *AD* 79; & perhaps Bible-inspired – *Deuteronomy*, 25. 15 **דברים** –
". . . A perfect and just weight shalt thou have; a perfect and just measure shalt thou have . . ."
". . .אבן שלמה וצדק יהיה-לך, איפה שלמה וצדק יהיה-לך. . .")

386 https://www.nist.gov/system/files/documents/pml/div684/fcdc/si_brochure_8.pdf
387 https://people.bu.edu/chamley/Ec765-18/MC.pdf
388 (*Mensa Ponderaria*, Pompei https://www.gutenberg.org/files/42715/42715-h/42715-h.htm [Page 94] https://www.gutenberg.org/files/42715/42715-h/images/fig034fs.jpg)
389 https://www.sefaria.org/Deuteronomy.25.19?lang=bi&aliyot=0
390 See, for example: A&D Instruction Manual, cited in a note above; pages 51, 52, 53. At page 53. https://cdn.shopify.com/s/files/1/2006/8997/files/FZ-i_FX-i_Manual.pdf?14788397102087296248
391 Amphetamine hydrochloride **hygroscopic:** [UK] Forensic Science Service – Methylamphetamine "FSS REPORT ON METHYLAMPHETAMINE CHEMISTRY, SEIZURE STATISTICS, ANALYSIS, SYNTHETIC ROUTES AND HISTORY OF ILLICIT MANUFACTURE IN THE UK AND THE USA FOR THE ACMD WORKING GROUP ON METHYLAMPHETAMINE," Prepared by: Dr. Mike White, Drugs Intelligence Unit, October 2004. At page 66.
https://assets.publishing.service.gov.uk/government/uploads/system/uploads/attachment_data/file/119161/ACMD-meth-annex-November-2005.pdf
392 *Controlled Drugs & Substances Act* SC 1996, c.19. http://laws-lois.justice.gc.ca/eng/acts/C-38.8/index.html
http://laws-lois.justice.gc.ca/eng/acts/C-38.8/index.html http://laws-lois.justice.gc.ca/eng/acts/C-38.8/page-9.html#h-19
393 *CD&S Act* s. 51.

[. . .]

**Designation of analysts**
**44** The Minister may designate, in accordance with the regulations made pursuant to paragraph 55(1)(o), any person as an analyst for the purposes of this Act and the regulations.

**Analysis**
**45(1)** A peace officer, inspector or prescribed person may submit to an analyst for analysis or examination any substance or sample of it taken by the peace officer, inspector or prescribed person.

**Report**
**(2)**An analyst who has made an analysis or examination under subsection (1) may prepare a certificate or report stating that the analyst has analysed or examined a substance or a sample thereof and setting out the results of the analysis or examination.

[. . .]

**Certificate of analyst**
**51(1)** Subject to this section, a certificate or report prepared by an analyst under subsection 45(2) is admissible in evidence in any prosecution for an offence under this Act or the regulations or any other Act of Parliament and, in the absence of evidence to the contrary, is proof of the statements set out in the certificate or report, without proof of the signature or official character of the person appearing to have signed it.

**Attendance of analyst**
**(2)** The party against whom a certificate or report of an analyst is produced under subsection (1) may, with leave of the court, require the attendance of the analyst for the purpose of cross-examination.

[. . .]

**394** Old City Hall, Toronto; 29 May 2018 (argument); 01 June (judgment). Before The Honourable Justice S.R. Clark.
**395** This judgment is apparently unreported in Canadian legal literature. It is copied, anonymized & slightly edited, unofficial for Canadian court purposes, in M.G., ***Forensic Chemistry*** *Fundamentals*, De Gruyter, Berlin; 2022; ISBN 978-3-11-071878-2; Appendix to Chapter 10, pages 412→419. https://www.degruyter.com/document/doi/10.1515/9783110718812/html
**396** RESPONSE OF THE ATTORNEY GENERAL OF CANADA ON BEHALF OF HEALTH CANADA, 23 May 2018; for Court proceedings, Toronto, 29 May 2018 (argument); 01 June (judgment).
**397** Paragraphs 17 and 18, pages 8 and 9.
**398**

⋮

17. There is no merit to the Applicant's assertion that accreditation standards (such as those promulgated by the International Organization for Standardization (ISO)) that apply to DAS laboratories mean that Police or PPSC "have obvious and complete access rights" to the information sought by the Applicant.[10] The DAS is accredited by the Standards Council of Canada and must meet both the requirements of ISO/IEC 17025 (*General Requirements for the Competence of Testing and Calibration*, as asserted by the Applicant), and, at the time these Certificates were produced, the requirements of Canadian Procedural Document CAN-P-1578 (*Guidelines for the Accreditation of Forensic Testing Laboratories*, and in particular Appendix 6 thereof, "Drug Chemistry"), which is an amplification of the ISO standard.[11] Section 5.10.1 of this latter standard states:

> Certificates of Analyst are legal documents. As such, information contained in these reports is directed by the appropriate laws of the land.[12]

18. This qualification supersedes any requirement which may arise under ISO/IEC 17025 to provide information beyond that statutorily required by the CDSA, namely, the Certificates.

[. . .]

---

[. . .]

[11] As stated in, for example, *Health Canada Internal Audit: Audit of Regional Laboratory Activities (July 2016)*, available at https://www.canada.ca/en/healthcanada/corporate/transparency/corporate-management-reporting/internalaudits/2016/audit-regional-laboratory-activities-july2016.html Note: CAN-P-1578 has since been replaced by the Standard Council of Canada's *Requirements and Guidance for the Accreditation for Forensic Testing Laboratories* (July 17, 2017) although, for the purposes of this application, the relevant sections remain materially unchanged.

[12] Standards Council of Canada's *Guidelines for the Accreditation of Forensic Testing Laboratories*, CAN-P-1578, Appendix 6, "Drug Chemistry", Respondent's Record, Tab E, section 5.10.1, available at:
https://www.scc.ca/sites/default/files/migrated_files/DLFE-433.pdf

[. . .]

**399** Guidelines for the Accreditation of Forensic Testing Laboratories CAN-P-1578 May 2009 Program Speciality Area – Forensic Testing Laboratories (PSA-FT)
http://www.jtc1sc34.org/en/c/document_library/get_file08cd.pdf page 91.

[. . .]

**5.10 Reporting the results**

5.10.1 Certificates of Analyst are legal documents. As such, information contained in these reports is directed by the appropriate laws of the land.

5.10.2 Where feasible, court testimony of Analysts shall be monitored on an annual basis.

[. . .]

**400** ISO/IEC 17025:2005, "General requirements for the competence of testing and calibration laboratories," International Organization for Standardization, Genève 20, Switzerland. www.iso.org

**401** [. . .]

5.10 Reporting the results

5.10.1 General

The results of each test, calibration, or series of tests or calibrations carried out by the laboratory shall be reported accurately, clearly, unambiguously and objectively, and in accordance with any specific instructions in the test or calibration methods.

The results shall be reported, usually in a test report or a calibration certificate . . ., and shall include all the information requested by the customer and necessary for the interpretation of the test or calibration results and all information required by the method used. This information is normally that required by 5.10.2, and 5.10.3 or 5.10.4.

[. . .]

**402** [. . .]

5.10.2 Test reports and calibration certificates

Each test report or calibration certificate shall include at least the following information, unless the laboratory has valid reasons for not doing so:

[. . .]

e) identification of the method used;

[. . .]

**403** International Organization for Standardization, Geneva, Switzerland. http://www.iso.org
**ISO/IEC 17025:2005** http://www.iso.org/iso/home/store/catalogue_tc/catalogue_detail.htm?csnumber=39883

**404** DAS – TORONTO LABORATORY, Standards Council of Canada Accredited Laboratory No.607, Toronto M1P 4R7 http://scc.ca/

**405** Text of SCOPE OF ACCREDIATION:

**406** **SCOPE OF ACCREDITATION**

**Health Canada**

**DRUG ANALYSIS SERVICE – TORONTO LABORATORY**

**2301 Midland Ave.**

**Scarborough, ON**

**M1P 4R7**

Accredited Laboratory No. 607

(Conforms with requirements of ISO/IEC 17025:2005, RG-FORENSIC)

CONTACT: V. Bell
TEL: (416) 973-1553
FAX: (416) 954-5923
EMAIL: das.manager.toronto@hc-sc.gc.ca

CLIENTS SERVED: Police officers, both RCMP and Municipal, in Ontario and other provinces, as well as members of other enforcement agencies; members of the legal profession; Canadian justice system

FIELDS OF TESTING: Forensic

FORENSICS DISCIPLINE(S): Forensic Drug Chemistry

PROGRAM SPECIALTY AREA: Forensic

INITIAL ACCREDITATION: 2006-08-29

MOST RECENT REACCREDITATION: 2018-11-04

ACCREDITATION VALID TO: 2022-08-29

**SCC Group Accreditation:**
This laboratory is a part of a Group Accreditation with the following facilities in accordance with SCC's policy on Group Accreditation documented in the Accreditation Services Accreditation

Program Overview.
- 15858 SANTÉ CANADA, SERVICE D'ANALYSE DES DROGUES – LABORATOIRE DE MONTRÉAL, 1001, rue Saint-Laurent ouest, Longueuil, QC, J4K 1C7, Accredited Laboratory No. 709
- 15885 Health Canada, Drug Analysis Service, 3155 Willingdon Green, Burnaby, BC, V5G 4P2, Accredited Laboratory No. 716

**FORENSICS**
**Forensic Drug Chemistry**
Description of Activities:
The Drug Analysis Service, Toronto Laboratory, carries out the following examinations / analyses:

Identification and quantification of controlled and non-controlled drugs, according to the Controlled Drugs and Substances Act and the Food and Drugs Act, and other related substances in or on powders, pharmaceuticals, plant materials and miscellaneous exhibits.

Techniques for which laboratory is accredited:
a. Gas chromatography
b. High performance liquid chromatography
c. Infra-red spectrometry
d. Macroscopic examination
e. Mass spectrometry
f. Micro-chemical tests
g. Nuclear magnetic resonance spectroscopy
h. Optical microscopy
i. Sample preparation, extraction, separation and general chemical and physical tests
j. Thin layer chromatography

**Notes:**
**RG-FORENSIC:** SCC Requirements and Guidance for the Accreditation for Forensic Testing Laboratories
**ISO/IEC 17025:2005:** General Requirements for the Competence of Testing and Calibration Laboratories

Elias Rafoul, Vice President
Accreditation Services

Date: 2020-08-18
Number of Forensic Techniques: 10
SCC 1003-15/747

Partner File #0
Partner: SCC

**407** https://www.scc.ca/en/accreditation/laboratories/health-canada-drug-analysis-service-toronto-laboratory
https://www.scc.ca/en/system/files/client-scopes/ASB_SOA_15747_Scope_v4_2020-08-13.pdf

**408** Joseph Heller, *CATCH-22*, Dell Publishing Co., Inc., New York City, 10017; 1955, 1961; New Dell Edition, 1985; ISBN 0-440-11120-X. For example, Chapter 39, at page 416.

**409** See in note above for definition of **mass**, **gram** and **weight**.

**410** The Hon. W. David Griffiths, Q.C., "Report to the Attorney General of Canada: Advice and Recommendations Respecting Certain Certificates of Analyst Issued by Health Canada," 23 May 2001.

**411** https://www.cbc.ca/news/canada/lab-errors-may-void-ontario-drug-convictions-1.283147
https://web.archive.org/web/20020220112655/ http://canada.justice.gc.ca/en/news/nr/2001/doc_26438.html

**412** See: Bruce A. MacFarlane, *DRUG OFFENCES IN CANADA*, CHAPTER 21, "DE MINIMIS NON CURAT LEX," pages 447→467; Canada Law Book Limited, Toronto, 1979. ISBN 0 88804 007 5.

**413** *de minimis non curat lex* = the law does not concern itself with trifles
https://www.merriam-webster.com/dictionary/de%20minimis%20non%20curat%20lex

**414** For example, *Certificates of Analyst* in a Toronto case informed only:

**415 ". . . Schedule I 18. Methamphetamine (N,α-dimethylbenzeneethanamine) . . ."**

**416 ". . . Schedule I 2. (2) Cocaine (benzoylmethylecgonine) . . ."**

**417** No indication of quantity (grams), concentration, proportion, ionization, chirality or hygroscopy. And, no molecular structural drawings or ACS CAS Registry Numbers.

**418** Agilent 5975 inert GC/MS System for Pharmaceutical Applications Data Sheet November 4, 2005 5989-3902EN
www.chem.agilent.com/

**419** Agilent 5975 Series MSD Operation Manual June 2012 G3170-90036 www.chem.agilent.com/

**420** This is in the context of the use of octafluoronaphthalene for the calibration of the Agilent instrumentation.

**421** . . . Manual . . . at pages 12 and 76: [1 pg/μL = $3.675 \times 10^{-9}$ mol/L OFN]

**422** Octafluoronaphthalene (OFN), $C_{10}F_8$, CAS 313-72-4, 272.09 g/mol (see Fig. 6.52).

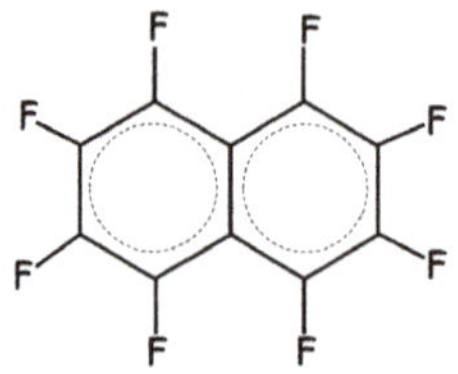


Fig. 6.52: Octafluoronaphthalene (OFN), CAS 313-72-4.

**423** Octafluoronaphthalene (OFN), CAS 313-72-4 abbreviations, conversions, *etc.*:
1 μL = 1 μL = 1 microlitre = $10^{-6}$ litres = .000001 litres
1 pg = 1 picogram = $10^{-12}$ g = $10^{-9}$ mg = $10^{-9}$ milligrams = $10^{-6}$ μg
$10^{-12}$ g / 272.09 g/mol = 3.675 × $10^{-15}$ mol
1 pg/μL = 3.675 × $10^{-9}$ mol/Ll
1 mol = 1 mole = mass/molar mass = g / (g/mol) = 6. 022 × $10^{23}$ molecules = Avogadro's number of particles.

**424** www.chem.agilent.com/
**Agilent 5975 inert GC/MS System for Pharmaceutical Applications**
Data Sheet **November 4, 2005**
**5989–3902EN**
[. . .]
**Installation checkout specifications**
EI scan sensitivity 100:1 S/N for 1 pg OFN scanning from 50–300 amu at nominal *m/z* 272 ion
[. . .] [page 2.]
[EI = electron impact]
[S/N = signal / noise.]
[1 picogram pg = $10^{-15}$ kg]

**425** www.chem.agilent.com/
**Agilent 5975**
**Series MSD Operation Manual** June 2012 G3170-90036
[. . .]
OFN Octafluoronaphthalene (calibrant)
[. . .] [page 12]
[. . .]
To verify system performance
Materials needed
– 1 pg/μL. (0.001 ppm (OFN sample (5188–5348)
[. . .] [page 76]
[1 pg/μL = 3.675 × $10^{-9}$ mol/L]

**426** https://www.agilent.com/store/en_US/LCat-SubCat2ECS_32500/GC-MS-Test-and-Performance-Samples
https://pubchem.ncbi.nlm.nih.gov/compound/Octafluoronaphthalene
https://www.sigmaaldrich.com/CA/en/product/aldrich/248061?lang=

**427** Octafluoronaphthalene OFN CAS 313-72-4.
1 :L = 1 microlitre = $10^{-6}$ L
1 picogram (pg) = $10^{-12}$g
$10^{-12}$ g/272.09g/mol = 3.675 × $10^{-15}$mol
1 pg/1:L = 3.675 × $10^{-9}$ mol/L

**428** https://scc-csc.lexum.com/scc-csc/scc-csc/en/item/808/index.do
https://www.lawnow.org/stinchcombe-and-crown-disclosure-of-criminal-evidence-2/#:~:text=Stinchcombe%20is%20a%20landmark%20Supreme,Crown%20will%20use%20in%20court
**429** (That would likely include a serial number. And maybe also various stickers or markings to alert to: inspection, calibration, maintenance, servicing, electrical safety, US PTO, *etc.*)
**430** https://chem.libretexts.org/Courses/Bellarmine_University/BU%3A_Chem_103_(Christianson)/Phase_3%3A_Atoms_and_Molecules_-_the_Underlying_Reality/9%3A_Chemical_Bonding/9.5%3A_Metallic_Bonding
**431** https://en.NBS]?>wikipedia.org/wiki/Metal_detector
**432** https://chem.libretexts.org/Courses/Saint_Marys_College_Notre_Dame_IN/CHEM_431%3A_Inorganic_Chemistry_(Haas)/CHEM_431_Readings/14%3A_Magnetism/14.02%3A_Magnetic_Properties_of_Materials
**433** https://www.silverinvestingguide.com/Silver-Canadian-Coins.html
**434** See below:
**435** Peter R. De Forest, "Foundations of Forensic Microscopy," as Chapter 9 in Richard Saferstein, ed., *Forensic Science Handbook* [Volume I], Prentice-Hall, Inc., Englewood Cliffs, New Jersey, 1982; pages 416→528, and references cited therein.
**436** Richard Saferstein, *CRIMINALISTICS/An Introduction to Forensic Science*, Seventh Edition, Prentice Hall, Upper Saddle River, New Jersey, 07458; 2001; ISBN 0-13-013827-4; "THE MICROSCOPE," Chapter 7, pages 164→188.
**437** Richard Saferstein, *CRIMINALISTICS/An Introduction to Forensic Science*, Seventh Edition, Prentice Hall, Upper Saddle River, New Jersey, 07458; 2001; ISBN 0-13-013827-4; page 64, et seq.
**438** Barry D. Gaudette, "The Forensic Aspects of Textile Fiber Examination," as Chapter 5 in Richard Saferstein, ed., *Forensic Science Handbook / Volume I*, Prentice Hall, Englewood Cliffs, New Jersey 07632, 1988; pages 209→272, and references cited therein.
**439** Elmer T. Miller, "Forensic Glass Comparisons," as Chapter 4 in Richard Saferstein, ed., *Forensic Science Handbook* [Volume I], Prentice-Hall, Inc., Englewood Cliffs, New Jersey, 1982.
**440** Thomas A. Kubic and Nicholas Petraco, in Stuart H. James and Jon J. Nordby, editors, *Forensic Science / An Introduction to Scientific and Investigative Techniques*, CRC Press, Boca Raton, 2003; ISBN 0-8493-1246-9; "Microanalysis and Examination of Trace Evidence," Chapter 14; pages 251→275. www.crcpress.com
**441** J.I. Thornton, "Forensic Paint Examination," as Chapter 10 in R. Saferstein, ed., *Forensic Science Handbook* [Volume I], Prentice-Hall, Inc., Englewood Cliffs, New Jersey, 1982, pages 547→549, and references cited therein.
**442** See also re use of computer technology and mathematical algorithms to recognize microscopic patterns: Tim Hornyak, "Forensic firm [Forensic Technology Inc., Montreal www.fti-ibis.com] aims to be No.1 with a bullet / Police from New York to Pretoria use identification technology to match spent ammunition with guns," *The Globe and Mail*, 29 July 1999, page T4.
**443** "Spectroscopy/Molecular energy levels"; Wikiversity; April 2020.
https://en.wikiversity.org/wiki/Wikiversity:Main_Page
https://en.wikiversity.org/wiki/Spectroscopy/Molecular_energy_levels
https://en.wikiversity.org/wiki/Spectroscopy/Rotational_spectroscopy https://en.wikiversity.org/wiki/Wikiversity:Welcome
**444** https://en.wikipedia.org/wiki/Microwave_spectroscopy
https://chem.libretexts.org/Bookshelves/Physical_and_Theoretical_Chemistry_Textbook_Maps/Supplemental_Modules_(Physical_and_Theoretical_Chemistry)/Spectroscopy/Rotational_Spectroscopy
https://www.sciencedirect.com/topics/chemistry/microwave-spectroscopy
**445** See: Walter J. Moore, *Physical Chemistry*, Prentice-Hall, Inc., Englewood Cliffs, New Jersey, 1962 and 1963, pages 595 and 596.

**446** Health effects of non-ionizing radiation would not be as obvious as for ionizing radiation, and are disputed and remain controversial.
**447** Adam Entous and John Lee Anderson, *HAVANA SYNDROME*, *The New Yorker*, 19 November 2018, pages 34→47.
https://longform.org/posts/the-mystery-of-the-havana-syndrome
**448** *Wikipedia*, April 2020, Havana syndrome https://en.wikipedia.org/wiki/Havana_syndrome
**449** https://en.wikipedia.org/wiki/Moscow_Signal
https://www.vanityfair.com/news/2019/01/the-real-story-behind-the-havana-embassy-mystery
**450** M.E. Bitran, Ph.D., D.E. Charron, P.Eng., J.M. Nishio, M.Sc., "Microwave Emissions and Operator Exposures from Traffic Radars used in Ontario," Non-Ionizing Radiation Section, Radiation Protection Service, Occupational Health And Safety Branch, Ontario Ministry of Labour, Weston, Ontario M9P 3T1; [1992].
**451** See W.D. Kinard and D.R. Lundy, "A Comparison of Neutron Activation Analysis and Atomic Absorption Spectroscopy on Gunshot Residue" as Chapter 11 in Geoffrey Davies, ed., *Forensic Science*, ACS Symposium Seeries 13, Washington, D.C., 1975; pages 97→107.
**452** John Joseph Fenton, "Forensic Toxicology," in Stuart H. James and Jon J. Nordby, editors, *Forensic Science / An Introduction to Scientific and Investigative Techniques*, CRC Press, Boca Raton, 2003; ISBN 0-8493-1246-9; Chapter 4. "Neutron Activation Analysis," at page 57. www.crcpress.com
**453** Richard Saferstein, *CRIMINALISTICS / An Introduction to Forensic Science*, Seventh Edition, Prentice Hall, Upper Saddle River, New Jersey, 07458; 2001; ISBN 0-13-013827-4; pages 157→160.
**454** https://en.wikipedia.org/wiki/Gamma_ray#Gamma_ray_production
**455** V.P. Guinn and R. Demiralp, "Arsenic in hair by INAA – A major recent murder case," *Journal of Radioanalytical and Nuclear Chemistry*, 1993 volume 168, pages 249→256. https://link.springer.com/article/10.1007/BF02040897
**456** Margaret Moffett, Associated Press, "Blanche Taylor Moore remains on death row after 25 years," *Washington Times*, 20 December 2015. ". . . secretly slipping arsenic into food – like her homemade banana pudding . . ."
https://www.washingtontimes.com/news/2015/dec/20/blanche-taylor-moore-remains-on-death-row-after-25/
**457** https://en.wikipedia.org/wiki/Blanche_Taylor_Moore https://murderpedia.org/female.M/m/moore-blanche-taylor.htm
**458** John Kenkel, *Analytical chemistry refresher manual*, Lewis Publishers, Chelsea, Michigan, 1992, "NUCLEAR MAGNETIC RESONANCE SPECTROSCOPY," Chapter 6.6, pages 155→162.
**459** https://www.acs.org/content/acs/en/education/whatischemistry/landmarks/mri.html https://en.wikipedia.org/wiki/Nuclear_magnetic_resonance_spectroscopy
**460** https://chem.libretexts.org/Bookshelves/Physical_and_Theoretical_Chemistry_Textbook_Maps/Supplemental_Modules_(Physical_and_Theoretical_Chemistry)/Spectroscopy/Magnetic_Resonance_Spectroscopies/Nuclear_Magnetic_Resonance/Nuclear_Magnetic_Resonance_II#:~:text=Nuclear%20Magnetic%20Resonance%20(NMR)%20is,structure%20determination%20of%20small%20molecules.
**461** J.O.S.H. Cleary, A.R. Guimarães, "Magnetic Resonance Imaging," *Pathobiology of Human Disease*, 2014; ScienceDirect, Elsevier.
https://www.sciencedirect.com/topics/social-sciences/magnetic-resonance-imaging
https://www.sciencedirect.com/referencework/9780123864574/pathobiology-of-human-disease
**462** See also:
https://www.nytimes.com/2025/07/17/nyregion/mri-chain-injury-long-island.html
https://www.nytimes.com/2025/07/18/health/mri-death-long-island.html
**463** https://en.wikipedia.org/wiki/Crystal_ball https://commons.wikimedia.org/wiki/File:3DCrystal_ball.png

**464** https://blogs.loc.gov/loc/2020/10/hollywood-houdini-and-the-halloween-seance-of-1936/
https://www.collectionscanada.gc.ca/obj/023017/f1/xx014595-vx.jpg
https://www.collectionscanada.gc.ca/laurier-house/023017-3220-e.html
https://en.wikipedia.org/wiki/Crystal_ball
https://en.wikipedia.org/wiki/Ouija#Scientific_investigation

**465** Before his death, 31 October 1926, Harry Houdini, the famous magician, had agreed with his wife to test the ability of spirits of the dead to communicate back. He (and presumably his wife) was known to not accept such spiritualism. The quote above came after a decade of annual Halloween sessions. The 1936 séance – radio broadcast – and likely prearranged – was to demonstrate that it didn't work.

**466** See also fictional TV – 1970s:
https://www.imdb.com/title/tt0641664/
https://www.imdb.com/review/rw3221337/?ref_=tt_ururv_c_1_hd
https://www.imdb.com/review/rw3071859/?ref_=tt_ururv_c_2_hd
https://en.wikipedia.org/wiki/Mannix

**467** See: Canadian Prime Minister Mackenzie King's crystal ball:

**468** https://www.elinorflorence.com/blog/mackenzie-king/

**469** Houdini Museum, Scranton, Pennsylvania.
http://houdini.org/
https://www.houdinimuseum.net/the-last-seance

**470** https://www.bac-lac.gc.ca/eng/news/podcasts/Pages/mackenzie-king-against-his-will.aspx

**471** http://houdini.org/MackenzieKingandSpiritualism.html

**472** https://www.collectionscanada.gc.ca/obj/023017/f1/xx014595-vx.jpg
https://www.canadashistory.ca/explore/religion-spirituality/speaker-startled-by-sparkling-sphere
https://en.wikipedia.org/wiki/Ouija#Scientific_investigation

**473** https://en.wikipedia.org/wiki/Crystal_ball https://commons.wikimedia.org/wiki/File:3DCrystal_ball.png

**474** https://www.sandiegoreader.com/news/1998/nov/12/did-houdini-contact-his-wife-after-he-died/#:~:text=Matthew%20Alice%3A,his%20wife%20AFTER%20he%20died.

**475** CBC ideas, "We believe in artificial intelligence the same way we believe in ghosts."

> [. . .]
>
> Hidden in the 1950 academic paper that launched the famous 'Turning Test' of machine intelligence, is a strange mystery. Cryptographer Alan Turning argued that humans might always be able to outsmart machines, because we have supernatural powers like ESP, telepathy, and telekinesis. His belief in the paranormal is just one part of the spooky side of artificial intelligence. Like hauntings or seances, AI is an exercise in self-deception; we imagine intelligence from computation and data, just like we imagine ghosts from strange lights and bumps in the night.
>
> [. . .]
>
> https://podcasts.apple.com/ca/podcast/we-believe-in-artificial-intelligence-the-same-way-we/id151485663?i=1000698951228

**476** See also: Cody Delistraty, "We're In a New Age of Techno-Spiritualism," *New York Times*, OPINION/GUEST ESSAY, 30 March 2025. https://www.nytimes.com/2025/03/30/opinion/grief-tech-ai-optimized.html

**477** Including in the movie industry: https://faroutmagazine.co.uk/the-10-greatest-occultist-movies/

**478** (Some of the clients might be better described as scam victims.)

**479** https://www.ojp.gov/ncjrs/virtual-library/abstracts/psychics-and-police-work
https://www.ojp.gov/ncjrs/virtual-library/abstracts/occult-crime-law-enforcement-primer
https://www.ojp.gov/ncjrs/virtual-library/abstracts/occult-criminal-investigation
https://www.gvsu.edu/cms4/asset/903124DF-BD7F-3286-FE3330AA44F994DE/occult_criminal_investiga

tion_2.pdf
https://en.wikipedia.org/wiki/Occult#Occultism

**480** See:

- Dr.H.Amsler, "Photography in the Service of Crime Detection / The Large-Format Camera – a Tool for Police and C.I.D.," pages 228→239, in Nikolaus Karpf, *Manual of Applied Photography* . . ., 2nd ed., Verlag Grossbild-Technik, Munich, 1966.
- H.J. Walls, *Forensic Science / An Introduction to Scientific Crime Detection*, Second Edition, Sweet and Maxwell, London, 1974, "Photography," Chapter 17, pages 2196225.
- Weston and Wells, cited in the note above, Recording the Crime Scene: Field Notes, Photographs, and Sketches, Chapter 5, pages 65→77.
- Richard A. Morton, ed., *Photography for the Scientist*, Second edition, Academic Press, London, . . ., 1984;
- Alfred A. Blaker, *Handbook for Scientific Photography*, Second Edition, Focal Press, Boston and London, 1989.

See also:

- *Ultraviolet and Fluorescence Photography / A Kodak Technical Publication, Kodak Publication No.M-27, Eastman Kodak Company,* Rochester, New York, 14,650, 1968 and 1972, and references cited therein.
- *Kodak Infrared Films, Kodak Publication No. N-17,* Eastman Kodak Company, Rochester, New York, 14650, 1971 [and 1972].
- *Kodak Professional Black-and-White Films*, Kodak Publication No. F-5, Eastman Kodak Company, Rochester, New York, 14650, 1990.
  http://www.kodak.com/
- https://en.wikipedia.org/wiki/George_Eastman

**481** See, for example: Raechel Huizinga and Iain Sherriff-Scott, "Victim testifies in poisoning case, sentencing marked for December / 'Like a scene from a horror movie,' victim tells court," *The Journal Queen's University*, 02 November 2018.
https://www.queensjournal.ca/story/2018-11-02/news/victim-testifies-in-poisoning-case-sentencing-marked-for-december/

**482** And nineteenth century very labour-intensive.

**483** Harry Asher, *Photographic Principles and Practices, Second Edition*, AMPHOTO American Photographic Book Publishing Co., Garden City New York, 1968; ISBN 0-8174-0507-0; pages 37→40.

**484** https://en.wikipedia.org/wiki/History_of_photography#:~:text=View%20from%20the%20Window%20at,the%20earliest%20surviving%20camera%20photograph

**485** This sketched diagram is to describe general concept; not to scale; not as an engineering plan or design; none of its parts are presented here as an engineering plan or design. The optical system here is represented as a simple thin convex lens with the approximate focal length shown from the plane of the photo-recording surface; in reality it is often a complex composite of different kinds of lenses, with variable focal length locations.

**486** https://www.canon.ge/pro/infobank/understanding-focal-length/

**487** f/d √2 C.B. Neblette, *PHOTOGRAPHY / ITS MATERIALS AND PROCESSES, SIXTH EDITION, 1962* (*First Edition, 1927*), D. Van Nostrand Company, Inc., New York City, 1962; page 57, et seq.

**488** f/d √2 Ralph E. Jacobson, Sidney F. Ray, Geoffrey G Attridge, *The Manual of Photography, Eighth Edition*, Focal Press, Butterworth-Heinemann, Surrey GU2 5BH, 1988, 1991; ISBN 0 240 51268 5; page 49, et seq.

**489** f/d √2 Harry Asher, *Photographic Principles and Practices, Second Edition*, AMPHOTO American Photographic Book Publishing Co., Garden City New York, 1968; ISBN 0-8174-0507-0; at page 10.

**490** https://www.dpreview.com/forums/post/53258728

**491** http://www.schoolphysics.co.uk/age11-14/Light/text/Camera_/index.html
**492** https://eportfolios.macaulay.cuny.edu/tenneriello15/2015/09/10/reading-response-to-every-portrait-tells-a-lie-and-how-john-singer-sargent-made-a-scene/
**493** https://expertphotography.com/how-does-a-camera-work/
**494** https://en.wikipedia.org/wiki/Camera
https://en.wikipedia.org/wiki/History_of_the_camera
https://en.wikipedia.org/wiki/Photography
https://en.wikipedia.org/wiki/Focal_length https://en.wikipedia.org/wiki/F-number
**495** Sometimes two-dimensional – for example to photograph documents.
**496** The *lens* is sketched here as a single thin lens, but is actually often/usually a *lens* system of substantial complexity.
**497** "/ " here is to indicate mathematical division.
**498** "/ " here is to indicate mathematical division.
**499** Sometimes f is represented as *f*
**500** "/ " here is **not** to indicate mathematical division;).
**501** "/ " here is **not** to indicate mathematical division;).
**502** – Sketched – not-to-scale.
**503** https://expertphotography.com/how-does-a-camera-work/
**504** $\sqrt{2} = 1.4142$
**505** Neblette, page 58.
**506** Sometimes: seconds, minutes, hours or longer.
**507** See also: Harold E. Edgerton & James R. Killian, Jr., *flash! / SEEING THE UNSEEN BY ULTRA HIGH-SPEED PHOTOGRAPHY*, Charles T. Branford Company, Boston, 1954.
**508** https://en.wikipedia.org/wiki/Lens_speed
**509** https://en.wikipedia.org/wiki/Film_speed
**510** Ernst Pawel, "*THE LABYRINTH OF EXILE / A Life of Theodor Herzl*," Collins Harvill, London W1, 1990; ISBN 0-00-272098-1; in photo pages, after text page 378; at text page 383.
**511** https://commons.wikimedia.org/wiki/File:THEODOR_HERZL_WITH_GERMAN_KAISER_WILHELM_II_IN_MIKVE_YISRAEL_IN_1898._ORIGINAL_PHOTOMONTAGE_OF_TWO_PHOTOS_BY_DAVID_WOLFSON,_DUE_TO_TECHNICAL_PROBLEM_IN_LAB_OR_CAMERA._%D7%AA%D7%90%D7%95%D7%93%D7%95%D7%A8_%D7%94.jpg
**512** https://www.amazon.ca/Labyrinth-Exile-Life-Theodor-Herzl/dp/0002720981
**513** https://www.amazon.ca/Labyrinth-Exile-Life-Theodor-Herzl/dp/0374523517
**514** https://www.haaretz.com/israel-news/.premium-the-first-zionist-fake-news-marks-its-120th-anniversary-1.6610695
**515** Haaretz.com, online edition of Haaretz Newspaper, Israel, 30 October 2018, "The First Zionist Fake News Marks 120th Anniversary, Starring Herzl and the Kaiser / When only the Zionist leader's foot appeared in a photo with Wilhelm II, it was time for a 19th-century 'Photoshop' effort."
**516**

[. . .]

The photomontage of Herzl and Kaiser Wilhelm II. Central Zionist Archives

The first Zionist fake news was a doctored photo meant to capture a meeting between Theodor Hertzl and Kaiser Wilhelm II – which really took place.

The meeting happened 120 years ago, on October 18, 1898, near the Mikveh Israel agricultural school near Jaffa. Herzl was trying to convince the Kaiser to grant a charter for the establishment of a Jewish state.

[. . .]

**517** http://www.zionistarchives.org.il/en/datelist/Pages/ZionistDelegation.aspx

**518** See:

- Cade Metz, "Spot the Deepfake. (It's Getting Harder.)," *The New York Times*, [25 November 2019], pages B1 and B4.

https://www.nytimes.com/2019/11/24/technology/tech-companies-deepfakes.html
https://www.brookings.edu/research/is-seeing-still-believing-the-deepfake-challenge-to-truth-in-politics/

- Stuart A. Thompson, "Which Parts of These Images Are A.I.-Generated? / Try our other A.I. quizzes: A.I.-generated images | A.I.-generated videos" *New York Times*, 26 December 2024. https://www.nytimes.com/interactive/2024/12/27/technology/artificial-intelligence-generative-fill-photoshop-openai.html
- Andre Rhoden-Paul, Joshua Cheetham & Jake Horton, "Princess of Wales: How might have Kate's photo been altered?" BBC Verify, 11 March 2024. https://www.bbc.com/news/uk-68534289

**519**

- Joshua Rothman, "Dept. of Technology / Afterimage / Now that everything can be faked, how do we know what's real?," *The New Yorker*, 12 November 2018, pages 34 →44.
- Kashmir Hill & Jeremy White, "It's Getting Easier to Create the Fake Faces in This Crowd," *The New York Times*, Science Times, 24 November 2020, page D2.
- *R v Nikolovski* [1996] 3 SCR 1197 (Supreme Court of Canada, on appeal from Court of Appeal for Ontario). https://scc-csc.lexum.com/scc-csc/scc-csc/en/item/1457/index.do

**520** https://en.wikipedia.org/wiki/Radar
**521** https://en.wikipedia.org/wiki/Luftwaffe
**522** See also: Robert Harris, *V2*; Knopf; 2020; ISBN 9780525656715.
https://www.penguinrandomhouse.com/books/607955/v2-by-robert-harris
**523** https://canadianarchaeology.com/caa/sites/default/files/page/technical_guide_final_april_11_sh.pdf
Canadian Archaeological Association Working Group on Unmarked Graves. 2023. Searching for Missing Children: A Guide to Remote Sensing Techniques.
**524** https://www.toronto.ca/services-payments/streets-parking-transportation/road-safety/vision-zero/safety-initiatives/automated-speed-enforcement/
**525** α = alpha particle = 2He4 ++ = helium + 2 ion.
β- = beta particle = $_1e^{0\,-}$ = electron.
γ = gamma ray, with energy determined by its wavelength λ. $E = h\,c / \lambda = h\nu$
$t_{1/2}$ = half-life = $\log_e 2$ / radiation decay constant = .69315 / radiation decay constant
= time for half of the radioactive isotope to disappear by decaying into becoming another isotope which might have decay kinetics of its own.
**526** https://en.wikipedia.org/wiki/Isotopes_of_lead
**527** http://www.rsc.org/periodic-table/
**528** http://www.rsc.org/periodic-table/element/86/radon
**529** http://10,043-92-2.msds-cas.com/
**530** http://en.wikipedia.org/wiki/Isotopes_of_radon
**531** Brookhaven National Laboratory – Chart of Nuclides http://www.nndc.bnl.gov/chart/
**532** M.G., *The Law of Occupational Health and Safety in Ontario, Second Edition*, Butterworths, Toronto and Vancouver, August 1994, ISBN 0-409-90414-7; APPENDIX II – radon.
**533** See: M.G., " CELA report on Radon," TORONTO WORKERS' HEALTH & SAFETY LEGAL CLINIC *newsletter*, January 2015. Vol. 23 No.1, pages 1, 2, 3. http://newsletter.workers-safety.ca.s3-website-us-east-1.amazonaws.com/#!/newsletters%2 F2015%2001%2F
**534** See also: "U of T engineering students' proposal for Ontario Building Code changes for Radon Gas," TORONTO WORKERS' HEALTH & SAFETY LEGAL CLINIC *newsletter*, Aug. 2016. Vol. 24 No.2, pages 1, 2, 3

August 2016 – Issue: Vol. 24, No. 2
https://s3.amazonaws.com/newsletter.workers-safety.ca/newsletters/Clinic + Newsletters/ 2010-present/ Vol + 24%2 C + No + 2%2C + August + 2016/2016%2B08.%2BVol.24%2BNo.2.pdf
**535** CAS 14859-67-7
**Radon-222**
$_{86}\mathbf{Rn}^{222}$
$t_{½}$ = 3.823 days [Xe] $4f^{14}$ $5d^{10}$ $6s^2$ $6p^6$
**536** http://www.epa.gov/radon/pubs/citguide.html
**537** http://en.wikipedia.org/wiki/Radon
**538** American Chemical Society – Periodic Table http://acswebcontent.acs.org/games/pt.html
**539** American Chemical Society – Periodic Table
www.acs.org/content/acs/en/education/whatischemistry/periodictable.html
**540** Radium-226
**88Ra226**
t½ = 1,600 years.
**541** Helen Mann and Chris Howden, CBC radio, *AS IT HAPPENS*, "Radium Girls," 30 December 2014 – interview of Deborah Blum.
https://www.cbc.ca/radio/asithappens/tuesday-ferry-boat-survivor-radium-girls-author-no-possum-drop-and-more-1.2941016/remembering-mae-keane-the-last-of-the-radium-girls-1.2941019#:~:text= News-,Remembering%20Mae%20Keane%2C%20the%20last%20of%20the%20%22Radium%20Girls%22, Last%20Updated:%20December%2029%2C%202014
**542** Deborah Blum, *The Poisoner's Handbook: Murder and the Birth of Forensic Medicine in Jazz Age New York*, Penguin Press, 2010; ISBN 978-1-101-52489-3.
http://deborahblum.com/The_Poisoners_Handbook.html
http://deborahblum.com/Home_Page.html
http://www.barnesandnoble.com/sample/read/9780143118824
http://www.publishersweekly.com/978-1-4001-1550-1
**543** William Yardley, "Mae Keane, 107, Whose Job Brought Radium to Her Lips / Many co-workers in a [Connecticut] watch factory died of cancer in their 20s," *The New York Times*, OBITUARIES, 14 March 2014, page B10.
**544** ". . . 1920s . . . paint . . . numbers on wristwatch dials: Put the tip of the tiny brush between your lips to shape the bristles into the finest of points. . . . Watchmakers liked it because it glowed in the dark. Later, it became clear that it killed. . . . radium girls . . . lost their teeth . . . jawbones . . ."
**545** https://www.nytimes.com/2014/03/14/nyregion/mae-keane-who-painted-watch-dials-with-a-radium -mixture-dies-at-107.html
**546** See, for example – regarding the medical-use isotopes, *molybdemun-99* & *technetium-99 m*, produced from *uranium-235*: M.G., "Chalk River & government inspection," Toronto Workers' Health & Safety Legal Clinic *newsletter*, 2008 04. Vol.16, No.2, pages 3 & 4.
2008 04_Vol_16_No_2_TWH_SLC newsletter_.doc
http://newsletter.workers-safety.ca.s3-website-us-east-1.amazonaws.com/#!/newsletters%2F2008%2F
http://newsletter.workers-safety.ca.s3-website-us-east-1.amazonaws.com/#!/newsletters%2F http://workers-safety.ca/
**547** Historically from the late nineteenth and early twentieth centuries – See:

- "*KELLY CLINIC CLOSES; Pioneered in Treating Cancer With X-Ray and Radium*," Special to *The New York Times*, 31 December 1952.
  https://www.nytimes.com/1952/12/31/archives/kelly-clinic-closes-pioneered-in-treating-cancer-with-xray-and.html
- https://en.wikipedia.org/wiki/Radium#: ~:text = Radium%20is%20a%20chemical%20ele ment,as %20the%20alkaline%20earth%20metals.

- https://en.wikipedia.org/wiki/Howard_Atwood_Kelly
- CITY OF BALTIMORE, ONE HUNDRED AND FORTIETH ANNUAL REPORT OF THE DEPARTMENT OF HEALTH, 1954; page 261: ". . . *Industrial Exposures* / The radium study at the Kelly Clinic at 1416–18 Eutaw Place was completed after the clinic's radium supply was removed along with laboratory and office equipment when the building was vacated after having been used for 50 years . . ." http://health.baltimorecity.gov/sites/default/files/City%20of%20Baltimore%20140th%20Annual%20Report%20of%20The%20Department% 20of%20Health.pdf
- Kelly Clinic, 1408 Eutaw Place, Baltimore, 21217 https://explore.baltimoreheritage.org/items/show/103
- https://www.hopkinsmedicine.org/gynecology_obstetrics/specialty_areas/gynecologic_oncology/
- https://www.heinzhistorycenter.org/magazine/Radium-City.pdf
- https://history.aip.org/exhibits/rutherford/sections/atop-physics-wave.html

**548** https://www.sciencedirect.com/topics/computer-science/radio-frequency-identification https://en.wikipedia.org/wiki/Radio-frequency_identification

**549** See: Walter J. Moore, *Physical Chemistry*, Prentice-Hall, Inc., Englewood Cliffs, New Jersey, 1962 & 1963, page 602, et seq.

**550** Peter Atkins, *Concepts in Physical Chemistry*, W.H. Freeman and Company, New York City 10,010; 1995; page 276, et seq.; ISBN 0-7167-2928-8.

**551** in Stuart H. James and Jon J. Nordby, editors, *Forensic Science / An Introduction to Scientific and Investigative Techniques*, CRC Press, Boca Raton, 2003; ISBN 0-8493-1246-9; at page 268, et seq. www.crcpress.com

**552** https://en.wikipedia.org/wiki/Raman_spectroscopy
https://www.horiba.com/int/scientific/technologies/raman-imaging-and-spectroscopy/raman-spectroscopy/
https://www.horiba.com/esp/scientific/resources/science-in-action/can-raman-microscopy-reveal-art-forgeries/
https://analyticalscience.wiley.com/do/10.1002/sepspec.1882education/full/ https://www.mt.com/ca/en/home/applications/L1_AutoChem_Applications/Raman-Spectroscopy.html

**553** Raman spectroscopy. Paint analysis. Chapters 3 & 6.

**554** See: Steven Weinberg, *DREAMS OF A FINAL THEORY*, Vintage Books, NewYork City, 1992, 1993, 1994. ISBN 978-0-679-74408-5.

**555** Weinberg, *DREAMS* . . ., page 202.

**556** See: W.H. Furry, E.M. Purcell, J.C. Street, *PHYSICS for Science and Engineering Students*, THE BLAKISON COMPANY, New York, Toronto, 1952:
- Chapter 31, "GENERAL BEHAVIOR AND NATURE OF LIGHT," pages 421→440;
- Chapter 32, "IMAGES," pages 441→453;
- Chapter 33, "LENSES," pages 454→466;
- Chapter 34, "OPTICAL INSTRUMENTS," pages 467→478.

**557** Or from one transparent medium to another.

**558** Furry, Purcell, Street, *PHYSICS* . . ., at page 427.

**559** Clarence Rainwater, Herbert S. Zim, Raymond Perlman; Vera R. Webster, ed.; *LIGHT AND COLOR*, GOLDEN PRESS SCIENCE GUIDE, GOLDEN PRESS, New York City, Western Publishing Company, Racine, Wisconsin; 1971; Library of Congress Catalog Card Number 69–11,967; "RE-FRACTION," at pages 42, 43, 44.
ISBN 0-307-63540-6
https://kupdf.net/download/light-and-color-a-golden-guide_5a1544c0e2b6f5f854f5e9c3_pdf

**560** Hazel Rossotti, *COLOUR / Why The World Isn't Grey*, Princeton University Press, Princeton, New Jersey, 08540; 1983; ISBN 0-691-02386-7; Chapter 2 "WHITE LIGHT ON CLEAR GLASS," page 26→34. https://press.princeton.edu/books/paperback/9780691023861/colour

**561** https://www.mathsisfun.com/sine-graph-exercise.html https://www.mathsisfun.com/algebra/trig-sine-law.html

**562** See: Howard A. Strobel, *Chemical Instrumentation / A Systematic Approach to Instrumental Analysis*, Addison-Wesley Publishing Company, Inc., Reading, Massachusetts & London, England, 1960; Chapter 9, REFRACTOMETRY, page 233→251.
See also: https://en.wikipedia.org/wiki/Refractive_index
n = speed of light in a vacuum / speed of light in a medium.

**563** Richard Saferstein, *CRIMINALISTICS / An Introduction to Forensic Science*, Seventh Edition, Prentice Hall, Upper Saddle River, New Jersey, 07458; 2001; ISBN 0-13-013827-4; page 94, et seq.

**564** *Gedankenexperiment*. Introduction 2.

**565** https://www.astm.org/DIGITAL_LIBRARY/JOURNALS/FORENSIC/PAGES/JFS10882J.htmhttps://www.documentcloud.org/documents/2801460-Spot-Tests-a-Color-Chart-Reference-for-Forensic.html
https://www.nist.gov/system/files/documents/2020/03/23/ASTM_WK70035_GEO_ color190806_OSACproposed.pdf

**566** Mike Nofi, "The Forensic Analysis of Color," *EVIDENCE TECHNOLOGY MAGAZINE*, Jan.-Feb. 2012. Vol. 10 No.1 Page 26, 27, 28.

**567** Definition as used here: *white light* = light of all wavelengths in the region of interest – here: UV and visible.

**568** See also:
https://www.ncbi.nlm.nih.gov/pmc/articles/PMC2841958/
https://www.cambridge.org/core/books/theory-of-reflectance-and-emittance-spectroscopy/reflectance-spectroscopy/C2B71AF4535899907ED5B8175188888A
https://www.sciencedirect.com/topics/agricultural-and-biological-sciences/reflectance-spectroscopy
https://en.wikipedia.org/wiki/Reflectance_difference_spectroscopy https://www.sas.upenn.edu/~rachelmr/labspec.html

**569** https://www.sciencedirect.com/science/article/pii/B9780128032244000194
https://www.sciencedirect.com/topics/agricultural-and-biological-sciences/reflectance-spectroscopy#:~:text=VIS https://www.sciencedirect.com/science/article/pii/B9780128032244000194

**570** https://www.filmetrics.com/technology

**571** https://www.britannica.com/science/color/The-psychology-of-colour
https://opg.optica.org/viewmedia.cfm?r=1&rwjcode=josa&uri=josa-63-5-615&html=true
https://en.wikipedia.org/wiki/Color_rendering_index
https://www.portrait.com/resource-center/visual-color-comparison/#:~:text=The%20dimensions%20for%20CIELAB%20color,dimension%20equals%201%20Delta%20E
https://colordesigner.io/compare-colors
https://hextoral.com/side-by-side/
https://plan-home.com/compare-colors
https://www.amnh.org/explore/ology/brain/seeing-color#:~:text=Special%20cells%20called%20rods%20and,%2C%20or%20colors%2C%20of%20light.

**572** See: W.H. Furry, E.M. Purcell, J.C. Street, *PHYSICS for Science and Engineering Students*, THE BLAKISON COMPANY, New York, Toronto, 1952; Chapter 37, "SPECTRA AND COLOR," pages 510→527.

**573** Clarence Rainwater, Herbert S. Zim, Raymond Perlman; Vera R. Webster, ed.; *LIGHT AND COLOR*, GOLDEN PRESS SCIENCE GUIDE, GOLDEN PRESS, New York City, Western Publishing Company, Racine, Wisconsin; 1971; Library of Congress Catalog Card Number 69-11967.
ISBN 0-307-63540-6
https://kupdf.net/download/light-and-color-a-golden-guide_5a1544c0e2b6f5f854f5e9c3_pdf

**574** Hazel Rossotti, *COLOUR / Why The World Isn't Grey*, Princeton University Press, Princeton, New Jersey, 08540; 1983; ISBN 0-691-02386-7.
https://press.princeton.edu/books/paperback/9780691023861/colour

**575** https://www.cbc.ca/radio/q/monday-dec-16-2019-michael-apted-ahmed-best-and-more-1.5395191/leatrice-eiseman-on-classic-blue-the-pantone-colour-of-the-year-for-2020-1.5395384
**576** See, for example: Raffi Khatchadourian, "PORTFOLIO / VITAL VESSELS / / PHOTOGRAPHS BY CHRISTOPHER PAYNE," *The New Yorker*, 07 December 2020, page 48→57.
https://www.newyorker.com/magazine/2020/12/07/the-race-to-make-vials-for-coronavirus-vaccines
http://www.chrispaynephoto.com/
**577** Rainwater, et al., at pages 50, 110, 111.
**578** Hazel Rossotti, at pages 60, 61.
**579** *KODAK WRATTEN FILTERS FOR SCIENTIFIC AND TECHNICAL USE,* 22nd edition; KODAK Publication No. B-3; 1965, 1962; Eastman Kodak Company, Rochester, New York, 14650.
**580** See:
https://www.brown.edu/Departments/Joukowsky_Institute/courses/greekpast/4929.html
https://www.sciencedirect.com/topics/biochemistry-genetics-and-molecular-biology/thermoluminescence
https://en.wikipedia.org/wiki/Thermoluminescence#:_09~:text_09=_09Thermoluminescence_is_a_form_of,upon_heating_of_the_material https://www.britannica.com/science/thermoluminescence
**581** See also: Patrick T. McCutcheon, "Thermoluminescent Analysis of Burned Bone: Assessing the Problems," as Chapter 22, page 311→326, in Mary Virginia Orna, ed., *Archaeological Chemistry / Organic, Inorganic Biochemical Analysis,* ACS Symposium Series 625, American Chemical Society, Washington, D.C., 1996.
**582** See also: Pamela S. Zuer, "Archaeological Chemistry / Physical Science helps to unravel human history," *Chemical & Engineering News*, SPECIAL REPORT, 21 February 1983, page 29.
**583** Farrington Daniels, J.W. Williams, Paul Bender, Robert A. Alberty, C.D. Cornwell, *Experimental Physical Chemistry, Sixth Edition*, McGraw-Hill Book Company, Inc., New York; 1962; pages 147→157, 456.
**584** Lejaren A. Hiller, Jr. & Rolfe H. Herber, *Principles of Chemistry*, McGraw-Hill Book Company, Inc., New York, . . ., 1960, pages 338→344.
https://pubs.acs.org/doi/10.1021/ed037p495.1
**585** Peter Atkins, *Concepts in Physical Chemistry*, W.H. Freeman and Company, New York City 10010; 1995; pages 346→348; ISBN 0-7167-2928-8.
**586** Alan Furst, *THE POLISH OFFICER*, A NOVEL, Random House, New York, 1995; pages 76-78; "The chemist and the commodity analyst." ISBN 978-0-375-75827-0.
https://www.fantasticfiction.com/f/alan-furst/polish-officer.htm
**587** In this novel, intelligence personnel of the Polish government-in-exile used viscosity data of oil analyzed from scavenged cloth rags used by the *Wehrmacht* when lubricating rifles and other equipment; to try to predict a winter offensive; too cold would require another oil grade.
**588** That modern analytical chemical methods largely involve instrumentation, rather than screening tests and wet chemistry, is reflected in this book's presentations.
**589** J.A.Siegel, "Forensic Identification of Controlled Substances," as Chapter 3 in R. Saferstein, ed., *Forensic Science Handbook* Volume II, Prentice Hall, Englewood Cliffs, New Jersey, 1988, pages 87→92.
**590** K. Bailey, "The value of the Duquénois Test for Cannabis – a Survey," *Journal of Forensic Sciences*, 1979, 24(4), pages 817→841.
**591** Daniel M. Perrine, *The Chemistry of Mind-Altering Drugs / History, Pharmacology, and Cultural Context*; American Chemical Society, Washington, D.C.; 1996; ISBN 0-8412-3253-9; page 347, et seq. – at page 347 7–10.
**592** http://en.wikipedia.org/wiki/Tetrahydrocannabinol
**593** *Duquénois-Levine*, Siegel, Saferstein, page 124.
**594** For example: "NIK® TEST G – COCAINE . . ."
https://forensicssource.com/products/nik-test-g-cocaine-crack-and-free-base-1006155?srsltid=Afm

BOorMdYnmbITLfZL_KGZIRpwOYFEFmZXHW_FXoJaFvlV8aAtow-W_
https://cdn.bfldr.com/TNZ0YWIS/at/8vp4r6976ssfv8g9pnnnpk6b/NIK_Test_G_Instructions_1004916_190-527_English.pdf
**595** https://www.safariland.com/products/forensics/field-drug-tests/nik-drug-tests-and-kits/
http://www.safariland.com/products/forensics/field-drug-tests/nik-test-g-cocaine-crack-and-free-base-1006155.html#start=1
https://www.safariland.com/resources_content/resource-landing-page.html
https://www.safariland.com/about-history.html
https://www.safariland.com/contact-us-new/contact-us-details.html
NIK®, Safariland, Jacksonville, Florida, 32218.
**596** https://en.wikipedia.org/wiki/Cobalt
**597** Cobalt(II)thiocyanate Co(SCN)2(H2O)3 cobalt thiocyanate test = Scott test. CAS 3017-60-5
https://www.oxfordreference.com/display/10.1093/acref/9780199594009.001.0001/acref-9780199594009-e-0225
**598** Alexander Aciman, "I Went Down a Lead-Testing Rabbit Hole. Here's What You Need to Know About Home Tests," 11 December 2024, *New York Times*, "HEALTH AND FITNESS / Wirecutter."
https://www.nytimes.com/wirecutter/reviews/home-lead-test-review/
https://www.amazon.ca/dp/B0CF6SV33S?ascsubtag=F0401JEVD9SPW03BT2W4KZGNPXB87&linkCode=gs2&tag=thewire0f-20&th=1
**599** However, within the link to **Lumetallix**, there was apparently no scientific indication of how this process works. No references to:

- Product specifications that would reveal the analytical chemistry.
- Links to technical literature from the manufacturer or supplier.
- Scientific literature.
- Safety data sheets for the chemical contents of the marketed kit, nor the UV light – as might be required by US cfr for hazard communication or Canada WHMIS.
- US or other patent numbers.

**600** "SHERLOCK HOLMES – The Naval Treaty – An Appeal to Holmes – By Sir Arthur Conan Doyle" Comic strip EPISODE 1, *The Boston Globe*, 05 December 1930, page 44. Illustrator = Leo O'Mealia. [Bell Syndicate (?)]
https://www.arthur-conan-doyle.com/index.php?title = The_Naval_Treaty_
https://www.arthur-conan-doyle.com/images/a/a9/The-boston-globe-1930-12-05-the-naval-treaty-p44-illu.jpg
https://en.wikipedia.org/wiki/Adaptations_of_Sherlock_Holmes#Comic_strip
**601** (The copyright provenance of this comic strip was too difficult to determine to be copied here.)
https://www.smithsonianmag.com/smart-news/sherlock-holmes-now-officially-copyright-and-open-business-180,951,794/
https://www.torontopubliclibrary.ca/detail.jsp?Entt = RDM364,990&R = 364,990
http://www.sherlockian-sherlock.com/
https://fourthgarrideb.com/2016/01/01/science-group-honors-sherlock-holmes-2002/
**602** Louise Haskett, "The Chemistry of Sherlock Holmes," [1983, 2003, 2025.].
http://surrey-shore.freeservers.com/HolmChem.htm
http://surrey-shore.freeservers.com/whats_new.html
http://surrey-shore.freeservers.com/index.html
**603** https://en.wikipedia.org/wiki/Wet_chemistry
**604** "nik® . . . TEST B – NITRIC ACID FOR: Opium Alkaloids.
https://www.copquest.com/nik-narcotic-test-kits-test-b-confirming-test_43-2101.htm
**605** instructions, SDS https://www.copquest.com/knowledgebase/index.htm#nik-instructions
**606** Nitric Acid HNO3 CAS 7697–37-2

**607** nik® PUBLIC SAFETY, Jacksonville, Florida, 32218.
**608** www.forensicssource.com
Home > Narcotics Identification > NIK® Drug Tests
**609** "nik® . . . TEST L FOR: Heroin.
https://forensicssource.com/products/nik-test-l-heroin-1006159?variant=34495759155245
**610** Instructions https://cdn.bfldr.com/TNZ0YWIS/at/nsm4ksx5wsmt9vwwggrcsvx/NIK_Test_L_Instructions_1004920_190-531_English.pdf
**611** Selenious Acid H2SeO3 CAS 7783–00-8 Sulfuric Acid H2SO4 CAS 7664–93-9
**612** "nik® . . . TEST F – ACID NEUTRALIZER . . ."
https://forensicssource.com/products/nik-test-f-acid-neutralizer-1006154?srsltid=AfmBOorz1-FlVvtTkT9taLZnYma0P7z_1CYWnaFTGCk4t1tIUyg2hWEA
**613** instructions https://cdn.bfldr.com/TNZ0YWIS/at/bvcr342t8ghcfvqgtmbc5fk8/NIK_Test_F_Instructions_1004915_190-526_English.pdf
**614** Soda Ash = sodium carbonate Na2CO3 CAS 497–19-8
**615** See: Richard Saferstein, *CRIMINALISTICS / An Introduction to Forensic Science*, Seventh Edition, Prentice Hall, Upper Saddle River, New Jersey, 07458; 2001; ISBN 0-13-013827-4; Chapter 6, pages 160, 161
**616** Thomas A. Kubic & Nicholas Pertaco, "Microanalysis and Examination of Trace Evidence," in Stuart H. James & Jon J. Nordby, editors, *Forensic Science / An Introduction to Scientific and Investigative Techniques*, CRC Press, Boca Raton, 2003; ISBN 0-8493-1246-9; Chapter 14, at pages 260, 261. www.crcpress.com
**617** H.L. Chen, D.W. Foreman & Kathryn A. Jakes, "X-Ray Diffractometric Analyses of Microstructure of Mineralized Plant Fibers," as Chapter 15, in Mary Virginia Orna, ed., *Archaeological Chemistry / Organic, Inorganic Biochemical Analysis,* ACS Symposium Series 625, American Chemical Society, Washington, D.C., 1996.
**618** Farrington Daniels, J.W. Williams, Paul Bender, Robert A. Alberty, C.D. Cornwell, *Experimental Physical Chemistry, Sixth Edition*, McGraw-Hill Book Company, Inc., New York; 1962; "*Diffraction,*" Chapter 12, pages 288→295; page 293, "Fig. 70. X-ray powder camera."
**619** Lejaren A. Hiller, Jr. & Rolfe H. Herber, *Principles of Chemistry*, McGraw-Hill Book Company, Inc., New York, . . ., 1960; page 271, et seq.
**620** https://www.britannica.com/science/crystal-defect
https://en.wikipedia.org/wiki/X-ray_crystallography

# Chapter 7
# When

[. . .][1]

. . . [stab] *wounds* . . . [1, 2, 3, 4]

[. . .]

. . . *pocket . . . watch . . . dented . . . hands pointed to . . . hour of the crime* . . . [5]

[. . .]

[Agatha Christie, *Murder on the Orient Express*, 1934] [6].

## 7.1 When – time: chemical kinetics

Time is of special concern in much of human culture [7, 8, 9, 10, 11, 12, 13, 14, 15, 16, 17, 18, 19, 20, 21].[2,3,4,5] Science, engineering, industrial manufacturing, government, commerce, and navigation on Earth and for space-travel, are all time-knowledge-dependent. Earth turns – time carefully observed –on its axis; geographic mapping requires the concept and practical application of longitude – which was established with accurately knowing anywhere on Earth the time at the Royal Observatory, Greenwich [22, 23]; and is related to the location of time zones. Arguably, the Royal Navy [24] was able to "rule the waves" with the eighteenth-century availability of John Harrison's accurate clock [25, 26].

The very concept of history is of time. Similar for forensic science.

Time is a special kind of quantification for aspects of the human stories that forensic science – particularly here as analytical chemistry – can help explain. This can be related to the *chemical kinetics* [27, 28] of processes studied by the forensic chemist. Chemical kinetics deals with the times taken for various quantities or concentrations of reactants, intermediates and final products to be depleted and created [29, 30]. This is related to the *reactions in progress* mentioned previously [31].[6,7]

Thus, the forensic chemist might be able to use some chemical reactions as a kind of "clockwork" [32, 33, 34] or "calendar" to help determine the *when* [35] of a human story. The face of the clock or calendar pages (to continue the metaphor) would be the quantities or concentrations of chemical substances measured as created or depleted, as at a time t=t, advancing from a start time of $t_0$=0.

---

1 Introductory quotes. Introduction 1.2.

2 When. Chemical kinetics. Reactions in progress. Chapters 2, 4 and 7

3 Time scale. Chapters 5 and 7.

4 When. Chapters 2, 4 and 7.

5 Time dilatation of relativity theory: $T = T_0 /\sqrt{(1 - (v/c)^2)}$. Chapters 2 and 7.

6 Chemical kinetics. Reactions in progress. Chapters 2, 4 and 7.

7 Energy. Thermodynamics. Entropy. "Time's Arrow" Chapters 4, 6 and 7.

https://doi.org/10.1515/9783111618555-008

However, a chemical kinetic theoretical accounting of forensic measurements can be very complex, so that empirical methods – estimated conclusions from other experimental observations – would usually be required. This includes biological changes, which might be regarded as complicated chemical reactions; for example postmortem interval for medical estimation of time of death. In this sense, all life – human, animal, plant, cellular, viral – may be thought of as complicated sets and series of chemical reactions – measurable, at least in concept, as to quantity and thermodynamic data as against time of chemical kinetics. In practice, difficulty would arise not only from the often unknown details of the complexities of the chemical reaction mechanisms, but also from various and changing conditions of temperature, humidity, concentration, isolation of the sample, catalysis, geometry, the container and contamination for the samples being examined.

However, a review of some of the very fundamentals of kinetic theory would be helpful for understanding.

## 7.2 Fundamentals of chemical kinetics

Consider a hypothetical molecule AB that, at appropriate conditions, undergoes a chemical reaction – a decomposition – to deplete the reactant AB and to create products A and B [36]:[8]

$$AB \rightarrow A + B$$

Assume that N molecules [37, 38][9] of AB as a gas are alone in a closed container at concentration and temperature high enough for the particles of that gas (the AB molecules) to collide with one another very frequently. Also assume that the container is large enough for reaction with and at the container walls to be neglected and that the temperature is high enough to allow breaches of vibration energy thresholds to break the A–B bond, so that the reaction proceeds. But low enough so that the AB does not decompose or oxidize or otherwise react than as written above. And, also assume that the reaction proceeds to near completion [39], so that eventually nearly all of AB is gone. These may not be unusual assumptions for a theoretical model, rather than for an actual forensic problem.

In such hypothetical circumstances, this reaction's progress over time to deplete AB and to subsequently create A and B may be expressed. At a particular temperature:

$$-(d[AB]/dt = +(d[A]/dt) = +(d[B]/dt) = k \times [AB]$$

8 Too much scientific technical detail here? Introduction 2.1.

9 Mole. Avogadro's number. Chapters 3, 4 and 7.

where the symbols here are given in Tab. 7.1.

Tab. 7.1: Chemical kinetic symbols.

| | | |
|---|---|---|
| d[ ]/dt | = | a calculus [40] notation for the incremental time rate of change of a chemical *concentration* (mol/L s) [41]. |
| [ ] | = | *concentration* (mol/L). (For the symbolism here, the empty space within the square brackets would contain the designation of the substance of interest – so that, for example, $[Cl^-]$ would represent the concentration of chloride ions measured in mol/L.) |
| t | = | *time* (seconds) [42] |
| k | = | *rate constant* (1/seconds) |

This simple differential equation [43] may be solved by exact or approximate methods; k may be determined by independent experiment. With k known, the times t it took for AB to be depleted and for A to be created might be calculated.

Such reactions can analogously occur with AB as solute molecules in liquid solution [44], or in more complex ways for AB as a solid [45].

Decompositions, such as above, are only one of very many of kinds of chemical reactions [46] possible, occurring in very many different ways, and often involving rather complex geometries, other reactants, catalysts [47] and multistep mechanisms. Their kinetic expressions may likewise be complex, often with multiple rate constants ($k_1$, $k_2$, $k_3$, . . ., $k_n$, . . .) and other factors, in complex mathematical expressions.

In practical terms, the complexities of chemical kinetics essentially mean that forensically usable systems – with mathematical expression – for which there is a detailed theoretical understanding would not be too often found. Fortunately, as indicated above, such details of theory would typically not be necessary to know, because an empirical (based on controlled and documented experimental experience) approach is sometimes possible, so that only a general understanding of kinetic concepts is needed – as, for example, when a pathologist makes a determination of time of death.[10] Unfortunately, for many, if not most, other forensic problems, reliable detailed kinetic determinations – neither theoretical nor empirical – would be known. Perhaps a most enticing – and elusive – forensic problem is dating when fingerprints were laid down.[11]

However, there is one forensically usable system for which theoretical details are not only well known, but are also relatively simple – presented as a further example to explain theoretical concept:

10 See below.

11 Fingerprint probability. Chapters 1 and 4. Fingerprint visualization. Fluorescence and phosphorescence. Chapters 6 and 9. Fingerprint when. Chapters 7 and 8.

## 7.3 Carbon-14 dating

The forensic method for which theoretical details are well known is carbon-14 dating [48, 49, 50, 51].[12,13,14] It is useful both because of ease of measurement, and because the nuclear processes on which it is based are generally independent of the forensic sample's temperature, humidity, concentration and contamination.

Carbon-14 dating is usually associated with archaeology[15] – rather than forensics – but the concepts are essentially the same. And, carbon-14 might be relevant to help test for fraudulently claimed antiques.

Carbon-14 dating is useful for things containing organic chemical substances – such as wood, paper, cloth of natural source, leather, and human and animal remains. "Organic" here means composed of the mostly carbon and hydrogen atoms of organic chemistry; but these organics would also have been related to living systems.

The Earth's atmosphere is always bombarded by solar and cosmic radiation (somewhat directed by Earth's magnetic field). One result is a nuclear reaction in the atmosphere [52]:[16,17,18,19,20]

$$ {}_0n^1 + {}_7N^{14} \rightarrow {}_6C^{14} + {}_1p^1 $$

Where the symbols here are defined in Tab. 7.2.

Tab. 7.2: Atomic symbols.

| | | |
|---|---|---|
| n | = | Neutron, atomic mass = 1 (g/mol), charge = 0. |
| N | = | **Nitrogen** atom, atomic mass = 14 (g/mol), with 7 protons and 7 neutrons – this is the nitrogen isotope of predominant natural abundance on the Earth. |
| C | = | Carbon atom, atomic mass = 14 (g/mol), with 6 protons and 8 neutrons – this **carbon-14** is not the carbon isotope of predominant natural abundance (which is carbon-12) – carbon-14 abundance is small on the Earth = $1 \times 10^{-10}$ % [53, 54].[21] |
| p | = | Proton, atomic mass = 1 (g/mol), charge = +1. |

12 Shroud of Turin. Carbon-14. Introduction 2; Chapters 3 and 7.
13 Too much scientific technical detail here? Introduction 2.1.
14 Isotope. Radioactive. Carbon-14. Chapters 4, 6 and 7.
15 Archeology. Carbon-14. Chapters 2 and 7.
16 Too much scientific technical detail here? Introduction 2.1.
17 Fundamental particles. Chapters 4 and 7.
18 submicroscopic particles: $p^+$, electrons $e^-$, neutrons n. Chapters 2, 4 and 7.
19 Neutron. Atomic Number. Chapters 2, 4 and 7.
20 Periodic Table. Chart of Nuclides. Chapters 4 and 7.
21 But see below where radioactive carbon-14 = ${}_6C^{14}$ = is also symbolized by C* (and about another possibly confusing notation).

The resulting carbon-14 combines chemically with oxygen in the atmosphere to form carbon dioxide $CO_2$, which is chemically essentially the same as for the more abundant – ~100% – carbon-12. $CO_2$, with both isotopic forms, mixed, interacts within the biosphere, so that C becomes incorporated into the biochemistry of plant tissue, by way of respiration. And, since animals eat plants, or eat other animals that have eaten plants, and there would also be exchanges with wastes and remains, both plants and animals have carbon-14-containing tissue – as a very small – but measurable – component of the total C.

But carbon-14 decays [55, 56]:

$$_{6}C^{14} \rightarrow {}_{7}N^{14} + {}_{-1}\beta^{0}$$

where: β = beta particle, atomic mass = 0, charge = –1 [57, 58, 59].

However, the decayed carbon-14 is always being replaced by fresh carbon-14 because living plants are always breathing, and living animals eating. Ideally, a steady-state concentration of carbon-14 is maintained – so long as the plant or animal lives. When it dies, there is no replacement carbon-14, and the remaining carbon-14 continues to decay. The carbon-14 concentration then decreases from the steady-state concentration that there was just before death.

Thus, kinetics for the above nuclear decay reaction can be used to determine the time since death. Since parts of previously living things are found archeologically, their time of death can help determine the *when* of a related story – of interest here, perhaps, for a human story.

A kinetic rate expression may be written and manipulated [60]:[22]

$$-(d[C^*]/dt) = k \times [C^*]$$

$$\int (d[C^*]/[C^*]) = -k \int dt$$

where [C*] is the carbon-14 concentration.[23]

Then

$$t = (1/k) \times \ln[C^*]_{t=0}/[C^*])\ [61]$$

where

$$[C^*]_{t=0} = [C^*]_0 \quad \text{at} \quad t = 0$$

22 Too much scientific technical detail here? Introduction 2.1.

23 But beware about C* symbolism – it has quite different definition/meaning in two unrelated contexts:

C* = an asymmetric carbon – indicating chirality.

C* = radioactivity of an atom – here carbon.

$[C^*]_0$ = a reference carbon-14 constant concentration over the time of the radiation bombardment of the Earth. (But $[C^*]_0$ is not really constant – see below.)

Define **half life** :

$t_{1/2}$ = the time it takes for half the radioactive material to disappear.

$t_{1/2}$ = **5730 years** = carbon-14 half-life determined by independent experiment.

$$t_{1/2} = \ln(2/1)/k = 0.6931/k = 5730 \text{ [62, 63]}$$

$$1/k = 5730/0.6931 = 8267.21$$

$$t = 8267 \times \ln([C^*]_0/[C^*]) \text{ (years) [64]}$$

Then: Knowing $[C^*]_0$ as a reference, and measuring $[C^*]$, t = time since death can be calculated.

In practice, $[C^*]_0$ and $[C^*]$ would not be known directly, but rather as proportions of the total carbon content of the samples for which the measurements would be made: $[C^*]_0/[C]_{total}$ and $[C^*]/[C]_{total}$ [65, 66]. An approximate sketch for carbon-14 exponential decay – for percentage of carbon-14 remaining – is shown in Fig. 7.1 [67].

To conveniently use carbon-14 for dating, $[C^*]_0/[C]_{total}$ should ideally be constant over the thousands of historical years for which the method is actually used. Reality is not so ideal – $[C^*]_0/[C]_{total}$ is not reliably constant, because the radiation flux and magnetic field may not be constant [68]. And, measurement is affected by the massive burning of fossil fuels since the later part of the eighteenth century (the industrial revolution's burning of coal would put much older carbon into the atmosphere; and more modern society also burns petroleum products) [69]; and because atomic bomb testing after WWII caused additional atmospheric radiation fluxes [70].

However, for particular time periods, $[C^*]_0/[C]_{total}$ can be set by calibration using artefacts of otherwise known human history (e.g. the binding or non-wood source [71] pages of an old book) or of otherwise known natural history (e.g. from identified tree rings [72, 73, 74]).

Although carbon-14 can be often be used to determine t with reasonable accuracy – perhaps within a few percent [75] – care must be taken not to confuse t with the time ago of the related human story – they are not necessarily the same. As a fictitious example [76], the carbon-14 dating of an ancient queen's wooden bowl might actually predate her majesty by decades, because, perhaps, wood of interior tree rings was used, or, perhaps, the wood – or the bowl itself – was somehow kept in storage for decades before her cipher was engraved into it. Care must also be taken to avoid errors that might be introduced into the measurement of t by the original craft processes – such as inlaying the bowl with wood of other source, or treating it with oils or colourants. Errors could also arise by environmental process, over the years, such as exposure to – and absorption of – the newer carbon of smoke.

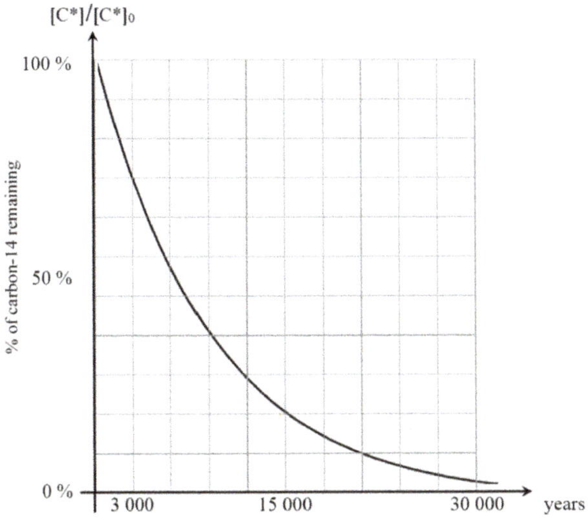


Fig. 7.1: Carbon-14 decay.

There is ongoing controversy about the reliability of the carbon-14 dating of the Shroud of Turin.[24]

## 7.4 About complex systems – empirically

As mentioned earlier, the complex detail of kinetic theory is fortunately not always needed for chemical forensic determinations of time, because an empirical (based directly on documented experimental experience) approach may be possible. Such an empirical approach would need to be based only on a general understanding of kinetic concepts. Two examples:

### 7.4.1 Age of bread

Most everybody finds that bread tastes best just after it is baked; a day later, it may still taste good – but not as good (hence day-old price reduction); within several days it becomes stale; and within several days more it may become mouldy [77]. It has aged with a sequence that could reveal an estimate of how old it is. In food science terms, a complex of biochemical reactions is occurring within the mixtures of chemical compounds that comprise the bread, and between those compounds and things of

**24** Shroud of Turin. Carbon-14. Introduction 2; Chapters 3 and 7.

the bread's environment (such as ambient moisture, oxygen, microorganisms (perhaps mould-causing) and photons from the sun). Also, the environment may influence these reactions in relation to catalytic factors, geometric shape, the container, barometric pressure and temperature, and insects and scavenging animals.

In principle, it is therefore possible to speak generally of bread ageing in terms of chemical kinetics. Such principle may be useful to the forensic scientist, even in the absence of chemical kinetic details. By an empirical approach, a forensic scientist (here, best, a food scientist), could experimentally produce a chart [78] of the some of the various stages of ageing of various kinds and brands of bread, with and without various preservatives, and under various conditions. Such a chart would also state defined protocols for scientific replication.

Also, if bread is still with its sale wrapper from time of manufacture – baking – some forensic data may be stated on the wrapper – including a best-before date.[25] And maybe a day-old or reduced-price label might help.

### 7.4.2 Pathology of time of death

More complex than bread, and compelling of more intense forensic attention, are dead human bodies. It is often possible to establish the time of death – *postmortem interval* – by applying the same general principles of chemical kinetic empirical observations, so that human [79] tissue decay, temperature [80] and parasite activity [81] may be indicators of time passage to a *forensic pathologist* [82]. There is a substantial literature on the complex of pathological possibilities [83, 84, 85, 86].[26,27]

## 7.5 Examples

It may be worthwhile to try to apply, from general concept, new approaches to forensic problems. As with the examples of bread and human bodies, the details of complex chemical kinetic mechanisms may not need to be known. For example:
- "A Microspectrophotometric Method for Dating Ballpoint Inks – A Feasibility Study" [87];[28]
- "Determining the relative age of ball-point ink using a single solvent extraction mass-independent approach" [88];
- "Ink dating – the state of the art" [89];

25 Bread. Best-before. Chapters 2.3.1.1 and 7.4.1.

26 Edmond Locard. Sherlock Holmes. Cocaine. Chapters 2, 3, 4 and 6.

27 Edmond Locard. Sir Arthur Conan Doyle. Forensics. CSI. Chapters 2.1.1 and 5.2.1.

28 When. Ink. Chapters 2 and 7.

- "A sequential multiple approach to determining the relative age of writing inks" [90];
- "Studies on Age Estimation Using Racemization of Aspartic Acid in Cementum" [91];
- "Kinetics of Ascorbic Acid Loss and Nonenzymatic Browning in Orange Juice Serum: Experimental Rate Constants" [92];
- "Initial Studies on Insect Succession on Carrion in Southwestern British Columbia" [93];
- "Estimating Maggot Age from Weight Using Inverse Prediction" [94];
- "Dating Flint Artifacts with Electron Spin Resonance: Problems and Prospects" [95, 96];
- Dating of ancient pottery by thermoluminescence [97];
- "Volume Traps – A New Retrospective Radon Monitor" [98];
- "Pollen key to discovering secret of mass grave in Magdenburg/New forensic method helps solve mystery of 32 young victims" [99].
- Use of dental x-ray to determine child age [100].
- Changes, over time, in materials used in art works [101, 102].[29,30]
- And in fictional literature too:[31] As in a story [103, 104] in which police forensics discovers a "nest of tiny spiders" in one of a pair of gloves – "Long time since a hand was in that glove" [105, 106].

But, some research that would look promising and worthwhile does not yield hoped-for comprehensive results [107] – for example, the efforts to date fingerprints reported by E.R. Menzel [108, 109, 110, 111, 112].[32] Although in principle, such dating, in some contexts, should be possible.

## 7.6 Chemical markers of *when* (other than chemical kinetics)

So far, this chapter has dealt with determining the *when* of the human story by assuming that a chemical reaction progresses after an incident has occurred, and that the passage of time might be determined by consideration of chemical kinetics. But there is another possibility, not of *chemical kinetics*, to use chemical analysis to determine time – or time ranges.[33] This relies on a known history of a chemical substance revealed by analysis – such substances might be called *chemical markers of when*.

---

**29** MALDI MS. art paint analysis. Chapter 7.

**30** Art analysis. Colour change. Chapters 3 and 7.

**31** Fictional literature. Ethylene glycol. CAS 107-21-1. Introduction 2.2; Chapters 4 and 7.

**32** Fingerprint probability. Chapters 1 and 4. Fingerprint visualization. Fluorescence and phosphorescence. Chapters 6 and 9. Fingerprint when. Chapters 7 and 8.

**33** Manufacture history when. Chapters 2 and 7.

As a fictitious example [113], if a purported corporate share certificate were found to be printed with ink that was known to have not been manufactured until after the shares' supposed issue date, a forensic chemist should conclude that that date is in contradiction – and a banker who allowed those shares as loan security might cry criminal fraud (– or maybe just simply weep).

In this example, the *chemical marker* was the ink. In principle, anything manufactured can be, or contain, such *markers*. In Chapter 2, it is mentioned about the manufacture history in the context of determining what an item is. That history might also be useful in determining the time range of manufacture – not before a certain date, and perhaps less likely after a certain date range.

Although the range might cover a lengthy period – for example, anything written with a ball point pen would likely have been done after about 1942, more probably after 1945, and more certainly not before 1939 [114] – it still might be a useful complement to other information.

To continue the example [115], if the ink manufacture and supply records show that the share certificate's ink was first formulated by the maker's chemists, during November 1995, and was first offered for sale to the printing industry the following month, any certificate with that ink purportedly issued before December 1995 should cause the loan banker very serious worry. If, however, in an alternative of the example, the records were to show that the ink was marketed between December 1932 and April 1933, and not afterwards, then the banker should still worry, but it would be possible for a certificate issued in 1996 to be genuine – printed from an old and until-then-forgotten container of that ink – further investigation would be appropriate.

Analogously, an environmental contamination might be dated [116].

## 7.7 Other than chemical markers of *when*

Chemical markers of when – other than of chemical kinetics – are mentioned above. The same concept of other markers might be considered even though not chemical. Examples:

- In the *Crippen* murder case, a label found on a pyjama top, used to wrap buried body parts, was presented as evidence in court, London, 1910, of a date range [117, 118, 119, 120, 121].[34]
- In a 2019 Ontario estate case, evidence of fraud was accepted when expert evidence showed that the font came into use too late for it to have been honestly used for the questioned document [122, 123].[35]

34 *Crippen* murder case. CAS 51-34-3. Marconi. Chapters 2 and 7.

35 Font. Chapters 1 and 7.

- For the theft of Yousuf Karsh's famous *Roaring Lion* photo of Winston Churchill, from the Château Laurier hotel: "Thanks to a January 2022 photo submitted by CBC reporter Paul Hunter, the window of disappearance was narrowed to the 12-day period between Christmas Day 2021 and Jan. 6, 2022." [124, 125]

## 7.8 Error

[. . .][36]

. . . "*. . . overturned when he fell. . . . time of the crime . . . .*"

[. . .]

. . . "*But that clock, . . . quarter of an hour fast?*"

[. . .]

[Agatha Christie, *Murder at the Vicarage*, 1930] [126].[37,38]

The indication in Chapter 2 that scientific quantification can sometimes be done with great accuracy, but often not, includes chemical kinetics, and error considerations dealt with in Chapter 8 should be applied.[39,40,41]

That forensically applied chemical kinetics are usually empirically based (see above), rather than directly linked to theory, is an extra reason to take care. Also, as indicated above, care must be taken not to confuse a chemical kinetically determined t with the time ago of the related human story – they may be closely related – but are not necessarily the same.

## Notes

**1** Agatha Christie, *Murder on the Orient Express*, 1934, Pocket Books, New York City, 1960.

**2** Agatha Christie, *Murder on the Orient Express*, 1934, WILLIAM MORROW/Harper Collins, New York City, 10022; 2011; ISBN 978-0-06-207349-5.
https://www.fantasticfiction.com/c/agatha-christie/murder-on-orient-express.htm

**3** . . . *Orient* . . . Harper Collins . . ., PART ONE/THE FACTS, Chapter 7, "The Body," at page 59.

**4** Hercule Poirot, page 59.

**5** Dr. Constantine, page 63.

**6** See also Agatha Christie, *Murder at the Vicarage.* [Murder victim stops a clock by falling on it.]

**7** In addition to the Agatha Christie quote above – *Orient Express* – and *Vicarage* – see also, for example:

---

**36** Introductory quotes. Introduction, top of the first page.

**37** Error. Clock fast. Agatha Christie. Chapters 7 and 8.

**38** Agatha Christie, *Orient Express*, 1934. *Vicarage*, 1930. Chapters 2, 7 and 8.

**39** When. Chemical kinetics. Reactions in progress. Chapters 2, 4 and 7.

**40** When. Systematic error. Chapters 7 and 8.

**41** Measurement. Error. Chapters 2, 3, 7 and 8.

**8** T.S. Eliot, "*Time present . . .,*" quoted by Edmond Cooper, *The Overman Culture*, CORONET BOOKS, Hodder Paperbacks Ltd., London, SCIENCE FICTION; 1971, 1974; ISBN 0 340 17860 4; at page [5].
**9** T. S. Eliot, Burnt Norton, 1935.
https://www.goodreads.com/quotes/716781-time-present-and-time-past-are-both-perhaps-present-in
https://en.wikiquote.org/wiki/Four_Quartets
**10** Salvador Dalí, *La persistència de la memòria (The Persistence of Memory)*, 1931, owned by Museum of Modern Art, New York City.
https://en.wikipedia.org/wiki/The_Persistence_of_Memory https://en.wikipedia.org/wiki/Museum_of_Modern_Art
https://www.dalipaintings.com/persistence-of-memory.jsp
https://www.dalipaintings.com/persistence-of-memory.jsp#prettyPhoto[image1]/0/
https:// www.moma.org/
https://en.wikipedia.org/wiki/The_Persistence_of_Memory
**11** Robert Silverberg, editor, *VOYAGERS IN TIME/twelve stories of science fiction*, TEMPO BOOKS, New York City, 10010, 1967, 1970. https://www.biblio.com/booksearch/author/silverberg-robert/title/voyagers-in-time
**12** "*Temponautic murder*" with a time machine – see Robert Silverberg, "MANY MANSIONS," republished, 1986, in *BEYOND THE SAFE ZONE/COLLECTED STORIES OF ROBERT SILVERBERG*, Warner Books, Inc., New York City, 10103; ISBN 0-446-30173-6; at pages 113 and 114.
**13** See also:
- Roz Chast, COMIC STRIP, "A CHEERY STORY" [end of the universe maybe], *The New Yorker,* 07 December 2020, page 41.
- Darrin Bell [cartoon – re relativity time dilation and Trump Presidency], *The New Yorker,* 07 December 2020, page 27.

**14** See also:
- Carlo Rovelli, "Time Does Not Exist," CBC Ideas, April 2021.
  https://www.cbc.ca/radio/ideas https://www.cbc.ca/player/play/1888683587729
- Carlo Rovelli, *The Order of Time*, Penguin, 2019; ISBN 9780141984964
  https://www.penguin.co.uk/books/301/301539/the-order-of-time/9780141984964.html
  https://www.nature.com/articles/d41586-018-04558-7
  https://www.theguardian.com/books/2018/apr/24/carlo-rovelli-the-order-of-time-review

**15** See also:
- David Rooney, *About Time / A History of Civilization in Twelve Clocks*, PENGUIN BOOKS, UK, 2021. ISBN 978-0-241-37051-3.

**16** – Dava Sobel, *Longitude / The True Story of a Lone Genius Who Solved the Greatest Scientific Problem of His Time*, Walker & Company, New York City 10011, 1995. ISBN 978-0-8027-1520-6
https://www.google.ca/books/edition/Longitude/lW8DYSPa6fEC?hl=en&gbpv=1&printsec=frontcover

**17** – Edwin Danson, *Drawing the Line/ How Mason and Dixon Surveyed the Most Famous Border in America*, Wiley, New York City, 10158-0012; 2001; ISBN 0-471-38502-6.
– https://www.amazon.ca/s?k=9780471385028&i=stripbooks&linkCode=qs

**18** https://en.wikipedia.org/wiki/Mason%E2%80%93Dixon_line
**19** US NIST US Naval Observatory https://time.gov/
**20** And, apparently for sometimes other-than-scientific process: Maya C. Miller,*"In Congress, a Day Can Last Months if Politics Demands It /* House Republicans are so intent on avoiding tough votes on President Trump's tariffs that they have rewritten the calendar," *New York Times*, CONGRESSIONAL MEMO, 10 April 2025.
https://www.nytimes.com/2025/04/10/us/congress-johnson-calendar.html

21 See also: Harold E. Edgerton & James R. Killian, Jr., *flash! / SEEING THE UNSEEN BY ULTRA HIGH-SPEED PHOTOGRAPHY*, Charles T. Branford Company, Boston, 1954; page 9, *et seq.*
22 Ship-board observations of the apparently-moving fixed stars as a kind-of-clock-face – because the earth-itself was steadily rotating – when calibrated with the time at Greenwich – allowed for determination of longitude.
23 https://www.rmg.co.uk/royal-observatory
24 https://digital-commons.usnwc.edu/cgi/viewcontent.cgi?article=2062&context=nwc-review
25 https://www.rmg.co.uk/stories/time/harrisons-clocks-longitude-problem
26 https://en.wikipedia.org/wiki/John_Harrison
27 For a more general reference that goes beyond the very limited treatment here, see Walter J. Moore, *Physical Chemistry*, Prentice-Hall, Inc., Englewood Cliffs, New Jersey, 1962 and 1963, Chapter 8 "Chemical Kinetics," pages 253→322.
28 See also"A general chemistry Libretexts Textmap organized around the textbook/Chemistry: The Central Science/by Brown, LeMay, Busten, Murphy, and Woodward," 2020, 2019.
https://chem.libretexts.org/Bookshelves/General_Chemistry/Map%3A_Chemistry_-_The_Central_Science_(Brown_et_al.)
https://libretexts.org/
"14: Chemical Kinetics"
https://chem.libretexts.org/Bookshelves/General_Chemistry/Map%3A_Chemistry_-_The_Central_Science_(Brown_et_al.)/14%3A_Chemical_Kinetics
"14.S: Chemical Kinetics (Summary)"
https://chem.libretexts.org/Bookshelves/General_Chemistry/Map%3A_Chemistry_-_The_Central_Science_(Brown_et_al.)/14%3A_Chemical_Kinetics/14.S%3A_Chemical_Kinetics_(Summary)
29 For example: a time fuse for which acid, released from a glass container, when crushed, would corrode a metal wire sufficiently to release a spring it was restraining under tension; to close an electrical circuit.
Perhaps:

$$Fe + H_2SO_4 \xrightarrow[H_2O]{} FeSO_4 + H_2$$

30 See: Roger Moorhouse, *Killing Hitler*, Bantam Books, New York City; 2007; ISBN 978-0-553-38255-6; at page 260.
31 In a philosophical sense "time's arrow" may also be related to the thermodynamic concept of entropy.
https://en.wikipedia.org/wiki/Entropy_(arrow_of_time)
32 Measures of time used by humans rely on some physical phenomenon; many of which are engineered rather better and more conveniently than set-ups to observe chemical kinetics. But when chemical substances are the forensic evidence, that is what has to be looked at.
33 As an example of human-regarded time and a physical phenomenon, consider stars in the night sky appearing as moving around the North Star as on a clock face – See Edwin Danson, *Drawing the Line/ How Mason and Dixon Surveyed the Most Famous Border in America*, Wiley, New York City, 10158-0012; 2001; ISBN 0-471-38502-6; at page 36.
https://www.amazon.ca/s?k=9780471385028&i=stripbooks&linkCode=qs
34 – But sometimes that "clockwork" can be more interesting than useful for forensics.
https://www.thoughtco.com/briggs-rauscher-oscillating-color-change-reaction-602057
https://en.wikipedia.org/wiki/Briggs%E2%80%93Rauscher_reaction
35 Usually that *when* is from a time past. However, chemical kinetics can be involved in the forensics of public policy decisions – as for example for estimations of the duration of human-caused damage to the atmospheric ozone layer – see Nigel J. Bunce, *Environmental Chemistry*, Wuerz Publishing Ltd., Winnipeg, 1991; Chapter 2, particularly Tab. 2.2 and Fig. 2.3.

36 For a more general reference that goes beyond the very limited treatment here; see Moore, cited above.
37 For convenience, N may be assumed for purposes here to be of the order of magnitude of Avogadro's number = $N_0$ = 6.022 × $10^{23}$ = 1 mol; for example = 18 g of water $H_2O$ ≈18 mL.
38 Symbol alert: The N and $N_0$ here refer to large numbers of atoms, molecules, particles; not to be confused with N for the nitrogen atom.
39 That is, the equilibrium constant = K = ([A] × [B])/([AB]) is very large and >1.
40 For further information, see William K. Morrill, *Calculus*, D. Van Nostrand Company, Inc., Princeton, New Jersey, 1956, page 48.
41 Concerning *moles*, see note above.
42 The unit of time here is seconds – others may also be convenient: min, h, day, year, *etc.*
43 For further information, see: William K. Morrill, *Calculus*, D. Van Nostrand Company, Inc., Princeton, New Jersey, 1956, page 465, et seq.
44 For further information on solution phase chemistry theory as analogue to gas phase, see Moore, cited above, pages 299 and 300.
45 For example, if AB were the decomposing substance of fingerprints on glass.
46 For further information, as a general reference, beyond the very limited treatment here, see Moore, cited in a note above, pages 257→273.
47 For further information, on catalysis, see Moore, cited above, pages 300→303.
48 https://www.acs.org/content/acs/en/education/whatischemistry/landmarks/radiocarbon-dating.html
49 See also:
- Isaac Asimov, "The Enemy Within," an essay published as Chapter 4 in *The Relativity of Wrong*, Pinnacle Books, Windsor Publishing Corporation, New York City, 1988; reprinted from *The Magazine of Fantasy and Science Fiction*, September 1986.
- Gerhart Friedlander, Joseph W. Kennedy & Julian Malcolm Miller, *Nuclear and Radiochemistry/ Second Edition*, John Wiley and Sons, Inc., New York, . . ., 1964, pages 505 and 506.
- Lejaren A. Hiller, Jr. and Rolfe H. Herber, *Principles of Chemistry*, McGraw-Hill Book Company, Inc., New York, . . ., 1960; Chapter 16.11 USES OF RADIOACTIVE ISOTOPES/radiocarbon dating, pages 691, et seq.; Chapter 15.3 THE ORDER OF A CHEMICAL REACTION/first-order reactions, pages 629 and 630.
- Pamela S. Zuer, "Archaeological Chemistry/Physical Science helps to unravel human history," *Chemical & Engineering News*, SPECIAL REPORT, 21 February 1983, pages 27→29.

50 See also:
- Brian H. Kaye, *Science and the Detective/Selected Reading in Forensic Science*, VCH Verlagsgesellschaft GmbH, Weinheim, Germany, 1995; ISBN-3-527-29252-7; pages 306 and 307, and references cited therein.
- Alan D. Adler,"Updating Recent Studies on the Shroud of Turin," as Chapter 17; D.A. Kouznetsov, A.A. Ivanov and P.R. Veletsky, "A Re-evaluation of the Radiocarbon Date of the Shroud of Turin Based on Biofractionation of Carbon Isotopes and a Fire-Simulating Model," as Chapter 18.
- A.J.T. Jull, D.J. Donahue and P.E. Damon, "Factors That Affect the Apparent Radiocarbon Age of Textiles," as Chapter 19; in Mary Virginia Orna, ed., *Archaeological Chemistry/Organic, Inorganic Biochemical Analysis*, ACS Symposium Series 625, American Chemical Society, Washington, D.C., 1996; and references cited therein.

51 https://en.wikipedia.org/wiki/Radiocarbon_dating

52

| $_6C^{12}$ | | |
|---|---|---|
| C | = | Carbon. |
| [He] $2s^22p^2$ | = | Electron configuration |
| 12 | = | Atomic mass number |
| 12.011 | = | Atomic mass (g/mol). |
| 6 | = | Atomic number. |

53 Brookhaven National Laboratory – Chart of Nuclides http://www.nndc.bnl.gov/chart/
54 https://en.wikipedia.org/wiki/Carbon-14
55 See: Hiller and Herber, 1960; equation (16.34), at page 691.
56 For chemical kinetic observations it is always important to monitor temperature; however, not necessary here because this is a nuclear process.
57 β is beta particle, atomic mass number is 0, charge is −1.
Often written as $\beta^-$ to show its minus 1 charge.
More completely written as $_{-1}\beta^0$ .
58 A beta particle is an electron: $m_e = 9.10938215 \times 10^{-31}$ kg $m_p = 1.672621637 \times 10^{-27}$ kg
59 https://en.wikipedia.org/wiki/Beta_particle
60 See: "A general chemistry Libretexts Textmap organized around the textbook/Chemistry: The Central Science/by Brown, LeMay, Busten, Murphy, and Woodward," 2020, 2019.
https://chem.libretexts.org/Bookshelves/General_Chemistry/Map%3A_Chemistry_-_The_Central_Science_(Brown_et_al.)
https://libretexts.org/
"14: Chemical Kinetics"
https://chem.libretexts.org/Bookshelves/General_Chemistry/Map%3A_Chemistry_-_The_Central_Science_(Brown_et_al.)
"14.4: The Change of Concentration with Time (Integrated Rate Laws)"
https://chem.libretexts.org/Bookshelves/General_Chemistry/Map%3A_Chemistry_-_The_Central_Science_(Brown_et_al.)/14%3A_Chemical_Kinetics/14.4%3A_The_Change_of_Concentration_with_Time_(Integrated_Rate_Laws)
61 See: Hiller and Herber, 1960; equation (15.12), at page 630.
62 See: Hiller and Herber, 1960; equation (16.15), at page 681.
63 See: Hiller and Herber, 1960; equation (16.16), at page 681.
64 See, in analogy: Hiller and Herber, 1960; Examples 16.6 and 16.7, at page 693.
65 $[C]_{total} = [C] + [C^*] \approx [C]$
66 Also, see, in analogy: Hiller and Herber, 1960; wherein Examples 16.6 and 16.7, at page 693, ancient and modern C-14 concentrations are taken as proportional to the measured rates of $\beta^-$ appearing from the C-14 disintegrations ((disintegrations / (minute × gram of carbon)). The ancient would be from the sample being examined; the modern from a reference standard; to date the ancient. The modern would be assumed = the concentration when the sample would have been a part of something live.
67 See also:
https://esrl.noaa.gov/gmd/infodata/isotopes/decay.html
https://study.com/academy/lesson/what-is-carbon-dating-definition-lesson-quiz.html
https://courses.lumenlearning.com/collegealgebra2017/chapter/putting-it-together-exponential-and-logarithmic-equations-and-models/
68 This would not be the major problem.
69 This would be the major problem.

70 Niraj Chokshi, "How Art-Forgery Sleuths Learned to Love the Bomb/The period of nuclear testing gave at least a temporary boost to radiocarbon dating," *The New York Times*, SCIENCE, 11 July 2019, page D4.
71 Wood pulp source might be inappropriate because it might contain varying tree ring woods.
72 *Dendrochronology* – see Pamela S. Zuer, "Archaelogical Chemistry/Physical Science helps to unravel human history," *Chemical and Engineering News*, SPECIAL REPORT, 21 February 1983, pages 28 and 29.
73 – Radiocarbon Tree-Ring Calibration, Beta Analytic TESTING LABORATORY, 2024, Miami, Florida, 33155
https://www.radiocarbon.com/tree-ring-calibration.htm
– Ronny Friedrich, Bernd Kromer, Frank Sirocko, Jan Esper, Susanne Lindauer, "Annual 14C Tree-Ring Data Around 400 AD: Mid- and High-Latitude Records Cambridge University Press, 30 April 2019.
https://www.cambridge.org/core/journals/radiocarbon/article/abs/annual-14c-treering-data-around-400-ad-mid-and-highlatitude-records/9CD950A4EAE0A1FDF4EF965892237420
Radiocarbon , Volume 61 , Issue 5: Radiocarbon 2018 Conference Proceedings Trondheim, Norway, June 17–22, 2018 Part 1 of 2 , October 2019 , pp. 1305 – 1316
https://doi.org/10.1017/RDC.2019.34
– Radiocarbon Calibration, [OxCal program] [2024]
https://c14.arch.ox.ac.uk/calibration.html
– https://en.wikipedia.org/wiki/Radiocarbon_calibration
74 Radiolab Podcast, see also: "The Fellowship of the Tree Rings."
https://www.youtube.com/watch?v=N4e-HMtsdOY https://www.radiolab.org/
75 But not without possibilities of controversy – see D.A. Kouznetsov, et al.; and A.J.T. Jull, et al.; cited above.
76 *Gedankenexperiment.* Introduction 2.2.
77 See:
– Karel Kulp and James Vetter, "Effect of Aging on Freshness of White Pan Bread," as Chapter 1 in George Charalambous, ed., *Handbook of Food and Beverage Stability/Chemical, Biochemical, Microbiological, and Nutritional Aspects*, Academic Press, Inc., Orlando, . . ., 1986, pages 1→31, and references cited therein;
– Ann-Charlotte Eliasson and Kåre Larsson, *Cereals in Breadmaking/A Molecular Colloidal Approach*, Marcel Dekker, Inc., New York, . . ., 1993, Chapter 7, IV, "Staling," pages 349→362, and references cited therein.
"Bread Staling," *LALLEMAND BAKING UPDATE*, VOLUME 3, NUMBER 9, Practical technology from Lallemand Inc., Montréal, Québec H1W 2N8, 2018. See, in particular, "MEASURING BREAD STALING," at page [2].
https://www.lallemandbaking.com/en/global/
– https://www.bbc.com/news/magazine-29701768
– https://bakerpedia.com/the-science-behind-bread-staling/#:~:text=Measuring%20Staling,Baker%20Compressimeter
– https://www.instron.com/en/resources/videos/2021/october/what-is-a-universal-testing-machine-tensile-testing-machine#:~:text=Instron%20universal%20testing%20machines%20are,an%20Instron%20universal%20testing%20machine.
– https://labomat.eu/gb/texture-faq/844-case-study-measuring-the-firmness-of-sliced-bread.html
– https://en.wikipedia.org/wiki/Bloom_(test)
– https://onlinelibrary.wiley.com/doi/abs/10.1111/j.1745-4603.1986.tb00716.x
– Toktam Mohammadi-Moghaddam, Mohammad Morshedi, Ramina Moalemzadeh Ansari, Amir Golmohammadi, Afsaneh Morshedi, Mohaddeseh Kariminejad, "Bread Staling Measurement

Techniques: A Review," *Journal of Food Chemistry & Nanotechnology*, 2024, *J Food Chem Nanotechnol* 10(2): 85-93.
https://foodchemistryjournal.com/jfcn/articles/v10n2/jfcn-178-toktam-mohammadi-moghaddam.pdf

- G.A.R. Sehn, C.J. Steel, 2020. Staling kinetics of whole wheat pan bread. *J Food Sci Technol* 57: 557→563.
  https://pubmed.ncbi.nlm.nih.gov/32116365/
  https://pmc.ncbi.nlm.nih.gov/articles/PMC7016051/

**78** See: "Bread Staling," *LALLEMAND* . . ., "MEASURING BREAD STALING," at page [2].
**79** Analogous observations might also be applied to animals and plants.
**80** See: Niels Lynnerup, "A Computer Program for the Estimation of Time of Death," *Journal of Forensic Sciences*, July 1993, pages 816–820; and follow-up letters to the editor, May 1994, pages 601 and 602.
**81** See:

- Tarek I. Tantaiwi and Bernard Greenberg, "The Effect of Killing and Preservative Solutions on Estimates of Maggot Age in Forensic Cases," *Journal of Forensic Sciences*, May 1993, pages 7026707.
- Gail S. Anderson and Serah L. Van Laerhoven, "Initial Studies on Insect Succession on Carrion in Southwestern British Columbia," *Journal of Forensic Sciences* 41(4), July 1996, pages 617–625.
- Jeffrey D. Wells and Lynne R. LaMotte, "Estimating Maggot Age from Weight Using Inverse Prediction," *Journal of Forensic Sciences* 40(4), July 1995, pages 585→590.

**82** [TOD not delt with in detail here.]
**83** See, for example:

- notes above, and –
- Marc S. Micozzi, *Postmortem Changes in Human and Animal Remains: a systematic approach*, Charles C Thomas, Publisher, Springfield, Illinois 62794–9265, 1991; ISBN 0398057478 (reviewed by William M. Bass, *Journal of Forensic Sciences*, January 1993, pages 225 and 226). (University of Toronto – Robarts Library CC 75 .7 M53 1991.)
- Arpad A. Vass, William M. Bass, Jeffery D. Wolt, John E. Foss and John T. Ammons, "Time Since Death Determinations of Human Cadavers Using Soil Solution," *Journal of Forensic Sciences*, Vol.37, No.5, September 1992, pages 1236→1253.
- F.O. Raasch, Jr., J.I. Hirvonen and C.J. Stahl, "Timing of Injury in Human Thermal Burns," *Journal of Forensic Sciences*, Vol.19, No.4, October 1974, pages 723→729.
- F.O. Raasch, Jr., J.I. Hirvonen and C.J. Stahl, "Timing of Injury in Human Thermal Burns," *Journal of Forensic Sciences*, Vol.19, No.4, October 1974, pages 723→729.
- David Montgomery, "LEARNING FROM DEATH/A place where all flesh is weak/University of Tennessee facility dedicated to the study of decaying corpses," *The Toronto Star*, Saturday, 24 July 1999, page K2.

**84** See also: Gil Reavill, *AFTERMATH, INC. / CLEANING UP AFTER CSI GOES HOME*, GOTHAM BOOKS Penguin, New York City, 1001; 2007. ISBN 978-1-592-40364-6.
**85** Included in Gil Reavill, *AFTERMATH* . . . – some commentary about related topics:

- Crime scene.
- CSI TV shows – pages 212→218.
- Forensics.
- Locard's principle Sir Arthur Conan Doyl, Sherlock Holmes – pages 215→218.
- postmortem interval (PMI) – time of death (TOD) – pages 50→58.

**86** – Time of death and why, and professional determinations thereof, are also popular in murder mysteries, and other fiction. See, for example:

- P.D. James, *A CERTAIN JUSTICE / AN ADAM DALGLIESH MYSTERY*, VINTAGE BOOKS. New York City, 1997, 2019; page 131, *et seq.*, [re forensic pathologist Miles Kynaston]. ISBN 978-0-593-08098-6.

- Agatha Christie, *The Moving Finger / A Miss Marple Mystery*, WILLIAM MORROW, New York City, 1942, 2022; pages 73 & 74 [re an inquest] ISBN 978-0-06-321403-3.
- Agatha Christie, 1929, *The Seven Dials Mystery*, Bantam Books, New York City, 1964, 1973; Chapter XV, "The Inquest," page 82, et seq.

**87** Valery N. Aginsky, *Journal of Forensic Sciences*, 40(3), May 1995, pages 475→477.
https://www.astm.org/DIGITAL_LIBRARY/JOURNALS/FORENSIC/PAGES/JFS13808J.htm

**88** R.L. Brunelle and H. Lee, *Journal of Forensic Sciences*, Vol.34, No.5, 1989, pages 1166→1182 (cited in Ellen Rohde, Annette C. McManus, Carla Vogt, William R. Heineman, "Separation and Comparison of Fountain Pen Inks by Capillary Zone Electrophoresis," *Journal of Forensic Sciences*, Vol.42, No. 6, November 1997, page 1004.).

**89** https://www.astm.org/DIGITAL_LIBRARY/JOURNALS/FORENSIC/PAGES/JFS12751J.htm
R.L. Brunelle, *Journal of Forensic Sciences*, Vol.37, No.1, 1992, pages 113→124 (cited in Ellen Rohde, Annette C. McManus, Carla Vogt, William R. Heineman, "Separation and Comparison of Fountain Pen Inks by Capillary Zone Electrophoresis," *Journal of Forensic Sciences*, Vol.42, No. 6, November 1997, page 1004.).
https://www.astm.org/DIGITAL_LIBRARY/JOURNALS/FORENSIC/PAGES/JFS13218J.htm

**90** https://www.astm.org/DIGITAL_LIBRARY/JOURNALS/FORENSIC/PAGES/JFS14252J.htm
R.L. Brunelle, *The International Journal of Forensic Document Examiners*, Vol.1, No.2, 1995, pages 94–98 (cited in Ellen Rohde, Annette C. McManus, Carla Vogt, William R. Heineman "Separation and Comparison of Fountain Pen Inks by Capillary Zone Electrophoresis," *Journal of Forensic Sciences*, Vol.42, No. 6, Nov. 1997, page 1004.).

**91** Susumu Ohtani, *Journal of Forensic Sciences*, 40(5), September 1995, pages 805→807.

**92** J.R. Johnson R.J. Braddock and C.S. Chen, *Journal of Food Science*, 60(3), 1995, pages 502→505.

**93** Gail S. Anderson and Serah L. Van Laerhoven, *Journal of Forensic Sciences* 41(4), July 1996, pages 617→625.

**94** Jeffrey D. Wells and Lynne R. LaMotte, *Journal of Forensic Sciences* 40(4), July 1995, pages 5856590.

**95** See: Anne F. Skinner and Mark N. Rudolph, "Dating Flint Artifacts with Electron Spin Resonance: Problems and Prospects," as Chapter 4; page 37, et seq.; in Mary Virginia Orna, ed., *Archaeological Chemistry/Organic, Inorganic Biochemical Analysis*, ACS Symposium Series 625, American Chemical Society, Washington, D.C., 1996; and references cited therein.

**96** ESR. Chapters 6 and 7.

**97** See: Pamela S. Zuer, "Archaeological Chemistry/Physical Science helps to unravel human history," *Chemical and Engineering News*, SPECIAL REPORT, 21 February 1983, page 29.
https://pubs.acs.org/toc/cenear/61/8
https://pubs.acs.org/doi/pdf/10.1021/cen-v061n008.p003
https://pubs.acs.org/doi/abs/10.1021/cen-v061n008.p026
https://pubs.acs.org/doi/pdf/10.1021/cen-v061n008.p026

**98** S. Oberstedt and H. Vanmarche, *Health Physics*, 70(2), pages 222→226, February 1996.
https://journals.lww.com/health-physics/Abstract/1996/02000/Volume_Traps___A_New_Retrospective_Radon_Monitor.10.aspx

**99** *Discover*, [reprinted in] *The Globe and Mail*, SCIENCE, 01 March 1999, page A8. The method reported here would allow for determination as to the time of year – for example, summer – of death, rather than the postmortem interval.

**100** "679: Save the Girl," This American Life, July 2019. https://www.thisamericanlife.org/679/transcript

**101** Sophie Haigney, "Time Robs 'The Scream' of its color/experts study how notable paintings lose pigment," *The New York Times*, Arts, 19 February 2020, pages C1 and C2.
https://www.nytimes.com/2020/02/07/arts/design/the-scream-edvard-munch-science.html

**102** See also:

- Jonathan Feakins, "Alba Álvarez-Martín is making mass spec more accessible in museums / The cultural heritage scientist is adapting a medical imaging technique to analyze paintings in Amsterdam's Rijksmuseum," *C&EN*, ART & ARTIFACTS, 02 January 2025.
- Alba Alvarez-Martin, Jusal Quanico, Teresa Scovacricchi, Ermanno Avranovich Clerici, Geert Baggerman, Koen Janssens, "Chemical Mapping of the Degradation of Geranium Lake in Paint Cross Sections by MALDI-MSI," *Analytical Chemistry*,2023, 95, 49, 18215–18223, 23 November 2023. https://doi.org/10.1021/acs.analchem.3c03992
- Jun Han, Teesha C Baker, Christoph H Borchers, "Recent advancements in matrix-assisted laser desorption/ionization mass spectrometry imaging," Current Opinion in Biotechnology, Elsevier, Volume 43, February 2017, Pages 62–69. https://doi.org/10.1016/j.copbio.2016.09.003

**103** Ian McEwan, *NUTSHELL*, VINTAGE CANADA, 2016, 2017; ISBN 978-0-345-81241-4. http://www.ianmcewan.com/books/nutshell.html http://www.ianmcewan.com/

**104** Wherein the murderers use an ethylene glycol-antifreeze [McEwan, *NUTSHELL* . . ., pages 49, 50] – poisoned smoothie [pages 93→99] to try to fake their victim's death as suicide. Among several mental distress indicating items planted at the death scene were also a pair of the deceased's gloves that he would have worn to conceal his psoriasis. And a cup and antifreeze bottle that he would have used. The gloves would account for no fingerprints on the cup and bottle [pages 100, 101.]; he would have removed the gloves before losing consciousness. But police forensics discovers the spiders [pages 181, 182].

**105** McEwan, *NUTSHELL* . . ., pages 181, 182.

**106** See also: https://miriamfeinbergvamosh.com/king-david-saw-it-on-the-web/
– story arising from 1 *Samuel* 24 שְׁמוּאֵל א https://mechon-mamre.org/p/pt/pt08a24.htm
https://mechon-mamre.org/p/pt/pt0.htm

**107** For example, if biofluid remains were the decomposing substance of fingerprints on glass, at known constant temperature and humidity, it might be worthwhile to try to estimate t. But, on other surfaces and environments, more difficult.

**108** See:

- E.R. Menzel, *Journal of Forensic Sciences*, letter to the editor, Vol.37, No.5, September 1992, pages 1212 and 1213. This relied on a change of laser-induced fluorescence emission wavelength for fresh → old prints (~550 nm – greenish yellow → ~ 580 nm – orange).
  https://www.astm.org/DIGITAL_LIBRARY/JOURNALS/FORENSIC/PAGES/JFS13307J.htm
  JFSCAS, 01 September 1992, Volume 37, Issue 5. Letters to the Editor

**109** – Ronny Merkel, *New Solutions for an Old Challenge: Chances and Limitations of Optical, Non-Invasive Acquisition and Digital Processing Techniques for the Age Estimation of Latent Fingerprints*, Logos, Berlin, 2014. ISBN 978-3-8325-3823-3.
[Ronny Merkel, Deutsche Bundesbank, Frankfurt am Main.
https://www.logos-verlag.de/cgi-bin/engbuchmid?isbn=3823&lng=eng&id=
https://www.researchgate.net/profile/Ronny-Merkel]

**110** See also:

- Y.S. Dikshitulu, L. Prasad, J.N. Pal and C.V.N. Rao, "Aging Studies on Fingerprint Residues Using Thin-Layer and High Performance Chromatography," *Forensic Science International*, Vol.31,1986, pages 261→266.
  https://pubmed.ncbi.nlm.nih.gov/3,744,217/
  https://www.sciencedirect.com/science/article/abs/pii/0379073886901659

Clay E. Allred, Tao Lin, and E. Roland Menzel "Lipid-Specific Latent Fingerprint Detection: Fingerprints on Currency," *Journal of Forensic Sciences*, Vol.42, No. 6, November 1997, page 997; and references cited therein.

111 – Kasey Wertheim, "Fingerprint age determination: Is there any hope?" *Journal of Forensic Identification*; Alameda Vol. 53, Iss. 1, (Jan/Feb 2003): 42–49.
https://www.proquest.com/openview/7c4633bc051bc63f31ae1f0e29d65dc2/1?pq-origsite=gscholar&cbl=29

112 – See also: K.M. Beesley, S. Damaskinos, A.E. Dixon, "Fingerprint Imaging with a Confocal Scanning Laser Macroscope," Journal of Forensic Sciences Volume:40 Issue: January 1995, pages:10→17.

113 https://asmedigitalcollection.asme.org/forensicsciences/article-abstract/40/1/10/1183411/Fingerprint-Imaging-with-a-Confocal-Scanning-Laser?redirectedFrom=fulltext
https://archive.org/details/sim_journal-of-forensic-sciences_1995_40_index
*Gedankenexperiment*. Introduction 2.2.

114 See: R.L. Brunelle in R. Saferstein, ed., *Forensic Science Handbook* [Volume I], Prentice-Hall, Inc., Englewood Cliffs, New Jersey:1982, Chapter 14, "Questioned Document Examination," pages 709→711, 716 and 717, and references cited therein.

115 *Gedankenexperiment*. Introduction 2.2.

116 See: Robert D. Morrison, "Forensic Techniques for Establishing the Origin and Timing of a Contaminant Release," as Chapter 10 in Carl Meyer, ed., *EXPERT WITNESSING/EXPLAINING AND UNDERSTANDING SCIENCE*, CRC Press LLC, Boca Raton, Florida, 33431, 1999; ISBN 0-8493-1197-7. (Derived from a symposium of the American Chemical Society, Division of Chemistry and the Law, ACS National Meeting, Las Vegas Nevada, 1997.) At page 149.

117 Dr. Hawley Harvey Crippen was convicted, London, 1910, of murdering his wife Cora, by poisoning her with hyoscine. Evidence included remnants of pyjamas and label, sold after 1909 by Jones Bros (Holloway) Ltd., Shirtmakers. Evidence was that parts of her remains, buried in the basement, were wrapped in the top, the bottom was found in the bedroom. Controversies about the case continue.
http://www.johnlewismemorystore.org.uk/page/dr_crippen_downed_by_jones_brothers_buyer
https://en.wikipedia.org/wiki/Hawley_Harvey_Crippen
http://www.pbs.org/wnet/secrets/hawley-crippen/199/
https://blablawriting.com/forensic-science-essay
https://www.bbc.com/news/magazine-10802059

118 See also: Erik Larson, *Thunderstruck*, CROWN, New York City, 2006. ISBN 978-1-4000-8067-0. **[**– about Hawley Harvey Crippen, whose arrest in Québec, 1910, was facilitated by the ship-shore wireless telegraphy of Guglielmo Marconi – the first use of such radio-transmitted data for law enforcement. Scotland Yard was able to send its officers on a faster Atlantic-crossing ship.**]**

119 Hyoscine, CAS CAS 51-34-3, $C_{17}H_{21}NO_4$, 303.35 g/mol (see Fig. 7.2).

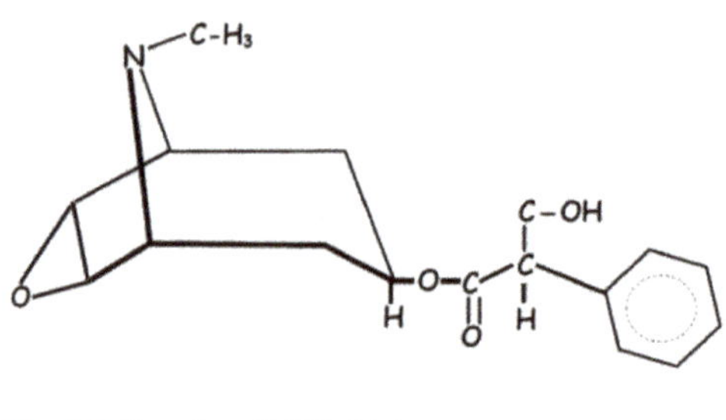


Fig. 7.2: Hyoscine, CAS 51–34-3.

**120** https://en.wikipedia.org/wiki/Hyoscine
https://www.drugbank.ca/drugs/DB00747
https://chem.nlm.nih.gov/chemidplus/rn/51-34-3
**121** https://www.shutterstock.com/editorial/image-editorial/'the-most-sensational-trial-many-years-'-7665061ow
https://en.wikipedia.org/wiki/Hawley_Harvey_Crippen
**122** Aidan Macnab, "Font expert exposes phony trust documents," *Canadian Lawyer Magazine*, 17 January 2019.
https://www.canadianlawyermag.com/practice-areas/trusts-and-estates/font-expert-exposes-phony-trust-documents/275788
**123** Thomas Phinney/The Font Detective https://thefontdetective.com/
**124** Lynn Saxberg, "Château Laurier welcomes return of the Roaring Lion — with extra security / Yousuf Karsh's iconic portrait of Winston Churchill, stolen from the hotel almost three years ago, will have a welcome home ceremony on Nov. 15," *Ottawa Citizen*, 13 November 2024.
https://ottawacitizen.com/news/chateau-laurier-roaring-lion-winston-churchill
**125** https://www.youtube.com/watch?v=T1s0dw_xCeQ
https://www.youtube.com/watch?v=9ZLKjEAYIrk
**126** Agatha Christie, *Murder at the Vicarage*/A MISS MARPLE MYSTERY, 1930, SIGNET, New York City, 10014; 2000; ISBN 0-451-20115-9; pages 38→41, at pages 39 and 40. (Murder victim stops a clock when falling on it.)

# Chapter 8
# Error

*All analytical measurements are wrong: it's just a question of how large the errors are, and whether they are acceptable* [1, 2, 3, 4].[1,2]

*Lab errors may void Ontario drug convictions* [Toronto, 2001] . . . [5, 6, 7][3]

[. . .]

*. . . The fibre evidence was contaminated within the Centre of Forensic Sciences. The timing and precise origin of the contamination cannot now be determined. However, it remains possible that this contamination tainted Ms. N . . . 's earliest findings. No inferences can safely be drawn from any alleged fibre similarities, given the existence of this in-house contamination.*

[. . .]

[*Report of the* [Ontario] *Kaufman Commission on Proceedings Involving Guy Paul Morin*, [1997]] [8][4]

[. . .]

*In March 1990, during the preparation for Guy Paul Morin's second trial, Crown attorney . . . learned that Sergeant M . . . had two notebooks for the . . . investigation, containing a number of divergent entries for the same events. Following an investigation by the Ontario Provincial Police, Sergeant M . . . was charged with perjury and attempting to obstruct justice in connection,* inter alia, [9][5] *with his evidence at the first trial relating to his notebook(s) and in relation to his evidence that a cigarette butt tendered as an exhibit at the first trial was the one found at the body site. The charges were judicially stayed in 1991 for reasons relating to Sergeant M . . . 's health. The Commissioner ruled, after receiving independent medical evidence, that Sergeant M . . . would not be compelled to testify at the Inquiry* [10, 11].

[. . .]

*Four months after Ms. S . . . was released from prison . . . the result of the hearing was overturned. The drug test was faulty . . .* *New York Times* 21 Nov. 2019 [12].

*After False Drug Test, He Was in Solitary Confinement for 120 Days / Hundreds of New York State prisoners were locked in cells, denied release or removed from programs when tests erroneously showed they had used narcotics, according to a lawsuit."*

*New York Times*, 20 Nov. 2019 [13].

---

**1** Introductory quotes. Introduction 1.2.

**2** Error. Sampling. Sampling plan. Chapters 3, 4 and 8.

**3** Fraud. Griffiths Report, 2001. QA failure. Gemma Ramlal, 2001. Annie Doukhan, 2012. Introduction 4.3; Chapters 8, 9 and 10.

**4** *Morin*. Fibre evidence contamination. Chapters 1, 3 and 8.

**5** = Among other things.

https://doi.org/10.1515/9783111618555-009

[. . .]
*To err is human; to forgive, divine.*
[. . .]
Alexander Pope, *An Essay on Criticism*, 1711 [14].[6,7]

## 8.1 Error: qualitative and quantitative

Analytical chemistry[8] includes both *qualitative* and *quantitative* evaluations [15].[9,10] Both can have error – the uncertainties mentioned previously.[11]

The first introductory quote above would appear to be for quantitative measurements for geological analyses. A more general recasting might be appropriate for forensic analytical chemistry:

- **Analytical chemistry comes with doubt**:
- Qualitative[12] determinations have at least a slight potential to be wrong as to what the examined substance is [16, 17, 18][13], or is not, or what sample impurities might or might not be; with questions as to what extent these doubts should be tolerated.
- Quantitative[14] measurements are, in a strict sense, inaccurate as to how much there is of whatever that substance is qualitatively accepted to be, with the question of how much, and what kinds of, numerical error should be tolerated.
- Sample selection can be problematic[15]; and sometimes so much so as to negate the entire analytical process.
- Forensic analytical chemistry sometimes (not too often) is confronted with arguments, requirements and prohibitions that are other than of science[16]; and further sometimes (hopefully hardly ever) ethically or unreasonably so.
- An error statement = an estimation of what the errors and error possibilities are or could be, a discussion of the contributing factors, and what it all is thought to mean for the results – is essential to any analytical chemical report.[17,18]

---

6 Error. Introduction 4.2; Chapter 8.

7 Error. Forgiveness. Chapter 8 intro. quote; Chapter 8.7.2.

8 Analytical chemistry. Expanded definition – more detail, and example. Introduction 2; Chapters 1, 2, 3, 5 and 8.

9 ***Qualitative*** = what it is. ***Quantitative*** = how much. Chapters 3.1.1 and 8.1.

10 Qualitative. Quantitative. Error Analytical chemistry. Chapters 2 and 8.

11 Measurement. Error. Chapters 2, 3, 7 and 8.

12 ***Qualitative*** = what it is. ***Quantitative*** = how much. Chapters 3.1.1 and 8.1.

13 Alice's Kitty's Looking-Glass milk. Chirality. Chapters 4 and 8.

14 ***Qualitative*** = what it is. ***Quantitative*** = how much; Chapters 3.1.1 and 8.1.

15 Error. Sampling. Sampling plan. Chapters 3, 4 and 8.

16 Legal/scientific truth. Chapters 1, 8 and 11.

17 Qualitative. Quantitative. Error. Chapters 2 and 8.

18 Error report. Chapters 8 and 9.

- **However, analytical chemistry can, with care, provide reasonably reliable and explainable data and conclusions – useful to help explain the forensic-related human story.**[19]

Also, within this, it is important to discriminate between a reasonable treatment of uncertainty that strives for reliable data and conclusions, and animosity to science – such as for creationism, climate change denial, and anti-vaccine advocacy.[20,21]

**This chapter deals with error, first, in the context of the kinds possible – the next several sections; and then how to try to limit and cope with error possibilities – in the remaining sections.**

## 8.2 Kinds of error: systematic errors – bias

***Systematic errors*** – sometimes called ***bias*** [19, 20, 21, 22][22] – alter the quantitative measurement from being correct in a repeatable way due to some aspect of the measuring system. Often this would be sufficiently known to be prevented, or to be corrected by formula – perhaps adjusted for by automated instrumentation.

Perhaps a simplest example [23]: A clock set as for 15 min ahead,[23,24] – correctable by simple formula:

$$\text{correct time (hour, minute)} = \text{time shown} - 15\,\text{min}$$

– or more simply, by comparison and resetting to https://time.gov/.

Another simplest example [24] – of common relevance to analytical chemistry: A mass[25] determination of a single item with an un-zeroed *balance* or *scale* [25], so that it is always off by a constant amount. This particular *bias* should be routinely noticed and prevented because a first step before using any balance or scale is to check the calibration, and to ensure a zero setting at zero. If however, the error is made and discovered later – the simple formula [26]:

$$\text{correct mass} = \text{uncorrected mass} + \text{c (kg)}$$

where c can be either a positive or negative constant number [27].

19 Analytical chemistry and forensically related human story. Chapters 1 and 8.
20 Creationism. Climate change denial. Antivaccine. Chapters 1, 2 and 8.
21 Science – good, bad & junk. Introduction 4; Chapters 1, 2, 8, 10 and 11.
22 Bias. Chapter 8.
23 Error. Clock fast. Agatha Christie. Chapters 7 and 8.
24 When. Systematic error. Chapters 7 and 8.
25 Mass, weight. Chapters 3 and 8.

Unfortunately, some *systematic* errors may be not so simple and obvious. For example: a gas chromatographic analysis[26] of a mixture for which some of its components react with the column packing could introduce a *bias* for both quantitative and qualitative determinations [28, 29].

## 8.3 Kinds of error: statistical errors – random

***Statistical***[27] – also called ***random*** [30, 31] – errors alter quantitative measurements in ways that are not repeated (except by chance) for any particular measurement. These *random fluctuations* – *scatter* – that occur for any set of quantitative measurements can be dealt with by statistical methods that address the extent of the fluctuations, along with decision-making as to the limits of fluctuation that would be tolerated.

The fluctuations may result from both the making of the measurements, and from what is being measured.[28]

Again, as a simplest example [32], repeated mass measurements[29] of the same item would yield variations that reflect the accuracy of the instrumentation. This would be dealt with by the mathematics of a *Gaussian* [33, 34, 35, 36, 37, 38, 39, 40, 41, 42, 43] *distribution* [44], also known as a *normal distribution* or a *bell curve*. [45, 46].[30]

As a fictitious example [47], five mass determinations repeated of that same single item can be characterized by a ***mean*** [48, 49, 50] (see Tab. 8.1):

Tab. 8.1: Mass measurement [51].

| | | | | |
|---|---|---|---|---|
| x = | 5.0031 | grams | 1 | |
| | 5.0027 | | 2 | |
| | 5.0035 | | 3 | n = 5 = the number of measurements. |
| | 5.0034 | | 4 | |
| | 5.0030 | | 5 | |
| | 25.0157 | = | Σ | |

$$\bar{x} = ((\Sigma x_n)/n) = 25.0157/5 = 5.0031$$

And deviations from the *mean* ($x - \bar{x}$) can be used to calculate a ***standard deviation*** [52][31]:

---

26 Gas chromatography. Chapters 6 and 8.
27 Too much scientific technical detail here? Introduction 2.1.
28 Measurement. Error. Chapters 2, 3 and 8.
29 War on drugs. *CD&A Act*. Mass (grams) measurement. *W&M Act*. Chapters 3, 6, 8 and 10.
30 *Gaussian* distribution. Chapters 8 and 11.
31 Too much scientific technical detail here? Introduction 2.1.

$$s = \surd\left(\left(\Sigma(x-\overline{X})^2\right)/(n-1)\right) = \surd(4.2\times10^{-7}/(5-1)) = \pm 0.00032\ [53, 54, 55, 56]$$

For such measurements, there is, in concept, a single true mass value that remains unchanged – the constant mass of the single item. The *scatter* results from the measuring process. The mathematical treatment tries to give a reliable number by a defined method that may be used for the measurement.

However, for many kinds of measurements, there may be many different true values for each of what is being measured, so that the *scatter* results from both measuring process and from actual variation in what is being measured. This might be regarded as a different kind of error. For example: a tabulation of heights and masses [57] of a population of people [58] – each individual person may have been measured accurately [59], but everyone would be at least a little different, so that there would be an error range when considering the whole group. The analysis would then provide numbers that could be used to study the group.

As a fictitious example [60], suppose, for a particular organic [61] solvent, used in the paint industry, in-blood concentration ([] mg/mL) [62] measurements are made for a random sampling of factory workers.[32] These might be important for a policy evaluation for occupational health and safety law [63]. Or for government inspectional enforcement. Or for union advocacy. Or for insurance coverage denial. Each worker surveyed would show a different number – because of *scatter* from the:
- measuring method,
- different exposures in the factory [64] and
- different personal metabolisms.

The resulting data [65] can be tabulated [66] – including for known control samples – and arranged to be analysed within a *Gaussian distribution* [67, 68, 69, 70, 71] (see Tab. 8.2).

Tab. 8.2: Worker # and [] (mg/mL) known control sample #1 known control sample#2 [72, 73, 74].

| Worker # = | In-blood concentration= [] (mg/mL) = | x – x̄ | (x – x̄)$^2$ |
|---|---|---|---|
| 1 | 5.6 | −0.79 | 0.62 |
| 2 | 7.2 | 0.81 | 0.66 |
| 3 | 4.0 | -2.39 | 5.71 |
| 4 | 8.5 | 2.11 | 4.45 |
| Known control sample #1 | 6.40 | | |
| 5 | 6.3 | -0.09 | 0.01 |
| 6 | 6.2 | -0.19 | 0.04 |
| 7 | 7.8 | 1.41 | 1.99 |

32 Error. Sampling. Sampling plan. Chapters 3, 4 and 8.

Tab. 8.2 (continued)

| Worker # = | In-blood concentration= [] (mg/mL) = | $x - \bar{x}$ | $(x - \bar{x})^2$ |
|---|---|---|---|
| 8 | 4.7 | −1.69 | 2.86 |
| 9 | 6.8 | 0.41 | 0.17 |
| 10 | 5.5 | −0.89 | 0.79 |
| 11 | 6.1 | −0.29 | 0.08 |
| Known control sample #2 | 6.38 | | |
| 12 | 7.1 | 0.71 | 0.50 |
| 13 | 5.1 | −1.29 | 1.66 |
| 14 | 8.6 | 2.21 | 4.88 |
| | Σ = 89.5 | | Σ = 24.42 |

From Tab. 8.2 [75], a ***mean*** and a ***standard deviation*** can be calculated[33]:

$$\bar{x} = 89.5/14 = 6.392857 \text{ [76, 77, 78]}$$

$$s = \sqrt{/}(24.42/13) = \pm\mathbf{1.371}$$

And the frequency of occurrence of conveniently chosen concentration ranges can be arranged graphically in Fig. 8.1 [79].

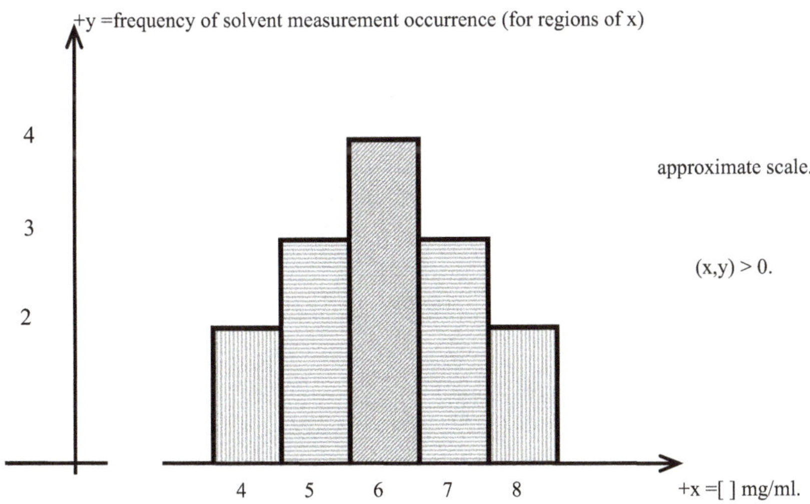


Fig. 8.1: Histogram.

33 Too much scientific technical detail here? Introduction 2.1.

Such a ***histogram*** [80] can then be idealized to a ***Gaussian distribution*** [81] also known as a ***normal distribution*** or a *bell curve* (see Fig. 8.2).

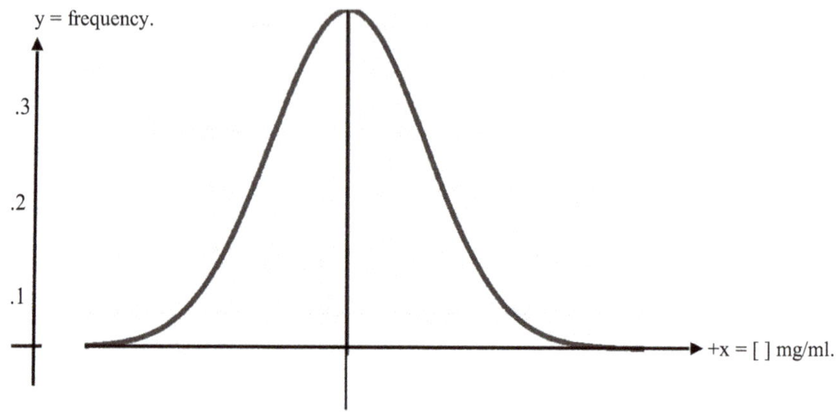


Fig. 8.2: *Gaussian* distribution.

A *Gaussian distribution* (Fig. 8.2) is appropriate for so many kinds of measurement that it is used as a part of a common method for experimental data statistics. Its x-axis can be recalibrated in terms of ***standard normal deviate*** [82, 83] (Tab. 8.3) numbers (Fig. 8.3).

Tab. 8.3: Standard normal deviate, known control sample #1 and known control sample #2 [84].[34]

| Worker # = | In-blood concentration = [](mg/mL) = x = | $x - \bar{x}$ = | $(x - \bar{x})^2$ = | $(x - \bar{x})$ /1.37= snd = |
|---|---|---|---|---|
| 1 | 5.6 | −0.79 | 0.62 | −0.58 |
| 2 | 7.2 | 0.81 | 0.66 | 0.59 |
| 3 | 4.0 | −2.39 | 5.71 | −1.75 |
| 4 | 8.5 | 2.11 | 4.45 | 1.54 |
| Known control sample #1 | 6.40 | | | 0.0073 |
| 5 | 6.3 | −0.09 | 0.01 | −0.07 |
| 6 | 6.2 | −0.19 | 0.04 | −0.141 |
| 7 | 7.8 | 1.41 | 1.99 | 1.03 |
| 8 | 4.7 | −1.69 | 2.86 | −1.24 |
| 9 | 6.8 | 0.41 | 0.17 | 0.2975 |
| 10 | 5.5 | −0.89 | 0.79 | −0.65 |
| 11 | 6.1 | −0.29 | 0.08 | −0.21 |
| Known control sample #2 | 6.38 | | | −0.0073. |

34 Too much scientific technical detail here? Introduction 2.1.

Tab. 8.3 (continued)

| Worker # = | In-blood concentration = [](mg/mL) = x = | x − x̄ = | (x − x̄)$^2$ = | (x − x̄) /1.37= snd = |
|---|---|---|---|---|
| 12 | 7.1 | 0.71 | 0.50 | 0.52 |
| 13 | 5.1 | −1.29 | 1.66 | −0.94 |
| 14 | 8.6 | 2.21 | 4.88 | 1.61 |
| New *mean* as snd. | = | (6.39 – 6.39)/1.37) | | = 0 |

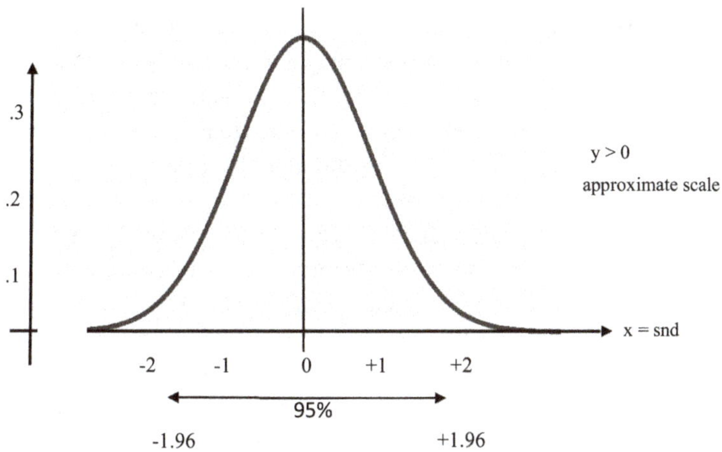


Fig. 8.3: *Gaussian* distribution snd.

The location of a particular data point can be related to the probability of finding such a value within the distribution. A 95% confidence interval is included within 1.96 standard normal deviate (snd) units [85] " on either side of the y-axis. The choice of 95% is somewhat arbitrary – for many analytical purposes it satisfies most people.

When such a choice is made, it signifies the range of fluctuation that would be tolerated without a further critical evaluation of the data point. It may also be used for a quantitative critical tracking of further incoming data. To continue the example [86], suppose the above data are charted, and then the next – the 15th measurement – is made and also charted in Tab. 8.4 [87].

Tab. 8.4: Fifteenth standard normal deviate (snd) measurement.

| Worker # = | In-blood concentration = [] (mg/mL) = x = | x – x = | (x – x̄)$^2$ = | (x – x̄) /1.37= snd = |
|---|---|---|---|---|
| 15 | 12.8 | 6.407143 | 41.05148 | 4.68 |

The enlarged chart – as **worker #** *versus* **snd** – would clearly indicate that this $15^{th}$ measurement should be examined for error possibilities, and perhaps rejected as a data point. Although it is possible that it is correct [88], it would likely be associated with a non-statistical error (e.g. a blunder – in instrument calibration) or an anomaly (*e.g.*, that worker had been using the organic paint solvent as a substance of abuse [89]; or that worker had been the sole-all-day occupant of a small and unventilated chemicals storage room, with leaking containers).

Taylor describes the use of various forms of *control charts* [90], which differ from Fig. 8.4 in that his are concerned with the tracking of *control samples* introduced to monitor the measuring system [91].[35]

The above statistical treatment is appropriate for *quantitative* measurements. But quite often a forensic determination is *qualitative* only. To alternatively continue the above example [92], an insurer might be interested only in the answer to the question: Is there *any* organic paint solvent in blood present? Or analogously, a criminal law judge might be interested only in the answer to the question: Is there *any* cocaine in the seized sample?

The yes or no answer to these questions is not statistically determined in the manner indicated above, although statistical methods of control may well be appropriate to try to assure that the answer is correct, particularly in a large busy lab. See below.

The above kinds of statistical considerations of error would often not be encountered directly in specific forensic science evaluations. But it is important to realize that they would help provide a theoretical underpinning for many analytical chemical results. Sometimes statistical analysis might arise when considering an evidential concept before a court. For example: When the evidence of a DRE[36,37] in an impaired-by-drug *Criminal Code* case would cite medical-type observations as not normal, an obvious next question from defence counsel should be about what is normal. The obvious answer should include reference to scientific literature of random studies evaluated with *Gaussian* mathematics. However, in Canadian criminal courts this would not likely happen[38]; but the concepts are still important.

**35** Too much scientific technical detail here? Introduction 2.1.

**36** *Gaussian* distribution. Chapters 8 and 11.

**37** DRE. Statistical. Chapters 8 and 11.

**38** Re criticism of Health Canada, Judges, et al. Introduction 4.1 and 4.3; Chapters 1, 10 and 11.

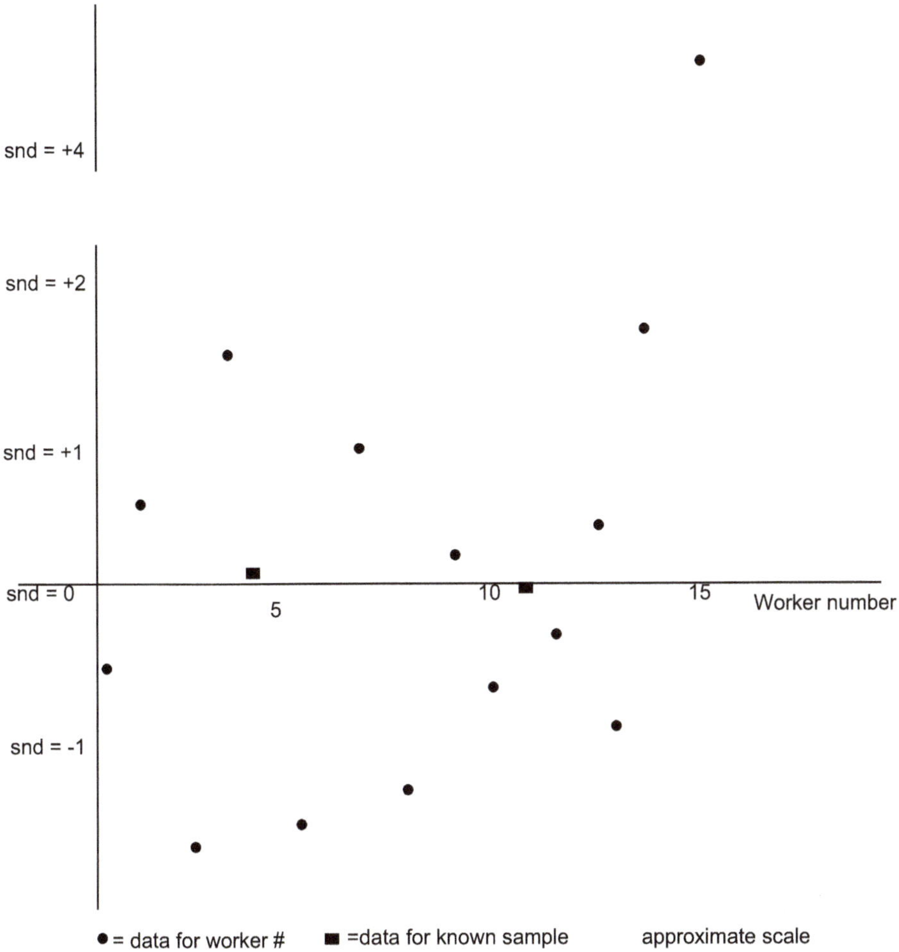


Fig. 8.4: Control chart.

## 8.4 If theory is wrong or method insufficient

### 8.4.1 If theory is wrong

In the considerations, so far, for both *systematic* and *statistical* kinds of errors, it has been assumed that the scientific theoretical basis of the measurement is correct. But if the theory is wrong, the results will be wrong, or useless, no matter how good the measurements otherwise [93]. For example: If in the ink analysis[39] example, the document had been written with a fountain pen containing ink of inorganic components in water as the solvent [94], then TLC with organic solvents would not be expected to lead to the most meaningful results.

It is common in science that theory is wrong, and the progress of science is often related to correction and refinement of such theories [95].[40] A wrong theory in a forensic investigation may not be a complete disaster; it might possibly be treatable as a kind of *bias*[41] and be subject to correction. But the forensic scientist should always be on the lookout for wrong theory problems.

### 8.4.2 If method is insufficient

It may be that the theory is correct-enough for forensic use, but that the method for applying it – including the instrumentation – is insufficient for the purpose. This requires making sure that the method and the instrumentation are valid for the purpose, either by the preparatory work of forensic lab itself, or by supporting literature. Such validation should be part of the lab's usual process.[42]

Examples of insufficiency might be found when commercially available instrumentation is not critically appraised before use.[43]

## 8.5 Graphical representation

### 8.5.1 Cartesian

Statistical data error is discussed above; now to consider how such data, along with the error, might be graphically represented [96][44].

First, about graphical representation more generally. Often it is convenient to display scientific results as a tabulation of numerical values. And, further convenient to plot some of those values on a *Cartesian* grid (Fig. 8.5) [97, 98][45] – as an x,y plot [99] – that might result in a straight or curved line.

---

**39** Variations of Ink analytical chemistry – as examples – *Gedankenexperiment* – continue throughout this book. Introduction 2.3 and 2.2.

**40** Naphthalene. Double bond? CAS 91-20-3. Chapters 4 and 8.

**41** Bias. Chapter 8.

**42** Method validation. Chapters 5, 8 and 11.

**43** Poppy. Too sensitive. Chapters 3 and 8.

**44** Too much scientific technical detail here? Introduction 2.1.

**45** *Cartesian* Grid. Chapters 3, 6 and 8.

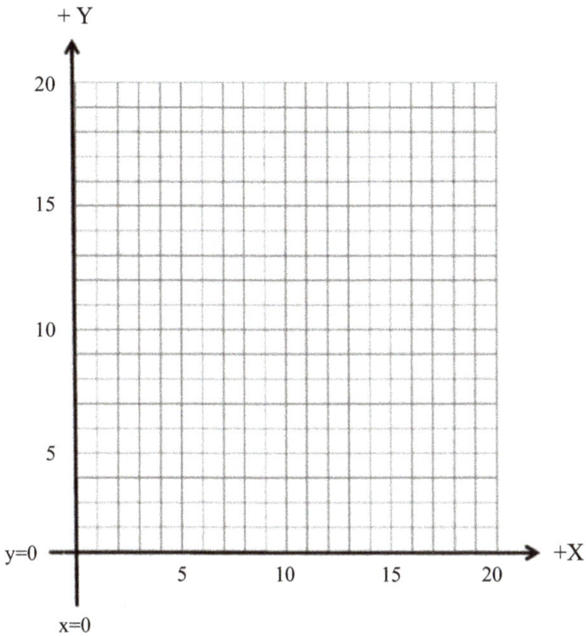


Fig. 8.5: *Cartesian* grid.

Often that resulting line can be found to have meaning, simply by visual inspection. For example [100]: a dose/response curve (Fig. 8.6) [101, 102, 103, 104][46] – with a gradient of increasing severity [105] between no effect and death that might be relevant to the acute toxicology of workplace chemical exposures, and might have a relevance in labour law litigation.

Often, that resulting line can be found to have meaning – as related to a mathematically supported theory – with arguments that the plotted line would "prove" the theory and allow for further data to be calculated. But also, sometimes, this seeking of a mathematically supported theory would fail.

Graphical representation considered as for general concept:

### 8.5.2 Straight line: $y = m \times x + b$

When observed data are to be tested against a theory with thought of mathematical expression, it is often convenient to look for algebraic rearrangement into a straight-

46 Dose/response. Chapters 1 and 8.

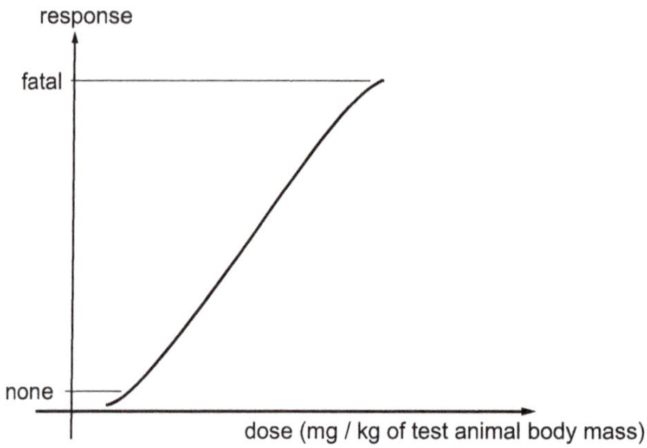


Fig. 8.6: Dose/response.

line equation on the x,y grid, in the general form: $y = m \times x + b$ (Fig. 8.7) – shown here for a specific example [106] as for $y = 0.5 \times x + 5$ [107, 108, 109, 110, 111, 112][47]:

In Fig. 8.7, x and y are values measured on the *abscissa* (horizontal) and *ordinate* (vertical) right-angled *axes*, m is the slope of the line = $\Delta y/\Delta x$ = *rise/run* = $\tan \phi$ (where $\phi$ is the angle with the x *axis*); and b is the intercept with the y *axis*.

### 8.5.3 Error representation

Although often not stated, data that are plotted carry error.[48] This may be indicated in the data tabulation, and carried on into the graphical representation (Fig. 8.8) – shown as an "error rectangle" – or better, ellipse [113, 114, 115, 116].

A fictitious example, sketched, is illustrated in Fig. 8.9a and b [117].

It may be sufficient to understand conclusions from a graphical representation by visual inspection – often obvious. But sometimes, not and even when obvious, formal mathematical method can be used. Thus, for straight-line plots – as above – *least squares* [118, 119] methods are often used to find a best fit for the data points. But care must be taken to be aware of outlier points that if not deleted from the calculations (but still recorded for the data record), would distort the conclusion. And further care must necessarily be stated for deletion decision-making.

Rod Pierce, *Math is Fun*, gives a formula for the slope for the ". . . the line of best fit for **N** points . . ." [120, 121]:

47 Too much scientific technical detail here? Introduction 2.1.

48 Too much scientific technical detail here? Introduction 2.1.

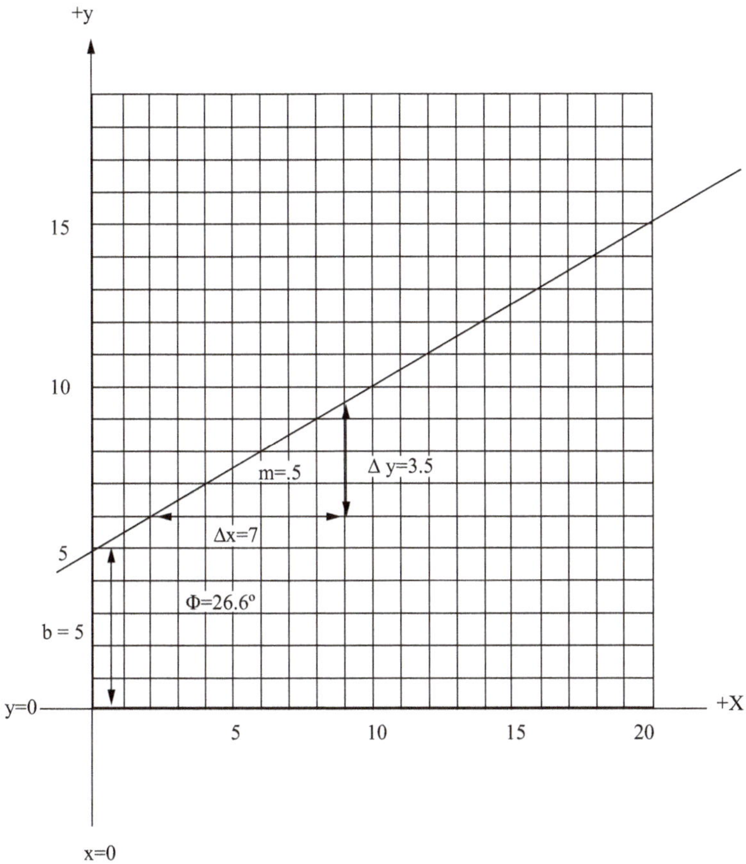


Fig. 8.7: y = m × x + b.

Fig. 8.8: Error rectangle and ellipse.

A)

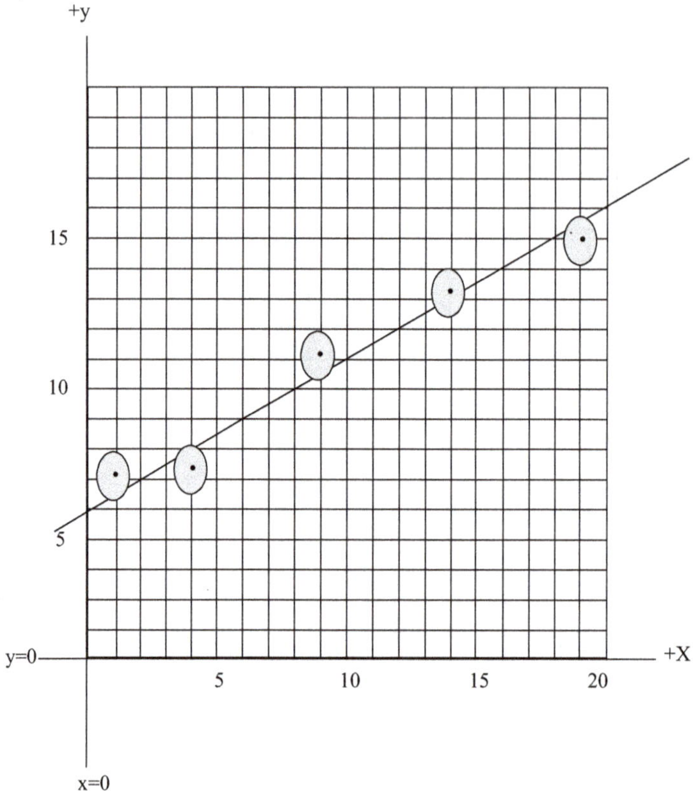


B)

$$y = m \times x + b$$
$$y = .5 \times x + 5$$

| x= | error ± | y= | error ± |
|---|---|---|---|
| 1.0 | .6 | 5.5 | .8 |
| 4.0 | .6 | 7.0 | .8 |
| 9.0 | .6 | 9.5 | .8 |
| 14.0 | .6 | 12.0 | .8 |
| 19.0 | .6 | 14.5 | .8 |

Fig. 8.9: (a) Error shown on Cartesian grid. (b) y = .5 × x + 5 x, y table.

$$m = (N\Sigma(xy) - \Sigma x\Sigma y)/N\Sigma(x^2) - (\Sigma x)^2$$

### 8.5.4 Further about graphical representation

The above treatments of graphical representations are for analyses related to straight-line geometry; the simplest and easiest to read and interpret. But other graphical geometries and related mathematics might be used; some graphical representation can continue to defy explanation; and some may convince the researchers that they got it wrong somewhere. There is also three-dimensional x, y and z mathematics.

Similar to the indication above, for statistical errors, these issues of graphical representations would usually not be encountered directly in specific forensic science evaluations. But again, it is important to realize that they can provide a theoretical underpinning for some analytical chemical results.

It should also be noted that some graphics may be so intuitively obvious as to be persuasive without showing the mathematics. For example: The patient data on the benefit of medical practitioner handwashing, presented by Ignaz Semmelweis (Fig. 8.10) [122, 123, 124].[49]

Some pictorial representations may similarly be intuitively obvious, without mathematical analysis. For example: The carpentry and wood grain matches in the Lindbergh kidnapping case, 1932 [125, 126].

## 8.6 Blunder

Even if the *systematic, statistical* and *theoretical* error possibilities are all under control, errors of *blunder*[50] can also happen. Possibilities abound, such as [127, 128]:
- sample mix-up so that the wrong sample is analyzed;
- the correct sample is compared against the wrong reference standard;
- the wrong method is applied; wrong instrumentation is used;
- the correct sample is routed to the wrong scientist who performs analyses irrelevant to the case at hand;
- the sample experiences the wrong handling conditions (*e.g.*, stored in an oven rather than a refrigerator, or is accidentally spilled on the floor);
- there is a labelling failure (a "time-sensitive – do it now" sticker falls off and is mistakenly reattached to the next container whose "incubate for 3 weeks" sticker also fell off);

**49** Semmelweis. Handwashing. Counting measurement. Introduction 4.1; Chapters 3 and 8.
**50** Blunder. Chapter 8.

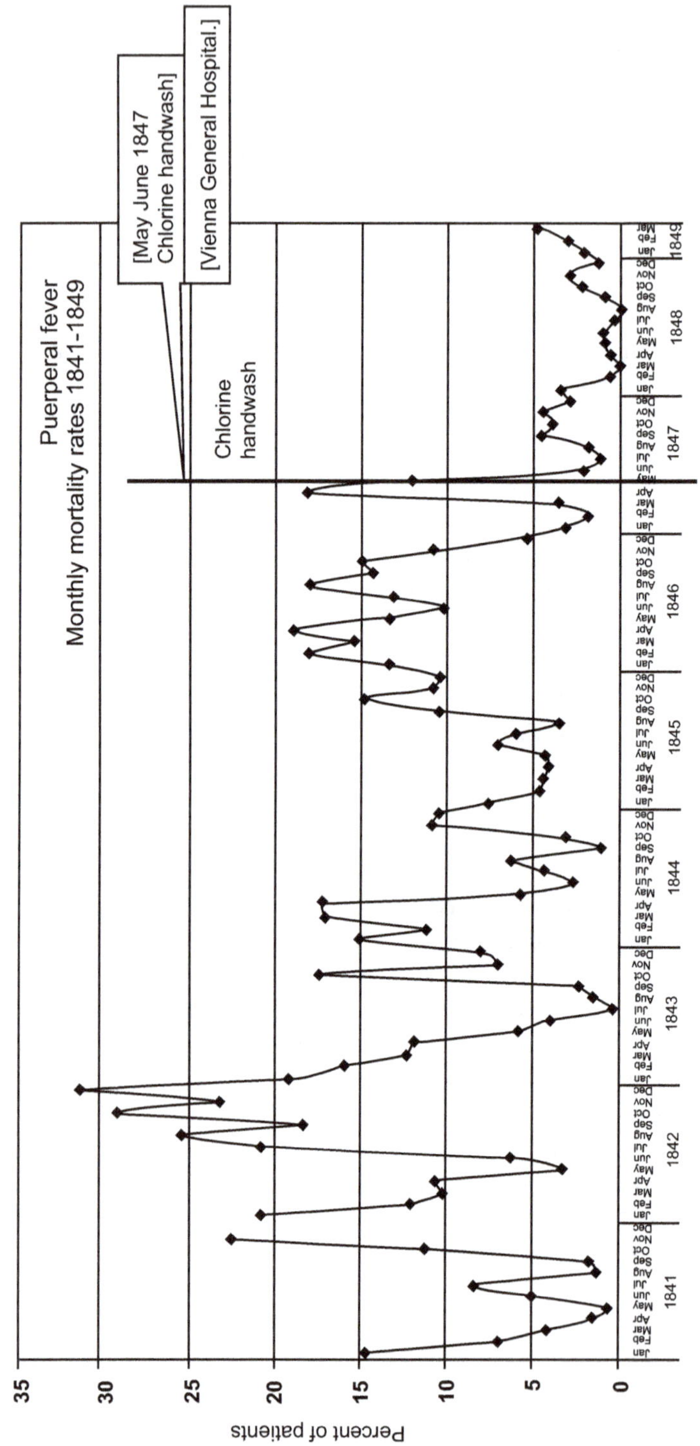


Fig. 8.10: Semmelweis (adapted from CBC Ideas, 2020).

- there is a paperwork mix-up so that the correct work on the correct sample is reported for some other unrelated sample;
- one or the other, or both, the sample and the paperwork are lost; or experience a break in chain-of-custody;
- the no-publication order is missed or misunderstood – sensitive data are published in error; [129
- on the court day, a wrong in-custody person is brought to appear before the court – and maybe it takes a while before anyone notices; and
- there are typographical/transcription errors in the final report (either originating there, or in prior paperwork).

Sometimes such errors are simple and obvious and can be quickly corrected and passed over with amusement [130, 131, 132, 133, 134], but sometimes not [135, 136].

Any of these kinds of blunders can render the forensic work wrong and useless; or worse. How embarrassing! In a legal context it can implicate or exonerate the wrong person. How tragic!

## 8.7 Ethical issues

How forensic evidence, including of experts, is brought before courts and tribunals was discussed in Chapter 1. There are not only error possibilities, but also ethical issues:[51]

### 8.7.1 Ethics

As indicated in Chapter 5, a hallmark of a professional is to be held to act within proper ethical and professional bounds [137, 138]. This may be defined and taught by professional organizations [139]. For some professions, their ethics are given an authority of law [140], which may to some extent be a trade-off for the state having granted a licensing regime that amounts to a monopoly [141] for commercial practice.

A theory behind these ethical regimes is that the professional's special knowledge and skill must be accompanied by a concept of public service and responsibility, which is linked to a professional obligation to always strive towards truth, and to otherwise try to be accountable to do a good job. Generally, for a forensic chemist, doing a good job includes producing and reporting reliably credible analytical results, in a proper way. It also includes proper accountability – especially when something goes wrong.

51 Ethics. Lawyers. Experts. *Etc.* Introduction intro. quote. Chapters 1, 5, 8 and 9.

In the author's opinion, the ethical obligation of this theory should transcend contrary personal requirements of livelihood, employers, and sometimes of the community and the law – with the result that, if need be, the professional must speak out, and/or resign position. In workplace cultures of the real work-a-day world this can be a tall order, particularly in years of a high unemployment economy that sometimes happen [142]. It may therefore be understandable if an operating person shirks aspects of ethical standard for fear of retribution; it is also, in some sense understandable and maybe forgivable [143].[52] But, in such circumstances, there should not be a self-deception of functioning as an ethical professional. However, such a charitable approach should not be so readily applied to leadership personnel.

Fortunately, most practices of personal livelihood, employers, the community and the law are well allied with professional ethics, so that the negative possibilities referred to above should be rare. But it is the possibility that gives definition to the concept of ethical obligation.

Another aspect of ethics requires realistic reporting by the forensic scientist, to all concerned, of the true capabilities and limitations of a method, and not to exaggerate to anyone's advantage. Although the scientist should be well aware of a method's capabilities, investigators, and the citizen (by self, or through legal counsel) whose case is at issue, may not; ethics requires that they should have ready access to the scientist's report.[53]

A forensic scientist should not be part of an exaggeration by others. For example [144], if a factory inspector employs a forensic scientist (an occupational hygienist) to monitor $CO_2$,[54] and then tells a worried worker that scientific air quality measurements prove there is no health hazard, that scientist is ethically bound to correct this, and should complain that the method is being seriously and wrongly overstated. This should be reported upward in the forensic agency's chain of authority, although it might be career limiting or damaging. This should also be of ethical concern to the factory inspectorate.

Ethical concerns about science-based subterfuge also extend to criminal investigations. If, in the ink example [145, 146],[55] a police officer tries to get the accused to panic into a confession by claiming that they can "prove the accused was there then because of scientific dating of fingerprints on the counter top where forgery was done," the forensic scientist, when/if learning of this, should complain that there is as yet no such method[56,57,58][147]. Even if the accused were actually guilty, and even if

52 Error. Forgiveness. Chapter 8 intro. quote; Chapter 8.7.1.
53 Proper scientific report. Chapters 5, 8, 9 and 11.
54 $CO_2$ CO. Chapters 2, 3, 6 and 8.
55 Ink analytical chemistry as examples. Introduction 2; Chapters 1, 2, 3, 8 and 9.
56 Police trickery allowed. Chapter 8.
57 Fingerprint probability. Chapters 1 and 4.
Fingerprint visualization. Fluorescence and phosphorescence. Chapters 6 and 9.
Fingerprint when. Chapters 7 and 8.
58 When. Chapters 2, 4, 7 and 8.

the “confession” would be an important economy in the constabulary’s “fight against crime,” the scientist should not compromise on ethics; and if otherwise, thinking that “fighting crime” is paramount, then should not claim to be an ethical scientist.

Another example of ethical concern is the use of “lie-detector” tests [148, 149] These require the subject to submit to an interrogative session that most lawyers would often advise against, simply on the basis that those with legal problems should only communicate with those of authority through counsel. This is of particular concern for criminal investigations where even truthful statements can be used against accused, who might even be innocent. Of further ethical concern is representation to accused of these tests as an error-free method.

**Some other aspects of ethical issues are given below.**

### 8.7.2 Ethics and economy

Not surprisingly, better, more reliable, forensic work may be expected if greater professional resources are expended, at greater cost; and conversely, less reliable if done on-the-cheap. This raises two ethical questions:[59]

Do cases involving more important or prominent:
- people [150]
- issues

– rate better, more costly and work?

The answer to the first question should be an obvious no [151, 152]. Actual practice might not reflect this.

The answer to the second question is more difficult. Surely, there are gradients of importance in real-world decision-making: For example an ink investigation for a commercial forgery case is trivial relative to a murder investigation. Everyone whose case would be under forensic study is surely entitled to a high standard of investigation, even at considerable expense to the state. But also, the investigation of a murder should take extra care, at extra cost, because of the extra seriousness. Phrased another way, the forgery case deserves no less than proper care and expense, but the murder investigation deserves more; this is not such a satisfying answer. However, a forensic scientist who would do sloppy work in the forgery case, thinking it unimportant, commits an ethical transgression, notwithstanding that the lab results would turn out to be correct, and the accused would be found guilty anyhow [153].[60]

---

**59** Ethics. Lawyers. Experts. *Etc.* Introduction 2.3; Chapters 1, 5, 8 and 9.

**60** Advocacy. Introduction 4.1; Chapters 1, 5 and 8.

An additional ethical concern here should be that the forensic scientist remains as blind as practicable to other reports of the case, public discussion and the accused's identity.

### 8.7.3 Ethical issues – forensic personnel

Even if all the error possibilities discussed above are under control, there are still possibilities that *ethical* issues could interfere with the reporting of correct forensic results.[61]

Thus, if the forensic scientist learns that the case being worked on involves someone or thing with a relationship, this must be reported as a *potential conflict of interest*, so that any of the parties (including that scientist) who wish to argue a *reasonable apprehension of bias* can do so [154].

Such *conflict* might include if the subject of the case is: a family member, friend, enemy, or business rival or partner; or is a business or other property in which there is a financial interest (*e.g.*, if the scientist is a shareholder in the ink manufacturing or sales company) [155].

These *conflicts* and *reasonable apprehensions* that occasionally do arise may be unavoidable surprises, and there is no *ethical* stigma for declaring them when discovered and withdrawing, if appropriate. Withdrawal is not always necessary. A typical *conflict* and *reasonable apprehension* would arise when the judge comes into the courtroom to start a criminal trial and notices that the accused used to be a client when the judge used to be defence counsel before appointment to the bench [156]. Sometimes Crown and Defence counsel will voice no objection, and sometimes otherwise. However, to be aware of a *conflict* and to not report it is *ethically* wrong.

While such *ethical* issues are not scientific errors, they have been dealt with here because if discovered too late they can, like scientific error, prevent forensic work from being used at law.[62]

### 8.7.4 Ethical behaviour of the state

Another *ethical* issue involves the behaviour of the state – when its criminal law enforcement activities involve intentionally misleading the citizenry [157]. In some circumstances, Canadian law clearly allows, if not encourages, such subterfuge [158, 159, 160, 161].[63]

The question here is not the use of subterfuge-source data at law – it can sometimes legally be used in Canada – but rather whether or not the forensic scientist

61 Ethics. Lawyers. Experts. *Etc.* Introduction 2.3; Chapters 1, 5, 8 and 9.

62 Ethical. Chapters 1, 5 and 8.

63 Police trickery allowed. Chapter 8.

should want to be in the employ of the state – and part of the process – when it does so.

More broadly, the question is whether or not the citizenry should have confidence in the scientific declaration of state agencies. For example, the author has argued that Health Canada – Canada's national health agency – has tainted health-science credibility throughout the entirety of its systems because of its involvement in criminal law enforcement.[64]

A corollary policy issue, that goes beyond forensic science ethics, is perhaps more important: If law enforcement personnel are allowed and encouraged to do subterfuge, then it becomes an effective part of their job description. If so, then there can develop a culture of deceit – where does it stop [162]?

## 8.8 Fraud

[. . .][65]

ז רַגְלֵיהֶם, לָרַע יָרֻצוּ, וִימַהֲרוּ, לִשְׁפֹּךְ דָּם נָקִי; מַחְשְׁבֹתֵיהֶם, מַחְשְׁבוֹת אָוֶן – שֹׁד וָשֶׁבֶר, בִּמְסִלּוֹתָם.

*7 Their feet run to evil, and they make haste to shed innocent blood; their thoughts are thoughts of iniquity, desolation and destruction are in their paths.*

[. . .]

[*Isaiah* יְשַׁעְיָהוּ] [163, 164, 165, 166, 167, 168]

In consideration of indications above, there are, regrettably, issues that do involve *ethical* wrongs.[66,67] Most regrettable of the various forms of scientific misconduct is fraud [169, 170, 171]. This is not unknown among scientists – forensic [172, 173, 174, 175, 176, 177, 178, 179, 180, 181, 182, 183, 184][68,69,70] and otherwise [185, 186, 187, 188, 189, 190, 191, 192]. And, lawyers' various misconducts [193, 194, 195, 196, 197, 198].[71,72] Nor is it unknown amongst other participants – for example, Colonel Hubert Henry – of the Dreyfus Affair.[73]

There are also stories of frauds/fakes:

– Theranos; blood data measuring device [199, 200, 201, 202, 203].

64 re criticism of Health Canada, Judges, et al. Introduction 4.1 and 4.3; Chapters 10 and 11.

65 Introductory quotes. Introduction.

66 Ethics. Lawyers. Experts. *Etc.* Introduction intro. quote. Introduction 2.3; Chapters 1, 5, 8 and 9.

67 Fraud. Griffiths Report, 2001. QA failure. Gemma Ramlal, 2001. Annie Doukhan, 2012. Introduction 4.3; Chapters 8, 9 and 10.

68 Bre-X. Chapters 8 and 9.

69 Fraud. Annie Doukhan, 2012. Gemma Ramlal, 2001. Chapters 8 and 9.

70 Error. Chapter 8.

71 Advocacy. Introduction 4.1; Chapters 1, 5 and 8.

72 Lawyer misconduct. Police subterfuge. *R v W*. . . Chapters 8.8 and 10.5.

73 Captain Alfred Dreyfus. *bordereau, petit bleu.* Colonel Hubert Henry. Introduction 2.3; Chapters 1 and 8.

- ADE 651; bomb detector. [204, 205]
- Annie Dookhan, Massachusetts Drug Lab. [206, 207][74]
- ExThera; with advertisement and purported treatment for cancer [208].
- And, in fictional literature, the "Infra-redioscope," 1953 [209].[75]

Because one of the main attributes of any profession is reputation for credibility and integrity, the dishonesty of even a small minority is very, very serious, and very, very damaging.

## 8.9 Near-fraud

The fraud discussed just above would involve such actions as the actual willful faking or altering of data [210, 211, 212, 213][76] or reports, either directly or indirectly, and cover-up therefor. But, there is also the acquiescence of others that might be described as-*near-fraud* (although in some sense still fraud, albeit it of a lesser nature). This would include not reporting upward in the chain of authority when actual fraud is observed. It may also include – if there is intention to mislead – taking a willfully cavalier or neglectful approach to the formalisms of science, its financing [214] and its record keeping.

Near-fraud might also include – again with intention to mislead – being unresponsive or intentionally unclear, or unfairly demanding of bureaucratic precision [215], to forestall legitimate requests of information [216, 217, 218]. Near-fraud might also include scientific negligence and sloppiness. For lawyers it might include what is called *sharp practice* [219][77]; and perhaps more-distantly related practices of politically inspired "lawfare" [220] or "weaponized justice system" [221, 222].[78]

Both fraud and near-fraud are characterized by some degree of willfulness. Near-fraud is less serious in a moral sense, but still involving correctable behaviour.

---

74 Fraud. Griffiths Report, 2001. QA failure. Gemma Ramlal, 2001. Annie Doukhan, 2012. Introduction 4.3; Chapters 8, 9 and 10.

75 Fictional literature. Christie. Poe. Introduction 2.2; Chapters 5 and 8.

76 *Leviticus*. *Deuteronomy*. Chapters 6 and 8.

77 Advocacy. Introduction 4.1; Chapters 1, 5 and 8.

78 *Sharp practice*; lawfare; weaponized justice system. Chapter 8.9.

## 8.10 How to try to limit and cope with error

### 8.10.1 Error control

It should first be noted that although careful work, including good organization and documentation, do facilitate good error control, they do not guarantee error-free work. It is always possible to miss something; that is why formal error control methods are so important. Good error control for quantitative measurements is always in the context of required-to-be-explained statistical uncertainty. This applies to forensic analytical chemistry measurements, as well as to theoretical studies [223].

It should also be noted that good error control does not remedy incorrect data to give it worthwhile meaning [224]. The scientist should, with avoidance in mind, be aware of the danger of *quod volimus credimus libenter* [225, 226, 227, 228, 229, 230].[79]

Good error control may be strived for by considering, with avoidance in mind:

## 8.11 Quality assurance

> . . . *making a list, checking it twice* . . .
>
> J. Fred Coots & Haven Gillespie, 1934,"Santa Claus Is Comin' To Town" [231].[80,81,82]

### 8.11.1 Introduction

As indicated in Chapter 5, modern commercial, industrial and laboratory practices should have a strong focus on producing quality products and services, and for assuring consumers of that quality.[83] Not only to look at the technology of the products and services, but also the organizational functions that produce them [232, 233].

Thus, a forensic lab must have a consistently applied and well-documented quality assurance programme, without which results reported by that lab should not be seen as worth much, and should be looked at with skepticism, by judges and paying contracting clients, including taxpayers when government is involved.

These chapter sections deal with quality assurance in the context of formalized processes.

---

**79** ≈ Believe what we want to believe.

**80** Introductory quotes. Introduction.

**81** Santa Claus. QA. Chapter 8.11.

**82** QA. ISO. Chapters 5.6; 8.11 and 9.1.1.

**83** Accreditation. QA ISO. ANAB. Chapters 5, 8, 10 and 11.

The service of a forensic lab is the provision of a report (see Chapter 9)[84] of forensic results and conclusions that is reliably accurate; and well-explained in writing, to the contracting client, and to whomever else is entitled to the report. For a forensic lab that whomever else would be the various other parties,[85] and since many legal processes are essentially public,[86] the forensic report may become a public document.

### 8.11.2 ISO 9002. ISO/IEC 17025:2005

The International Organization for Standardization [234] publishes its 9000 series of standards for quality assurance [235]. **ISO 9002**, also published by the Canadian Standards Association as CAN/CSA-ISO 9002-94 [236], deals, in general terms, with the organizational structure and functions of a quality assurance programme.

**ISO/IEC 17025:2005** [237, 238] applies specifically for laboratories that do testing.

ISO/IEC 17025:2005 and ISO 9002 are used here to show what a quality standard looks like, and what should go into a quality assurance regime. They are chosen here for convenience because they are so widely known. There is nothing especially magical about their particular wording – their essential features are found elsewhere [239]. What is important is the content [240]. Generally, some important concepts:
- Defined and documented quality policy – by management with executive responsibility and authority.
- Functions of quality-directed personnel are defined and documented.
- Management with executive responsibility and authority regularly reviews the quality system.
- There must be a quality manual.
- Documented procedures.
- Definition and documentation for quality planning.
- Requirements of work to be done for clients adequately defined and documented, and for the forensic lab to be capable of delivery.
- Procedures to control documents and data; and to be documented.
- Control of sample identification, movement and processing.
- Documented procedures for validation, inspection and testing[87].
- Results of analytical chemistry work inspected and tested.
- Documented procedures for what to do when the inspection and testing reveal problems with the analytical chemistry work; and for prevention.
- Documented procedures for the records related to all of this; and such record keeping is to be done.

---

**84** QA. Quality assurance. Report. Chapters 8 and 9.
**85** Advocacy. Introduction 4.1; Chapters 1 and 8.
**86** Confidential *verses* public. Introduction 4; Chapers 5.8; and 8.
**87** Method validation. Chapters 5, 8 and 11.

- Internal and external quality audits.
- Documented procedures for personnel training and qualification, and accreditation.
- Statistical techniques for controlling the quality processes; and documented.

Generally, the above is reflected in the ISO/IEC 17025:2005 headings [241].[88] Generally, it would be similar to other ISO standards, such as ISO 9002. Such provisions would try to assure quality work by careful consideration of the processes that are accessible by comprehensive documentation and document control – to allow for scrutiny and correction. Much of these would seem intuitive to a science team dedicated to trying to do good work, and while stating so may seem repetitive and tedious, it is also important to state it formally.

Considerations of ISO/IEC 17025:2005 and ISO 9002 lead to some other matters that should also be dealt with:

### 8.11.3 Public

Public scrutiny should be regarded as a potential method of error control.[89] However, this should be viewed in contrast to some needs for confidentiality, and there should be a reasonable approach to government secrecy – see Chapter 5.

As indicated in Chapter 5, the forensic report may sometimes become a public document. Depending on the nature of the case, not only this report, but also other aspects of the forensic process may become public. ISO/IEC 17025:2005 and ISO 9002 do not appear to deal with this; their background assumptions would appear to be that the report and supporting information are the private business of the contracting parties – the forensic lab and its client – and that the lab and its personnel would hold all confidential at the pleasure of the client. The client may do anything it likes [242] with the report and supporting information – from publishing it by paid advertisement in *The New York Times* or posting it on the Internet, to trying to hide it forever.

But, since, as mentioned in Chapter 5, many government and legal processes are/or should be essentially public, it may happen that others than the forensic lab and its client may be able to have access. While many litigants may wish otherwise, this is an important feature of the justice systems of Canada, the United Kingdom and the United States [243].[90]

**88** Method validation. Chapters 5, 8 and 11.

**89** Public. Introduction 4; Chapters 5 and 8.

**90** Public process. Introduction 4.1; Chapters 5, 8, 10.

Another cause for the forensic report and supporting information to become public, as a matter of public policy, is because of state treasure in the process – either directly as a part of the government structure, or indirectly as government-sponsored work, or even more indirectly where even a small amount of government money is in a larger project of mostly non-government money. In the United States, the involvement of federal government money has additional *qui tam* possibilities.[91]

Public access to such reports and information may come quite easily and routinely – by:

- publication as public documents,
- releases to the media, including Internet,
- publication in scientific journals,
- grants of patents of invention,
- parliamentary statements, and
- replies to citizens' enquiries – either by direct process; or under freedom of information legislation.[92]

But when it is sometimes not so easy,[93] it is reasonable and responsibly democratic for the citizen to speculate that "Something is rotten . . .," [244] and to investigate and raise a fuss – by:

- freedom of information process,
- litigation – as for disclosure to be provided to defence counsel,
- criticism in the media.
- political process.

However, when the controlling authorities do not want to disclose, the eventual disclosure, if it comes at all, would likely not be timely, and maybe disappointing anyhow.

### 8.11.4 Peer review

Peer review is an important part of the scientific process. It is usually the critical scrutiny by other scientists, knowledgeable in the field, that is required before publication in a scientific journal [245, 246], or other publications, or the granting of research funds. It is typically a formalized process; sometimes uncomfortable to those being reviewed. While editors and granters would both have the final say for publishing or granting, and the author may reply, the process of referee comment is essential. How-

91 US *Qui tam*. Chapters 1, 8 and 9.

92 Freedom of info. legislation. Chapters 5, 8 and 10.

93 Health Canada. Introduction 4.3; Chapters 8 and 10.

ever, it can be time-consuming – and sometimes exasperating – and sometimes perhaps with ugly scenes (see Fig. 8.11) [247].[94,95]

Fig. 8.11: Peer review.
Most scientists regarded the new streamlined peer-review process as "quite an improvement" [cartoon by Dr. Nick Kim, Wellington, New Zealand].

Peer review should neither be used for political censorship, nor be part of personal animosity, but is a reasonable speculation that sometimes it is.

Essentially, peer review is another opinion in a hopefully organized way. The concept is consistent with ISO 9002's quality audits.[96]

While peer review would mostly be associated with science publications and research grants, it may also find a place in the approval process for forensic reports – most particularly for method evaluation, and reports that are non-routine.

94 Introductory quotes. Introduction 1.2.
95 Dr. Nick Kim. Cartoons. Copyright acknowledgement. Chapters 2, 6 and 8.
96 Audit. Chapter 8.

### 8.11.5 Audits and scrutiny

Audits and scrutiny, internally, and done by an external agency, can be important to a lab's credibility [248].[97] These may be covered by effective inclusion in other error control methods. But they can also be done in addition, and independently.

### 8.11.6 Accreditation

Also consistent with ISO quality audits, is the accreditation process for analytical labs [249, 250, 251, 252, 253, 254, 255, 256, 257, 258].[98] In this process, an external agency [259] scrutinizes the lab and its personnel and subsequently accredits or certifies that lab, or not, or conditionally. Such accreditation is according to the terms and criteria of the accrediting agency, and care should be taken not to confuse the terms of one accreditation with another. For example, ISO 9000 series registration is not the same as accreditation of a lab for specific forensic analytical purposes.

There can also be accreditation processes of lab personnel.[99]

### 8.11.7 Document trail

A most essential aspect of any quality assurance programme is its documentation – see above, and in Chapter 5.[100] While that documentation does not guarantee quality, nor is it absolutely tamper-proof evidence, the absence of even part of it, or defects, clearly raise questions of credibility. Any inadequately explained break in the document trail – or even a break in its transparency – should be addressed.

### 8.11.8 Method validation

As indicated above, an important quality aspect is the assurance that the analytical chemistry method and the related instrumental systems actually work – that they can be relied on to produce correct results.[101,102] If there is some flaw to stand in the way, then the method is not valid and should not be used, or perhaps only used with stated

---

**97** Audit. Chapter 8.

**98** Accreditation. QA ISO. ANAB. Chapters 5, 8, 10 and 11.

**99** Personnel – scientific qualification. Automated instrumentation. Chapters 5, 6 and 8.

**100** Lab core documents. SOP. Scientific literature. Chapters 3, 5 and 8.

**101** Method validation. Chapters 5 and 8.

**102** Validation. DRE. Chapters 8 and 11.

reservations. Perhaps the method might be good enough to give approximations, but if so, that limitation must be stated.

Part of any quality assurance regime is a checking of the method. This should be by formal processes, documentation of which forms part of the record of the analysis. Method validation failure is a serious reason to not accept the analytical results, as is a failure to show the documentation.

## 8.12 Replication and multiple-blind check samples

The ideal concept of blind determinations is associated with law-related process. Ancient and subsequent imagery of *Justitia* survives – with blindfold, scales and sword [260, 261, 262, 263].[103] By analogy, then, forensic analyses should be with the analysts being as blind as practicable to the related human story.

Applying this further: ***Blind replication*** of lab analyses and ***multiple-blind check samples*** can be important to avoid some kinds of error. These are based on the idea that such errors would be more likely caught if the same work is done again (perhaps several times) as remotely as practicable from the first time. These analyses-again would be for actual samples analyses and for created check samples.

Thus, for the forensic ink analysis used[104] as examples in previous chapters, the results from the procedure set out in Chapter 3, would have added reliability if two separate TLC plates were used. And more reliability still, if those second TLC procedures were done by a second chemist, blind to the work of the original chemist. And, even more reliability still could be achieved by having the second experiments done yet again by another chemist in another lab. And so on.

How far the so on should go becomes not as much a question of science, but rather a question of administrative feasibility as to time, cost and how much of the sample is left; and is ultimately a question of judicial interpretation as to necessity; and maybe judicial patience.

For a single case of only moderate importance [264], such as the ink example, with well-documented lab work, in a reputable lab, by a reputable chemist, one set of experiments is arguably enough (and many chemists would routinely have done more than one spotting on the TLC plates, anyhow). It might well be reasonable for a judge to tell defence counsel that the once-only TLC work would stand, unless the defence itself arranges for additional work, or has acceptable specific reasons otherwise [265].

**103** Justitia. Introduction 4.1; Chapters 5 and 8.

**104** Variations of Ink analytical chemistry – as examples – *Gedankenexperiment* – continue throughout this book. Introduction 2.3 and 2.2.

An extended kind of *replication* is for the forensic process to be repeated independently at several other labs [266]. In the ink example, this would not be too practicable because the destructive testing[105] method used by the first lab might also be used by subsequent labs; the sample would go to being used-up.

However, for some processes, replication might be replaced by of some kind of proficiency monitoring. As for labs that do large volumes of routine analyses.

As a fictitious example [267], suppose a region economically depends on its paint industry to such an extent that to assure quality its government inspectors take dozens of samples daily from producers, wholesalers and retailers. At the end of each day, these samples are delivered to a government forensic chemistry lab (totaling more than 15,000 samples last year). For those samples not meeting the criteria set out in the legislated regulations, a *Certificate of Failure to Meet Standard* is forwarded to the Crown Attorney for prosecution for an absolute liability offence.[106]

Thus, teams of chemists at the government lab continually do analyses on very many samples. It is dreary routine work, and no chemist can remember which sample was done by whom, or when – there are so many – it seems to go on forever. Therefore, the evaluation and legal process relies absolutely on the documentation.

The reasonable question arises: For such a volume of routine work, with scant possibility of personal recollection, what are the chances of error – that something sometimes goes wrong – for example: a sample mix-up *blunder* (see above)?

In the author's view, the answer is: not zero; any claim to the contrary, in the real world, where absolute perfection is elusive, is absurd; perhaps in a fantasized alternative universe, but not here. Readers should examine this statement in the context of their own experiences with large institutions and their bureaucracies. Who has never heard of such an error – in analogy: at school, at work, with a public utility, with a government agency [268], in news reports [269, 270, 271, 272, 273, 274] with a large retailer, with professionals who serve the public, or as typos [275] in a reputably published book? [276]

Rather, for such a high-volume lab situation, a statistical gauge[107] should be used, to try to obtain a reasonable assessment of what the error rate is, so that it can be reasonably coped-with. This may be called proficiency monitoring.

Such a statistical gauge could be a system of *multiple-blind check sampling*, by which known samples are prepared external to the group doing the analyses. These would then be distributed as though they were real, and the analytical results subsequently compared to the preparation records. Such a procedure might be regarded as a kind-of statistical substitute for *replicate* analyses.

---

**105** Destructive and non-destructive testing. Chapters 3, 5 and 8.

**106** *Absolute liability offence.* Chapters 1 and 8.

**107** Referred to in Chapter 8.3 with respect to yes-or-no analyses.

*Multiple-blind* here means that although the group of analysts – the chemists – would know there would be checking, they would not know of – be blind to – some various aspects of the checking procedure. This would include as blind to some or all of:
- Which samples are for checks? [277]
- What would be in any check sample?
- How frequently check sampling is done?
- For how many check samples, and when?
- How the check samples get into the analysis stream?
- What supervisory personnel, at what levels, administer the check sampling?
- To what extent supervisory personnel are blind to the details of the checking processes?
- Who gets to see the results of the check sampling, and when?
- Who determines the failure that the check sampling would be seeking; and the consequences?

*Single-blind* lab analyses here would mean that the analysts would know which samples are for checks, but would not know what is in them. (The term *double-blind* is often used for pharmaceutical trials for when neither the medical researchers nor the patients know, until revealed later, whether it is an active drug or *placebo*; not the situation here although similar in general concept, with similar terminology.)

To continue this example [278], if, in due-course, the statistical report were to reveal an error rate of 5% (that is, the analysis is quantitatively off beyond ± defined limits, or is qualitatively wrong), then everyone involved should have serious concern about the lab's credibility, including industry owners, judges, insurance carriers and taxpayers, or others who are paying for the work. Product liability lawyers may start taking an interest.

But an error rate of, to alternatively extend the example [279], 0.05% is not so easy to decide about. Perhaps a judge would allow the lab's work as credible, provided that there would be an adequate ISO 9002-type document trail (see above) for defence counsel and others to scrutinize. In the absence of the document trail, it is the author's view that no work of the lab should be accepted as credible; and likewise if there is no *multiple-blind* statistical gauge of error, or *replicate* analysis.

It should be noted that *single-blind check sampling* is not sufficient to gauge the error rate.[108] Although single-blind sampling scrutinizes the proficiency of the chemist under certain conditions, it is not adequate to scrutinize the lab's *in-process* performance.

Generally, the use of replication and multiple-blind check samples is a realistic way to reduce the chance for error.

**108** Error. Sampling. Sampling plan. Chapters 3, 4 and 8.

## 8.13 Some general advice

For some kinds of errors discussed above – ***systematic, wrong theory, blunder***, and ***ethical issues***, there is little effective advice other than to take care, to follow the quality assurance procedures – including SOPs and documentation referred to in Chapter 5 – and to do *replicate* analyses and *multiple blind check samples* referred to in this chapter, when possible. And not to be arrogant that errors cannot happen, but rather to be prepared to discuss probabilities.

**Fraud** and **near-fraud** (see above) are more difficult to deal with because both involve some elements of dishonesty/uncooperativeness on the part of someone who might well be trying to cover-up. A sometimes effective method to discover fraud and near-fraud is to look for irregularities in the document trail (see in Chapter 5), that might have been overlooked in a cover-up attempt, or be beyond its reach.

Perhaps a most newsworthy exposure of fraud and near-fraud is because of whistle-blowers, particularly in the United States under strengthened Civil War–era bounty hunter – *qui tam* – legislation.[109] Whistle blowers would often be disgruntled former employees, or soon-to-be former employees.

***Statistical error*** (see above) can be dealt with by clearly defined decision-making as to the tolerable limits of fluctuation, and by control chart tracking.

## 8.14 Critical appraisal – *nullius in verba*

As written above the service of a forensic lab is the provision of a report (Chapter 9)[110] of reliably accurate results and conclusions – well-explained in writing [280, 281, 282].[111]

Perhaps, a good indicator of quality processes with good error control is the appearance of that report. If the report is concisely written to convey meaning, presenting and interpreting the relevant data in a meaningful way, with convenient reference to other supporting and explanatory literature, and with an error discussion, then it is likely reliable. These criteria are not absolutely sure tests for reliability; but any of these criteria failure would be a good indicator of unreliability.

The above is consistent with the concept that an understandable meaningful report would allow for a meaningful critical appraisal – so that the reader can apply the concepts, as stated in Introduction 4, that science is *nullius in verba*, and that *ipse dixit* is not good enough.

---

**109** US *Qui tam*. Chapters 1, 8 and 9.

**110** Report. Chapter 9.

**111** *Nullius in verba*. Introduction 4.1; Chapters 2, 5 and 8.

The above is also consistent with the concept implied by the Yiddish expression שֵׂכֶל *séchel* [283] – that what is presented is to be thought about and seems to make good sense [284, 285, 286].

And, also consistent with the concept expressed by the fictional Prince Kinnall, quoted in Chapter 9.[112]

Another indicator of forensic quality processes are the responses of the report's providers when asked reasonable questions that are reasonably answerable. Non-answers are perhaps suggestive of unreliability, as is arrogance or an anger response for asking [287].

## Notes

**1** "All analytical measurements are wrong: it's just a question of how large the errors are, and whether they are acceptable." M. Thompson, *Analytical quality control in theory and practice, Proceedings of the 3rd International Symposium on the Harmonization of Quality Assurance Systems in Analytical Chemistry*, 19–20 Apr., 1989, Washington, DC (ISO/REMCO 184): 183–189 (1989) – as quoted by W.M. Johnson in "Quality Control and Quality Assurance" – Chapter 8 in C. Riddle, ed., *Analysis of Geological Materials*, Marcel Decker, Inc., New York City, 1993, page 343.

**2** https://books.google.ca/books/about/The_Analysis_of_Geological_Materials.html?id=leJMMQAACAAJ&redir_esc=y

**3** Also quoted elsewhere: Virginia T. McLemore, Kathleen S. Smith, Carol C. Russell, editors, *Sampling and Monitoring for the Mine Life Cycle*, Society for Mining, Metallurgy, and Exploration, 28 May 2014; page 79.

**4** Also quoted elsewhere: *Hydrometallurgy 2008: Proceedings of the Sixth International Symposium* https://books.google.ca/books/about/Hydrometallurgy_2008.html?id=1etfSdk55SYC&redir_esc=y

**5** CBC News · Posted: Jun 24, 2001 10:38 PM ET | Last Updated: June 24, 2001
https://www.cbc.ca/news/canada/lab-errors-may-void-ontario-drug-convictions-1.283147

> **Lab errors may void Ontario drug convictions** [Toronto, 2001]
>
> Botched paperwork at a federal lab may void thousands of drug convictions in Ontario.
>
> Anyone convicted of a drug offence in Ontario during the past 12 years, even those now serving time, may be off the hook because of bungled paperwork at a federal drug lab.
>
> The problem involves thousands of drug analysis certificates that Health Canada says might be not be accurate.
>
> The certificates are issued in every case where suspected illegal drugs are found. They are used by courts to identify and determine the purity of various drugs.
>
> "The chances are very good that if you just get in touch with a lawyer any lawyer that you will be out of jail very quickly," said Alan Gold of the Criminal Lawyers Association.

---

**112** Prince Kinnall. Chapters 8 and 9.

> The analyst at the centre of the controversy is Gemma Ramlal, who resigned from her Health Canada lab job two months ago. She now works at a grocery store. She denies any wrongdoing.
>
> "I want to know how it took my supervisors so many years to figure out my work was wrong," Ramlal said.
>
> Dann Michols, assistant deputy minister at Health Canada, told CBC TV: "We trust . . . the professionalism and training of the individuals to follow these procedures and only in this one case have we been disappointed."
>
> Ramlal had been doing the work for 12 years. Earlier this year, she went on vacation and another analyst took over and discovered some things were wrong with the certificates.
>
> A review of Ramlal's work discovered substances misidentified, data entry errors, wrong dates and instances when Health Canada procedures were not followed. The discovery casts doubt on some 16,000 certificates issued over the past 12 years.

**6** See also: *The Globe and Mail,* NATIONAL NEWS/NATIONAL REPORT, 28 September 1998, page A6; CP. ". . .VANCOUVER. The overdose death of an employee of the [Health Canada] federal drug-testing lab has cast doubt on 20 years of drug convictions, lawyers say. . . ."
7 See also: Anne McIlroy, Parliamentary Bureau, *The Globe and Mail,* Tuesday, 13 July 1999. "Safety defect found at high-risk laboratory/Winnipeg waste-water leak a 'wake-up call' for Ottawa [1999] . . ."
**8** Ontario Ministry of the Attorney General, *Report of the Kaufman Commission on Proceedings Involving Guy Paul Morin* [1997], Executive Summary, page 6. https://librarysearch.library.utoronto.ca/discovery/fulldisplay/alma991106708569506196/01UTORONTO_INST:UTORONTO
https://en.wikipedia.org/wiki/Kaufman_Report
**9** [Latin] *inter alia* = among other things. https://dictionary.law.com/Default.aspx?selected=996
**10** Ontario Ministry of the Attorney General, *Report of the Kaufman Commission on Proceedings Involving Guy Paul Morin/Executive Summary*; Chapter V: The Investigation by Durham Regional Police and the Prosecution of Guy Paul Morin/"The Body Site."
**11** http://netk.net.au/Canada/Morin2.asp
**12** Jan Ransom, "Inmates Received Harsh Punishments Over False Positives for Drugs," *The New York Times*, 21 November 2019, page A22.
**13** Jan Ransom, "*After False Drug Test, He Was in Solitary Confinement for 120 Days /* Hundreds of New York State prisoners were locked in cells, denied release or removed from programs when tests erroneously showed they had used narcotics, according to a lawsuit," *The New York Times*, 20 November 2019.
https://www.nytimes.com/2019/11/20/nyregion/prison-inmate-drug-testing-lawsuit.html
**14** "To err is human . . ."
https://en.wikipedia.org/wiki/An_Essay_on_Criticism
https://www.eighteenthcenturypoetry.org/works/o3675-w0010.shtml
https://en.wikipedia.org/wiki/To_err_is_human
Pat Croskerry, "To err is human – and let's not forget it," *Canadian Medical Association Journal*, 23 Mar 2010.
https://pmc.ncbi.nlm.nih.gov/articles/PMC2842843/#:~:text=Alexander%20Pope%2C%20poet%20of%20the,safety%3A%20To%20Err%20is%20Human.&text=The%20remainder%20of%20the%20line,further%20reinforced%20the%20report's%20message
https://www.cmaj.ca/content/182/5/524
**15** Analytical chemistry is the branch of chemical science for the chemical determination of substances, as to what they are (*qualitative*), how much (*quantitative*), and the uncertainty. . . .

**16** The potential to wrongly identify a substance could be related to very subtle molecular aspects – for example chirality. Or otherwise – for example: Iron pyrite = iron disulphide = fool's gold ≠ gold. $Fe^{+2}$ $S_2^{-2}$ ~~Au~~

**17** https://en.wikipedia.org/wiki/Pyrite

**18** "Substance" here might be a single chemical compound – or a mixture of several compounds. And there might be phase changes with temperature and pressure changes (for example liquid water to ice).

**19** See also: Julian Tyson, *ANALYSIS/What Analytical Chemists Do*, Royal Society of Chemistry Paperbacks, Cambridge CB4 4WF; 1988, 1994; ISBN 0-85186-463-5; Chapter 5 "Tools of the Trade," at pages 117→133.

**20** See also: Robert L. Peesok, Kenneth Chapman, Wade W. Ponder, *et al.*, *CHEMICAL TECHNOLOGY HANDBOOK/Guidebook for Industrial Chemical Technologists and Technicians*, American Chemical Society, Washington, D.C.; 1975, 1980; SBN 0-8412-0578-7; Chapter 12 "Use and Interpretation of Data." pages 111→125.

**21** See John Keenan Taylor, *Quality Assurance of Chemical Measurements*, Lewis Publishers, Inc., 1987; page 7.

**22** Farrington Daniels, J.W. Williams, Paul Bender, Robert A. Alberty, C.D. Cornwell, *Experimental Physical Chemistry, Sixth Edition*, McGraw-Hill Book Company, Inc., New York; 1962; Chapter 18, "Treatment of Experimental Data," page 393, et seq.

**23** *Gedankenexperiment*. Introduction 2.2.

**24** *Gedankenexperiment*. Introduction 2.2.

**25** Generally, a *balance* is an analytical device to measure *mass* by balancing the *weight* of what is being measured against standardized weights that are labelled as to their mass (grams) – all done within the earth's gravitational field. Or balanced against a calibrated magnetic system with electronic read-out. Or both. Often balances are very precise and accurate – and expensive. Generally, for a *scale* the deformation of a stiff metal coiled spring is used – based on Hooke's equation – $f = -k\ x$. https://www.westlab.com/blog/hookes-law-understanding-definition-formula-and-graph. The terminology is often used loosely.

**26** This systematic error would typically be simple; but possibly not – for example, it might change with ambient temperature or altitude.

**27** And may be easily and quickly discovered by calibration experiment.

**28** And for such an example, the "constant" may not actually be constant. Perhaps it might be a function of such properties as: temperature, column length, carrier gas pressure, how many previous times the column has been used.

**29** And progressively render the column as not usable.

**30** See John Keenan Taylor, *Quality Assurance of Chemical Measurements*, Lewis Publishers, Inc., 1987; page 7.

**31** Farrington Daniels, J.W. Williams, Paul Bender, Robert A. Alberty, C.D. Cornwell, *Experimental Physical Chemistry, Sixth Edition*, McGraw-Hill Book Company, Inc., New York; 1962; Chapter 18, "Treatment of Experimental Data," page 393, et seq.

**32** *Gedankenexperiment*. Introduction 2.2.

**33** See: Betty R. Kirkwood, *ESSENTIALS OF Medical Statistics*, Blackwell Scientific Publications, Oxford, OX2 0EL; 1988; ISBN 0-632-01052-5; pages 10 and 11; page 21, et seq.

**34** – after the mathematician Carl Fredr. Gauß, 1777–1855.

**35** DEUTCHE BUNDESBANK, Frankfurt am Main, Zehn – 10 – Deutche Mark, 1999, Carl Friedrich Gauss.
https://commons.wikimedia.org/wiki/File:10_DM_Serie4_Vorderseite.jpg
https://en.numista.com/catalogue/note203955.html
https://www.ma-shops.com/cdma/item.php?id=214688

**36** $y = (1/\sigma(\sqrt{(2\pi)})e^{-(x-\mu)^2/2\sigma^2)}$   $e = 2.71828$

**37** https://www.mathsisfun.com/numbers/e-eulers-number.html
https://www.mathsisfun.com/algebra/logarithms.html

**38** e = 2.71828. . . = natural or Napierian log base. See William K. Morrill, *Calculus*, D. Van Nostrand Company, Inc., Princeton, N.J. . . ., 1956, page 125 et seq.

**39** Generally expressed:

$y = (1/(\sqrt{(2\pi)})e^{-x^2/2}$

$y = (1/ (\sqrt{(2\pi)})\exp(-x^2/2)$

**40** Carl Friedrich Gauss http://en.wikipedia.org/wiki/Normal_distribution

**41** Jane Muir, "Carl Friedrich Gauss, 1777–1885," in *OF MEN AND NUMBERS/THE STORY OF GREAT MATHEMATICIANS*, DOVER PUBLICATIONS, INC., New York, pages 157→183, 1996; ISBN-13: 978-0-486-28,973-1; ISBN-10: 0-486-28,973-7.

**42** http://www.storyofmathematics.com/19th_gauss.html https://en.wikipedia.org/wiki/Carl_Friedrich_Gauss

**43** https://en.wikipedia.org/wiki/Normal_distribution https://en.wikipedia.org/wiki/Observational_error

**44** For purposes here: $y = (1/(s\ (/(2\pi))))\exp\ [-(x - \bar{x})^2/2s^2]$
See John Keenan Taylor, *Statistical Techniques for Data Analysis*, Lewis Publishers, Inc., CRC Press, Inc., Boca Raton, Florida, 33421; 1990; ISBN 0-87371-250-1; 1990; pages 25 and 26.
In the treatment here $\bar{x}$ is used to approximate μ used by Taylor, and s is used to approximate his σ.
Applying the data of the fictitious example here: $y = (.29)\ \exp\ [-(x - \bar{x})^2 / 3.74]$

**45** It, kind-of, looks like a bell.

**46** Not to be confused with Richard J. Herrnstein and Charles Murray, *The Bell Curve / Intelligence and Class Structure in American Life*; Free Press, 1994; ISBN 0-02-914673-9. Nor with Charles Murray, *Human Diversity: The Biology of Gender, Race, and Class*, Twelve/Hatchette Book Group, New York City, 10104. January 2020. Neither recommended here.
https://www.amazon.com/Charles-Murray/e/B000AP5UJQ
https://www.nytimes.com/2006/07/23/magazine/23wwln_idealab.html
https://en.wikipedia.org/wiki/The_Bell_Curve

**47** *Gedankenexperiment*. Introduction 2.2.

**48** *Mean* =
$\bar{x} = (3x) / n$ Kirkwood, page 12.
$\bar{x}$ = : Kirkwood, page 22.

**49** $\bar{x}$ ". . . spoken 'x bar'. . .," Kirkwood, page 12.

**50** Also called *average*. https://www.mathsisfun.com/definitions/average.html
There is also a similar concept *root mean square* – sometimes used for electrical measurements.
$rms = \sqrt{}((x_1^2 + x_2^2 + x_3^2 + x_4^2 + \ldots + x_n^2)/n) = \sqrt{}(\Sigma x_n^2/n) = (\Sigma x_n^2/n)^{1/2}$https://www.mathsisfun.com/definitions/root-mean-square.html#:~:text=The%20square%20root%20of%20the,are%20(%3D%20the%20arithmetic%20mean)
https://www.sciencedirect.com/topics/engineering/root-mean-square
https://www.britannica.com/technology/root-mean-square-voltage

**51** Σ = math symbol *sigma* for a summation of a list of numbers. https://www.mathsisfun.com/algebra/sigma-notation.html https://www.mathsisfun.com/definitions/letter-s.html

**52** *Standard deviation* =
$s = \sqrt{}((\Sigma(x - \bar{x})^2) / (n-1))$ Kirkwood, page 14.
s = Φ Kirkwood, page 22.

**53** $\sqrt{}$ = math symbol for square root of a number. $= ()^{1/2}$
https://www.mathsisfun.com/definitions/square-root.html
https://www.mathsisfun.com/definitions/letter-s.html

**54** See Tab. 8.5.

Tab. 8.5: Standard deviation.

| x = | | $\bar{x}$ = | | $(x - \bar{x})$ = | $(x - \bar{x})^2$ = |
|---|---|---|---|---|---|
| 5.0031 | - | 5.0031 | = | 0 | 0 |
| 5.0027 | - | 5.0031 | = | −0.0004 | 0.00000016 |
| 5.0035 | - | 5.0031 | = | 0.0004 | 0.00000016 |
| 5.0034 | - | 5.0031 | = | 0.0003 | 0.00000009 |
| 5.0030 | - | 5.0031 | = | −0.0001 | 0.00000001 |
| Σ = 25.0157 | | | | Σ = | 0.00000042 |

**55** $= \sqrt{((0.00000042)/\ (5-1))} = \sqrt{((0.00000042)/\ (4))} = \sqrt{(0.000000105)} = \pm 0.\mathbf{00032}$
**56** ± is the mathematical representation for a number that is either positive or negative, or maybe both; spoken as "plus or minus." Often one or the other would be ignored as impractical.
**57** Metres and kilograms.
**58** See, for example: Kirkwood, page 8, *et seq.*, for haemoglobin measure; and page 21 et seq., for heights.
**59** For purposes here "accurate" should be taken to mean as sufficiently so that no one is arguing about it. It is not intended to mean here as absolutely accurate, or that absolute accuracy can actually be surely known.
**60** *Gedankenexperiment.* Introduction 2.2.
**61** "Organic" here means a hydrocarbon composed mostly of C and H atoms, often with some other atoms. This is a chemical science definition and is not a now popularly used meaning for natural and safe. Many, if not most organics by the chemical science meaning would not be safe in the popular meaning; and many would not be natural.
**62** [] is used to indicate a concentration – $[C_x H_y]$ – here an organic solvent in blood. Thus, for worker #1 tabulated below – $[C_x H_y]$ = 5.6 milligrams of the organic solvent per millilitre of blood (which, by volume, is mostly water).
**63** Such as in Ontario for lead in blood – R.R.O. 1990 Reg. 843 – under the *Occupational Health and Safety Act*, R.S.O. 1990 c. O.1. See M.G., *The Law of Occupational Health and Safety in Ontario*, 2nd ed., Butterworths, Toronto, 1994, Appendix II.
**64** But some of these differences would not be random and should be tabulated in appropriate groupings, as best they are known. For example, different measurements for different work locations and exposure patterns in the factory (e.g. a sprayer *vs* a clerk in the front office *vs* a welder working on sprayed items) are not fully attributable to random scatter and are therefore not appropriate for a purely statistical treatment together. It would be a complicating factor that these may not all be known.
**65** *Gedankenexperiment.* Introduction 2.2.
**66** These data are concocted for illustrative purposes here; and in real life, a much larger sampling would be appropriate.
**67** See: Betty R. Kirkwood, *ESSENTIALS OF Medical Statistics*, Blackwell Scientific Publications, Oxford, OX2 0EL; 1988; ISBN 0-632-01052-5; Chapters 2, 3 and 4, page 5, et seq.
**68** See also: Julian Tyson, *ANALYSIS / What Analytical Chemists Do*, Royal Society of Chemistry Paperbacks, Cambridge CB4 4WF; 1988, 1994; ISBN 0-85186-463-5; pages 117→133.
**69** See also: Robert L. Peesok, Kenneth Chapman, Wade W. Ponder, *et al.*, *CHEMICAL TECHNOLOGY HANDBOOK/Guidebook for Industrial Chemical Technologists and Technicians*, American Chemical Society, Washington, D.C.; 1975, 1980; ISBN 0-8412-0578-7; pages 122→125

70 Assuming – for purposes of the example here – to illustrate statistical treatment – that most of the measured differences are from random scatter, and that the measuring method is very accurate.
71 *Gaussian* distribution
https://www.mathsisfun.com/data/standard-normal-distribution.html
https://www.mathsisfun.com/definitions/letter-s.html
72 Both the known control samples #1 and #2 are prepared to monitor the measuring instrumentation, and are not from the worker blood samples.
73 Since both of the known control samples #1 and #2 are set here for concentration = 6.390 mg/mL – their variation is entirely due to scatter in the measuring method. The results of 6.40 and 6.38 suggest an error of the measurement system of ± ~ 0.01/6.390 ≈ ±0.2%.
74 Control samples #1 and #2 not included to calculate Σ = 89.5 and Σ = 24.42.
75 Not including the known control samples.
76 With the advent of modern calculators, it is most convenient to carry more decimal places than needed through the calculations, and to defer round-off decision-making for the extent of significant figures in the decimal string until the conclusion. The extra decimal places do not necessarily mean extra precision. See John Keenan Taylor, *Quality Assurance of Chemical Measurements*, Lewis Publishers, Inc., 1987; page 10.
77 For purposes of this fictitious example a two decimal place round-off is used.
78 https://brownmath.com/swt/symbol.htm
https://en.wikipedia.org/wiki/Normal_distribution
https://www.statlect.com/probability-distributions/normal-distribution
79 https://www.mathsisfun.com/data/standard-normal-distribution.html
80 See also: Larry J. Goldstein and David L Schneider, *Finite Mathematics and Its Applications*, Prentice-Hall, Inc., Englewood Cliffs, New Jersey, 07632; 1980; ISBN 0-13-317263-5; at page 271, et seq.
81 Gaussian distribution – see notes above.
82 *standard normal deviate* =

$$\text{snd} = (x - \bar{x})/s$$

$$\text{snd} = (x - \mu)/\sigma$$

$$\text{snd} = z$$ Kirkwood, pages 22 and 23.

83 Such a recalibration results in the *mean* moving to $\bar{x} = 0$.
84 Since both of the known control samples #1 and #2 are set here for concentration = 6.390 mg/ml – their variation is entirely due to scatter in the measuring method. The results of 6.40 and 6.38 suggest an error of the measurement system of ± ~ .01/6.390. ≈ ±.2%.
85 Kirkwood, 25 and 26. See also John Keenan Taylor, *Statistical Techniques for Data Analysis*, Lewis Publishers, Inc., 1990, page 66→75.
86 *Gedankenexperiment*. Introduction 2.2.
87 For convenience here, this new 15th measurement is charted without calculating a new mean (which would be = $\bar{x}$ = 6.82), and therefore without recalculating the table above. In real life, beyond this example, such a recalculation should be done.
88 Fictitiously, within the context of this *Gedankenexperiment*.
89 For such substance abuse, the measurement would be anomalous only if the abuse were rare; such widespread abuse might increase the *mean*, but not necessarily the *scatter*. As an alternative example, if the survey were intended to monitor a suspected substance abusing population, a non-abusing but occupationally caused surveyed might present the anomaly.
90 See:

- John Keenan Taylor, *Quality Assurance of Chemical Measurements*, Lewis Publishers, Inc., 1987; Chapter 14, "Control Charts," pages 129→146.
- John Keenan Taylor, *Statistical Techniques for Data Analysis*, Lewis Publishers, Inc., 1990, pages 151→154.

- W.M. Johnson in "Quality Control and Quality Assurance" as Chapter 8 in C. Riddle, ed., *Analysis of Geological Materials*, Marcel Decker, Inc., New York City, 1993, pages 351→356.

**91** For a general reference, see also: Howard A. Strobel, *Chemical Instrumentation / A Systematic Approach to Instrumental Analysis*, Addison-Wesley Publishing Company, Inc., Reading, Massachusetts and London, England, 1960; Chapter 2, ERRORS OF MEASUREMENT, pages 14→34.

**92** *Gedankenexperiment.* Introduction 2.2.

**93** ". . . No Matter How Carefully Incorrect Data Are Analysed The Results Are Meaningless . . ."
John Keenan Taylor, *Statistical Techniques for Data Analysis*, Lewis Publishers, Inc., 1990, page 20.

**94** See, for example: Ellen Rohde, Annette C. McManus, Carla Vogt, William R. Heineman "Separation and Comparison of Fountain Pen Inks by Capillary Zone Electrophoresis," *Journal Of Forensic Sciences*, Vol.42, No. 6, November 1997, page 1004. https://www.astm.org/DIGITAL_LIBRARY/JOURNALS/FORENSIC/PAGES/JFS14252J.htm

**95** See, for example: "This IMPOSSIBLE Theory Led to the Discovery of Oxygen"
https://www.youtube.com/watch?v=xqsqiOTY1t8

**96** https://www.mathsisfun.com/data/graphs-index.html

**97** https://mathinsight.org/cartesian_coordinates https://en.wikipedia.org/wiki/Cartesian_coordinate_system

**98** https://www.mathsisfun.com/data/cartesian-coordinates.html

**99** Also x, y, z for three dimensions (Fig. 8.12).

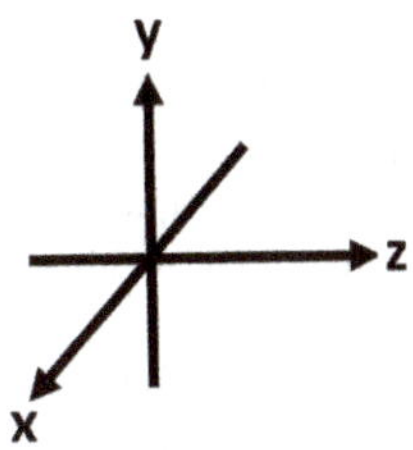


Fig. 8.12: x,y,z.

**100** *Gedankenexperiment.* Introduction 2.2.

**101** See: Michael A. Kamrin, *Toxicology/A Primer on Toxicology Principles and Applications/•Indoor and Outdoor Air/•Drinking Water/•Food/•Workplace Environment*, Lewis Publishers, Inc., Chelsea, Michigan, 48118; 1988; ISBN 0-87371-133-5; Chapter 2, "Chemical Individuality and Toxicology," 2.5 "Dose–response," page 12, et seq., page 13, Figure 2.3.

**102** https://en.wikipedia.org/wiki/Dose%E2%80%93response_relationship
https://www.sciencedirect.com/topics/engineering/doseresponse-curve

**103** Hypothetically sketched here.

**104** *Gedankenexperiment.* Introduction 2.2.

**105** The concept of severity of response remains numerically undefined here – not needed for the limited purpose of this illustration.

**106** *Gedankenexperiment.* Introduction 2.2.

**107** y = mx + b See William K. Morrill, *Calculus*, D. Van Nostrand Company, Inc., Princeton, N.J. . . ., 1956, APPENDIX, page 496.

**108** See: Rod Pierce, *Math Is Fun*, "Equation of a Straight Line," November 2019.
y = m x + b https://www.mathsisfun.com/equation_of_line.html
https://www.mathsisfun.com/ https://www.mathsisfun.com/aboutmathsisfun.html
https://www.mathsisfun.com/equation_of_line.html

**109** See also: Larry J. Goldstein and David L Schneider, *FINITE MATHEMATICS AND ITS APPLICATIONS*, PRENTICE-HALL, INC., Englewood Cliffs, New Jersey, 07632; 1980; ISBN 0-13-317263-5; at page 27, et seq.

**110** See also: https://www.nku.edu/~whitsonma/Bio150LSite/Lab%202%20Water/SpectrophotometryII.html

**111** See, for example: [Peter Scott], "LEAST SQUARES: FITTING A CURVE TO DATA POINTS," Chapter 3, equations 3.1 and 3.2 – wherein for a physics experiment an oscillating mass M hung on a spring of mass m has a period = $T = 2\pi[(M + m)/k]^{1/2}$. When rearranged as $T^2 = (4\pi^2/k)M + (4\pi^2/k)m$, it is in the general form of a straight line equation. Actual data do plot as on a straight line – as illustrated.
https://scott.physics.ucsc.edu/pdf/133_draftman.pdf
[PHYSICS 133 LAB MANUAL / Rumple Stiltskin / Summer, 2015]
[Chapter 3 LEAST SQUARES: FITTING A CURVE TO DATA POINTS] [page 26, et seq.]
https://scott.physics.ucsc.edu/
https://library.ucsc.edu/reg-hist/peter-scott-professor-of-physics
https://www.santacruzsentinel.com/2025/01/13/chris-krohn-santa-cruz-renaissance-man-peter-scott-dies-at-91/

**112** See also: Harry F. Meiners, Walter Eppenstein, Kenneth H. Moore, John P. Nickol, *Analytical Laboratory Physics/Second Edition*, Rensselaer Polytechnic Institute, Troy, New York; J.W. Edwards, Publisher, Inc., Ann Arbor, Michigan; 1956; Chapter III, "Graphing – Tabular Differences," pages 45→64; at page 54. Library of Congress Catalog Card Number 59–15 190.
https://www.amazon.com/Analytical-Laboratory-Physics-Harry-Meiners/dp/B001ESMLWC

**113** See: Harry F. Meiners, Walter Eppenstein, Kenneth H. Moore, John P. Nickol, *Analytical Laboratory Physics/Second Edition*, Rensselaer Polytechnic Institute, Troy, New York; J.W. Edwards, Publisher, Inc., Ann Arbor, Michigan; 1956; Chapter III, "Graphing – Tabular Differences," pages 45→64; at page 50. Library of Congress Catalog Card Number 59–15 190. https://www.amazon.com/Analytical-Laboratory-Physics-Harry-Meiners/dp/B001ESMLWC

**114** See also: John Keenan Taylor, *Quality Assurance of Chemical Measurements*, Lewis Publishers, Inc., an imprint of CRC Press, Inc., Boca Raton, Florida, 33431; 1987; ISBN 0-87371-097-5; Chapter 10, "Principles of Calibration," at page 102.

**115** See also: UNSW Sydney, NSW, 2052, Australia, School of Physics, "Errors and Error Estimation/ First Year Physics Laboratory Manual" [2021].
https://www.animations.physics.unsw.edu.au/sf/toolkits/Errors_and_Error_Estimation.pdf
https://www.physics.unsw.edu.au/

**116** https://www.mathsisfun.com/measure/error-measurement.html

**117** *Gedankenexperiment*. Introduction 2.2.

**118** – Rod Pierce, "Least Squares Regression," *Math Is Fun*, 2019.
http://www.mathsisfun.com/data/least-squares-regression.html
https://www.mathsisfun.com/
https://www.mathsisfun.com/citation.php
https://en.wikipedia.org/wiki/Least_squares
https://en.wikipedia.org/wiki/Ordinary_least_squares

**119** http://www.storyofmathematics.com/19th_gauss.html
https://en.wikipedia.org/wiki/Carl_Friedrich_Gauss

**120** Rod Pierce, "Least Squares Regression," *Math Is Fun*, 2024.

**121** https://www.mathsisfun.com/data/least-squares-regression.html

**122** See, for example – "Semmelweis Reflex": Nahlah Ayed, CBC, Ideas, "The dirt on handwashing: the tragic death behind a life-saving act," May 2020, "Even with impeccable data, [Ignaz Semmelweis] the 'saviour of mothers' was shunned."
https://www.cbc.ca/radio/ideas

https://www.cbc.ca/radio/ideas/the-dirt-on-handwashing-the-tragic-death-behind-a-life-saving-act-1.5587319

**123** https://u.osu.edu/korzen.1/2016/11/26/the-semmelweis-reflex-or-semmelweis-effect-is-a-metaphor-for-the-reflex-like-tendency-to-reject-new-evidence-or-new-knowledge-because-it-contradicts-established-norms-beliefs-or-paradigms/
https://en.wikipedia.org/wiki/Semmelweis_reflex
https://en.wikipedia.org/wiki/Ignaz_Semmelweis

**124** https://commons.wikimedia.org/wiki/Category:Ignaz_Semmelweis
https://commons.wikimedia.org/wiki/Category:Ignaz_Semmelweis#/media/File:Monthly_-mortality_rates_1841-1849.png
https://commons.wikimedia.org/wiki/File:Monthly_mortality_rates_1841-1849.png

**125** Richard Saferstein, *Criminalistics/An Introduction to Forensic Science*, Seventh Edition, Prentice Hall, Upper Saddle River, New Jersey, 07458; 2001; ISBN 0-13-013827-4.

**126** Saferstein, *Criminalistics*. . ., 2001; pages 181→188.

**127** *Gedankenexperiment*. Introduction 2.2.

**128** See: John Keenan Taylor, *Quality Assurance of Chemical Measurements*, Lewis Publishers, Inc., 1987, page 9.

**129** See, for example, in a different context: Stefanos Chen & Benjamin Weiser, "*U.S. Sidelines Lawyers Who Doubted Their Own Case on Congestion Pricing /* The Trump administration replaced lawyers who had exposed flaws in its legal battle over New York City's congestion pricing program," *New York Times*, 24 April 2025.
https://www.nytimes.com/2025/04/24/nyregion/nyc-congestion-pricing-duffy-lawyers.html

[. . .]

> . . . U.S. attorney's office . . . said it mistakenly filed in federal court in Manhattan . . . a confidential memo that questioned the department's legal strategy and urged a new approach.

[. . .]

> It's not unusual for lawyers to advise their clients confidentially in this way. But the filing telegraphed the government's legal weaknesses in the middle of a tense fight with Gov. Kathy Hochul and transit leaders who have vowed to keep the tolling program running.

[. . .]

**130** Niraj Chokshi, "A Tiny Typo Repeated 46 Million Times Is Circulating in Australia," *The New York Times*, BUSINESS, 11 March 2019, page B4. ". . . new $50 dollar bills . . . spelling error . . ."

**131** Engraved in stone, years ago, around the main hall, Union Station, Toronto, near the ceiling, are the Canadian cities historically served by the railway systems. Including "SAULT ST. MARIE" – likely better spelled "Sault Ste. Marie."

**132** https://commons.wikimedia.org/wiki/File:Typo_in_the_carvings_at_Union_Station.jpg
http://torontothenandnow.blogspot.com/2011/03/21-union-station-then-and-now.html

**133** Sault Ste. Marie, Ontario. https://en.wikipedia.org/wiki/Sault_Ste._Marie,_Ontario

**134** Lynsey Chutel, "*Westminster Abbey's Brontë Plaque Had a Typo for 85 Years. It's Fixed Now. /* Punctuation delayed, but not denied: A memorial to Charlotte, Emily and Anne Brontë at Poets' Corner in the celebrated London church finally gets its accent marks," New York Times, 27 Sept. 2024.
https://www.nytimes.com/2024/09/27/world/europe/westminster-abbey-bronte-plaque-accent-marks.html
https://www.babbel.com/en/magazine/the-history-of-the-umlaut-and-the-diaeresis-and-how-to-pronounce-them#:~:text=Whereas%20the%20umlaut%20represents%20a,O%20and%20the%20E%20separately

**135** See: Ali Watkins & Alan Feuer, "The U.S. says a Maryland man was deported because of an 'administrative error.'" *New York Times*, 01 April 2025.
https://www.nytimes.com/live/2025/04/01/us/trump-news-updates#maryland-man-deportation-error-el-salvador
Lindsay Whitehurst & Regina Garcia Cano, "Judge questions if Trump admin ignored order to turn around deportation flights," Associated Press, *Baltimore Banner*, 18 March 2025.
Alan Feuer, "Judge Calls Mistaken Deportation of Maryland Man a 'Grievous Error' / Judge Paula Xinis, who has ordered that the Trump administration return the man, Kilmar Armando Abrego Garcia, by Monday, also rejected a request to pause that order," *New York Times*, 06 April 2025.
https://www.nytimes.com/2025/04/06/us/politics/trump-deportation-el-salvador-judge-xinis.html
https://www.thebaltimorebanner.com/politics-power/national-politics/trump-administration-deportation-flights-LLAMD3RB4BEKPCM3BAAGKHRKTA/
**136** See also, in a different context:
https://mx-schroeder.medium.com/that-one-the-1895-paris-montparnasse-france-train-derailment-f0f382715383
https://www.cbc.ca/listen/live-radio/1-77-bookends-with-mattea-roach/clip/16135326-emma-donoghue-boards-train-destined-disaster
**137** See: Oliver C. Schroeder, Jr., "Ethical and Moral Dimensions in Forensic Science," as Chapter 3, pages 27→48, in Geoffrey Davies, Editor, *Forensic Science*, Second Edition, ACS Symposium Series, Volume 13, American Chemical Society; 1986; ISBN 0-8412-0918-9.
https://books.google.ca/books/about/Forensic_Science.html?id=5MXyAAAAMAAJ&source=kp_book_description&redir_esc=y
**138** Brent Ostrum, 1aSCa2 "A Canadian perspective on forensic science versus pseudoscience," ICA 2013 Montreal Montreal, Canada 2–7 June 2013 Speech Communication Session 1aSCa: Distinguishing Between Science and Pseudoscience in Forensic Acoustics I. https://asa.scitation.org/doi/pdf/10.1121/1.4800649
**139** See:
- Patricia Hunter and George Atkinson, "The Science Ethics Forum at the University of Waterloo," *Canadian Chemical News*, June 1995, pages 13 and 14;
- "The Chemical Institute of Canada Code of Ethics / Code d'Éthique. . .," *Canadian Chemical News*, June 1995, page 14;
- "ACS Strategic Plan 2025–2029"; approved by ACS Board of Directors 06 December 2024; American Chemical Society, Washington, DC, 20036.
  https://www.acs.org/about/strategicplan.html
- Martin P.J. Kratz, "Ethics, Law and Moonlighting," *Canadian Chemical News*, June 1995, pages 15 and 16.
- Deborah Illman, "Courses Address Ethical Research Practices In The Chemical Sciences /#Chemistry Departments at Iowa and Maryland offer courses on scientific fraud; at Tennessee, casebook spurs ethics discussion," EDUCATION, *Chemical and Engineering News*, 24 October 1994, pages 38 and 39.
- "SYMPOSIUM: ETHICAL CONFLICTS IN THE FORENSIC SCIENCES," *Journal of Forensic Sciences*, Vol.34, No.3, May 1989: Joseph L. Peterson, "Introduction," pages 717 and 718;
- Douglas M. Lucas, "The Ethical Responsibilities of the Forensic Scientist: Exploring the Limits," pages 719→729;
- Paul C. Giannelli, "Evidentary and Procedural Rules Governing Expert Testimony," pages 730→748;
- Joseph L. Peterson and John E. Murdock, "Forensic Science Ethics: Developing an Integrated System of Support and Enforcement." pages 749→762;

- Mark S. Frankel, “Ethics and the Forensic Sciences: Professional Autonomy in the Criminal Justice System,” pages 7636771;
- Michael J. Saks, “Prevalence and Impact of Ethical Problems in Forensic Science,” pages 772→793.
- Special Communication: Papers Delivered at the Plenary Session of the 49th Annual Meeting, American Academy of Forensic Sciences, New York NY on Ethics in Forensic Science, *Journal of Forensic Sciences*, Vol.42, No.6, November 1997:
- Robert Weinstock, “Ethical Practice in the Forensic Sciences An Introduction,” page 1189; Richard Rosner, “Foundations of Ethical Practice in the Forensic Sciences,” page 1191;
- Jon J. Nordby, “A Member of the Roy Rogers Riders Club Is Expected to Follow the Rules Faithfully” page 1195;
- F. M. Kamm, “Physician-Assisted Suicide and the Doctrine of Double Effect,” page 1198; Ruth Macklin, “Ethics and Value Bias in the Forensic Sciences,” page 1203;
- Don Harper Mills, “Comments from the Perspective of the AAFS Ethics Committee Chairman,” page 1207.

**140** For example: for practice in Ontario under the *Medicine Act, 1991*, SO 1991c.30; and the *Regulated Health Professions Act, 1991*, SO 1991c.18, as am. Concerning ethics, see . . .*Professions Act*, Schedule 2, s.3(1)#2, #3, #5, and s.51(c). https://www.gov.on.ca/MBS/english/publications/statregs/contents.html

**141** For example, . . . *Professions Act*. . ., s.27.

**142** See Michael *Heylin, Chemical and Engineering News*, pages 10→16; and Madeline Jacobs, editor's page, page 3.

**143** See: Giacomo Puccini, *LA FANCIULLA DEL WEST* (*THE GIRL OF THE GOLDEN WEST*), 1910.
https://www.metopera.org/season/in-cinemas/synopsiscast/2018-19/la-fanciulla-del-west/ https://www.jhiblog.org/2018/12/17/variations-on-a-theme-by-puccini-theologizing-la-fanciulla-del-west/
https://archive.org/stream/lafanciulladelwe00pucc/lafanciulladelwe00pucc_djvu.txt

> Minnie:
> Ciò vuol dire, ragazzi, che non v'è, al mondo, peccatore cui non s'apra una via di redenzione. Sappia ognuno di voi chiudere in se questa suprema verità d'amore.
> Which is to say, boys, that there is no sinner in the world for whom a path to redemption is not open. May each of you learn how to hold this supreme truth of love within you.

Apparently inspired from Biblical Psalms Chapter 51 תְּהִלִּים
https://www.biblestudytools.com/psalms/51.html
https://www.mechon-mamre.org/p/pt/pt2651.htm

**144** *Gedankenexperiment*. Introduction 2.2.

**145** *Gedankenexperiment*. Introduction 2.2.

**146** Variations of this example – *Gedankenexperiment* – continue throughout this book.

**147** See:

- John J. Furedy, Professor of Psychology, University of Toronto, “The detection of lies,” letter to the editor, *The Globe and Mail*, 4 September 1996, page A11;
- Russell Wodell, British Columbia Civil Liberties Association, “Test only measures stress,” letter to the editor, *The Globe and Mail*, 31 August 1996, page D7;
- Jane Gadd, “Lie-detector tests seen as inaccurate” *The Globe and Mail*, 11 September 1996, page A6.
- Harry Hollien, Laura Geison and James W. Hicks, Jr., “Voice Stress Evaluators and Lie Detection,” *Journal of Forensic Sciences*, Vol.32, No.2, March 1987, pages 4056418;
- Frank Horvath, “Detecting Deception: The Promise and the Reality of Voice Stress Analysis,” *Ibid.*, Vol.27, No.2, April 1982, pages 340–351.

- Peter Hermann, "City Police buy high-tech device to help detect lies," *Baltimore Sun*, 14 July 1995, page 1A; and Jeff Stein, "If the truth be known, polygraphs not reliable / Law-abiding citizens are wrongly being denied federal jobs," *ibid.*, 05 October 1997, page 5M (*Baltimore Sun* Archive – http://www.sunspot.net/archive/search/ – as of 07 March 1998).

**148** Malcolm Gladwell, "Annals of Psychology/The Naked Face/*Can you read people's thoughts just by looking at them?*," *The New Yorker*, 05 August 2002, pages 38→49. But see also: Julian Bleecker, New York City, The mail/"deceptive faces," *The New Yorker*, 2 September 2002, page 9; and *ibid.*, Thomas Strong, San Francisco.

**149** See also: Adam Goldman, "*Unease at F.B.I. Intensifies as Patel Ousts Top Officials* / Senior executives are being pushed out and the director, Kash Patel, is more freely using polygraph tests to tamp down on news leaks about leadership decisions and behavior," *New York Times*, 01 June 2025. https://www.nytimes.com/2025/06/01/us/politics/fbi-patel.html

**150** See, for example:

- Knowlton Nash, "The premier and the pocketful of pot he said wasn't his / RCMP found a 35-gram bag of marijuana, but Richard Hatfield said he didn't know anything about it / . . . / Pot on the royal plane," *The National*, 22 Oct. 1984, CBC Archives, 22 Oct. 2019, 22 Oct,2020; ". . . The pot had been discovered while Hatfield was travelling with the Queen . . ."
  https://www.cbc.ca/archives/the-premier-and-the-pocketful-of-pot-he-said-wasn-t-his-1.5313671
- "*PREMIER CLEARED IN DRUG CASE,*" *New York Times* [Archive], 30 January 1985, Section A, Page 8. https://www.nytimes.com/1985/01/30/world/premier-cleared-in-drug-case.html
- https://en.wikipedia.org/wiki/Richard_Hatfield
  See also:
- Richard Hatfield, Premier of New Brunswick, cover letter, 04 Feb. 1974, *DRUGS CAN TURN YOU INTO A SHADOW OF YOURSELF / NARCOTIC AWARENESS PROGRAM / Fact Manual 1974 edition / Fredericton, N.B.*; John Howard Society.

**151** "*Canadian Charter of Rights and Freedoms* [being Schedule B, Part I of the *Canada Act* 1982, including the *Constitution Act*, 1982 (1982 c. 11 (United Kingdom)] . . . [s.]15(1) Every individual is equal before and under the law and has the right to the equal protection and equal benefit of the law without discrimination. . ."

**152** "In Congress, July 4, 1776. A Declaration [of Independence] By the Representatives of the United States of America, In General Congress assembled ". . . We hold these Truths to be self-evident, that all Men are created equal, that they are endowed by their Creator with certain unalienable Rights, that among these are Life, Liberty, and the Pursuit of Happiness . . ."

**153** Although perhaps still vigorously protesting innocence; and firing the lawyer for even suggesting otherwise.

**154** Tara Parker-Pope and Anahad O'Connor, "Researcher in Meat Study Had Ties to Trade Group," *The New York Times*, 05 October 2019, page A19.

**155** See also, in another context: Meg Kinnard, "What is the Emoluments Clause? And how might it apply to Qatar giving Trump a plane?" ASSOCIATED PRESS, 12 May 2025.
https://apnews.com/article/emoluments-clause-trump-jet-qatar-56f995acb57eb741b0214aba5e7e8220

**156** In the author's experience as defence counsel, there was once a client – as accused, in custody – who had not only been previously represented by the judge, when that judge was a defence lawyer, but also prosecuted by that same judge when that judge had even more previously been a Crown counsel. In the courtroom, the accused and the judge exchanged courteous greetings – but the matter had to go to another judge.

**157** Analogous to criminal law enforcement issues is the ethical behaviour of agents of the state for foreign policy purposes. For example, The Gulf of Tonkin Resolution was presented to Congress on 5 August 1964, by then President Lyndon Johnson, in response to two allegedly unprovoked attacks, on 2 and 4 August, by North Vietnamese torpedo boats, on two ships of the US Navy. Approved by over-

whelming majorities in both houses of Congress, the Resolution became a basis for subsequent escalation of the Vietnam War. The issue of the correctness of the evidence presented to Congress has been raised. In 1995, Vietnam's military commander during the war acknowledged the first attack on a US naval ship, but denied the second. See: "Gulf of Tonkin Resolution," Encyclopaedia Britannica.
https://www.britannica.com/event/Gulf-of-Tonkin-incident
https://www.britannica.com
**158** *R v Liew*, Supreme Court of Canada, 1999.
https://scc-csc.lexum.com/scc-csc/scc-csc/en/item/1731/index.do
**159** See also: Timothy Appleby, "Under interrogation/How far can the police go with questioning suspects and witnesses? Two murder charges dropped this week put the spotlight on investigators' tactics" *The Globe and Mail*, Friday, 12 November 1999:
TORONTO . . .

> [. . .]
>
> Years ago, when . . . was a member of the homicide squad, he had a murder suspect in his sights who he was sure – really sure – was guilty. But the man wouldn't crack.
>
> . . .Go fetch the shoes you were wearing . . . Our crime lab has got some new fangled laser thing they want to use [emphasis added]. . . .
>
> On getting home, the man scurried inside . . . did exactly what . . . had hoped he would do. He crept out and tossed his shoes into a nearby dumpster. Pounced on by waiting police . . . swiftly confessed.
>
> There was, of course, no laser. The story was an out-and-out ruse – a wholly legitimate device in the complexities of police interrogations [emphasis added]. Nothing anywhere prevents law enforcers from deceiving a suspect to obtain information.
>
> '. . .we can use trickery,' said Detective . . .
>
> [. . .]
>
> At the same time, however, two court cases this week illustrate some of the boundaries in a police interrogation.
>
> . . .the Crown dropped a murder charge in what was almost certainly a drug-related execution. Among other things, it emerged that the prosecution's star witness had initially exonerated the accused, but then incriminated him after learning from investigators that a $25,000 reward in the killing had been raised to $50,000.
>
> [. . .]

**160** See: *R v Oickle* – Kirk Makin, JUSTICE REPORTER, "Deceptive police tactics justified, top court says / Inflated importance of failed polygraph test to get confession from suspected arsonist," *The Globe and Mail*, Saturday, 30 September 2000, page A23.http://www.theglobeandmail.com/ – Her Majesty The Queen *Appellant v* Richard Floyd Oickle *Respondent* and The Attorney General for Ontario and the Criminal Lawyers' Association (Ontario) *Interveners*, 2000 SCC 38. File No.: 26535. 29 September 2000. https://scc-csc.lexum.com/scc-csc/scc-csc/en/item/1801/index.do
**161** See also: Michael S. Schmidt and Tariq Panja, *"A Doping Feud Almost Cost Salt Lake City the Olympics. It Still Might. / A* [US] *Justice Department investigation into whether antidoping authorities covered up positive tests for elite Chinese swimmers set off a power play to kill the inquiry," New York Times*, 1 December 2024.
https://www.nytimes.com/2024/12/31/us/politics/wada-doping-salt-lake-city-2034-olympics.html
**162** See Kirk Makin, Justice Reporter, "Appeal court handcuffs defence bar / Police witnesses can no longer be confronted with previous judicial findings that they lied," *The Globe and Mail*, Wednesday 17 Nov. 1999, page A8. https://www.theglobeandmail.com/.
**163** Isaiah Chapter 59 יְשַׁעְיָהוּ

**164** A Hebrew – English Bible According to the Masoretic Text and the JPS 1917 Edition www.mechon-mamre.org/p/pt/pt1059.htm
**165** Also:
**166** Proverbs Chapter 1 מִשְׁלֵי
**167**

[. . .]

טז כִּי רַגְלֵיהֶם לָרַע יָרוּצוּ; וִימַהֲרוּ לִשְׁפָּךְ-דָּם.
16 For their feet run to evil, and they make haste to shed blood.

[. . .]

**168** A Hebrew – English Bible According to the Masoretic Text and the JPS 1917 Edition https://www.mechon-mamre.org/p/pt/pt2801.htm
**169** See Bette Hileman, "White House defines scientific misconduct," *C&EN*, 25 October 1999, page 12 See also: [*Code of Federal Regulations*][Title 45, Volume 3, Parts 500 to 1199][Revised as of October 1, 1998]From the U.S. Government Printing Office via GPO Access[CITE: 45*CFR*689] [Pages 229–233] TITLE 45 – PUBLIC WELFARE CHAPTER VI – NATIONAL SCIENCE FOUNDATION PART 689 – MISCONDUCT IN SCIENCE AND ENGINEERING.
**170** See also:
- Patricia Meisol, "Scientists clash in battle over honesty in research / . . .," *Baltimore Sun*, 25 July 1991, pages 1A and 12A.
- Brian Sullam, "Scientist files lawsuit against colleague, Sun," *Baltimore Sun*, 25 July 1991, page 12A.

**171** https://www.baltimoresun.com/news/bs-xpm-1991-07-25-1991206043-story.html
**172** James E. Starrs, "Mountebanks among Forensic Scientists," as Chapter 1 in Richard Saferstein, ed., *Forensic Science Handbook*, Prentice Hall, Englewood Cliffs, New Jersey, 1988, pages 1→37.
**173** Jean Pagel, Associated Press, Lubbock, Texas, " 'Made-to-order' autopsies return to haunt Texas doctor / Investigations have revealed that a pathologist fabricated or tampered with evidence for prosecutors. In the ensuing uproar, a dozen convicts on death row have appealed, and one's conviction has been overturned." *The Globe and Mail*, 10 June 1995, page A13.
**174** Samuel Spencer, "Where Is Annie Dookhan Now? Chemist from 'How to Fix a Drug Scandal' Didn't Serve Her Whole Prison Sentence," *Newsweek*, Culture, 01 APRIL 2020.
**175** https://www.newsweek.com/how-fix-drug-scandal-netflix-annie-dookhan-now-2020-prison-free-1495521
**176** https://en.wikipedia.org/wiki/Annie_Dookhan
**177** Michael Gold, "A Fake Psychologist Treated Troubled Children for Years, Prosecutors Say," *The New York Times*, 30 September 2019, page A18.
**178** Jacques Gallant, "Expert's prior cases could draw scrutiny / Psychologist who misrepresented credentials gave evidence in child protection hearings," *Toronto Star*, Star Investigation, 01August 2019, pages A1 and A8.
**179** http://www.abebooks.co.uk/book-search/title/bre-x/
**180** http://en.wikipedia.org/wiki/National_Instrument_43-101
**181** https://minestarters.com/reports/
**182** https://en.wikipedia.org/wiki/Ben_Johnson_(Canadian_sprinter)
**183** Charles Dubin, Royal Commission to Inquire into the Use of Drugs and Banned Practices Intended to Increase Athletic Performance (1988).
**184** http://en.wikipedia.org/wiki/Charles_Dubin
**185** See:

- Kenneth Chang, "On Scientific Fakery and the Systems to Catch It / At Bell Labs, Lots of Blame Goes Around," *The New York Times / Science Times*, 15 October 2002, pages D1 and D4.
- George Johnson, "At Lawrence Berkeley, Physicists Say Colleague Took Them for a Ride," pages D1 and D4.

**186** Editorial, "What Happens When Scientists Cheat," *The New York Times*, 1 June 2015, page A16.
**187** Bethany Halford, Andrea L. Widener, "MISCONDUCT / Harvard chemist Charles Lieber charged with fraud / Feds allege that he failed to disclose Chinese funding to US agencies and his university," *C&EN*, VOLUME 98, ISSUE 5, 28 January 2020. https://cen.acs.org/topics/research-integrity.html#misconduct
2 CFR § 180.800 https://www.govinfo.gov/app/details/CFR-2008-title2-vol1/CFR-2008-title2-vol1-sec180-800
**188** See also: Helena Champion, "Data fabrication," letter to the editor, *C&EN*, 1 March 2021, page 3.
**189** See also: National Academies of Sciences, Engineering, and Medicine, *Reproducibility and Replicability in Science*, The National Academies Press, Washington, DC; 2019; 978-0-309-48,616-3.
https://nap.nationalacademies.org/catalog/25303/reproducibility-and-replicability-in-science
**190** See:

- Kenneth Chang, "On Scientific Fakery and the Systems to Catch It / At Bell Labs, Lots of Blame Goes Around," *The New York Times / Science Times*, 15 October 2002, pages D1 and D4.
- George Johnson, "At Lawrence Berkeley, Physicists Say Colleague Took Them for a Ride," pages D1 and D4.

**191** ASSOCIATED PRESS, "Fraud taints 5 major leukemia studies / Junior researcher admits fabricating test results," *Baltimore Sun*, 30 October 1996, page 8A.
**192** Carl Zimmer, *"Fraudulent Scientific Papers Are Rapidly Increasing, Study Finds /* A statistical analysis found that the number of fake journal articles being churned out by "paper mills" is doubling every year and a half," *New York Times*, .04 Aug. 2025.
https://www.nytimes.com/2025/08/04/science/04hs-science-papers-fraud-research-paper-mills.html
**193** Philip Slayton, "Lawyers Gone Bad / Money, Sex, and Madness in Canada's Legal Profession," Penguin Canada, Toronto M4P 2Y3; 2007, 2008; ISBN 978-0-14-305610-2.
https://www.penguinrandomhouse.ca/books/391688/lawyers-gone-bad-by-philip-slayton/9780143179658
**194** Slayton, *"Lawyers Gone Bad* . . ., 2007, 2008; re Harry Kopyto, page 290, et seq.; page 327.
**195** *Law Society of Ontario v Harry Kopyto*, 2020 ONSC 35, COURT FILE NO.: CV-19-617826, 20 Dec. 2019, 07 Jan. 2020, Justice Edward P. Belobaba; "Application for Permanent Injunction under section 26.3 of the *Law Society Act*."
**196** [. . .] [pages 16 and 17]

**Disposition**

[78] The Law Society of Ontario's application for a permanent injunction prohibiting Mr. Kopyto from contravening s. 26.1 of the *Act*, and in particular:

i. prohibiting Mr. Kopyto, and any business or business entity that he controls, from practicing law or providing legal services in Ontario, contrary to s. 26.1(1) of the *Act*, and

ii. prohibiting Mr. Kopyto from holding himself out as, or representing himself to be, a person who can practice law or provide legal services in Ontario, contrary to s. 26.1(2) of the *Act* is **granted**.

[. . .]

**197** https://globalnews.ca/news/6433275/judge-permanently-retires-harry-kopyto/
https://en.wikipedia.org/wiki/Harry_Kopyto
**198** See also: *R v Wijesinha*, [1995] 3 SCR 422.
https://decisions.scc-csc.ca/scc-csc/scc-csc/en/item/1294/index.do
**199** Theranos. Elizabeth Holmes.

- CBC aih Theranos https://www.cbc.ca/player/play/1239137347658
- John Carreyrou, *The Wall Street Journal*, October 2015.
- John Carreyrou, "How One Company Scammed Silicon Valley. And How It Got Caught," The *New York Times* 21 May 2018. ISSN 0362-4331 – Erin Griffith & Erin Woo, "Schemer or Naïf? Theranos Chief Is Going to Trial / Elizabeth Holmes, the blood-testing start-up founder who lured billionaires and dignitaries, has much to answer for and some predictable defenses," *The New York Times*, Business, 31 August 2021, pages B1 & B4.

**200** https://en.wikipedia.org/wiki/Theranos
https://en.wikipedia.org/wiki/John_Carreyrou
https://www.cbc.ca/news/business/theranos-elizabeth-holmes-fraud-1.4575955
https://www.cbc.ca/radio/asithappens/as-it-happens-monday-full-episode-1.4671937/new-book-documents-the-rise-and-fall-of-silicon-valley-wunderkind-elizabeth-holmes-1.4671941
https://www.esquire.com/entertainment/tv/a26810723/elizabeth-holmes-now/
https://www.gatesnotes.com/bad-blood

**201** Connie Roff, "Everything you need to know about the Theranos scandal / Your questions about Elizabeth Holmes and the Theranos scandal answered," PAN MACMILLAN, UK, 25 March 2024.
https://www.panmacmillan.com/about-pan-macmillan

**202** https://www.nytimes.com/2025/05/10/business/elizabeth-holmes-partner-blood-testing-startup.html

**203** Curiously, some frauds of purported science and engineering share an aspect of *Ponzi* schemes: obviously defective explanation of how it actually works. For example: the financial investment "split-strike conversion strategy." See: Andrew Kirtzman, *BETRAYAL / The Life and Lies of Bernie Madoff*; Harper, New York City, 10022; 2008, 2010; Page 222; 266, et seq. ISBN 978-0-06-187074.

**204** Caroline Hawley, World Affairs Correspondent, "The story of the fake bomb detectors," BBC, October 2014.
https://www.bbc.com/news/uk-29459896

**205** https://en.wikipedia.org/wiki/ADE_651
https://www.vanityfair.com/news/2015/06/fake-bomb-detectors-iraq
https://www.foreignaffairs.com/articles/iraq/2016-08-05/iraq-and-big-swindle
https://en.wikipedia.org/wiki/ADE_651
https://www.bbc.com/news/uk-29459896

**206** See also: Andrea Widener and Carmen Drahl, Washington, DC, "FORCING CHANGE IN FORENSIC SCIENCE / SCANDALS IN LABS across the country illustrate the need to reform the discipline, but change has been slow in coming," *C&EN*, 92(19), 12 May 2014, pages 10→14.
https://cen.acs.org/articles/92/i19/Forcing-Change-Forensic-Science.html

**207** https://en.wikipedia.org/wiki/Annie_Dookhan#:~:text=Dookhan%2C%20Massachusetts%20Department%20of%20Correction,was%20subsequently%20released%20from%20prison .

**208** See: John Carreyrou, *"A Start-Up Claimed Its Device Could Cure Cancer. Then Patients Began Dying.* / Two U.S. companies teamed up to treat cancer patients using an unproven blood filter in Antigua, out of reach of American regulators," *New York Times*, 23 January 2025.
https://www.nytimes.com/2025/01/23/business/exthera-cancer-blood-filtering-device.html

**209** Bertrand Russell, *Satan in the Suburbs and Other Stories*, Simon and Schuster, New York City, 1953. "The Infra-redioscope," pages 87→117 (see regarding fraud at pages 94, 104, 106, 108).
tpl TRL Call number = 823.912 R79.3 https://www.torontopubliclibrary.ca/detail.jsp?Entt=RDM462659&R=462659
https://www.amazon.com/Satan-Suburbs-Stories-Bertrand-Russell/dp/1852900024
https://books.google.ca/books/about/Satan_in_the_suburbs.html?id=yD9JAAAAMAAJ&redir_esc=y
https://www.goodreads.com/book/show/1101287.Satan_in_the_Suburbs_and_Other_Stories https://www.amazon.com/Satan-Suburbs-Bertrand-Russell/dp/0851246281

https://en.wikipedia.org/wiki/Bertrand_Russell
https://library.mcmaster.ca/archives/r/russell.200.htm

**210** As perhaps happened with certain mine samples: "A so-called fabulous gold find in Mali that sparked a spectacular runup in the shares of Timbuktu Gold Corp. went bust yesterday when officials disclosed its assays are bogus and were probably faked." See: Allan Robinson, Mining Reporter, "Timbuktu reveals assays are bogus / Mali gold samples suggest tampering," *The Globe and Mail*, 11 June 1996, pages B1 and B6.

**211** Paul Waldie and Allan Robinson, "Timbuktu tampering preceded test / Samples altered before reaching lab," *The Globe and Mail*, 12 June 1996, pages B1 and B5.

**212** John Saunders and Paul Waldie, "Fool's Gold / FICTIONAL FIND / Oliver Reese's Calgary-based mining firm, Timbuktu, capitalized on a market mad with fear of missing out on the next great bonanza with one of the oldest tricks in the book." *The Globe and Mail*, 15 June 1996, pages B1 and B5.

**213**
- Michael Grange and Karen Howlett, "VSE probes runup in Borneo Gold share prices / Halts trading after stock soars on surface sample assessment," *The Globe and Mail*, 19 September 1996, page B3.
- Allan Robinson, "Mineral analysts group urges lab accreditation / Investors stung by incompetent testing," *The Globe and Mail*, 19 September 1996, page B3.

**214** Or perhaps the less serious ". . .inappropriate accounting procedures. . .," see "Lawrence Livermore Lab, government settle on audit results," *Chemical and Engineering News*, government concentrates, 28 October 1996, page 21: "The University of California will pay $2 million to settle claims stemming from inappropriate accounting procedures at Lawrence Livermore National Laboratory. . . ."

**215** Amongst lawyers, taking ". . . advantage of . . . slips, irregularities or mistakes . . . not going to the merits. . ." of the issues is called "sharp practice," – which "[t]he lawyer should avoid" – see *Professional Conduct Handbook*, The Law Society of Upper Canada, Toronto, 1983, Rule 16 (page 44), Commentary 4, and footnote 5.

**216** For example: See Paul Koring and Jeff Sallot, "Top brass fostered secrecy, probe told / Papers renamed, colonel testifies / Access law scorned by senior officials, inquiry told," *The Globe and Mail*, 30 August 1996, pages A1 and A4.

**217** See:
- David Folkenflik, "Academic career unravels / # Misconduct: Young scientist leaves [Johns] Hopkins after he is found to have misrepresented findings in an effort to win a [US] federal grant." *Baltimore Sun*, 24 July 1996, pages 1A, 6A and 7A;
- and response by Walter Ehrlich, professor emeritus, Johns Hopkins School of Hygiene and Public Health, "Falsified data hurts honest scientists," letter to the editor, *Baltimore Sun*, 07 August 1996, page 18A.

See also:
- Donn Devine, letter to the editor, *Chemical and Engineering News*, 26 August 1996, pages 4 and 95.
- Rebecca Rawls, "news of the week / Imanishi-Kari Cleared / [US Department of] H[ealth] and H[uman] S[ervices] panel rejects misconduct charges," *C&EN*, 01 July 1996, pages 6 and 7;
- Daniel J. Kevles, "annals of science / The Assault on David Baltimore / *Congress and the media hailed the case of a whistle-blower . . .*," *The New Yorker*, 27 May 1996, pages 95→109;
- Pamela Zurer, "science insights / In scientific misconduct cases, justice isn't always blind," *C&EN*, 24 June 1996, page 31.
- Brian H. Kaye, *Science and the Detective / Selected Reading in Forensic Science*, VCH Verlagsgesellschaft GmbH, Weinheim, Germany; 1995; ISBN-3-527-29252-7; pages 326→332, and references cited therein.

**218** Brian H. Kaye, *Science and the Detective / Selected Reading in Forensic Science*, VCH Verlagsgesellschaft GmbH, Weinheim, Germany; 1995; ISBN-3-527-29252-7.
https://books.google.ca/books?id=kNyPkc2ADTsC&pg=PR4&lpg=PR4&dq=ISBN-3-527-29252-7&source=bl&ots=w_ZILdlgpS&sig=ACfU3U2VXDlNAjHLYno36nd3SslrbBGSoA&hl=en&-sa=X&ved=2ahUKEwimg7SJ3o3pAhXic98KHbV3CMgQ6AEwAXoECA8QAQ#v=onepage&-q=ISBN-3-527-29252-7&f=false
https://pubs.rsc.org/en/content/articlelanding/1996/an/an996210005n/unauth

**219** https://en.wikipedia.org/wiki/Sharp_practice

**220** https://www.collinsdictionary.com/us/dictionary/english/lawfare

**221** Ben Shpigel, "*Trump Pardons Reality-Show Couple Convicted of $36 Million Fraud* / The clemency for Todd and Julie Chrisley, the stars of "Chrisley Knows Best," is Mr. Trump's latest gift to people he calls victims of a weaponized justice system," *New York Times*, 27 May 2025. https://www.nytimes.com/2025/05/27/us/politics/trump-chrisley-pardon.html

**222** See also: re *willful blindness / deliberate ignorance* – "Appeals / What is "Wilful Blindness" in Criminal Law? / Written on behalf of Barrison & Manitius."
https://criminallawoshawa.com/what-is-wilful-blindness-in-criminal-law/
https://www.canlii.org/en/on/onca/doc/2017/2017onca789/2017onca789.html?resultIndex=1

**223** See: CBC Radio, *Ideas*, November 2019, "Psychologists confront impossible finding, triggering a revolution in the field" ". . . study proved the ability to sense the future is real, inspiring fundamental research changes" "Replication crisis" https://www.cbc.ca/radio/ideas/psychologists-confront-impossible-finding-triggering-a-revolution-in-the-field-1.5344467

**224** See: John Keenan Taylor, *Statistical Techniques for Data Analysis*, Lewis Publishers, Inc., 1990, page 20.

**225** *Quod volimus credimus libenter* ≈ believe what we want to believe.

**226** *Quod volimus credimus libenter* = we always believe what we want to believe. Robert Harris, *Archangel*, Berkley Publishing, New York City, 1998; Arrow, 2009; ISBN-10: 0099527936, ISBN-13: 978-0099527930.
https://www.amazon.ca/Archangel-Robert-Harris/dp/0099527936
https://www.goodreads.com/quotes/377660-quod-volimus-credimus-libenter-we-always-believe-what-we-want

**227** Robert Harris, *Archangel*, Arrow Books, The Random House Group Limited, London SW1V 2SA; 1999; ISBN 0-09-28241-0; at page 48: " . . . '*Quod volimus credimus libenter*' . . . 'we always believe what we want to believe' . . ."

**228** https://en.wikiquote.org/wiki/Julius_Caesar: Julius Caesar, *Commentarii de Bello Gallico* [= *Commentaries on the Gallic War*], Book I, Ch. 14, translated by W.A. McDevitte and W.S. Bohn: "*Fere libenter homines id quod volunt credunt*" = In most cases men willingly believe what they wish.

**229** Latin. Introduction 4.1.

**230** Introductory quotes. Introduction.

**231** https://en.wikipedia.org/wiki/Santa_Claus_Is_Comin%27_to_Town
https://www.azlyrics.com/lyrics/lonestar/santaclausiscomintotown.html

**232** See:

- Brian Rothery, *ISO 9000*, Second Edition, Gower Press, 1993.
- Ronald J. Cottman, *A Guidebook to ISO 9000 and ANSI/ASQC Q90*, American Society for Quality Control, Quality Press, 611 East Wisconsin Avenue Milwaukee 53202, 1993.
- Dwayne D. Mathers, Robert T. Marshall, Pierre F. Callibot and Malcolm J. Phipps, *The ISO 9000 Essentials*,Canadian Standards Association, 1994.
- John Keenan Taylor, *Quality Assurance of Chemical Measurements*, Lewis Publishers, Inc., 1987.

**233** See "Preventing Fatal Medical Errors," *The New York Times*, EDITORIAL, 01 December 1999, page A30.

**234** ISO Central Secretariat address: International Organization for Standardization (ISO), 1, rue de Varembé, Case postale 56, CH-1211 Genève 20, Switzerland; 41 22 749 01 11; fax 41 22 733 34 30; central@-iso.ch. ISO Online: https://www.iso.org/home.html
**235** See Rothery; Cottman; Mathers, *et al.* – cited above.
**236** CAN/CSA- ISO 9002-94 (ISO 9002:1994), *Quality systems – Model for quality assurance in production, installation and servicing*, 2nd ed. 1994-07-01 (cancelling and replacing 1st ed. – ISO 9002:1987); CSA, 178 Rexdale Boulevard, Rexdale (Toronto), Ontario M9W 1R3; (416)-747-4000. According to introductory information included with this CSA printing, the CSA is a major participant in ISO affairs, and in regard to the ISO 9000 family of Standards.
**237** ISO/IEC 17025:2005 International Organization for Standardization, Geneva, Switzerland. https://www.iso.org/standard/66912.html
**238** ISO/IEC 17025:2005 revised by ISO/IEC 17025:2017 General requirements for the competence of testing and calibration laboratories. https://www.iso.org/standard/66912.html
**239** See:
- good laboratory practice for nonclinical laboratory studies (*Food, Drug, and Cosmetic Act*), 21 US Code of Federal Regulations ' 58 (as of 01 April1995 (consolidated from the *Federal Register*));
- good laboratory practice standards (*Insecticide, Fungicide and Rodenticide Act)* 40 CFR ' 160, *Federal Register*, 17 August 1989 ( reprinted in *Good Laboratory Practice Standards*, ACS, Washington, D.C., 1992; Appendix A, pages 433→444);
- national voluntary laboratory accreditation program procedures, 15 CFR ' 7 (as of 01 January 1992 (consolidated from the *Federal Register*)).
- ISO/IEC Guide 25: 1990, distributed by the Standards Council of Canada, 45 O'Connor, Suite 1200, Ottawa K1P 6N7, (613)-238-3222. This Guide, which is particularly directed to quality systems in laboratories, is published by the ISO, Genève – see note above
- Stuart H. James and Jon J. Nordby, editors, *Forensic Science / An Introduction to Scientific and Investigative Techniques*, "DNA Advisory Board (DAB) Guidelines," CRC Press, Boca Raton, 2003; ISBN 0-8493-1246-9; Appendix A, page 597, et seq.

**240** – Selectively summarized here.
**241** – Clause headings selected from ISO/IEC 17025:2005 contents pages: clause

4 Management requirements
4.1 Organization
4.2 Management system
4.3 Document control
4.3.2 Document approval and issue
4.3.3 Document changes
4.4 Review of requests, tenders and contracts
4.5 Subcontracting of tests and calibrations
4.6 Purchasing services and supplies
4.7 Service to the customer
4.8 Complaints
4.9 Control of nonconforming testing and/or calibration work
4.10 Improvement
4.11 Corrective action
4.11.2 Cause analysis
4.11.3 Selection and implementation of corrective actions
4.11.4 Monitoring of corrective actions
4.11.5 Additional audits

4.12 Preventive action
4.13 Control of records
4.13.2 Technical records
4.14 Internal audits
4.15 Management reviews
5 Technical requirements
5.2 Personnel
5.3 Accommodation and environmental conditions
5.4 Test and calibration methods and method validation
5.4.6 Estimation of uncertainty of measurement
5.4.7 Control of data
5.5 Equipment
5.6 Measurement traceability
5.6.2 Specific requirements
5.6.3 Reference standards and reference materials
5.7 Sampling
5.8 Handling of test and calibration items
5.9 Assuring the quality of test and calibration results
5.10 Reporting the results
5.10.2 Test reports and calibration certificates
5.10.3 Test reports
5.10.4 Calibration certificates
5.10.5 Opinions and interpretations
5.10.6 Testing and calibration results obtained from subcontractors
5.10.7 Electronic transmission of results
5.10.8 Format of reports and certificates
5.10.9 Amendments to test reports and calibration certificates

**242** Subject to its own contractual and legal commitments otherwise.
**243** *Sixth Amendment*, US Constitution.
**244** "Something is rotten in the state of Denmark," – to borrow a phrase from William Shakespeare – Hamlet, Act I, Scene 5.
**245** See: Peter J. Stang, "Peer Review at the *Journal of the American Chemical Society*," *J. Am. Chem. Soc.* 2017, 139, 46, 16431→16432. https://pubs.acs.org/doi/full/10.1021/jacs.7b11541
**246** See: *J. Am. Chem. Soc.* 114 (1) (1992) page 8A; and *Can. J. Chem.* 71 (1) 1993 page (viii).
**247** See: Fig. 8.12. https://theanalyticalscientist.com/fields-applications/cartoon
Dr. Nick Kim, *the Analytical Scientist*.
**248** See: J Hurst, K Nickel, LH Hilborne, "Are Physicians' Office Laboratory Results of Comparable Quality to Those Produced in Other Laboratory Settings?," *Journal of the American Medical Association*, 1998; 279(6):468–471.
doi:10.1001/jama.279.6.468.
https://jamanetwork.com/journals/jama/fullarticle/187234
Reported by associated press, "Study finds test errors more likely at doctor's offices / 91% accuracy, compared with 97% at hospitals, labs," *Baltimore Sun*, 11 February 1998, page 4A.
**249** ANSI National Accreditation Board / ANAB; Cary, North Carolina, 27518. SCOPE OF ACCREDITATION TO ISO/IEC 17025:2017, FORENSIC TESTING, Certificate Number: FT-0124, Valid to 31 May 2023
http://search.anab.org/public/organization_files/Centre-of-Forensic-Sciences-Toronto-Cert-and-Scope-File-04-18-2019_1555609155.pdf

**250** [previously] ANAB = ANSI-ASQ National Accreditation Board, Milwaukee, Wisconsin, 53203. http://www.anab.org/forensic-accreditation (previously: ASCLD/LAB = American Society of Crime Laboratory Directors/Laboratory Accreditation Board, Garner, North Carolina, 27529. https://www.ascld.org/
**251** ANAB = ANSI-ASQ National Accreditation Board. http://www.anab.org/about-anab/definitions
ANSI = American National Standards Institute.
ASQ = American Society for Quality.
**252** http://search.anab.org/public/organization_files/Centre-of-Forensic-Sciences-Toronto-Certand-Scope-File-04-18-2019_1555609155.pdf
**253** **ISO/IEC 17025:2017** General requirements for the competence of testing and calibration laboratories.
**254** **ISO/IEC 17025:2017** revises and replaces **ISO/IEC 17025:2005**.
**255** International Organization for Standardization, Geneva, Switzerland. http://www.iso.org
**ISO/IEC 17025:2017** https://www.iso.org/standard/66912.html
**256** International Organization for Standardization, Geneva, Switzerland. http://www.iso.org
**ISO/IEC 17025:2005** http://www.iso.org/iso/home/store/catalogue_tc/catalogue_detail.htm?csnumber=39883
**257** ANSI National Accreditation Board / ANAB; Cary, North Carolina, 27518. SCOPE OF ACCREDITATION TO ISO/IEC 17025:2017, FORENSIC TESTING, Certificate Number: FT-0124, Valid to 31 May 2023.
http://search.anab.org/public/organization_files/Centre-of-Forensic-Sciences-Toronto-Certand-Scope-File-04-18-2019_1555609155.pdf
**258** Previous SCOPE of ACCREDITATION ALI-087-T
http://search.anab.org/public/organization_files/Centre-of-Forensic-Sciences-Toronto-Cert-and-Scope-File-04-20-2017_1492714555.pdf
**259** See:
- accreditation in Canada to ISO/IEC Guide 25 – the Standards Council of Canada
- Kenneth J. Mintz, “Accreditation of a Testing Laboratory,” *Canadian Chemical News*, May 1995, pages 37→39.
- Louisa Newbury, society news / president = s message, *Canadian Society of Forensic Science Journal*, Vol.28, No.3, September 1995, page 229
- Glenn R. Carroll, president’s message, *Canadian Society of Forensic Science Journal*, Vol.29, No.1, March 1996, page 34.
- Allan Robinson, “Mineral analysts group urges lab accreditation / Investors stung by incompetent testing,” *The Globe and Mail*, 19 September 1996, page B3.

**260** Lady Justice – from Roman goddess Justitia – with scales, blindfold and sword – as in arms of Ilshofen, Baden-Württemberg.
**261** https://en.wikipedia.org/wiki/Lady_Justice#/media/File:DEU_Ilshofen_COA.svg
https://en.wikipedia.org/wiki/Lady_Justice#:~:text=Since%20the%2016th%20century%2C%20Lady,%2C%20power%2C%20or%20other%20status
https://www.heraldry-wiki.com/heraldrywiki/index.php?title=Ilshofen
https://commons.wikimedia.org/wiki/File:DEU_Ilshofen_COA.svg
**262** Justitia – at Vancouver, with scales and blindfold, but with a scroll instead of her usual sword. “Scales of Justice, Vancouver Law Courts. The blindfolded Lady Justice symbolizes the impartial manner in which our laws are administered: blind to all considerations but the facts”
https://www.canada.ca/en/immigration-refugees-citizenship/corporate/publications-manuals/discover-canada/read-online/justice-system-images.html
**263** http://www.inquiriesjournal.com/articles/896/the-visual-rhetoric-of-lady-justice-understanding-jurisprudence-through-metonymic-tokens
**264** The individual litigants may continue to vigorously disagree as to their cases’ importance.
**265** For example, if the lab accreditation had expired.

266 For example: for the Carbon-14 analysis of the Shroud of Turin. Introduction 2, Chapters 3 & 7.
267 *Gedankenexperiment*. Introduction 2.2.
268 Ivan Penn, "Steps taken to combat inmate release errors / Officials upgrading case review policies, computer tracking," *Baltimore Sun*, 24 November 1997, pages 1B and 6B: "In response to concerns about mistaken release of four Maryland inmates, prison officials say they have established new policies . . ."
269 Mark Quinn, "2nd story of babies switched at birth– same year, same Come By Chance [Newfoundland] hospital / Muriel and Cecil Stringer were handed the wrong baby boy in 1962 – but noticed the mix-up almost immediately," CBC News, 17 February 2020.
https://www.cbc.ca/news/canada/newfoundland-labrador/hodge-s-cove-wrong-baby-1.5459421
270 Mark Quinn, "By Chance: Switched at birth, they were brought together again by coincidence / 2 lives took a wrong turn early on. Some 50 years later, the N.L. men are struggling with the fallout," CBC News, 11 December 2019. https://www.cbc.ca/news/canada/newfoundland-labrador/switched-at-birth-newfoundland-1.5366043
271 https://en.wikipedia.org/wiki/Babies_switched_at_birth
272 https://www.netflix.com/ca/title/81627034#:~:text=Two%20sets%20of%20identical%20twins,identities%20in%20this%20captivating%20documentary
273 https://www.disneyplus.com/en-ca/browse/entity-1d79681c-c3c7-4983-b9f1-a832a7714741
274 See also: Kevin Peraino, "Separated as Toddlers, Raised on Opposite Sides of the World / In her entrancing, disturbing "Daughters of the Bamboo Grove," Barbara Demick traces the wildly divergent paths of twins born in China under the one-child rule," *New York Times*, 24 & 28 May 2025. Book review of Barbara Demick, *DAUGHTERS OF THE BAMBOO GROVE: From China to America, a True Story of Abduction, Adoption, and Separated Twins.*
https://www.nytimes.com/2025/05/24/books/review/daughters-of-the-bamboo-grove-barbara-demick.html
275 Hopefully not too many here.
276 See "Completely off the Track an afterword by David Garnett," 1992, 1998.
http://www.infinityplus.co.uk/stories/offtrack.htm
277 However, see the caution given by John Keenan Taylor, *Quality Assurance of Chemical Measurements*, Lewis Publishers, Inc., 1987, page 146.
278 *Gedankenexperiment*. Introduction 2.2.
279 *Gedankenexperiment*. Introduction 2.2.
280 *Nullius in verba* ≈ We don't take anybody's word for it.
281 http://royalsociety.org/ "The Royal Society's motto 'Nullius in verba' roughly translates as 'take nobody's word for it'. It is an expression of the determination of Fellows to withstand the domination of authority and to verify all statements by an appeal to facts determined by experiment."
282 Edward Neville da Costa Andrade, *Sir Isaac Newton / His Life and Work*, DOUBLEDAY ANCHOR, Garden City, New York; SCIENCE STUDIES SERIES, S42; reprinted from Macmillan Company, New York, and W. Collins Sons and Co, Ltd., London, first published 1954; page 60.
283 https://en.wiktionary.org/w/index.php?title=%D7%A9%D7%82%D7%9B%D7%9C&oldid=46239729
284 See also: ". . . if something doesn't make sense, it doesn't make sense. . . ."
285 Mark Carney, Lecture 2: "From Credit Crisis to Resilience," December 2020, BBC Radio 4. Also, CBC Ideas. "THE REITH LECTURES 2020: HOW WE GET WHAT WE VALUE Reith Lecturer: Dr. Mark Carney, former Governor of the Bank of England Lecture 2: From Credit Crisis to Resilience TX: 09.12.2020 at 9am, BBC Radio 4"
". . .Bob Hurst . . . then one of the partners at Goldman Sachs. Bob's rule was if something doesn't make sense, it doesn't make sense . . . ask the person to repeat the rationale, and if that response still doesn't make sense, you should run . . ."
https://downloads.bbc.co.uk/radio4/reith2020/Reith_2020_Lecture_2_transcript.pdf

https://www.bbc.co.uk/programmes/m000q3sp
https://www.bbc.co.uk/programmes/articles/43GjCh72bxWVSqSB84ZDJw0/reith-lectures-2020-how-we-get-what-we-value

**286** https://en.wikipedia.org/wiki/Mark_Carney

**287** See also: William Shakespeare, *Hamlet*, Act 3, Scene 2; ~1601; Queen Gertrude: "The lady doth protest too much, methinks."
https://dictionary.cambridge.org/dictionary/english/protest-too-much
https://literarydevices.net/lady-doth-protest-too-much/
https://en.wikipedia.org/wiki/The_lady_doth_protest_too_much,_methinks
https://www.sciencedirect.com/science/article/abs/pii/S0271530920300264
https://interestingliterature.com/2020/11/methinks-the-lady-doth-protest-too-much-meaning-origin/

# Chapter 9
# Report

[. . .][1]

*The White Rabbit put on his spectacles. "Where shall I begin, please your Majesty?" he asked.*

*"Begin at the beginning," the King said, very gravely, "and go on till you come to the end: then stop."*

[. . .] **[**Alice at a court of justice**]**. [1]

[. . .]

*Meanwhile in my office at the Port Judiciary I struggled to do the job my bondfather had given me. Each day a formidable stack of reports and memoranda reached my desk; each day I tried to decide which must go before the High Justice and which were to be ignored. At first, naturally, I had no grounds for judgment. Sevgord helped me, though, as did several of the senior officials . . .*

*Most of the material submitted for the guidance of the High Justice was nonsense. I learned swiftly to detect that sort by a quick scanning, often by looking at a single page. The style in which it was written told me much: I found that a man who cannot phrase his thoughts cleanly on paper probably has no thoughts worth notice. The style is the man. If the prose is heavy-footed and sluggish, so too, in all likelihood, is the mind of its author, and then what are his insights into the operations of the Port Judiciary worth? . . .*

[. . .]

**[**Fictional Prince Kinnall, as private secretary to the High Justice of the Port Judiciary**]**. [2, 3, 4, 5, 6, 7, 8][2,3]

*Доверя́й, но проверя́й*

**[**Russian proverb = *Trust, but verify* – quoted by President Ronald Reagan to General Secretary Mikhail Gorbachev**]**. [9, 10, 11]

## 9.1 Report

The forensic scientist's report[4,5] is a very important component of both science and law. It is the formal and explained end-product of the scientific work done in a legal context.

That report is the professional written statement of findings and conclusions, and speaks both for the scientist and for the scientist's employer [12]. It should contain, or refer to, all that is relevant to expressing the theory and conclusions of the science – as far as the scientist could know – including doubts and wonderings about unknowns.

---

**1** Introductory quotes. Introduction 1.2.

**2** Prince Kinnall. Chapters 8 and 9.

**3** Robert Silverberg, *A Time of Changes*. Copyright acknowledgement; Chapter 9.

**4** Proper scientific report. *CD&S Act*. Health Canada. Introduction 4.3; Chapters 5, 8, 9, 10 and 11.

**5** Forensic report. Chapters 5 and 9.

https://doi.org/10.1515/9783111618555-010

The forensic science report would be used by prosecutors, police, defence lawyers, judges and others, to help try to construct a theory of the human story of what happened. And maybe used by historians, long afterwards.

The report should be as comprehensive as possible, clear, cite all relevant documentation and state its limitations. It should deal with scientific matters in a proper scientific way. It should clearly indicate its authorship, its authorship responsibility and contact data. It should also identify the key participants in the forensic work.

The report should provide its readers not only with a clear idea of the issues to be addressed, but also ample information for a reader's critical appraisal that might suggest other conclusions to be argued about.

The report would be a formal written narrative with sufficient explanation so that a responsible reader, with some knowledge of science, but not necessarily closely familiar with the detailed science theory and facts, could gain a decent understanding of what is going on. That reader should be expected to have at least a high school level appreciation of science and mathematics, and should not profess animosity to science and mathematics.

The report must also be appropriate for an expert reader who would be closely familiar with the detailed science of concern.

A forensic science report would look like any proper science report[6] – generally – as to content and organization. Although there are many possible formats, all science reports would be organized to explain what is being studied, in such a way that a critical reader or reviewer can know what challenging questions to ask, and could have those questions addressed. Perhaps, it is relevant to speculate that all scientific reports, at various levels – from high school science projects – to research reported in the *Journal of The American Chemical Society* – to meeting Swedish royalty [13] – should share some concepts of explanation.

A forensic science report should address, and neither hide, obscure nor obfuscate inconvenient relevancies, if such exist.

But, if in the midst of a public controversy, the report's authors should tread carefully – for example: In a report on the health physics of a particular nuclear power proposal, the report's authors should stick to just that. That report should not be seen as for advocacy – those are the roles of the public policy advocates and their adversaries. The report should not have a commercial advertising kind of tone; the reporting agency's website should not give the appearance of an advertising platform or of brand promotion.

The concepts of science report writing are well known. Examples abound – including from the American Chemical Society [14, 15], from the US *Code of Federal Regulations* for the EPA [16], from high schools [17], *etc* [18, 19, 20, 21].

---

6 *Proper scientific report. CD&S Act.* Polemic. Health Canada. Introduction 4; Chapters 9 and 10.

As a specific further example: A textbook experiment to measure enthalpy of combustion with a bomb calorimeter [22, 23, 24].

And, further examples: As indicated in quality assurance standard ISO/IEC 17025:2005, including clauses 5.10.1, 5.10.2 and 5.10.5 [25].[7]

And, further examples: reports as for cases that would appear before the Ontario Court of Justice – *FIRE INVESTIGATION REPORT*, and CFS *REPORT*.[8] But not from Health Canada,[9] under the *CD&S Act*, nor from the *Criminal Code* DRE.[10,11]

Individual report writers and their science-related agencies and corporate entities should be expected to be science-report-literate. Documented scientific reporting goes back at least more than three centuries to *Principia*.[12]

As a science report, a forensic science report would include several necessary parts. Those parts would each be reasonably well-complete explanations, either by their own narrative, or by way of reference to reasonably readily well-available other documents. Descriptions of those parts follow in this chapter.

A forensic science report with missing parts should be accepted only provisionally, if at all, until it would become complete. But, report parts, by way of reference, if reasonably and readily available, should not be considered as missing.

However, reports for many purposes may be ok as in abbreviated or summarized format, so long as a fully proper scientific report exists somewhere and is reasonably readily available. For example: Most medical patients, and their physicians, are satisfied with the quality of lab results (often analytical chemistry) as brief data printout, because there is sufficient confidence in the lab's and the physician's professionalism, in the physician's explanation, and in the government's regulatory supervision. But, not always is such confidence warranted – for example: "Motherisk" laboratory at Toronto Hospital for Sick Children.[13]

And, again, in a medical context, there may be questions of confidence with data security.[14]

As another example: It may be reasonably speculated that investors interested in Bre-X [26, 27, 28, 29, 30][15,16] gold Au mining assays (also analytical chemistry) should have (now in retrospect of the exposed fraud) insisted on proper scientific reports [31].[17]

7 QA. ISO. Chapters 5.6, 8.11 and 9.1.1.

8 Fire Marshall. CFS. Chapters 9 and 10.

9 Health Canada DAS *Certificate of Analyst*. Chapters 9 and 10.

10 DRE. Chapters 9 and 11.

11 re criticism of Health Canada, Judges, et al. Introduction 4.1 and 4.2; Chapters 10 and 11.

12 Newton. Haley. *Principia*. Chapters 9, 10 and 11.

13 "Motherisk" Chapter 9 [see also other notes.].

14 LifeLabs. Chapters 5 and 9.

15 Bre-X, the gold mining fraud – not Brexit, the EU-leaving – which sounds kind-of-similar, but is not.

16 Bre-X. Chapters 8 and 9.

17 Sample selection. Chapters 3, 5 and 9.

Some reports that are in the administrative context of an issuing agency – perhaps set out in law – would not necessarily qualify as a proper scientific report. Those agency-directed reports would be used for whatever the agency's bureaucracy – or the law – requires, but are not necessarily reflective of proper science. For example: Health Canada's *Certificates of Analyst*, as applicable at law, should not be confused with properly reported science.[18,19,20]

**Science and conclusions that would be conveyed by a forensic report that is not a proper scientific report should not be seen as scientifically acceptable, for that reason alone. A forensic report not in writing (as print-on-paper or as permanent electronic record) should not be seen as scientifically acceptable, for that reason alone. A mere presentation of recorded data is not an acceptable scientific report; it must come with reporting explanation and formalism.**

That Canadian courts and legal process are sometimes entitled at law to proceed without a proper scientific report, and often do, is a matter for public policy debate. For example: Health Canada's *Certificates of Analyst* – see above.

Sometimes, there are publicly revealed tragic happenings that would/should have involved considerations of analytical chemistry, and for which there are government enquiries. And, for which journalists take notice. Within the resulting government-published enquiries, there would/should have been presented and discussed issues of proper scientific reports, references to them and mention of at least some science concepts (See Table 5.1).[21,22]

After a proper scientific report would be issued, requests for relevant explanations are always in order, and the report's writer and/or the issuing agency should always be willing to further explain, or argue about it; although within the chain of authority, and sequence, of the legal process. In this, it should be remembered that good science requires good faith explanation.[23]

Often a forensic report is presented to a court or tribunal by way of expert[24] testimony. While the oral testimony of an expert witness is often appropriate and helpful for a court or tribunal, and required; if that expert refers to a specific examination or

---

**18** Health Canada DAS *Certificate of Analyst*. Chapters 9 and 10.

**19** Polemic. Health Canada. Introduction 4; Chapters 10 and 11.

**20** Re criticism of Health Canada, Judges, et al. Introduction 4.1 and 4.3; Chapters 10 and 11.

**21** Government enquiries: Introduction 4; Chapters 8 and 9.
- Grange re Susan Nelles, 1984;
- Griffiths, re Gemma Ramlal, Health Canada, 2001;
- Goudge re Dr. Smith, Hospital for Sick Children, 2008;
- Lang re "Motherisk," Dr. Koren Hospital for Sick Children, 2015;
- Gillese re Ontario Long-Term Care Homes murders, 2019.

**22** Proper scientific report. *CD&S Act*. Health Canada. Introduction 4.3; Chapters 5, 8, 9, 10 and 11.

**23** Good faith explanation. Chapters 2 and 9.

**24** Expert. Chapters 1 and 9.

Tab. 9.1: Tragic public happening enquiries.

| Enquiry | | | | Comment [32] | |
|---|---|---|---|---|---|
| – Grange Royal Commission. Ontario. | 1994 | Re | Susan Nelles | An innocent nurse was accused of murder-by-poisoning of babies. Eventually exonerated. For an issue where analytical chemistry would seem to be important, scant mention is made. | [33, 34, 35, 36, 37] |
| – Griffiths report. Canada federal. | 2001 | Re | Gemma Ramlal | A government chemist was involved with serious quality assurance failure. For an issue where analytical chemistry and QA would seem to be of fundamental and paramount importance, there was essentially no mention.<br>Mention within the Griffiths report of a to-follow-up QA report has apparently never been published, if it was ever written.<br>Later comparison with a Massachusetts case, 2012, is appropriate. | [38, 39, 40, 41, 42, 43, 44, 45, 46, 47, 48, 49, 50, 51, 52, 53, 54][25] |
| – Goudge Inquiry. Ontario | 2008 | Re | Dr. Smith | An MD, acting in pathology, apparently made serious errors amid disorganized administrative process. Apparently, no one of senior authority within the Ministry of the Attorney General actively intervened. | [55, 56, 57, 58] |

25 Fraud. Griffiths Report, 2001. QA failure. Gemma Ramlal, 2001. Annie Doukhan, 2012. Introduction 4.3; Chapters 8, 9 and 10.

Tab. 9.1 (continued)

| Enquiry | | | | Comment [32] | |
|---|---|---|---|---|---|
| – Lang report. Ontario | 2015 | Re | "Motherisk" Dr. Koren | An MD, in charge of a hospital clinic used questionable tests and conclusions, with legal consequences for patients. Concepts of analytical chemistry and QA appear to be well dealt with in this report. | [59, 60, 61, 62, 63, 64, 65, 66, 67, 68, 69, 70, 71, 72, 73, 74][26] |
| – Gillese report. Ontario | 2019 | Re | Long-term care homes murders | A nurse is convicted of a series of murders for which there is little specific direct physical evidence, other than to confirm the deaths, with an opportunity to murder by-way-of misuse of medication. Essentially, conviction was based on her confession in a psychiatric facility. There is scant mention of analytical chemistry. | [75, 76, 77, 78] |

observations, done by or on behalf of that expert, a written report would also be necessary. Without the written report, the expert's testimony should not be seen as acceptable, for that reason alone.

Some of the above descriptions may seem somewhat too fussy, redundant and tedious. But, it is very important.

## 9.2 Report parts

Descriptions of each of the report parts that must appear – by way of direct narrative or reference – for purposes of a forensic analytical chemistry proper scientific report as contemplated in this book.

### 9.2.1 Title, …

The report must declare at its beginning – or, if not at the actual beginning, referenced from the beginning, or otherwise obviously displayed [79]:

26 "Motherisk" Chapter 9 [see also other notes].

- A meaningful title.
- The author(s).
- Their government agency or corporate entity; the supporting or financing agency or entity.
- The key participants in the forensic work.
- An effective date of report issue; and indications as to when the work reported on was actually done.
- Various identifying codes, classifications, and cataloguing numbers.
- If acronyms are used – defined.
- An indication (often obvious) that the report is of a prepared form or is *de novo* [80].[27]

### 9.2.2 Signature

Also:
- The report should be signed by its authors and others of answerable responsible authority involved in its work.[28] In this regard, those in responsible authority must be clearly defined and identified; a nebulous team with no one clearly designated to meaningfully respond to enquiries or criticism does not satisfy this. The person of responsibility would typically be the analytical chemist who did or directed the work and authored the report.

Some comments are appropriate about the mechanics and meaning of a signature: A personally handwritten name, initials, or mark – to affirm as true the document being signed [81, 82, 83]. Although sometimes of great historical significance [84, 85, 86, 87, 88, 89], document signing is often routine business, but still of significance in its own context.

The signature would be inscribed near words to indicate: The place, jurisdiction, date and time of signing; and that it's original or an official copy. Impliedly, if not explicitly, it would be for the entire document, including boilerplate [90]. As a practical matter, typically, if the document is type-set (black), the ink signature would be in colour (blue). The signature might be witnessed and attested to with another's signature; sometimes in affidavit form, sworn under oath.

For a forensic report, the signature would mean that the signer, having the authority to sign, had examined the report, mindful of its meaning, including at the moment

27 = From the beginning.

28 Signed report. Chapter 9.

just before signing. The subsequent signing – just after – would represent personal approval, attestation, affirmation, responsibility, commitment, *etc.* This symbolic formality – of a human hand, directed by a mindful human brain – is quite important; casual, *pro forma* [91],[29] meaningless signing should not happen; nor for purposes here should there be signing-in-blank, nor other-than-actual dating. Signing must not be delegated to a robot [92, 93] or artificial intelligence.

In more ordinary, everyday language, the signing process described above would be to assure that someone of responsible authority was actually "minding the store" [94].

In our modern electronic document era, there are equivalent alternatives to signing by a brain-directed human hand using a physical pen exuding chemical-substance ink [95]. Such equivalents must include that the electronic signing be of analogous criteria as above. There are analogies for signings for commercial transactions [96, 97].

The above – perhaps too fussy and idealized – commentary might be of little remarkable necessity. But then there are *facsimile* signatures – with perhaps neither explanations nor indications on the document or obviously elsewhere of the criteria above. If the forensic report is with a facsimile signature, it should be evaluated and proven with the same criteria as above: caused to be by a human hand, directed by a mindful human brain, and taking-on responsibilities. Then too, in the modern world, there are robots – and now advancing artificial intelligence (AI) [98, 99, 100] with similar problematic issues.[30]

If the facsimile signature is otherwise than as recommended above – perhaps somehow found to be automatically applied, electronically, with the computer production of the report – the report should not have been issued. This is a problem to be addressed when computer systems are involved in the forensics process.[31]

Sometimes, Canadian law bypasses these concepts.[32,33]

Related to this are autopen signatures [101] which may be valid but should require the same criteria as for a signature with-a-human-hand – to be human-mind-directed.

---

**29** = As a matter of form.

**30** Robots. Artificial intelligence. AI. Chapter 9.2.2.

**31** Signature. Electronic. Facsimile Autopen. Chapters 5 and 9.

**32** Health Canada DAS *Certificate of Analyst. CD&S* **Act** s. 51. Chapters 9 and 10.

**33** Re criticism of Health Canada, Judges, and others. Introduction 4.1 and 4.2; Chapters 10 and 11.

### 9.2.3 For example – title

For example [102], for the forensic ink analysis used[34] as examples in previous chapters, and being followed through in this chapter: Title:

**FORENSIC REPORT for Police File [][][][][][].**
**DETERMINATION OF INK ON A DOCUMENT & IN A PEN.**

This report issued [][] [][][][] 202[], [][]:[][] []m. EST [103].
For forensic science work done

[][] [][][][] 202[] → [][] [][][][] 202[], [][]:[][] []m. EST.
Authorizing signature: ______________________
(as at the issue date above, as for this entire report. Σpages=[][].)
(The signer self-declares to be a person of answerable responsible authority.)
[][][][][] [][][][][], Analytical Chemist. Employee id # = [][][][][].

Personnel involved as key participants in this forensic work:
✓ Analytical chemist signing above.
☐& others:______________________

Forensic Service Agency,
Ministry of the Solicitor General.
[][][][] [][][][][][][][][], [104]
[][][][][][][][][][], [][][][][][][][], [][][][][][][]. [][][] [][][] [105]
This lab's file number for this work = [][][][][][][][].

## 9.3 Abstract

An abstract is a very brief summary near the beginning of the report. It is usual in science; in it is analogous to a "head note" in published law reports.

For example, for the forensic ink analysis used[35] as examples in previous chapters, and being followed through in this chapter [106]:

> Abstract: From items provided by police, TLC and IR analyses were done for ink and ink-writing, with comparison for light blue ball point pen ink to standard reference materials. It was concluded that all three (ink, writing and reference) are the same. Statistical data also indicates that the ink is not usually found.

34 Variations of Ink analytical chemistry – as examples – *Gedankenexperiment* – continue throughout this book. Introduction 2.3 and 2.2.

35 Variations of Ink analytical chemistry – as examples – *Gedankenexperiment* – continue throughout this book. Introduction 2.3 and 2.2.

## 9.4 Contents outline

A contents outline, somewhere near the beginning of the report, allows for a convenience of reading.[36] For example: Selected from the paragraph headings of this chapter – for the forensic ink analysis used as examples in previous chapters, and being followed through in this chapter [107]:

Contents outline:

- Problem to be addressed<<<<
- Theory of the human story being investigated
- Science theory
- Items for analysis
- Equipment and reagents
- Analysis
- Error statement
- SOPs
- Conclusion<<<<
- Quality assurance
- Environmental and workplace health and safety
- Legal and disclaimer
- Control of distribution
- Data archival
- Finance
- Language

## 9.5 Problem to be addressed: blind?

The problem to be addressed should be very briefly stated. It involves the human story that is being sought, for which forensic science is being applied. It is usually and mostly not a treatment of the scientific issues. And, importantly, the forensic scientist should try to be blind, as best as practicable, to the human story, to relieve apprehension of bias.

For example, for the forensic ink analysis used[37] as examples in previous chapters, and being followed through in this chapter [108]:

> Problem to be addressed: An ink-writing – alleged as a forgery, and a pen containing ink, were analyzed as to identify the ink as from the pen, or not; and to determine the popularity of that pen and ink in the region of the alleged offence.

**36** With emphasis <<<< indicated here for where most readers would likely be most interested.

**37** Variations of Ink analytical chemistry – as examples – *Gedankenexperiment* – continue throughout this book. Introduction 2.3 and 2.2.

> The forensic analytical chemist, although having knowledge that an alleged forgery was being investigated, in a criminal or civil law process, had no other involvement in that investigation; nor contact with police personnel other than to receive the physical evidence; nor contact with Crown office personnel other than to receive notice of a trial date for testimony.

## 9.6 Theory of the human story being investigated

The human story being investigated is often very interesting. But that story – aside from how it might inform about the scientific problem – essentially should not be of concern to the forensic scientist. And, as just indicated, the forensic scientist should try to be blind, as best as practicable, to it, to relieve apprehension of bias.

For example, for the forensic ink analysis used as examples in previous chapters, and being followed through in this chapter [109]:

> Re Theory of the human story being investigated: Personnel at this lab, although aware that a criminal or civil legal issue (more likely criminal) is being investigated, do not have detailed knowledge of the allegations of incident; nor express any opinion therefor. Nor have any known contact with anyone connected with the incident, other than police & Crown office personnel as noted above in "Problem to be addressed."

## 9.7 Science theory

The theoretical scientific basis for the forensic work reported – and hence for the forensic conclusion – must be stated. It would usually be concisely summarized in the report, with references as found in chemical literature – textbooks or journal articles, or government or corporate scientific reports, or in US patents. Or, the theory might be found by reference to the examining lab's own SOPs that would refer to the scientific literature.

For the fictitious example being followed through in this chapter [110]:

> Science theory: The conclusions are based on analysis by TLC,[38] with comparisons to reference standards, and the US International Ink Library,[39] as reported by [][][][][] & [][][][][], "[][][][][][][] [][]," *[][][][][][][][][]*, Vol.[][][], No.[][], pages [][][]→[][][], [][] [][][][] 20[][]; [111] and as presented in this lab's SOP[][][]. Some IR spectra were also observed for conformation. Assumptions were made to estimate the occurrence probability of the pen of interest in the geographical region, based on industry-provided sales data.

**38** TLC. Chapters 6 and 9.

**39** Ink reference standards. US International Ink Library. ATF. Chapters 2, 3, 4 and 9.

## 9.8 Items for analysis

The source, handling, storage and otherwise care of the items being analysed must be stated.[40] This would include registry within the lab's numbering/cataloguing/recording systems, and would refer to police systems. Reference would also be made to the lab notebook.[41] Photographs would also be included.

For the fictitious example being followed through in this chapter [112]:

Items for analysis:
- 1 single sheet of white paper [].[][] x [].[][] $cm^2$ with light blue writing apparently from a ball point pen.
- 1 ball point pen. Markings on the pen: [][][][][][][][][][][] [pen company] [][][][][] [= product identification numbers] engineering standards reference = [][][][][][42]

(Item identification numbers: [][][], [][][].) (See photos as figs. [][][], [][][] – below)
(Lab notebook ref.: pages [][][], *et seq.*)

## 9.9 Equipment and reagents

Much of the equipment and chemicals used would be more or less routine for most chemistry labs. Only those specific to the analysis reported need be mentioned.

For the fictitious example being followed through in this chapter [113]:

Equipment and reagents: Surgical-type clean sharp scalpels were used to scrape from the ink lines.

Commercially prepared TLC plates, of [][][][][][][][] [lab supply company], catalogue number [][][] [][], were used.

TLC solvents of [][][][][][][][] [chemical supply company], catalogue numbers [][][][], [][][][][], [][][][], were used. Production and lot numbers: [][][][], [][][][][]. All were of "spectral grade." None contained any preservatives that would be of concern for this analysis. All were used as supplied without further treatment.

The IR instrument used was of [][][][][][][][] [instrument manufacturing company], catalogue number [][][][]; modified in this lab as described in SOP [][][].

K Br pellets were pressed as set out in SOP [][][].

Refer to SOPs [][][], [][][], [][][] for more details.

40 Polymer envelopes. Chapters 3, 5 and 9.
41 Lab notebook. Chapters 5 and 9.
42 ASTM. Crayola label. Chapters 1, 5 and 9.

## 9.10 Analysis

The analysis that led to the conclusion would be concisely described in the report, so that others would be able to critically appraise what was done, and repeat it, notwithstanding that such repetition would usually not be necessary.

For the fictitious example being followed through in this chapter [114]:

Analysis: The items for analysis were photographed – see figs. [][] – of the ink-writing on the paper, and from the pen.

The pen and ink appeared to be of a standard ball point type – see figs.: [][] – as confirmed by the analysis here.

The analysis of the ink taken from the paper followed SOP no.[][][] – TLC (using SOP no. [][][] – solvent mix [][][]); and whole ink IR (using SOP nos. [][[] & [][][]). TLC plate photos and IR spectra are shown as figs. [][].

The analysis of the ink taken form the pen was the same as just above.

All inks from the items for analysis were found to be identical – by both TLC and IR; to each-other, and to US International Ink Library, reference standard sample: [][][] – light blue. No matches were found with any other ink as reported in the US International Ink Library.

These findings were consistent with data from the pen manufacturing company identified from markings on the pen.

The data described above are set-out as tables [][], [][], [][].

Ink industry production and sales statistics, bank-industry-published (ref. [][]), for various inks are also presented as table [][]. These statistical data indicate that the light blue ink is rarely found in the region.

Care was taken to preserve other-than ink evidence – such as handwriting, fingerprints (on paper & pen), and pen ball microscopic markings. Before starting any ink work, close-up colour photographs and photomicrographs were made to best preserve such evidence (as indicated in the conclusion), should further work be required. The fingerprint photography used a laser light source ($\lambda$=[][][] nm) to try to activate a fluorescence response range ($\lambda$=[][][]→[][][] nm) (also as indicated in the conclusion). That no fingerprint images were observed by this process does not exclude fingerprint possibilities that might be revealed by other process not attempted in this lab. (Lab notebook ref.: page [][][], *et seq.*)

## 9.11 Error statement

As indicated in Chapter 8,[43] consideration of error in scientific evaluations is important. This should be reflected in the forensic report.

For the fictitious example being followed through in this chapter [115, 116]:

**43** Error report. Chapters 8 and 9.

Error statement: As appropriate for this work, the chemical analyses were all of a qualitative nature, without numerically precise determination of ink component quantities. The validity of the analyses methods for use here is supported by reports in the scientific literature – cited in SOP no.[][][].

The ink and bank industry-published statistical data are only indirectly related to pen use quantities, so that the probability of light blue ink use should be regarded as only an approximate estimate.

## 9.12 SOPs, ...

Full copies of all SOPs,[44] and as appropriate other of the lab's core documents, referred to in the report must be readily available, either as appended to the report, or as separately available, perhaps in an SOP manual, or other of the lab's manuals. Although much of this would be of little interest to the report's readers, it should be available to whomever would want to refer to it, for whatever reason. This should be regarded as routine. SOP and document unavailability or secrecy, without explanation, should be seen as detracting from the credibility of the report.

For the fictitious example being followed through in this chapter [117][45]:

SOPs: The full texts of all SOPs referred to in this report are either as Appendix [] to this report, or are downloadable from the lab's public-access website www.[][][][][][][] None of these SOPs contain sensitive state or personal information; all are public documents, with routine copyright requirements. All of the lab work reported herein conformed to these SOPs, without variation.

## 9.13 Conclusion

If not the most important part of the report, the conclusion would likely be the most attractive to readers who would want to know what happened. (Some of those readers might be hard-pressed, anyhow, to thread through what might to them be the tedious technical detail of the report's other parts.)

The report-interested would include: Crown and defence counsel, and law enforcement, insurance investigators, *et al.* The forensic report does not necessarily say what happened for the theory of the human story, but it offers to help determine how that story might have happened, or not, or somewhat.

Crown and defence counsel would use the forensic conclusion in pursuing their arguments, or dropping them, or looking elsewhere still seeking to understand unknowns. Defence counsel, after seeing the report, and sharing it with client, might

44 Core documents. SOPs. Chapters 5 and 9.

45 Variations of Ink analytical chemistry – as examples – *Gedankenexperiment* – continue throughout this book. Introduction 2.3 and 2.2.

want to more carefully explain the advantages of a plea bargain; or alternatively, the wonders and risks of a contested trial. Crown counsel, after seeing the report, might want to drop the case; or alternatively confidently proceed to trial.

A report example, from US Senate hearings, 1989, stating conclusions[46]:

> [. . .]
>
> Q11 through Q13 – Three handwritten letters bearing questioned ink entries
>
> RESULTS OF EXAMINATION:
> Chemical and physical examinations were performed on the questioned ink entries appearing on exhibits Q11 through Q13 and the results were compared with each other and with those from inks in our standard ink library.
>
> One class of ink prepared all the black non-ball point ink entries. Representative samples of this ink taken from each of Exhibits Q11 through Q13 contained characteristics which matched those of standard inks available prior to 1981. The blue ball point ink entry ("him") on the back of Exhibit Q12 matched a standard ink available prior to 1981. Hence no conclusion can be made regarding the date or dates of preparation of Exhibits Q11 through Q13. [118][47]
>
> [. . .]

For the fictitious example being followed through in this chapter [119]:

> Conclusion: The ink of the writing on the questioned document, and the ink in the pen, are the same, and are identified as a light blue ink of [][][][][][] [ink manufacturing company], catalogue number [][][][][]; [][] [][][] 19[][], lot [][][][]. And further identified as an ink in the US International Ink Library – reference identification [][][][][] – with no other US International Ink Library find.
>
> From published sales data (see attached table [][]), it may be inferred that this light blue ink is rather rarely found in the region – having an occurrence probability expectation of <½% (but this number would have only an approximate and tenuous relationship to actual occurrence probability expectation and should be regarded with care).[48]
>
> The pen was of [][][][][][] [pen manufacturing company], catalogue number [][][][][]; assembled containing the here-identified ink, [][] [][][] 20[][], lot [][][][].
>
> No forensic conclusions are drawn here concerning any other factor – most particularly concerning handwriting, fingerprints and pen ball microscopic markings. However, before starting any other work at this lab, close-up colour photographs and photomicrographs were made to best preserve such evidence (see photos of fig. [][]), should further work be required.
>
> No forensic ink indications were sought as to when the document was made, although ink and pen manufacture historical data (see attached table [][]), might be suggested as for limiting *not-before* date ranges.

46 US Senate Hastings Impeachment hearings, 1989. Chapters 5 and 9.

47 Ink reference standards. Chapters 2 and 9.

48 Occurrence probability – ink example. Find expectation. Chapters 2, 4 and 9.

The fingerprint photography (of paper and pen) was done to help avoid damaging possible fingerprint evidence while analysing ink samples. That photography used laser and other light sources to try to activate fluorescence response, at λ as noted above; according to the methods of E. Roland Menzel, [120, 121] and Keith Beesly, Savvas Damaskinos and A.E. Dixon [122].[49] No treatment was done to the items for analysis to try to enhance fingerprint observation. As indicated above, that no fingerprint images were observed by this process does not exclude fingerprint possibilities that might be revealed by other process not attempted in this lab.

It should be noted in regard to forensics other than for ink, that this lab does not perform such further work. Only the ink analysis as described above was done.

## 9.14 Quality assurance

For the fictitious example being followed through in this chapter [123][50]:

> QA: The formal quality assurance system in place for this report, and for the work done for this report, is described in SOP [][][][], *et seq.* This lab's quarter-yearly QA evaluation reports are public documents available at www.[][][][][][].

## 9.15 Environmental and workplace health and safety

For the fictitious example being followed through in this chapter [124][51]:

> Environmental and workplace health and safety: Environmental and workplace health and safety practices of this lab, in-place for this report, and for the work done for this report, are described in SOP [][][], *et seq.* and SOP [][][], *et seq.*; which include applicable requirements of law, and management/trade union collective agreements, as indicated in those SOPs. There is no related litigation going-on; nor grievance proceedings.

## 9.16 Legal and disclaimer

[. . .][52]
*the large print giveth*
*and the small print taketh away*
[. . .] [125, 126, 127]

---

49 Fingerprint visualization. Fluorescence and phosphorescence. Chapters 6, 7 and 9.
Fingerprint probability. Chapters 1 and 4. Fingerprint visualization. Fluorescence and phosphorescence.
Chapters 6 and 9. Fingerprint when. Chapters 7 and 8.

50 QA. Quality assurance. Report. Chapters 8 and 9.

51 Regulatory law. Environmental and workplace health and safety. Chapters 1, 5, 9 and 11.

52 Introductory quotes. Introduction 1.2.

**Legal and disclaimer** concepts may apply to forensic science reports. But first, some **general commentary:**

Books [128, 129],[53] other publications [130, 131],[54] legal documents [132, 133, 134], *etc.* [135, 136] often contain various legal disclaimers that would appear to try to limit the responsibility and liability of the writer/publisher/issuer; and would make various declarations of claimed fact, intent and legal rights, *etc.*; including government science publications [137, 138].[55] There may be a legitimate purpose to this, in that it would avoid misleading readers and users of the publication [139].[56] But, also it can obfuscate and be used as a less-than-legitimate excuse or ruse to avoid or evade responsibility and liability. Sometimes, similarly, in another context – as for agreeing to website terms and conditions – the reader might be given what might appear to be a choice, but in view of a vastly superior bureaucratic advantage – perhaps including massive and complex contract wording – is not really fairly so.

Scientific instrument manufacturers may also have their own legal disclaimer. For example [SCIEX]: "For Research Use Only. Not for use in Diagnostic Procedures" [140, 141, 142, 143, 144, 145]. The consequences of a forensic lab violating this are not indicated. This would appear to have a possible relevance for measurements made in an impaired by drug case – with equipment reserved in the commercial market place for pharmaceutical research [146].

Sometimes declarations and disclaimers might be placed by a publisher, on legal advice, beyond the wishes or concerns of the writer, perhaps added in a final stage without the prior knowledge of the writer. Some may see this as just legal boilerplate [147] that does not matter too much – no need to worry – which might be true unless/until – hopefully rarely – someone of senior authority thinks it does matter – and then the legal advisors might take over, to the chagrin of others.

**Applying these concepts to forensic science reports:** Care needs to be taken that the declarations and disclaimers do not negate, or muddle, what is being reported. For example [148], "E&OE" [149] (usual for Canadian lottery tickets [150]; unheard of for a forensic report) [151] at the end of a report finding a criminalized molecule, with a prison term possibility, would raise questions of what the report is actually saying, and who is responsible to finally determine the correctness, or not, of the finding. Such a forensic report, with "E&OE" un-explained, should not be used for criminal court purposes.

For the fictitious example being followed through in this chapter [152]:

---

**53** Disclaimers. Error [. . .], Not legal advice [. . .] Commercial products, [. . .] Introduction 2 and 4.1; Chapters 1, 5, 6 and 9.

**54** $LD_{50}$ fragrance CCOHS Disclaimer Chapters 1, 6 and 9.

**55** Health Canada Guide. Chapters 5 and 9.

**56** Ethics. Lawyers. Experts. Introduction.; Chapters 1, 5, 8 and 9.

Legal: This report is, at its time of issue, according to its context, the final determination of this agency's investigation, results & conclusions; *functus officio*,[57] unless otherwise court-ordered, for example as when an author of the report is testifying in court. Otherwise, revisitation by this agency of anything within this report would be as investigation *de novo*, [153] and must be by formal written request, stating reasons, and indicating budgetary allowance.

©202[] Ministry of the Solicitor General. Licence *gratis*, [154] for all purposes of legal process. Confidential to the litigants until legal processes and appeal periods would be completed. Then, unless otherwise labelled, and when anonymized as to the people referred-to within it, this report may be considered as a public document.

## 9.17 Control of distribution: public document? Confidential? Freedom of information

Often, best practice might be similar to the example [155] being followed through in this chapter [156]:

Distribution: As originally signed,[58] this report first goes to Crown counsel (by-way-of the police agency that delivered the items investigated to this lab), to convey copies to other litigants, and then to the court or tribunal, if and when appropriate. This report is confidential, depending on court orders; until litigation and appeal periods would be completed; and afterwards as a public document when anonymized as to the people referred-to within it.

## 9.18 Data archival

As indicated in Chapter 5, the scientific data should be retained – in its raw forms and, as appropriate, at its various stages of processing, even though much of those might not eventually appear in the report, or perhaps anywhere else. Those data should be referenced in the report as archived and accessible.[59]

For the fictitious example being followed through in this chapter [157]:

Data archival: The scientific data upon which this report is based, including data not actually appearing herein, have been archived and are available as data archival ref. [][][][], to be retained there until [][] [][][][][] 20[][].

(Lab notebook ref.: page [][][], *et seq.*)

**57** *Functus officio*. Chapters 1 and9.

**58** Signed report. Chapter 9.

**59** Data archive. Chapters 5 and 9.

## 9.19 Finance: US *Qui tam*

The financing of the report's issuing agency, and more specifically, of how the report came to be, should be stated.[60] This may be very brief, but should be sufficient to allow critics to be able to otherwise access relevant financial statements, contracts with supporting external agencies, and with participating contractors. This should include government contract reference numbers, if they exist. Critics should be entitled to be legitimately interested in considering various legal processes, including US *Qui tam*.

For the fictitious example being followed through in this chapter [158]:

> Financial: This report is produced by the Forensic Service Agency, Ministry of the Solicitor General, from 20[][] budgetary provision [][][][][][]. No other funding is involved. All forensic work was done at government facilities, by regular government employees – of civil service designation; no private contractors were involved.

## 9.20 Clarity and quality of writing

As indicated above, in this chapter, a forensic science report should give a clear idea of the issues addressed, concisely written, with clear and ample referencing. Clarity and quality of writing takes practice that sometimes is not taught well-enough along with the rest of a science and engineering education; but is very important. Perhaps, the quote of the fictional Prince Kinnall,[61] is worthwhile for some guidance. The popular science writings[62] of Isaac Asimov [159] and George Gamow [160, 161] are worthwhile to consider.

## 9.21 Language

> *turn the box around*
>
> [Pierre Trudeau, 1979(?), to a heckler complaining about French on his cereal box**]**. [162]

Generally, the report would be in English in most of Canada's provincial jurisdictions, in French in Québec, and both languages federally and New Brunswick. Other local practices might provide for various languages, either routinely, or on request. In modern Canada this usually seems to be no longer contentious [163].

---

**60** US *Qui tam*. Chapters 1, 8 and 9.

**61** Prince Kinnall. Chapters 8 and 9.

**62** Asimov. Gamow. Introduction 4; Chapter 9.

For the example being followed through in this chapter [164]:

Language: Disponibles en français.
Traducción al español disponible.
Verfügbar in Deutsch und Englisch

## 9.22 Format and short form

### 9.22.1 Format

Any number of formats and styles for forensic reports may be appropriate – depending on the standards and practices of the issuing agencies and their writers. What is important is that, in principle, the elements of the above discussions are addressed.

### 9.22.2 Short form

For some forensic work, where very many reports would be routinely issued for very many of the same kind of analyses, a short form [165] format, relying on overall reference documents, might be appropriate, so long as those references, and other, documents, are easily and readily available; and so long as the essential criteria of this chapter, as indicated above, are met by the references.

For the example [166] for the drug analysis referred to in Chapter 5 [167].[63]

---

FORENSIC REPORT – **SHORT FORM** – for Police File □□□□□□.
**DETERMINATION OF MARIHUANA,** ***etc.*****; or not.**

---

This report issued **□□ □□□□ 202□**, □□:□□ □m. EST. [168]
For forensic science work done □□ □□□□ 202□; concluded □□:□□ □m. EST.
Authorizing signature: ______________________________
(as at the issue date above, as for this entire report. Σpages = 1.)
□□□□□ □□□□□, Analytical Chemist. Employee id # = □□□□□.

Forensic Service Agency,
Ministry of the Solicitor General.
□□□□ □□□□□□□□□, [169]
□□□□□□□□□□, □□□□□□□, □□□□□□. □□□ □□□ [170]

This lab's file number for this work = □□□□□□□.

---

63 Report – Short Form. Chapters 5 and 9.

(continued)

---

- General appearance of the item: □□□□□□□□□□□□□.
- □□□.□□ g
- □□□.□□cm x □□□.□□cm x □□□.□□cm.
- Other: □□□□□□□□□□□□□□□□□.

---

Conclusion: ✔ ☐ is ☐ is not **marihuana** – plant material.
☐ does ☐ does not contain **THC**.

---

Tetrahydrocannabinol THC CAS 1972-08-3 314.469 g/mol. )$^9$-THC. [see Fig. 9.1][64]

---

This **SHORT FORM REPORT:**

- Is dependent on the LAB NOTES PROCESS SHEET for this lab's file number indicated above. Copy available to legal counsel on request to Crown counsel.
- Is for analysis that was done & conformed exactly to the forensic analytical chemistry procedures, requirements & other criteria, as set out in detail in this lab's SOP □□□ = "DETAILS OF FORENSIC ANALYTICAL CHEMISTRY TO DETERMINE MARIHUANA, *etc.*; or not." SOP available as a public document & downloadable from this lab's public-access website www.□□□□□□□□.

---

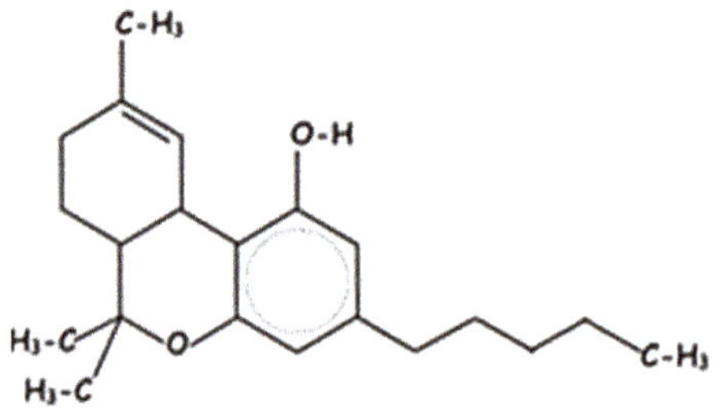


Fig. 9.1: Short form – THC.

## Notes

**1** Lewis Carroll, *Alice's Adventures in Wonderland* and *Through the Looking-Glass*, Signet Classics, The New American Library, New York City, 1960 & 1962; from . . . *Adventures* . . ., "Alice's Evidence", Chapter XII, page 111.
**2** Robert Silverberg, *A Time of Changes*, Warner Books, Inc., New York City, 10103; 1971; ISBN 0-446-34061-8 (an enlarged version from *Galaxy Science Fiction*, Universal Publishing and Distribution Corporation, 1971).
**3** Robert Silverberg, *A Time of Changes*, copyright 1971, Agberg, Ltd., Oakland, California.
**4** Silverberg, Chapter 27, at page 89. Kinnall Darival, tells part of his story:
**5** He is a fugitive prince of the province of Salla, on the continent Velada Borthan, on the planet Borthan, far from Earth and settled from there long ago. He left Salla eventually arriving in Manneran, a city in a province of that same name, and the most important port and commercial centre of Velada Borthan. He became private secretary to his bondfather, Segvord Helalam, High Justice of the Port Judiciary, and in that very important position prospers, before having a great fall.
**6** Silverberg, Chapter 27, pages 89 and 90.

---

**64** Tetrahydrocannabinol THC CAS 1972-08-3 Chapters 6 and 9; Figs. 6.20 and 9.1.

7 https://en.wikipedia.org/wiki/Robert_Silverberg_bibliography
https://en.wikiquote.org/wiki/Robert_Silverberg https://en.wikipedia.org/wiki/Robert_Silverberg

**8** Robert Silverberg. Copyright acknowledgement; Chapter 9.

9 Доверя́й, но проверя́й = *Doveryáy, no proveryáy* = Trust, but verify.
Russian proverb. https://en.m.wikipedia.org/wiki/Trust,_but_verify

**10** Suzanne Massie and President Ronald Reagan. President Reagan and General Secretary Mikhail Gorbachev.

**11** https://www.washingtonpost.com/opinions/trust-but-verify-an-untrustworthy-political-phrase/2016/03/11/da32fb08-db3b-11e5-891a-4ed04f4213e8_story.html
http://www.suzannemassie.com/
https://www.nytimes.com/2019/12/02/books/robert-k-massie-dead.html

**12** = A government Ministry, an independent or quasi-independent agency of the state, or a charitable institution, that would serve government or public processes; or a private corporation or private business that would serve private business or corporate purposes, or serve government Ministries or state agencies by contract.

**13** https://www.nobelprize.org/ceremony/the-nobel-prize-award-ceremonies-and-banquets/

**14** Anne M. Coghill and Lorin R. Garson, Editors, *The ACS Style Guide / Effective Communication of Scientific Information*, Third Edition; American Chemical Society, Washington, DC; Oxford University Press, New York City, 10016; 2006; Chapter 2, "Scientific Papers," "Standard Format for Reporting Original Research," page 19, *et seq.*

**15** "Preparing a Research Report," ACS Committee on Professional Training, American Chemical Society, Washington, DC; August 2015.
https://www.acs.org/content/dam/acsorg/about/governance/committees/training/acsapproved/degreeprogram/preparing-a-research-report.pdf

**16** 40 US CFR §160.185 EPA PESTICIDE PROGRAMS – GOOD LABORATORY PRACTICE STANDARDS – Reporting of study results. https://www.law.cornell.edu/cfr/text/40/160.185

**17** https://pmc.ncbi.nlm.nih.gov/articles/PMC3528086/
https://science-fair.org/students-parents/writing-abstracts/
https://www.thoughtco.com/how-to-write-a-lab-report-606052
https://panchbhaya.weebly.com/uploads/1/3/7/0/13701351/grade_9-10_formal_lab_report_guidelines.pdf

**18** Dr. Rami El Hayek, Analytical Chemistry Lab (CH 313),UMass. Boston, CH 313, Spring 2012.
http://alpha.chem.umb.edu/chemistry/ch313 http://alpha.chem.umb.edu/chemistry/ch313/lecture0.pdf
http://alpha.chem.umb.edu/chemistry/ch313/How%20to%20write%20a%20lab%20report.pdf

**19** Academia.edu, San Francisco [April 2020].
https://www.academia.edu/7307995/How_to_Write_a_Good_Lab_Report_Analytical_Chemistry_Lab_
https://www.academia.edu/ https://www.academia.edu/copyright
https://www.academia.edu/about https://www.academia.edu/hiring https://www.academia.edu/press

**20** https://www.elsevier.com/connect/11-steps-to-structuring-a-science-paper-editors-will-take-seriously

**21** See also: John Keenan Taylor, *Quality Assurance of Chemical Measurements*, Lewis Publishers, Inc., CRC Press; Boca Raton, Florida, 33431; 1987; ISBN 087371-097-5; Chapter 22, "Reporting Analytical Data," pages 197→207; at "REPORT FORMAT," pages 205, 206, 207.

**22** Farrington Daniels, J.W. Williams, Paul Bender, Robert A. Alberty, C.D. Cornwell, *Experimental Physical Chemistry, Sixth Edition*, McGraw-Hill Book Company, Inc., New York; 1962.
https://datapdf.com/experimental-physical-chemistry-daniels-farrington0b82eabb506c0b6b514d5a6ddebf69dd93482.html

**23** Daniels, et al.; Chapter 2, "Thermochemistry," page 15, *et seq.*: Using an experiment to measure enthalpy of combustion with a bomb calorimeter as an example description of a lab process.

24 R.S. Jessup, *Precise Measurement of Heat of Combustion with a Bomb Calorimeter*, US Department of Commerce, National Bureau of Standards, Monograph 7, 26 February 1960; Superintendent of Documents, US Government Printing Office, Washington 25, D.C.; 25c. https://nvlpubs.nist.gov/nistpubs/Legacy/MONO/nbsmonograph7.pdf
25 General requirements for the competence of testing and calibration laboratories
ISO/IEC 17025:2005(E)

[. . .] [page 20]

**5.10 Reporting the results**

**5.10.1 General**

The results of each test, calibration, or series of tests or calibrations carried out by the laboratory shall be reported accurately, clearly, unambiguously and objectively, and in accordance with any specific instructions in the test or calibration methods.

The results shall be reported, usually in a test report or a calibration certificate (see Note 1), and shall include all the information requested by the customer and necessary for the interpretation of the test or calibration results and all information required by the method used. This information is normally that required by 5.10.2, and 5.10.3 or 5.10.4.

In the case of tests or calibrations performed for internal customers, or in the case of a written agreement with the customer, the results may be reported in a simplified way. Any information listed in 5.10.2 to 5.10.4 which is not reported to the customer shall be readily available in the laboratory which carried out the tests and/or calibrations.

[. . .]

**5.10.2 Test reports and calibration certificates**

Each test report or calibration certificate shall include at least the following information, unless the laboratory has valid reasons for not doing so:

a) a title (e.g., "Test Report" or "Calibration Certificate");
b) the name and address of the laboratory, and the location where the tests and/or calibrations were carried out, if different from the address of the laboratory;
c) unique identification of the test report or calibration certificate (such as the serial number), and on each page an identification in order to ensure that the page is recognized as a part of the test report or calibration certificate, and a clear identification of the end of the test report or calibration certificate;
d) the name and address of the customer;
e) identification of the method used;
f) a description of, the condition of, and unambiguous identification of the item(s) tested or calibrated;
g) the date of receipt of the test or calibration item(s) where this is critical to the validity and application of the results, and the date(s) of performance of the test or calibration;
h) reference to the sampling plan and procedures used by the laboratory or other bodies where these are relevant to the validity or application of the results;
i) the test or calibration results with, where appropriate, the units of measurement;
j) the name(s), function(s) and signature(s) or equivalent identification of person(s) authorizing the test report or calibration certificate;
k) where relevant, a statement to the effect that the results relate only to the items tested or calibrated.

[. . .] [page 22]

**5.10.5 Opinions and interpretations**

When opinions and interpretations are included, the laboratory shall document the basis upon which the opinions and interpretations have been made. Opinions and interpretations shall be clearly marked as such in a test report.

[. . .]

**26** See: Douglas Goold and Andrew Willis, *The Bre-X Fraud,* McClelland and Stewart, 1998; ISBN 10:0771033346 & ISBN 13: 9780771033346.
https://www.abebooks.com/book-search/isbn/0771033346/

**27** http://en.wikipedia.org/wiki/Bre-X

**28** Peter Brieger, "Red Flags Missed at BRE-X / Felderhoff Cleared / Geologist not the only one fooled by Indonesian gold mine hoax, says judge," *Financial Post,* 1 August 2007, page FP1, FP3.

**29** Charles Lewis, "De Guzman Story Has Shades of Cooper / 'Vanishing' Standard," *Financial Post,* 1 August 2007, page FP3.

**30** Diane Francis, Comment, *"Bre-x fiasco shows flaw in Canadian law," Financial Post,* 1 August 2007, page FP3.

**31** As to sample selection and preparation – which included fraud – before the samples went for actual chemical analysis. See: Goold and Willis at page 254, *et seq.*

**32** This synoptic commentary is opined by M.G.

**33** Grange Royal Commission – re Susan Nelles:

**34** The Honourable Mr. Justice Samuel G. M. Grange, Supreme Court of Ontario, Commissioner; *Report of the Royal Commission of Inquiry into Certain Deaths at the Hospital for Sick Children and Related Matters,* Ontario Ministry of the Attorney General; Toronto, 1984; ISBN 0-7743-9968-6.
https://archive.org/details/reportofsickkids00onta

**35** Gavin Hamilton, M.D. *The Nurses are Innocent – The Digoxin Poisoning Fallacy.* Dundurn Press, 2011. ISBN 978-1459700574. https://en.wikipedia.org/wiki/Special:BookSources/978-1459700574.

**36** http://en.wikipedia.org/wiki/Susan_Nelles

**37** See also: Megan Specia, *"Letby's Lawyers Seek Appeal After Lead* [Pb] *Expert' Changed Mind' on Murders /* Lawyers for Lucy Letby, a British nurse found guilty of the murders of seven babies, said that new evidence about the prosecution's lead witness undermined the case against her." *New York Times,* 16 December 2024.
https://www.nytimes.com/2024/12/16/world/europe/letby-murders-appeal.html

**38** Griffiths Report, 2001 – re Gemma Ramlal:

**39** CBC News · Posted: Jun 24, 2001 10:38 PM ET | Last Updated: June 24, 2001
https://www.cbc.ca/news/canada/lab-errors-may-void-ontario-drug-convictions-1.283147

Lab errors may void Ontario drug convictions

Botched paperwork at a federal lab may void thousands of drug convictions in Ontario.

Anyone convicted of a drug offence in Ontario during the past 12 years, even those now serving time, may be off the hook because of bungled paperwork at a federal drug lab.

The problem involves thousands of drug analysis certificates that Health Canada says might be not be accurate.

The certificates are issued in every case where suspected illegal drugs are found. They are used by courts to identify and determine the purity of various drugs.

"The chances are very good that if you just get in touch with a lawyer any lawyer that you will be out of jail very quickly," said Alan Gold of the Criminal Lawyers Association.

> The analyst at the centre of the controversy is Gemma Ramlal, who resigned from her Health Canada lab job two months ago. She now works at a grocery store. She denies any wrongdoing.
>
> "I want to know how it took my supervisors so many years to figure out my work was wrong," Ramlal said.
>
> Dann Michols, assistant deputy minister at Health Canada, told CBC TV: "We trust . the professionalism and training of the individuals to follow these procedures and only in this one case have we been disappointed."
>
> Ramlal had been doing the work for 12 years. Earlier this year, she went on vacation and another analyst took over and discovered some things were wrong with the certificates.
>
> A review of Ramlal's work discovered substances misidentified, data entry errors, wrong dates and instances when Health Canada procedures were not followed. The discovery casts doubt on some 16,000 certificates issued over the past 12 years.

**40** www.hc-sc.gc.ca/english/archives/releases/2001/2001_40e.htm

**41** Drug Analysis Service, within Health Canada.
https://health-infobase.canada.ca/drug-analysis-service/analyzed-drug-report.html

**42** "Report to the Attorney General of Canada: Advice and Recommendations Respecting Certain Certificates of Analyst Issued by Health Canada," by the Honourable W. David Griffiths, Q.C., May 23, 2001. http://www.canada.justice.gc.ca/en/news/nr/2001/doc_26438.html.

**43** Health Canada / News Release / 2001–40 April 19, 2001.
"Announcement concerning the Toronto Drug Analysis Service Laboratory" http://www.hcsc.gc.ca/ahc-asc/media/nr-cp/2001/2001_40_e.html

**44** *The Globe and Mail*, National News / National Report, 28 September 1998, page A6.

**45** Drug evidence questioned
VANCOUVER. The overdose death of an employee of the [Health Canada] federal drug testing lab has cast doubt on 20 years of drug convictions, lawyers say.
The federal Justice Department is reviewing drug cases where Henry Sadowski handled the evidence, after the lab chemist died in May, 1997, of a drug overdose. Tests showed Mr. Sadowski, 51, had snorted a lethal mixture of cocaine and heroin, called a speedball. An autopsy revealed he had been using the drugs for some time.
Now lawyers who defend people on drug charges are wondering whether Mr. Sadowski fudged his reports. "How do we know what he analyzed and what he put up his nose?" defence lawyer Ken Young said. CP

**46** https://www.worldcat.org/title/report-to-the-attorney-general-of-canada-advice-and-recommendations-respecting-certain-certificates-of-analyst-issued-by-health-canada/oclc/48790343?referer=di&ht=edition

**47** https://web.archive.org/web/20020111071110/http:/www.canada.justice.gc.ca/en/news/nr/2001/doc_26436.html

> RELEASE OF REPORT TO THE ATTORNEY GENERAL OF CANADA CONCERNING CERTIFICATES OF ANALYST ISSUED BY HEALTH CANADA
>
> OTTAWA, June 7, 2001 – Today, the Honourable Anne McLellan, Minister of Justice and Attorney General of Canada, released a report by the Honourable W. David Griffiths containing advice and recommendations respecting certain Certificates of Analyst issued by Health Canada. The Department of Justice is moving to implement the recommendations in this report.
>
> Health Canada recently identified concerns with some of the Certificates of Analyst issued by one analyst in the Toronto Drug Analysis Service Laboratory. Certificates of Analyst are used to prove the nature of substances in drug prosecutions.

From the time that the problem was identified, no further certificates were issued by this analyst. Prosecutors were instructed not to use certificates issued by this analyst in ongoing cases. On April 19, 2001, officials from the Department of Justice and Health Canada issued a media release and held a technical briefing in Toronto to ensure that those who might be affected could be made aware of the issue.

At that time, the Honourable Mr. Griffiths, a retired judge of the Ontario Court of Appeal, was asked to provide advice and recommendations to the Attorney General of Canada on the steps to be taken to deal fairly and expeditiously with those who may be affected. In preparing his report, Mr. Griffiths consulted with members of the defence bar and prosecutors. Prominent criminal lawyer Mark J. Sandler assisted Mr. Griffiths with his report.

Health Canada retained the services of Dr. Joel Mayer, Deputy Director of Scientific Affairs at the Ontario Centre for Forensic Sciences. Dr. Mayer was asked to verify the methodology used by Health Canada to identify questionable certificates and will also advise Health Canada on changes and improvements to its Drug Analysis quality assurance systems.

Dr. Mayer's report will be made public shortly.

Health Canada reviewed the work of the analyst and determined that there were problems with 192 certificates regarding the nature of the substance and 13 certificates regarding the purity of the substance. Dr. Mayer found the methodology used by Health Canada to review the work of the analyst acceptable. The Griffiths report makes recommendations with respect to those and other certificates issued by the analyst.

In keeping with the recommendations made by Mr. Griffiths, the Department of Justice Canada will notify each accused found guilty of an offence to which one of the identified certificates relates. Further steps would be taken if a request for relief is made.

For those who have been found guilty of an offence where the case involved a certificate issued by the analyst, the Griffiths report sets out a range of possible remedies that may be available depending on the circumstances of the case. These include: an extension of time in which to file an appeal; an appeal to set aside a conviction and enter an acquittal or a stay of proceedings or order a new trial; a direction from the Minister of Justice for a new trial or appellate hearing under s.690 of the Criminal Code; a pardon, repayment of any fine paid or release from custody, where appropriate.

With respect to cases currently before the courts, the Department of Justice has been identifying those cases involving certificates issued by the analyst in question, making disclosure to the defence, and taking steps to deal with such cases based on the individual circumstances. Mr. Griffiths has approved the steps undertaken by the Department of Justice in ongoing cases.

All persons who have been found guilty of an offence pertaining to an illicit drug between June 1, 1988 and May 31, 2001, and who wish to determine whether their Certificate of Analyst is one of those in question, should access the following web site: www.hc-sc.gc.ca/expo. This site lists the numbers of all certificates prepared by this analyst. Individuals without Internet access can contact the Department of Justice by phone (416) 973–0957 or by facsimile (416) 973–8253.

A copy of the report by Mr. Griffiths can be found on the Department of Justice website at canada.justice.gc.ca/en/news/nr/2001/doc_26438.html.The Department of Justice is moving to implement the recommendations in this report and remains committed to taking all necessary steps to ensure that public confidence in the integrity of the justice system is maintained.

[. . .]

48 "Report to the Attorney General of Canada: Advice and Recommendations Respecting Certain Certificates of Analyst Issued by Health Canada by the Honourable W. David Griffiths, Q.C."; 23 May 2001.

[. . .]

**OVERVIEW**

Most drug prosecutions involve, as a substantive component, proof of analysis and identification of a substance forming the subject matter of the Crown's allegations. To establish this element, the Crown generally tenders a Certificate of Analyst which states that a designated analyst has examined the substance and found it to contain a controlled substance. A sample copy of such certificate is attached as "Appendix A". Analysts who prepare these certificates are designated by the Minister of Health and are employed in various Health Canada Drug Analysis laboratories. Recently, Health Canada identified concerns with some of the certificates issued by one analyst in the Toronto Drug Analysis Service laboratory. This analyst was employed by Health Canada since June 1982, and was working for the Drug Analysis Service laboratory since June 1988. After March 8, 2001, when the concerns were identified, no further certificates were issued by this analyst.

An internal review by Health Canada of the certificates issued by this analyst and any available supporting documentation was recently completed. As I shall point out later in this report, this review process has identified a number of problematic certificates.

As well, Dr. Joel Mayer, Deputy Director, Scientific Affairs at the Ontario Centre of Forensic Sciences has been retained by Health Canada to review the methodologies used by Health Canada to assess the work of the individual in the Toronto laboratory, analyze the processes used by Health Canada to review the work of all other analysts in the Toronto and all other laboratories across Canada, advise Health Canada on changes and improvements that must be made to its Drug Analysis operations and quality assurances systems and oversee the implementation of the plan to prepare the Drug Analysis Laboratories to seek and obtain accreditation from international bodies.[1] I am advised that, while Dr. Mayer has not completed his entire work for Health Canada, he has reviewed and favourably passed upon the criteria used by Health Canada in evaluating the acceptability or unacceptability of certificates issued by the subject analyst.

[. . .]

**FOOTNOTES**

1 Government of Canada News Release dated April 19, 2001.

[. . .]

49 https://web.archive.org/web/20020220112655/http:/canada.justice.gc.ca/en/news/nr/2001/doc_26438.html

50 A report by Dr. Mayer to ". . . be made public shortly . . ." – from 7 June 2001 – has not been found in researching for this book.

51 See also: https://criminallawyers.ca/

- Alan Gold, CLA Letter, *FOR THE DEFENCE*, Criminal Lawyers Association Newsletter, Toronto, Vol. 22, No. 2, March / April 2001; pages 19 → 21.
- M.G., "Cheating on Drug Tests / Health Canada's involvement in the war on drugs has tainted the credibility of purported science everywhere in its systems. This is serious and long term damage.," Letter to the Editor, *FOR THE DEFENCE*, Criminal Lawyers Association Newsletter, Toronto, Vol. 22, No. 3, May/June 2001; pages 14 & 15.

52 [slightly edited]

*Earlier in May*
*we posted information regarding Federal Drug Agent Gemma Ramlal. This agent has been suspended for improper work practices.*
*Alan Gold, Frank Addario and Leslie Pringle met with Justice Griffiths and others to discuss this issue. Justice Griffiths has been retained by the Justice Department to investigate this issue with stakeholders and make recommendations.*
*Here is the letter written from CLA following this meeting.*

**The Honourable Mr. Justice P. Griffiths**
*Dear Justice Griffiths;*
The Criminal Lawyers Association is pleased to offer its assistance and advice concerning this unprecedented situation in the criminal justice system. By way of summary of, and addenda to the comments made at our meeting this past Tuesday, we offer the following.
We reiterate the general observation that how the criminal justice system responds is itself important as a matter of principle. At issue is a state employee, hired and supervised by the state, for whose malfeasance the state must accept full responsibility. The state's laboratory facilities, including staffing, are a crucial component of the criminal justice system. If one of the employees becomes 'tainted', the system must demonstrate in no uncertain terms that it accepts full responsibility and will completely and exhaustively rectify the wrong-doing. This appears to be the first time such a situation has arisen; consequently it is especially vital to set a definitive precedent that will send a message to state institutions as to how seriously such an episode is to be viewed.
We also restate the fundamental premise that this analyst's certificates are now to be viewed as blank pieces of paper. They cannot be seen as having any valid content whatsoever. In on-going cases, the state will simply have to carry on the case, if it can, by means of another, untainted analyst.
In closed cases, the situation must be viewed as one where a "hole" or "gap" in the Crown's proof has now materialized. It is important that "begging the question" not occur by any assumption, express or hidden, that the substance in any case is in fact an illicit drug based upon the conduct of the defence or accused. In any case involving this analyst, at least *prima facie* there is no proof of the illicit drug. It follows that:
1. In any extant appeal case the Crown must agree the gap exists so that prima facie the appeal must be allowed, and the argument must then be over whether a) the now-absent certificate matters, or the proviso can be applied, or b) whether an acquittal or merely a new trial is the appropriate remedy;
2. In any unappealed past case, concluded outside the appeal period, it should be turned into an extant appeal by the Crown's agreement to an extension of time to appeal. Thereafter the matter comes within 1.
3. It should be irrelevant whether the accused's plea was not guilty or guilty at trial, so long as this analyst's certificate was part of the trial record, or the accused can show the analyst's certificate was part of their case.

A guilty plea in our view does not affect the applicable approach because a nature of the substance in a drug case is an essential element of the offence and it is rare in our caselaw for a drug to be proved by circumstantial evidence without any certificate. Even in guilty pleas judges rely on the certificate in accepting the accused's guilty plea. The fact that the nature of a drug is almost inevitably proved by certificate evidence is a reflection of the trust and respect that the laboratory services were held by all involved in the criminal justice system, as well as defence counsel's appropriate standard of behaviour, encouraged by judges and prosecutors, not to waste time litigating such issues. When the trust turns out to have been betrayed, it is completely indefensible to use defence counsel's cooperation to try to minimize the effect of the certificates turning "blank". Fairness and logic dictate that we simply now do not know what the substance was. It should not matter whether the plea was guilty or not guilty.

4. Steps should be taken to bring the situation to the attention of any person who may have been affected by this analyst's certificates as follows:

a. all current cases should have full disclosure made, and a promise of Department of justice cooperation;
b. all possible efforts to identify accused and past cases involving this analyst should be carried out, and any accused so located should be treated as in a.
c. the entire situation should be publicized as widely as possible so that any person that may be able affected is made aware of the situation. A contact person should be set up so that anyone making inquiries could be given whatever assistance is possible to ascertain if the analyst was involved in their case.
d. any person able to show that the analyst was involved in their case should then be assisted to obtain an appellate remedy as in 2. above.

5. As indicated, regarding cases that have concluded at the trial level, whether last week, last year, or a decade ago, we regard the appellate remedy as the most convenient and appropriate. Once completed (at the trial level) cases have reached the point where this analyst is identified as involved, an extension of time should be consented to by the Department of Justice and an appeal launched, either in the Court of Appeal or the Summary Conviction Appeal Court.

6. Once an appeal is launched then the Department of Justice can consider its response. It seems to us the possibilities break down into several categories.

a. consent to admission of fresh evidence of certificate invalidity and consent to acquittal being entered;
b. consent to admission of fresh evidence of certificate invalidity and consent to new trial being ordered (whether or not the accused chooses to go further and seek an acquittal on appeal)
c. contested appeals where the Crown opposes fresh evidence only on the basis it would not make any difference because of Crown's responding fresh evidence of new, independent analysis or examination of available material allows acceptance of this analyst's results, or based upon trial record conviction was inevitable and proviso should be applied.

For reasons we stated, we would urge the Crown to be generous, rather than niggardly in its approach to these cases, and allow acquittals to go in all but the clearest cases where the public interest is definite in requiring a new trial or opposition to a successful appeal.

For category b. cases, we submit the Crown should only take this position where it is clear there is a reasonable prospect of conviction on the same theory as at the first trial. In other words, we would be against new trial sought on the basis that a "trafficking in marihuana" case could be retried as a "trafficking in a substance held out to be marihuana". That is an unfair second bite for the Crown and an unfair second trial for an accused who is in no way at fault for this analyst's malfeasance.

For category c. cases, we submit the Crown should only try to finalize the matter in the appeal court in the clearest of cases. A *post-facto* assessment of the tainted analyst's work claiming confirmation is inherently suspect, and would require substantial evidentiary procedures at the appeal level. Even an independent confirmatory analysis, given the circumstances, could reasonably be viewed as warranting testing by the defence at trial. We urge that the trial forum would be more appropriate for exploration of such matters and so the Crown should not in general rely on these factors in seeking to defeat an appeal, though there would be a basis for proceeding with a new trial.

Finally, we reiterate our position that all costs, including transcripts of past cases, occasioned by this forensic disaster, should be borne by the federal government. We are especially concerned that the Ontario Legal Aid plan not become charged with the expense occasioned as that would be an unfair and undeserved charge on an already tightly stretched budget.

Thank you for the opportunity to express our views. The Criminal Lawyers Association wishes you well in carrying out your challenging and highly important mandate and assures you of our continuing cooperation.

Yours truly,

*Alan D. Gold*

**53** [Slightly edited] [re Griffiths Report, *etc.*]

***Letters to the Editor***

**Cheating on Drug Tests**

About 10 years ago, I defended a client in Toronto, charged with possession of cocaine. Resulting from a subpoena I had sought, the Health Canada Analyst who did purported analytical chemistry for that case testified at trial. Despite my efforts, the judge found my client guilty.

The Analyst who testified was Gemma Ramlal.

I mostly agree with Alan Gold's letter, for the CLA, copied in *For The Defence.*[1] But unfortunately, the Health Canada lab did not maintain my ". . . trust and respect . . ."[2] As a chemist, from before becoming a lawyer, the word pair that would be increasingly appropriate would be "shocked and appalled." And, after a few years of my professional encounters with the federal government's purported forensics, no longer shocked.

There may be some implication that Ms. Ramlal should take most of the blame. She should not. Whatever she may have done, or not, she was a very minor player.

What she was allegedly involved in should hardly have been possible in a modem analytical chemistry lab, with a proper quality assurance programme. The quality assurance failure of the purported analytical chemistry of Health Canada's unaccredited, secrecy shrouded, forensic labs, invited, if not made inevitable, whatever Ms. Ramlal is accused of.

The major responsibility is with the lab supervisors, and the federal Ministers of the Crown on whose watches it happened. Ms. Ramlal is apparently a convenient fall guy in a very serious system failure.

The government position[3] would minimize the damage – that this was a

**Health Canada's involvement in the war on drugs has tainted the credibility of purported science everywhere in its systems. This is serious and long term damage.**

specific one-person problem at a specific site that will be attended to . . . .[4] Whatever Ms. Ramlal is alleged to have done would likely be only a small part of major and general quality assurance failure. And, such failure in the forensics system taints the credibility of purported science everywhere in Health Canada's systems.[5]

A recent survey in *The Economist*[6] opines that justice in the USA has been a major victim, as collateral damage,[7] in the war on drugs. I argue that in Canada there is another major victim: the scientific credibility of the national health ministry.

This raises the question of why Health Canada is purporting to do analytical chemistry in the war on drugs in the first place. The US FDA is not so involved.

This ethical curiosity for Health Canada has been government policy for years.[8] In my opinion, a health agency has no business being involved, even remotely, in criminal law enforcement that includes police subterfuge; and subterfuge in criminal law enforcement is also lawful government policy in Canada.[9]

As criminal defence lawyers, we naturally, and properly, address the issues surrounding Ms. Ramlal in a criminal law context, for our clients. But, we must also recognize when these issues have broader implications. I argue here that the issues of quality assurance failure in a national health ministry, and that ministry's ethical mis-direction, are actually of competing importance with the criminal law issues, if not more important.

A national health ministry with tainted scientific credibility is a national disaster. Why risk relying on any of its purported scientific pronouncements? Canadians would be well advised the seek their answers to health science questions elsewhere – the US FDAwould be a good place to start.

In another context, the importance of credible behaviour by health agencies was exemplified by President Clinton, when apologizing to the Tuskegee victims.[10] That tragedy of deception continued for decades as a tragedy of failure of trust in professional health providers.[11]

For a health agency, everything it is accountable for must be governed by health agency standards –

even non-health lab activities. Otherwise, its credibility everywhere in its systems – including its health activities – suffers. Health Canada's involvement in the war on drugs has tainted the credibility of purported science everywhere in its systems. This is serious and long term damage.

*M. Grossman,*

Barrister & Solicitor

**Notes**

1. Vol. 22, No. 2, March/April 2001; pages 19–21.
2. letter page 20; middle column, top.
3. www.hcsc.gc.ca/english/archives/releases/2001/2001_40e.htm
4. to borrow a phrase: – John Wayne. *BIG JAKE*, 1971. http://www.reelclassics.com/Gallery/quotes.htm
5. I have argued this before, elsewhere: M. Grossman, "Some Recent Canadian Judicial Decisions Relevant to Forensic Chemistry," presented at a Chemistry and the Law session, American Chemical Society national meeting, Anaheim, California, 22 March 1999.
6. Volume 360, Number 8232, 28 July-03 August 2001, between pages 48 & 49: "A survey of illegal drugs," 28 July 2001, sixteen pages;
7. "Collateral damage," survey pages 12 & 13.
8. MEMORANDUM OF UNDERSTANDING Between the DEPARTMENT OF NATIONAL HEALTH & WELFARE . and the ROYAL CANADIAN MOUNTED POLICE . CONCERNING THE PROVI SION OF AN ANALYSIS SERVICE OF SEIZED DRUGS AND DRUG PARAPHERNALIA, Ottawa, 18 February 1985 & 5 November 1984, five pages.
9. In *R v Oickle*, paragraph 65, quoting *Rothman v. The Queen* [1981] 1 S.C.R. 640, it is clear that the Supreme Court of Canada allows, within limits, police trickery and deceit against citizens, in criminal investigations. https://ca.vlex.com/vid/r-v-oickle-r-680829197

10. See Alison Mitchell, Washington, 16 May 1997, "Survivors of Tuskegee Study Get Apology From Clinton", *The New York Times*, 17 May 1997, page 9.
11. Ibid.; and see Lynda Richardson, "An Old Experiment's Legacy: Distrust of AIDS Treatment", *The New York Times*, 21 April, 1997, pages A1 & A9.

**54** See also: M.G., "What's Up with Canadian Science?," *C&EN*, letter, Vol. 92, Issue 2, 13 January 2014. https://cen.acs.org/articles/92/i2/s-Canadian-Science.html .
**55** Goudge Inquiry – Dr. Smith.
**56** Inquiry into Pediatric Forensic Pathology in Ontario, 2007–2008.
**57** http://en.wikipedia.org/wiki/Charles_Randal_Smith
**58** "Public register provides no explanations," *Toronto Star*, 28 February 2015, pages A25.
**59** See: Rachel Mendleson, *Toronto Star* [many reports re Motherisk]
http://www.thestar.com/authors.mendleson_rachel.html
**60** http://en.wikipedia.org/wiki/Motherisk
**61** *R v Broomfield*, 2014 ONCA 725. 21 October 2014. DOCKET: C52434.
https://www.canlii.org/en/on/onca/doc/2014/2014onca725/2014onca725.html?searchUrlHash=AAAAAQAJbW90aGVyaXNrAAAAAAE
**62** Rachel Mendleson, "Lawyers push for answers on drug test / Some refuse to allow clients to agree to hair analysis until concerns about lab addressed," *Toronto Star*, 18 November 2014, pages A1 and A18.
**63** Rachel Mendleson, *Toronto Star*.
http://www.thestar.com/authors.mendleson_rachel.html [30 November 2014.]

- Ontario's child advocate praises probe of drug hair tests 28 November 2014
- Ontario launches review of hair drug tests performed at Sick Kids 28 November 2014
- Health Minister Eric Hoskins punts to attorney general on use of Motherisk hair tests 24 November 2014
- Expert concedes limits to Motherisk's hair test for drugs 21 November 2014
- Sick Kids breaks silence about cocaine hair tests 20 November 2014–
- Reliability of hair drug tests up for debate 18 November 2014
- Motherisk concerns a 'wake-up call' for family lawyers 18 November 2014
- Trio of ministries defend Motherisk amid mounting concerns 13 November 2014
- Sick Kids says it will not review Motherisk hair analysis 11 November 2014
- Sick Kids defends reliability of tests amid calls for review 8 November 2014
- Sick Kids quiet amid growing concerns about hair tests 6 November 2014
- Lawyer seeks 'reassurance' on Motherisk drug tests 5 November 2014.

**64** Rachel Mendleson, "Patrol Targets Sick Kids Hair Tests / Provinces launches review of controversial Motherisk lab program after Star stories," *Toronto Star*, 28 November 2014, pages A1 and A19.
**65** Rachel Mendleson, "Motherisk review could affect cases across country / Child protection agencies in other provinces have relied on drug tests now under review," *Toronto Star*, 12 March 2015, pages A1 and A4.
**66** Rachel Mendleson, "Hair tests halted at controversial Sick Kids lab / Lawyer whose case sparked Star probe calls decision to suspend Motherisk program ' a relief' / Hospital lab standards not as strict as forensic lab," *Toronto Star*, 6 March 2015, pages A1 and A14.
**67** The Honourable Susan E. Lang, Independent Reviewer, Report of the Motherisk Hair Analysis Independent Review, Ontario Ministry of the Attorney General, 2015; ISBN 978-1-4606-7157-3 (PDF); ISBN 978-1-4606-7156-6 (HTM).
http://m-hair.ca/docs/default-source/default-document-library/motherisk_enbfb30b45b7f266c

c881aff0000960f99.pdf
https://www.ppsc-sppc.gc.ca/eng/pub/is-ip/ch6.html
**68** André Picard, "Motherisk Report / Sick Kids in need of some serious self-reflection," *The Globe and Mail*, 12 January 2016, page A11.
https://www.theglobeandmail.com/opinion/after-motherisk-report-sick-kids-needs-some-serious-self-reflection/article28117696/
**69** The Honourable Susan E. Lang, Independent Reviewer, Report of the Motherisk Hair Analysis Independent Review, Ontario Ministry of the Attorney General, 2015; ISBN 978-1-4606-7157-3 (PDF); ISBN 978-1-4606-7156-6 (HTM).
https://www.archives.gov.on.ca/en/e_records/motheriskcommission/the_motherisk_independent_review.html
**70** Eric S. Lander, "Fix the Flaws in Forensic Science / We are often dazzled by scientific 'evidence.' We shouldn't be," *The New York Times*, 21 April 2015, page A23.
**71** **"Motherisk" laboratory at Toronto Hospital for Sick Children:**
**72** http://en.wikipedia.org/wiki/Motherisk https://en.wikipedia.org/wiki/Gideon_Koren
**73** The Honourable Susan E. Lang, Independent Reviewer, Report of the Motherisk Hair Analysis Independent Review, Ontario Ministry of the Attorney General, 2015; ISBN 978-1-4606-7157-3 (PDF); ISBN 978-1-4606-7156-6 (HTM).
https://www.archives.gov.on.ca/en/e_records/motheriskcommission/the_motherisk_independent_review.html
**74** The Honourable Judith C. Beaman, Commissioner, *Harmful Impacts: The Reliance on Hair Testing in Child Protection Report of the Motherisk Commission*, February 2018.
https://www.ontario.ca/page/harmful-impacts-reliance-hair-testing-child-protection
**75** The Honourable Eileen E. Gillese, Commissioner, *Public Inquiry into the Safety and Security of Residents in the Long-Term Care Homes System REPORT*, Queen's Printer for Ontario, Toronto, 31 July 2019. ISBN 978-1-4868-3585-0 (PDF); ISBN 978-1-4868-3581-2 (Print)
Volume 1 – Executive Summary and Consolidated Recommendations
Volume 2 – A Systemic Inquiry into the Offences
Volume 3 – A Strategy for Safety
Volume 4 – The Inquiry Process.
**76** Volume 2 – A Systemic Inquiry into the Offences.
**77** https://wayback.archive-it.org/17275/20220405150521/https:/longtermcareinquiry.ca/en/
**78** Sandro Contenta, "Province vows 'immediate action' to make nursing homes safer after report into how Elizabeth Wettlaufer's killing spree went undetected / 'Your pain, loss and grief are not in vain'," *Toronto Star*, 1 August 2019, pages A1 and A9.
**79** Or, if not at the actual beginning, referenced from the beginning, or otherwise obviously displayed.
**80** Latin *de novo* = from the beginning. https://www.dictionary.com/browse/de-novo
**81** There are various definitions, descriptions, uses, practices and meanings of the concept of a signature. The description here represents the author's observation and interpretation of generally commonly used practice.
**82** See: Treasury Board of Canada Secretariat, "Government of Canada Guidance on Using Electronic Signatures," July 2019, August 2020, February 2021.
https://www.canada.ca/en/government/system/digital-government/online-security-privacy/government-canada-guidance-using-electronic-signatures.html#toc1

[. . .]

Whether a signature is paper-based or electronic, the fundamental purpose of the signature is the same. A signature links a person to a document (or transaction) and typically provides evidence of that person's intent to approve or to be legally bound by its contents. The primary function of a signature is to provide evidence of the signatory's:

- identity
- intent to sign
- agreement to be bound by the contents of the document

The requirement for a signature can be:

- imposed by an act of Parliament
- imposed by policy
- a customary practice

[. . .]

**83** See also below:

**84** See: Erin Schaff and Carl Hulse, "Insider's View of a Long Day / Drama and debate outside the hearing room doors as the House considers impeachment," *The New York Times*, 14 December 2019, pages A12 and A13.
https://www.nytimes.com/2019/12/12/us/politics/house-impeachment-vote.html

**85** https://en.wikipedia.org/wiki/Signature
https://legalbeagle.com/6283095-reference-dba-name-contract.html
https://legal-dictionary.thefreedictionary.com/signature
For example: US H. RES. "(Original Signature of Member)" – *The New York Times*, 11 December 2019, page A15.

**86** At page A13, bottom right: "After Democrats reviewed the proposed articles of impeachment one last time on Tuesday morning, Mr. Nadler officially signed off on them."

**87 House Judiciary Committee Chairman Jerrold Nadler (D-N.Y.) signs the articles of impeachment in his office on Capitol Hill on Tuesday, December 10, 2019. (Erin Schaff/The New York Times)**
EDS.: TWENTY-THIRD IN A SERIES OF 36 PHOTOS DEPICTING BEHIND THE SCENES OF THE IMPEACHMENT PROCESS – House Judiciary Committee Chairman Jerrold Nadler (D-N.Y.) signs the articles of impeachment in his office on Capitol Hill on Tuesday, December 10, 2019. Members of the House granted The New York Times rare access to photograph their impeachment preparations.
https://archive.reduxpictures.com/id/15301231

**88** https://prologue.blogs.archives.gov/2019/09/12/john-hancock-and-his-signature/ https://en.wikipedia.org/wiki/John_Hancock#:~:text=He%20is%20remembered%20for%20his%20large%20and,become%20a%20colloquialism%20for%20a%20person's%20signature.&text = He%20is%20primarily%20remembered%20by%20Americans%20for,United%20States%2C%20an%20informal%20synonym%20for%20signature

**89** https://prologue.blogs.archives.gov/2019/09/12/john-hancock-and-his-signature/
https://en.wikipedia.org/wiki/John_Hancock#:~:text=Signing%20the%20Declaration,-Hancock's%20signature%20on&text = Hancock%20was%20president%20of%20Congress,an%20informal%20synonym%20for%20signature.

**90** https://en.m.wikipedia.org/wiki/Boilerplate_text

**91** = as a matter of form https://en.wikipedia.org/wiki/Pro_forma

**92** https://niryo.com/tasks-your-company-delegate-to-a-robot/

**93** https://en.wikipedia.org/wiki/Three_Laws_of_Robotics
https://en.wikipedia.org/wiki/I,_Robot

**94** https://www.yourdictionary.com/mind-the-store
https://dictionary.cambridge.org/us/dictionary/english/mind-the-store
https://www.merriam-webster.com/dictionary/mind%20the%20store
**95** For example: The AUTHOR CONTRACT for this book's first edition, DE GRUYTER, Berlin, 23 July 2020, was by-way of DocuSign. No paper; no pen; no ink. https://www.docusign.ca/.
**96** See, for example: Ontario *Electronic Commerce Act, 2000,* S.O. 2000, c. 17
https://www.ontario.ca/laws/statute/00e17#act-verion
**97** See also: **Pete Mitton,** CBC "*The Sunday Magazine* with Piya Chattopadhyay / Old fashioned signatures are disappearing. What are we signing away?"
https://www.cbc.ca/listen/live-radio/1-57-the-sunday-magazine/clip/16156356-old-fashioned-signatures-disappearing.-what-signing-away
**98** See, for example: Leif Weatherby, "A.I. Killed the Math Brain," *New York Times*, OPINION / GUEST ESSAY, 2 June 2025. https://www.nytimes.com/2025/06/02/opinion/ai-coders-jobs.html
**99** Yoshua Bengio, *LawZero / Safe AI for Humanity*, 3 June 2025. https://yoshuabengio.org/
**100** Muriel Draaisma, "An Ontario judge tossed a court filing seemingly written with AI. Experts say it's a growing problem / Artificial intelligence can produce false information that can make its way into court submissions," CBC News, Toronto, 03 June 03 2025. ". . . Fake cases, known as AI hallucinations, can make their way into legal submissions if a lawyer doesn't take additional steps to make sure the cases actually exist, says Amy Salyzyn, an associate professor at the University of Ottawa's faculty of law. . . ."
https://www.cbc.ca/news/canada/toronto/artificial-intelligence-legal-research-problems-1.7550358
https://www.youtube.com/watch?v=QA2zBW2rUJc Amy Salyzyn, YouTube
**101** See, for example: Charlie Savage &Tyler Pager, "*Biden Says He Made the Clemency Decisions That Were Recorded With Autopen* / Donald J. Trump and his allies have begun investigations to support their claims that Joseph R. Biden Jr. was incapacitated and his staff conspired to take presidential actions in his name," *New York Times*, 13 July 2025.
https://www.nytimes.com/2025/07/13/us/politics/biden-pardon-autopen-trump.html
**102** *Gedankenexperiment*. Introduction 2.2.
**103** [][] [][][][] 202[], [][]:[][] []m. EST.
day month year hour minute am. or pm. time zone.
**104** Municipal address: [][][][] [][][][][][][][][],
**105** [][][][][][][][][][][], [][][][][][][][][], [][][][][][][]. [][][] [][][]
Geopolitical data: City or local jurisdiction. Province or State. Dominion, Kingdom or Republic. Postal code or ZIP. (ZIP = Zone Improvement Plan; USA postal code, 1963. https://en.wikipedia.org/wiki/ZIP_Code)
**106** *Gedankenexperiment*. Introduction 2.2.
**107** *Gedankenexperiment*. Introduction 2.2.
**108** *Gedankenexperiment*. Introduction 2.2.
**109** *Gedankenexperiment*. Introduction 2.2.
**110** *Gedankenexperiment*. Introduction 2.2.
**111** [][][][][] & [][][][][], "[][][][][][][][][][][]," *[][][][][][][][][]*, Vol.[][][], No.[][], pages [][][]→[][][], [][] [][][] 19[][]. = authors article title journal date.
**112** *Gedankenexperiment*. Introduction 2.2.
**113** *Gedankenexperiment*. Introduction 2.2.
**114** *Gedankenexperiment*. Introduction 2.2.
**115** *Gedankenexperiment*. Introduction 2.2.
**116** The error report here is of a qualitative process. Quantitative reports would also require some statistical analysis.
**117** *Gedankenexperiment*. Introduction 2.2.

**118** “prepared” – typo?? Should have been some version of “compared” ?
**119** *Gedankenexperiment.* Introduction 2.2.
**120** E R Menzel, “Recent advances in photoluminescence detection of fingerprints,” *Scientific World Journal*, 2001 Oct 2;1:498–509. doi: 10.1100/tsw.2001.76. https://pubmed.ncbi.nlm.nih.gov/12805842/
**121** Menzel, *Fingerprint Detection with Lasers*, Second Edition, CRC Press, Mar. 3, 1999 - https://scholar.google.ca/scholar?q=Menzel,+Fingerprint+Detection+with+Lasers,+Second+Edition,+CRC+Press,+Mar.+3,+1999&hl=en&as_sdt=0&as_vis=1&oi=scholart
**122** K.M. Beesley, S. Damaskinos, A.E. Dixon, “Fingerprint Imaging with a Confocal Scanning Laser Macroscope,” Journal of Forensic Sciences Volume:40 Issue: January 1995, pages:10→17. https://asmedigitalcollection.asme.org/forensicsciences/article-abstract/40/1/10/1183411/Fingerprint-Imaging-with-a-Confocal-Scanning-Laser?redirectedFrom=fulltext
**123** *Gedankenexperiment.* Introduction 2.2.
**124** *Gedankenexperiment.* Introduction 2.2.
**125** Tom Waits, Benjamin Heyward Allen, Elijah Abbott Jones, *Used Songs* [album], 1973–1980, “Step Right Up.” http://www.tomwaits.com/songs/song/322/Step_Right_Up/
**126** *Job*, 1:21, Bible, King James Version: “The Lord giveth and the Lord taketh away.”
**127** *Job* Chapter 1
**21** And he said; naked came I out of my mother’s womb, and naked shall I return thither; the LORD gave, and the LORD hath taken away; blessed be the name of the LORD.
https://www.mechon-mamre.org/p/pt/pt2701.htm
**128** Typically, published books variously declare and restrict re –

- copyright,
- copying re-sale, *etc.* including in other binding or cover,
- fictional characters not being actual people,
- declaration of authorship,
- recycling from wood grown in sustainable forests,
- if purchased without a cover the book is stolen,
- cataloguing, ISBN,
- manufactured in USA,
- *etc.*

-- each likely important to someone; often not the book’s reader.
**129** See reverse of the title page in this book too.
**130** Chemicals and Materials, “What is a $LD_{50}$ and $LC_{50}$?” Canadian Centre for Occupational Health and Safety, Hamilton, Ontario, 10 May 2024. https://www.ccohs.ca/oshanswers/chemicals/ld50.html

[. . .]

> **Disclaimer**
> Although every effort is made to ensure the accuracy, currency and completeness of the information, CCOHS does not guarantee, warrant, represent or undertake that the information provided is correct, accurate or current. CCOHS is not liable for any loss, claim, or demand arising directly or indirectly from any use or reliance upon the information.

[. . .]

**131** Health and Safety Programs, “Scent-Free Policy for the Workplace” Canadian Centre for Occupational Health and Safety, Hamilton, Ontario, 29 August 2024.
https://www.ccohs.ca/oshanswers/hsprograms/scent_free.html
https://www.ccohs.ca/products/posters/scentfree

[. . .]

**Disclaimer**

Although every effort is made to ensure the accuracy, currency and completeness of the information, CCOHS does not guarantee, warrant, represent or undertake that the information provided is correct, accurate or current. CCOHS is not liable for any loss, claim, or demand arising directly or indirectly from any use or reliance upon the information.

[. . .]

132

[. . .]

**2018 Mortgage Statement**
**Summary of transactions from**
**January 1, 2018 to December 31, 2018**

[. . .]

E&OE

[. . .]

**133** E&OE = Errors and Omissions Excepted.
https://en.wikipedia.org/wiki/Errors_and_omissions_excepted https://dictionary.cambridge.org/dictionary/english/e-oe.
**134** See notes below:
**135** Food wrappers, dry cleaner tickets, websites, . . .
**136** Legal disclaimer. MD. Chapters 9 and 11.
**137** *GUIDE TO THE DRUGS DIRECTORATE LABORATORY ACTIVTIES QUALITY ASSURANCE PROGRAM*; Minister of National Health & Welfare Canada, Minister of Supply & Services Canada; February 1991; Catalogue No. H42-2/26-1990; ISBN 0-662-57250-5.
**138** **[**inside front cover page – reverse:**]**
This report was prepared by the Drugs Directorate, Health Protection Branch, Department of National Health and Welfare, for the guidance of work in its own laboratories. The views and recommendations contained herein are those of the author(s) and do not necessarily constitute endorsement by the Department.
**139** E&OE. Chapter 9.
**140** QTRAP® 4500 LC-MS/MS System

[. . .]

The SCIEX clinical diagnostic portfolio is For *In Vitro* Diagnostic Use. Rx Only. Product(s) not available in all countries. For information on availability, please contact your local sales representative or refer to https://sciex.com/diagnostics. All other products are For Research Use Only. Not for use in Diagnostic Procedures.

Trademarks and/or registered trademarks mentioned herein, including associated logos, are the property of AB Sciex Pte. Ltd. or their respective owners in the United States and/or certain other countries.

© 2020 DH Tech. Dev. Pte. Ltd.

[. . .]

**141** https://sciex.com/products/mass-spectrometers/qtrap-systems/qtrap-4500-system

**142** SCIEX 3200 QTRAP® System
https://sciex.com/products/mass-spectrometers/qtrap-systems/qtrap-3200-system
https://sciex.com/search-results?term=3200
**143** SCIEX, Framingham, Massachusetts, 01701.
https://sciex.com/products/mass-spectrometers/qtrap-systems/qtrap-5500-system
**144** https://sciex.com/products/mass-spectrometers/qtrap-systems/qtrap-5500-system
AB Sciex is doing business as SCIEX. © 2010–2017 AB Sciex. The trademarks mentioned herein are the property of the AB Sciex Pte. Ltd. or their respective owners. AB SCIEX™ is being used under license. Beckman Coulter® is being used under license. For research use only. Not for use in diagnostic procedures.
**145** US Food and Drug Administration, Silver Spring, Maryland, 20993.
https://www.fda.gov/medicaldevices/deviceregulationandguidance/overview/classifyyourdevice/ucm051512.htm
**146** Arguably, a forensic lab analysing for drugs in blood drawn by medical personnel for an impaired by drug case would be doing a kind of medical diagnosis; and forensic labs would usually not be doing research, within the meaning of how SCIEX instrumentation would usually be used. If instrumentation with such a requirement would be used, that should be stated in the forensic report, along with an explanation of what it means for the forensic study, and the consequences.
**147** https://en.m.wikipedia.org/wiki/Boilerplate_text
**148** *Gedankenexperiment.* Introduction 2.2.
**149** See note above.
**150** https://www.olg.ca/content/dam/olg/web/product/instants/product/chrome-2154/Chrome_2154_Actuals.pdf
https://www.olg.ca/en/instants/play-the-bigger-spin-3148/how-to-play.html
https://www.olg.ca/en/instants/play-wild-side-2239.html .
**151** E&OE never seen by the author in a forensic report.
**152** *Gedankenexperiment.* Introduction 2.2.
**153** Latin – *de novo* = from new; from the beginning; anew; starting from the beginning.
**154** Middle English, Latin – *gratis* = without monetary charge; free; no payment required. Here = no copyright restriction, except as indicated.
**155** *Gedankenexperiment.* Introduction 2.2.
**156** *Gedankenexperiment.* Introduction 2.2.
**157** *Gedankenexperiment.* Introduction 2.2.
**158** *Gedankenexperiment.* Introduction 2.2.
**159** Notwithstanding his humorous suggestion, in his own biographical synopsis, *Asimov's Biographical Encyclopedia of Science and Technology / Second Revised Edition,* Doubleday & Company, Inc., Garden City, New York, 1982; ISBN 0-385-17771-2: ABOUT THE AUTHOR, Isaac Asimov; at page 900: "**[. . .]** autobiography – the self-indulgent product of a man who cheerfully admits that nothing much has ever happened to him, but insists he can hide that fact by clever writing. . . .".
**160** George Gamow, *One Two Three . . . Infinity / Facts and Speculations of Science*, Bantam Books, New York City, 10016, April 1967; copyright 1947, 1961 by the author; page 142, *et seq.*
**161** George Gamow, *Thirty Years that Shook Physics / The Story of Quantum Theory*, Anchor Books, Doubleday & Company, Inc., Garden City, New York, 1966; pages 13, 14, 15.
**162** Jane Koustas, *Robert Lepage on the Toronto Stage: Language, Identity, Nation*, McGill-Queen's Press, 2016; ISBN 978-0-7735-4674-5; "Notes / INTRODUCTION . . . 8 Indeed, some may recall Pierre Trudeau's quip during a campaign in British Columbia. When a heckler complained that he was tired of seeing French on his cereal box every morning, Trudeau advised him to "turn the box around," suggesting yet another way of redeploying the box! . . ."
https://books.google.ca/books?id = 0UVyDAAAQBAJ&pg = PA153&lpg = PA153&dq = trudeau + turn + the

+ cereal + box + around&source = bl&ots = NeEUrGZg7w&sig = ACfU3U1IoYx96s88xdMekCgQ_27xquB9-Q&hl = en&sa = X&ved = 2ahUKEwjmtM6lqsHpAhUZK80KHR39DZIQ6AEwAXoECAoQAQ#v = onepage&q = trudeau%20turn%20the%20cereal%20box%20around&f = false
https://archive.macleans.ca/article/1979/12/3/rip-a-flaming-failure-fuelled-by-a-cerealbox

**163** https://www.theglobeandmail.com/news/national/new-bilingualism-taking-hold-in-canada/article4650408/
https://nationalpost.com/opinion/conrad-black-pierre-trudeau-was-a-dilettante-but-a-successful-one
https://en.wikipedia.org/wiki/Official_bilingualism_in_Canada
https://www.amazon.ca/Oh-Canada-Quebec-Requiem-Divided/dp/0140168176.

**164** *Gedankenexperiment.* Introduction 2.2.

**165** *The Adventures of Buckaroo Banzai Across The 8th Dimension*; 1984; W.D. Richter, Director; with Peter Weller, Ellen Barkin, John Lithgow, Jeff Goldblum, Christopher Lloyd, et al.; Sherwood Productions and Twentieth Century Fox; USA.
https://en.wikipedia.org/wiki/The_Adventures_of_Buckaroo_Banzai_Across_the_8th_Dimension
http://img.stanleylieber.com/src/14530/img/1346293164.png
". . . DECLARATION OF WAR: The Short Form . . . Date ____ . . . . Name of Enemy _______ . . ."

**166** *Gedankenexperiment.* Introduction 2.2.

**167** Readers are cautioned that this is a creation of the author, for illustration purposes, that has not been reality- or lab-tested. Some of it might be regarded as too fanciful.

**168** [][] [][][][] 202[], [][]:[][] []m. EST.
day month year hour minute am.or pm. time zone.

**169** Municipal address: [][][][] [][][][][][][][][],

**170** [][][][][][][][][][], [][][][][][][][], [][][][][][]. [][][] [][][]
Geopolitical data: City or local jurisdiction. Province or State. Dominion, Kingdom or Republic. Postal code or ZIP.

# Chapter 10
# Essay – biased opinion – polemic: Canada *Controlled Drugs & Substances Act – ipse dixit*

This chapter[1] addresses some forensic issues of the war on drugs [1, 2][2,3,4,5,6,7,8]:

[. . .][9]

*It is* [US] *constitutional black letter law. To obtain a criminal conviction, the prosecution must prove every element of the offense, by proof beyond a reasonable doubt. The Constitution* [Sixth Amendment [3]] *entitles a defendant to confront and cross-examine all witnesses against him. Yet, for the past thirty years, state legislatures have quietly approved laws that cheat the Constitution. These laws fly, undetected, beneath the constitutional radar, violating fundamental constitutional rights.*

*. . . these . . .* [are characterized] *as forensic* ipse dixit *statutes, because the bare assertion of an uncrossexamined state witness becomes,* ipse dixit, *an adjudicated fact. . . . The forensic* ipse dixit *statutes deprive defendants of the right to confrontation and relieve the government of its burden of proof. . . . also discourage vigorous defense advocacy, promote carelessness and fraud in crime laboratories, and increase the likelihood of wrongful convictions and sentences . . . .*

[. . .]

[Pamela R. Metzger, Tulane University, 2006]. [4, 5, 6, 7][10,11,12]

[. . .]

*Confrontation is designed to weed out not only the fraudulent analyst, but the incompetent one as well. Serious deficiencies have been found in the forensic evidence used in criminal trials. One commentator asserts that "[t]he legal community now concedes, with varying degrees of urgency, that our system produces erroneous convictions based on discredited forensics". Metzger . . . One study of cases in*

---

**1** Essay – biased opinion – polemic. Introduction 4.3; Chapters 10 and 11.

**2** *CD&S Act*. Chapters 1 and 10.

**3** Polemic. Health Canada. Introduction 4; Chapters 9 and 10.

**4** Author's biases. Introduction 1.2, 4.1 and 4.3; Chapters 10 and 11.

**5** Legislative process – policy-making. Introduction 4.3; Chapters 1, 10 and 11.

**6** Ontario Court proceedings, May and June 2018. Chapter 10; Chapter 10 Appendix.

**7** War on drugs. *CD&A Act*. Mass (grams) measurement. *W&M Act*. Chapters 3, 6, 8 and 10.

**8** *CD&S Act – ipse dixit*. Chapter 10; Chapter 10 Appendix.

**9** Introductory quotes. Introduction 1.2.

**10** *Melendez-Diaz*. Justice Scalia. US Constitution confrontation clause. *ipse dixit*. Tulane professor Metzger. Chapters 10 and 11.

**11** *Criminal law. Regulatory/administrative law*. Proof *beyond a reasonable doubt* test. Chapters 1.4.2.1 and 1.4.2.2; Chapters 10 and 11.

**12** Professor Pamela R. Metzger, *Vanderbilt Law Review*. Copyright acknowledgment. Chapter 10.

https://doi.org/10.1515/9783111618555-011

*which exonerating evidence resulted in the overturning of criminal convictions concluded that invalid forensic testimony contributed to the convictions in 60% of the cases. Garrett & Neufeld . . .*

[. . .]

[Justice Antonin Scalia, *Melendez-Diaz v Massachusetts*, US Supreme Court, *2009*.] [8][13]

[. . .]

*"When I use a word," Humpty Dumpty said, in a rather scornful tone, "it means just what I chose it to mean – neither more nor less."*

*"The question is," said Alice, "whether you* can *make words mean so many different things."*

*"The question is," said Humpty Dumpty, "which is to be the master – that's all."*

*". . . They've a temper, some of them – particularly verbs: they're the proudest adjectives you can do anything with, but not verbs – however* I *can manage the whole lot of them! Impenetrability! That's what* I *say!"* (see Fig. 10.1) [9, 10, 11, 12, 13][14]

[. . .]

Fig. 10.1: Alice and Humpty Dumpty.

[. . .]

*Where, in proceedings under subsection (1), a court concludes that evidence was obtained in a manner that infringed or denied any rights or freedoms guaranteed by this* Charter, *the evidence*

13 *Melendez-Diaz*. Justice Scalia. US Constitution confrontation clause. *ipse dixit*. Tulane professor Metzger. Chapters 10 and 11.

14 *Alice and Humpty Dumpty*. Chapter 10.

*shall be excluded if it is established that, having regard to all the circumstances, the admission of it in the proceedings would bring the administration of justice into disrepute.*

[. . .]

[Canadian *Charter* s. 24(2).] [14]

## 10.1 War on drugs

In the United States, prior to the beginning of 2025, the criminal law war on drugs of abuse and recreation that still continued, even as politics and policies to which it is related – from the Nixon, Rockefeller and Reagan eras – slowly seemed to become, in a bipartisan way, somewhat, sometimes, going towards an apparent decline. There seemed to have been, amongst some in government, a belated realization that a healthcare treatment approach might be better. Both the decriminalization of marijuana in a few states (federal prohibition remaining) and the opioid crisis affected attitudes [15, 16, 17, 18]. But, at the start of the second Trump administration, 2025, such cautious generalizations become lost [19]. Most particularly for fentanyl [20, 21, 22, 23, 24, 25, 26].

The situation is similar in Canada. At least, prior to the beginning of 2025, notwithstanding the previous decriminalization of marijuana, and a serious opioid crisis too, Canada retained, in policy and practice, *criminalized molecular offences* [27].[15] Perhaps, sometimes, pursued somewhat more mildly, and with drug treatment-courts, and supervised injection sites; but criminal law enforcement, court judgement and imprisonment still continued. For example, it remains a crime to possess cocaine[16,17] and a worse crime to trade in it [28]. Police stings, and police arrangements with criminals [29], can still result in convictions. And, in 2025, in apparent response to President Tump – a Canadian *fentanyl czar* [30, 31, 32] was named. *Czar* – apparently previously-not-too-well-known terminology of Canadian legal administration.

The criminal law of concern for this polemic is the Canada federal ***Controlled Drugs & Substances Act*** [33, 34]; its relevant case law and government practice.

But there is an aspect of Canadian government policy, beyond the law as it now is, that should be considered. It is in that obscure corner – the law of evidence and forensic analytical chemistry – for argument here.[18]

Lawyers' defence offerings should include evaluating the analytical chemistry of the government's lab process. This does not happen (". . . What, never?/No, never!/What, never?/Well, hardly ever! . . .") [35]; the law severely discourages it.

**15** Statutes. Regulations. Criminalized molecular offences. Chapters 1 and 10.

**16** Cocaine. CAS 53-21-4 CAS 50-36-2. Chapters 4, 6 and 10.

**17** Edmond Locard. Sherlock Holmes. Cocaine. Chapters 2, 3, 4, 6 and 10.

**18** Evidence. Section 1.5; Chapter 10.

Even accepting, for the sake of hypothetical argument, that the war on drugs[19] is wise – that it should be enforced and prosecuted vigorously, and harshly for deterrence, and for moral purpose – it should not be done, as Canada does, with bad and unethical science. Nor should it be done without protections to which US accused are *Sixth Amendment*-entitled – the right to confront your accuser, including the chemist who says it's a criminally forbidden drug [36, 37, 38, 39, 40, 41].[20,21]

Mostly of concern in this polemic is the defective way that the purported scientific proof of criminally forbidden substance is reported by the government to the accused, and presented in the law courts. So defective that it would be better described as *not* reported under legally provided guise. A *proper scientific report* is not available and might not even exist.[22]

## 10.2 Quest for a *proper scientific report*

The importance of a *proper scientific report* is presented in Chapter 9 [42].[23]

**That a *proper scientific report*, related to enforcement of the *CD&S Act*, is not available – or even existent – and maybe never has been, is the central issue of this polemic and is the essential reason to opine that the government of Canada, in this regard, does bad science.[24] This is mostly an issue of policy rather than law.** And largely of science rather than crime.

That a *proper scientific report* cannot be found by defence counsel and writer of this polemic requires explanation as to how it was sought. The *Certificate of Analyst*, produced by Health Canada – the national health agency – and presented by federal Crown counsel to the court as proof of substance, is not a *proper scientific report*. It could and should be, but it is not – apparently by government intention.

The *CD&S Act* has been used as Canada's criminal law war on drugs instrument [43]. *CD&S Act* s. 51 [44, 45][25] alters the normal law of evidence [46][26] so that the Crown's mere presentation of a *Certificate of Analyst* [47, 48][27] is sufficient to support conviction:

19 War on drugs – Canada. Chapters 5.3.4.1 and 10.1.

20 US *Sixth Amendment*. Chapter 10.

21 *Melendez-Diaz*. Justice Scalia. US Constitution confrontation clause. *ipse dixit*. *Tulane professor Metzger*. Chapters 10 and 11.

22 Some science controversy remains. Introduction 4; Chapters 1, 2, 10 and 11.

23 *Proper scientific report*. *CD&S Act*. Polemic. Health Canada. Introduction 4; Chapters 9 and 10.

24 Science – good, bad and junk. Introduction 4; Chapters 1, 2, 5, 8 and 10.

25 Drug law. Chapters 1 and 10.

26 Identification of molecules. *CD&S Act*. Chapters 4 and 10.

27 Health Canada DAS Certificate of *Analyst*. *CD&S Act* s. 51. Chapters 9 and 10.

> [. . .]
>
> . . . a certificate . . . by an analyst . . . is admissible in evidence . . . and, in the absence of evidence to the contrary, is proof of the statements set out . . ., without proof of the signature or official character of the person appearing to have signed it.
>
> [. . .]

No explanation of science is required at Canadian law, and it is government practice and apparent policy to not provide it. No *Sixth Amendment* jurisdiction here.[28] **Parliament is apparently constitutionally, and at law otherwise, entitled to enact anti-science legislation – and has done so – s. 51; and the government continues to take-up on this with bad science.**[29]

*CD&S Act* s. 51 relieves judges in Canadian courts of some jurisdiction in matters of fact and law, in favour of government-appointed *Analysts*; Parliament intended it so. It is also apparent federal government policy to remove quality-of-science issues from *CD&S Act* considerations. The *CD&S Act* [49] and regulations [50] effectively allow that whatever the *Analyst* might know of analytical chemistry, or do, need not be revealed.

It should be noticed that nothing that would be kept secret by way of *CD&S Act* s. 51 would either involve national security nor trade secret concerns. Nor would any documentation be expected to reveal other than routine activity of a modern analytical chemistry laboratory.

Non-disclosure of science, and science-based procedure, and non-explanation, are achieved by:

- The *Certificate of Analyst* does not describe the analytical chemistry beyond declaring a result.[30] Neither CAS numbers nor molecular structural diagrams ever appear on any *Certificate.*
- The Crown sometimes does not respond to data requests.
- There is never revelation of a *proper scientific report*, and no indication that such a report would exist (in contrast with a *FIRE INVESTIGATION REPORT* [51] and a CFS *REPORT* [52],[31,32] both routinely disclosed in other kinds of cases). It is a reasonable speculation that for the *CD&S Act* a *proper scientific report* does not exist; and perhaps never has in years of previous prosecutions.
- There is sometimes no revelation of forensic data or documents, after being requested. However, occasionally, there would be some, which although somewhat helpful (for guessing), does not tell a full-enough, or sufficiently coherent, story to be a *proper scientific report.*

---

**28** US *Sixth Amendment.* Chapter 10.

**29** Science – good, bad and junk. Introduction 4; Chapters 1, 2, 5, 8, 10 and 11.

**30** Health Canada DAS Certificate of *Analyst.* Chapters 9 and 10.

**31** Health Canada DAS Certificate of *Analyst.* Chapters 9 and 10.

**32** Fire Marshall. CFS. Chapters 9 and 10.

- Legislation [53] is used in a way that effectively denies access, encouraging opaqueness. Notwithstanding its public statement [54], Health Canada seems to be a pioneer in opaqueness [55, 56, 57].[33]
- The federal freedom-of-information regime [58] is not designed for such disclosure. FOI sometimes may be used to obfuscate, delay, and for the government to hide behind as a ruse [59, 60, 61, 62, 63, 64, 65].[34,35,36]

The quest for a *proper scientific report*, by pursuing each of these non-disclosure achievements, is an adventure – perhaps something like a kindergarten treasure hunt – with obstacles more difficult than in *Pet Shop* [66] or *Cheese Shop* [67].

Invoking the *Canadian Charter of Rights and Freedoms* [68, 69, 70] invites further adventure:

- A *Charter* s. 7 "security of the person" [71, 72] application would be problematic because of the very nature of *Charter* litigation. Citizens' purported rights can be in convoluted language, with escape clauses that hinder enforcement; and the not-withstanding clause. Humpty Dumpty might well be consulted.[37] There also appears to be government and court reluctance to allow the *Charter* for positive rights advancement. And, *Charter* litigation is expensive and protracted – it is neither for the poor nor for those in a hurry to be somewhere else, and people detained pre-trial in jail without bail tend to be both.

Another adventure:

- Obtaining, now, a copy of the *Griffiths* report[38]; that followed a quality assurance calamity of some years-ago, but did not actually delve much into QA.

The quest for a *proper scientific report*, and related data, should not be such adventures, effectively blocked by law apparently intended to block.

[. . .][39]

וְלִפְנֵי עִוֵּר לֹא תִתֵּן מִכְשֹׁל וְיָרֵאתָ מֵּאֱלֹהֶיךָ . . . Before the blind, do not put a stumbling block . . .

[. . .]

[*Leviticus* Chapter 19 [73, 74] [וַיִּקְרָא]

33 Transparency. Opaqueness. Chapters 2, 5 and 10.

34 Obfuscation. Introduction 4. Chapter 10.

35 Health Canada. Introduction 4.3; Chapters 8 and 10.

36 Freedom of info. legislation. Chapters 5, 8 and 10.

37 *Alice and Humpty Dumpty*. Chapter 10.

38 Fraud. Griffiths Report, 2001. QA failure. Gemma Ramlal, 2001. Annie Doukhan, 2012. Introduction 4.3; Chapters 8, 9 and 10.

39 Introductory quotes. Introduction top of first page.

## 10.3 Bad science – *ipse dixit*

Parliament and the government have something of a tradition with bad science [75, 76, 77, 78, 79, 80, 81][40,41,42,43,44,45,46,47] – perhaps pioneered with the Health Canada Drug Analysis Service under the *CD&S Act*. With *CD&S Act* s. 51, DAS need not reveal its analytical chemistry, and apparent federal government policy is to not reveal it in a meaningful and coherent way. Nor does the government seem to care.[48] *CD&S Act Certificates of Analyst*, science unexplained, and ISO out-of-compliance[49,50] are the essential evidence for conviction and imprisonment. Even the guilty deserve better.

Tulane University law professor Pamela R. Metzger[51] in "Cheating the Constitution", [82] develops an analysis of "forensic *ipse dixit* statutes" – in her opinion *Sixth Amendment* [83] unconstitutional. (Her paper is referred to by Justice Antonin Scalia in *Melendez-Diaz v Massachusetts* [84].[52]) Canadian *CD&S Act* s. 51 is *ipse dixit* [85], without *Sixth Amendment* remedy.

Included with the bad science tradition was a serious quality assurance failure that came to public attention some years ago[53] – within the activities of a DAS *Analyst*. While that *Analyst* apparently then ceased government employment, there were apparently no publicly reported serious consequences for senior officials on whose watches it happened; perhaps the ministers of the Crown might have tried to explain why not resigning. The resulting *Griffiths* report inadequately addressed QA. Comparison with a similar occurrence, years later, in the Commonwealth of Massachusetts is appropriate – the chemist there eventually served prison time, and the lab was shut down. With the secrecy-enabling *CD&S Act* s. 51 still operative, the same conditions that lead to *Griffiths* would appear to remain today.

---

**40** Author's biases. Introduction 1.2, 2, 4.1 and 4.3.
**41** Science – good, bad and junk. Introduction 4, Chapters 1, 2, 5, 8, 10 and 11.
**42** Re criticism of Health Canada, Judges et al. Introduction 4.1 and 4.3, Chapters 10 and 11.
**43** US *Sixth* & *Fourteenth* Amendment. Confrontation of witness. Chemist. *Melendez-Diaz*. Section 10.3.
**44** US *First* Amendment. Introduction top of first page; Chapter 10.3.
**45** *Grundgesetz für die Bundesrepublik Deutschland* Artikel 5. Introduction top of first page; Chapter 10.3.
**46** DAS and bad science. Introduction 4.1; Chapters 10 and 11.
**47** Science – good, bad and junk. Introduction 4; Chapters 1, 2, 8, 10 and 11.
**48** Health Canada. DAS. Ethics. Introduction 4.3; Chapters 5 and 10.
**49** **Alert**: But see Catch-22 – below.
**50** DAS. Catch-22. Chapters 10 and 11.
**51** DAS. Catch-22. Chapters 10 and 11.
**52** *Melendez-Diaz*. Justice Scalia. US Constitution confrontation clause. ipse dixit. *Tulane professor Metzger*. Chapters 10 and 11.
**53** Fraud. Griffiths Report, 2001. QA failure. Gemma Ramlal, 2001. Annie Doukhan, 2012. Introduction 4.3; Chapters 8, 9 and 10.

A *CD&S Act*-accused person, in modern Canada, remains in jeopardy of imprisonment because of defective lab process, with a reasonable apprehension of at least some defect somewhere:

There are anecdotal reports of serious forensic lab flaws from Ontario [86, 87, 88, 89], British Columbia [90, 91] and Massachusetts [92, 93, 94] in addition to commentary [95] and cited references [96, 97], in *Melendez-Diaz v Massachusetts* [98], with the opinion of Justice Antonin Scalia [99] for the US Supreme Court [100, 101, 102, 103, 104, 105, 106, 107].

Justice Scalia refers [108] to Pamela R. Metzger, "Cheating the Constitution".[54] Although not specifically mentioned by Justice Scalia, Tulane University law professor Metzger [109] developed an analysis of "forensic *ipse dixit* statutes" – in her opinion *Sixth Amendment* is unconstitutional [110] in the United States.[55] *CD&S Act* s. 51, as Canadian forensic *ipse dixit* legislation, without *Sixth Amendment* remedy, should be unconstitutional too – although for different reasons in the context of Canadian law.

An example of the related "principles of fundamental justice", of *Canadian Charter* s. 7, may be found in *Melendez-Diaz v Massachusetts*,[56] reaffirmed and extended in *Bullcoming v New Mexico* [111, 112], wherein the US Supreme Court interprets the *Sixth* [113] (and *Fourteenth*) [114] *Amendment* – "[i]n all criminal prosecutions, the accused shall enjoy the right . . . to be confronted with the witnesses against him" [115][57] – to include the government analytical chemist. Thus, an accused has, in effect, a right to seek reports, data and explanation by the analytical chemist who determined the molecule of concern. That right might be achieved by *Sixth Amendment* confrontation in the United States; there is apparently no Canadian equivalent.

To be scientifically accepted, a scientific declaration is published so that it that can be questioned, corrected and improved-on or maybe abandoned.[58] This has been usual practice for at least the more than three centuries since the publication of *Principia*, by Isaac Newton, encouraged by Edmond Halley [116].[59] The Parliament of Canada seems to have missed this.

*CD&S Act* s. 51 allows the Canadian federal state to "prove" drugs, merely on the signature of a shrouded government appointee, by way of a Health Canada *Certificate of Analyst*, without explaining the forensic analytical chemistry. This is bad science –

54 *Melendez-Diaz*. Justice Scalia. US Constitution confrontation clause. *ipse dixit*. Tulane professor Metzger. Chapters 10 and 11.

55 *Melendez-Diaz*. Justice Scalia. US Constitution confrontation clause. *ipse dixit*. Tulane professor Metzger. Chapters 10 and 11.

56 *Melendez-Diaz*. Justice Scalia. US Constitution confrontation clause. *ipse dixit*. Tulane professor Metzger. Chapters 10 and 11.

57 US *Sixth* & *Fourteenth* Amendment. Confrontation of witness. *Melendez-Diaz*. Chemist. Introduction; Chapter 10.3.

58 Science. DRE. Chapters 2 and 10.

59 Newton. Haley. *Principia*. Chapters 9–11.

of a policy that is anti-science – because science (like democracy) requires explanation to be presented for argument by everyone.[60]

That the *Certificate* might usually turn out to offer some correct data does not fix the problem. It is not a *proper scientific report*; it does not explain what was done (although further data that is sometimes provided can be helpful for guessing).

**The absence of explanation is of itself the bad science**. The Canadian government should not do bad science, and Parliament should not allow and encourage it; especially for a national health agency. That people are sent to prison on this basis makes it worse.

The normal law of evidence is short-circuited by *CD&S Act* s. 51. S. 51 should be repealed.[61]

Such repeal might be expensive, but, by analogy, comment by Justice Scalia, in *Melendez-Diaz*, is instructive: "The [US Constitution] confrontation clause may make the prosecution of criminals more burdensome, but that is equally true of the right to trial by jury and the privilege against self-incrimination."

## 10.4 ISO out of compliance? and Catch-22

### 10.4.1 ISO out of compliance?

Every *Certificate of Analyst* issued by Health Canada would appear as obviously out-of-compliance,[62] as a "test report", with its lab's accreditation [117, 118, 119, 120, 121], as according to ISO/IEC 17025:2005 [122] Clause 5.10.2 e – "identification of the method used" [123, 124] is missing.

And, would be further out of compliance in that neither Health Canada, nor the federal Crown office as its *CD&S Act* gatekeeper, are willing to engage in science-based conversation to explain – as would appear required by Clause 5.10.1 [125].

That the Standards Council of Canada – the "national accreditation body" [126] – would not agree with these assessments is a further problem that should be examined. It is a rather serious problem because engineering and laboratory standards credibility is so important to the national economy.

**60** Explanation. Method validation. Introduction 4.1; Chapters 5.3.2 and 10.3.

**61** Evidence. Chapters 1 and 10.

**62** **Alert**: But see below.

### 10.4.2 Catch-22

However, in regard to the *Certificates*, as apparently ISO out-of-compliance, there is effectively a Catch-22 = CAN-P-1578, Section 5.10.1 [127, 128, 129, 130, 131, 132].[63]

In relation to Ontario Court proceedings, May and June 2018 [133][64] federal Crown counsel [134] informed that [135, 136, 137, 138][65] "Canadian Procedural Document CAN-P-1578 . . . Section 5.10.1 . . . states: 'Certificates of Analyst are legal documents. As such, information contained in these reports is directed by the appropriate laws of the land.' . . . supersedes any requirement which may arise under ISO/IEC 17025 to provide information beyond that statutorily required by the CDSA, namely, the Certificates".

CAN-P-1578 [139] "is one of several [documents] issued by the Standards Council of Canada to define the policies, plans, and procedures established by the Council to help achieve its mandate" [140]. As the accrediting agency, SCC would appear to have some authority to supersede ISO/IEC 17025. But, *CD&S Act* s. 51(1) as the "laws of the land" does not require secrecy. It allows secrecy. It is a government policy that invokes s. 51 that imposes the secrecy.

The government could reveal a *proper scientific report*, or "test report", if it wanted to; it does not with *CD&S Act* s. 51. A reasonable speculation is that no such report exists. Why else would the government strive so hard to keep secret such a routine document? Is government policy of secrecy for a *proper scientific report*, or "test report", "laws of the land"? Apparently yes. Catch-22.

Resulting from Ontario Court proceedings, May and June 2018,[66] the presiding Justice [141] agreed entirely [142,143] with the federal Crown [144]. The Justice effectively found that adequate science was done and was sufficiently reported to the court. The *Certificates of Analyst* that would later be considered by the court would be safe from challenge; unassailable evidence of a criminally forbidden molecule.

Neither a *proper scientific report*, nor a "test report", was ever revealed. Effectively, the Justice, May and June 2018 [145], created case law that the government policy of secrecy is within "laws of the land". Catch-22. Other courts may later differ, but that is how it remains now.

As indicated above,[67] the central issue of this polemic, and the essential reason to opine that the government of Canada does bad science, is that a *proper scientific report* is not available. And that this is mostly a matter of policy rather than law. And, of science policy, rather than criminal law policy. These issues might be argued in

63 DAS. Catch-22. Chapters 10 and 11.
64 Ontario Court proceedings, May and June 2018. Chapter 10; Chapter 10 Appendix.
65 ". . . no merit to the Applicant's/defence assertion . . ." *O'Connor*. Introduction 4.1; Chapter 10.
66 Ontario Court proceedings, May and June 2018. Chapter 10; Chapter 10 Appendix.
67 Chapter 10.2.

court again, but the public political forum[68] is the most important – Parliament should repeal *CD&S Act* s. 51.

## 10.5 Unethical science

The unethical science mentioned above is from the mere fact of Health Canada's involvement in the criminalized war on drugs.[69]

The *CD&S Act* is essentially of the criminal law rather than of health law. Quite simply, a national health agency should not be involved in criminal law enforcement with police subterfuge[70] – stings – even indirectly, and even in a remote and obscure corner of its systems. To do so is a serious ethical wrong that taints everything else it might stand for. The Supreme Court of Canada allows [146], if not encourages, stings; but that should be the work of a police agency, not a health agency. The US FDA and DEA are separate agencies.

Health Canada should be removed from involvement in the criminal law war on drugs.[71]

Another ethical issue: Neither Health Canada nor its DAS explains the science it purportedly does. Opined here as a general proposition: State treasure should be allotted neither to any state agency nor to any private entity, that does not, when asked, willingly explain the purported science being financed; except for very good reason. Such good reason is not found within the war on drugs.[72]

Health Canada should, when asked, always willingly explain its purported science.[73]

## 10.6 Lawyers, judges and parliamentarians

In a Canada that claims world-class status as a modern industrial nation, it should not be too much to ask that lawyers and judges who would be involved in a process that sends people to prison for *criminalized molecular offences* have some idea about molecular science. So too for the parliamentarians who enacted the legislation.[74]

They should have at least a high school level appreciation of science and mathematics. Experience in Ontario Court suggests they do not. A result of *CD&S Act* s. 51 is

68 Public process. Introduction 4.1; Chapters 5, 8 and 10.
69 RCMP mou. Chapters 5 and 10.
70 Lawyer misconduct. Police subterfuge. *R v W*. . . Chapters 8.8 and 10.5.
71 re criticism of Health Canada, Judges, *et al.* Introduction 4.1 and 4.3; Chapters 10 and 11.
72 Health Canada. DAS. Ethics. Introduction 4.3; Chapters 5 and 10.
73 re criticism of Health Canada, Judges, *et al.* Introduction 4.1 and 4.3; Chapters 10 and 11.
74 re criticism of Health Canada, Judges, *et al.* Introduction 4.1 and 4.3; Chapters 10 and 11.

that a shrouded government appointee can effectively make decisions of law and fact to which most everyone involved, including the trial judge, is oblivious. However, the severest criticism here is for defence lawyers who should pursue these matters more vigorously.

## 10.7 Recommendations: repeal *CD&S Act* s. 51 because it is anti-science – don't allow Catch-22

**Parliament is constitutionally – and at law otherwise – entitled to enable bad and unethical science, and the government is lawfully entitled to follow its lead.**[75,76,77] **Both do so. The courts are entitled to allow this to go on and to shy-away from understanding the science. They do both.**[78]

Perhaps the most serious policy problem is that Canadian legislators, judges, prosecutors and defence lawyers – and the government itself [147·148][79,80] – seem content that inadequately considered science comes into law and is accepted on authority, without apparent care for understanding – for concepts that would require high school level appreciation of science and mathematics. Maybe they are simply ignorant – science is too-alien a culture, or genuinely anti-science, or *chemknowphobic* [149] or do not care.

Public regard for science is important public policy; Canadian legislators, judges, prosecutors and defence lawyers – and the government itself – should try to do better. Even for the guilty, criminal conviction and jail should at least come with the courtesy of the judge having some clue of the science.

Both law and practice should be changed:

- *CD&S Act* s. 51 should be repealed. Because it is anti-science, it effectively prevents providing a *proper scientific report* to give explanation. And because it unfairly distorts the law of evidence.[81] And because it can be replaced by process that is more careful about science. Such replacement policy can be economical too.
- As already opined above[82]: Health Canada should, when asked, always willingly explain its purported science.

---

75 Author's biases. Introduction 1.2, 2, 4.1 and 4.2.

76 re criticism of Health Canada, Judges, *et al.* Introduction 4.1 and 4.3; Chapters 10 and 11.

77 DAS. Catch-22. Chapters 10 and 11.

78 Re criticism of Health Canada, Judges, *et al.* Introduction 4.1 and 4.3; Chapters 10 and 11.

79 Astrology and occult. Crystal ball. Quantum mechanics. Governor-General. Prime Minister Mackenzie King. Harry Houdini. Chapter 4; Chapters 6.26 and 10.7.

80 Governor General. Chapters 2, 10 and 11.

81 Evidence. Chapters 1 and 10.

82 Health Canada. DAS. Ethics. Introduction 4.3; Chapters 5 and 10.

- Health Canada should be removed from criminal law involvement.
- The Standards Council of Canada accreditation of the Health Canada DRUG ANALYSIS SERVICE – TORONTO LABORATORY should be re-examined by an external engineering or analytical chemistry consultant. Perhaps overseeing ISO standards for Canadian agencies should be redirected from SCC to ANAB, Cary, North Carolina. (Ontario CFS is accredited to ANAB, that binds CFS to ISO/IEC 17025:2017, with a published SCOPE OF ACCREDIATION.)[83]
- ISO/IEC 17025 should not be used with a Catch-22.
- Law education should include some science.
- There should be generous federal government financing for local school boards to allow judges, parliamentarians and Crown and defence counsel, to attend high school science and math courses.[84]

## 10.8 Some further biased opinion – raising a difficult question

The essay – biased opinion – polemic, above,[85,86,87] implies an additional question: Should *CD&S Act*, s. 51 – in its effective removing of scientific explanation – and thereby imposing bad science for everyone – be regarded, *ipso facto* [150],[88] as imposing a *reasonable doubt* – of the *criminal law test*[89] – for whatever specific case relied on s. 51. Not such an easy question. But, the author's biased opinion: maybe yes. The issue here is that actual scientific explanation has not been adequately legislated by Parliament.

Opined in other words: Although Parliament's anti-science legislation may be constitutionally and otherwise valid at law, does its content, of itself, raise a criminal law test of *reasonable doubt*, when it would be applied?

---

**83** Accreditation. QA ISO. ANAB. Chapters 5, 8, 10 and 11.
**84** *Daubert.* Judicial unfamiliarity with science. Introduction 4.1; Chapters 1 and 10.
**85** *CD&S Act*, s.51. reasonable doubt. Chapter 10.7.
**86** *Criminal Code* s.320.12 (d). DRE. legislated method validation. Chapters 11.8 and 11.9.
**87** This paragraph is very similar for both Chapters 10.8 and 11.11.
**88** *Ipso facto* = by that very fact or act : as an inevitable result.
**89** Reasonable doubt. Science explanation. Introduction 4.1; Chapters 1.1.1, 10.7, 10.8 and 11.11.

## Appendix Law case – Old City Hall, Toronto, 2018

(see Fig. 10.2) [151][90,91]

Fig. 10.2: City Hall, 1898.

[. . .][92]

. . . *the production of a* [CD&S Act, s. 51(1)] *certificate in and of itself establishes the scientific underpinnings of the conclusions set out therein* . . . .

[. . .]

[From this judicial ruling] [152].[93]

[. . .]

. . . *There is no merit to the defence assertion . . . that accreditation standards . . . that apply to DAS laboratories mean that police or the Federal Crown have obvious and complete access rights to the information sought by the defence. . . .*

[. . .]

[From this judicial ruling].[94]

[. . .]

. . . *It is important for the effective administration of justice that criminal trials remain focused on the issues to be tried, and that scarce judicial resources not be squandered on what has been referred to as, "fishing expeditions" for irrelevant evidence. The likely relevance threshold reflects this gatekeeper function. In* McNeil . . . *paraphrased, is that courts should screen these applica-*

**90** Introductory quotes. Introduction top of fist page.

**91** Introductory quotes. Introduction top of fist page; Introduction 4.1.

**92** Introductory quotes. Introduction 1.2.

**93** Appendix: Law case – Old City Hall, Toronto, 2018. Chapter 10; Chapter 10 Appendix.

**94** ". . . no merit to the Applicant's / defence assertion . . ." *O'Connor*. Introduction 4.1; Chapter 10; Chapter 10 Appendix.

> *tions in an effort to prevent the defence from engaging in either speculative, fanciful, disruptive, unmeritorious, obstructive and time-consuming requests . . . . The importance of preventing unnecessary applications for production from consuming scarce judicial resources cannot be overstated. In this case, the defence adduces no credible grounds for the likely relevance of the records sought. Rather, the defence advances two fundamentally flawed propositions according to the Crown.*
>
> [. . .]
>
> [From this judicial ruling].[95,96,97]

> [. . .]
>
> . . . *There is no merit to the Applicant's assertion that accreditation standards . . . that apply to DAS laboratories mean that police or PPSC "have obvious and complete access rights" to the information sought by the Applicant . . .*
>
> [. . .]
>
> [From AG Response, 2018] [153]

In this chapter,[98] a critical appraisal is made of a case, Toronto, Old City Hall, May and June 2018, that effectively determines how a particular scientific process is regarded at Canadian law. That ruling [154] states the law – at least for the time-being – maybe a very long time, if not forever. Although the ruling is from a lower court, it is comprehensive – and well-presented.[99,100,101]

Parliament's and the government's intentions are interpreted by courts so as to make the law. But that is not the same as good science policy.[102] The ruling [155, 156]:

ONTARIO COURT OF JUSTICE
HER MAJESTY THE QUEEN
V
□□□□□□□
BEFORE THE HONOURABLE JUSTICE S.R. CLARK
on 01 June 2018, at TORONTO, Ontario [157]

M. Lockner, Counsel for the Crown.

---

**95** Appendix: Law case – Old City Hall, Toronto, 2018. Introduction 4.1; Chapter 10; Chapter 10 Appendix.

**96** ". . . no merit to the Applicant's / defence assertion . . ." *O'Connor*. Introduction 4.1; Chapter 10; Chapter 10 Appendix.

**97** Other descriptions: "fishing expeditions", "speculative", "fanciful", "disruptive", "unmeritorious", "obstructive", "time-consuming", "unnecessary", "judicial resources", "no credible grounds", "relevance" and "fundamentally flawed".

**98** *CD&S Act – ipse dixit*. Chapter 10; Chapter 10 Appendix.

**99** Forensic Chemistry. Chapter 10. Appendix. A law case. Notes:

**100** Ontario Court proceedings, May and June 2018. Chapter 10; Chapter 10 Appendix.

**101** Introductory quotes. Introduction 1.2.

**102** Re criticism of Health Canada, Judges et al. Introduction 4.1 and 4.3; Chapters 10 and 11.

[Presentation for the Federal Crown by Roger Flaim, Senior Counsel, Department of Justice Canada, Ontario Regional Office, Toronto M5H 1T1; as counsel for Health Canada.]

M. Grossman, Counsel for □□□□□□□

FRIDAY, 01 JUNE 2018
R U L I N G
CLARK, J. (Orally):

The defendants □□□□□□□ and □□□□□□□ are charged with a number of drug related offences on or about □□□□□ □□, 2017. I believe there are companion provincial charges as well. The trial is scheduled to commence on □□□□□ □□, 2018. Three weeks have been set aside for this matter.

Counsel for □□□□□□□, Mr. Grossman, has brought a pre-trial motion, originally dated 16 October 2017, seeking a ruling in the form of a *Charter*-based declaration that the certificate of analyst not be allowed as sufficient, by itself, evidence of criminal molecular presence in a police-seized item, and that the Crown provide a proper scientific report that tells a sufficiently coherent story of the analytical chemistry, or to inform the court that such a report does not exist.

There are other motions as well set out in the defence *factum* dated 25 April 2018 that are not the subject of today's proceedings such as an application seeking to cross-examine the analyst. This may well be entertained and dealt with on a separate and subsequent date.

More simply put, the defence seeks disclosure of some further evidentiary or scientific basis for the certificates of analyst prepared by Health Canada's Drug Analysis Service, . . . dated 1 May 2017, which provide or confirm that samples provided by the Toronto Police Service to the DAS were analysed and found to contain cannabis. Counsel for the co-accused and the provincial Crown attorney's office have been properly served with this application. Counsel for the co-accused has chosen not to participate in these proceedings. Mr. Lockner, the provincial Crown, is present to at least accept this ruling today. The following is the defence position on this motion as this court understands it.

The defence essentially acknowledges that s. 51 of the *Controlled Drugs and Substances Act*, . . ., provides a rebuttable presumption that the certificate of analysis upon presentation to the court is sufficient to support the conviction and that no explanation of science is required. That it is the government practice and policy not to have to provide same. However, there are situations where there should be a full or sufficiently coherent story or report or records relating to the analysis of the substances and a position on the analytical chemistry. The legislation effectively denies access to this in the view of the defence. The defence, therefore, seeks an order requiring the Crown and Health Canada to provide an analytical chemistry of the government's lab process pursuant to s. 7 of the *Charter of Rights*, relating to the security of the person, and to be able to determine if there have been any defective lab processes. The defence argues, essentially, that any *bona fide* and properly functioning lab would have this documentation, that it would be readily available and that it is not out of the ordinary to expect same. The defence counsel has requested this and received it on other occasions the court has been told. Defence further submits that this should be routinely provided, and, in its absence, unexplained science is bad science. The defence therefore has a right to a proper scientific report or data and an explanation by the analyst who has determined the molecules of concern. If not, the certificate of analysis should not be allowed or admitted as sufficient, by itself, evidence. Furthermore, on a s. 24(2) *Charter* analysis, the manner of obtaining the evidence and denial of the *Charter* right to production by non-disclosure of reports or data or any explanation should result in the certificate of analysis being inadmissible in the trial.

In essence the defence challenges the accuracy of the certificate of analysis, which may possibly contain substances misidentified or data entry errors or wrong dates, or that certain Health Canada procedures were not followed. All of this is being sought pursuant to what the defence submits is the Crown *Stinchcombe* disclosure obligation as first-party records. What this should include, therefore, is copies of test reports, or identification of methods used, copies of related data records of the analyses of the police seized samples, copies of the related standard operating procedures or SOP and a review worksheet.

The . . . [Public] Prosecution Service of Canada (. . . Federal Crown) opposes the application, substantially, on two grounds. First, s. 51(1) of the *CDSA* does not require production of same. The defence has received the certificates as part of the Crown's *Stinchcombe* disclosure obligations and has been advised of its intention to rely on same at trial. They will presumably be then admitted into evidence pursuant to s. 51, which provides a certificate or report prepared by an analyst under s. 45(2), is admissible in evidence in any prosecution for an offence under this act or any other act of Parliament, and in the absence of evidence to the contrary is proof of statements set out in the certificate or report, without proof of the signature or official character of the person appearing to have signed it. Notwithstanding this express statutory authority, the defence now applies, without any evidentiary basis according to the prosecution, for an order for further disclosure to allow for, as they claim, evaluation and decisions about further proceedings.

The Crown contends, however, that speculation about the outcome of evaluation or the possibility of further proceedings does not amount to evidence to the contrary required by the *CDSA*. Therefore, to order such production would be to effectively nullify s. 51(1). The Crown submits further that if the defence position is that no evidence is required or needed to obtain disclosure regarding the science that underlies the certificates, the defence ought to be challenging the constitutionality of the *CDSA*, itself.

The second position advanced in response to the defence application is that in any event, any application for disclosure by the defence, can or must only be done through a third-party records application, as this is not part of the Crown's *Stinchcombe* disclosure obligations. This should be properly the subject of what is more commonly referred to as an *O'Connor* application[103] in which case the defence is required to serve and file the appropriate materials on Health Canada. The court made an endorsement in this matter on 7 May 2018 to the effect that Health Canada should be served with this particular application, resulting in this ruling today.

On 29 May 2018, Mr. Flaim appeared as counsel for the Attorney General of Canada, representing Health Canada as the third-party record holder when this motion was argued. Bound, written materials, titled, *Response of the Attorney General of Canada on Behalf of Health Canada*, have been filed for the court's consideration. It should be noted that Mr. Flaim is in attendance in court today. It should also be noted that on a strictly voluntary and without prejudice basis, given that the trial is imminent, commencing . . . [][], Mr. Flaim, has taken a pragmatic approach in an effort to stave off any delay in the trial proceedings by disclosing the notes created by the analyst in this case setting out the scientific tests conducted, underscoring that this is, in the court's language, a "one-off" situation.

In the analysis of this issue the court will deal with the two issues in reverse order. The first issue is whether the records sought are subject to *Stinchcombe* or *O'Connor*, that is, first-party or third-party records. This is a third-party records matter for the following reasons. In the context of disclosures from Health Canada's . . . DAS, a determination of this distinction is critical. This court has been advised that to date, this rather novel issue has not been considered by any other courts in Ontario. It has, however, been recently decided in the British Columbia Supreme Court, in the case of *R. v. Koumoutsidis*, 2017 BCSC 2129. Paragraph 18, provides as follows:

> In this case, the records are held by Health Canada. The records sought are clearly held and maintained outside of the control of the investigative agency involved, [namely, in that case the Surrey RCMP]. On the hearing of this application counsel for the accused appeared to acknowledge this position. Thus, the records are third party records and the *O'Connor* rules apply.

This court substantially agrees with the written submissions set out by counsel, Mr. Flaim, in the response materials. Thus, a proper application of the principles that distinguish first-party disclosure, or *Stinchcombe*, from third-party disclosure, *O'Connor*, clearly establishes that records held by the DAS are held by a third party and thus their disclosure is subject to the *O'Connor* regime. An *O'Connor* application imposes a significant, but not onerous burden on an applicant to establish that the records sought are likely relevant. This common law standard has been amended by the re-

103 ". . . no merit to the Applicant's / defence assertion . . ." *O'Connor*. Introduction 4.1; Chapter 10; Chapter 10 Appendix.

quirement of s. 51(1) of the *CDSA* for an evidentiary basis on which disclosure should be ordered. Moreover, in contrast to an application for further Stinchcombe disclosure, an *O'Connor* application must be served on the third-party record holder, in this case, Health Canada. This requirement grants Health Canada standing to advance important policy issues expressed in the *CDSA* that should or would inform any decision to order further disclosure. As confirmed by the Ontario Court of Appeal in the case of *R. v. Jackson*, 2015, ONCA, 832, at paragraph 80 and 82:

> For the purposes of first party [. . .] disclosure, the term "the Crown" refers to the prosecuting Crown only, not to all Crown entities, [such as departments and agencies of the Federal government].
>
> [. . .] *Stinchcombe* disclosure [. . .] extends only to material relating to the accused's, [or the defendant's case] in the possession or control of the prosecuting Crown entity. This material is commonly described as the "fruits of the investigation", that is to say, material gathered during the investigation of the offence [for] which [an] accused is charged.

The documents the defence seeks those setting out some scientific basis for the certificates are not in the possession or control of the prosecuting Crown entity. They are not "fruits of the investigation", insofar as they were not gathered during the investigation of the offence with which the defendant is charged. Indeed, s. 45(2) of the *CDSA* only authorizes the analyst to produce a certificate or a report, each of which are in a prescribed form and do not encompass or contemplate working notes made in the preparation of either document. The "fruits of the investigation" that have been gathered and are in the possession and control of the Federal Crown are limited to the certificates. These have been disclosed pursuant to *Stinchcombe*. Any documentation held exclusively by the DAS is thus only available by way of the *O'Connor* application. There is no merit[104] to the defence assertion, therefore, that accreditation standards, such as those promulgated by the International Organization for Standardization or ISO, that apply to DAS laboratories mean that police or the Federal Crown have obvious and complete access rights to the information sought by the defence. The DAS is accredited by the Standards Council of Canada and must meet both the requirements of ISO/IEC 17025 regarding general requirements for the competence of testing and calibration as asserted by the defence, and at the same time these certificates were produced the requirements of Canadian procedural document certificates were produced. The requirements of Canadian procedural document or CAN-P-1578, regarding guidelines for the accreditation of forensic testing laboratories, and in particular Appendix 6, dealing with drug chemistry, which is an amplification of the ISO standard, s. 5.10.1 of this standard provides that:

> [As Read] Certificates of Analyst are legal documents, as such, information contained in these reports is directed by the appropriate laws of the land.

**104** ". . . no merit to the Applicant's / defence assertion . . ." *O'Connor*. Introduction 4.1; Chapter 10.

This qualification supersedes any requirement which may arise under ISO/IEC 17025 to provide information beyond that statutorily required by the *CDSA*, namely, the certificates.

The requirement for applicants or defendants to bring an *O'Connor* application is consistent with two critical features that govern the work of the DAS to the CDSA. First, the need to meet a threshold of likely relevance, as opposed to mere relevance. This is consistent with the requirement in s. 51(1) for an evidentiary basis that justifies going behind or beyond the certificates. The requirement of s. 51(1) amends the common law requirement for likely relevance, insofar as it must be founded on an evidentiary basis. In any event, in both cases more is required than a demonstration of bare relevance that would apply to Stinchcombe disclosure. Second, one of the objectives of the *CDSA* is to create a management scheme that permits proof of the results of the analysis of seized substances, without the necessity of requiring Health Canada to take steps beyond what is required by the *CDSA* in every case where an accused or counsel makes a document request. There is also a logical or practical reason for this position.

The Crown, Mr. Flaim, has submitted to the court that there are two further underpinnings. One relates to efficiency. From paragraph 38 of the *Koumoutsidis* case, it is stated that, "[Every] year, Health Canada receives approximately 120,000 drug samples for analysis". This would equate to approximately 450 such requests per juridical day. The second underpinning relates to affordability.

In support of this, Justice Ramsay, in the case of *R. v. Buffone and Kompon*, 1 June 2017, decision of the Superior Court of Justice in Welland, Ontario, at page 2, line 10, stated:

> [As Read] The purpose of s. 51 of the *CDSA* is to allow the Federal government to prove drug offences with an affordable number of analysts which would not be possible if they were liable to be compelled to testify in any and every case at the request of the accused.

Mr. Flaim submits, therefore, that it is a science lab and should not have to account for its work by going to court. This significant policy concern is reflected in s. 5.10.1 of the CAN-P-1578, which amends what information must otherwise be provided to those under applicable accreditation standards. Moreover, this policy concern has long been recognized by the courts, albeit in the context of considering whether leave should be granted to cross-examine DAS analysts. Mr. Flaim has cited in his written materials, which this court also endorses, various passages from the case of *R. v. Howden* [1997] O.J. No. 4727, at Tab F of the respondents record, at paragraphs 8–10:

> [As Read] It is apparent from the authorities that evidence to the contrary must be evidence which calls into question the correctness of the very analysis which is being tendered by the Crown.

The defence in certain cases may wish to retain their own defence expert. Another example is if the defence suspects or knows that the substance might not test because it was either some form of cutting agent such as baking powder. This might provide an evidentiary basis or informational basis at least on which to bring such an application. In conclusion, this court rules that disclosure from DAS is subject to the third-party *O'Connor* disclosure regime.

Turning to issue two, whether this motion meets the requirements of any such *O'Connor* application, the procedure for same has been summarized in the case of *R. v. McNeil* [2009] 1 SCR 66, 2009 SCC 3, at paragraph 27, as follows:

1) The accused first obtains a subpoena *duces tecum* identifying the documents sought and serves it on the third-party record holder.
2) The accused then brings an application, served on the Crown and the third-party record holder, supported by appropriate affidavit evidence, showing that the records sought are likely to be relevant in his or her trial.
3) The judge addresses any questions of privilege and, if no privilege is a bar release, proceeds to determine whether production should be compelled in accordance with the two-stage test established in *O'Connor*.
4) At the first stage, the judge rules on whether the documents are likely relevant based on the applicable test. If so, the judge proceeds to the second stage to resolve whether, and to what extent, production should be ordered to the accused in order to protect any privacy rights or privileges that may exist in the documents.

The present application complies with none of the foregoing requirements. It does not contain a subpoena by which specific documents or classes of documents are identified. It is unsupported by any affidavit evidence on which the statutory requirement for evidence to the contrary might be satisfied, that is, evidence of an issue with the certificate. Even if the court were to dispense with the need for an application record that meets the formal requirements of an *O'Connor* application, the court must not overlook the lack of evidentiary basis for the disclosure sought in this case. Under the first stage of the *O'Connor* regime, the applicant has the onus of satisfying the court that the third-party records are likely relevant to the proceedings. In *O'Connor*, the court set out at paragraph 140, as paraphrased, that this must be accompanied by affidavit evidence which establishes to the court's satisfaction that the information sought is likely to be relevant. Likewise, in the same case, at paragraph 20, again paraphrased, the accused must bring a formal written application supported by affidavit, setting out the specific grounds for production. Similarly, in the *McNeil* case at paragraph 28, also paraphrased, one of the important considerations is that the applicant must, therefore, justify to the court the use of state power to compel their production. It is important for the effective administration of justice that criminal trials remain focused on the issues to be tried and that scarce judicial resources not be squandered on what has been referred to as, "fishing expeditions" for irrelevant evidence. The likely relevance threshold reflects this gatekeeper function. In *McNeil*, at paragraph 29, also paraphrased, is that courts should screen these applications in an effort to prevent the defence from engaging in either speculative, fanciful, disruptive, unmeritorious, obstructive and time-consuming requests for production. The importance of preventing unnecessary applications for production from consuming scarce judicial resources cannot be overstated. In this case, the defence adduces no credible grounds for the likely relevance of the records sought. Rather, the defence advances two fundamentally flawed propositions according to the Crown.

First, the defence states that there is no evidence before the court of science involvement. The *CDSA* states otherwise. Pursuant to ss. 44 and 45, the minister delegates individuals with scientific qualifications to act as analysts and those analysts carry out analyses or examinations at DAS laboratories which operate according to the previously referenced accreditation standards applicable to drug chemistry. Pursuant to s. 51(1) of the *Act*, a certificate or report prepared is admissible in evidence for the proof of the statements set out therein. Accordingly, the production of a certificate in and of itself establishes the scientific underpinnings of the conclusions set out therein. Second, the defence alleges that production is required for evaluation and decisions about further proceedings. This is tantamount to an admission by counsel that the request for documents is premised on purely speculative grounds. The defence speculates that the evaluation of documents supporting the certificate may yield some unspecified information of some unspecified value in relation to some unspecified issue that may or may not inform some unspecified further proceedings. This falls short of establishing relevance much less likely relevance. It is, in the words of Mr. Flaim, and indeed, supported by the court, "a prototypical or proverbial fishing expedition". The Crown points out that in *Koumoutsidis*, the court overlooked the threshold that applicants must meet in establishing likely relevance before ordering the disclosure of the notes created by an analyst during analysis of a sample submitted by police. In *Koumoutsidis*, the court noted, at para. 34, that it would be very difficult for the defence to verify the accuracy or determine the nature of the analysis carried out because of the limited disclosure the defence receives in a drug prosecution. In the present case this court is not unsympathetic to the position of the defence. It might be referred to, without simplifying matters too much, as a "chicken or egg" type of argument. One could ask, rhetorically, how does one know what to challenge unless given the necessary information through disclosure? This position, however, undermines the express statutory presumption that Parliament has imposed in s. 51(1). The Crown submits, therefore, that to no[t] follow this provision would effectively allow an applicant to dispense with the requirement to articulate a credible basis of likely relevance. This is not a precedent that should be followed in Ontario. This court agrees.

In conclusion, this court rules that the certificates provided under the authority of the *CDSA* are documents in the possession or control of a third party and are thus subject to disclosure based only a properly constituted *O'Connor* application. This requires an evidentiary basis on which the court may order disclosure. In the result, this defence application, as presently constituted is dismissed, without prejudice to the defence to bring a further third party records application.

Thank you very much for your attendance today. As I say, if I had more time, I think my ruling could have been much more concise and tighter for lack of a better expression but I hope that that is of some assistance to the parties moving forward. Thank you!

# Notes

**1** Chapters 10 and 11 are composed and presented as "Essay – Biased Opinion – Polemic".
**2** *Ipse dixit* ≈ he himself said it – an assertion without proof. See notes below.
**3** US *Sixth Amendment* https://constitution.congress.gov/constitution/amendment-6/
**4** Pamela R. Metzger, Tulane University, 2006.
**5** Pamela R. Metzger, "Cheating the Constitution," *Vanderbilt Law Review*, 59 475 (2006).
https://scholar.smu.edu/law_faculty/580/
https://scholarship.law.vanderbilt.edu/cgi/viewcontent.cgi?article=1589&context=vlr
**6** Metzger, at page 476.
**7** Professor Pamela R. Metzger Copyright acknowledgement. Chapter 10.
**8** Justice Scalia, *Melendez-Diaz*, page 13.
https://supreme.justia.com/cases/federal/us/557/305/
**9** Lewis Carroll, *Alice's Adventures in Wonderland* and *Through the Looking-Glass* [1865], Signet Classics, The New American Library, New York City, 1960 and 1962.
**10** Carroll, . . . *-Glass* . . ., "*Humpty Dumpty*", Chapter VI, at page 183.
**11** https://www.alice-in-wonderland.net/resources/pictures/tweedledum-tweedledee/
https://www.alice-in-wonderland.net/resources/pictures/through-the-looking-glass/
https://www.alice-in-wonderland.net/resources/faq/#copyright
**12** re Humpty Dumpty – US courts – see also: *Cheryl Hogue,* "Court Castigation/Fairy Tales and Judicial Opinions," Vol. 86, Issue 18, at page 36, 5 May 2008.
https://cen.acs.org/articles/86/i18/Court-Castigation.html
*New York v EPA*, US Court of Appeals, DC Circuit, 2006 443 F.3D 880 (D.C. CIR. 2006)
*TVA v Hill*, 1978.
https://earthjustice.org/news/press/2007/supreme-court-puts-final-nail-in-coffin-in-administration-plan-to-gut-key-clean-air-safeguard
https://www.latimes.com/archives/la-xpm-2006-mar-18-na-epa18-story.html
https://www.reuters.com/article/us-usa-environment-court/top-court-wont-hear-powerplant-pollution-rule-idUSWBT00690120070430
https://thehill.com/blogs/congress-blog/energy-environment/275921-the-road-from-massachusetts-v-epa-through-the-looking/
https://thehill.com/blogs/congress-blog/politics/24922-mass-v-epa-decision-is-a-step-in-right-direction/
http://www.law.harvard.edu/faculty/freeman/vermeule.freeman.paper.pdf
**13** . . . Carroll, . . . *-Glass* . . ., "*Humpty Dumpty*", Chapter VI, page 186.
**14** *Canadian Charter* s.24(2) See note below.
**15** See: Wm. F. Buckley Jr., Ethan A. Nadelmann, Kurt Schmoke, Joseph D. McNamara, Robert W. Sweet, Thomas Szasz and Steven B. Duke, "The War on Drugs Is Lost," *National Review*, 12 February 1996, pages 34–48
https://www.nationalreview.com/2014/07/war-drugs-lost-nro-staff/ www.nationalreview.com
**16** A. Robert Kaufman, "Attack the profit, not the fact, of drugs," OPINION-COMMENTARY, *The Baltimore Sun*, 10 December 1996, page 21A. www.sunspot.net.
**17** But see: Rod J. Rosenstein (Deputy Attorney General of the United States), "Fight Drug Abuse, Don't Subsidize It/Americans struggling with addiction need treatment and reduced access to deadly drugs. They do not need a taxpayer-sponsored haven to shoot up," *The New York Times*, Opinion [OP-ED], 27 August 2018.
https://www.nytimes.com/2018/08/27/opinion/opioids-heroin-injection-sites.html
**18** Josh Katz, Margot Sanger-Katz & Nick Thieme, "Overdoses plague a generation of Black men in Baltimore and cities across America," *Baltimore Banner*, 20December 2024.

https://www.thebaltimorebanner.com/community/public-health/us-overdose-data-black-men-SE3ZYFVJHZFW5LQR4T5TA2N6E4/
https://www.thebaltimorebanner.com/
**19** Jan Hoffman, *"Trump's Focus on Punishing Drug Dealers May Hurt Drug Users Trying to Quit/*Public health and addiction experts fear millions could lose access to treatment and prevention services if the administration's proposed cuts are enacted," *New York Times*, 13 May 2025. https://www.nytimes.com/2025/05/13/health/trump-drugs-treatment.html
**20** Fentanyl, CAS 437-38-7, $C_{22}H_{28}N_2O$, 336.47 g/mol (see Figs. 10.3a, 10.3b, 10.3c).

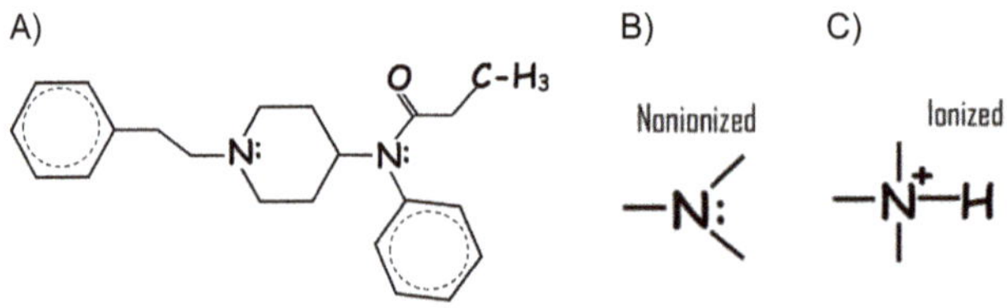


Fig. 10.3: (a) Fentanyl. (b) Nonionized. (c) Ionized.

**21** Fentanyl citrate, CAS 990-73-8, $C_{22}H_{28}N_2O \times C_6H_8O_7$, 528.59 g/mol (see Fig. 10.4).

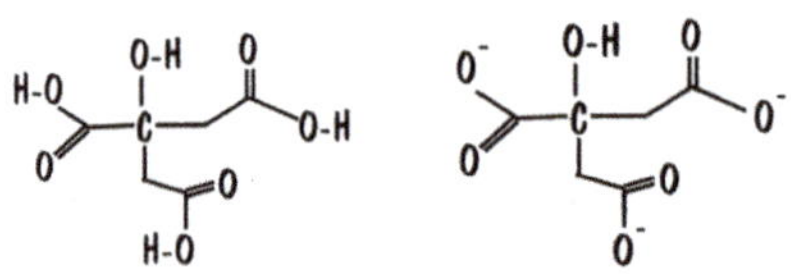

Fig. 10.4: Citric acid, CAS 77-92-9, 192.12 g/mol.
Citrate – if for $Na^+$
CAS 68-04-2
CAS 6,132-04-3 (dihydrate)
CAS 6,858-44-2 (pentahydrate)

**22** Fentanyl hydrochloride, $C_{22}H_{28}N_2O \times HCl$, 372.9 g/mol, CAS 1443-54-5.
**23** 336.47 + 192.12 = 528.59 g/mol.
**24** **See:** Daniel M. Perrine, *The Chemistry of Mind-Altering Drugs/History, Pharmacology, and Cultural Context*, American Chemical Society, Washington, DC, 1996; ISBN 0-8412-3253-9; pages 82 and 83.
**25** Carfentanil, CAS 59708-52-0, $C_{24}H_{30}N_2O_3$, 394.515 g/mol (see Fig. 10.5).

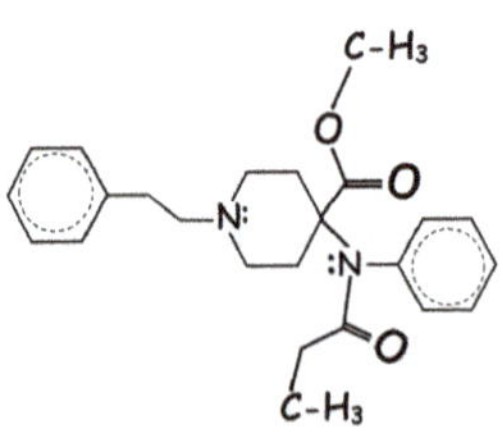

Fig. 10.5: Carfentanil.

**26** https://en.wikipedia.org/wiki/Fentanyl
https://chem.nlm.nih.gov/chemidplus/rn/437-38-7
http://webbook.nist.gov/cgi/cbook.cgi?ID=437-38-7
https://www.spectrumchemical.com/media/sds/F1147_AGHS.pdf?srsltid=AfmBOookHiIUfcmC59m_9ehNI1DNM3FClA05D8Uh_hF1tkoNO7RW8KYd
https://www.spectrumchemical.com/cas/990-73-8?srsltid=AfmBOooUnnoFbp-7RPcOkmnAd2Mr1tY0piaYmumE7rMbTDWG17AwOt4e
https://www.spectrumchemical.com/fentanyl-citrate-cii-usp-f1147
https://en.wikipedia.org/wiki/Citric_acid
https://en.wikipedia.org/wiki/Trisodium_citrate
https://pubchem.ncbi.nlm.nih.gov/compound/Fentanyl-Hydrochloride
https://en.wikipedia.org/wiki/Carfentanil
https://www.nytimes.com/2025/03/07/well/what-is-fentanyl.html
**27** See: CBC, As It Happens, "Why this former Harper adviser [=Benjamin Perrin] had a change of heart about supervised injection sites," 27 January 2020.
https://www.cbc.ca/radio/asithappens/why-this-former-harper-adviser-had-a-change-ofheart-about-supervised-injection-sites-1.5442142
**28** *CD&S Act* SC 1996, c. 19. s.5 Schedule I, item 2. http://laws-lois.justice.gc.ca/eng/acts/C-38.8/index.html
**29** – that is – other criminals than the target arrested – who would be serving the police as informants – perhaps for milder treatment for their own drug crimes – and also perhaps for hire.
**30** Vjosa Isai, *"What to Know About Canada's Role in the Fentanyl Crisis/*Responding to threatened tariffs by President Trump, Canada has unveiled asecurity plan aimed at tackling the trafficking of fentanyl, which is also killing many Canadians," *New York Times*, 06 February 2025.
https://www.nytimes.com/2025/02/06/world/canada/canada-fentanyl-trump.html
**31** Catherine Lévesque, "Former RCMP deputy commissioner Kevin Brosseau is Canada's new 'fentanyl czar'/Brosseau has had an 'over 20 year career in public safety and national security including tackling drug trafficking and organized crime,' Trudeau said in a news release," 11 February 2025, *NATIONAL POST*.
https://nationalpost.com/news/kevin-brosseau-canada-fentanyl-czar
**32**

[. . .]

> **Canada Gazette, Part I, Volume 158, Number 47: GOVERNMENT NOTICES**
> 23 November 2024

[. . .]

> **DEPARTMENT OF INDUSTRY**
> **OFFICE OF THE REGISTRAR GENERAL**
> ***Appointments***

[. . .]

> Deputy National Security and Intelligence Advisor to the Prime Minister and Deputy Secretary to the Cabinet (Emergency Preparedness), Privy Council Office
> Brosseau, Kevin, Order in Council 2024-1082

[. . .]

https://canadagazette.gc.ca/rp-pr/p1/2024/2024-11-23/html/notice-avis-eng.html
**33** – and regulations thereunder.
**34** Cited below.

**35** To borrow from Gilbert and Sullivan, *H.M.S. Pinafore*, 1878; "My Gallant Crew/I Am the Captain of the Pinafore". https://genius.com/Gilbert-and-sullivan-my-gallant-crew-i-am-the-captain-of-the-pinafore-annotated
https://casetext.com/analysis/what-never-no-never-what-never-well-hardly-ever-the-supreme-court-rule-23f-and-gilbert-sullivan
**36** ". . . [i]n all criminal prosecutions, the accused shall enjoy the right . . . to be confronted with the witnesses against him . .".
**37** *Melendez-Diaz v Massachusetts*, reaffirmed and extended by *Bullcoming v New Mexico* – in the US Supreme Court.
**38** *Melendez-Diaz v Massachusetts*, 2009. https://www.law.cornell.edu/supct/html/07-591.ZS. html
https://www.law.cornell.edu/supct/pdf/07-591P.ZS
http://www.supremecourt.gov/
**39** *Bullcoming v New Mexico*, 2011.
https://www.law.cornell.edu/supct/html/9-10876.ZS.html
https://www.law.cornell.edu/supct/pdf/9-10876P.ZS
http://www.supremecourt.gov/
**40** Adam Liptak, "Supreme Court Ruling Accepts No Substitutes in Crime Lab Testimony," *The New York Times*, 24 June 2011, page A16.
**41** *CD&S Act* s.51(2) does not meet such a confrontation concept. *CD&S Act* s.51(2) is more-or-less useless.
**42** Some of this is reworked from court pleading documents from the author's law practice.
**43** *CD&S Act* replaces and continues from the previous *Narcotics Control Act* – which for purposes here is effectively the same. http://www.cfdp.ca/drug.htm.
**44** *Controlled Drugs and Substances Act* SC 1996, c.19.
http://laws-lois.justice.gc.ca/eng/acts/C-38.8/index.html
http://laws-lois.justice.gc.ca/eng/acts/C-38.8/page-9.html#h-19
**45** *CD&S Act* s. 51.

[. . .]

**Designation of analysts**

**44** The Minister may designate, in accordance with the regulations made pursuant to paragraph 55(1)(o), any person as an analyst for the purposes of this Act and the regulations.

**Analysis**

**45(1)** A peace officer, inspector or prescribed person may submit to an analyst for analysis or examination any substance or sample of it taken by the peace officer, inspector or prescribed person.

**Report**

**(2)** An analyst who has made an analysis or examination under subsection (1) may prepare a certificate or report stating that the analyst has analysed or examined a substance or a sample thereof and setting out the results of the analysis or examination.

[. . .]

**Certificate of analyst**

**51(1)** Subject to this section, a certificate or report prepared by an analyst under subsection 45(2) is admissible in evidence in any prosecution for an offence under this Act or the regulations or any other Act of Parliament and, in the absence of evidence to the contrary, is proof of the statements set out in the certificate or report, without proof of the signature or official character of the person appearing to have signed it.

**Attendance of analyst**

(2) The party against whom a certificate or report of an analyst is produced under subsection (1) may, with leave of the court, require the attendance of the analyst for the purpose of cross-examination.

[. . .]

**46** *R v* □□□□□□□, 2017, Ontario Court, Justice Kearney, Brampton L6W 4T1, 02 February 2017 (copied in full below), at paragraph [7]:

[. . .]

**[7]** Parliament has provided an evidentiary shortcut that enables the Crown to put before the court expert opinion evidence as to the nature of the substance analyzed.

[. . .]

**47** For example: English language text from a Health Canada DAS *Certificate of Analyst* used for a *CD&S Act* case:

**48** Health Canada – Drug Analysis Service No.-N° □□□□□□□

2301 Midland Avenue

Toronto Ontario M1P 4R7

**Certificate of Analyst**

I □□□□□□□ being a person on the staff of the Department of Health duly designated as an Analyst under the Controlled Drugs and Substances Act, and also duly designated as an Analyst under the Food and Drugs Act, do hereby certify:

1. That at the Health Canada laboratory in Toronto in the province of Ontario on the □□ day of □□□□□ 2009 there was delivered by Depository box from Toronto Police Service a sealed and unopened package which bore the following identification marks, initials or numbers: □□□□ □□□□
2. That I did take possession of the sealed and unopened package on the □□ day of □□□□□ 2009.
3. That I did open the said package and did remove therefrom a sample for analysis.
4. That I duly analysed and examined the said sample and I found it to contain a controlled substance within the meaning of the Controlled Drugs and Substances Act, to wit:
   **Methamphetamine (N, α-dimethylbenzeneethanamine)**
5. That this certificate is true to the best of my knowledge and skill.
   Dated at Toronto, Ontario
   This □□ day of □□□□□ 2009
   □□□□
   Revision: DAS-SAD 03-2009a Analyst – Department of Health

**49** *CD&S Act* s. 44. http://laws-lois.justice.gc.ca/eng/acts/C-38.8/page-9.html#h-19

**50** *Qualifications for Designations as Analysts Regulations* SOR/98–594
http://laws-lois.justice.gc.ca/eng/regulations/SOR-98-594/page-1.html#h-1
http://laws-lois.justice.gc.ca/eng/acts/C-38.8/index.html

51 Ontario Office of the Fire Marshall.
https://one.ofmem.ca/Login
52 Ontario Ministry of Community Safety and Correctional Services, Centre of Forensic Sciences, Toronto M3M 0B1.
53 *CD&S Act* s.51.
54 Opaque here, notwithstanding the government's transparency claims otherwise:
http://hc-sc.gc.ca/home-accueil/rto-tor/index-eng.php
http://hc-sc.gc.ca/home-accueil/alt_formats/pdf/rto-tor/2015-18-fap-cpa-eng.pdf

[. . .]

> About Health Canada
>
> Regulatory transparency and openness
>
> The Government of Canada is making more data and information available to Canadians than ever before. Canadians are also being offered more opportunities to participate in discussions on government policies and priorities.
>
> As a regulator, Health Canada plays an important role in protecting the health and safety of Canadians and is committed to greater transparency and openness to further strengthen trust in our regulatory decisions.

[. . .]

55 See also: Steve Chapman, "[Canada] Controlled Drugs and Substances Act/Definitions and interpretations," Isomer Design, Toronto M5B 2R3; 02 December 2015, January 2021: [e-mail exchanges with Health Canada – extracts].
http://isomerdesign.com/Cdsa/
http://isomerdesign.com/Cdsa/definitions.php
http://isomerdesign.com/Home/index.php
https://rocketreach.co/isomer-design-profile_b5cd7f63f42e0b04
56 See also: M.G., *SOME RECENT CANADIAN JUDICIAL DECISIONS RELEVANT TO FORENSIC CHEMISTRY*, presented at a Chemistry and the Law session, 217th American Chemical Society National Meeting, Anaheim, California, 22 March 1999. (CHAL: http://membership.acs.org/C/CHAL/ . ACS: www.acs.org . *Chemical & Engineering News*, 01 March 1999, page 94.
57 See also: M.G., "POLICY/What's Up with Canadian Science?," letter to the Editor, *C&EN*, Volume 92, Issue 2, 13 January 2014.
https://cen.acs.org/articles/92/i2/s-Canadian-Science.html
https://cen.acs.org/articles/91/i43/Acrimony-Canadian-Science.html
58 *Access to Information Act* RSC 1985 c. A-1 as amended.
http://laws-lois.justice.gc.ca/eng/acts/A-1/ http://laws-lois.justice.gc.ca/eng/acts/A-1/page-1.html
59 M.G., "[*R v Filsinger's Organic Foods*] – Ministry of Labour inspections – Clinic's *Freedom of Information* request –" Toronto Workers' Health & Safety Legal Clinic *newsletter*, August 2012, Vol. 20, No. 4, Pages 1–4. http://workers-safety.ca/features/newsletters/ http://workers-safety.ca/index.html
60 At page 4:

> **Some comment on the *FOI* process.**
>
> The analysis and commentary above is in the context of problems with using *FOI*-source data, and some explanation is appropriate.
>
> Many *FOI*-data-seekers are simply looking for copies of some of their government-held personal data. Much of this is quite routine and practical, and confidentiality requirements quite acceptable. Others are looking for specific uncontroversial historical data; this too may not be much of a problem, if the data still exists. And, others are looking for data for purposes of some

legal process. These *FOI*-data-seekers would typically know what documents they are looking for and when-from.

However, *FOI*-data such as for our Clinic's request about Filsinger's is in a different category. We are not actually that much interested in released documents. Rather we seek answers to reasonably relevant questions in a public policy conversation. For example: How many Raven tanks were in use at Filsinger's at the time of the tragedy? [1] . . .. These are questions from citizens who may be publicly criticizing government policy. In a free and democratic society,[2] . . . official persons of the Crown should make reasonable efforts to try to give answers. The reality of modern Canada is that they sometimes don't. And they sometimes obfuscate and sometimes hide behind *FOI* as a ruse.

But *FOI* personnel are not in the business of answering questions. They are in the business of providing document copies or declining to provide them for statutory reasons. They would not normally be expected to delve into what the documents might mean in a public policy conversation.

In this context, *FOI*-released documents are requested by citizens to search for answers that the documents might reveal or might imply. But for citizens to use the *FOI* process, it helps to know which documents to request. But if they knew that – if government officials simply answered reasonably relevant questions – citizen activists wouldn't need *FOI*. The result is that to get actual answers the citizens may have to go on fishing expeditions – and perhaps play Monty Python's *Pet Shop* [3] . . . or *Cheese Shop* [4] . . . with government officials – often for unknown documents and then hope those documents would reveal answers. The results are often unsatisfactory.

So, for the worker tragedy at Filsinger's, our Clinic made our *FOI* request; we got what we got; and we have to try to make the best of it. In this context, the analysis and commentary above, hopefully, makes a reasonable contribution to the public conversation on workplace health and safety. So, perhaps that secret MOL ". . . hygienist report . . ." [5] . . . would reveal something interesting, but we will have to make do without it.

**MOL** = Ontario **Ministry of Labour**.

***FOI*** = Ontario ***Freedom of Information*** *Act*.

**61** . . . [1] See above at page 2.

**62** . . . [2] to borrow a phrase – *Canadian Charter of Rights and Freedoms* C being Schedule B, Part I of the *Canada Act* 1982, including the *Constitution Act*, 1982 (1982 c. 11 (United Kingdom), s. 1.

**63** . . . [3] Monty Python http://www.montypython.net/scripts/petshop.php

**64** . . . [4] Monty Python http://www.montypython.net/scripts/cheese.php

**65** . . . [5] See in this *newsletter*, mid-July, Vol.20, No.3, page 4.

**66** Monty Python https://www.youtube.com/watch?v=4vuW6tQ0218
http://montypython.50webs.com/scripts/Series_1/53.htm

**67** Monty Python http://montypython.50webs.com/scripts/Series_3/61.htm

**68** *Constitution Act, 1982*, SCHEDULE B, PART I, *CANADIAN CHARTER OF RIGHTS AND FREEDOMS.*

**69** *Constitution Act*, 1867 (originally called The *British North America Act*, 1867; [United Kingdom] 30 and 31 Victoria, c. 3.). http://www.solon.org/Constitutions/Canada/English/
http://www.solon.org/Constitutions/Canada/English/ca_1867.html

**70** *CANADIAN CHARTER OF RIGHTS AND FREEDOMS* http://laws-lois.justice.gc.ca/eng/Const/page-15.html

**71** "7. Everyone has the right to life, liberty and security of the person and the right not to be deprived there of except in accordance with the principles of fundamental justice".

**72** *Canadian Charter of Rights and Freedoms,* http://laws-lois.justice.gc.ca/eng/Const/page-15.html

[. . .]

Enforcement of guaranteed rights and freedoms

**24. (1)** Anyone whose rights or freedoms, as guaranteed by this Charter, have been infringed or denied may apply to a court of competent jurisdiction to obtain such remedy as the court considers appropriate and just in the circumstances.

Exclusion of evidence bringing administration of justice into disrepute

**(2)** Where, in proceedings under subsection (1), a court concludes that evidence was obtained in a manner that infringed or denied any rights or freedoms guaranteed by this Charter, the evidence shall be excluded if it is established that, having regard to all the circumstances, the admission of it in the proceedings would bring the administration of justice into disrepute.

[. . .]

73 וְלִפְנֵי עִוֵּר לֹא תִתֵּן מִכְשֹׁל וְיָרֵאתָ מֵּאֱלֹהֶיךָ Before the blind, do not put a stumbling block

74 A Hebrew – English Bible According to the Masoretic Text and the JPS 1917 Edition. https://www.mechon-mamre.org/p/pt/pt0.htm https://www.mechon-mamre.org/p/pt/pt0319.htm

75 Some of this is reworked from: *R v* . . ., APPLICATION, Toronto, . . . 2017.

76 *ipse dixit* ≈ he himself said it – an assertion without proof.

77 http://legal-dictionary.thefreedictionary.com/Ipse + Dixit = ". . . [Latin, He himself said it.] *An unsupported statement that rests solely on the authority of the individual who makes it* . . .".

78 Latin. Introduction 4.1.

79 https://en.wikipedia.org/wiki/Ipse_dixit = ". . . Ipse dixit (Latin for 'he himself said it') is an assertion without proof; or a dogmatic expression of opinion . . . The fallacy of defending a proposition by baldly asserting that it is 'just how it is' distorts the argument by opting out of it entirely: the claimant declares an issue to be intrinsic, and not changeable . . . Cicero (106–43 BC) . . . De Natura Deorum . . .".

80 Pamela R. Metzger, "Cheating the Constitution," Vanderbilt Law Review, 59 V and. L. Rev. 475 (2006).
https://scholarship.law.vanderbilt.edu/cgi/viewcontent.cgi?article=1589&context=vlr

81 Pamela R. Metzger, Robert A. Ainsworth Professor in the Courts, Tulane University Law School, New Orleans, Louisiana, 70118.

82 Pamela R. Metzger, "Cheating the Constitution," *Vanderbilt Law Review*, 59 Vand. L. Rev. 475 (2006).

83 http://www.archives.gov/index.html
http://www.archives.gov/exhibits/charters/constitution.html

84 ***Melendez-Diaz v Massachusetts***, US Supreme Court, 2009; at page 13, last paragraph.
https://www.law.cornell.edu/supct/html/07-591.ZS. html https://www.law.cornell.edu/supct/pdf/07-591P.ZS http://www.supremecourt.gov/

85 https://www.law.cornell.edu/wex/ipse_dixit#:~:text=Ipse%20dixit%20is%20a%20Latin,any%20supporting%20evidence%20or%20proof ". . . *Ipse dixit* (Latin for 'he himself said it') is an assertion without proof; or a dogmatic expression of opinion . . . The fallacy of defending a proposition by baldly asserting that it is 'just how it is' distorts the argument by opting out of it entirely: the claimant declares an issue to be intrinsic, and not changeable . . . Cicero (106–43 BC) . . . *De Natura Deorum*. . .".

86 "Report to the Attorney General of Canada: Advice and Recommendations Respecting Certain Certificates of Analyst Issued by Health Canada," by the Honourable W. David Griffiths, Q.C., May 23, 2001.

87 2001 http://www.cbc.ca/news/canada/lab-errors-may-void-ontario-drug-convictions-1.283147

88 **CBCnews | Canada**

**Lab errors may void Ontario drug convictions**

CBC News Posted: Jun 24, 2001 10:38 PM ET | Last Updated: Jun 24, 2001 10:38 PM ET

Anyone convicted of a drug offence in Ontario during the past 12 years, even those now

serving time, may be off the hook because of bungled paperwork at a federal drug lab.
. . . thousands of drug analysis certificates . . ..
[. . .]

89 There are also examples, in other contexts, of problematic results from insufficiently scrutinized analytical chemistry. See for example: Rachel Mendleson, *Toronto Star* [many reports re Motherisk] http://www.thestar.com/authors.mendleson_rachel.html. Or, if gold mining is to be scrutinized by government financial regulators: Douglas Goold and Andrew Willis, *The Bre-X Fraud*, McClelland and Stewart, 1998; ISBN 10: 0771033346 & ISBN 13: 9780771033346.

90 *The Globe and Mail*, National News/National Report, 28 September 1998, page A6.

91 Drug evidence questioned:

> VANCOUVER. The overdose death of an employee of the [Health Canada] federal drug testing lab has cast doubt on 20 years of drug convictions, lawyers say.
>
> The federal Justice Department is reviewing drug cases where Henry Sadowski handled the evidence, after the lab chemist died in May, 1997, of a drug overdose. Tests showed Mr. Sadowski, 51, had snorted a lethal mixture of cocaine and heroin, called a speedball.
>
> An autopsy revealed he had been using the drugs for some time.
>
> Now lawyers who defend people on drug charges are wondering whether Mr. Sadowski fudged his reports. "How do we know what he analyzed and what he put up his nose?" defence lawyer Ken Young said. CP

92 Carmen Drahl, "DOOKHAN NAMED 'SOLE BAD ACTOR'/FORENSIC SCIENCE: Inspector general faults management neglect for Massachusetts crime lab scandal," *Chemical and Engineering News*, 92(10), American Chemical Society, Washington, DC 20036, 10 March 2014, page 6.

93 See also: Andrea Widener and Carmen Drahl, Washington, DC, "FORCING CHANGE IN FORENSIC SCIENCE/SCANDALS IN LABS across the country illustrate the need to reform the discipline, but change has been slow in coming," *C&EN*, 92(19), 12 May 2014, pages 10–14.
https://cen.acs.org/articles/92/i19/Forcing-Change-Forensic-Science.html

> [. . .]
>
> **Pinpointing Red Flags In Massachusetts Drug Lab Scandal**
>
> Disgraced forensic chemist . . . was sentenced to prison more than five months ago. Yet her saga continues to roil both her home state . . . and the forensics community, providing a very public example of the need for reform in forensic science.
>
> . . . fabricated drug test data and lied under oath. . . cases involving more than 40,000 individuals . . . Massachusetts has set aside $30 million to review.
>
> . . . report . . . Inspector General . . . revealed a rash of management failures at the . . . now-shuttered drug lab . . . failed to provide proper chemistry training, and they ignored the concerns . . .
>
> [. . .]

94 Jess Bidgood, "*Massachusetts Justices Clear Way for New Trials in Cases Chemist May Have Tainted*," *The New York Times*, 19 May 2015, page A11.

95 *Melendez-Diaz v Massachusetts* Pages 13–15.
https://supreme.justia.com/cases/federal/us/557/305/

96 *Pamela R. Metzger, "Cheating the Constitution," Vanderbilt Law Review*, 59 Vand. L. Rev. 475 (2006).
https://scholarship.law.vanderbilt.edu/cgi/viewcontent.cgi?article=1589&context=vlr
http://www.law.tulane.edu/tlsfaculty/profiles.aspx?id=450&vpubcat=Articles

**97** Brandon L. Garrett and Peter J. Neufeld, "Invalid Forensic Science Testimony and Wrongful Convictions," *Virginia Law Review*, Volume 95 March 2009 Number 1.
https://papers.ssrn.com/sol3/papers.cfm?abstract_id=1354604
**98** *Melendez-Diaz v Massachusetts*, US Supreme Court, 2009.
https://www.law.cornell.edu/supct/html/07-591.ZS.html
https://www.law.cornell.edu/supct/pdf/07-591P.ZS
http://www.supremecourt.gov/opinions/08pdf/07-591.pdf
http://www.supremecourt.gov/
**99** Pages 1–23.
**100** *Melendez-Diaz v Massachusetts*, No. 07-591.
**101** – see Adam Liptak, " Justices Rule Crime Analysts Must Testify on Lab Results," *The New York Times*, 26 June 2009, pages A1 and A16. See also Linda Greenhouse, ". . . Crime Lab Reports," *The New York Times*, 18 March 2008, page A19.
**102** http://www.supremecourtus.gov/
**103**

> [. . .]
>
> The Sixth Amendment to the United States Constitution,[* **] made applicable to the States via the Fourteenth Amendment, . . . provides that "[i]n all criminal prosecutions, the accused shall enjoy the right . . . to be confronted with the witnesses against him." In *Crawford*, after reviewing the Clause's historical underpinnings, we held that it guarantees a defendant's right to confront those "who 'bear testimony'" against him.[***]
>
> . . .
>
> In short, under our decision in *Crawford* the analysts' affidavits were testimonial statements, and the analysts were "witnesses" for purposes of the Sixth Amendment. Absent a showing that the analysts were unavailable to testify at trial *and* that petitioner had a prior opportunity to cross-examine them, petitioner was entitled to "'be confronted with'" the analysts at trial.
> [. . .] [****]
>
> . . .

**104** * http://www.archives.gov/http://www.archives.gov/exhibit_hall/index.html http://www.archives.gov/exhibits/charters/ .
**105** ** https://www.archives.gov/founding-docs/constitution .
**106** *** *Melendez-Diaz* at page 3.
**107** **** *Melendez-Diaz* at page 5.
**108** At page 13, last paragraph.
**109** Pamela R. Metzger, then Professor, Tulane University Law School, New Orleans, Louisiana, 70,118; now Professor, Dedman School of Law, and Director of the Deason Family Criminal Justice Reform Center, Southern Methodist University, Dallas, Texas, 75275.
**110** http://www.archives.gov/index.html
http://www.archives.gov/exhibits/charters/constitution.html
**111** *Bullcoming v New Mexico*, 2011. https://www.law.cornell.edu/supct/html/9-10876.ZS.html
https://www.law.cornell.edu/supct/pdf/9-10876P.ZS http://www.supremecourt.gov/opinions/10pdf/09-10876.pdf http://www.supremecourt.gov/
**112** Adam Liptak, "Supreme Court Ruling Accepts No Substitutes In Crime Lab Testimony," *The New York Times*, 24 June 2011, page A16.
**113** US *Sixth Amendment*
https://constitution.congress.gov/constitution/amendment-6/
**114** US *Fourteenth Amendment*
https://constitution.congress.gov/constitution/amendment-14/

115 http://www.archives.gov/index.html http://www.archives.gov/exhibits/charters/constitution.html
116 Sir Isaac Newton, *Philosophiae Naturalis Principia Mathematica* (Mathematical Principles of Natural Philosophy), Royal Society, London, 1687. See Edward Neville da Costa Andrade, *Sir Isaac Newton/ His Life and Work*, DOUBLEDAY ANCHOR, Garden City, New York; SCIENCE STUDIES SERIES, S42; reprinted from Macmillan Company, New York, and W. Collins Sons and Co, Ltd., London, first published 1954; page 70, et seq.
117 Regarding ISO/IEC 17025:2005, Clauses 5.10.1 and 5.10.2 e – that require that there be a "test report" that includes ". . . identification of the method used". That "test report" is not provided to Defence counsel, and perhaps it does not exist. Defence counsel, as ". . . members of the legal profession; Canadian justice system . . .," are "CLIENTS SERVED," according to the DAS Lab's SCOPE OF ACCREDITATION, and thereby a "customer" in ISO Clause 5.10.1.
118 DAS – Toronto Laboratory, Standards Council of Canada Accredited Laboratory No.607, Toronto M1P 4R7
http://scc.ca/
119 Text of SCOPE OF ACCREDIATION:
**120 SCOPE OF ACCREDITATION**
**Health Canada**
**DRUG ANALYSIS SERVICE – TORONTO LABORATORY**
**2301 Midland Ave.**
**Scarborough, ON**
**M1P 4R7**

Accredited Laboratory No. 607
(Conforms with requirements of ISO/IEC 17025:2005, RG-FORENSIC)
CONTACT: V. Bell
TEL: (416) 973-1553
FAX: (416) 954-5,923
EMAIL: das.manager.toronto@hc-sc.gc.ca
CLIENTS SERVED: Police officers, both RCMP and Municipal, in Ontario and other provinces, as well as members of other enforcement agencies; members of the legal profession; Canadian justice system
FIELDS OF TESTING: Forensic
FORENSICS DISCIPLINE(S): Forensic Drug Chemistry
PROGRAM SPECIALTY AREA: Forensic
INITIAL ACCREDITATION: 2006-08–29
MOST RECENT REACCREDITATION: 2018-11–04
ACCREDITATION VALID TO: 2022-08–29
**SCC Group Accreditation:**
This laboratory is a part of a Group Accreditation with the following facilities in accordance with SCC's policy on Group Accreditation documented in the Accreditation Services
Accreditation
Program Overview.
– 15858 SANTÉ CANADA, SERVICE D'ANALYSE DES DROGUES – LABORATOIRE DE MONTRÉAL, 1,001, rue Saint-Laurent ouest, Longueuil, QC, J4K 1C7, Accredited Laboratory
No. 709
– 15885 Health Canada, DRUG ANALYSIS SERVICE, 3,155 Willingdon Green, Burnaby, BC, V5G 4P2, Accredited Laboratory No. 716
**FORENSICS**
**Forensic Drug Chemistry**
Description of Activities:

The Drug Analysis Service, Toronto Laboratory, carries out the following examinations/analyses: Identification and quantification of controlled and non-controlled drugs, according to the Controlled Drugs and Substances Act and the Food and Drugs Act, and other related substances in or on powders, pharmaceuticals, plant materials and miscellaneous exhibits.
Techniques for which laboratory is accredited:

a. Gas chromatography
b. High performance liquid chromatography
c. Infra-red spectrometry
d. Macroscopic examination
e. Mass spectrometry
f. Micro-chemical tests
g. Nuclear magnetic resonance spectroscopy
h. Optical microscopy
i. Sample preparation, extraction, separation and general chemical and physical tests
j. Thin layer chromatography

**Notes:**
**RG-FORENSIC:** SCC Requirements and Guidance for the Accreditation for Forensic Testing Laboratories
**ISO/IEC 17025:2005:** General Requirements for the Competence of Testing and Calibration Laboratories
Elias Rafoul, Vice President
Accreditation Services
Date: 2020-08–18
. . .
Number of Forensic Techniques: 10
SCC 1,003-15/747
Partner File #0
Partner: SCC

**121** https://www.scc.ca/en/accreditation/laboratories/health-canada-drug-analysis-service-toronto-laboratory
https://www.scc.ca/en/system/files/client-scopes/ASB_SOA_15747_Scope_v4_2020-08-13.pdf

**122** ISO/IEC 17025:2005, "General requirements for the competence of testing and calibration laboratories," International Organization for Standardization, Genève 20, Switzerland. www.iso.org

**123**

> [. . .]
> 5.10.2 Test reports and calibration certificates
> Each test report or calibration certificate shall include at least the following information, unless the laboratory has valid reasons for not doing so:
> [. . .]
> (e) identification of the method used;
> [. . .]

**124** International Organization for Standardization, Geneva, Switzerland. http://www.iso.org
**ISO/IEC 17025:2005** http://www.iso.org/iso/home/store/catalogue_tc/catalogue_detail.htm?csnumber=39883

**125**

[. . .]

5.10 Reporting the results

5.10.1 General

The results of each test, calibration or series of tests or calibrations carried out by the laboratory shall be reported accurately, clearly, unambiguously and objectively and in accordance with any specific instructions in the test or calibration methods.

[. . .]

The results shall be reported, usually in a test report or a calibration certificate . . ., and shall include all the information requested by the customer and necessary for the interpretation of the test or calibration results and all information required by the method used. This information is normally that required by 5.10.2, and 5.10.3 or 5.10.4.

[. . .]

**126** Standards Council of Canada, Ottawa. K1P 6L5 http://www.scc.ca/en/accreditation http://www.scc.ca/en/accreditation/laboratories http://www.scc.ca/en/forensic http://www.scc.ca/en

**127** Catch-22. An exception to a rule or arrangement that the would-be-beneficiaries are informed of later-on, so that they do not actually benefit. The disappointed would-be-beneficiaries are not entitled to explanation.

**128** Joseph Heller, *CATCH-22*, Dell Publishing Co., Inc., New York City, 10017; 1955, 1961; New Dell Edition, 1985; ISBN 0-440-11120-X.

**129** Joseph Heller, *CATCH-22*, Chapter 39, page 416:

[. . .]

"They don't have to show us Catch-22", . . .

"The law says they don't have to."

"What law . . .?"

"Catch-22."

[. . .]

**130** https://dictionary.cambridge.org/dictionary/english/catch-22 https://en.wikipedia.org/wiki/Catch-22_(logic) https://en.wikiquote.org/wiki/Catch-22

**131** *Catch-22*, Paramount Pictures, 1970. https://en.wikipedia.org/wiki/Catch-22_(film)

**132** https://www.scc.ca/en/system/files/publications/ASB_RG_Forensic-Laboratories_v1_2017-07-17.pdf

**133** *R v* □□□□□□□, Ontario Court proceedings, Old City Hall, Toronto; 29 May 2018 (argument); 01 June (judgment).

**134** For the case then being litigated, counsel for Health Canada, acting for Health Canada, was effectively also acting for the federal Crown.

**135** Response of the Attorney General of Canada on Behalf of Health Canada, 23 May 2018; for Court proceedings, Toronto, 29 May 2018 (argument); 01 June (judgment).

**136** Paragraphs 17 and 18, pages 8 and 9.

**137**

[. . .]

17. There is no merit to the Applicant's assertion that accreditation standards (such as those promulgated by the International Organization for Standardization (ISO)) that apply to DAS laboratories mean that police or PPSC "have obvious and complete access rights" to the information sought by the Applicant. . .. The DAS is accredited by the Standards Council of Canada and must

meet both the requirements of ISO/IEC 17025 (*General Requirements for the Competence of Testing and Calibration*, as asserted by the Applicant), and, at the time these Certificates were produced, the requirements of Canadian Procedural Document CAN-P-1578 (*Guidelines for the Accreditation of Forensic Testing Laboratories*, and in particular Appendix 6 thereof, "Drug Chemistry"), which is an amplification of the ISO standard.[11] Section 5.10.1 of this latter standard states:

Certificates of Analyst are legal documents. As such, information contained in these reports is directed by the appropriate laws of the land.[12]

18. This qualification supersedes any requirement which may arise under ISO/IEC 17025 to provide information beyond that statutorily required by the CDSA, namely, the Certificates.

[. . .]

[11] As stated in, for example, *Health Canada Internal Audit: Audit of Regional Laboratory Activities (July 2016)*, available at

[https://publications.gc.ca/collections/collection_2024/sc-hc/H14-558-2016-eng.pdf]

. . . Note: CAN-P-1578 has since been replaced by the Standard Council of Canada's *Requirements and Guidance for the Accreditation for Forensic Testing Laboratories* (July 17, 2017) although, for the purposes of this application, the relevant sections remain materially unchanged.

[12] Standards Council of Canada's *Guidelines for the Accreditation of Forensic Testing Laboratories*, CAN-P-1578, Appendix 6, "Drug Chemistry", Respondent's Record, Tab E, section 5.10.1, available at:

[ https://www.jtc1sc34.org/en/c/document_library/get_file08cd.pdf ] . . .

[. . .]

**138** Government of Canada Home → Departments and agencies → Health Canada → Health Canada's transparency → Health Canada's corporate management reporting → Health Canada internal audits → 2016
Final Report: Audit of Regional Laboratory Activities – July 2016
https://www.canada.ca/en/health-canada/corporate/transparency/corporate-management-reporting/internal-audits/2016/audit-regional-laboratory-activities-july-2016.html
Accreditation Services SCC Requirements and Guidance for the Accreditation for Forensic Testing Laboratories 2017-07-17
https://www.jtc1sc34.org/en/c/document_library/get_file08cd.pdf
Guidelines for the Accreditation of Forensic Testing Laboratories CAN-P-1578 May 2009
Program Speciality Area – Forensic Testing Laboratories (PSA-FT)
http://www.jtc1sc34.org/en/c/document_library/get_file08cd.pdf
SCC Requirements and Guidance for the Accreditation for Forensic Testing Laboratories
This document replaces CAN-P-1578.
https://www.scc.ca/en/about-scc/publications/scc-requirements-and-guidance-for-accreditation-for-forensic-testing-laboratories
Accreditation Services SCC Requirements and Guidance for the Accreditation for Forensic Testing Laboratories 2017–07–17
ASB_RG_Forensic-Laboratories_v1_2017-07–17.pdf
Accreditation Services SCC Requirements and Guidance for the Accreditation for Forensic Testing Laboratories 2017–07–17
https://scc-ccn.ca/search/site?fulltext=en%20about%20scc%20publications%20scc%20requirements%20guidance%20for%20accreditation%20for%20forensic%20testing%20laboratories

**139** Guidelines for the Accreditation of Forensic Testing Laboratories CAN-P-1578 May 2009 Program Speciality Area – Forensic Testing Laboratories (PSA-FT)

https://www.jtc1sc34.org/en/c/document_library/get_file08cd.pdf page 91
**5.10 Reporting the results**
5.10.1 Certificates of Analyst are legal documents. As such, information contained in these reports is directed by the appropriate laws of the land.
5.10.2 Where feasible, court testimony of Analysts shall be monitored on an annual basis.
**140** CAN-P-1578, May 2009, FORWARD.
**141** Judges of the Ontario Court (OCJ) are referred to as "Justice".
**142** Defence arguments not accepted by the Judge: Ruling by The Honourable Justice S.R. Clark, 01 June 2018, Old City Hall, Toronto, Ontario Court Of Justice, *Her Majesty The Queen v M . . .*; Court File No. 17–15007031.
**143** Text copied as Chapter 10 Appendix.
**144** For the case then being litigated, counsel for Health Canada, acting for Health Canada, was effectively also acting for the federal Crown.
**145** See note above.
**146** *Rothman v The Queen* [1981] 1 SCR 640 at page 697.
http://scc-csc.lexum.com/scc-csc/scc-csc/en/item/2514/index.do
**147** But apparently not including the former Governor General herself – on other science issues – although there were criticisms concerning issues of vice-regal protocol.
**148** https://www.nationalobserver.com/2017/11/06/opinion/science-v-religion-and-new-governor-general-under-fire
https://www.nationalobserver.com/2017/11/04/news/payette-takes-climate-change-deniers-and-horoscopes-science-conference
https://torontosun.com/opinion/editorials/editorial-julie-payette-speech-oversteps-her-role
https://www.cbc.ca/news/opinion/governor-general-speech-julie-payette-climate-change-1.4384481
https://www.gg.ca/en
https://en.wikipedia.org/wiki/Governor_General_of_Canada
**149** *chem-know-phobia* = a fear of a knowledge of chemistry.
**150** *ipso facto* = by that very fact or act : as an inevitable result.
https://www.merriam-webster.com/dictionary/ipso%20facto
**151** Copied from a copy of a part of an 1898 drawing.
**152** [From the judicial ruling copied here, third from last paragraph, second and third sentences.]
**153** Response of the Attorney General of Canada on Behalf of Health Canada, 23 May 2018; for Court proceedings, Toronto, 29 May 2018 (argument); 01 June (judgment).
**154** Canadian and Ontario criminal law cases may be titled as *HER MAJESTY THE QUEEN verses* [accused defendant], abbreviated as *R v* []. *R* = *Regina* = Queen. Sometimes phrased as Her Majesty The Queen in right of Canada, or as Her Majesty The Queen in right of Ontario. In other eras, and now: *R* = *Rex* = King.
**155** Copied here with minor editorial adjustments as to format, *etc.*; for academic and public discussion purposes. **This copy is not for official Ontario court purposes**. Official copies must be obtained from the official court reporter or otherwise as according to accepted court practice (which can be confusing); fees may apply.
**156** The copyright claim here is for the commentary and arrangement only, and does not include the ruling itself.
**157** Court File No. 17-15007031.

# Chapter 11
# Essay – biased opinion – polemic: Canada *Criminal Code* – driving impaired by drug – science?

*Quod volimus credimus libenter* [1, 2, 3, 4][1,2,3]

## 11.1 Forensic question

The Canada *Criminal Code* criminalizes [5, 6, 7, 8, 9, 10, 11, 12, 13, 14, 15, 16, 17, 18, 19, 20, 21, 22, 23, 24, 25, 26][4,5,6] drug-impaired driving.[7] This is distinct from regulatory law.[8] For purposes here, this may be rephrased: <<driving while ability to drive is impaired to any degree by a drug is a crime.>>.

The forensic question to be addressed, in the Canadian *criminal law* context, with its *proof beyond a reasonable doubt* test:[9] <<**How to determine ability to drive is impaired to any degree by a drug?**>> [27][10,11,12,13]

---

1 ≈ Believe what we want to believe.

2 *Quod volimus credimus libenter*. Latin. Introduction 4.1; Chapter 11, top of the first page.

3 Introductory quotes. Introduction 1.2.

4 ***Criminal Code***, as amended, **s. 320.14(1)** https://laws-lois.justice.gc.ca/eng/acts/C-46/page-46.html#h-121277

[. . .]

> **Operation while impaired**
> **320.14 (1)** Everyone commits an offence who
> **(a)** operates a conveyance while the person's ability to operate it is impaired to any degree by alcohol or a drug or by a combination of alcohol and a drug;

[. . .]

5 Toxicology. Driving impaired. Chapters 1 and 11.

6 Chart – legislation. Introduction 4.3; Chapter 11.

7 **Alert**: "impairment" is to be tediously defined and argued about in this polemic.

8 Regulatory law. Environmental and workplace health and safety. Chapters 1, 5, 9 and 11.

9 *Criminal law. Regulatory/administrative law*. Proof *beyond a reasonable doubt* test. Chapters 1.4.2.1 and 1.4.2.2; Chapters 10 and 11.

10 Author's biases. Introduction 1.2, 4.1 and 4.3; Chapters 10 and 11.

11 Polemic. Health Canada. Introduction 4; Chapters 9 and 10.

12 Legislative process – policy making. Introduction 4.3; Chapters 1, 10 and 11.

13 Drug-impaired driving. Chapter 11.

https://doi.org/10.1515/9783111618555-012

Within the framework of the DRE regime – the law process, procedure and judgement discussed here,[14,15,16,17,18] no other question or issue need be considered in detail for this polemic.[19]

## 11.2 Newly enforced law

Although drug-impaired driving legislation had been in place for several years, law enforcement and prosecution in Ontario seemed to become more newly popular – with perhaps an increase when recreational marijuana became legalized – to remain until automation of motor vehicle operation. And: *Bingley*, 2018 – a new and important Supreme Court of Canada decision.

When police investigate – at a routine road or street stop, or for bad driving or at a crash scene – the possibility of impairment is a top agenda item – and should be. Until recent years the impairment was typically regarded as ethanolic [28]; but impairment was usually not sought to be actually literally proven scientifically or medically. Rather, a machine-measured amount of over 80 [20,21] was criminalized by definition, by operation of law, so that the rather more difficult court judgement of impaired need not be addressed. While this procedure is well-accepted, for ethanol-impaired driving,[22] albeit with some dissent, and with complexities of heavy-duty litigation [29]; for drug-impaired it is not.

Procedures for drug-impaired driving impose a newer and more controversial regime, and this is what this polemic is about.[23] The simplicity of finding criminal guilt, machine-measured, does not apply as it does for ethanol [30]. For this, Parliament is legally and constitutionally entitled to criminalize highway traffic safety, notwithstanding that a federal-provincial regulatory approach might be more sensible, effective and life-saving and less damaging to the concept of *mens rea*[24] and less expensive.

---

**14 The process of law, procedure and judgement discussed here is referred to here as the DRE process. And, depending on the context, DRE may also refer to the police officer who would conduct parts of the process.**

15 DRE. Introduction 4.3; Chapter 11.

**16 DRE = drug recognition expert.**

17 **"DRE"** as used in this Essay – biased opinion – polemic – should usually be regarded as "DRE [sic erat scriptum]" – see Chapter 11.6.

18 *Criminal Code* ethanol in blood. SOP. Chapters 5 and 11.

19 Essay – biased opinion – polemic. Introduction 4.3; Chapters 10 and 11.

20 Over-80 ≥ 80 mg ethanol/100 mL in blood. Chapter 11.

21 80 mg/100 mL = 0.8 g/L 0.8 g/L/46.07 g/mol = 0.0173648795311 . . . mol/L.

22 *Criminal Code* ethanol in blood. SOP. Chapters 5 and 11.

23 Some science controversy remains. Introduction 4; Chapters 1, 10 and 11.

24 *mens rea*. Chapters 1 and 11.

Parliament is also entitled to distort the law of evidence, for ease of conviction, facilitated with bad science[25] to be applied to determine criminality, and the Supreme Court of Canada is entitled to allow this. Both Parliament and the Court have done so.

(During the COVID-19 pandemic, the face-to-face, up-front person-to-person, DRE examinations for a strictly disciplined actual health sciences context did not seem to affect DRE law enforcement and prosecutions. Perhaps it should have – but it seemed not.)

> *The "Red Death" had long devastated the country. No pestilence had ever been so fatal, or so hideous . . . .* [Edgar Allan Poe, 1842]. [31, 32][26]

## 11.3 Science literature

From a forensic science perspective, the answer to <<How to determine ability to drive is impaired to any degree by a drug>> should include some reference, within the course of the legal process, that is sufficiently relevant to science literature.[27] But it does not. This is problematic for Canadian *criminal* law.

To be sure, impaired driving is a very serious public safety issue – an obvious menace – calling for prompt and definitive police action, and follow-up, whenever and wherever noticed. For *regulatory law* purposes – that does not require proof of *mens rea* – to determine, restrict and forbid drug-impaired drivers – in addition to obvious and intuitive observations – there is ample scientific and reliable popular literature. But while this should lead to prompt regulatory law action, science-based support is apparently not available to satisfy the *criminal law* test of *proof beyond a reasonable doubt*, that should be applied most strictly [33],[28] including proving *mens rea*.

Arguably, *criminal* law proof beyond a reasonable doubt,[29] and *mens rea* is intended as a difficult barrier – to be extra careful to protect the innocent, and secure their freedom. While it may have practical regulatory aspects, "criminal law" is essentially a moral-based construct. A crime is a serious intentional transgression against society and its morals.

Significant here: The "*criminal law* test" of *proof beyond a reasonable doubt*.[30] Not merely strong indicators for *regulatory* purpose. Not the balance of probabilities of civil litigation.

---

25 Science – good, bad and junk. Introduction 4; Chapters 1, 2, 8, 10 and 11.

26 Introductory quotes. Introduction, top of the first page.

27 Impairment and science literature. Chapters 5 and 11.

28 *Criminal law. Regulatory/administrative law.* Proof *beyond a reasonable doubt* test. Chapters 1.4.2.1 and 1.4.2.2; Chapters 10 and 11.

29 *Criminal law. Regulatory/administrative law.* Proof *beyond a reasonable doubt* test. Chapters 1.4.2.1 and 1.4.2.2; Chapters 10 and 11.

30 *Criminal law. Regulatory/administrative law.* Proof *beyond a reasonable doubt* test. Chapters 1.4.2.1 and 1.4.2.2; Chapters 10 and 11.

Science literature – reported support of the DRE process – is not found to satisfy the *criminal* law test. Or, if it does, it remains unrevealed by Crown prosecution counsel. Although there are science references within the DRE process[31] – cited in a US NHTSA *Manual* [34] – that should be reliably considered as for real negative effects of drugs on human performance – this is not the same as *proof beyond a reasonable doubt*, at *criminal* law, of <<ability to drive is impaired to any degree by a drug>>. Negative effects, measured and interpreted scientifically, may be well on the way to convince a judge to convict of *crime* – but not yet – some more is needed.

That needed more:

## 11.4 How to determine <<ability to drive>> and <<impairment>>?

How to scientifically determine <<ability to drive>> and <<impairment>>? The needed more would require assuming definitions [35] of what <<ability to drive>> is, what <<impaired>> means and how they would be measured. This is a difficult problem in toxicology.[32]

This problem is apparently not addressed in statute, regulation or case law; nor in the DRE literature [36]. This problem, not only elusive to frame as a question, is even more elusive to answer. An answer would have to include the prior-agreed-to definitions and assumptions.

Although indications of human impairment are often obvious, actual measurable determination is difficult.[33] It is difficult even with ethanol [37, 38, 39, 40],[34,35] for which judgement about impairment is accepted – consistent with cultural historical experience – including as personal to many people. That would appear to be why breath measuring technology [41][36] and over-80 [42, 43, 44][37] legislation [45, 46, 47, 48, 49] were invented. The technical measurements are held to be sufficient on their own, at law, without the judge actually having to decide on <<ability to drive>> and <<impairment>>.

But with drugs of abuse and recreation it is so much more difficult to decide. Analytical chemistry may easily be able to determine the presence and quantity of drugs, in blood, but not so easily in other target organs. Observed altered body appearance and performance and sloppy behaviour can indicate likely impairment – but with problematic amounts of doubt.

**31** DRE. Introduction 4.3; Chapter 11.

**32** Toxicology. Hazard communication. Chapters 1 and 11.

**33** Toxicology. Chapters 1 and 11.

**34** *Ethanol*. $CH_3$-$CH_2$-OH CAS 64–17-5. Chapters 1, 4 and 11.

**35** Toxicology. Chapters 1 and 11.

**36** Concentration measurement – IR. Chapters 6 and 11.

**37** Over-80 = 80 mg ethanol/100 mL in blood. Chapter 11.

There is no scientifically reliable measure for impairment found within the course of the legal process because impairment is not well-enough defined; and such definition would have to include prior unimpaired performance measures – perhaps with a motor vehicle driving virtual simulator [50, 51, 52]. There is apparently no adequate reference to the science literature – found within the legislation, case law or DRE literature for the measurement of impairment in this context.

Sometimes doubt about impairment can be minimized, within the court's other evidence processes – by stark observations, common sense and admissions of fact of the accused. But often not. It is a judgement call for the judge and that is what the judge is for. However, in many cases the problematic amounts of doubt remain; and, again, that is what this polemic is about.

By this polemic, impairment is not, by the DRE process,[38] sufficiently well-enough scientifically measured – in the context of *criminal* law's *beyond a reasonable doubt* test – see below.

## 11.5 Bad science

The DRE process is bad science.[39] That bad science is the determination of <<ability to drive is impaired to any degree by a drug>> – a *crime* – by police officers who are government-appointed for the purpose of pronouncing opinions without explanation supported by sufficient scientific literature, using methods beyond the bounds of science practice, and who do not have the education and years of supervised clinical practice of real health science professionals (such as MDs and RNs).

The DRE evaluation, as a kind of analytical chemistry and health science clinical diagnosis, is also bad science simply because of – and if only – the severely hindered opportunity for scrutiny. Quite simply, science should be seen as badly done unless there is a formal proper scientific report[40] and the opportunity for scientific conversation with explanation.[41] This is not the DRE process.

Notwithstanding the more than three centuries after publication of *Principia*, by Isaac Newton, encouraged by Edmond Halley [53],[42] the Parliament of Canada, the Governor-in-Council and the Supreme Court are entitled in law and constitution to ignore usual scientific practice of reporting and explanation. And do so ignore.

**38** DRE. Scientific measurement. Introduction 4.3; Chapter 11.

**39** Science – good, bad and junk. Introduction 4; Chapters 1, 2, 5, 8, 10 and 11.

**40** Proper scientific report. CD&S Act. Health Canada. Introduction 4.3; Chapters 5–11.

**41** Science – good, bad and junk. Introduction 4; Chapters 1, 2, 5, 8, 10 and 11.

**42** Newton. Haley. *Principia*. Chapters 9–11.

## 11.6 DRE opinion

The Supreme Court of Canada, 2018, in *R v Bingley* [54, 55],[43] unfortunately gives the DRE process legal credibility so that DRE decisions support the criminal conviction and imprisonment of citizens without sufficient scientific reporting or otherwise scientific, clinical and courtroom controls.[44, 45 46]

Crown counsel seeks the criminal law proof in court for impaired by drug in two stages:

- **First**, the opinion of a *Criminal Code*-created "Drug Recognition Expert", a police officer accredited by the INTERNATIONAL ASSOCIATION of CHIEFS OF POLICE [56],[47] is taken to be a presenter of reliable science by operation of law. Actual expertise need not be shown to the court. Nor actual science.

  That this "Expert" [*sic erat scriptum*] [57, 58, 59][48] might, or not, actually be knowledgeable would remain unrevealed.

  The DRE [*sic*] has a special status in court of an expert witness – entitled to give opinion evidence. That opinion establishes – at law – drug impairment [60, 61, 62].

  The accused person will have been subjected to personal performance tests and examinations, as listed in regulation [63] under the *Criminal Code*.[49] (But not with a vehicle driving simulator.) The DRE opinion would lead to demanding a urine or blood sample.
- **Second**, an opinion is given by an actual scientist from the Ontario Centre of Forensic Sciences, with CFS-done analytical chemistry of urine or blood. Work by CFS appears as well-done and with professional integrity that importantly includes a willingness to engage in scientific conversation including discussing disagreements with some of its conclusions. The work of CFS is not challenged here.

However, once the presence of drug is CFS-confirmed, it is largely the DRE opinion from DRE observations that prevails. And that leads to the central question of this polemic: How to determine <<ability to drive is impaired to any degree by a drug>>? In court, this is essentially done by the DRE; and not CFS; and effectively not by the Judge, after hearing from the DRE.

---

**43** *R v Bingley*. Chapters 1 and 11.

**44** DRE = drug recognition expert.

**45** This process of law, procedure and judgement discussed here is referred to as the DRE process. And, depending on the context, DRE may also refer to the police officer who would conduct parts of the process.

**46** DRE. Introduction 4.3; Chapter 11.

**47** IACP, Alexandria, Virginia, 22314

**48** = thus it had been written.

**49** DRE *Criminal Code* provisions. Chapter 11.

But, in reality, toxicology can be difficult.

Analytical chemistry to determine blood system drug is not too difficult – with modern automated GC- or LC-MS.[50] In contrast, the interpretation of how to define and interpret <<impaired>> is exceptionally difficult and problematic. This is a problem in toxicology.

In a Canadian legal context, the answer to <<How to determine ability to drive is impaired to any degree by a drug?>> – as a policy matter – is dictated by Parliament (statute = *Criminal Code*), the Governor-in-Council (regulation under statute) [64] and the Supreme Court of Canada (*Bingley*). They are entitled by law and constitution to be science-avoidant, if not science-ignorant; and possibly anti-science; and they effectively direct Judges to accept the DRE opinion as legal truth.[51]

## 11.7 Result of the DRE opinion

The *Criminal Code*, its regulation and Supreme Court of Canada decision in *R v Bingley*, when read together,[52] allow the DRE opinion its special legal status. That opinion can support conviction and imprisonment of a citizen.

The DRE report and testimony has its complexities, but it is essentially the single phrase in the written report and testimony [65]: **"opinion that [accused] is impaired by drug and not fit to drive"** – that effectively requires the court to find criminal guilt [66, 67].

With the *Bingley*-determined special status as expert [*sic*] by operation of law, without actual meaningful qualification (as otherwise should be usual for expert witnesses),[53] Defence counsel may be reduced to asking about scraps of data, of uncertain meaning, that in an other-than-law-court context would seem foolish.

*Bingley* allows the DRE as expert [*sic*] not only to conduct the tests but also for the interpretation of the results [68], notwithstanding that the scientific meanings are missing in the legislation. *Bingley* also effectively recognizes that Parliament [69] and the Governor-in-Council [70] are entitled to legislate, notwithstanding that they might get the science wrong, and that *Bingley* enforcement of even wrong science is law in Canada. That the DRE process is bad science has been argued above[54]; and it is also unwise law.

---

**50** GC-MS. Chapters 6 and 11.

**51** Legal/scientific truth. Chapters 1, 8 and 11.

**52** DRE *Criminal Code* provisions. Chapter 11.

**53** DRE. Expert witness. Chapters 1 and 11.

**54** Science – good, bad and junk. Introduction 4; Chapters 1, 2, 8, 10 and 11.

## 11.8 Criticisms of the DRE process

The arguments here, while in the context of criminal law, are essentially of policy-directed issues. As argued in this polemic,[55,56,57] impairment cannot be adequately scientifically measured by the DRE process – and, in principle, the DRE report should not be scientifically relied on because [58,59]

- For each of the tests and examinations set out in the *Criminal Code* regulation [71], and for the apparent DRE practice of combining the results for a concluding opinion, adequate supporting references to the health sciences literature are missing – for the Canadian *criminal law* test of *proof beyond a reasonable doubt.*[60] In a forensic science context this is a very serious failing. Parliament, the government and the Supreme Court seem oblivious to this [72, 73, 74, 75, 76, 77, 78, 79, 80, 81, 82, 83, 84, 85, 86, 87, 88].[61]
- Canada *Criminal Code* s. 320.12 (d)[62] is inadequate for scientific method validation.[63]
- "Impairment", however defined, cannot be reliably measured, within the DRE process, because there are scant prior basis measurements – as for comparison with (at least one of):
  - The accused in the known absence of drugs.
  - Published scientific studies as to what is normal.
  - At least some medical history – taken more adequately than can be known from the DRE process. And, taken and interpreted by someone trained professionally – and licensed – for a health science (MD, RN, *etc.*)
- <<Ability to drive>> is not measured by the DRE; neither before nor after. This would require some kind of simulator machinery [89].
- The DRE training and literature do not adequately address that the test and examination results are difficult to interpret because observations and measurements of humans can have a broad range of non-drug possibilities. There are scant DRE statistical discussions; and none of *Gaussian* (normal (bell curve)) or other distributions.[64,65]

---

55 Author's biases. Introduction 1.2, 4.1 and 4.3; Chapters 10 and 11.

56 Re criticism of Health Canada, Judges et al. Introduction 4.1 and 4.3; Chapters 10 and 11.

57 DRE. Scientific measurement. Introduction 4.3; Chapter 11.

58 *R v Bingley*. Chapters 1 and 11.

59 DRE *Criminal Code* provisions. Chapter 11.

60 *Criminal law. Regulatory/administrative law.* proof *beyond a reasonable doubt* test. Chapters 1.4.2.1 and 1.4.2.2; Chapters 10 and 11.

61 Re criticism of Health Canada, Judges et al. Introduction 4.1 and 4.3; Chapters 10 and 11.

62 Validation. DRE. Chapters 8 and 11.

63 Method validation. Chapters 5, 8 and 11.

64 *Gaussian* distribution. Chapters 8 and 11.

65 DRE. Statistical. Chapters 8 and 11.

- How to interpret the results meanings of the tests and examinations listed in the *Criminal Code* regulation [90] is missing. That interpretation role is simply given to the DRE.
- The DRE police officers who act as expert [*sic*] do so devoid of the education and years of supervised clinical practice – and licensing – of real health science professionals.
- The validation of the DRE method, as apparently implied in the IACP DRE *International Standards* [91, 92, 93], with a specific literature reference [94, 95], appears to be problematic [96, 97, 98, 99, 100].
- The DRE process is not externally scrutinized with a quality assurance regime. In contrast, CFS is accredited to ANAB for ISO/IEC 17025:2017.[66]
- The invasion of the federal *Criminal Code* into the regimes of the provincial health sciences professions presents a constitutional problem. What the DRE does to the accused looks something like a medical examination that should not be done except with Ontario-legislated licensure [101, 102, 103]. And, generally, only a licensed Medical Doctor is entitled to diagnose [104, 105, 106, 107, 108].[67] Perhaps this raises *Constitution Act* s. 91 and s. 92 issues? [109, 110]. And otherwise, parts of an apparent evaluating medical examination should not be done by anyone without actual health science education.
- The DRE process presents something of a public health risk; even aside from the COVID-19 pandemic. (For example, if the DRE does not use the medical equipment – such as the blood pressure cuff – in a proper hygienic way.)
- The DRE training and literature do not appear to sufficiently qualify the DRE to distinguish human illness from drug impairment – as for the *criminal law* test of *beyond a reasonable doubt.*
- The DRE, as a police officer, investigating a person known to be charged with a crime, is not a neutral observer. There is a reasonable apprehension of inherent built-in bias. *Quod volimus credimus libenter.* [111, 112].[68] Someone else should do the evaluation, with bias-control protocols, and as multi-blind as practicable.
- The DRE observations and measurements are made by only one person – in only one session – by a process that is inherently difficult to repeat.
- The IACP DRE "12-step" [113, 114, 115, 116, 117, 118, 119, 120] mentioned frequently in *Bingley* should be scientifically explained; along with its provenance, in more detail than: ". . . Nothing in or about the DRE protocol is new or novel. The DRE protocol is a compilation of tests that physicians have used for decades to identify and assess alcohol- and/or drug-induced impairment" [121].

---

**66** Accreditation. QA ISO. ANAB. Chapters 5, 8, 10 and 11.

**67** Legal disclaimer. MD. Chapters 9 and 11.

**68** ≈ Believe what we want to believe.

- The DRE process of performance tests, and even the few invasive health history questions, as a criminal law adventure, is insulting to the dignity of innocent citizens. And, even the guilty deserve better.
- As an ethical issue, hospital emergency care facilities should be used for professional healthcare providers to minister to patients – not for criminal law enforcement. Professional healthcare providers should not be involved in active law enforcement evidence gathering – to do so raises questions of healthcare conflict of interest.

## 11.9 Parliament and government, and bad science – a tradition

[. . .][69]

*It is recognized and declared that . . . an evaluation conducted by an evaluating officer* [=DRE] *is a reliable method of determining whether a person's ability to operate a conveyance is impaired by a drug or by a combination of alcohol and a drug.*

[. . .]

**[**Canada *Criminal Code* s. 320.12 (d)**]** [122][70]

As opined in Chapter 10, Parliament and the government have something of a tradition with bad science. Professor Metzger's analysis of "forensic *ipse dixit* statutes" should also be considered as for the *Criminal Code* drug-impaired driving provisions.[71,72]

The DRE process is likely not junk science[73]; the personal performance and health science-appearing examinations are likely able to give some – if also problematic – indications of a drug-impaired condition. But without adequate science-based explanation, and meaningfully science-relevant scrutiny by cross-examination, it is reasonable to speculate about junk; *Bingley* encourages that speculation.

## 11.10 Most-serious policy issues

The policy issues above are in the context of the question posed at the beginning of this polemic: <<**How to determine ability to drive is impaired to any degree by a drug?**>>

- Without adequate reference to the scientific literature, and observations and measurements done properly scientifically, there is not a satisfactorily scientific answer for this question in the Canadian criminal law process.

69 Introductory quotes. Introduction 1.2.

70 Validation. DRE. Chapters 8 and 11.

71 Re criticism of Health Canada, Judges et al. Introduction 4.1 and 4.3; Chapters 10 and 11.

72 DAS and bad science. Introduction 4.1; Chapters 10 and 11.

73 Science – good, bad and junk. Introduction 4; Chapters 1, 2, 5, 8, 10 and 11.

- Without observations and measurements done by actually provincially licensed health science professionals, there is not a satisfactory answer for this question in the Canadian criminal law process.
- But, perhaps the most serious policy problem is that Canadian legislators, judges, prosecutors and defence lawyers – and the government itself [123, 124][74] – seem content that inadequately considered science can come into law, and be accepted on authority, without apparent care for understanding – for concepts that would require high school level appreciation of science and mathematics.[75] Public regard for science is important public policy; Canadian legislators, judges, prosecutors and defence lawyers – and the government itself – should try to do better. Even for the guilty, criminal conviction and jail should at least come with the courtesy of the judge having some clue of the science.
- Impaired driving, surely a public safety menace, is obviously *malum in se* [125, 126, 127].[76] But if Parliament and the Governor-in-Council are intent on making it *malum prohibitum*,[77] at *criminal* law, then they should do so with science-compatible legislation. But the science issues for impaired by drug would be so difficult that, for most situations provincial *regulatory* law – rather than federal *criminal* – would be so much better and would avoid many of the problems argued above and maybe more economical too.

## 11.11 Some further biased opinion – raising a difficult question

The Essay – biased opinion – polemic, above,[78,79,80] implies an additional question: Should *Criminal Code* s. 320.12 (d) – in its legislated method validation [81,82] and thereby imposing bad science for everyone – be regarded, *ipso facto* [128],[83] as imposing a *reasonable doubt* – of the *criminal law test* – for whatever specific case relied on s. 320.12 (d)? Not such an easy question. But the author's biased opinion: maybe yes. The issue here is that actual scientific method validation has not been adequately legislated by Parliament.

---

74 Governor General. Chapters 2, 10 and 11.
75 Re criticism of Health Canada, Judges et al. Introduction 4.1 and 4.3; Chapters 10 and 11.
76 *Malum in se* = wrong of itself; naturally wrong.
77 *Malum prohibitum* = wrong because it's prohibited by law.
78 *CD&S Act*, s. 51. reasonable doubt. Chapter 10.7
79 *Criminal Code* s. 320.12 (d). DRE. Legislated method validation. Chapters 11.8 and 11.9.
80 This paragraph is very similar for both Chapters.
81 Method validation. Chapters 5 and 8.
82 Reasonable doubt. Science explanation. Introduction 4.1; Chapters 1.1.1, 5.3.1, 10.7, 10.8 and 11.11.
83 = By that very fact or act: as an inevitable result.

Opined in other words: Although Parliament's anti-science legislation may be constitutionally and otherwise valid at law, does its content, of itself, raise a criminal law test of *reasonable doubt*, when it would be applied?

## Notes

**1** *Quod volimus credimus libenter* ≈ believe what we want to believe.
**2** *Quod volimus credimus libenter* = we always believe what we want to believe. Robert Harris, *Archangel*, Berkley Publishing, New York City, 1998; Arrow, 2009; ISBN-10: 0099527936, ISBN-13:978-0099527930.
https://www.amazon.ca/Archangel-Robert-Harris/dp/0099527936
https://www.goodreads.com/quotes/377660-quod-volimus-credimus-libenter-we-alwaysbelieve-what-we-want
**3** Robert Harris, *Archangel*, Arrow Books, The Random House Group Limited, London SW1V 2SA; 1999; ISBN 0-09-28241-0; at page 48: ". . . '*Quod volimus credimus libenter*' . . . 'we always believe what we want to believe'. . ."
**4** https://en.wikiquote.org/wiki/Julius_Caesar Julius Caesar, *Commentarii de Bello Gallico* [=*Commentaries on the Gallic War*], Book I, Ch. 14, translated by W.A. McDevitte and W.S. Bohn: "*Fere libenter homines id quod volunt credunt*" = In most cases men willingly believe what they wish.
**5** ***Criminal Code***, as amended **s. 320.14(1)** https://laws-lois.justice.gc.ca/eng/acts/C-46/page-46.html#h-121277
https://laws-lois.justice.gc.ca/eng/annualstatutes/2018_21/page-1.html#h-2

[. . .]

> **Operation while impaired**
>
> **320.14 (1)** Everyone commits an offence who
>
> **(a)** operates a conveyance while the person's ability to operate it is impaired to any degree by alcohol or a drug or by a combination of alcohol and a drug;

[. . .]

**6** *Criminal Code* amended so that: s. 253(1)(a), *et seq.* → s. 320.11, *et seq.* – coming into force as of 21 June 2018.
**7** https://en.wikipedia.org/wiki/Impaired_driving_in_Canada
**8** Canadian legislation and case law for prosecution and judgment of driving impaired by drug – when read together, allow the DRE opinion special legal status in a process that would give the appearance of health science observation and decision making.
**9** *Her Majesty The Queen v* ***Bingley*** 2017 SCC 12 Supreme Court of Canada, 23 February 2017.
https://scc-csc.lexum.com/scc-csc/scc-csc/en/item/16417/index.do?r=AAAAAQAHYmluZ2xleQE
**10** https://www.justice.gc.ca/eng/cj-jp/sidl-rlcfa/
http://gazette.gc.ca/rp-pr/p2/2018/2018-07-11/html/sor-dors148-eng.html
**11** Blood drug concentration regulations: **SOR/2018-148**
http://gazette.gc.ca/rp-pr/p2/2018/2018-07-11/html/sor-dors148-eng.html (see Tab. 11.1).

[. . .].

**Blood Alcohol Concentration and Blood Drug Concentration**

**Summary offence**

**1** For the purpose of paragraph 253(3)(b) of the *Criminal Code*, the prescribed blood drug concentration for tetrahydrocannabinol (THC) is 2 ng of THC per mL of blood.

**Hybrid offence – drugs**

**2** For the purpose of paragraph 253(3)(a) of the *Criminal Code*, the prescribed blood drug concentration for each drug set out in column 1 of the table to this section is set out in column 2 (see Tab. 11.1).

[. . .]

12 *Criminal Code* as newly amended **s. 320.27 (1) (a) and (c)**

[. . .]

Tab. 11.1: Any detectable level.

| Item | Column 1<br>Drug | Column 2<br>Concentration |
|---|---|---|
| 1 | Tetrahydrocannabinol (THC) | 5 ng/mL of blood |
| 2 | Lysergic acid diethylamide (LSD) | Any detectable level |
| 3 | Psilocybin | Any detectable level |
| 4 | Psilocin | Any detectable level |
| 5 | Phencyclidine (PCP) | Any detectable level |
| 6 | 6-Monoacetylmorphine | Any detectable level |
| 7 | Ketamine | Any detectable level |
| 8 | Cocaine | Any detectable level |
| 9 | Gamma hydroxybutyrate (GHB) | 5 mg/L of blood |
| 10 | Methamphetamine | Any detectable level |

[. . .]

13 [. . .]

**Investigative Matters**

**Testing for presence of alcohol or drug**

**320.27 (1)** If a peace officer has reasonable grounds to suspect that a person has alcohol or a drug in their body and that the person has, within the preceding three hours, operated a conveyance, the peace officer may, by demand, require the person to comply with the requirements of either or both of paragraphs (a) and (b) in the case of alcohol or with the requirements of either or both of paragraphs (a) and (c) in the case of a drug:

(a) to immediately perform the physical coordination tests prescribed by regulation and to accompany the peace officer for that purpose;

(b) to immediately provide the samples of breath that, in the peace officer's opinion, are necessary to enable a proper analysis to be made by means of an approved screening device and to accompany the peace officer for that purpose;

(c) to immediately provide the samples of a bodily substance that, in the peace officer's opinion, are necessary to enable a proper analysis to be made by means of approved drug screening equipment and to accompany the peace officer for that purpose.

[. . .]

14 *Criminal Code* as amended **s. 320.28 (2)**
15 [320.28]

[. . .]

**Evaluation and samples of blood – drugs**

**(2)** If a peace officer has reasonable grounds to believe that a person has operated a conveyance while the person's ability to operate it was impaired to any degree by a drug or by a combination of alcohol and a drug, or has committed an offence under paragraph 320.14(1)(c) or (d) or subsection 320.14(4), the peace officer may, by demand, made as soon as practicable, require the person to comply with the requirements of either or both of paragraphs (a) and (b):

(a) to submit, as soon as practicable, to an evaluation conducted by an evaluating officer to determine whether the person's ability to operate a conveyance is impaired by a drug or by a combination of alcohol and a drug, and to accompany the peace officer for that purpose or

(b) to provide, as soon as practicable, the samples of blood that, in the opinion of the qualified medical practitioner or qualified technician taking the samples, are necessary to enable a proper analysis to be made to determine the person's blood drug concentration, or the person's blood drug concentration and blood alcohol concentration, as the case may be, and to accompany the peace officer for that purpose.

[. . .]

16 *Criminal Code* as newly amended **s. 320.28 (4)**
17 [320.28]

[. . .]

**Samples of bodily substances**

**(4)** If, on completion of the evaluation, the evaluating officer has reasonable grounds to believe that one or more of the types of drugs set out in subsection (5) – or that a combination of alcohol and one or more of those types of drugs – is impairing the person's ability to operate a conveyance, the evaluating officer shall identify the type or types of drugs in question and may, by demand made as soon as practicable, require the person to provide as soon as practicable,

(a) a sample of oral fluid or urine that, in the evaluating officer's opinion, is necessary to enable a proper analysis to be made to ascertain the presence in the person's body of one or more of the types of drugs set out in subsection (5); or

(b) the samples of blood that, in the opinion of the qualified medical practitioner or qualified technician taking the samples, are necessary to enable a proper analysis to be made to ascertain the presence in the person's body of one or more of the types of drugs set out in subsection (5) or to determine the person's blood drug concentration for one or more of those types of drugs.

[. . .]

18 *Criminal Code* as amended **s. 320.38.**

19 [. . .]

Regulations

320.38 The Governor in Council may make regulations:

(a) prescribing the qualifications required for a peace officer to act as an evaluating officer and respecting the training of evaluating officers;

(b) prescribing the blood drug concentration for a drug for the purpose of paragraph 320.14 (1)(c);

(c) prescribing a blood alcohol concentration and a blood drug concentration for a drug for the purposes of paragraph 320.14(1)(d);

(d) prescribing the blood drug concentration for a drug for the purpose of subsection 320.14(4);

(e) prescribing the physical coordination tests to be conducted under paragraph 320.27(1)(a) and

(f) prescribing the tests to be conducted and procedures to be followed during an evaluation under paragraph 320.28(2)(a) and the forms to be used in recording the results of the evaluation.

[. . .]

20 *Criminal Code* as amended **s. 320.11**

21 [. . .]

**PART VIII.1**

**Offences Relating to Conveyances**

**Interpretation**

**Definitions**

**320.11** The following definitions apply in this Part

***conveyance*** means a motor vehicle, a vessel, an aircraft or railway equipment. (*moyen de transport*)

***evaluating officer*** means a peace officer who has the qualifications prescribed by regulation that are required in order to act as an evaluating officer. (*agent évaluateur*)

[. . .]

22 *Criminal Code* as newly amended **s. 320.12**

[. . .]

**Recognition and Declaration**

Recognition and declaration

320.12 It is recognized and declared that

(a) operating a conveyance is a privilege that is subject to certain limits in the interests of public safety that include licensing, the observance of rules and sobriety;

(b) the protection of society is well served by deterring persons from operating conveyances dangerously or while their ability to operate them is impaired by alcohol or a drug because that conduct poses a threat to the life, health and safety of Canadians;

(c) the analysis of a sample of a person's breath by means of an approved instrument produces reliable and accurate readings of blood alcohol concentration and

(d) an evaluation conducted by an evaluating officer is a reliable method of determining whether a person's ability to operate a conveyance is impaired by a drug or by a combination of alcohol and a drug.

[. . .]

23 *Criminal Code* regulation **SOR/2008-196**
http://laws-lois.justice.gc.ca/eng/regulations/SOR-2008-196/page-1.html#s-3

**24 Evaluation of impaired operation (drugs and alcohol) regulations**
**SOR/2008-196**

CRIMINAL CODE
Registration 2008–06–11
Evaluation of Impaired Operation (Drugs and Alcohol) Regulations
P.C. 2008-1033 2008-06-11
Her Excellency the Governor General in Council, on the recommendation of the Minister of Justice, pursuant to section 254.1 of the *Criminal Code*, hereby makes the annexed *Evaluation of Impaired Operation (Drugs and Alcohol) Regulations*.

**Qualification Required of Evaluating Officer**
**1** An evaluating officer must be a certified drug recognition expert accredited by the International Association of Chiefs of Police.

**Physical Coordination Tests**
**2** The physical coordination tests to be conducted under paragraph 254(2)(a) of the *Criminal Code* are the following standard field sobriety tests:
(a) the horizontal gaze nystagmus test;
(b) the walk-and-turn test; and
(c) the one-leg stand test.

**Evaluation Tests and Procedures**
**3** The tests to be conducted and the procedures to be followed during an evaluation under subsection 254(3.1) of the *Criminal Code* are
(a) a preliminary examination, which consists of measuring the pulse and determining that the pupils are the same size and that the eyes track an object equally;
(b) eye examinations, which consist of
(i) the horizontal gaze nystagmus test,
(ii) the vertical gaze nystagmus test and
(iii) the lack-of-convergence test;
(c) divided-attention tests, which consist of
(i) the Romberg balance test,
(ii) the walk-and-turn test referred to in paragraph 2(b),
(iii) the one-leg stand test referred to in paragraph 2(c), and
(iv) the finger-to-nose test, which includes the test subject tilting the head back and touching the tip of their index finger to the tip of their nose in a specified manner while keeping their eyes closed;
(d) an examination, which consists of measuring the blood pressure, temperature and pulse;
(e) an examination of pupil sizes under light levels of ambient light, near total darkness and direct light and an examination of the nasal and oral cavities;
(f) an examination, which consists of checking the muscle tone and pulse and
(g) a visual examination of the arms, neck and, if exposed, the legs for evidence of injection sites.

**Coming into Force**
**4** These regulations come into force on 2 July 2008.

**25** ***R v Bingley*** = *Her Majesty The Queen v Bingley* 2017 SCC 12 Supreme Court of Canada, 23 Feb. 2017. https://scc-csc.lexum.com/scc-csc/scc-csc/en/item/16417/index.do?r=AAAAAQAHYmluZ2xleQE

**26** See also SOR/2018-148.

27 *Criminal Code*, as amended **s. 320.14(1)**

[. . .]

**Operation while impaired**

**320.14 (1)** Everyone commits an offence who

**(a)** operates a conveyance while the person's ability to operate it is impaired to any degree by alcohol or a drug or by a combination of alcohol and a drug;

[. . .]

**28** For whatever reason, the Canada *Criminal Code* uses "alcohol" a chemically general term, rather than "ethanol", which specifically refers to the intoxicating substance intended for legal application.
**29** See: Joseph F. Kenkel, *Impaired Driving in Canada*, 5th Edition, LexisNexis Canada
https://store.lexisnexis.com/en-ca/products/impaired-driving-in-canada-the-charter-cases-5th-edition.html
**30** See: Ernesto Londoño, *"As Marijuana Use Grows, Effect on Road Safety Remains a Blind Spot* / Scientists want to create breathalyzers that show how recently a driver used marijuana. Researchers also are studying how cannabis impairs motor skills and reflexes for habitual and occasional users," New York Times, 01 January 2025.
https://www.nytimes.com/2025/01/01/us/marijuana-driving-impairment.html
**31** The Works of Edgar Allan Poe, THE MASQUE OF THE RED DEATH
https://etc.usf.edu/lit2go/147/the-works-of-edgar-allan-poe/5383/the-masque-of-the-reddeath/
https://etc.usf.edu/lit2go/147/the-works-of-edgar-allan-poe/
**32** See also: Mary Shelley, *The Last Man*, 1826.
https://mary-shelley.fandom.com/wiki/The_Last_Man_(1826) https://en.wikipedia.org/wiki/The_Last_Man
**33** The Canadian *criminal law* test is proof, by the state, *beyond a reasonable doubt*. Analogously, the same test applies to *regulatory/administrative law*, but perhaps, in practice, not quite as strictly considered. Arguably, the strictness of consideration would be related to the general theoretical proposition that *crime* accusations are for serious moral transitions of individual persons whose personal freedom may be in the balance. In contrast, *regulatory/administrative law* is as for methods of community improvement, encouraged by legal penalties; important in its own right, but dealt with differently.
**34** References cited in: NHTSA, *Participant Manual/Drug Recognition Expert Course*, Revised: 10/2015.
www.nhtsa.gov
https://www.nhtsa.gov/sites/nhtsa.dot.gov/files/documents/dre_7-day_participant_manual-tag.pdf
https://www.nhtsa.gov/sites/nhtsa.gov/files/documents/dre_7-day_participant_manual-tag.pdf
https://www.nhtsa.gov/search?keywords=participant+manual
https://www.nhtsa.gov/search?keywords=participant+manual#gsc.tab=0
**35** – and perhaps some of those definitions would be somewhat arbitrary, but should be well-defined and explained in scientific and public literature.
**36** The DEC Program Technical Advisory Panel of the IACP Highway Safety Committee, *The International Standards of the Drug Evaluation and Classification Program*, Revised October 2017.
https://www.theiacp.org/sites/default/files/all/i-j/International%20Standards%20of%20the%20DECP%20October%202017.pdf
https://www.theiacp.org/projects/the-international-drug-evaluation-classification-program
**37** M.G., *The Law of Occupational Health and Safety in Ontario, Second Edition*, Butterworths, Toronto and Vancouver, August 1994, ISBN 0-409-90414-7; Section 10.3.f.
**38** Toronto Workers Health & Safety Legal Clinic *newsletter*, 2010 10. Vol. 18, No. 4, pages 5 and 6.

39 Stacey Cowley & Jessica Silver-Greenberg, "Paying Dearly for a Defective Breath Test/A Linchpin of the Criminal Justice System Is Often Unreliable," *The New York Times*, 04 November 2019, pages A1, A14, A15, A16.
https://www.nytimes.com/2019/11/03/business/drunk-driving-breathalyzer.html "*These Machines Can Put You in Jail. Don't Trust Them.* / Alcohol breath tests, a linchpin of the criminal justice system, are often unreliable, a Times investigation found."
40 *Ethanol.* $CH_3$–$CH_2$–OH, CAS 64-17-5 46.07 g/mol (see Fig. 11.1) (Molecular structural drawing. CAS. Chapters 4.4 and 4.6).

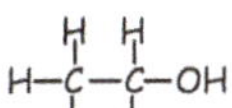

Fig. 11.1: Ethanol, CAS 64-17-5.

https://pubchem.ncbi.nlm.nih.gov/compound/Ethanol
41 Such as listed in Canada *Approved Breath Analysis Instruments Order*, SI/85-201 – see note below.
42 Measured as 80 mg ethanol/100 mL in blood.
43 80 mg/100 mL = 0.8 g/L 0.8 g/L/46.07 g/mol = 0.0173648795311. . . mol/L.
44 https://en.wikipedia.org/wiki/Blood#:~:text=Blood%20is%20composed%20of%20blood,to%20resist%20infections%20and%20parasites
45 *Criminal Code* as amended **s. 320.14(1)**
https://laws-lois.justice.gc.ca/eng/acts/C-46/page-46.html#h-121277
https://laws-lois.justice.gc.ca/eng/annualstatutes/2018_21/page-1.html#h-2

[. . .]

**Operation while impaired**

**320.14 (1)** Everyone commits an offence who

**(a)** operates a conveyance while the person's ability to operate it is impaired to any degree by alcohol or a drug or by a combination of alcohol and a drug [or]

**(b)** subject to subsection (5), has, within two hours after ceasing to operate a conveyance, a blood alcohol concentration that is equal to or exceeds 80 mg of alcohol in 100 mL of blood.

[. . .]

46 *Approved Breath Analysis Instruments Order* SI/85-201 *Criminal Code*
https://laws-lois.justice.gc.ca/eng/regulations/SI-85-201/page-1.html

**Approved Breath Analysis Instruments Order**

**SI/85–201**

*CRIMINAL CODE*

**Short Title**

1 This Order may be cited as the *Approved Breath Analysis Instruments Order.*

**Approved Instruments**

2 The following instruments, each being an instrument of a kind that is designed to receive and make an analysis of a sample of the breath of a person in order to measure the concentration of alcohol in the blood of that person, are hereby approved as suitable for the purposes of section 258 of the *Criminal Code:*

**(a) to (c)** [Repealed, SOR/2012-237, s. 1]

**(d) to (g)** [Repealed, SOR/2013-107, s. 1]

**(h)** Intoxilyzer® 5,000 C;
**(i)** [Repealed, SOR/2012-237, s. 1]
**(j)** [Repealed, SOR/2013-107, s. 1]
**(k)** BAC Datamaster C;
**(l)** Alco-Sensor IV-RBT IV;
**(m)** [Repealed, SOR/2013-107, s. 1]
**(n)** Alco-Sensor IV/RBT IV-K;
**(o)** Alcotest 7,110 MKIII Dual C;
**(p)** Intoxilyzer® 8000 C;
**(q)** DataMaster DMT-C;
**(r)** Intox EC/IR II; and
**(s)** Intoxilyzer® 9000.

47 *Approved Screening Devices Order* SI/85-200 *Criminal Code*
https://laws-lois.justice.gc.ca/eng/regulations/SI-85-200/page-1.html
48 *Approved Drug Screening Equipment Order* SOR/2018-179 *Criminal Code*
https://laws-lois.justice.gc.ca/eng/regulations/SOR-2018-179/page-1.html
49 https://en.wikipedia.org/wiki/Impaired_driving_in_Canada
50 https://www.nhtsa.gov/research-data/driver-simulation
51 https://www.ckas.com.au/driver_training_simulators_43.html?gclid=EAIaIQobChMIvMWvcGB7gIVdOW1Ch3QFADeEAAYASAAEgK5vvDBwE
52 *The Sunday Edition*, "Michael Enright piloted a Boeing 777 and landed it in Lake Ontario," CBC Radio, 14 December 2018.
https://www.cbc.ca/radio/thesundayedition/the-sunday-edition-december-16-2018-1.4928642/michael-enright-piloted-a-boeing-777-and-landed-it-in-lake-ontario-1.4928645
53 Sir Isaac Newton, *Philosophiae Naturalis Principia Mathematica* (Mathematical Principles of Natural Philosophy), Royal Society, London, 1687. See Edward Neville da Costa Andrade, *Sir Isaac Newton / His Life and Work*, DOUBLEDAY ANCHOR, Garden City, New York; SCIENCE STUDIES SERIES, S42; reprinted from Macmillan Company, New York, & W. Collins Sons & Co, Ltd., London, first published 1954; page 70, et seq.
54 *Her Majesty The Queen v* ***Bingley*** 2017 SCC 12 Supreme Court of Canada, 23 February 2017.
https://scc-csc.lexum.com/scc-csc/scc-csc/en/item/16417/index.do?r=AAAAAQAHYmluZ2xleQE
55 See: Ronald Joseph Delisle, Don Stuart, David M. Tanovich, Lisa Dufraimont; *EVIDENCE / PRINCIPLES AND PROBLEMS* / Twelfth Edition; CARSWELL / THOMSON REUTERS CANADA, Toronto M1T 3V4; 2018. ISBN 978-0-7798-8655-5. At pages 606→614 – *R v Bingley*.
https://digitalcommons.osgoode.yorku.ca/faculty_books/362/
56 **IACP** = INTERNATIONAL ASSOCIATION of CHIEFS OF POLICE, Alexandria, Virginia, 22314.
http://www.theiacp.org/ http://www.theiacp.org/Drug-Recognition-Expert-Section
https://www.theiacp.org/sites/default/files/all/i-j/International%20Standards%20of%20the%20DECP%20October%202017.pdf
57 *sic erat scriptum* = thus it had been written.
https://etymologyotd.wordpress.com/2017/10/02/sic-short-for-sic-erat-scriptum-latin-for-thus-it-had-been-written-read-more/
https://en.wikipedia.org/wiki/Sichttps://www.theguardian.com/notesandqueries/query/0,5753,-23558,00.html https://en.wiktionary.org/wiki/sic#Etymology_1http://mymemory.translated.net/en/Latin/English/sic-erat-scriptum
58 *Sic erat scriptum* is used here in an ironic-critical sense to imply that the government-created "expert" would be lacking in actual expertise – other than as imposed by operation of Canadian law –

such Canadian law being the topic of this Essay – biased opinion – polemic. The author opines that the DRE [*sic*] process is not actually worthy for use other than for seeking criminal convictions of citizens (who may be imprisoned therefor). The author further opines that while the DRE [*sic*] process is lawful before Canadian law courts, it is otherwise unwise in concept, and employs bad science.

**59** [**DRE** = Drug Recognition Expert = a designation of a police officer by operation of law. See in *R v Bingley*, at paragraphs [20] and [21]:

[. . .]

> [20] The DRE, literally, is a "drug recognition expert", certified as such for the purposes of the scheme. It is undisputed that the DRE receives special training in how to administer the 12-step drug recognition evaluation and in what inferences may be drawn from the factual data he or she notes. It is for this limited purpose that a DRE can assist the court by offering expert opinion evidence.
>
> [21] While a DRE's evaluation certainly has an investigative purpose, their application of the 12-step drug recognition evaluation and determination of impairment is relevant evidence and can assist the trier of fact. The DRE's opinion is based on his or her specialized training and experience in conducting the evaluation. By reason of this training and experience, all DREs undoubtedly possess expertise on determining drug impairment that is outside the experience and knowledge of the trier of fact.

[. . .]

**60** – and the category of drug.

**61** *DRUGS THAT IMPAIR DRIVING / PARTICIPANT'S MANUAL* [US DEPARTMENT OF TRANSPORTATION, Transportation Safety Institute, National Highway Traffic Safety Administration, February 2006], HS178A R2/06.; "SESSION VI / DRUG CATEGORIES AND THEIR OBSERVABLE EFFECTS."
https://www.wsp.wa.gov/breathtest/docs/webdms/DRE_Forms/Manuals/drugs/Instructor%20Manual%20(Drugs)%20-%20February%202006.pdf

**62** *DWI Detection and Standardized Field Sobriety Testing / March 2013 Edition / Guide*, International Association of Chiefs of Police & US National Highway Traffic Safety Administration; "C. Drug Categories and Their Observable Effects / Seven Categories of 'Drugs;'" HS 178 R5/13.
https://www.nhtsa.gov/sites/nhtsa.gov/files/documents/sfst_ig_full_manual.pdf

**63** *Criminal Code* regulation **SOR/2008-196** http://laws-lois.justice.gc.ca/eng/regulations/SOR-2008-196/page-1.html#s-3

**64** In the Canadian system, Parliament (= Commons, Senate & Monarch (represented by the Governor General)) enact statutes. The Governor-in-council – effectively the cabinet – brings regulations into effect. Both statutes and regulations (to specifically implement parts of the statutes) are considered as legislation. The use of the term legislation should be considered here in this context.

**65** – effectively, in these or similar words:

**66** DC #, Traffic Services, Toronto Police.

**67** **9. DRE opinion**

[. . .]

> In the opinion of DC . . ., the evaluating DRE Officer, at the time of the evaluation, subject's ability to operate a motor vehicle was impaired by a combination of drugs from Cannabis and Central Nervous System Depressants categories.

[. . .]

**68** *Bingley* paragraphs [20] and [21]:

[. . .]

> [20] The DRE, literally, is a "drug recognition expert", certified as such for the purposes of the scheme. It is undisputed that the DRE receives special training in how to administer the 12-step drug recognition evaluation and in what inferences may be drawn from the factual data he or she notes. It is for this limited purpose that a DRE can assist the court by offering expert opinion evidence.
>
> [21] While a DRE's evaluation certainly has an investigative purpose, their application of the 12-step drug recognition evaluation and determination of impairment is relevant evidence and can assist the trier of fact. The DRE's opinion is based on his or her specialized training and experience in conducting the evaluation. By reason of this training and experience, all DREs undoubtedly possess expertise on determining drug impairment that is outside the experience and knowledge of the trier of fact.
>
> [. . .]

69 *Criminal Code* as amended **s. 320.14(1)**
https://laws-lois.justice.gc.ca/eng/acts/C-46/page-46.html#h-121277
https://laws-lois.justice.gc.ca/eng/annualstatutes/2018_21/page-1.html#h-2

> [. . .]
>
> **Operation while impaired**
>
> **320.14 (1)** Everyone commits an offence who
>
> **(a)** operates a conveyance while the person's ability to operate it is impaired to any degree by alcohol or a drug or by a combination of alcohol and a drug; [or]
>
> **(b)** subject to subsection (5), has, within two hours after ceasing to operate a conveyance, a blood alcohol concentration that is equal to or exceeds 80 mg of alcohol in 100 mL of blood;
>
> [. . .]

70 *Criminal Code* regulation SOR/2008-196
http://laws-lois.justice.gc.ca/eng/regulations/SOR-2008-196/page-1.html#s-3
71 *Criminal Code* regulation SOR/2008-196.
72 It should be noticed from the INTERNATIONAL ASSOCIATION of CHIEFS OF POLICE statement (with apparently no health science status of its own), in its International Standards, that: "DRE training has been validated through both laboratory and field studies conducted by Johns Hopkins University . . .". This is inconsistent with Bigelow et al., Johns Hopkins School of Medicine, 1985, who apparently conclude otherwise. Nor does Heishman et al. (1996) appear to fulfil the IACP claim.
73 The DEC Program Technical Advisory Panel of the IACP Highway Safety Committee, *The International Standards of the Drug Evaluation and Classification Program*, Revised October 2017.
https://www.theiacp.org/sites/default/files/all/i-j/International%20Standards%20of%20the%20DECP%20October%202017.pdf
74 *International Standards*, at page 4.

> [. . .]
>
> Since 1984, the National Highway Traffic Safety Administration (NHTSA) has supported the Drug Evaluation and Classification (DEC) Program, often referred to as the Drug Recognition Expert (DRE) Training Program. Initially developed by the Los Angeles, California, Police Department, DRE training has been validated through both laboratory and field studies conducted by Johns Hopkins University.
>
> [. . .]

75 George E. Bigelow, Warren E. Bickel, John D. Roache, Ira A. Liebson, Pat Nowowieski, "Identifying Types of Drug Intoxication: Laboratory Evaluation of a Subject-Examination Procedure," Johns Hopkins School of Medicine, Baltimore, US DOT HS 806 753, May 1985.
http://www.decp.us/DREpdfs.htm
http://www.decp.us/pdfs/Bigelow_1985_DRE_validation_study.pdf
76 Stephen J. Heishman, Edward G. Singleton & Dennis J. Crouch, "Laboratory Validation Study of Drug Evaluation and Classification Program: Ethanol, Cocaine and Marijauna," *Journal of Analytical Toxicology*, Vol. 20, Oct. 1996, pages 468–483.
77
- Stephen J. Heishman, Addiction Research Center, National Institute on Drug Abuse, Baltimore.
- Edward G. Singleton, Department of Psychiatry & Behavioral Sciences, Johns Hopkins University School of Medicine, Baltimore; & Department of Psychology, Morgan State University, Baltimore.
- Dennis J. Crouch, Center for Human Toxicology, University of Utah, Salt Lake City.

78 Further, while the DRE PARTICIPANT'S MANUAL does refer to three of the tests as having ". . . been scientifically validated . . .", for alcohol, no actual health science literature is cited, for either alcohol or drugs.

When read together, legislation and case law allow the DRE opinion special legal status in a process that would give the appearance of health science observation and decision making. But, nowhere within these, nor apparently in related published DRE literature, is found reference to verified scientific validation of the tests and procedures to rely on. Nor health science meanings of the tests and procedures. Nor personnel of actual health science expertise or licensure. The science is either actually absent, or unrevealed in such a way that the court should regard it as absent.
79 Also, judgment in *Bingley*: ". . . Parliament has established the reliability of the 12-step drug evaluation by statute. . . ." But, *Bingley* appears unaware that actual scientific validation, meaning, expertise and licensure are absent; and *Bingley* is not *Charter* reliant. The cross-examination suggested by *Bingley* would allow questioning about data scraps of limited and uncertain meaning without challenge to the very basis of the DRE process.
**80** ["12-step" not explicitly mentioned in the regulation, but apparently that is what the regulation procedure is called, as in *Bingley*.]
**81** [Perhaps better described as by the Governor-in-Council by regulation, as authorized by statute.]
82 ***Bingley***

> [. . .]
>
> [32] That Parliament has established the reliability of the 12-step drug evaluation by statute does not hinder the trier of fact's ability to critically assess a DRE's conclusion of impairment or an accused person's right to test that evidence. Cross-examination of the DRE may undermine his or her conclusion. Evidence of bias may raise doubt about the officer's conclusion. The officer may fail to conduct the drug recognition evaluation in accordance with his or her training. A DRE may draw questionable inferences from his or her observations. Bodily sample evidence obtained under s. 254(3.4) may refute the DRE's assessment, as may evidence of bystanders or other experts. It will always be for the trier of fact to determine what weight to give a DRE's opinion. Any weight given to a DRE's evidence will necessarily respect the scope of the DRE's expertise and the fact that it is not conclusive of impairment.
>
> [. . .]

**83** ***Bingley*** at paragraph [32].
**84** Within the Supreme Court of Canada's judgment in *Bingley*, there is an especially difficult interpretation issue. Although frequent reference is made to the "12-step," as a central feature of the DRE process, actually finding its provenance is problematic. The "12-step" is not mentioned in the legislation'

and hardly appears in the DRE literature; neither explanation nor reference.

The "12-step" does not appear as disclosure to Defence counsel, other than where it is found is on a printed card provided by police DRE. It is also found, in another version, on the IACP website. Neither version has verified reference to scientific validation or literature.

While the "12-step" has similarities to the text of the regulation, they are not congruent. Correlating to what is in the regulation with the "12-step" is difficult; as is knowing which "12-step" version should be regarded as primary. If the "12-step" featured in *Bingley* is to have meaning, it should be unambiguously identified in the legislation, and its provenance declared.

**85** There is certainly a possible prospect of the wrong scientific determinations. In a different context, it is argued by US Supreme Court Justice Antonin Scalia.

**86** *Melendez-Diaz v Massachusetts*, pages 1–23.

**87** In *R v* . . ., 2017, Old City Hall, Her Honour Justice C. A. Nelson, in view of *Bingley*, did not allow Defence questions about scientific basis in cross-examination of the DRE. *Charter* issues were not raised in that case. In *R v*. . ., 2018, Old City Hall, Her Honour Justice M. Greene, with *Charter* issues raised, would not allow Defence questions about scientific basis because of the lack of evidence brought by Defence, but would have allowed re-opening the issue with a proper evidentiary foundation.

Also, in *R v* . . ., 2018, Old City Hall, Her Honour, Justice M. Greene, giving judgment orally, would allow Defence to ask questions about scientific basis in cross-examination of the DRE in relation to weight, but this is not repeated in her written Reasons for judgment (27 November 2018). Her Honour, orally, did not elaborate, but such cross-examination of the DRE in relation to weight would appear to reasonably extend what is allowed by *Bingley*. As it actually turned out in . . ., no questions were asked because Crown counsel eventually did not call the DRE to give any evidence.

**88** *Bingley* at paragraph [32].

**89** (– or a road test – not recommended in these circumstances.)

**90** *Criminal Code* regulation SOR/2008-196 – cited & quoted in a note below.

**91** The DEC Program Technical Advisory Panel of the IACP Highway Safety Committee, *The International Standards of the Drug Evaluation and Classification Program*, Revised October 2017.
https://www.theiacp.org/sites/default/files/all/i-j/International%20Standards%20of%20the%20DECP%20October%202017.pdf

**92** **IACP** = INTERNATIONAL ASSOCIATION of CHIEFS OF POLICE, Alexandria, Virginia, 22314.
http://www.theiacp.org/
http://www.theiacp.org/Drug-Recognition-Expert-Section http://www.theiacp.org/Portals/0/documents/DRE/International%20DEC%20Program%20Standards%20October%202016.pdf

**93** NHTSA, *Participant Manual / Drug Recognition Expert Course*, Revised: 10/2015.
www.nhtsa.gov
https://wsp.wa.gov/breathtest/docs/dre/manuals/pre-school_dre/2015_pre_dre/student_pre_oct2015.pdf
https://www.nhtsa.gov/search?keywords=participant+manual

**94** *International Standards* at page 4.

**95** Bigelow et al.

**96** The apparently implied validation of the DRE method appears defective. According to the IACP DRE International Standards:

> . . . Initially developed by the Los Angeles, California, Police Department, DRE training has been validated through both laboratory and field studies conducted by Johns Hopkins University.
>
> [. . .]

But, Bigelow et al., Johns Hopkins School of Medicine, Baltimore, 1985, apparently conclude otherwise. Nor do Heishman et al., Baltimore, 1995, satisfy this.

97 George E. Bigelow, Warren E. Bickel, John D. Roache, Ira A. Liebson, Pat Nowowieski, "Identifying Types of Drug Intoxication: Laboratory Evaluation of a Subject-Examination Procedure," Johns Hopkins School of Medicine, Baltimore, US DOT HS 806 753, May 1985.
http://www.decp.us/DREpdfs.htm
http://www.decp.us/pdfs/Bigelow_1985_DRE_validation_study.pdf
98 Bigelow et al. At page 16:
99 CONCLUSIONS

> This laboratory simulation study does not represent a direct test of the validity of these or related behavioral examination procedures for detecting and identifying drug intoxication in field situations. It does, however, provide valuable scientific information concerning the potential accuracy and utility of such procedures.

100 Stephen J. Heishman, Edward G. Singleton, & Dennis J. Crouch, "Laboratory Validation Study of Drug Evaluation and Classification Program: Ethanol, Cocaine, and Marijuana," Journal of Analytical Toxicology, Vol. 20, October 1996; pages 468–483. [1995 SOFT meeting, Baltimore.]
101 (– assuming Ontario here for purposes of example.)
102 Ontario *Regulated Health Professions Act, 1991*

> [. . .]
>
> PROHIBITIONS
>
> **Controlled acts restricted**
>
> **27. (1)** No person shall perform a controlled act set out in subsection (2) in the course of providing health care services to an individual unless
>
> (a) the person is a member authorized by a health profession Act to perform the controlled act or
>
> (b) the performance of the controlled act has been delegated to the person by a member described in clause (a).
>
> **Controlled acts**
>
> **(2)** A "controlled act" is any one of the following done with respect to an individual:
>
> **1.** Communicating to the individual or his or her personal representative a diagnosis identifying a disease or disorder as the cause of symptoms of the individual in circumstances in which it is reasonably foreseeable that the individual or his or her personal representative will rely on the diagnosis.
>
> [. . .]
>
> **Offences**
>
> **40. (1)** Every person who contravenes subsection 27(1) . . . is guilty of an offence . . .
>
> [. . .]

https://www.ontario.ca/laws/statute/91r18
103 Ontario *Medicine Act, 1991*

> [. . .]
>
> **Scope of practice**
>
> **3.** The practice of medicine is the assessment of the physical or mental condition of an individual and the diagnosis, treatment and prevention of any disease, disorder or dysfunction.
>
> **Authorized acts**
>
> **4.** In the course of engaging in the practice of medicine, a member is authorized, subject to the terms, conditions and limitations imposed on his or her certificate of registration, to perform the

following:

**1.** Communicating a diagnosis identifying a disease or disorder as the cause of a person's symptoms.

[. . .]

https://www.ontario.ca/laws/statute/91m30

**104** And, when an MD would be involved, as sometimes happens, to sign-off for the taking of a blood sample – as lawfully demanded by the DRE – there would be an apparent potential ethical conflict of interest. This would arise because if the MD would have been ministering to the accused in a hospital emergency room as a patient. That same MD should not then act effectively in the service of law enforcement with regard to that patient. If police need a sign-off for the blood taking, they should hire their own MD.

**105** English text from CFS form, 2015:

**106** Ontario Centre of Forensic Sciences . . . Certificate of a Qualified Medical Practitioner (S.258(1) (h) (i) and (ii))

[. . .]

I _________a person duly qualified by provincial law to practice medicine in the Province of Ontario and being therefore a qualified medical practitioner as defined in subsection 254(1) of the Criminal Code.

**DO HEREBY CERTIFY**:

[. . .]

. . . **THAT** I caused samples of blood to be taken under my direction by a qualified technician (blood samples) as defined in subsection 254(1) from a person identified to me as _____________

. . . **THAT**, before the said samples were taken and at the times the said samples were taken, I was of the opinion that the taking of blood samples from the said person would not endanger the life or the health of the said person;

[. . .]

Dated . . . _________ at . . . ________ . . . Ontario ________________

[. . .]

CFS 6038 (06/03)

**107** From a medical report of an MD, 2015: ". . . The police approached me and asked for me to sign a form stating that the they could get blood from . . . They stated that . . . had consented . . . I asked . . . about this . . . [understood & consented] . . . I . . . [signed] the form]".

**108** See also:

https://www.ontario.ca/page/toxicology-unit-centre-forensic-sciences

https://www.canlii.org/en/commentary/doc/1998CanLIIDocs408#!fragment/zoupio-_Tocpdf_bk_1_2/BQCwhgziBcwMYgK4DsDWszIQewE4BUBTADwBdoAvbRABwEtsBaAfX2zhoBMAzZgI1TMAjMwBMASgA0ybKUIQAiokK4AntADkGyREJhcCJSvVadegyADKeUgCF1AJQCiAGScA1AIIA5AMJPJUjA+aFJ2cXEgA

**109** *Constitution Act*, s. 91 and 92. https://laws-lois.justice.gc.ca/eng/const/page-3.html
**110** *Constitution Act*, 1982 (1982 c. 11 (United Kingdom)). https://laws-lois.justice.gc.ca/pdf/const_e.pdf
**111** *Quod volimus credimus libenter* = we always believe what we want to believe. Robert Harris, *Archangel*, Berkley Publishing, New York City, 1998; Arrow, 2009; ISBN-10: 0099527936, ISBN-13:978–0099527930.
https://www.amazon.ca/Archangel-Robert-Harris/dp/0099527936
https://www.goodreads.com/quotes/377660-quod-volimus-credimus-libenter-we-alwaysbelieve-what-we-want
**112** https://en.wikiquote.org/wiki/Julius_Caesar:
Julius Caesar, *Commentarii de Bello Gallico* [=*Commentaries on the Gallic War*], Book I, Ch. 14, translated by W.A. McDevitte and W.S. Bohn: "*Fere libenter homines id quod volunt credunt*" = In most cases men willingly believe what they wish.
**113** [From a Toronto law case.] [Not to scale here.] (see Fig. 11.2).

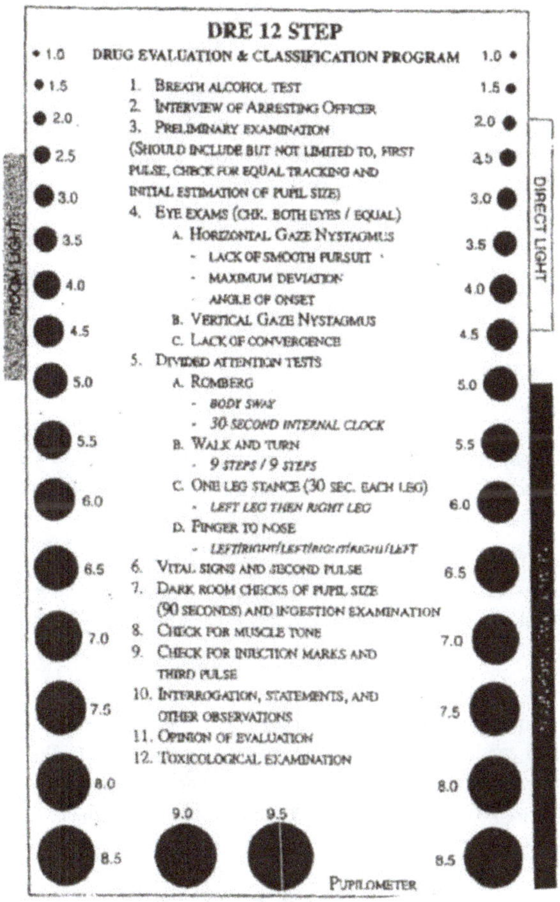


Fig. 11.2: Pupilometer.

**114** (That printed card (see again Fig. 11.2) also doubling as a "PUPILOMETER" is apparently without US FDA CFR reference, nor with manufacture specifications, as a medical device; apparently without provenance; apparently without health science literature reference.)
**115** See also: https://store.iptm.org/products/12-step-drug-influence-evaluation-checklist-card
**116** See also:
https://en.wikipedia.org/wiki/Pupilometer#:~:text=Pupillometer%2C%20also%20spelled%20pupilometer%2C%20is,to%20characterize%20pupillary%20light%20reflex
**117** See also: Robert G. Kerr, MD, PhD; Andrea M. Bacon, RN, BSN; Laura L. Baker, MS, ARNP, FNP-BC; Janelle S. Gehrke, RN, BSN, CCRC; K. Danielle Hahn, BS; Cheryl L. Lillegraven, MSN, ARNP, ACNS-BC; Catherine Hackett Renner, PhD; Sarah K. Spilman, MA; "Underestimation of Pupil Size by Critical Care and Neurosurgical Nurses" *Am J Crit Care* (2016) 25(3): 213–219.
https://doi.org/10.4037/ajcc2016554
**118** **IACP** = INTERNATIONAL ASSOCIATION of CHIEFS OF POLICE, Alexandria, Virginia, 22314. http://www.theiacp.org/ http://www.theiacp.org/Drug-Recognition-Expert-Section
**119** [IACP 12 STEP; February 2021, & previous] https://www.theiacp.org/12-step-process
**120** [. . .]

**12 Step Process**

The DRE protocol is a standardized and systematic method of examining a Driving Under the Influence of Drugs (DUID) suspect to determine the following: (1) whether or not the suspect is impaired; if so, (2) whether the impairment relates to drugs or a medical condition; and if drugs, (3) what category or combination of categories of drugs are the likely cause of the impairment. The process is *systematic* because it is based on a complete set of observable signs and symptoms that are known to be reliable indicators of drug impairment.

The DRE evaluation is *standardized* because it is conducted the same way, by every drug recognition expert, for every suspect whenever possible.

**The 12-Step DRE Protocol**

The DREs utilize a 12-step process to assess DUID suspects:

1. *Breath Alcohol Test*

The arresting officer reviews the subject's breath alcohol concentration (BrAC) test results and determines if the subject's apparent impairment is consistent with the subject's BrAC. If the impairment is *not* explained by the BrAC, the officer requests a DRE evaluation.

2. *Interview of the Arresting Officer*

The DRE begins the investigation by reviewing the BrAC test results and discussing the circumstances of the arrest with the arresting officer. The DRE asks about the subject's behavior, appearance, and driving.

3. *Preliminary Examination and First Pulse*

The DRE conducts a preliminary examination, in large part, to ascertain whether the subject may be suffering from an injury or other condition unrelated to drugs. Accordingly, the DRE asks the subject a series of standard questions relating to the subject's health and recent ingestion of food, alcohol, and drugs, including prescribed medications. The DRE observes the subject's attitude, coordination, speech, breath and face. The DRE also determines if the subject's pupils are of equal size and if the subject's eyes can follow a moving stimulus and track equally. The DRE also looks for horizontal gaze nystagmus (HGN) and takes the subject's pulse for the first of three times. If the DRE believes that the subject *may* be suffering from a significant medical condition, the DRE will seek medical assistance immediately. If the DRE believes that the subject's condition is drug-related, the evaluation continues.

4. *Eye Examination*
The DRE examines the subject for HGN, vertical gaze Nystagmus (VGN), and a lack of convergence.

5. *Divided Attention Psychophysical Tests*
The DRE administers four psychophysical tests: the Modified Romberg Balance, the Walk and Turn, the One Leg Stand, and the Finger to Nose test.

6. *Vital Signs and Second Pulse*
The DRE takes the subject's blood pressure, temperature, and pulse.

7. *Dark Room Examinations*
The DRE estimates the subject's pupil sizes under three different lighting conditions with a measuring device called a pupilometer. The device will assist the DRE in determining whether the subject's pupils are dilated, constricted or normal.

8. *Examination for Muscle Tone*
The DRE examines the subject's skeletal muscle tone. Certain categories of drugs may cause the muscles to become rigid. Other categories may cause the muscles to become very loose and flaccid.

9. *Check for Injection Sites and Third Pulse*
The DRE examines the subject for injection sites, which may indicate recent use of certain types of drugs. The DRE also takes the subject's pulse for the third and final time.

10. *Subject's Statements and Other Observations*
The DRE typically reads *Miranda*, if not done so previously, and asks the subject a series of questions regarding the subject's drug use.

11. *Analysis and Opinions of the Evaluator*
Based on the totality of the evaluation, the DRE forms an opinion as to whether or not the subject is impaired. If the DRE determines that the subject is impaired, the DRE will indicate what category or categories of drugs may have contributed to the subject's impairment.

12. *Toxicological Examination*
The toxicological examination is a chemical test or tests that provide additional scientific, admissible evidence to support the DRE's opinion.

Nothing in or about the DRE protocol is new or novel. The DRE protocol is a compilation of tests that physicians have used for decades to identify and assess alcohol- and/or drug induced impairment.

[. . .]

**121** [IACP 12 STEP; February 2021, and previous] https://www.theiacp.org/12-step-process [at last paragraph]

**122** Canada *Criminal Code* s. 320.12 (d) – see notes above.

**123** – but apparently not including the then Governor General herself – on other science issues – although there were criticisms concerning issues of vice-regal protocol.

**124** https://www.nationalobserver.com/2017/11/06/opinion/science-v-religion-and-new-governor-general-under-fire
https://www.nationalobserver.com/2017/11/04/news/payette-takes-climate-change-deniers-and-horoscopes-science-conference
https://torontosun.com/opinion/editorials/editorial-julie-payette-speech-oversteps-her-role
https://www.cbc.ca/news/opinion/governor-general-speech-julie-payette-climate-change-1.4384481

https://www.gg.ca/en
https://en.wikipedia.org/wiki/Governor_General_of_Canada
**125** Latin. Introduction 4.1.
**126** *Malum in se* = wrong of itself; naturally wrong.
*Malum prohibitum* = wrong because it's prohibited by law.
https://en.wikipedia.org/wiki/Malum_in_se https://en.wikipedia.org/wiki/Malum_prohibitum
**127** See: Joel Coen, *True Grit*, Paramount Pictures, 2010; Jeff Bridges, Hailee Steinfeld, Matt Damon, Josh Brolin. https://www.scripts.com/script/true_grit_22307 Script pdf page 32: [Mattie Ross explains this concept to US Marshal Rooster Cogburn.]
**128** *ipso facto* = by that very fact or act : as an inevitable result.
https://www.merriam-webster.com/dictionary/ipso%20facto

# Index



https://doi.org/10.1515/9783111618555-013

www.ingramcontent.com/pod-product-compliance
Lightning Source LLC
Chambersburg PA
CBHW061150240926
51846CB00006B/971
* 9 7 8 3 1 1 9 1 4 9 7 0 9 *